The Mycota

Edited by
K. Esser

The Mycota

The Mycota

A Comprehensive Treatise on Fungi as Experimental Systems for Basic and Applied Research

Edited by K. Esser

III

Biochemistry and Molecular Biology
3rd Edition

Volume Editor:
D. Hoffmeister

Springer

Series Editor

Professor Dr. Dr. h.c. mult. Karl Esser
Allgemeine Botanik
Ruhr-Universität
44780 Bochum, Germany

Tel.: +49 (234)32-22211
Fax.: +49 (234)32-14211
e-mail: Karl.Esser@rub.de

Volume Editor

Professor Dr. Dirk Hoffmeister
Friedrich-Schiller-Universität
Pharmazeutische Mikrobiologie
Winzerlaer Straße 2
07745 Jena
Germany

Tel.: +49 (3641) 949851
Fax: +49 (3641) 949852
e-mail: dirk.hoffmeister@leibniz-hki.de

ISBN 978-3-319-27788-2 ISBN 978-3-319-27790-5 (eBook)
DOI 10.1007/978-3-319-27790-5

Library of Congress Control Number: 2016932310

Springer Cham Heidelberg New York Dordrecht London

Printed on acid-free paper

Springer International Publishing AG Switzerland is part of Springer Science+Business Media (www.springer.com)

Karl Esser
(born 1924) is retired Professor of General Botany and Director of the Botanical Garden at the Ruhr-Universität Bochum (Germany). His scientific work focused on basic research in classical and molecular genetics in relation to practical application. His studies were carried out mostly on fungi. Together with his collaborators he was the first to detect plasmids in higher fungi. This has led to the integration of fungal genetics in biotechnology. His scientific work was distinguished by many national and international honors, especially three honorary doctoral degrees.

Dirk Hoffmeister
(born 1972) is a mycologist and Full Professor for Pharmaceutical Microbiology at the Friedrich-Schiller-University Jena (Germany). He is also affiliated with the Leibniz Institute for Natural Product Research and Infection Biology—Hans Knöll Institute. Particularly emphasizing the basidiomycetes, research in his group is focused on the biochemical and genetic basis underlying fungal natural product biosyntheses.

Series Preface

Mycology, the study of fungi, originated as a sub discipline of botany and was a descriptive discipline, largely neglected as an experimental science until the early years of this century. A seminal paper by Blakeslee in 1904 provided evidence for self-incompatibility, termed "heterothallism," and stimulated interest in studies related to the control of sexual reproduction in fungi by mating-type specificities. Soon to follow was the demonstration that sexually reproducing fungi exhibit Mendelian inheritance and that it was possible to conduct formal genetic analysis with fungi. The names Burgeff, Kniep and Lindegren are all associated with this early period of fungal genetics research.

These studies and the discovery of penicillin by Fleming, who shared a Nobel Prize in 1945, provided further impetus for experimental research with fungi. Thus, began a period of interest in mutation induction and analysis of mutants for biochemical traits. Such fundamental research, conducted largely with Neurospora crassa, led to the one gene:one enzyme hypothesis and to a second Nobel Prize for fungal research awarded to Beadle and Tatum in 1958. Fundamental research in biochemical genetics was extended to other fungi, especially to *Saccharomyces cerevisiae*, and by the mid-1960s fungal systems were much favored for studies in eukaryotic molecular biology and were soon able to compete with bacterial systems in the molecular arena.

The experimental achievements in research on the genetics and molecular biology of fungi have benefited more generally studies in the related fields of fungal biochemistry, plant pathology, medical mycology, and systematics. Today, there is much interest in the genetic manipulation of fungi for applied research. This current interest in biotechnical genetics has been augmented by the development of DNA-mediated transformation systems in fungi and by an understanding of gene expression and regulation at the molecular level. Applied research initiatives involving fungi extend broadly to areas of interest not only to industry but to agricultural and environmental sciences as well.

It is this burgeoning interest in fungi as experimental systems for applied as well as basic research that has prompted publication of this series of books under the title The Mycota. This title knowingly relegates fungi into a separate realm, distinct from that of either plants, animals, or protozoa. For consistency throughout this series of volumes, the names adopted for major groups of fungi (representative genera in parentheses) are as follows:

Pseudomycota

Division:	Oomycota (*Achlya, Phytophthora, Pythium*)
Division:	Hyphochytriomycota

Eumycota

Division:	Chytridiomycota (*Allomyces*)
Division:	Zygomycota (*Mucor, Phycomyces, Blakeslea*)
Division:	Dikaryomycota
Subdivision:	Ascomycotina
Class:	Saccharomycetes (*Saccharomyces, Schizosaccharomyces*)
Class:	Ascomycetes (*Neurospora, Podospora, Aspergillus*)
Subdivision:	Basidiomycotina
Class:	Heterobasidiomycetes (*Ustilago, Tremella*)
Class:	Homobasidiomycetes (*Schizophyllum, Coprinus*)

We have made the decision to exclude from The Mycota the slime molds which, although they have traditional and strong ties to mycology, truly represent nonfungal forms insofar as they ingest nutrients by phagocytosis, lack a cell wall during the assimilative phase, and clearly show affinities with certain protozoan taxa.

The series throughout will address three basic questions: what are the fungi, what do they do, and what is their relevance to human affairs? Such a focused and comprehensive treatment of the fungi is long overdue in the opinion of the editors.

A volume devoted to systematics would ordinarily have been the first to appear in this series. However, the scope of such a volume, coupled with the need to give serious and sustained consideration to any reclassification of major fungal groups, has delayed early publication. We wish, however, to provide a preamble on the nature of fungi, to acquaint readers who are unfamiliar with fungi with certain characteristics that are representative of these organisms and which make them attractive subjects for experimentation.

The fungi represent a heterogeneous assemblage of eukaryotic microorganisms. Fungal metabolism is characteristically heterotrophic or assimilative for organic carbon and some nonelemental source of nitrogen. Fungal cells characteristically imbibe or absorb, rather than ingest, nutrients and they have rigid cell walls. The vast majority of fungi are haploid organisms reproducing either sexually or asexually through spores. The spore forms and details on their method of production have been used to delineate most fungal taxa. Although there is a multitude of spore forms, fungal spores are basically only of two types: (1) asexual spores are formed following mitosis (mitospores) and culminate vegetative growth, and (2) sexual spores are formed following meiosis (meiospores) and are borne in or upon specialized generative structures, the latter frequently clustered in a fruit body. The vegetative forms of fungi are either unicellular, yeasts are an example, or hyphal; the latter may be branched to form an extensive mycelium.

Regardless of these details, it is the accessibility of spores, especially the direct recovery of meiospores coupled with extended vegetative haploidy, that have made fungi especially attractive as objects for experimental research.

The ability of fungi, especially the saprobic fungi, to absorb and grow on rather simple and defined substrates and to convert these substances, not only into essential metabolites but into important secondary metabolites, is also noteworthy. The metabolic capacities of fungi have attracted much interest in natural products chemistry and in the production of antibiotics and other bioactive compounds. Fungi, especially yeasts, are important in fermentation processes. Other fungi are important in the production of enzymes, citric acid, and other organic compounds as well as in the fermentation of foods.

Fungi have invaded every conceivable ecological niche. Saprobic forms abound, especially in the decay of organic debris. Pathogenic forms exist with both plant and animal hosts. Fungi even grow on other fungi. They are found in aquatic as well as soil environments, and their spores may pollute the air. Some are edible; others are poisonous. Many are variously associated with plants as copartners in the formation of lichens and mycorrhizae, as symbiotic endophytes or as overt pathogens. Association with animal systems varies; examples include the predaceous fungi that trap nematodes, the microfungi that grow in the anaerobic environment of the rumen, the many insect associated fungi, and the medically important pathogens afflicting humans. Yes, fungi are ubiquitous and important. There are many fungi, conservative estimates are in the order of 100,000 species, and there are many ways to study them, from descriptive accounts of organisms found in nature to laboratory experimentation at the cellular and molecular level. All such studies expand our knowledge of fungi and of fungal processes and improve our ability to utilize and to control fungi for the benefit of humankind.

We have invited leading research specialists in the field of mycology to contribute to this series. We are especially indebted and grateful for the initiative and leadership shown by the Volume Editors in selecting topics and assembling the experts. We have all been a bit ambitious in producing these volumes on a timely basis and therein lies the possibility of mistakes and oversights in this first edition. We encourage the readership to draw our attention to any error, omission, or inconsistency in this series in order that improvements can be made in any subsequent edition.

Finally, we wish to acknowledge the willingness of Springer-Verlag to host this project, which is envisioned to require more than 5 years of effort and the publication of at least nine volumes.

Bochum, Germany
Auburn, AL, USA
April 1994

Karl Esser
Paul A. Lemke
Series Editors

Volume Preface

The years in between the previous and the current third edition of The Mycota III witnessed a dramatic change in how we address scholarly questions in fungal biology. Above all, the rapidly advancing technologies to generate massive amounts of nucleic acid sequence data that enabled the exponential growth of the number of sequenced genomes and the increased quality of transcriptomes have provided insight into the function and evolution of fungi to an extent we would probably not have even dreamed of 10 years ago. Consequently, results reviewed in the newly written or updated chapters of this volume underscore the copious availability of genomes, transcriptomes, or proteomes, which has become the expected norm.

To truly advance fungal biology, we need to be careful in order not to get sidetracked and drowned in an ocean of computer-produced data. "We are at the start of what will be one of the most exciting periods of advance and discovery in the history of our field." These words, written by Professors Robert Brambl and George A. Marzluf in the preface of the second edition of The Mycota III, should encourage us, more than ever, to be thoughtful in asking the right questions and to test hypotheses—beyond the in silico level—experimentally through wet-bench work.

To produce the third edition of The Mycota III, the Editor was privileged to work with diligent and enthusiastic experts as both recurring and first-time chapter contributors. The Editor is both pleased and thankful of the five returning authors/author teams and additionally excited to have many new authors. It was the aim of the Editor to cover fungi as broadly and comprehensively as possible. Accordingly, chapters highlighting the aspects of zygomycete and basidiomycete biology appear to complement contributions that focus on precious model species, such as Aspergilli. Also, primary metabolism, gene regulation, and signal transduction were further emphasized in the new volume, e.g., with chapters on major metabolic routes, RNA interference, regulation in plant pathogenic fungi, but also on global regulation of *Aspergillus* natural product biosyntheses, the latter to reflect the markedly increased interest in fungal secondary metabolites.

The Editor cordially thanks Dr. Andrea Schlitzberger of Springer Publishers for excellent and competent guidance throughout the production process and, last but not least, the Series Editor, Professor Emeritus Karl Esser, for precious advice and encouragement that was always combined with a fine sense of humor.

Jena, Germany
October 2015

Dirk Hoffmeister

Contents

Molecular Aspects of Biochemical Pathways

List of Contributors

Özgür Bayram
Department of Biology, Maynooth University, National University of Ireland Maynooth, Maynooth Co. Kildare, Ireland

Jin Woo Bok
Department of Medical Microbiology and Immunology, University of Wisconsin-Madison, Madison, WI, USA

Katherine A. Borkovich
Department of Plant Pathology and Microbiology, University of California, Riverside, CA, USA

Matthias Brock
Microbial Biochemistry and Physiology, Leibniz Institute for Natural Product Research and Infection Biology e. V., -Hans Knöll Institute, Jena, Germany

Institute for Microbiology, Friedrich Schiller University Jena, Jena, Germany

Fungal Genetics and Biology Group, School of Life Sciences, University of Nottingham, Nottingham, UK

Rekha Deka
Department of Plant Pathology and Microbiology, University of California, Riverside, CA, USA

Candace E. Elliott
School of Biosciences, The University of Melbourne, Melbourne, VIC, Australia

Elena Geib
Microbial Biochemistry and Physiology, Leibniz Institute for Natural Product Research and Infection Biology e. V., -Hans Knöll Institute, Jena, Germany

Fungal Genetics and Biology Group, School of Life Sciences, University of Nottingham, Nottingham, UK

Arit Ghosh
Department of Plant Pathology and Microbiology, University of California, Riverside, CA, USA

MATTHIAS GUBE
Soil Science of Temperate Ecosystems, Georg August University Göttingen, Göttingen, Germany

Microbial Communication, Friedrich Schiller University Jena, Jena, Germany

STEFAN IRNIGER
Department of Molecular Microbiology and Genetics, Georg August University, Göttingen, Germany

KERSTIN KAERGER
National Reference Center for Invasive Mycoses, Leibniz Institute for Natural Product Research and Infection Biology - Hans-Knöll-Institute, Jena, Germany

NANCY P. KELLER
Department of Medical Microbiology and Immunology, University of Wisconsin-Madison, Madison, WI, USA

JOAN M. KELLY
Department of Genetics and Evolution, The University of Adelaide, Adelaide, Australia

MARK J. LEE
Department of Microbiology and Immunology, McGill University, Montréal, QC, Canada

GÁBOR NAGY
Faculty of Science and Informatics, Department of Microbiology, University of Szeged, Szeged, Hungary

KATRIN OCHSENREITER
Karlsruhe Institute of Technology (KIT), Institute of Process Engineering in Life Sciences, Section II: Technical Biology, Karlsruhe, Germany

ÅKE OLSON
Department of Forest Mycology and Plant Pathology, Swedish University of Agricultural Sciences, Uppsala, Sweden

JOHN V. PAIETTA
Department of Biochemistry and Molecular Biology, Wright State University, Dayton, OH, USA

TAMÁS PAPP
Faculty of Science and Informatics, Department of Microbiology, University of Szeged, Szeged, Hungary

HEE-SOO PARK
Department of Bacteriology, University of Wisconsin-Madison, Madison, WI, USA

Cesar Roncero
Departamento de Microbiologia y Genetica, IBFG, CSIC, Universidad de Salamanca, Salamanca, Spain

Marta Rubio-Texeira
Laboratory of Molecular Cell Biology, Institute of Botany and Microbiology, KU Leuven, Leuven, Belgium

Department of Molecular Microbiology, VIB, Leuven-Heverlee, Flanders, Belgium

Alberto Sanchez-Diaz
Instituto de Biomedicina y Biotecnologia de Cantabria (IBBTEC), Universidad de Cantabria, CSIC, Santander, Spain

Departamento de Biologia Molecular. Facultad de Medicina, Universidad de Cantabria, Santander, Spain

Özlem Sarikaya-Bayram
Department of Biology, Maynooth University, National University of Ireland Maynooth, Maynooth Co. Kildare, Ireland

Ekaterina Shelest
Systems Biology/Bioinformatics, Leibniz Institute for Natural Product Research and Infection Biology – Hans-Knöll-Institute, Jena, Germany

Donald C. Sheppard
Department of Microbiology and Immunology, McGill University, Montréal, QC, Canada

Department of Medicine, McGill University, Montréal, QC, Canada

Department of Microbiology & Immunology, McGill University, Montréal, Qc, Canada

Ranjan Tamuli
Department of Biosciences and Bioengineering, Indian Institute of Technology Guwahati, Guwahati, Assam, India

Johan M. Thevelein
Laboratory of Molecular Cell Biology, Institute of Botany and Microbiology, KU Leuven, Leuven, Belgium

Department of Molecular Microbiology, VIB, Leuven-Heverlee, Flanders, Belgium

Richard B. Todd
Department of Plant Pathology, Kansas State University, Manhattan, KS, USA

M.-Henar Valdivieso
Departamento de Microbiologia y Genetica, IBFG, CSIC, Universidad de Salamanca, Salamanca, Spain

Kerstin Voigt
Jena Microbial Resource Collection, Leibniz Institute for Natural Product Research and Infection Biology – Hans-Knöll-Institute, Jena, Germany

Department of Microbiology and Molecular Biology, Institute of Microbiology, University of Jena, Jena, Germany

Thomas Wolf
Systems Biology/Bioinformatics, Leibniz Institute for Natural Product Research and Infection Biology – Hans-Knöll-Institute, Jena, Germany

Jae-Hyuk Yu
Department of Bacteriology, University of Wisconsin-Madison, Madison, WI, USA

Department of Genetics, University of Wisconsin-Madison, Madison, WI, USA

Griet Van Zeebroeck
Laboratory of Molecular Cell Biology, Institute of Botany and Microbiology, KU Leuven, Leuven, Belgium

Department of Molecular Microbiology, VIB, Leuven-Heverlee, Flanders, Belgium

Regulation of Gene Expression

1 Molecular Biology of Asexual Sporulation in Filamentous Fungi

Hee-Soo Park[1], Jae-Hyuk Yu[1,2]

CONTENTS

I. Introduction

Filamentous fungi are of great importance to humankind as pathogens, environmental recyclers, industrial producers, and agricultural aids. Fungi produce spores as the primary means of reproduction, propagation, and infectivity (Ebbole 2010). Fungi use asexual sporulation as a main reproductive mode (Adams et al. 1998; Ni et al. 2010). Asexual nonmotile spores, called conidia (conidiospores), are formed on the specialized developmental structure called conidiophore. The entire processes of the initiation, progression, and completion of asexual sporulation are regulated by various positive and negative genetic elements that direct the expression of genes required for the proper assembly of the conidiophore and the formation and maturation of conidiospores (Adams et al. 1998; Park and Yu 2012). This chapter presents a snapshot of the up-to-date molecular biology of conidiation in the three model fungi *Aspergillus nidulans*, *Penicillium marneffei*, and *Fusarium graminearum*.

II. Asexual Sporulation in *Aspergillus nidulans*

A. Morphology of Asexual Structure

A conidiophore is a specialized asexual structure, which is the most remarkable characteristic of a specific fungal species (Adams et al. 1998; Yu 2010). The formation of conidiophores in *A. nidulans* is highly sophisticated and can be divided into several differential stages (Fig. 1.1) (Mims et al. 1988; Timberlake 1990). First, conidiation starts from the foot cells, thick-walled cells, which elongate into the air to produce aerial stalks. After the extension of stalks ceases, the tip of stalks then starts to swell and form apical vesicles, which contain multiple nuclei. Through a budding-like process, 60 metulae are formed on the surface of the vesicle and then each metula produces 2 buds that develop into uninucleate sterigmata, termed phialides. Phialides undergo repeated asymmetric mitotic divisions to generate chains of conidia (about 120 conidia per

[1]Department of Bacteriology, University of Wisconsin-Madison, Madison, WI 53706, USA; e-mail: jyu1@wisc.edu
[2]Department of Genetics, University of Wisconsin-Madison, 1550 Linden Drive, Madison, WI 53706, USA; e-mail: jyu1@wisc.edu

Biochemistry and Molecular Biology, 3rd Edition
The Mycota III
D. Hoffmeister (Ed.)

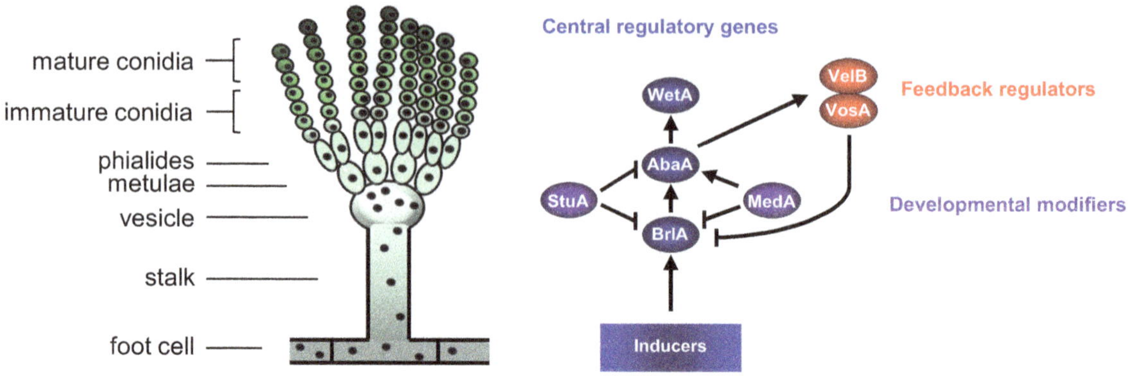

Fig. 1.1 Conidiophore and the central regulators of conidiation in *A. nidulans*. Simplified illustration of the conidiophore (*left*) and a genetic model of the central regulatory pathway of conidiation (*right*) are shown

phialide) (Adams et al. 1998; Mims et al. 1988). Conidia are haploid cells and capable of forming a new colony under appropriate conditions. Soon after conidia are produced, they undergo a maturation process, involving the modification of the conidial wall and trehalose biogenesis for the establishment of conidial dormancy (Ni et al. 2010; Sewall et al. 1990b; Timberlake 1990). The whole process of conidiation is genetically programmed, and various regulators are responsible for the progression of each stage (Adams et al. 1998).

B. Regulators of Asexual Development

1. Central Regulators of Conidiation

a) BrlA

An essential initiation step of conidiation is the activation of *brlA* encoding a C_2H_2 zinc-finger transcription factor (TF) (Adams et al. 1988, 1990). The conidiophores of *brlA* null mutants fail to form a vesicle and other asexual structures and continue to grow conidiophore stalks (thus termed "bristle") (Clutterbuck 1969). Contrarily, the overexpression of *brlA* causes termination of hyphal growth coupled with the formation of spores directly from the hyphal tips (Adams et al. 1988), suggesting that *brlA* is sufficient to induce the transition from polar hyphal growth to asexual sporulation.

As a key TF, BrlA is necessary for activating the expression of development-related genes during the early phase of conidiation (Adams et al. 1988; Boylan et al. 1987; Clutterbuck 1990; Mirabito et al. 1989; Stringer et al. 1991; Timberlake 1991). The *brlA* mutants fail to accumulate transcripts of certain development-specific genes, including *abaA*, *wetA*, *rodA*, and *yA* (Birse and Clutterbuck 1991; Boylan et al. 1987; Mayorga and Timberlake 1990; Stringer et al. 1991). BrlA contains two C_2H_2 zinc-finger motifs (Adams et al. 1988), which are important for DNA binding and BrlA TF activity (Adams et al. 1990). Mutation in either one of the two motifs results in a complete loss of BrlA activity (Adams et al. 1990). This central developmental TF interacts with the **BrlA response elements** (BREs; 5′-(C/A)(G/A)AGGG(G/A)-3′) and activates mRNA expression of multiple developmental genes (Chang and Timberlake 1993). Chang and Timberlake demonstrated that *brlA* expression in *Saccharomyces cerevisiae* caused activation of the *brlA*-dependent gene, and multiple BREs were present in the promoter regions of several developmentally regulated genes in *A. nidulans* (Chang and Timberlake 1993).

Expression of *brlA* is subject to complex regulation. The *brlA* gene consists of two overlapping transcripts, designated *brlA*α and *brlA*β, which are regulated at different levels and are individually required for proper development (Han and Adams 2001; Han et al. 1993; Prade and Timberlake 1993). The *brlA*β transcript encodes two open reading frames (ORFs), *brlA*β ORF, and a short upstream ORF (*brlA*β μORF) (Han et al. 1993). Expression of *brlA*α is only regulated at the transcriptional level and requires both BrlA and AbaA (Han et al. 1993). Expression of *brlA*β is regulated at both the transcriptional and translational levels. For transcriptional control, there are multiple *cis*-acting regulatory sequences involved in the activation or repression of *brlA*β mRNA (Han and Adams 2001). In vegetative cells, translation of the *brlA*β μORF represses BrlA translation and thereby prevents premature development (Han and Adams 2001; Han et al. 1993).

b) AbaA

AbaA is a presumed TF that is induced by BrlA during the middle phases of asexual development after differentiation of metulae and is required for proper formation and function of phialides (Boylan et al. 1987; Sewall et al. 1990a). The *abaA* null mutants bear an abacus-like structure with swellings at intervals in place of chains of conidia, suggesting its possible role in cytokinesis (Clutterbuck 1969; Sewall et al. 1990a). Non-sporulating conidiophores of *abaA* mutants are differentiated from sterigmata but do not form conidiogenous phialides, suggesting that *abaA* is required for proper function of phialides as sporogenous cells. The overexpression of *abaA* in vegetative cells results in growth termination and cellular vacuolization but no spore formation (Adams and Timberlake 1990; Mirabito et al. 1989).

AbaA is a member of TEF-1 (Transcriptional Enhancer Factor-1) family that contains an ATTS (AbaA, TEC1p, TEF-1 sequence)/TEA DNA-binding domain with a potential leucine zipper for dimerization (Andrianopoulos and Timberlake 1991, 1994; Mirabito et al. 1989). Results from a gel shift assay showed that AbaA interacts with the consensus sequence 5′-CATTCY-3′ (**AbaA response element** (ARE), where Y is a T or C) (Andrianopoulos and Timberlake 1994). One or multiple AREs are present in the promoter regions of several developmental genes such as *yA*, *rodA*, *wA*, and *wetA*, the upstream gene *brlA* and *abaA* itself (Andrianopoulos and Timberlake 1994) (Aramayo and Timberlake 1993; Mayorga and Timberlake 1990; Mirabito et al. 1989; Stringer et al. 1991). Further studies demonstrated that AbaA regulates the chitin synthase gene *chsC* (Ichinomiya et al. 2005; Park et al. 2003), the *velvet* genes *vosA* and *velB* (Ni and Yu 2007; Park et al. 2012b), and a component of the axial bud site marker Axl2 (Si et al. 2012) during conidiation.

Genetic interaction between *abaA* and *brlA* is complicated (Aguirre 1993; Aguirre et al. 1990). The overexpression of *abaA* in hyphae induces expression of *brlA*, whereas the *abaA* null mutant also activates *brlA* expression (Aguirre 1993), indicating that AbaA acts as both an activator and a repressor of *brlA*. Further studies proposed that AbaA exerts its repressive role in *brlA* expression via activating VosA, a key feedback negative regulator of *brlA* (Han and Adams 2001; Ni and Yu 2007).

c) WetA

The *wetA* gene is activated by sequential expression of *brlA* and *abaA* during asexual sporulation, and WetA is essential for the modification of the conidial wall resulting in the impermeable and resilient conidia (Marshall and Timberlake 1991; Sewall et al. 1990b). The phenotype of *wetA* null mutants is described as "wet-white" because these mutants produce colorless and autolytic conidia (Clutterbuck 1969; Sewall et al. 1990b). Conidia of the *wetA* null mutant lack both the condensation of the C2 wall layer and the formation of C3 and C4 layers (Sewall et al. 1990b). The *wetA* gene is sufficient for the activation of the conidium-specific gene, including *wA*, and the *wetA* null mutant fails to accumulate mRNAs of conidium-specific genes, indicating that WetA may act as a (transcriptional) regulator of conidium-specific genes (Marshall and Timberlake 1991; Sewall et al. 1990b). The overexpression of *wetA* in hyphae inhibits hyphal growth and causes excessive branching without causing the activation of *brlA* and *abaA* expression or the induction of precocious conidiation (Marshall and Timberlake 1991). Overall, WetA is required for spore maturation and is proposed to be an activator of a set of conidium-specific genes.

d) StuA and MedA

StuA (stunted) and MedA (medusa) function as developmental modifiers that are necessary for the precise organization of conidiophores (Adams et al. 1998). The StuA is a sequence-specific TF, which contains the APSES domain (Dutton et al. 1997). Mutational inactivation of *stuA* causes production of abnormal conidiophores with the lack of metulae and phialides.

However, some *stuA* mutants can produce viable conidia directly from vesicle (Clutterbuck 1969). The *stuA* gene encodes two transcription units, *stuAα* and *stuAβ*, and transcription of both mRNAs dramatically increases during the establishment of developmental competence (Miller et al. 1991, 1992). StuA is required for proper activation of *brlA* and repression of *abaA* (Dutton et al. 1997). Apposite *stuA* expression is required for the proper conidiophore morphology, but StuA may not affect the temporal development of cell types (Wu and Miller 1997). Mutational inactivation of *medA* results in aberrant conidiophores with multiple layers of sterigmata before conidia formation (Clutterbuck 1969). While StuA is essential for spatial expression of *brlA*, MedA is required for proper temporal expression of *brlAα* and *brlAβ*. In addition, MedA acts as a coactivator required for *abaA* expression (Busby et al. 1996).

2. Controllers of the Central Regulators

a) FluG-Mediated Signaling Pathway

Developmental transition from vegetative growth is highly complex and involves various positive elements in response to internal and external cues (Fig. 1.2) (Adams et al. 1998). Two decades ago, Wieser et al. carried out genetic analyses of recessive mutations and identified six upstream developmental activators, including *fluG*, *flbA*, *flbB*, *flbC*, *flbD*, and *flbE* (Wieser et al. 1994). The deletion of any one of these genes leads to the fluffy phenotypes (cotton-like colonies) coupled with decreased *brlA* expression (Adams et al. 1998; Wieser et al. 1994). A series of follow-up studies has revealed the molecular mechanisms and genetic relationships of these genes and proposed a model for the roles of the upstream developmental activators in conidiation (Adams et al. 1998; Etxebeste et al. 2010; Park and Yu 2012; Yu 2010).

FluG contains a glutamine synthetase I like domain and is a key activator of asexual development in *A. nidulans* (Lee and Adams 1994). The *fluG* deletion mutant cannot produce normal conidiophores under standard culture conditions, whereas the overexpression of *fluG* causes conidiophore formation and *brlA* activation in liquid submerged culture (D'Souza et al. 2001; Lee and Adams 1996). Lee and Adams proposed that FluG is required for the synthesis of an extracellular sporulation inducing factor (ESIF, also known as the FluG factor), which signals the activation of conidiophore development (Lee and Adams 1994). Later studies by Rodriguez-Urra and colleagues have revealed that the FluG factor is a diorcinol-dehydroaustinol adduct, which can rescue the conidiation defects caused by the deletion of *fluG* (Rodriguez-Urra et al. 2012). The FluG-initiated developmental signaling then leads to two independent regulatory processes: (1) inhibition of hyphal growth via activation of FlbA and (2) activation of developmental-specific regulatory cascades (FlbB–FlbE) (Adams et al. 1998). Further genetic studies have revealed that both FluG-mediated processes occur via the removal of repression of conidiation imposed by SfgA (Seo et al. 2003, 2006). SfgA is an upstream repressor of conidiation with a Gal4-type $Zn(II)_2Cys_6$ binuclear DNA-binding domain. Double mutant analyses indicate that SfgA functions downstream of FluG but upstream of FlbA–FlbD and BrlA (Seo et al. 2006).

FluG-initiated developmental-specific regulatory cascades (FlbB–FlbE) consist of at least two separated pathways, FlbC and FlbE/FlbB/FlbD, for activation of *brlA* (Etxebeste et al. 2010; Park and Yu 2012). FlbC is a C_2H_2 zinc-finger protein that directly binds to the promoter region of *brlA* and activates *brlA* expression (Kwon et al. 2010a). Kwon et al. also demonstrated that FlbC also binds to the *cis*-element of *abaA* and *vosA* and activates their expression (Kwon et al. 2010a). FlbE contains two uncharacterized conserved domains and co-localizes with FlbB, a basic leucine zipper (b-zip) TF, at the hyphal tip (Etxebeste et al. 2009; Garzia et al. 2009; Kwon et al. 2010b). FlbE and FlbB together induce *flbD* expression interdependently (Garzia et al. 2009; Kwon et al. 2010b). Then, FlbB cooperates with

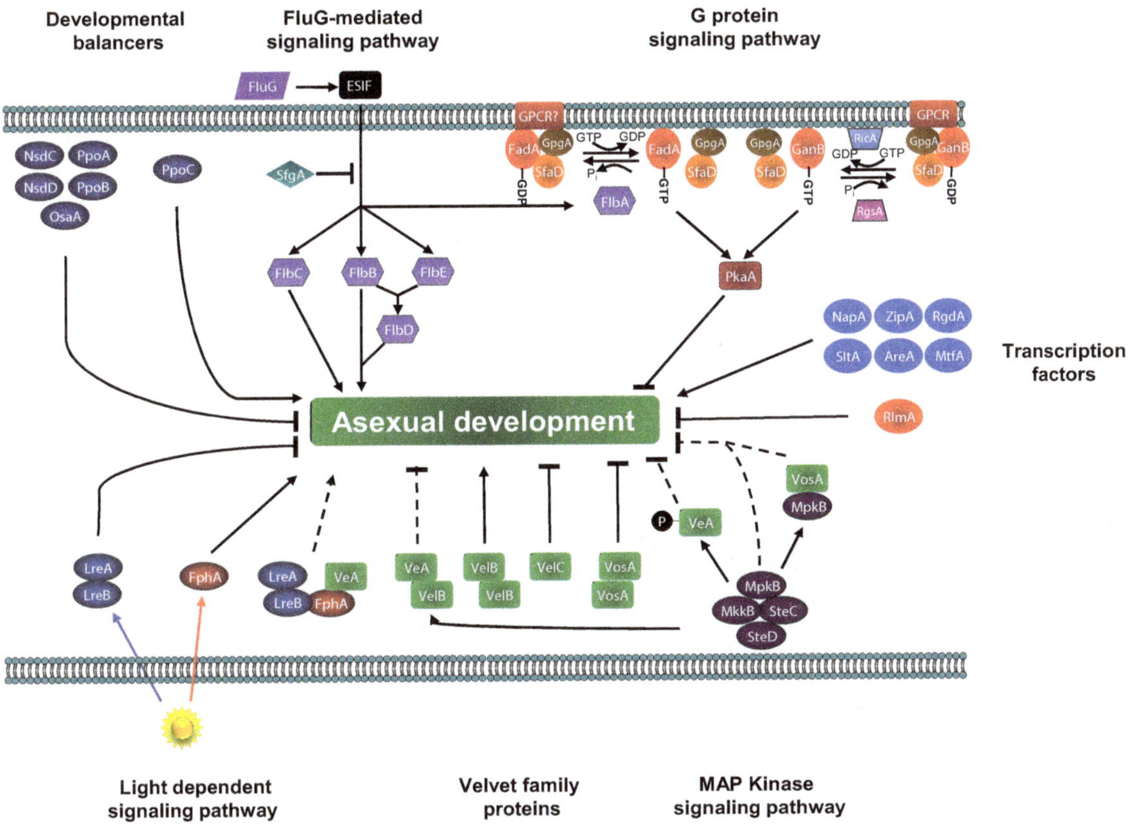

Fig. 1.2 A comprehensive model for the regulation of asexual sporulation in *A. nidulans*. This model includes almost all elements that influence conidiation

FlbD (a cMyb-type transcription factor) for *brlA* activation (Garzia et al. 2010). Overall, FlbB~E play independent or interdependent roles and are necessary for full activation of *brlA*.

In filamentous fungi, FlbA is the first studied regulator of G-protein signaling (RGS) protein (Lee and Adams 1994; Yu et al. 1996). RGS proteins are a group of proteins containing a conserved ~130 amino acid RGS box that interacts with an activated GTP-Gα subunit and increases its intrinsic GTPase activity, thereby facilitating the hydrolysis of GTP-Gα (active form) to GDP-Gα (inactive form) (Chidiac and Roy 2003; McCudden et al. 2005). FlbA is the cognate RGS of the FadA Gα protein, and together they govern the upstream regulation of vegetative growth, development, and biosynthesis of secondary metabolites (Hicks et al. 1997; Tag et al. 2000; Yu and Keller 2005; Yu et al. 1996). As an RGS, FlbA is presumed to enhance the intrinsic GTPase activity of FadA (Yu et al. 1996, 1999). Loss of *flbA* function results in a similar fluffy autolytic phenotype without sporulation caused by dominant mutations in FadA (Hicks et al. 1997; Lee and Adams 1994; Yu et al. 1996).

b) Heterotrimeric G Protein Signaling Pathways

In *A. nidulans*, two main G heterotrimeric G protein signaling pathways involving FadA and GanB Gα subunits play a predominant role in controlling growth, development, stress response, and secondary metabolism (Yu 2006). For more information on G protein signaling pathways, please refer to Chap. 7. FadA (fluffy autolytic dominant A) was identified and characterized by studying dominant activating mutants (G42R, R178L, G183S, R178C, and Q204L) that exhibit the fluffy autolytic phenotype (Wieser et al. 1994; Yu et al. 1996, 1999). Dominant interfering mutation (G203R) in FadA caused hyperactive sporulation and reduced hyphal growth (Hicks et al. 1997; Yu et al. 1996). This FadA-mediated signaling needs to be attenuated by FlbA (RGS) for proper asexual and sexual development to occur (Hicks et al. 1997; Wieser et al. 1997). Similar to FadA, GanB (G protein alpha subunit in *A. nidulans*) negatively controls asexual development (Chang et al. 2004). While the *ganB* null or dominant interfering (G207R) mutants produce conidiophores in liquid submerged culture, constitutively active GanB mutations (Q208L and R182L) cause severe defects in asexual development (Chang et al. 2004).

GanB-mediated signaling is in part activated by the putative GDP/GTP exchange factor, RicA (Kwon et al. 2012), and is negatively controlled by RgsA (Han et al. 2004a). These two Gα subunits work with the SfaA(Gβ): GpgA(Gγ) dimer (Lafon et al. 2005; Rosen et al. 1999; Seo et al. 2005) and inhibit asexual development in part via the cyclic AMP (cAMP)-dependent protein kinase PkaA (Shimizu and Keller 2001). Recently, Kong and colleagues characterized the Gβ-like protein B CpcB that is required for proper conidiation and *brlA* expression in *A. nidulans* (Kong et al. 2013). Overall, these results indicate that individual G protein components are associated with asexual development and may play differential roles in conidiation (Yu 2006).

c) MAP Kinase Signaling Pathways

Mitogen-activated protein kinase (MAPK) pathways respond to various environmental cues and amplify the signals, leading to appropriate cellular responses in fungi (Gustin et al. 1998). *A. nidulans* contains four MAPK genes: *mpkA*, *mpkB*, *mpkC*, and *hogA* (Galagan et al. 2005). Among these, MpkB, a homologue of Fus3p in *S. cerevisiae*, is associated with conidiation, sexual development, and secondary metabolism (Atoui et al. 2008; Bayram et al. 2012; Jun et al. 2011; Kang et al. 2013; Paoletti et al. 2007). The MAPK kinase kinase SteC, a component of the MAP kinase module Ste11–Ste7–Fus3 complex, is also required for proper formation of conidiophore (Wei et al. 2003). The deletion of *mpkB* causes increased *brlA* expression and the formation of conidiophores in liquid submerged cultures (Kang et al. 2013). Moreover, the Δ*mpkB* mutant produces abnormal conidiophores and exhibited a decreased number of conidia (Bayram et al. 2012; Jun et al. 2011; Kang et al. 2013). Interestingly, the deletion of *mpkB* results in decreased VeA phosphorylation and formation of the VeA–VelB heterodimer, which acts as a key and essential activator of sexual development (Bayram et al. 2012). MpkB also interacts with VosA in the nucleus. These results suggest that MpkB is also involved in the regulation of the *velvet* proteins' activities, thereby drastically affecting fungal development (Bayram et al. 2012). In-depth information on MAP kinase pathways is found in Chap. 6.

d) The *Velvet* Family Proteins

The *velvet* family proteins, mainly VeA, VelB, VelC, and VosA, play a central and global role in coordinating fungal growth, development, virulence, and metabolism in ascomycetes and basidiomycetes (Bayram and Braus 2012; Ni and Yu 2007). These proteins define a new fungi-specific TF class that contains the *velvet* domain with DNA-binding activity (Ahmed et al. 2013; Ni and Yu 2007). Interestingly, the *velvet* regulators form diverse complexes that play differential roles in coordinating various biological processes in fungi (Bayram et al. 2008; Park et al. 2012b, 2014; Sarikaya Bayram et al. 2010). VeA is the founding member of the *velvet* regulators playing a key role in the light-dependent developmental control. The *veA1* mutation abolishes the light affected developmental changes and causes severely restricted sexual development with increased conidiation (Kafer 1965; Mooney and Yager 1990). Further studies demonstrated that the nuclear localization and complex formation of VeA are important for the regulation of light-controlled development (Bayram et al. 2008; Stinnett et al. 2007). In the dark, VeA predominantly localizes in the nucleus and forms the VeA–VelB heterodimer, which controls sexual development or the VelB–VeA–LaeA heterotrimeric complex that regulates secondary metabolism (Bayram et al. 2008). VeA also interacts with the light complex components, which balance asexual and sexual development (Purschwitz et al. 2008, 2009). Recently, Bayram et al. demonstrated that phosphorylation of VeA is regulated by the MAP kinase MpkB (AnFus3) (Bayram et al. 2012). Similar to VeA, VelB is required for sexual development and production of certain secondary metabolites (Bayram et al. 2008). In addition, VelB functions as a positive regulator of conidiation (Park et al. 2012b). The *velB* deletion mutant exhibits the decreased conidiospore number, whereas the overexpression of *velB* results in enhanced conidial production (Park et al. 2012b). Another multifunctional regulator VosA functions as a key repressor of conidiophore formation and an essential activator of trehalose biogenesis in conidia (Ni and Yu 2007). VosA directly binds to the promoter of *brlA* and represses *brlA* expression in vegetative cells (Ahmed et al. 2013; Ni and Yu 2007). VosA physically interacts with VelB, VosA, or VelC, and each complex is presumed to control different sets of genes thereby regulating various aspects of fungal development (Park et al. 2012b, 2014; Sarikaya Bayram et al. 2010). VelC is important for balancing asexual and sexual development, and acts as an activator of sexual development, likely by binding to VosA during the early stages of sexual development thereby leading to the increased formation of VelB–VeA that is essential for the initiation of sexual fruiting (Park et al. 2014). As a consequence, VelC indirectly inhibits asexual development in *A. nidulans*.

e) Light and Signals

Light is a key environmental cue affecting diverse cellular processes, including fungal development and secondary metabolism in *A. nidulans* (Bayram et al. 2010; Mooney and Yager 1990). For example, light induces conidiation but represses sexual development and ST production (Bayram et al. 2010). In fungi, several photoreceptors play a crucial role in light-response processes (Bayram et al. 2010). The red-light sensor FphA is a **fungal phytochrome** that stimulates asexual development (Blumenstein et al. 2005). The deletion of *fphA* causes reduced mRNA accumulation of the *brlA* and *fluffy* genes and decreased conidial production (Blumenstein et al. 2005; Ruger-Herreros et al. 2011). LreA and LreB are main components of the blue light-sensing system, and they form complexes with FphA (Purschwitz et al. 2008). The deletion of *lreA* or *lreB* causes slightly increased conidiospore production in light and dark conditions, suggesting that LreA and LreB repress asexual development (Purschwitz et al. 2008). These sensors form the photoreceptor complex LreA/LreB/FphA/VeA, which activates the accumulation of the *brlA* and *fluffy* genes (Ruger-Herreros et al. 2011).

f) Developmental Balancers

Balance between asexual and sexual development is regulated by various factors. First, certain **sexual developmental activators** negatively control *brlA* expression and asexual development. For instance, NsdC and NsdD were initially identified as sexual activators with DNA-binding domains (Han et al. 2001; Kim et al. 2009; Lee et al. 2014). The deletion of one of them results in the absence of sexual fruiting bodies, whereas the overexpression of these genes causes enhanced formation of Hűlle cells (Han et al. 2001; Kim et al. 2009; Lee et al. 2014). Further studies have revealed that NsdC and NsdD repress *brlA* expression and conidiophore formation (Kim et al. 2009; Lee et al. 2014). Genetic analyses indicate that NsdD functions downstream of FlbE/B/D/C and upstream of *brlA*. Second, **Psi factors** (precocious sexual inducers), derived from oleic, linoleic, and linolenic acids, control the balance between asexual and sexual development (Bayram and Braus 2012; Dyer and O'Gorman 2012). In *A. nidulans*, the three oxylipin biosynthetic genes *ppoA*, *ppoB*, and *ppoC* (psi factor producing oxygenase) are present (Brodhun and Feussner 2011; Tsitsigiannis and Keller 2007; Tsitsigiannis et al. 2004, 2005). The deletion of *ppoA* or *ppoB* causes increased conidial production, suggesting that PpoA and PpoB negatively affect asexual development (Tsitsigiannis et al. 2004, 2005). However, the deletion of *ppoC* leads to decreased asexual sporulation, suggesting that PpoC positively influences conidiation, and has antagonistic activity to PpoA and PpoB (Tsitsigiannis et al. 2004, 2005). Third, OsaA, an ortholog of Wor1 in *Candida albicans*, has been identified by a gain-of-function genetic screen and is an orchestrator of sexual and asexual development (Ni and Yu

2007). The deletion of *osaA* causes enhanced sexual fruiting with reduced conidiation, suggesting that OsaA acts as a repressor of sexual development and indirectly affects asexual development in a positive way.

g) Other Transcription Factors

A diverse range of other TFs have been shown to influence asexual development in *A. nidulans* (Fig. 1.2). RgdA (retarded growth and development) is a putative APSES TF, which is required for proper conidial production and *brlA* expression. The *rgdA* deletion mutant exhibits a reduced number of conidia, and conidiophores of this mutant show irregular shaped phialides. In addition, the deletion of *rgdA* results in decreased mRNA accumulation of *brlA* and *abaA*, suggesting that RgdA may act as an upstream regulator of central regulatory genes (Lee et al. 2013). MtfA with a C_2H_2 zinc-finger domain is termed as a master TF regulating secondary metabolism and differentiation (Ramamoorthy et al. 2013). The absence of *mtfA* results in a drastic reduction of conidiophore formation, conidial production, and expression of *brlA*. The zinc-finger TF SltA mainly plays a role in cation homeostasis (Shantappa et al. 2013). Conidia production and *brlA* expression were reduced in the *sltA* deletion mutant, suggesting that *sltA* is required for normal conidiation. RlmA containing a MADS-box domain is a major MpkA-dependent TF, which regulates cell wall integrity signaling (Fujioka et al. 2007; Kovacs et al. 2013). The deletion of *rlmA* results in an increased number of conidiospores and *brlA* accumulation in surface cultures. In addition, the *rlmA* deletion mutant produces conidiophore in liquid submerged culture, where WT strains do not. Kovács et al. proposed that *sfgA*, *flbB*, and *flbE*, upstream regulators of *brlA*, are putative target genes of RlmA, indicating that RlmA indirectly regulates *brlA* expression (Kovacs et al. 2013). Two basic leucine zipper (bZIP) transcription factors, NapA and ZipA, affect asexual and sexual development. The overexpression of *napA* or *zipA* causes increased conidial production but decreased sexual spore production (Yin et al. 2013). AreB is a putative GATA zinc-finger TF containing a leucine zipper motif at N-terminal region and acts as a negative regulator of nitrogen catabolism (Conlon et al. 2001). The *areB* deletion mutant exhibits reduced growth and conidiation, suggesting that AreB is crucial for normal growth and conidiation (Wong et al. 2009).

3. Feedback Regulators of Conidiation

Along with the formation of conidiospores, expression of the key developmental activator *brlA* is repressed, and many other spore-specific genes are induced (Adams et al. 1998). As mentioned, AbaA is required for the repression of *brlA* through the control of *brlA*β (Han and Adams 2001). However, AbaA indirectly represses *brlA* expression, because the promoter of *brlA* does not contain AbaA response element (Han and Adams 2001). Recent genetic analyses demonstrated that the VosA–VelB complex is involved in AbaA-mediated repression of *brlA* (Ni and Yu 2007; Park et al. 2012b). During the conidiophore development, the VosA and VelB proteins are completely degraded. Then, in phialides, AbaA directly binds to the promoter regions of *vosA* and *velB* and induces their expression, thereby a large amount of the VosA and VelB proteins are newly synthesized in the developing cells (Park et al. 2012b). VosA, then, interacts with VelB, and the VosA–VelB heterocomplex binds to the *brlA* promoter and represses *brlA* expression in conidia (Ahmed et al. 2013). Moreover, the VosA–VelB complex regulates spore-specific genes, suggesting that the VosA–VelB dimeric complex acts as a key functional unit regulating spore maturation (trehalose biogenesis and spore wall synthesis), viability, and attenuating conidial germination in conidiospores (Ahmed et al. 2013; Park et al. 2012b).

III. Asexual Sporulation in *Penicillum marneffei*

A. Morphology of Asexual Structure

Penicillium marneffei is an emerging fungal pathogen and a dimorphic fungus, which

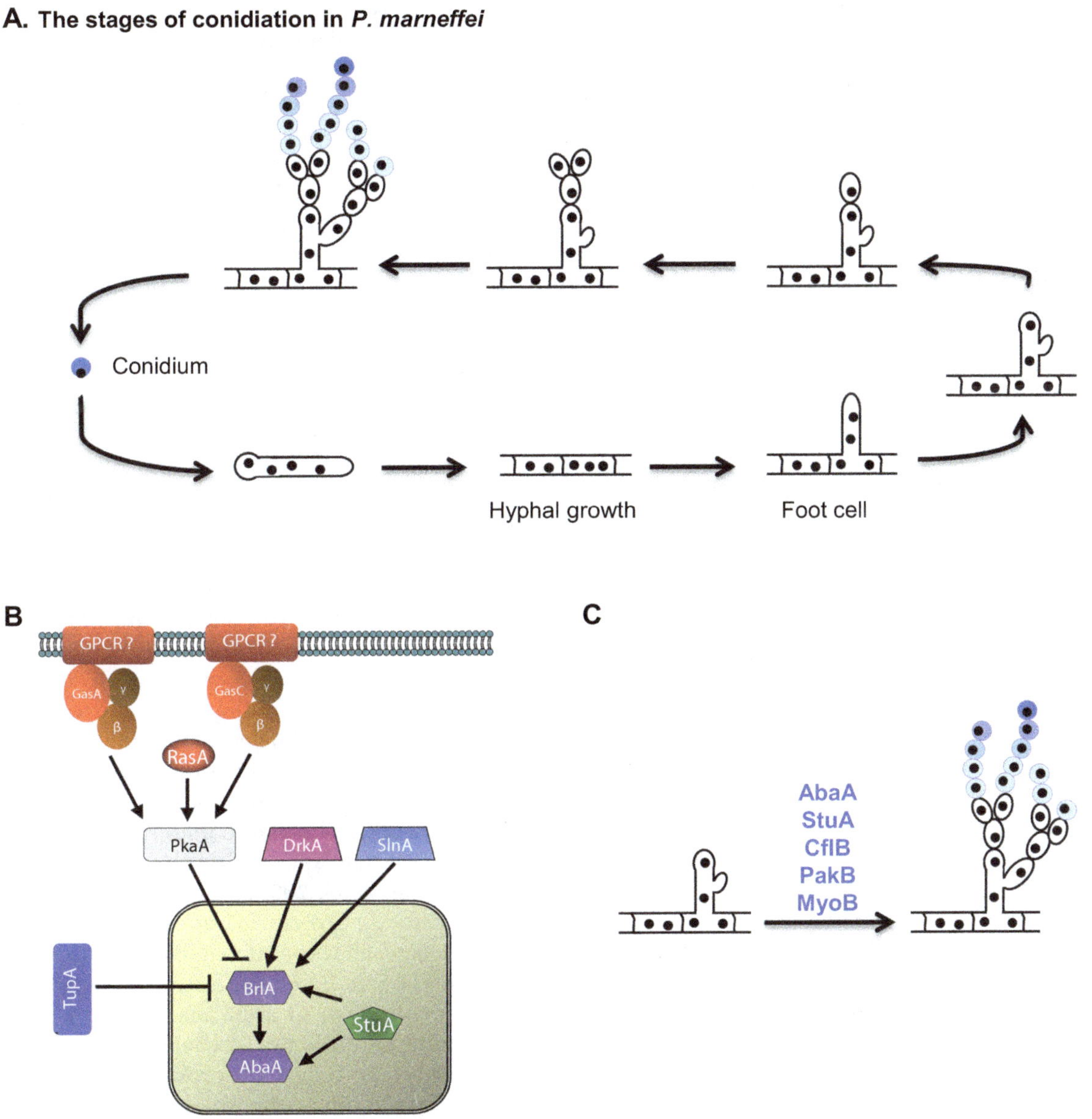

Fig. 1.3 Stages and regulation of conidiation in *P. marneffei*. (**a**) The stages of conidiation in *P. marneffei*. (**b**) Simplified model for the genetic regulation of conidiation in *P. marneffei*. (**c**) Genes involved in the conidiophore morphogenesis

grows in the yeast form at 37 °C and the filamentous form at 25 °C. In the filamentous phase, *P. marneffei* undergoes the complex asexual development and forms conidiophores (Andrianopoulos 2002; Pasricha et al. 2013). In response to inducing environmental signals, hyphal growth is arrested and the fungus starts to form asexual structures. Multinucleate stalk cells, formed from foot cells, are subsequently septated and form the subapical branches called rama. Two sterigmata, metulae and phialides, are formed from the apical cells. Finally, phialides produce conidia, resulting in the formation of brush-like structures called penicillin (Fig. 1.3a) (Andrianopoulos 2002; Roncal and Ugalde 2003; Vanittanakom et al. 2006).

B. Regulators of Asexual Development

The role of two central regulatory genes, *brlA* and *abaA*, is conserved in *A. nidulans* and *P. marneffei* (Borneman et al. 2000). The *brlA* deletion mutant in *P. marneffei* produces only conidiophore stalks but not conidia (Borneman et al. 2000; Boyce and Andrianopoulos 2013). AbaA in *P. marneffei* plays a similar role in asexual development (Borneman et al. 2000). The *abaA* deletion mutant can produce conidiophore stalks, rama and metulae, but not phialides and conidia. In addition, ectopic overexpression of *abaA* results in the formation of aberrant apical and multinucleate hyphal cells. These results indicate that AbaA is required for controlling cell cycle events during conidiation (Borneman et al. 2000).

As in *A. nidulans*, the central developmental pathway in *P. marneffei* is controlled by various regulatory inputs (Fig. 1.3b). First, **G protein signal pathways** are involved in the activation of conidiation. GasA, the homologue of FadA in *A. nidulans*, negatively regulates asexual development in *P. marneffei* (Zuber et al. 2002). The *gasA* deletion and dominant interfering $gasA^{G203R}$ mutant strains show elevated and precocious asexual development and *brlA* expression, whereas the dominant activating *gasA* mutant ($gasA^{G42R}$) fails to display conidiation and *brlA* expression in developmental cultures, suggesting a conserved cellular response to this particular G protein signal transduction. GasC, another G-protein alpha subunit, also plays a crucial role in germination and negatively controlling asexual development (Zuber et al. 2003). The small GTPase RasA acts downstream of heterotrimeric G proteins and negatively regulates the onset of conidiation (Boyce et al. 2005). The dominant-negative allele $RasA^{D125A}$ causes precocious initiation of conidiation, whereas the dominant positive $RasA^{G19V}$ allele causes reduced conidiation. The GasA-, GasC-, and RasA-mediated regulation of asexual development occurs via activation of the PKA signaling pathway (Boyce and Andrianopoulos 2007). Second, **TupA**, an ortholog of Tup1p in *S. cerevisiae*, acts as a **repressor of asexual development** (Todd et al. 2003). The deletion of *tupA* leads to premature conidiation and *brlA* expression, suggesting that TupA is required for proper control of the *brlA*-dependent conidiation. Third, the two-component histidine kinases DrkA and SlnA also play a crucial role in *brlA*-dependent asexual development (Boyce et al. 2011). The deletion of *slnA* or *drkA* results in delayed and reduced conidiation. The *drkA* deletion mutant also exhibits decreased expression of *brlA* mRNA. Interestingly, mRNA levels of *drkA* are increased in the Δ*brlA* mutant but reduced in the Δ*abaA* mutant, suggesting a potential feedback regulatory circuit (Boyce and Andrianopoulos 2013; Boyce et al. 2011).

Proper morphogenesis of conidiophore in *P. marneffei* is governed by multiple genes (Fig. 1.3c). StuA, a member of the APSES protein, is required for the formation of metula and phialide during conidiation (Borneman et al. 2002). The *stuA* deletion mutant fails to produce sterigmata cells, but can elaborate spores directly from the tips of stalks. The *cflB* gene, which encodes a Rho GTPase, is required for the polarized growth and cell division in both hyphal growth and conidiogenesis (Boyce et al. 2003, 2005). The *cflB* deletion mutant produces abnormal conidiophores with swollen and malformed cells. The dominant-negative $cflB^{D123A}$ mutant displays aberrant conidiophores with a single, terminal, and multinucleate conidium (Boyce et al. 2003, 2005). RfxA, an RFX (regulatory factor X) protein, is important for the cell division and checkpoint regulation with morphogenesis. Decreased expression of *rfxA* leads to the production of abnormal conidiophores containing multiple nuclei (Bugeja et al. 2010). MyoB is a type II myosin protein required for chitin deposition at the sites of cell division. The *myoB* deletion strain produces defective conidiophores, which lack clearly defined cell types due to malformed septa and faulty nuclear division (Canovas et al. 2011).

IV. Asexual Sporulation in *Fusarium graminearum*

A. Morphology of Asexual Structure

Fusarium graminearum (teleomorph *Gibberella zeae*) is a major plant pathogen that causes

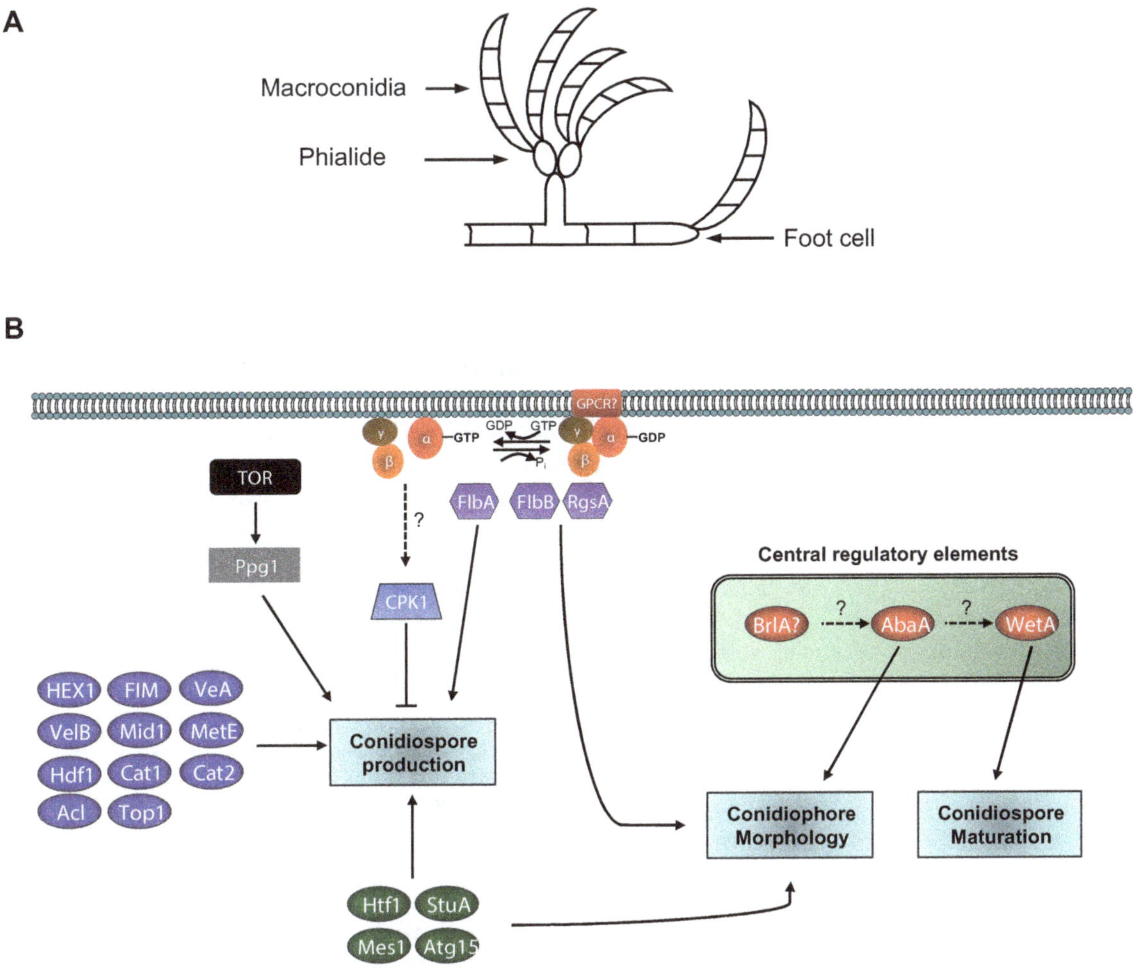

Fig. 1.4 Morphology and regulation of conidiation in *F. graminearum*. (**a**) Morphology of the asexual developmental structure in *F. graminearum*. (**b**) A simplified model for the genetic regulation of conidiation in *F. graminearum*

Fusarium head blight (FHB) in cereal crops, such as wheat, rice, and oats (Fernando et al. 1997; Goswami and Kistler 2004; Parry et al. 1995). This pathogenic fungus produces both asexual (conidia) and sexual (ascospore) spores, which can be major propagules causing FHB (Markell and Francl 2003). Different from *A. nidulans* and *P. marneffei*, *F. graminearum* directly produces phialides from the hyphae and continuously generates macroconidia or microconidia from phialides through a basipetal division (Leslie and Summerell 2006; Zheng et al. 2012) (Fig. 1.4a). Macroconidia are moderately curved containing multiple septa, whereas microconidia are formed on the simple conidiophores (Harris 2005; Leslie and Summerell 2006).

B. Regulators of Asexual Development

A number of regulators are associated with the asexual development in *F. graminearum* (Fig. 1.4b). First, *abaA* and *wetA*, the orthologs of those central regulators in *A. nidulans*, are required for conidiogenesis (Son et al. 2013a, 2014). The Δ*abaA* mutant strains produce abnormally shape phialides. The overexpression of *abaA* causes a reduction in the number of conidia and the formation of abacus-like phialides. Genomic studies have revealed that AbaA regulates several genes required for conidiation, suggesting a pivotal role of AbaA in asexual sporulation (Son et al. 2013a). WetA in *F. graminearum* is essential for conidiogenesis and maturation of conidia (Son et al. 2014). The deletion of *wetA* results in decreased number of conidia and the formation of abnormal conidiospores with longer and fewer septa. Conidia of the *wetA* deletion mutant are sensitive to various stresses and exhibit reduced long-term viability. In addition, the *wetA* deletion mutant contains numerous autophagic bodies in the conidium. These results indicate that WetA plays a crucial role in conidial dormancy by suppressing microcycle conidiation (Son et al. 2014).

Three RGS proteins are required for conidia morphology or production in *F. graminearum* (Park et al. 2012a). The deletion of *flbA* results in a significant reduction of conidia production. The *flbB* deletion mutant produces thinner and short conidia with few septa, whereas conidia of the *rgsA* deletion mutant show wider and longer conidia. The cAMP-PKA pathway plays a crucial role in growth and differentiation in *F. graminearum*. The deletion of *cpk1* encoding the main catalytic subunit of PKA, or the adenylate cyclase encoding gene *fac1* causes growth defects, indicating that the cAMP-PKA pathway is required for proper hyphal growth. In addition, the *cpk1* mutant produces phialides and conidia earlier than WT strain and the deletion of *cpk1* causes elevated mRNA expression of genes related to conidiation, suggesting that *cpk1* is negatively associated with conidiation (Hu et al. 2014). The target of rapamycin (TOR) signaling pathway also plays an important role in vegetative growth and differentiation in *F. graminearum* (Yu et al. 2014). Rapamycin has an inhibitory effect on fungal growth and asexual development. Moreover, the deletion of *ppg1* encoding a component of TOR signaling pathway causes decreased conidiophore production and impaired septum formation.

Several other proteins are required for proper production of asexual spores in *F. graminearum*. Both the deletion and overexpression of *hex1* encoding a hexagonal peroxisome protein result in the reduced production of conidia, suggesting that appropriate expression of HEX1 is important for controlling conidiogenesis (Son et al. 2013b). The actin bundling protein, Fim, plays a vital role in various cellular processes, and the *fim* deletion mutant exhibits reduced conidiation (Zheng et al. 2014). The *velvet* genes, *veA* and *velB*, act as repressors of conidia production (Jiang et al. 2011, 2012; Lee et al. 2012). Both the *veA* and *velB* deletion mutants show increased conidial production, and their conidia contain a large number of bulky lipid droplets. Additional proteins involved in conidiation include Mid1 (Stretch-activated ion channel) (Cavinder et al. 2011), MetE (Homoserine *O*-acetyltransferase) (Han et al. 2004b), HDF1 (Histone Deacetylase) (Li et al. 2011), CATs (Carnitine AcetylTransferases; CAT1 and CAT2) (Son et al. 2012), Acl (ATP Citrate Lyase) (Son et al. 2011), and Top1 (Topoisomerase I) (Baldwin et al. 2010).

A number of other regulators function in the production and morphogenesis of conidia. For instance, the homeobox TF Htf1 is required for phialidegenesis, conidiogenesis, and macroconidia basal cell division (Zheng et al. 2012). The deletion of *htf1* results in reduced conidia production and abolished macroconidia development, and this function is conserved in other *Fusarium* species. StuA, an APSES protein, acts as a master regulator controlling diverse processes in *F. graminearum*. The Δ*stuA* mutant fails to form conidiophores or phialides and produces aberrant macroconidia directly from the hyphae. The autophagy-related lipase Atg15 is also important for conidia formation and morphogenesis (Nguyen et al. 2011). The deletion of *atg15* leads to a reduced number of conidia and production of aberrantly shaped conidia. Mes1, a homologue of MesA in *A. nidulans* required for the formation of stable polarity axes, is also necessary for conidiogenesis in *F. graminearum*. The deletion of *mes1* leads to a reduction of asexual production and causes production of abnormal macroconidia (Rittenour and Harris 2008).

V. Conclusions

In this chapter, we have summarized current molecular biology of asexual sporulation in the three important filamentous fungal species. As discussed, several regulatory and signaling elements are conserved in three major fungal genera, and they play vital roles in various aspects of conidial production and morphogenesis of conidiophore. Further studies aimed at revealing the detailed molecular mechanisms of asexual sporulation in diverse fungal species will illuminate the common and distinct regulators and signaling cascades governing growth and development in fungi.

Acknowledgements We thank Dr. Ellin Doyle for critically reviewing the manuscript. This work was supported by the Intelligent Synthetic Biology Center of Global Frontier Project (2011-0031955) funded by the Ministry of Education, Science and Technology grants to JHY.

References

Adams TH, Timberlake WE (1990) Developmental repression of growth and gene expression in *Aspergillus*. Proc Natl Acad Sci USA 87:5405–5409

Adams TH et al (1988) *brlA* is necessary and sufficient to direct conidiophore development in *Aspergillus nidulans*. Cell 54:353–362

Adams TH et al (1990) *brlA* requires both zinc fingers to induce development. Mol Cell Biol 10:1815–1817

Adams TH et al (1998) Asexual sporulation in *Aspergillus nidulans*. Microbiol Mol Biol Rev 62:35–54

Aguirre J (1993) Spatial and temporal controls of the *Aspergillus brlA* developmental regulatory gene. Mol Microbiol 8:211–218

Aguirre J et al (1990) Spatial control of developmental regulatory genes in *Aspergillus nidulans*. Exp Mycol 14:290–293

Ahmed YL et al (2013) The *velvet* family of fungal regulators contains a DNA-binding domain structurally similar to NF-kappaB. PLoS Biol 11: e1001750

Andrianopoulos A (2002) Control of morphogenesis in the human fungal pathogen *Penicillium marneffei*. Int J Med Microbiol 292:331–347

Andrianopoulos A, Timberlake WE (1991) ATTS, a new and conserved DNA binding domain. Plant Cell 3:747–748

Andrianopoulos A, Timberlake WE (1994) The *Aspergillus nidulans abaA* gene encodes a transcriptional activator that acts as a genetic switch to control development. Mol Cell Biol 14:2503–2515

Aramayo R, Timberlake WE (1993) The *Aspergillus nidulans yA* gene is regulated by *abaA*. EMBO J 12:2039–2048

Atoui A et al (2008) *Aspergillus nidulans* natural product biosynthesis is regulated by *mpkB*, a putative pheromone response mitogen-activated protein kinase. Appl Environ Microbiol 74: 3596–3600

Baldwin TK et al (2010) A role for topoisomerase I in *Fusarium graminearum* and *F. culmorum* pathogenesis and sporulation. Mol Plant Microbe Interact 23:566–577

Bayram O, Braus GH (2012) Coordination of secondary metabolism and development in fungi: the *velvet* family of regulatory proteins. FEMS Microbiol Rev 36:1–24

Bayram O et al (2008) VelB/VeA/LaeA complex coordinates light signal with fungal development and secondary metabolism. Science 320:1504–1506

Bayram O et al (2010) Spotlight on *Aspergillus nidulans* photosensory systems. Fungal Genet Biol 47:900–908

Bayram O et al (2012) The *Aspergillus nidulans* MAPK module AnSte11-Ste50-Ste7-Fus3 controls development and secondary metabolism. PLoS Genet 8:e1002816

Birse CE, Clutterbuck AJ (1991) Isolation and developmentally regulated expression of an *Aspergillus nidulans* phenol oxidase-encoding gene, *ivoB*. Gene 98:69–76

Blumenstein A et al (2005) The *Aspergillus nidulans* phytochrome FphA represses sexual development in red light. Curr Biol 15:1833–1838

Borneman AR et al (2000) The *abaA* homologue of *Penicillium marneffei* participates in two developmental programmes: conidiation and dimorphic growth. Mol Microbiol 38:1034–1047

Borneman AR et al (2002) A basic helix-loop-helix protein with similarity to the fungal morphological regulators, Phd1p, Efg1p and StuA, controls conidiation but not dimorphic growth in *Penicillium marneffei*. Mol Microbiol 44:621–631

Boyce KJ, Andrianopoulos A (2007) A p21-activated kinase is required for conidial germination in *Penicillium marneffei*. PLoS Pathog 3:e162

Boyce KJ, Andrianopoulos A (2013) Morphogenetic circuitry regulating growth and development in the dimorphic pathogen *Penicillium marneffei*. Eukaryot Cell 12:154–160

Boyce KJ et al (2003) Control of morphogenesis and actin localization by the *Penicillium marneffei* RAC homolog. J Cell Sci 116:1249–1260

Boyce KJ et al (2005) The Ras and Rho GTPases genetically interact to co-ordinately regulate cell polarity during development in *Penicillium marneffei*. Mol Microbiol 55:1487–1501

Boyce KJ et al (2011) The two-component histidine kinases DrkA and SlnA are required for in vivo growth in the human pathogen *Penicillium marneffei*. Mol Microbiol 82:1164–1184

Boylan MT et al (1987) Isolation and physical characterization of three essential conidiation genes from *Aspergillus nidulans*. Mol Cell Biol 7:3113–3118

Brodhun F, Feussner I (2011) Oxylipins in fungi. FEBS J 278:1047–1063

Bugeja HE et al (2010) The RFX protein RfxA is an essential regulator of growth and morphogenesis in *Penicillium marneffei*. Eukaryot Cell 9: 578–591

Busby TM et al (1996) Suppression and enhancement of the *Aspergillus nidulans medusa* mutation by altered dosage of the bristle and stunted genes. Genetics 143:155–163

Canovas D et al (2011) The fungal type II myosin in *Penicillium marneffei*, MyoB, is essential for chitin deposition at nascent septation sites but not actin localization. Eukaryot Cell 10:302–312

Cavinder B et al (2011) Mid1, a mechanosensitive calcium ion channel, affects growth, development, and ascospore discharge in the filamentous fungus *Gibberella zeae*. Eukaryot Cell 10:832–841

Chang YC, Timberlake WE (1993) Identification of *Aspergillus brlA* response elements (BREs) by genetic selection in yeast. Genetics 133:29–38

Chang MH et al (2004) The GanB Galpha-protein negatively regulates asexual sporulation and plays a positive role in conidial germination in *Aspergillus nidulans*. Genetics 167:1305–1315

Chidiac P, Roy AA (2003) Activity, regulation, and intracellular localization of RGS proteins. Receptors Channels 9:135–147

Clutterbuck AJ (1969) A mutational analysis of conidial development in *Aspergillus nidulans*. Genetics 63:317–327

Clutterbuck AJ (1990) The genetics of conidiophore pigmentation in *Aspergillus nidulans*. J Gen Microbiol 136:1731–1738

Conlon H et al (2001) The *Aspergillus nidulans* GATA transcription factor gene *areB* encodes at least three proteins and features three classes of mutation. Mol Microbiol 40:361–375

D'Souza CA et al (2001) Characterization of the role of the FluG protein in asexual development of *Aspergillus nidulans*. Genetics 158:1027–1036

Dutton JR et al (1997) StuAp is a sequence-specific transcription factor that regulates developmental complexity in *Aspergillus nidulans*. EMBO J 16:5710–5721

Dyer PS, O'Gorman CM (2012) Sexual development and cryptic sexuality in fungi: insights from *Aspergillus* species. FEMS Microbiol Rev 36:165–192

Ebbole DJ (2010) The conidium. In: Borkovich KA, Ebbole DJ (eds) Cellular and molecular biology of filamentous fungi. ACM Press, Washington, pp 577–590

Etxebeste O et al (2009) The bZIP-type transcription factor FlbB regulates distinct morphogenetic stages of colony formation in *Aspergillus nidulans*. Mol Microbiol 73:775–789

Etxebeste O et al (2010) *Aspergillus nidulans* asexual development: making the most of cellular modules. Trends Microbiol 18:569–576

Fernando WG et al (1997) Head blight gradients caused by *Gibberella zeae* from area sources of inoculum in wheat field plots. Phytopathology 87:414–421

Fujioka T et al (2007) MpkA-Dependent and -independent cell wall integrity signaling in *Aspergillus nidulans*. Eukaryot Cell 6:1497–1510

Galagan JE et al (2005) Sequencing of *Aspergillus nidulans* and comparative analysis with *A. fumigatus* and *A. oryzae*. Nature 438:1105–1115

Garzia A et al (2009) *Aspergillus nidulans* FlbE is an upstream developmental activator of conidiation functionally associated with the putative transcription factor FlbB. Mol Microbiol 71:172–184

Garzia A et al (2010) The concerted action of bZip and cMyb transcription factors FlbB and FlbD induces *brlA* expression and asexual development in *Aspergillus nidulans*. Mol Microbiol 75:1314–1324

Goswami RS, Kistler HC (2004) Heading for disaster: *Fusarium graminearum* on cereal crops. Mol Plant Pathol 5:515–525

Gustin MC et al (1998) MAP kinase pathways in the yeast *Saccharomyces cerevisiae*. Microbiol Mol Biol Rev 62:1264–1300

Han S, Adams TH (2001) Complex control of the developmental regulatory locus *brlA* in *Aspergillus nidulans*. Mol Genet Genomics 266:260–270

Han S et al (1993) Translational repression of *brlA* expression prevents premature development in *Aspergillus*. EMBO J 12:2449–2457

Han KH et al (2001) The *nsdD* gene encodes a putative GATA-type transcription factor necessary for sexual development of *Aspergillus nidulans*. Mol Microbiol 41:299–309

Han KH et al (2004a) Regulators of G-protein signalling in *Aspergillus nidulans*: RgsA downregulates stress response and stimulates asexual sporulation through attenuation of GanB (Galpha) signalling. Mol Microbiol 53:529–540

Han YK et al (2004b) Functional analysis of the homoserine O-acetyltransferase gene and its identification as a selectable marker in *Gibberella zeae*. Curr Genet 46:205–212

Harris SD (2005) Morphogenesis in germinating *Fusarium graminearum* macroconidia. Mycologia 97:880–887

Hicks JK et al (1997) *Aspergillus* sporulation and mycotoxin production both require inactivation of the FadA G alpha protein-dependent signaling pathway. EMBO J 16:4916–4923

Hu S et al (2014) The cAMP-PKA pathway regulates growth, sexual and asexual differentiation, and pathogenesis in *Fusarium graminearum*. Mol Plant Microbe Interact 27:557–566

Ichinomiya M et al (2005) Expression of asexual developmental regulator gene *abaA* is affected in the double mutants of classes I and II chitin synthase genes, *chsC* and *chsA*, of *Aspergillus nidulans*. Curr Genet 48:171–183

Jiang J et al (2011) Involvement of a *velvet* protein FgVeA in the regulation of asexual development, lipid and secondary metabolisms and virulence in *Fusarium graminearum*. PLoS One 6:e28291

Jiang J et al (2012) *FgVELB* is associated with vegetative differentiation, secondary metabolism and virulence in *Fusarium graminearum*. Fungal Genet Biol 49:653–662

Jun SC et al (2011) The MpkB MAP kinase plays a role in post-karyogamy processes as well as in hyphal anastomosis during sexual development in *Aspergillus nidulans*. J Microbiol 49:418–430

Kafer E (1965) Origins of translocations in *Aspergillus nidulans*. Genetics 52:217–232

Kang JY et al (2013) The MpkB MAP kinase plays a role in autolysis and conidiation of *Aspergillus nidulans*. Fungal Genet Biol 61:42–49

Kim HR et al (2009) The *nsdC* gene encoding a putative C2H2-type transcription factor is a key activator of sexual development in *Aspergillus nidulans*. Genetics 182:771–783

Kong Q et al (2013) Gbeta-like CpcB plays a crucial role for growth and development of *Aspergillus nidulans* and *Aspergillus fumigatus*. PLoS One 8:e70355

Kovacs Z et al (2013) Effect of cell wall integrity stress and RlmA transcription factor on asexual development and autolysis in *Aspergillus nidulans*. Fungal Genet Biol 54:1–14

Kwon N-J et al (2010a) FlbC is a putative nuclear C2H2 transcription factor regulating development in *Aspergillus nidulans*. Mol Microbiol 77:1203–1219

Kwon N-J et al (2010b) Characterization of the developmental regulator FlbE in *Aspergillus fumigatus* and *Aspergillus nidulans*. Fungal Genet Biol 47:981–993

Kwon N-J et al (2012) The putative guanine nucleotide exchange factor RicA mediates upstream signaling for growth and development in *Aspergillus*. Eukaryot Cell 11:1399–1412

Lafon A et al (2005) The heterotrimeric G-protein GanB(alpha)-SfaD(beta)-GpgA(gamma) is a carbon source sensor involved in early cAMP-dependent germination in *Aspergillus nidulans*. Genetics 171:71–80

Lee BN, Adams TH (1994) The *Aspergillus nidulans fluG* gene is required for production of an extracellular developmental signal and is related to prokaryotic glutamine synthetase I. Genes Dev 8:641–651

Lee BN, Adams TH (1996) FluG and *flbA* function interdependently to initiate conidiophore development in *Aspergillus nidulans* through *brlA* beta activation. EMBO J 15:299–309

Lee J et al (2012) *FgVelB* globally regulates sexual reproduction, mycotoxin production and pathogenicity in the cereal pathogen *Fusarium graminearum*. Microbiology 158:1723–1733

Lee JY et al (2013) A putative APSES transcription factor is necessary for normal growth and development of *Aspergillus nidulans*. J Microbiol 51:800–806

Lee MK et al (2014) NsdD is a key repressor of asexual development in *Aspergillus nidulans*. Genetics 197:159–173

Leslie JF, Summerell BA (2006) *Fusarium* laboratory workshops-A recent history. Mycotoxin Res 22:73–74

Li Y et al (2011) The HDF1 histone deacetylase gene is important for conidiation, sexual reproduction, and pathogenesis in *Fusarium graminearum*. Mol Plant Microbe Interact 24:487–496

Markell SG, Francl LJ (2003) *Fusarium* head blight inoculum: species prevalence and *Gibberella zeae* spore type. Plant Dis 87:814–820

Marshall MA, Timberlake WE (1991) *Aspergillus nidulans wetA* activates spore-specific gene expression. Mol Cell Biol 11:55–62

Mayorga ME, Timberlake WE (1990) Isolation and molecular characterization of the *Aspergillus nidulans wA* gene. Genetics 126:73–79

McCudden CR et al (2005) G-protein signaling: back to the future. Cell Mol Life Sci 62:551–577

Miller KY et al (1991) Isolation and transcriptional characterization of a morphological modifier: the *Aspergillus nidulans* stunted (*stuA*) gene. Mol Gen Genet 227:285–292

Miller KY et al (1992) StuA is required for cell pattern formation in *Aspergillus*. Genes Dev 6: 1770–1782

Mims CW et al (1988) Ultrastructural analysis of conidiophore development in the fungus *Aspergillus nidulans* using freeze-substitution. Protoplasma 144:132–141

Mirabito PM et al (1989) Interactions of three sequentially expressed genes control temporal and spatial specificity in *Aspergillus* development. Cell 57:859–868

Mooney JL, Yager LN (1990) Light is required for conidiation in *Aspergillus nidulans*. Genes Dev 4:1473–1482

Nguyen LN et al (2011) Autophagy-related lipase FgATG15 of *Fusarium graminearum* is important for lipid turnover and plant infection. Fungal Genet Biol 48:217–224

Ni M, Yu J-H (2007) A novel regulator couples sporogenesis and trehalose biogenesis in *Aspergillus nidulans*. PLoS One 2:e970

Ni M et al (2010) Regulation of *Aspergillus* Conidiation. In: Borkovich K, Ebbole D (eds) Cellular and molecular biology of filamentous fungi. ACM Press, Washington, pp 559–576

Paoletti M et al (2007) Mating type and the genetic basis of self-fertility in the model fungus *Aspergillus nidulans*. Curr Biol 17:1384–1389

Park H-S, Yu J-H (2012) Genetic control of asexual sporulation in filamentous fungi. Curr Opin Microbiol 15:669–677

Park BC et al (2003) Activation of *chsC* transcription by AbaA during asexual development of *Aspergillus nidulans*. FEMS Microbiol Lett 220:241–246

Park AR et al (2012a) Functional analyses of regulators of G protein signaling in *Gibberella zeae*. Fungal Genet Biol 49:511–520

Park H-S et al (2012b) The role, interaction and regulation of the *velvet* regulator VelB in *Aspergillus nidulans*. PLoS One 7:e45935

Park H-S et al (2014) VelC positively controls sexual development in *Aspergillus nidulans*. PLoS One 9: e89883

Parry DW et al (1995) *Fusarium* ear blight (scab) in small-grain cereals – a review. Plant Pathol 44:207–238

Pasricha S et al (2013) Cell-type-specific transcriptional profiles of the dimorphic pathogen Penicillium marneffei reflect distinct reproductive, morphological, and environmental demands. G3 (Bethesda) 3:1997–2014

Prade RA, Timberlake WE (1993) The *Aspergillus nidulans brlA* regulatory locus consists of overlapping transcription units that are individually required for conidiophore development. EMBO J 12:2439–2447

Purschwitz J et al (2008) Functional and physical interaction of blue- and red-light sensors in *Aspergillus nidulans*. Curr Biol 18:255–259

Purschwitz J et al (2009) Mapping the interaction sites of *Aspergillus nidulans* phytochrome FphA with the global regulator VeA and the White Collar protein LreB. Mol Genet Genomics 281:35–42

Ramamoorthy V et al (2013) The putative C2H2 transcription factor MtfA is a novel regulator of secondary metabolism and morphogenesis in *Aspergillus nidulans*. PLoS One 8:e74122

Rittenour WR, Harris SD (2008) Characterization of *Fusarium graminearum* Mes1 reveals roles in cell-surface organization and virulence. Fungal Genet Biol 45:933–946

Rodriguez-Urra AB et al (2012) Signaling the induction of sporulation involves the interaction of two secondary metabolites in *Aspergillus nidulans*. ACS Chem Biol 7:599–606

Roncal T, Ugalde U (2003) Conidiation induction in *Penicillium*. Res Microbiol 154:539–546

Rosen S et al (1999) The *Aspergillus nidulans sfaD* gene encodes a G protein beta subunit that is required for normal growth and repression of sporulation. EMBO J 18:5592–5600

Ruger-Herreros C et al (2011) Regulation of conidiation by light in *Aspergillus nidulans*. Genetics 188:809–822

Sarikaya Bayram O et al (2010) LaeA control of *velvet* family regulatory proteins for light-dependent development and fungal cell-type specificity. PLoS Genet 6:e1001226

Seo JA et al (2003) Suppressor mutations bypass the requirement of *fluG* for asexual sporulation and sterigmatocystin production in *Aspergillus nidulans*. Genetics 165:1083–1093

Seo JA et al (2005) Multiple roles of a heterotrimeric G-protein gamma-subunit in governing growth and development of *Aspergillus nidulans*. Genetics 171:81–89

Seo JA et al (2006) FluG-dependent asexual development in *Aspergillus nidulans* occurs via derepression. Genetics 172:1535–1544

Sewall TC et al (1990a) *abaA* controls phialide differentiation in *Aspergillus nidulans*. Plant Cell 2:731–739

Sewall TC et al (1990b) Conidium differentiation in *Aspergillus nidulans* wild-type and *wet-white* (*wetA*) mutant strains. Dev Biol 138:499–508

Shantappa S et al (2013) Role of the zinc finger transcription factor SltA in morphogenesis and sterigmatocystin biosynthesis in the fungus *Aspergillus nidulans*. PLoS One 8:e68492

Shimizu K, Keller NP (2001) Genetic involvement of a cAMP-dependent protein kinase in a G protein signaling pathway regulating morphological and chemical transitions in *Aspergillus nidulans*. Genetics 157:591–600

Si H et al (2012) Morphogenetic and developmental functions of the *Aspergillus nidulans* homologues of the yeast bud site selection proteins Bud4 and Axl2. Mol Microbiol 85:252–270

Son H et al (2011) ATP citrate lyase is required for normal sexual and asexual development in *Gibberella zeae*. Fungal Genet Biol 48:408–417

Son H et al (2012) Mitochondrial carnitine-dependent acetyl coenzyme A transport is required for normal sexual and asexual development of the ascomycete *Gibberella zeae*. Eukaryot Cell 11:1143–1153

Son H et al (2013a) AbaA regulates conidiogenesis in the ascomycete fungus *Fusarium graminearum*. PLoS One 8:e72915

Son M et al (2013b) The *HEX1* gene of *Fusarium graminearum* is required for fungal asexual reproduction and pathogenesis and for efficient viral RNA accumulation of *Fusarium graminearum* virus 1. J Virol 87:10356–10367

Son H et al (2014) WetA is required for conidiogenesis and conidium maturation in the ascomycete fungus *Fusarium graminearum*. Eukaryot Cell 13:87–98

Stinnett SM et al (2007) *Aspergillus nidulans* VeA subcellular localization is dependent on the importin alpha carrier and on light. Mol Microbiol 63:242–255

Stringer MA et al (1991) Rodletless, a new *Aspergillus* developmental mutant induced by directed gene inactivation. Genes Dev 5:1161–1171

Tag A et al (2000) G-protein signalling mediates differential production of toxic secondary metabolites. Mol Microbiol 38:658–665

Timberlake WE (1990) Molecular genetics of *Aspergillus* development. Annu Rev Genet 24:5–36

Timberlake WE (1991) Temporal and spatial controls of *Aspergillus* development. Curr Opin Genet Dev 1:351–357

Todd RB et al (2003) TupA, the *Penicillium marneffei* Tup1p homologue, represses both yeast and spore development. Mol Microbiol 48:85–94

Tsitsigiannis DI, Keller NP (2007) Oxylipins as developmental and host-fungal communication signals. Trends Microbiol 15:109–118

Tsitsigiannis DI et al (2004) The lipid body protein, PpoA, coordinates sexual and asexual sporulation in *Aspergillus nidulans*. J Biol Chem 279:11344–11353

Tsitsigiannis DI et al (2005) Three putative oxylipin biosynthetic genes integrate sexual and asexual development in *Aspergillus nidulans*. Microbiology 151:1809–1821

Vanittanakom N et al (2006) *Penicillium marneffei* infection and recent advances in the epidemiology and molecular biology aspects. Clin Microbiol Rev 19:95–110

Wei H et al (2003) The MAPKK kinase SteC regulates conidiophore morphology and is essential for heterokaryon formation and sexual development in the homothallic fungus *Aspergillus nidulans*. Mol Microbiol 47:1577–1588

Wieser J et al (1994) Genetic requirements for initiating asexual development in *Aspergillus nidulans*. Curr Genet 27:62–69

Wieser J et al (1997) Dominant mutations affecting both sporulation and sterigmatocystin biosynthesis in *Aspergillus nidulans*. Curr Genet 32:218–224

Wong KH et al (2009) Deletion and overexpression of the *Aspergillus nidulans* GATA factor AreB reveals unexpected pleiotropy. Microbiology 155:3868–3880

Wu J, Miller BL (1997) *Aspergillus* asexual reproduction and sexual reproduction are differentially affected by transcriptional and translational mechanisms regulating stunted gene expression. Mol Cell Biol 17:6191–6201

Yin WB et al (2013) bZIP transcription factors affecting secondary metabolism, sexual development and stress responses in *Aspergillus nidulans*. Microbiology 159:77–88

Yu J-H (2006) Heterotrimeric G protein signaling and RGSs in *Aspergillus nidulans*. J Microbiol 44:145–154

Yu J-H (2010) Regulation of Development in *Aspergillus nidulans* and *Aspergillus fumigatus*. Mycobiology 38:229–237

Yu J-H, Keller N (2005) Regulation of secondary metabolism in filamentous fungi. Annu Rev Phytopathol 43:437–458

Yu J-H et al (1996) The *Aspergillus* FlbA RGS domain protein antagonizes G protein signaling to block proliferation and allow development. EMBO J 15:5184–5190

Yu J-H et al (1999) Extragenic suppressors of loss-of-function mutations in the aspergillus FlbA regulator of G-protein signaling domain protein. Genetics 151:97–105

Yu F et al (2014) The TOR signaling pathway regulates vegetative development and virulence in *Fusarium graminearum*. New Phytol 203:219–232

Zheng W et al (2012) A conserved homeobox transcription factor Htf1 is required for phialide development and conidiogenesis in *Fusarium* species. PLoS One 7:e45432

Zheng Z et al (2014) FgFim, a key protein regulating resistance to the fungicide JS399-19, asexual and sexual development, stress responses and virulence in *Fusarium graminearum*. Mol Plant Pathol 15:488–499

Zuber S et al (2002) G-protein signaling mediates asexual development at 25 degrees C but has no effect on yeast-like growth at 37 degrees C in the dimorphic fungus *Penicillium mameffei*. Eukaryot Cell 1:440–447

Zuber S et al (2003) The G-protein alpha-subunit GasC plays a major role in germination in the dimorphic fungus *Penicillium marneffei*. Genetics 164:487–499

2 Insight into Fungal Secondary Metabolism from Ten Years of LaeA Research

Jin Woo Bok[1], Nancy P. Keller[1]

CONTENTS

I. Introduction

Fungi are well known for their ability to produce copious numbers of bioactive small molecules known as natural products or secondary metabolites (SMs), the moniker used in this chapter. Since the discovery of penicillin in 1928 by Alexander Fleming, the number of partially or fully characterized fungal SMs has risen exponentially. The interest in fungal SMs lies primarily in their useful antibiotic and pharmaceutical activities, although several of these metabolites are also potent phytotoxins or mycotoxins, contributing adversely to plant, animal, and/or human health (Leitão and Enquita 2014). A literature survey of fungal metabolites, covering 1500 compounds that were isolated and characterized between 1993 and 2001, showed that more than half of the molecules had antibacterial, antifungal, or antitumor activity (Pelaez et al. 2005). In particular, certain members of the Ascomycetes and Basidiomycetes encode a large wealth of SMs that—as observed from genomes of sequenced fungi—remain largely untapped.

The first genetically characterized fungal SMs, the β-lactam antibiotics—penicillin and cephalosporins (Martin 1992) and the mycotoxins—aflatoxin and sterigmatocystin (Brown et al. 1996; Yu et al. 1995; Trail et al. 1995), revealed the near-universal clustered arrangement of genes involved in the production of a single SM. This clustering of fungal SM genes (reviewed in Hoffmeister and Keller 2007) has accelerated the ability to identify SM clusters in fungal genomes and led to the development of various bioinformatic algorithms, such as SMURF, antiSMASH, or MIDDAS-M (Khaldi et al. 2010; Medema et al. 2011; Umemura et al. 2013). While unable to predict intertwined superclusters containing genes for more than one SM (Wiemann et al. 2013) or account for genes outside of the cluster (Sanchez et al. 2011), these programs have greatly assisted in initial predictions of fungal SM gene clusters.

A major goal of studying SM is to understand how SM cluster genes are regulated. Some of the clusters contain cluster-specific transcription factors (e.g., AflR regulating expression of aflatoxin and sterigmatocystin clusters, Fernandes et al. 1998; Woloshuk et al. 1994) that, when activated naturally or through genetic manipulations, induce expression of other genes within the cluster (examples in Hoffmeister and Keller 2007; Brakhage 2013). Rarely, these types of in-cluster transcription

[1]Department of Medical Microbiology and Immunology, University of Wisconsin-Madison, Madison, WI, USA; e-mail: jwbok@wisc.edu; npkeller@wisc.edu

Biochemistry and Molecular Biology, 3rd Edition
The Mycota III
D. Hoffmeister (Ed.)

factors have been reported to regulate another SM cluster such as AflR regulation of the asperthecin gene cluster (Yin et al. 2012). Thus, the discovery of LaeA, capable of regulating multiple SM clusters simultaneously, was remarkable and recognized early on as a useful tool in SM sleuthing (Bok and Keller 2004; Bok et al. 2006a). Here, we present an overview of LaeA function in SM production in fungi, and in so doing, compile a list of all SMs currently known to be regulated by this protein. For a more thorough review of LaeA impact on other aspects of fungal development, we refer the reader to Jain and Keller (2013).

II. LaeA Mechanism

LaeA was identified through a chemical mutagenesis of an *Aspergillus nidulans* norsolorinic acid-producing strain. This compound is a visible orange precursor of both sterigmatocystin and aflatoxin, and loss of its production is easy to screen (Butchko et al. 1999). Twenty-three single gene mutants were obtained with LaeA representing one of three mutants showing loss of *aflR* expression. Chemical characterization of *ΔlaeA* showed a decrease not only in sterigmatocystin production but also in multiple secondary metabolites (Bok and Keller 2004). The number and types of LaeA regulated SMs in *A. nidulans* and other fungi are described in the next section.

A. Methyltransferase

LaeA contains an *S*-adenosyl methionine (SAM)-binding site that when mutated yields a null-LaeA phenotype (Bok and Keller 2004), presumably indicative of methyltransferase activity. However, to date, other than demonstrating automethylation at a methionine residue near the SAM-binding site, a modification which is not required for in vivo function (Patananan et al. 2013), no substrate-specific methyltransferase activity has been found for LaeA. Interestingly, microarray analysis of the *A. fumigatus ΔlaeA* mutant shows it to be down-regulated in the sulfur/methionine regulon (Perrin et al. 2007); however, no mechanistic connection between LaeA and this metabolic pathway has been established.

B. Epigenetics

Due to LaeA's similarities to methyltransferases, its localization in the nucleus, and its often precise regulation of SM clusters (Bouhired et al. 2007), it has been suggested that LaeA regulates transcription by protein lysine or protein arginine methyltransferase functions (Bok and Keller 2004; Bok et al. 2006b; Fox and Howlett 2008). Although no direct biochemical studies have demonstrated such a role, this protein has been linked to changes in chromatin structure in SM gene clusters where loss of LaeA leads to increased heterochromatin marks (Reyes-Dominguez et al. 2010). Several papers have indicated a role for LaeA in interactions with canonical histone-modifying enzymes, including HdaA, HstD, and CclA (Kawauchi et al. 2013; Bok et al. 2009; Shwab et al. 2007).

C. Velvet Complex Member

A clue to how LaeA works also came from the finding that it is a member of a nuclear complex known as the Velvet Complex composed of LaeA, VeA, and VelB (Bayram et al. 2008). Although first noted for its role in SM regulation, LaeA also has a profound effect on both asexual and sexual spore development, as do both VeA and VelB (Sarikaya Bayram et al. 2010; Bayram and Braus 2012). Thus, the Velvet Complex as a unit links morphological development with chemical development in all fungi examined so far (Wiemann et al. 2010; Lopez-Berges et al. 2013; Wu et al. 2012; Kosalková et al. 2009; Amaike and Keller 2009; Baba et al. 2012). When described, the phenotypes of deletants of these three genes are not equivalent but overlapping in some regulatory aspects of SM and morphological development.

III. Secondary Metabolites Regulated by LaeA

The initial characterization of LaeA in *A. nidulans* reported LaeA as positively regulating two well characterized endogenous SMs (sterigmatocystin and penicillin) as well as the heterologous lovastatin SM cluster genetically engineered into *A. nidulans* (Bok and Keller 2004). A second study of *A. nidulans* LaeA using microarray analysis identified additional uncharacterized SM clusters positively regulated by LaeA where one was characterized as producing terrequinone A (Bok et al. 2006a). Many more *A. nidulans* SMs have been discovered since these papers, and it is likely that LaeA regulates some, perhaps a majority, of these newly characterized SMs (Yaegashi et al. 2014). Microarray studies of at least four additional species (*A. fumigatus*, *A. flavus*, *Fusarium fujikuroi*, and *Trichoderma reesei*) show that many unknown and known SM clusters are regulated by LaeA; however, here we will only focus on those assigned to a metabolite (Bok et al. 2006a; Perrin et al. 2007; Georgianna et al. 2010; Karimi-Aghcheh et al. 2013; Wiemann et al. 2010 and Table 2.1). The reader should note that Table 2.1 represents only a small fraction of SMs regulated by LaeA, as many papers report an association of SM with LaeA without reporting what these metabolites are (Perrin et al. 2007; Georgianna et al. 2010; Karimi-Aghcheh et al. 2013; Wiemann et al. 2010; Rachmawati et al. 2013). Below, NRPS indicates a non-ribosomal peptide synthase derived SM, PKS a polyketide derived SM, and DMATS a dimethylallyl tryptophan synthase derived SM.

A. *Aspergillus* species

LaeA regulated SMs have been partially characterized in five *Aspergillus spp.*, including *A. nidulans*, *A. fumigatus*, *A. flavus*, *A. oryzae*, *and A. carbonarius*. In *A. fumigatus*, LaeA regulated SMs include gliotoxin (NRPS, cluster size: 25 kb), fumitremorgin (NRPS, cluster size: 25 kb), pseurotin (PKS/NRPS hybrid, part of an intertwined cluster with fumagillin, cluster size: 50 kb), fumagillin (PKS/terpene hybrid), endocrocin (PKS, cluster size: 15 kb), festuclavine (DMATS), elymoclavine (DMATS), fumigaclavines (DMATS), helvolic acid (terpene, cluster size 17kbref), fumiquinazolines (NRPS cluster size: 15 kb), and hexadehydroastechrome (NRPS, cluster size: 25 kb). Several of these metabolites have been implicated as playing a role in virulence in this human pathogen (Abad et al. 2010).

LaeA in *A. flavus* regulates aflatoxin (PKS, cluster size: 80 kb), diastereomeric piperazines (two duplicated clusters encoding NRPS-like adenylating reductases, cluster sizes each: 13 and 20 kb), morpholine (NRPS), pyrazines (NRPS), cyclopiazonic acid (PKS/NRPS), 3-(p-hydroxyphenyl)-1,2-propanediol (NRPS), kojic acid (simple organic acid from glucose), aspergillic acid (NRPS), paspaline (DMATS), paspalinine (DMATS), aflatrem (DMATS, cluster size:10 kb), and aflavinines (DMATS). LaeA in *A. oryzae* regulates kojic acid and the heterologously expressed terrequinone A (NRPS, cluster size: 10 kb) and monacolin K (PKS/NRPS) clusters. LaeA in *A. carbonarius* regulates ochratoxin A (NRPS).

B. Other Genera

LaeA orthologs have been identified in other fungal genera. LaeA has been characterized in several *Fusarium* species including *F. oxysporum* where it regulates beauvericin (NRPS, cluster size: 10 kb), ferricrocin (NRPS), and triacetylfusarinine C (NRPS). Lae1 in *F. verticillioides* regulates bikaverin (PKS, cluster size: 12 kb), fumonisin (PKS, cluster size 43 kb), fusaric acid (PKS, cluster size: 13 kb), and fusarins (PKS/NRPS). FfLae1 in *F. fujikuroi* regulates gibberellin (terpene, cluster size: 15 kb), fumonisin (PKS, cluster size: 42 kb), fusarin C (PKS/NRPS, cluster size: 25 kb), and bikaverin (PKS, cluster size: 12 kb). FgLaeA in *F. graminearum* regulates trichothecenes (terpene, cluster size: 25 kb) and zearalenone (PKS, cluster size: 22 kb).

ChLae1 in *Cochliobolus heterostrophus* regulates T-toxin (PKS) and melanin (PKS). LaeA in *Monascus pilosus* regulates monacolin K

Table 2.1 LaeA linked secondary metabolite regulation in filamentous fungi

Gene name	Species	Secondary metabolites	References
LaeA	*Aspergillus nidulans*	Sterigmatocystin, penicillin, lovastatin	Bok and Keller (2004), Bok et al. (2006b)
		Hyphal pigments	Sarikaya Bayram et al. (2010)
		Terrequinone A	Bok et al. (2006a), Bouhired et al. (2007)
		Monodictyphenone, F9775A , F9775B	Bok et al. (2009)
LaeA	*A. fumigatus*	Gliotoxin	Bok et al. (2005), Bok and Keller (2004), Sugui et al. (2007), Ben-Ami et al. (2009), Perrin et al. (2007)
		Fumagillin	Dhingra et al. (2013)
		Fumitremorgin, pseurotin	Wiemann et al. (2013), Perrin et al. (2007)
		Endocrocin	Lim et al. 2012
		Festuclavine, elymoclavine, fumigaclavines	Perrin et al. (2007)
		Hexadehydroastechrome	Yin et al. (2013)
		Helvolbic acid	Lodeiro et al. (2009)
		Fumiquinazolines	Lim et al. (2014)
LaeA	*A. flavus*	Aflatoxin	Amaike and Keller (2011), Kale et al. (2008), Georgianna et al. (2010)
		Diastereomeric piperazines, morpholine, pyrazines, 3-(p-hydroxyphenyl)-1,2-propanediol	Forseth et al. (2013)
		Cyclopiazonic acid	Kale et al. (2008), Georgianna et al. (2010)
		Aspergillic acid, paspaline, paspalinine, aflatrem, aflavinines, kojic acid	Kale et al. (2008)
LaeA	*A. oryzae*	Kojic acid	Oda et al. (2011)
		Terrequinone A, monacolin K	Sakai et al. (2012)
LaeA	*A. carbonarius*	Ochratoxin A	Crespo-Sempere et al. (2013)
ChLae1	*Cochliobolus heterostrophus*	T-toxin, melanin	Wu et al. (2012)
LaeA	*Fusarium oxysporum*	Beauvericin	López-Berges et al. (2014)
		Triacetylfusarinine C , ferricrocin	Lopez-Berges et al. (2013)
Lae1	*F. verticillioides*	Bikaverin, fumonisins, fusaric acid, fusarins	Butchko et al. (2012)
FfLae1	*F. fujikuroi*	Gibberellin, fumonisins, fusarin C, bikaverin	Wiemann et al. (2010)
		Fusarin C	Niehaus et al. (2013)
FgLaeA	*F. graminearum*	Trichothecenes, zearalenone	Kim et al. (2013)
LaeA	*Monascus pilosus*	Monacolin K, pigments	Lee et al. (2013), Zhang and Miyake (2009)
LaeA	*Penicillium citrinum*	ML236B	Baba et al. (2012)
PcLaeA	*P. chrysogenum*	Penicillin	Kosalková et al. (2009), Kopke et al. (2013), Hoff et al. (2010), Martín et al. (2012), Veiga et al. (2012)
		Pigments	Kosalková et al. (2009)
Lae1	*Trichoderma reesei*	Sterigmatocystin, siderophore	Karimi-Aghcheh et al. (2013)

(PKS, cluster size: 42 kb) and various pigments. LaeA in *Penicillium citrinum* regulates ML236B (PKS/NRPS, cluster size: 20 kb). PcLaeA in *P. chrysogenum* regulates penicillin (NRPS, cluster size: 15 kb) and pigments. Lae1 in *Trichoderma reesei* controls siderophore (NRPS) and the heterologously expressed sterigmatocystin cluster (PKS, 60 kb).

IV. Processes Identified Through LaeA Microarrays

As mentioned above, several microarray studies have led to characterization of several SMs, including but not limited to terrequinone A (Bok et al. 2006b), piperazines (Forseth et al. 2013), pseurotin (Wiemann et al. 2013), fumagillin (Wiemann et al. 2013), endocrocin (Lim et al. 2012), fumiquinazoline (Lim et al. 2014), and hexadehydroastechrome (Yin et al. 2012). However, other non-SM genes regulated by LaeA also may impact SM production. Characterization of LaeA regulated transcription factors include the sporulation specific regulatory protein BrlA as mediating LaeA regulation of spore-specific SMs (Berthier et al. 2013; Lim et al. 2014), NosA as mediating germination defects of the *ΔlaeA* mutant (Soukup et al. 2012b), and MeaB, a bZIP protein, enhancing virulence in *A. flavus* (Amaike et al. 2013). Details of BrlA are discussed in Chap. 1.

Both BrlA and MeaB affect SM production. BrlA is required for transcription and production of several spore-specific SMs, including endocrocin, fumiquinazoline, fumigaclavines, trypacidin, and various uncharacterized SMs in *A. fumigatus* (Berthier et al. 2013; Lim et al. 2014; Twumasi-Boateng et al. 2009; Coyle et al. 2007; Gauthier et al. 2012). Currently, it is not known if LaeA regulation of spore SMs is also mediated by BrlA—or the appropriate sporulation transcription factor in non-Aspergilli—in other fungal spp. Although not reported to be through BrlA, one study suggested that LaeA regulation of aflatoxin in *A. flavus* might be mediated through alterations in conidial development (Chang et al. 2012), and it was noted that *laeA* loss also impacted hydrophobin content in *A. fumigatus* spores (Dagenais et al. 2010). MeaB had a regulatory impact on aflatoxin synthesis in *A. flavus* where loss of MeaB greatly reduced production of this mycotoxin (Amaike et al. 2013).

A microarray analysis of *Trichoderma reesei* showed that *lae1* loss in this species resulted in complete loss of enzymes (CAZymes) responsible for lignocellulose degradation. On the other hand, overexpression of *lae1* led to enhanced CAZyme gene transcription (Seiboth et al. 2012). Another study, this one in *P. chrysogenum*, resulted in the identification of 62 genes co-regulated by both PcVelA and PcLaeA. One gene positively regulated by both proteins was *PcchiB1* encoding a class V chitinase required for cell wall integrity and pellet formation in *P. chrysogenum* (Kamerewerd et al. 2011). These two studies did not examine if there was relationship between SM production and these enzymes.

V. Processes Identified Through LaeA Mutagenesis

A multicopy suppressor screen looking for restoration of secondary metabolism in an *A. nidulans ΔlaeA* background has resulted in the identification of several novel regulators of SM. RsmA (remediation of secondary metabolism A) is a bZIP protein that directly regulates the sterigmatocystin gene cluster by binding to the intergenic region of AflR and AflJ (Shaaban et al. 2010; Yin et al. 2012, 2013). Asperthecin was also regulated by RsmA, apparently through transactivation by AflR (Yin et al. 2012). Overexpression of RsmA partially restored sterigmatocystin synthesis but not sporulation defects in both *ΔlaeA* and *ΔveA* backgrounds. The RsmA ortholog in *A. fumigatus* positively regulates gliotoxin in that species (Sekonyela et al. 2013).

The same screen also found EsaA, a histone acetyltransferase, to be a global regulator of SM. Like RsmA, overexpression of EsaA partially restored sterigmatocystin synthesis (and again, not sporulation defects) in *ΔlaeA* (Soukup et al. 2012a). Moreover, EsaA was determined to increase transcript levels of multiple SM cluster genes; this increase was associated with an increase in total H4 acetylation and specifically H4K12 acetylation of SM gene promoters. As mentioned earlier, several histone-modifying enzymes have been found to be important in SM regulation, often in relation with LaeA functionality.

VI. Conclusion

Since its discovery in 2004, LaeA has provided the research community with a new paradigm of regulation of SM gene clusters in fungi. The global nature of SM regulation by LaeA, presumably as part of the Velvet Complex, suggests an evolved requirement for production of certain SM in concert with morphological development, possibly as part of a stress response in protecting fungi from both abiotic and biotic stresses (Hong et al. 2013). Although present in most Ascomycetes, LaeA and other members of the Velvet complex are conspicuously missing in *Saccharomyces cerevisiae* and *Schizosaccharomyces pombe*. Recently, putative VeA and VelB orthologs have been found in the Basidiomycete *Ustilago maydis* (Karakkat et al. 2013), and it remains to be seen if LaeA also exists in this fungus.

Considering the large number of sequenced fungi and unknown SM clusters, LaeA is likely to continue to be a valuable tool in natural product studies, both as a means to activate endogenous SM clusters and also, increasingly, as a tool to activate heterologously expressed clusters. This was recently demonstrated where *laeA* overexpression in *A. oryzae* activated transcription of the monacolin K gene cluster from *M. pilosus* and the terrequinone A gene cluster from *A. nidulans* (Sakai et al. 2012). In another embodiment, *A. nidulans laeA* was overexpressed in *Cordyceps militaris* to awaken silent secondary metabolite clusters in that fungus (Rachmawati et al. 2013). An alternative approach in utilizing LaeA as a SM enhancer was recently demonstrated in *P. chrysogenum* where 1,3-diaminopropane and spermidine were found to enhance *laeA* transcript levels and, thus, increase penicillin production (Martín et al. 2012; Pfeifer and Khosla 2001).

Acknowledgements This work was funded by NIH 1 R01 AI065728-01 and NIH P01 GM084077 to N.P.K.

References

Abad A, Fernandez-Molina JV, Bikandi J, Ramirez A, Margareto J, Sendino J, Hernando FL, Ponton J, Garaizar J, Rementeria A (2010) What makes *Aspergillus fumigatus* a successful pathogen? Genes and molecules involved in invasive aspergillosis. Rev Iberoam Micol 27(4):155–182

Amaike S, Keller NP (2009) Distinct roles for VeA and LaeA in development and pathogenesis of *Aspergillus flavus*. Eukaryot Cell 8(7):1051–1060

Amaike S, Keller NP (2011) Aspergillus flavus. Annu Rev Phytopathol 49:107–133

Amaike S, Affeldt K, Yin WB, Franke S, Choithani A, Affeldt KJ, Keller NP (2013) The bZIP protein MeaB mediating virulence attributes in *Aspergillus flavus*. PLoS One 8(9), e74030

Baba S, Kinoshita H, Nihira T (2012) Identification and characterization of *Penicillium citrinum* VeA and LaeA as global regulators for ML-236B production. Curr Genet 58(1):1–11

Bayram Ö, Braus GH (2012) Coordination of secondary metabolism and development in fungi: the velvet family of regulatory proteins. FEMS Microbiol Rev 36(1):1–24

Bayram Ö, Krappmann S, Ni M, Bok JW, Helmstaedt K, Valerius O, Braus-Stromeyer S, Kwon NJ, Keller NP, Yu JH, Braus GH (2008) VelB/VeA/LaeA complex coordinates light signal with fungal development and secondary metabolism. Science 320 (5882):1504–1506

Ben-Ami R, Lewis RE, Leventakos K, Kontoyiannis DP (2009) *Aspergillus fumigatus* inhibits angiogenesis through the production of gliotoxin and other secondary metabolites. Blood 114(26):5393–5399

Berthier E, Lim FY, Deng Q, Guo CJ, Kontoyiannis DP, Wang CCC, Rindy J, Beebe DJ, Huttenlocher A, Keller NP (2013) Low-volume toolbox for the discovery of immunosuppressive fungal secondary metabolites. PLoS Pathog 9(4):e1003289

Bok JW, Keller NP (2004) LaeA, a regulator of secondary metabolism in *Aspergillus* spp. Eukaryot Cell 3 (2):527–535

Bok JW, Balajee SA, Marr KA, Andes D, Nielsen KF, Frisvad JC, Keller NP (2005) LaeA, a regulator of morphogenetic fungal virulence factors. Eukaryot Cell 4(9):1574–1582

Bok JW, Hoffmeister D, Maggio-Hall LA, Murillo R, Glasner JD, Keller NP (2006a) Genomic mining for *Aspergillus* natural products. Chem Biol 13(1):31–37

Bok JW, Noordermeer D, Kale SP, Keller NP (2006b) Secondary metabolic gene cluster silencing in *Aspergillus nidulans*. Mol Microbiol 61(6):1636–1645

Bok JW, Chiang YM, Szewczyk E, Reyes-Dominguez Y, Davidson AD, Sanchez JF, Lo HC, Watanabe K, Strauss J, Oakley BR, Wang CC, Keller NP (2009) Chromatin-level regulation of biosynthetic gene clusters. Nat Chem Biol 5(7):462–464

Bouhired S, Weber M, Kempf-Sontag A, Keller NP, Hoffmeister D (2007) Accurate prediction of *Aspergillus* natural product gene cluster boundaries using the transcriptional regulator LaeA. Fungal Genet Biol 44(11):1134–1145

Brakhage AA (2013) Regulation of fungal secondary metabolism. Nat Rev Microbiol 11(1):21–32

Brown DW, Yu JH, Kelkar HS, Fernandes M, Nesbitt TC, Keller NP, Adams TH, Leonard TJ (1996) Twenty-five coregulated transcripts define a sterigmatocystin gene cluster in *Aspergillus nidulans*. Proc Natl Acad Sci USA 93(4):1418–1422

Butchko RAE, Adams TH, Keller NP (1999) *Aspergillus nidulans* mutants defective in *stc* gene cluster regulation. Genetics 153(2):715–720

Butchko RA, Brown DW, Busman M, Tudzynski B, Wiemann P (2012) Lae1 regulates expression of multiple secondary metabolite gene clusters in *Fusarium verticillioides*. Fungal Genet Biol 49 (8):602–612

Chang PK, Scharfenstein LL, Ehrlich KC, Wei Q, Bhatnagar D, Ingber BF (2012) Effects of *laeA* deletion on *Aspergillus flavus* conidial development and hydrophobicity may contribute to loss of aflatoxin production. Fungal Biol 116(2):298–307

Coyle CM, Kenaley SC, Rittenour WR, Panaccione DG (2007) Association of ergot alkaloids with conidiation in *Aspergillus fumigatus*. Mycologia 99 (6):804–811

Crespo-Sempere A, Marín S, Sanchis V, Ramos AJ (2013) VeA and LaeA transcriptional factors regulate ochratoxin A biosynthesis in Aspergillus carbonarius. Int J Food Microbiol 166(3):479–486

Dagenais TR, Giles SS, Aimanianda V, Latge JP, Hull CM, Keller NP (2010) *Aspergillus fumigatus* LaeA-mediated phagocytosis is associated with a decreased hydrophobin layer. Infect Immun 78 (2):823–829

Dhingra S, Lind AL, Lin H-C, Tang Y, Rokas A, Calvo AM (2013) The fumagillin gene cluster, an example of hundreds of genes under *veA* control in *Aspergillus fumigatus*. PLoS One 8(10):e77147

Fernandes M, Keller NP, Adams TH (1998) Sequence-specific binding by *Aspergillus nidulans* AflR, a C6 zinc cluster protein regulating mycotoxin biosynthesis. Mol Microbiol 28(6):1355–1365

Forseth RR, Amaike S, Schwenk D, Affeldt KJ, Hoffmeister D, Schroeder FC, Keller NP (2013) Homologous NRPS-like gene clusters mediate redundant small-molecule biosynthesis in *Aspergillus flavus*. Angew Chem Int Ed Engl 52(5):1590–1594

Fox EM, Howlett BJ (2008) Secondary metabolism: regulation and role in fungal biology. Curr Opin Microbiol 11(6):481–487

Gauthier T, Wang X, Sifuentes Dos Santos J, Fysikopoulos A, Tadrist S, Canlet C, Artigot MP, Loiseau N, Oswald IP, Puel O (2012) Trypacidin, a spore-borne toxin from *Aspergillus fumigatus*, is cytotoxic to lung cells. PLoS One 7(2), e29906

Georgianna DR, Fedorova ND, Burroughs JL, Dolezal AL, Bok JW, Horowitz-Brown S, Woloshuk CP, Yu J, Keller NP, Payne GA (2010) Beyond aflatoxin: four distinct expression patterns and functional roles associated with *Aspergillus flavus* secondary metabolism gene clusters. Mol Plant Pathol 11 (2):213–226

Hoff B, Kamerewerd J, Sigl C, Mitterbauer R, Zadra I, Kürnsteiner H, Kück U (2010) Two components of a velvet-like complex control hyphal morphogenesis, conidiophore development, and penicillin biosynthesis in *Penicillium chrysogenum*. Eukaryot Cell 9(8):1236–1250

Hoffmeister D, Keller NP (2007) Natural products of filamentous fungi: enzymes, genes, and their regulation. Nat Prod Rep 24(2):393–416

Hong SY, Roze LV, Wee J, Linz JE (2013) Evidence that a transcription factor regulatory network coordinates oxidative stress response and secondary metabolism in aspergilli. Microbiologyopen 2 (1):144–160

Jain S, Keller NP (2013) Insights to fungal biology through LaeA sleuthing. Fungal Biol 27(2):51–59

Kale SP, Milde L, Trapp MK, Frisvad JC, Keller NP, Bok JW (2008) Requirement of LaeA for secondary metabolism and sclerotial production in *Aspergillus flavus*. Fungal Genet Biol 45(10): 1422–1429

Kamerewerd J, Zadra I, Kürnsteiner H, Kück U (2011) PcchiB1, encoding a class V chitinase, is affected by PcVelA and PcLaeA, and is responsible for cell wall integrity in *Penicillium chrysogenum*. Microbiology 157(Pt 11):3036–3048

Karakkat BB, Gold SE, Covert SF (2013) Two members of the *Ustilago maydis* velvet family influence teliospore development and virulence on maize seedlings. Fungal Genet Biol 61:111–119

Karimi-Aghcheh R, Bok JW, Phatale PA, Smith KM, Baker SE, Lichius A, Omann M, Zeilinger S, Seiboth B, Rhee C, Keller NP, Freitag M, Kubicek CP (2013) Functional analyses of *Trichoderma reesei* LAE1 reveal conserved and contrasting roles of this regulator. G3 (Bethesda) 3(2):369–378

Kawauchi M, Nishiura M, Iwashita K (2013) Fungus-specific sirtuin HstD coordinates secondary metabolism and development through control of LaeA. Eukaryot Cell 12(8):1087–1096

Khaldi N, Seifuddin FT, Turner G, Haft D, Nierman WC, Wolfe KH, Fedorova ND (2010) SMURF: genomic mapping of fungal secondary metabolite clusters. Fungal Genet Biol 47(9):736–741

Kim HK, Lee S, Jo SM, McCormick SP, Butchko RA, Proctor RH, Yun SH (2013) Functional roles of FgLaeA in controlling secondary metabolism, sexual development, and virulence in Fusarium graminearum. PLoS One 8(7):e68441

Kopke K, Hoff B, Bloemendal S, Katschorowski A, Kamerewerd J, Kück U (2013) Members of the *Penicillium chrysogenum* velvet complex play functionally opposing roles in the regulation of penicillin biosynthesis and conidiation. Eukaryot Cell 12(2):299–310

Kosalková K, García-Estrada C, Ullán RV, Godio RP, Feltrer R, Teijeira F, Mauriz E, Martín JF (2009) The global regulator LaeA controls penicillin biosynthesis, pigmentation and sporulation, but not roquefortine C synthesis in *Penicillium chrysogenum*. Biochimie 91(2):214–225

Lee SS, Lee JH, Lee I (2013) Strain improvement by overexpression of the *laeA* gene in *Monascus pilosus* for the production of monascus-fermented rice. J Microbiol Biotechnol 23(7):959–965

Leitão AL, Enquita FJ (2014) Fungal extrolites as a new source for therapeutic compounds and as building blocks for applications in synthetic biology. Microbiol Res. doi:10.1016/j.micres.2014.02.007

Lim FY, Hou Y, Chen Y, Oh JH, Lee I, Bugni TS, Keller NP (2012) Genome-based cluster deletion reveals an endocrocin biosynthetic pathway in *Aspergillus fumigatus*. Appl Environ Microbiol 78(12):4117–4125

Lim FY, Ames B, Walsh CT, Keller NP (2014) Co-ordination between BrlA regulation and secretion of the oxidoreductase FmqD directs selective accumulation of fumiquinazoline C to conidial tissues in *Aspergillus fumigatus*. Cell Microbiol. doi:10.1111/cmi.12284

Lodeiro S, Xiong Q, Wilson WK, Ivanova Y, Smith ML, May GS, Matsuda SP (2009) Protostadienol biosynthesis and metabolism in the pathogenic fungus *Aspergillus fumigatus*. Org Lett 11(6):1241–1244

Lopez-Berges MS, Hera C, Sulyok M, Schäfer K, Capilla J, Guarro J, Di Pietro A (2013) The velvet complex governs mycotoxin production and virulence of *Fusarium oxysporum* on plant and mammalian hosts. Mol Microbiol 87(1):49–65

López-Berges MS, Schäfer K, Hera C, Di Pietro A (2014) Combinatorial function of velvet and AreA in transcriptional regulation of nitrate utilization and secondary metabolism. Fungal Genet Biol 62:78–84

Martin JF (1992) Clusters of genes for the biosynthesis of antibiotics: regulatory genes and overproduction of pharmaceuticals. J Ind Microbiol 9(2):73–90

Martín J, García-Estrada C, Kosalková K, Ullán RV, Albillos SM, Martín JF (2012) The inducers 1,3-diaminopropane and spermidine produce a drastic increase in the expression of the penicillin biosynthetic genes for prolonged time, mediated by the *laeA* regulator. Fungal Genet Biol 49 (12):1004–1013

Medema MH, Blin K, Cimermancic P, de Jager V, Zakrzewski P, Fischbach MA, Weber T, Takano E, Breitling R (2011) antiSMASH: rapid identification, annotation and analysis of secondary metabolite biosynthesis gene clusters in bacterial and fungal genome sequences. Nucl Acids Res 39 (Web server issue):W339–346

Niehaus EM, Kleigrewe K, Wiemann P, Studt L, Sieber CM, Connolly LR, Freitag M, Güldener U, Tudzynski B, Humpf HU (2013) Genetic manipulation of the *Fusarium fujikuroi* fusarin gene cluster yields insight into the complex regulation and fusarin biosynthetic pathway. Chem Biol 20 (8):1055–1066

Oda K, Kobayashi A, Ohashi S, Sano M (2011) *Aspergillus oryzae laeA* regulates kojic acid synthesis genes. Biosci Biotechnol Biochem 75(9):1832–1834

Patananan AN, Palmer JM, Garvey GS, Keller NP, Clarke SG (2013) A novel automethylation reaction in the *Aspergillus nidulans* LaeA protein generates S-methylmethionine. J Biol Chem 288 (20):14032–14045

Pelaez T, Alcala L, Alonso R, Martin-Lopez A, Garcia-Arias V, Marin M, Bouza E (2005) In vitro activity of ramoplanin against *Clostridium difficile*, including strains with reduced susceptibility to vancomycin or with resistance to metronidazole. Antimicrob Agents Chemother 49(3):1157–1159

Perrin RM, Fedorova ND, Bok JW, Cramer RA, Wortman JR, Kim HS, Nierman WC, Keller NP (2007) Transcriptional regulation of chemical diversity in *Aspergillus fumigatus* by LaeA. PLoS Pathog 3(4), e50

Pfeifer BA, Khosla C (2001) Biosynthesis of polyketides in heterologous hosts. Microbiol Mol Biol Rev 65 (1):106–118

Rachmawati R, Kinoshita H, Nihira T (2013) Establishment of transformation system in *Cordyceps militaris* by using integration vector with benomyl resistance gene. Proc Environ Sci 17:142–149

Reyes-Dominguez Y, Bok JW, Berger H, Shwab EK, Basheer A, Gallmetzer A, Scazzocchio C, Keller NP, Strauss J (2010) Heterochromatic marks are associated with the repression of secondary metabolism clusters in *Aspergillus nidulans*. Mol Microbiol 76(6):1376–1386

Sakai K, Kinoshita H, Nihira T (2012) Heterologous expression system in *Aspergillus oryzae* for fungal biosynthetic gene clusters of secondary metabolites. Appl Microbiol Biotechnol 93(5):2011–2022

Sanchez JF, Entwistle R, Hung JH, Yaegashi J, Jain S, Chiang YM, Wang CC, Oakley BR (2011) Genome-based deletion analysis reveals the prenyl xanthone biosynthesis pathway in *Aspergillus nidulans*. J Am Chem Soc 133(11):4010–4017

Sarikaya Bayram O, Bayram O, Valerius O, Park HS, Irniger S, Gerke J, Ni M, Han KH, Yu JH, Braus GH (2010) LaeA control of velvet family regulatory proteins for light-dependent development and fungal cell-type specificity. PLoS Genet 6(12), e1001226

Seiboth B, Karimi RA, Phatale PA, Linke R, Hartl L, Sauer DG, Smith KM, Baker SE, Freitag M, Kubicek CP (2012) The putative protein methyltransferase LAE1 controls cellulase gene expression in *Trichoderma reesei*. Mol Microbiol 84(6):1150–1164

Sekonyela R, Palmer JM, Bok JW, Jain S, Berthier E, Forseth R, Schroeder F, Keller NP (2013) RsmA regulates *Aspergillus fumigatus* gliotoxin cluster metabolites including cyclo(L-Phe-L-Ser), a potential new diagnostic marker for invasive aspergillosis. PLoS One 8(5), e62591

Shaaban MI, Bok JW, Lauer C, Keller NP (2010) Suppressor mutagenesis identifies a velvet complex remediator of *Aspergillus nidulans* secondary metabolism. Eukaryot Cell 9(12):1816–1824

Shwab EK, Bok JW, Tribus M, Galehr J, Graessle S, Keller NP (2007) Histone deacetylase activity regulates chemical diversity in *Aspergillus*. Eukaryot Cell 6(9):1656–1664

Soukup AA, Chiang YM, Bok JW, Reyes-Dominguez Y, Oakley BR, Wang CC, Strauss J, Keller NP (2012a) Overexpression of the *Aspergillus nidulans* histone 4 acetyltransferase EsaA increases activation of secondary metabolite production. Mol Microbiol 86(2):314–330

Soukup AA, Farnoodian M, Berthier E, Keller NP (2012b) NosA, a transcription factor important in *Aspergillus fumigatus* stress and developmental response, rescues the germination defect of a laeA deletion. Fungal Genet Biol 49(11):857–865

Sugui JA, Pardo J, Chang YC, Mullbacher A, Zarember KA, Galvez EM, Brinster L, Zerfas P, Gallin JI, Simon MM, Kwon-Chung KJ (2007) Role of laeA in the Regulation of alb1, gliP, Conidial Morphology, and Virulence in *Aspergillus fumigatus*. Eukaryot Cell 6(9):1552–1561

Trail F, Mahanti N, Rarick M, Mehigh R, Liang SH, Zhou R, Linz JE (1995) Physrical and transcriptional map of an aflatoxin gene cluster in Aspergillus parasiticus and functional disruption of a gene involved early in the aflatoxin pathway. Appl Environ Microbiol 61(7):2665–2673

Twumasi-Boateng K, Yu Y, Chen D, Gravelat FN, Nierman WC, Sheppard DC (2009) Transcriptional profiling identifies a role for BrlA in the response to nitrogen depletion and for StuA in the regulation of secondary metabolite clusters in *Aspergillus fumigatus*. Eukaryot Cell 8(1):104–115

Umemura M, Koike H, Nagano N, Ishii T, Kawano J, Yamane N, Kozone I, Horimoto K, Shin-ya K, Asai K, Yu J, Bennett JW, Machida M (2013) MIDDAS-M: motif-independent de novo detection of secondary metabolite gene clusters through the integration of genome sequencing and transcriptome data. PLoS One 8(12):e84028

Veiga T, Nijland JG, Driessen AJ, Bovenberg RA, Touw H, van den Berg MA, Pronk JT, Daran JM (2012) Impact of velvet complex on transcriptome and penicillin G production in glucose-limited chemostat cultures of a beta-lactam high-producing *Penicillium chrysogenum* strain. OMICS 16(6):320–333

Wiemann P, Brown DW, Kleigrewe K, Bok JW, Keller NP, Humpf HU, Tudzynski B (2010) FfVel1 and FfLae1, components of a velvet-like complex in *Fusarium fujikuroi*, affect differentiation, secondary metabolism and virulence. Mol Microbiol 77:972–994

Wiemann P, Guo CJ, Palmer JM, Sekonyela R, Wang CC, Keller NP (2013) Prototype of an intertwined secondary-metabolite supercluster. Proc Natl Acad Sci USA 110(42):17065–17070

Woloshuk CP, Foutz KR, Brewer JF, Bhatnagar D, Cleveland TE, Payne GA (1994) Molecular characterization of *aflR*, a regulatory locus for aflatoxin biosynthesis. Appl Environ Microbiol 60(7):2408–2414

Wu D, Oide S, Zhang N, Choi MY, Turgeon BG (2012) ChLae1 and ChVel1 regulate T-toxin production, virulence, oxidative stress response, and development of the maize pathogen *Cochliobolus heterostrophus*. PLoS Pathog 8(2):e1002542

Yaegashi J, Oakley BR, Wang CC (2014) Recent advances in genome mining of secondary metabolite biosynthetic gene clusters and the development of heterologous expression systems in *Aspergillus nidulans*. J Ind Microbiol Biotechnol 41(2):433–442

Yin WB, Amaike S, Wohlbach DJ, Gasch AP, Chiang YM, Wang CC, Bok JW, Rohlfs M, Keller NP (2012) An *Aspergillus nidulans* bZIP response pathway hardwired for defensive secondary metabolism operates through aflR. Mol Microbiol 83(5):1024–1034

Yin WB, Baccile JA, Bok JW, Chen Y, Keller NP, Schroeder FC (2013) A nonribosomal peptide synthetase-derived iron(III) complex from the pathogenic fungus *Aspergillus fumigatus*. J Am Chem Soc 135(6):2064–2067

Yu J, Chang PK, Cary JW, Wright M, Bhatnagar D, Cleveland TE, Payne GA, Linz JE (1995) Comparative mapping of aflatoxin pathway gene clusters in *Aspergillus parasiticus* and *Aspergillus flavus*. Appl Environ Microbiol 61(6):2365–2371

Zhang MY, Miyake T (2009) Development and media regulate alternative splicing of a methyltransferase pre-mRNA in *Monascus pilosus*. J Agric Food Chem 57(10):4162–4167

3 RNAi Function and Diversity in Fungi

Åke Olson[1]

CONTENTS

Abbreviations

aRNA	Aberrant RNA
DdRP	DNA-dependent RNA polymerase
dsRNA	Double-stranded DNA
miRNA	microRNA
milRNAs	microRNA-like small RNAs
MSUD	Meiotic Silencing by Unpaired DNA
PTGS	Posttranscriptional gene silencing
qiRNA	QDE-2-interacting small RNAs
RdRP	RNA-dependent RNA polymerase
RISC	RNA-induced silencing complex
RNAi	RNA interference
SIS	Sex-induced silencing
siRNAs	small interfering RNAs
snRNA	small nuclear RNA

[1]Department of Forest Mycology and Plant Pathology, Swedish University of Agricultural Sciences, Almas alle 5, 756 51 Uppsala, Sweden; e-mail: ake.olson@slu.se

I. Introduction

The development of high-throughput DNA-sequencing over the last decades has revealed that the vast majority of the eukaryotic genome is transcribed. In the ENCODE project (Kellis et al. 2014), it is suggested that around three-quarters of the human genome is transcribed, while the corresponding levels for the budding yeast (*Saccharomyces cerevisiae*) and the fission yeast (*Schizosaccharomyces pombe*) genomes are 85 and 94 %, respectively (David et al. 2006; Wilhelm et al. 2008). Transcripts represent protein-coding mRNAs, as well as transcripts which appear to have no coding capacity. These noncoding RNAs (ncRNAs) include housekeeping RNAs, which are required for cell viability, like transfer RNA (tRNA), ribosomal RNA (rRNA), small nuclear RNA (snRNA), and spliceosomal RNAs, but also regulatory ones which are expressed during specific developmental stages, or in response to nutritional or other external stimuli.

The regulatory RNAs can be classified based on their size into **small ncRNA** and **long ncRNA**. The long ncRNAs are defined as transcripts being more than 200 nucleotides in length and transcribed by RNA polymerase II. The known mechanisms for long ncRNA gene regulation include transcriptional interference, chromatin remodeling, and double-stranded RNA (dsRNA) formation and has been reviewed elsewhere (Donaldson and Saville 2012). Small ncRNAs are more diverse than long ncRNAs and with many different names and features, such as **small interfering RNAs** (siRNAs), **microRNAs** (miRNAs), **microRNA-**

Biochemistry and Molecular Biology, 3rd Edition
The Mycota III
D. Hoffmeister (Ed.)

like small RNA (milRNA), **PIWI-interacting RNAs** (piRNAs), and **primal small RNAs** (priRNAs) depending on their precursor structure, biogenesis pathways, and modes of action (Billmyre et al. 2013). All small ncRNAs are about 20–30 nucleotides in length and involved in a process known as RNA interference (RNAi), a conserved homology-dependent mechanism for genome defense against viral infections, heterochromatin formation, and gene regulation. This chapter's focus is the diversity and function of components and pathways of RNAi in fungi.

RNAi was first described in *Caenorhabditis elegans* in 1998 as a dsRNA-induced phenomenon which regulates the silencing of genes complementary to the dsRNA (Fire et al. 1998). This process was first thought to be a mechanism to defend the genome from invasive transposons and viruses. In mammals, RNAi was first observed in mouse oocytes where microinjection of dsRNA resulted in the downregulation of gene expression via mRNA degradation (Wianny and Zernicka-Goetz 2000). However, already in 1992, a **post-transcriptional gene-silencing** (PTGS) phenomenon in *Neurospora crassa* was described by Romano and Macino (1992), later to be discovered as RNAi-related. Since then, RNAi silencing mechanisms and components thereof have been identified in a large set of fungal species. As the fungal kingdom is diverse, the variation of RNAi pathways and processes is also diverse. The best understood system for RNAi-mediated transcriptional gene silencing is quelling and meiotic silencing of unpaired DNA (MSUD) in *N. crassa*. Another well-studied system where RNAi has been shown to be involved is in heterochromatin formation and gene regulation at the transcriptional level in the fission yeast *S. pombe* (Volpe et al. 2002). Today, the RNAi gene regulation phenomenon is known from many fungal species all over the kingdom.

Dicer, **Argonaute,** and **RNA-dependent RNA polymerase** (RdRP) are core proteins of the RNAi machinery in most of the biogenesis pathways that generates the various regulatory small interfering RNAs (siRNAs). Dicer, an RNase III endonuclease, processes the pre-RNA to form the siRNA. The mature siRNA is then incorporated into a multiprotein complex cell, the **RNA-induced Silencing Complex** (RISC) in which Argonaute is the core protein. Argonaute functions as a siRNA-guided endonuclease in the RISC. RISC recognizes mRNAs with complementary sequence to the engaged siRNA. An RdRP then generates dsRNA from single-stranded transcripts either by de novo, primer-independent second-strand synthesis or by the use of siRNA as primers to synthesize RNA complementary to the target mRNA resulting in the production and amplification of dsRNA. Although this is the general picture of RNAi-mediated silencing, the variation and species-specific differences seem at the moment to be unlimited and the more systems that are explored the more mechanisms are discovered.

A. Evolution of RNAi and Its Protein Components

The RNAi phenomenon is taxonomically widely distributed among eukaryotes and has been found in fungi, plants, and animals, suggesting that it was already present very early in eukaryotic evolution (Shabalina and Koonin 2008). The fungal kingdom is one of the major branches among eukaryotic organisms and estimated to embrace more than one million species with enormous diversity in morphology, ecology, and life strategies. Today, RNAi has been found in all the major groups of fungi, e.g., Ascomycota (Cogoni and Macino 1999), Basidiomycota (Wang et al. 2010), and Zygomycota (Nicolás et al. 2003). However, in some fungal species a functional RNAi mechanism involving the components known today has been lost. For instance, the top model organism *S. cerevisiae* has lost its RNAi, although the pathway can be introduced by transforming the main components Dicer and Argonaute genes from the closely related species *Saccharomyces castellii* into it (Drinnenberg et al. 2009). Also the human pathogen *Cryptococcus gatti*, responsible for an ongoing disease outbreak in the Vancouver area, Canada, has lost its RNAi machinery, while other *Cryptococcus* species has retained the RNAi mechanism (Wang et al. 2010). Interestingly, the fungal species that lost the RNAi mechanism are spread out in the fungal phylogeny and have probably lost the mechanism in relatively recent time, within 10 million years ago (MYA) (Billmyre et al. 2013). This poses an interesting question: When is RNAi not essential or even disadvantageous for an organism? It would be unlikely that the loss could be random and neutral considering how few species have lost their RNAi mechanism. Rather, it has to be a selective advantage under certain condition to discard the RNAi mechanism for the species where it has been lost. In their review, Billmyre and colleagues (2013) discuss two possible scenarios where it could be favorable to lose the RNAi mechanism, viral defense, and hypermutability. While RNAi usually is considered a defense mechanism against invasive viral infection, there are at least a few cases where viral infections are considered to be of advantage for the organism. In *S. cerevisiae*, the infection by dsRNA viruses known as killer viruses is considered such a trait. The killer viruses are transmitted either vertically from mother cell to daughter cell or horizontally through the fusion of two zygotic cells during mating. Killer virus-infected strains have a distinct phenotype: killer-positive strains inhibit the growth of neighboring negative strains. This provides an advantage in competitive environments and is incompatible with the RNAi mechanism which would silence the killer virus. The presence of killer-like viruses in fungi correlates inversely with the presence of RNAi in four of nine documented cases, suggesting that this could at least partially explain the loss of RNAi (Drinnenberg et al. 2011).

The other mechanism discussed by Billmyre et al. (2013) is hypermutability. Since one role of RNAi is to control transposable elements, the loss of RNAi allows an increased movement of transposons. This has been observed in *C. gatti* (Janbon et al. 2010; Wang et al. 2010, 2013). Loss of RNAi could conceivably contribute to a hypermutator phenotype which is advantageous in organisms required to adapt rapidly to new environments. Mutators tend to be beneficial only in the short term. There are attempts to explore if such hypermutator phenotypes can explain the molecular type (VGII) *C. gattii* outbreak in the Pacific Northwest since this is normally considered a tropical or subtropical pathogen (Billmyre et al. 2013).

During evolution, the RNAi pathway has undergone adaption, leading to diversification and differentiation of the biogenesis mechanisms as well as which process and genes that are regulated, although the major components (Dicer, Argonaute, and RdRP) have remained the same. In fungi, the number of genes for the components varies from zero to four for Dicer and up to nine for Argonaute, and in general, basidiomycetes have more gene copies than ascomycetes (Hu et al. 2013). Several Ascomycota yeast species as well as the Basidiomycota yeast species *C. gattii* and *Ustilago maydis* seem to lack Argonaute or Dicer orthologs (Billmyre et al. 2013; Hu et al. 2013). Reconciliation analysis suggests that the ancestor of fungi had only one copy of each gene and that one duplication seem to have taken place before the development of dikarya (Hu et al. 2013). Several duplications and losses have then taken place in individual species or lineages, but dating of these events have proven difficult, probably because the lack of information of gene copy numbers in many species. Phylogenetic analyses divide fungal Argonaute genes into two subgroups and the Dicer genes into three, but no functional differences between the groups have been shown (Hu et al. 2013). Although it is clear that the number of gene copies of both Dicer and Argonaute are elevated in agaricomycetes, no hypothesis about what drives the expansion has been presented at the moment (Hu et al. 2013).

II. RNAi Function

The biological functions in which RNAi are involved can be classified into three main categories: genome defense, heterochromatin formation, and gene regulation. It has been suggested that the ancestral RNAi primarily had a function in defense against viruses and transposons (Shabalina and Koonin 2008), but more and more data accumulates that suggest RNAi as an important gene regulation mechanism.

A. Quelling

Quelling is an RNAi mechanism that is induced by repetitive transgenic sequences in *N. crassa*. It was the first transgene-induced gene silencing phenomenon discovered in fungi (Romano and Macino 1992). The phenomenon was found through transformation of wild type with the *al-1* and *al-3* (albino 1 and 3) genes, required for the biosynthesis of carotenoids, to wild-type isolates of *N. crassa* (Romano and Macino 1992). The transformation resulted in white/pale isolates as a consequence of silencing of the endogenous *al-1* and *al-3* genes (Romano and Macino 1992), and efficiency of the silencing seemed to be correlated to high copy number of tandem repeats of the transgene (Cogoni and Macino 1997). Mutations of the *al* genes are usually recessive but when the *al* transgene was introduced and silenced by quelling the silenced nuclei dominated over the wild type in heterokaryons, which indicates a diffusible and transacting molecule responsible for the silencing (Cogoni et al. 1996). Since quelling did not affect the levels of mRNA precursors, this observation led to the hypothesis that production of aberrant RNA (aRNA) in the presence of transgenes causes PTGS (Cogoni et al. 1996).

Later, quelling-deficient (*qde*) mutants were isolated and divided into three complementation groups: *qde-1*, *qde-2*, and *qde-3* (Cogoni and Macino 1997). The corresponding genes were identified and found to correspond to key components in PTGS pathway. The first such component to be identified was RdRP, encoded by the *qde-1* gene. The verification that RdRP is important for quelling demonstrates that dsRNA is an essential step in the silencing mechanism. Argonaute was the second component to be identified; it is encoded by the *qde-2* gene and show homology with the *qde-1* gene of *C. elegans*, suggesting that PTGS and RNAi share common mechanisms and components. QDE-3 is a RecQ helicase which is involved in homologous recombination, DNA repair, and DNA replication. The requirement of QDE-3 in quelling raises the possibility that repetitive DNA may form aberrant DNA (aDNA) structures that are recognized by QDE-3 to promote aRNA and siRNA formation. Expression of inverted repeats containing transgenes can bypass *qde-1* and *qde-3* and trigger silencing, indicating that indeed QDE-3 and QDE-1 products are involved in the production of aRNA and siRNA (Catalanotto et al. 2004). It also suggests that QDE-1 is not involved in siRNA amplification. How QDE-1 on the other hand recognizes aRNA and produces siRNA is not known. Recently, it was demonstrated that QDE-1 is not only an RdRP but also a DNA-dependent RNA polymerase (DdRP) and uses both single-stranded DNA and RNA as template to form DNA/RNA hybrids or dsRNA (Lee et al. 2010a, b). Together with a subunit of the replication protein A (RPA-1), QDE-1 associates with ssDNA-binding complex involved in repair, recombination, and DNA-replication (Lee et al. 2010a, b). However, the interaction between QDE-1 and RPA-1 requires QDE-3. Probably, RPA-1 has two roles in the process: it recruits QDE-1 to ssDNA with the aid of QDE-3 and it blocks the formation of DNA/RNA hybrids. This mechanism explains how aRNA are produced and are specifically recognized by an RdRP in the quelling pathway.

Quelling also requires the activity of two partially redundant Dicer proteins (DCL-1 and DCL-2) which process dsRNA into 25 nt long siRNA (Catalanotto et al. 2004). The nuclease activity where the siRNA molecules are produced by Dicer is an energy-demanding process which requires ATP. The deletion of both Dicer genes in *N. crassa* completely abolishes quelling, while deletion of just one results in reduced quelling activity (Catalanotto et al. 2004). The *dcl-2* gene product associates with siRNA duplexes to form an RISC. An active RISC requires a single-stranded RNA molecule, which QDE-2 has the required slicer activity to

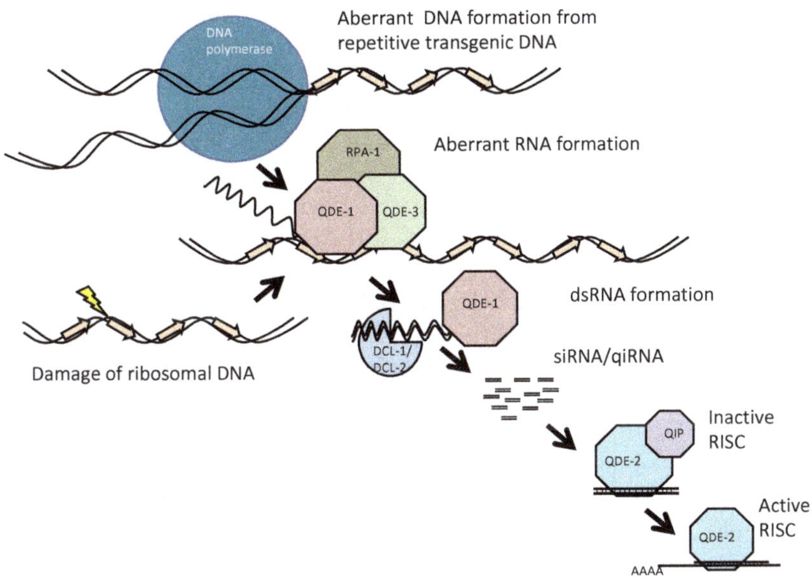

Fig. 1 A model for quelling and DNA damage-induced RNAi gene silencing in *Neurospora crassa*. During vegetative growth, both repetitive transgenes and damaged ribosomal DNA lead to aberrant RNA formation by QDE-1 and QDE-3 together with RPA-1 that is part of the DNA replication machinery. Single stranded aRNA is converted to dsRNA by QDE-1 and processed to siRNA/qiRNA by Dicer, DCL-1 or DCL-2. These siRNAs/qiRNAs are then loaded to the RISC consisting of QDE-2 and QIP. This complex is activated through the removal of one strand of the siRNA/qiRNA, where upon it bind to complementary mRNA and silence the gene. Abbreviations: RNAi, RNA interferens; QDE-1/2/3, Quelling defective 1/2/3; RPA-1, Replication protein A; aRNA, aberrant RNA; dsRNA, double stranded RNA; siRNA, small interfering RNA; qiRNA, QDE-2-interacting small RNAs; DCL-1, Dicer-like protein 1; RISC, RNA-induced silencing complex; QIP, QDE-2 interacting protein

produce (Maiti et al. 2007), providing evidence that the Argonaute protein is involved in production of the single strand siRNA. However, QDE-2 cannot do this by itself but interacts with a QDE-2 interacting protein (QIP), which has exonuclease activity required for the removal of the passenger strand (Maiti et al. 2007).

A model for the quelling pathway in *N. crassa*, based on the current knowledge, has been proposed by Chang and colleagues (2012). During replication, aDNA structures are formed from repetitive DNA sequences. The aDNA is then recognized by a complex of QDE-3 and RPA-1, which recruits QDE-1 to the ssDNA area (Nolan et al. 2008). Through its DdRP activity, QDE-1 produces aRNAs which are further converted to long dsRNAs through the RdRP activity of QDE-1. These long dsRNAs are then processed into siRNA duplexes by Dicer proteins. The siRNAs are loaded onto the RISC where they are cleaved by QDE-2, and the passenger strand is removed by the help of the QIP to form an active RISC. With guiding of the single-stranded siRNA, the activated RISC can now target complementary mRNA for degradation and silencing (Fig. 1).

Quelling is also triggered by DNA damage in vegetative tissue. Small RNAs that are induced upon DNA damage differ slightly from other siRNAs, e.g., they are 21–23 nt in length and have a 5′ uridine. These damage-induced siRNAs are most often generated from the rDNA locus and have been named qiRNA for QDE-2-interacting small RNAs (Lee et al. 2009). The biogenesis of qiRNAs requires QDE-1, QDE-3, and Dicers which show that their synthesis is dependent on the RNAi machinery (Lee et al. 2009). For instance, isolates mutated in *qde-3* are unable to produce aRNA which suggest that the Rec-Q helicase is involved in the damage-induced aRNA production. In addition, the RdRP QDE-1 is essential for the aRNA production after DNA damage. Furthermore, *dcl* and *qde-1* mutants have an increased sensitivity to DNA damage. Both transgene repeat silencing (quelling) and DNA damage-induced silencing (qiRNA) seem to use the same basic components and probably they share the similar molecular machinery for silencing (Lee et al. 2009, 2010a, b). Thus, it have been speculated that quelling also could be a result of DNA damage at the transgene integration site triggered by repetitive DNA (Chang et al. 2012).

B. Meiotic Silencing by Unpaired DNA

A process related to quelling known as **meiotic silencing of unpaired DNA** (MSUD) takes place during meiosis in the sexual cycle in *N. crassa* (Aramayo and Metzenberg 1996; Shiu and Glass 2000; Shiu and Metzenberg 2002). Ascomycetes like *N. crassa* are haploid during the vegetative state and become transiently diploid during meiosis after fertilization and prior to the formation of new ascospores. MSUD takes place in the prophase I of meiosis at the stage where unpaired DNA is present, which leads to the silencing of homologous genes in the diploid cell (Shiu and Metzenberg 2002). The recognition of unpaired DNA and the silencing are two distinct steps in MSUD, where the former occurs in the nucleus and the latter, where aRNA is converted to dsRNA and siRNA, is believed to at least partly take place in the cytoplasm. So far, very little is known about the recognition phase although DNA methylation appears to play a role (Pratt et al. 2004), while the silencing has much in common with quelling.

Two genes, *sad-1* and *sad-2* (suppression of ascus dominance 1 and 2), required for MSUD have been identified (Shiu and Metzenberg 2002; Shiu et al. 2006). *Sad-1* encodes RdRP that is a paralog of QDE-1 while the *sad-2* encoded protein has no conserved domains and no homology to any protein with known function. Together, SAD-1 and SAD-2 are localized to the perinuclear region and probably it is SAD-2 that recruits SAD-1 there, since a proper localization of SAD-1 requires SAD-2 (Shiu et al. 2006). In addition, *sms-2* (suppressor of meiotic silencing 2) which encodes an Argonaute homolog is required for MSUD (Lee et al. 2003). In contrast, QDE-2 is not required for MSUD which suggests that there are two parallel RNAi pathways functioning in the vegetative and meiotic phases in *N. crassa*. The major Dicer protein in quelling DCL-2 is not required for MSUD (Catalanotto et al. 2004). On the other hand, DCL-1 is colocalized with the other MSUD components, SAD-1, SAD-2, and SMS-2 (Alexander et al. 2008; Shiu et al. 2006). In addition, the QIP exonuclease, a component of the RISC of quelling, is also required for MSUD (Xiao et al. 2010). Furthermore, a putative RNA/DNA helicase, SAD-3, has been shown to be required for MSUD (Hammond et al. 2011). This is a homolog to *Hrr1* of *S. pombe*, which is a component of the RNAi-mediated heterochromatin formation for transcriptional silencing (see below). This suggests that this process and MSUD use similar components and might be conceptually similar. Also, SAD-3 associates with other MSUD components at the perinuclear region (Hammond et al. 2011). In addition, Hammond and colleagues (2013) identified two other partners in the MSUD process, SAD-4 and SAD-5. Although none of them shows sequence homology with proteins of known function, both are involved in the silencing. SAD-4 colocalizes with the known proteins at the perinuclear region, while SAD-5 seems to be the first protein identified which function within the nucleus (Hammond et al. 2013). This implies that the components assemble in a large complex that together converts aRNA to dsRNA and later to siRNA to achieve gene silencing (Fig. 2).

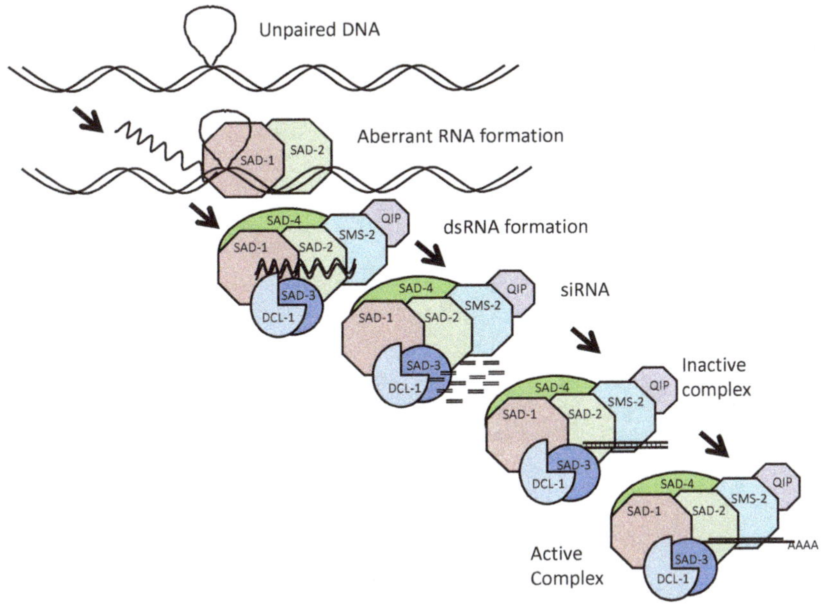

Fig. 2 A model for meiotic silencing in *Neurospora crassa*. During the meiotic prophase, unpaired DNA might trigger the transcription of aRNA by the two proteins SAD-1 and SAD-2. Then several proteins, SAD-1, SAD-2, SAD-3, SAD-4, DCL-1, SMS-2, and QIP, co-localize at the nuclear periphery to form a complex that first produce dsRNA, then siRNA, and then an active silencing complex. Abbreviations: aRNA, aberrant RNA; dsRNA, double stranded RNA; siRNA small interfering RNA; DCL-1, Dicer-like protein 1; SAD-1/2/3/4/5, suppression of ascus dominance; SMS-2 suppressor of meiotic silencing-2

C. Heterochromatin Formation and Transcriptional Gene Silencing

RNAi is a well-established conserved eukaryotic gene regulation mechanism that uses small noncoding RNAs to mediate posttranscriptional/transcriptional gene silencing (Chang et al. 2012). In filamentous fungi, RNAi mostly results in posttranscriptional gene silencing and has no role in transcriptional gene silencing (Dang et al. 2011). However, in the fission yeast, *S. pombe* RNAi mediates gene silencing at the transcriptional level by heterochromatin formation. Heterochromatin in *S. pombe* is distributed in three different loci: telomers, centromers, and mating type loci. Heterochromatin loci are highly condensed part of the chromosomes that harbor repetitive DNA, like transposable elements. These regions are hypoacetylated and coated with histone H3 that are methylated at lysin 9. These modifications create recognition sites for chromodomain proteins that maintain the transcriptionally silent status of heterochromatin. The formation of heterochromatin prevents expression of transposons, improper recombination of repetitive regions, and missegregation of chromosomes during mitosis and meiosis, thus maintaining genome stability (Grewal and Jia 2007).

In 2002, Volpe and colleagues (2002) found the first evidence for a direct link between heterochromatin formation and an RNAi pathway in the fission yeast when they showed that RNAi components are required for heterochromatin formation in the centromeric regions. Heterochromatin was shown to accumulate in isolates mutated in RNAi components, including RNA-dependent RNA polymerase (*rdp1*), Dicer (*dicer1*), and Argonaute (*ago1*) (Volpe et al. 2002). In heterochromatic regions, DNA repeats are transiently transcribed and the tran-

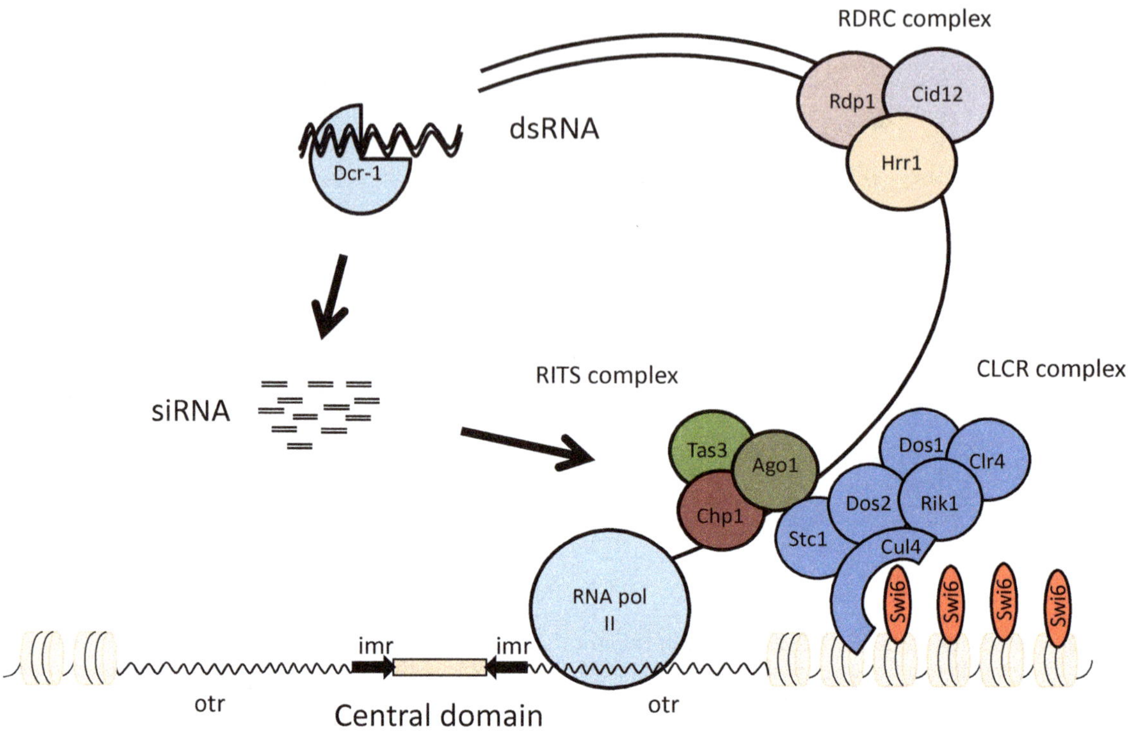

Fig. 3 A model for RNAi-mediated heterochromatin formation in *Schizoscaccharomyces pombe*. RNA pol II synthesizes a transcript from the centromeric region. From this transcript, the RDRC complex (Rdp1, Cid12, and Hrr1) generates dsRNA which is further processed by Dcr-1 to siRNA. These small RNAs are then guiding the RITS complex (Ago1, Tas3, and Chp1) to the nascent transcript. Now the CLCR complex (Dos1, Dos2, Cul4, Rik1, and Clr4) associates with the RITS complex through the Stc1 protein to come in close proximity to the chromatin. Here the Clr4 methylate the histone and allows Swi6 to dock at the modified chromatin and thereby form heterochromatin. This creates a positive feedback loop aiding the formation of heterochromatin. Abbreviation: imr, innermost repeats; otr, outermost repeats; RDRC, RNA-dependent RNA polymerase complex; Rdp1, RNA-dependent polymerase 1; Cid12, caffeine-induced death resistant 12; Hrr1, helicase required for RNAi-mediated heterochromatin assembly; dsRNA, double stranded RNA; Dcr-1, Dicer 1; siRNA, small interfering RNA; RITS, RNA-induced transcriptional silencing; Ago1, Argonaute 1; Tas3, tyrosine auxotrophy suppressor 3; Chp1, chromodomain-containing protein 1; CLCR, Clr4-Rik1-Cul4 complex; Clr4, Cryptic loci regulator 4; Rik1 RS2-interacting KH domain protein 1; Swi6, Switch/sucrose nonfermentable 6

scripts are processed by the RNAi machinery into siRNA (Fig. 3). These siRNAs originate from both DNA strands of the heterochromatin region indicating that they are made from a dsRNA precursor (Volpe et al. 2002). Nascent transcripts are made during the S-phase of the cell cycle by RNA-polymerase II and further processed by Ago1. They are subsequently reverse-transcribed by the RNA-directed RNA polymerase complex (RDRC) consisting of Rdp1, Cid12, and Hrr1 to produce double-stranded RNA molecules (Kato et al. 2005). The dsRNA is further processed to siRNAs by the Dcr1 which are loaded onto the Argonaute siRNA chaperone complex and then onto the **RNA-induced transcriptional silencing complex** (RITS) (Motamedi et al. 2004). The RITS complex consisting of Ago1, Tas3, and Chp1 is targeted to the repeat region through base paring between the siRNA and the nascent transcripts. The RITS complex then recruits another protein complex called CLRC (Clr4-Rik1-Cul4), which includes the H3K9 methyltransferase. It is the Stc1 protein that links RITS and the CLRC complexes together and thereby couple RNAi to chromatin modification (Bayne et al. 2010). Together, they allow Swi6 to dock at the modified chromatin to compact the chromatin to heterochromatin (Verdel et al. 2004). These interactions place the RITS complex in a central role, integrating transcription and chromatin modification. The interactions also create a positive feedback loop between siRNA, RITS complex localization, and histone 3 methylation.

There is growing evidence that nuclear RNAi may co-transcriptionally regulate loci outside heterochromatin in *S. pombe*. Nuclear RNAi is suggested to have a role in preventing read-through transcription at convergently transcribed genes, presumably through RNA polymerase release (Gullerova et al. 2011). In addition, Dcr1 physically interacts with genes in euchromatin regions, suggesting a role in gene regulation without interacting with histone modification (Woolcock et al. 2011). For instance, Dcr1 is involved in regulating heat stress-responsive genes through a thermo-switch. In unstressed cells, Dcr1 is located in the nucleus to repress stress-responsive genes but under heat stress Dcr1 is exported out of the nucleus where after the stress-responsive genes are activated (Woolcock et al. 2012).

D. Sex-Induced Silencing

A silencing process related to RNAi operates during the sexual cycle of *Cryptococcus neoformans* (Wang et al. 2010). This human pathogen undergoes a dimorphic transition from yeast to hyphae during opposite-sex (a-α) mating and during unisexual reproduction (Idnurm et al. 2005). Silencing of a tandem insertion during opposite sex reproduction is higher than during unisex mating and about 250-fold higher than during asexual mitotic vegetative growth (Wang et al. 2013). This silencing process was called **sex-induced silencing** (SIS) although it also operates during vegetative growth. This homology-based silencing process, which requires multicopy transgenes to be active, functions to limit the transposon activity during the sexual cycle, thus serving as a genome defense mechanism during meiosis (Wang et al. 2010). Small RNAs characterized from *C. neoformans* under both mating and vegetative growth map at a high frequency to transposable elements support a role in transposon control (Wang et al. 2010). Transposon activity is higher during sexual reproduction of RNAi mutants, but the RNAi machinery components are more abundant during mating as well. The diminishing of these siRNAs in the *rdp1* mutant supports the fact that SIS is related to RNAi and dependent on dsRNA (Wang et al. 2010). Another quelling-like co-suppression phenomenon is present in *C. neoformans*, named mitotic-induced silencing (Wang et al. 2012). Transformation of *cpa1* resulted in isolates with decreased mRNA levels of CPA1 and CPA2 genes. This silencing is dependent on RNAi components Rdp1, Ago1, and Dcr2 (Wang et al. 2012).

III. MicroRNAs

MicroRNA (miRNA) is a class of snRNAs with a biogenesis different from other snRNAs since they are generated through transcription from miRNA genes. The miRNA are small noncoding RNAs of 18–25 nucleotides that regulate gene expression by binding to the target mRNA. They were discovered in *C. elegans*, the first in the early 1990s (Lee et al. 1993) and the second one in 2000 (Reinhart et al. 2000). Soon thereafter, orthologous RNAs were found in mammals and have now been found in plants, green algae, and viruses as well (Llave et al. 2002; Ambros 2004; Bartel 2004; Molnar et al. 2007; Zhao et al. 2007). The number of confidently identified miRNAs in *C. elegans*, *Drosophila melanogaster*, and human has now surpassed 110, 140, and 400, respectively (Ruby et al. 2006, 2007; Landgraf et al. 2007). These numbers approaches 1–2 % of the total number of protein-coding genes in the respective organisms, indicating their importance in gene regulation. With advanced molecular and bioinformatics tools, numerous

Table 1 Biogenesis pathways of milRNA in *Neurospora crassa*

milRNA	Dicer	Argonaute	Exonuclease	RNAse III homolog
milR-1	DCL-1/2	QDE-2	QIP	MRPL3[a]
milR-2		QDE-2	Unknown enzyme	
milR-3	DCL-1/2			
milR-4	DCL-1/2		Unknown enzyme	MRPL3[a]

[a]Homolog of *Saccharomyces cerevisiae* mitochondrial ribosomal protein MRPL3

miRNAs have now been identified in all kinds of organisms, and presently >25,000 are available from the miRNA database, miRBase release 19.0 (Griffiths-Jones et al. 2006). Although small RNA pathways and many different small RNAs are known from fungi, it was not until early 2000 that the miRNA and its role in fungi began to be investigated.

miRNAs are today known as one of the most abundant regulator molecules for gene expression and have shown to play various roles in processes ranging from cell development, proliferation, differentiation, apoptosis, and carcinogenesis to immunity in animals (Ulitsky and Bartel 2013). In plants, miRNAs are involved in development, stress response, and antibacterial resistance (Allen and Howell 2010), but its function in fungi is less understood. Mature miRNA are produced from primary miRNA transcripts (pri-miRNA) by the endonuclease Drosha, producing a precursor hairpin structure of 60–70 nucleotides, termed pre-miRNA. These pre-miRNAs are then transported from the nucleus to the cytoplasm where they are cleaved by a Dicer to yield mature miRNAs (He and Hannon 2004). Studies of *N. crassa* have led to the discovery of miRNA-like RNAs (milRNAs) in filamentous fungi (Lee et al. 2010a, b). They discovered at least 25 potential milRNA loci which share many similarities with conventional miRNAs from animals; they originate from highly specific stem-loop RNA precursors; most of the milRNAs require Dicer for the biogenesis; and milRNAs may silence endogenous targets with imperfect complementarity (Lee et al. 2010a, b). These milRNAs have a clear preference for an uracil base at their 5′ termini and more heterogeneity at the 3′ end. The analysis revealed that there are at least four different biogenesis pathways, milR-1 to milR-4, for milRNAs in *N. crassa* (Table 1) (Lee et al. 2010a, b).

The milR-1 pathway is completely dependent on Dicer for the pre-milRNA and mature milRNA production, and the QDE-2 is required for the mature milRNA and the exonuclease activity of QIP. A biogenesis model have been put forward where the pri-milRNA is produced by Dicer to generate double-stranded pre-milRNA, to which QDE-2 binds and recruits the exonuclease QIP to produce the mature milRNA. milR-2 on the other hand is Dicer-independent but requires QDE-2. The milR-2 pri-milRNAs form a hairpin structure to which QDE-2 binds and cleaves one of the strands. Afterwards, an unknown exonuclease further cuts the pre-milRNA to generate the mature milRNA. The biogenesis of milRNAs through the milR-2 pathway is the first known Dicer-independent but Argonaute-dependent pathway to generate siRNAs. The milR-3 pathway is similar to the milRNA pathway in plants and requires Dicer for pre-milRNA and mature milRNAs production. milRNAs produced through the milR-4 pathway is only partially dependent on Dicer, which suggests that other nucleases might be involved in siRNA production. Lee and colleagues (2010a, b) identified a putative ribonuclease with homology to *S. cerevisiae* mitochondrial ribosomal protein L3 (MRPL3) to be involved in the biogenesis of milR-1 and milR-4 production.

IV. RNAi During Plant–Microbe Interaction

The plant pathogenic fungus, *Botrytis cinerea*, is the causative agent of gray mold and has a wide host range, infecting more than 200 plant species (Dean et al. 2012). During infection, *B. cinerea* takes advantages of the fact that the machinery for RNA interference is conserved within its plant hosts. Weiberg and colleagues (2013) identified three *B. cinerea* siRNA (Bc-siRNA) with homology to protein-coding genes with a likely function in the plant immunity. The Bc-siRNAs targets are mitogen-activated protein kinases 1 and 2 (MPK1 and MPK2), an oxidative stress-related gene, peroxiredoxin (PRXiiF), and a cell wall-associated kinase (WAK). During infection of *Arabidopsis*, both target genes are downregulated while typical defense marker genes are upregulated (Weiberg et al. 2013). The mechanism for this downregulation is that *B. cinerea* hijacks the plant RNAi machinery when the Bc-siRNA binds to the *Arabidopsis* Argonaut 1 (AGO1) protein, which in turn suppresses the host immunity by downregulation of the target genes (Weiberg et al. 2013). These findings raise a lot of interesting questions about the evolution of these host-related pathogen siRNAs. Did they arise through selection in the pathogen or are they gained through horizontal gene transfer? How are these Bc-siRNA delivered into the plant cell cytoplasm? Questions about how general this mechanism functions and if it is used in other interactions, not only in pathogenic ones, are to be addressed by future research.

V. RNAi as a Biotechnology Tool

Since RNA interference is a naturally occurring PTGS process in which endogenous gene expression is reduced, it has been exploited as a tool in fungal research and in diverse biotech applications. Here, selected illustrative examples are highlighted. The readers are referred to the paper by Salame and colleagues (2011) for a more detailed summary on particular systems where RNAi has been successfully applied. In their paper, they listed examples from more than 40 species in which RNAi have proven to modify gene expression and the number of example is constantly growing. The technology has the possibility not only to downregulate the expression of one gene but also to target several homologous genes. This provides a fantastic opportunity since it opens up possibilities to modify gene expression in pathways with functional redundant genes, but it also makes the technology susceptible for undesired non-target effects which have to be considered when designing an experiment.

Aspergillus oryzae, the fungus most widely used in fermentation industries in Japan and China to produce sake and soy sauce, is also used for large scale heterologous production of enzymes and proteins (Machida et al. 2008). The yield of heterologously produced protein is generally low, and one important factor for this limitation is the degradation of secreted proteins by native fungal proteases present in the culture medium (Archer et al. 1992). To circumvent this problem Zheng et al. (1998) used RNAi to downregulate the expression of a serine type carboxypeptidase (CPase) gene to produce a strain with low extracellular peptidase activity. The strain produced about 30 % of the peptidase of the untransformed strain (Zheng et al. 1998). The transformed strain also showed a reduced activity of other CPases indicating that the construct not only targeted a specific gene but were able to control the activity of CPases as a group.

In *Mortierella alpina,* Takeno and colleagues (2005) used RNAi to silence a delta 12-desaturase, a key enzyme in the arachidonic acid biosynthetic pathway, which desaturates oleic acid (18:1$n-9$) to linoleic acid (18:2$n-6$). Silencing of the desaturase not only lead to the accumulation of 18:2$n-9$ but also to accumulation of the fatty acids 20:2$n-9$ and 20:3$n-9$ (Mead acid), which had not been detected in the original strain (Takeno et al. 2005). This shows that RNAi silencing technology can be used to quantitatively as well as qualitatively alter the fatty acid production of commercial interesting fungi.

Genes in basidiomycetes have also been targeted for silencing through RNAi. For instance, Kemppainen et al. (2009) have implemented RNAi as a tool in the investigation of nitrogen nutrition of the symbiotic mushroom

Laccaria bicolor. This ectomycorrhizal fungus is believed to deliver nitrogen and other nutrients to its tree host in exchange for carbon, and understanding its nitrogen metabolism is important to understand the symbiotic interaction. Knockdown of a nitrate reductase gene resulted in a strain of *L. bicolor* that was strongly affected in its ability to grow with nitrate as its sole nitrogen source (Kemppainen et al. 2009). The phenotype correlated with a clear reduction of mRNA levels of the target gene. Furthermore, Kemppainen and Pardo (2010) have developed a versatile vector system for gene silencing in *L. bicolor* and other hygromycin-sensitive homobasidiomycetes.

Salame et al. (2010) have used a reverse genetics approach to silence the *Pleurotus ostreatus mnp3* gene using RNAi. *mnp3* encodes one of the nine Mn-peroxidases present in the genome of *P. ostreatus*, enzymes most commonly known for their capacity to degrade lignin. RNAi silencing and the subsequent analysis of the gene expression of all MnP gene family members and the fungus' capacity to decolorize Orange II dye demonstrated sequence-specific characteristics in efficiency to reduce the expression of different genes in the family (Salame et al. 2010). This shows that RNAi can efficiently be used to analyze functional redundant gene families when members share high sequence similarities, and their gene expression can be altered simultaneously. In the review by Salame et al. (2011), an algorithm was compiled for predicting the effect of particular RNAi constructs on various target and nontarget genes. This suggests that the target properties of an RNAi construct could be predicted for a rational design, especially when gene family members and the genome sequence of the host are known.

This target-independent tool show high potential in a wide range of fungal species, especially filamentous fungi that are multinucleated or lack available DNA-based systems. Although almost all investigated species have the basic components for the RNAi silencing machinery, it is important to investigate the possibilities in the target host species before designing an experiment (Hu et al. 2013).

VI. Conclusion

Fungal genome sequencing has demonstrated that RNA interference is most likely a widespread phenomenon in the fungal kingdom. The central components Dicer, Argonaute, and RNA-dependent RNA polymerases have been found in almost all species examined, although the mechanisms and details of the silencing systems are not known. RNAi participates in gene silencing in both the vegetative and sexual phases of fungal life cycles as a genome defense mechanism. It also regulates gene expression at the transcriptional and post-transcriptional level. A few systems are studied in detail from which we learned much of the mechanisms of regulation, specificity, and amplification but also many species-specific differences have been discovered indicating a highly adaptable mechanism with a wide diversity in function. In the future, we will probably encounter RNAi as a common mechanism for gene regulation in a wide variety of processes and new variations of the mechanism that control the silencing process.

Acknowledgment The author would like to thank The Swedish Foundation for Strategic Research for financial support. Dr M. Lind is acknowledged for his constructive comments on this chapter.

References

Alexander WG, Raju NB, Xiao H, Hammond TM, Predue TD et al (2008) DCL-1 colocalizes with other components of the MSUD machinery and is required for silencing. Fungal Genet Biol 45:719–727

Allen E, Howell MD (2010) miRNAs in the biogenesis of *trans*-acting siRNAs in higher plants. Semin Cell Dev Biol 21:798–804

Ambros V (2004) The function of animal microRNAs. Nature 431:350–355

Aramayo R, Metzenberg RL (1996) Meiotic transvection in fungi. Cell 86:103–113

Archer DB, Mackenzie DA, Jeenes DJ, Roberts IN (1992) Proteolytic degradation of heterologous protein expressed in *Aspergillus niger*. Biotechnol Lett 14:357–362

Bartel DP (2004) MicroRNAs: genomics, biogenesis, mechanism, and function. Cell 116:203–207

Bayne EH, White SA, Kagansky A, Bijos D, Sanchez-Pulido L, Hoe K-L, Kim D-U, Park H-O, Ponting CP, Rappsilber J, Allshire RC (2010) Stc1: A critical link between RNAi and chromatin modification required for heterochromatin integrity. Cell 140:666–677

Billmyre RB, Calo S, Feretzaki M, Wang X, Heitman J (2013) RNAi function, diversity, and loss in the fungal kingdom. Chromosome Res 21:561–572

Catalanotto C, Pallotta M, ReFalo P, Sachs MS, Vayssie L et al (2004) Redundancy of the two Dicer genes in transgene-induced posttranscriptional gene silencing in *Neurospora crassa.* Mol Biol 24:2536–2545

Chang S, Zhang Z, Liu Y (2012) RNA interference pathways in fungi: mechanisms and function. Annu Rev Microbiol 66:305–323

Cogoni C, Macino G (1997) Conservation of transgene-induced post-transcriptional gene silencing in plants and fungi. Trends Plant Sci 2:205–443

Cogoni C, Macino G (1999) Gene silencing in *Neurospora crassa* requires a protein homologous to RNA-dependent RNA polymerase. Nature 399:166–169

Cogoni C, Irelan JT, Schumacher M, Schmidhauser TJ, Selker EU, Macino G (1996) Transgene silencing of the *al-1* gene in vegetative cells of *Neurospora* is mediated by a cytoplasmic effector and does not depend on DNA-DNA interactions or DNA methylation. EMBO J 15:3153–3163

Dang Y, Yang Q, Xue Z, Liu Y (2011) RNA interference in fungi: pathways, functions, and applications. Eukaryotic Cell 10:1148–1155

David L, Huber W, Granovskaia M, Toedling J, Palm CJ, Bofkin L et al (2006) A high-resolution map of transcription in the yeast genome. Proc Natl Acad Sci USA 103:5320–5325

Dean R, Van Kan JAL, Pretorius ZA, Hammond-Kosack KE, Di Pietro A, Spanu PD, Rudd JJ, Dickman M, Kahmann R, Ellis J, Foster GD (2012) The top 10 fungal pathogens in molecular plant pathology. Mol Plant Pathol 13:414–430

Donaldson ME, Saville BJ (2012) Natural antisense transcripts in fungi. Mol Microbiol 85:405–417

Drinnenberg IA, Weinberg DE, Xie KT, Mower JP, Wolf KH, Fink GR, Bartel DP (2009) RNAi in budding yeast. Science 333:544–550

Drinnenberg IA, Fink GR, Bartel DP (2011) Compatibility with killer explains the rise of RNAi-deficient fungi. Science 333:1592

Fire A, Xu S, Montgomery MK, Kostas SA, Driver SE, Mello CC (1998) Potent and specific genetic interference by double-stranded RNA in *Caenorhabditis elegans.* Nature 391:806–811

Grewal SIS, Jia S (2007) Heterochromatin revisited. Nat Rev Genet 8:35–46

Griffiths-Jones S, Grocock RJ, van Dongen S, Bateman A, Enright AJ (2006) miRBase: microRNA sequences, targets and gene nomenclature. Nucleic Acids Res 34(Database issue):D140–D144

Gullerova M, Moazed D, Proudfoot NJ (2011) Autoregulation of convergent RNAi genes in fission yeast. Genes Dev 25:556–568

Hammond TM, Xiao H, Boone EC, Predue TD, Pukkila PJ, Shiu PK (2011) SAD-3, a putative helicase required for meiotic silencing by unpaired DNA, interacts with the other components of the silencing machinery. Genes Genomes Genet 1:369–376

Hammond TM, Xiao H, Boone EC, Decker LM, Lee SA, Predue TD, Pukkila PJ, Shiu PK (2013) Novel proteins required for meiotic silencing by unpaired DNA and siRNA generation in *Neurospora crassa.* Genetics 194:91–100

He L, Hannon GJ (2004) MicroRNAs: small RNAs with a big role in gene regulation. Nat Rev Genet 5:522–531

Hu Y, Stenlid J, Elfstrand M, Olson Å (2013) Evolution of RNA interferance proteins dicer and argonaute in Basidiomycota. Mycologia 105:1489–1498

Idnurm AYS, Bahn K, Nielsen K, Lin X, Fraser JA et al (2005) Deciphering the model pathogenic fungus *Cryptococcus neoformans.* Nat Rev Microbiol 3:753–764

Janbon G, Maeng S, Yang D-H, Ko Y-J, Jung K-W, Moyrand F, Floyd A, Heitman J, Bahn Y-S (2010) Characterizing the role of RNA silencing components in *Cryptococcus neoformans.* Fungal Genet Biol 47:1070–1080

Kato H, Goto DB, Martienssen RA, Urano T, Furukawa K, Marakami Y (2005) RNA polymerase II is required for RNAi-dependent heterochromatin assembly. Science 309:467–469

Kellis M, Wold B, Snyder MP, Berstein BE, Kundaje A (2014) Defining functional DNA elements in the human genome. PNAS 111:6131–6138

Kemppainen MJ, Pardo AG (2010) pHg/pSILBAgamma vector system for efficient gene silencing in homobasidiomycetes: optimization of ihpRNA-triggering in mycorrhizal fungus Laccaria bicolor. Microb Biotechnol 3:178–200

Kemppainen MJ, Duplessis S, Martin F, Pardo AG (2009) RNA silencing in the model mycorrhizal fungus *Laccaria bicolor*: gene knock-down of nitrate reductase in inhibition of symbiosis with *Populus.* Environ Microbiol 11:1878–1896

Landgraf P, Rusu M, Sheridan R, Sewer A, Iovino N, Aravin A, Pfeffer S, Rice A, Kamphorst AO, Landthaler M, Lin C, Socci ND, Hermida L, Fulci V, Chiaretti S, Foà R, Schliwka J, Fuchs U, Novosel A, Müller RU, Schermer B, Bissels U, Inman J, Phan Q, Chien M, Weir DB, Choksi R, De Vita G, Frezzetti D, Trompeter HI, Hornung V, Teng G, Hartmann G, Palkovits M, Di Lauro R, Wernet P, Macino G, Rogler CE, Nagle JW, Ju J, Papavasiliou FN, Benzing T, Lichter P, Tam W, Brownstein MJ, Bosio A, Borkhardt A, Russo JJ, Sander C, Zavolan M, Tuschl T (2007) A mammalian microRNA expression atlas based on small RNA library sequencing. Cell 129:1401–1414

Lee RC, Feinbaum RL, Ambros V (1993) The *C. elegans* heterochronic gene lin-4 encodes small RNAs with antisense complementarity to lin-14. Cell 75:843–854

Lee DW, Pratt RJ, McLaughlin M, Aramayo R (2003) An Argonaute-like protein is required for meiotic silencing. Genetics 164:821–828

Lee HC, Chang SS, Choudhary S, Aalto AP, Maiti M et al (2009) qiRNA is a new type of small interfering RNA induced by DNA damage. Nature 459:274–277

Lee HC, Aalto AP, Yang Q, Chang SS, Huang G et al (2010a) The DNA/RNA-dependent RNA polymerase QDE-1 generates aberrant RNA and dsRNA for RNAi in a process requiring replication protein A and a DNA helicase. PLoS Biol 8:e1000496

Lee HC, Li L, Gu W, Xue Z, Crosthwaite SK, Pertsemlidis A, Lewis ZA, Freitag M, Selker EU, Mello CC, Liu Y (2010b) Diverse pathways generate MicroRNA-like RNAs and dicer-independent small interfering RNAs in fungi. Mol Cell 38:803–814

Llave C, Kasschau KD, Rector MA, Carrington JC (2002) Endogenous and silencing-associated small RNAs in plants. Plant Cell 14:16051619

Machida M, Yamada O, Gomi K (2008) Genomics of *Aspargillus oryzae*: learning from the history of koji mold and exploration of its future. DNA Res 15:173–183

Maiti M, Lee HC, Liu Y (2007) QIP, a putative exonuclease, interacts with the *Neurospora* Argonaute protein and facilitates conversion of duplex siRNA into single strands. Genes Dev 21:590–600

Molnar A, Schwach F, Studholme DJ, Thuenemann EC, Baulcombe DC (2007) miRNAs control gene expression in the single-cell alga *Chlamydomonas reinhardtii*. Nature 447:1126–1129

Motamedi MR, Verdel A, Colmenares SU, Gerber SA, Gygi SP, Moazed D (2004) Two RNAi complexes, RITS and RDRC, physically interact and localize to noncoding centromeric RNAs. Cell 119:789–802

Nicolás FE, Torres-Martinez S, Ruiz-Vazquez RM (2003) Two classes of small antisense RNAs in fungal RNA silencing triggered by non-integrative transgenes. EMBO J 22:3983–3991

Nolan T, Cecere G, Mancone C, Alonzi T, Tripodi M, Catalanotto C, Cogoni C (2008) The RNA-dependent RNA polymerase essential for post-transcriptional gene silencing in *Neurospora crassa* interacts with replication protein A. Nucleic Acids Res 36:532–538

Pratt RJ, Lee DW, Aramayo R (2004) DNA methylation affects meiotic trans-sensing, not meiotic silencing, in *Neurospora*. Genetics 168:1925–1935

Reinhart BJ, Slack FJ, Basson M, Pasquinelli AE, Bettinger JC et al (2000) The 21-nuleotide *let-7* RNA regulates development timing in *Caenorhabditis elegans*. Nature 403:901–906

Romano N, Macino G (1992) Quelling: transient inactivation of gene expression in *Neurospora crassa* by transformation with homologous sequences. Mol Microbiol 6:3343–3353

Ruby JG, Jan C, Player C, Axtell MJ, Lee W, Nusbaum C, Ge H, Bartel DP (2006) Large-scale sequencing reveals 21U-RNAs and additional microRNAs and endogenous siRNAs in *C. elegans*. Cell 127:1193–1207

Ruby JG, Stark A, Johnston WK, Kellis M, Bartel DP, Lai EC (2007) Evolution, biogenesis, expression, and target predictions of a substantially expanded set of *Drosophila* microRNAs. Genome Res 17:1850–1864

Salame TM, Yarden O, Hadar Y (2010) *Pleurotus ostreatus* manganese-dependent peroxidase silencing impairs decolourization of Orange II. Microb Biotechnol 3:93–106

Salame TM, Ziv C, Hadar Y, Yarden O (2011) RNAi as a potential tool for biotechnological applications in fungi. Appl Microbiol Biotechnol 89:501–512

Shabalina SA, Koonin EV (2008) Origins and evolution of eukaryotic RNA interference. Trends Ecol Evol 23:578–587

Shiu PK, Glass NL (2000) Cell and nuclear recognition mechanisms mediated by mating type in filamentous ascomycetes. Curr Opin Microbiol 3:183–188

Shiu PK, Metzenberg RL (2002) Meiotic silencing by unpaired DNA: properties, regulation and suppression. Genetics 161:1483–1495

Shiu PK, Zickler D, Raju NB, Ruprich-Robert G, Metzenberg RL (2006) SAD-2 is required for meiotic silencing by unpaired DNA and perinuclear localization of SAD-1 RNA-directed RNA polymerase. Proc Natl Acad Sci USA 103:2243–2248

Takeno S, Sakuradani E, Tomi A, Inohara-Ochiai M, Kawashima H, Ashikari T, Shimizu S (2005) Improvement of the fatty acid composition of an oil-producing filamentous fungus, *Mortierella alpina* 1S-4, through RNA interference with delta 12-desaturase gene expression. Appl Environ Microbiol 71:5124–5128

Ulitsky I, Bartel DP (2013) lincRNAs: genomics, evolution, and mechanisms. Cell 154:26–46

Verdel A, Jia S, Gerber S, Sugiyama T, Gygi S et al (2004) RNAi-mediated targeting of heterochromatin by the RITS complex. Science 303:672–676

Volpe TA, Kidner C, Hall IM, Teng G, Grewal SIS, Martienssen RA (2002) Regulation of heterochromatic silencing and histone H3 lysine-9 methylation by RNAi. Science 297:1833–1837

Wang X, Hsueh Y-P, Li W, Floyd A, Skalsky R, Heitman J (2010) Sex-induced silencing defends the genome of *Cryptococcus neoformans* via RNAi. Genes Dev 24:2566–2582

Wang X, Wang P, Sun S, Darwiche S, Idnurm A, Heitman J (2012) Transgen induced co-suppression during vegetative growth in *Cryptococcus neoformans*. PLos Genet 8:e1002885

Wang X, Darwiche S, Heitman J (2013) Sex-induced silencing operates during opposite-sex and unisexual reproduction in *Cryptococcus neoformans*. Genetics 193:1163–1174

Weiberg A, Wang M, Lin F-M, Zhao H, Zhang Z, Kaloshian I, Huang H-D, Jin H (2013) Fungal small RNAs suppress plant immunity by hijacking host RNA interference pathways. Science 342:118–123

Wianny F, Zernicka-Goetz M (2000) Specific interference with gene function by double-stranded RNA in early mouse development. Nat Cell Biol 2:70–75

Wilhelm BT, Marguerat S, Watt S, Schubert F, Wood V, Goodhead I et al (2008) Dynamic repertoire of a eukaryotic transcriptome surveyed at single-nucleotide resolution. Nature 453:1239–1243

Woolcock KJ, Gaidatzis D, Punga T, Bühler M (2011) Dicer associates with chromatin to repress genome activity in *Schizosaccharomyces pombe*. Nat Struct Mol Biol 18:94–99

Woolcock KJ et al (2012) RNAi keeps Atf1-bound stress response genes in check at nuclear pores. Genes Dev 26:683–692

Xiao H, Alexander WG, Hammond TM, Boone EC, Perdue TD et al (2010) QIP, a protein that converts duplex siRNA into single strands, is required for meiotic silencing by unpaired DNA. Genetics 186:119–126

Zhao T, Li G, Mi S, Li S, Hannon GJ, Wang XJ, Qi Y (2007) A complex system of small RNAs in the unicellular green alga Chlamydomonas reinhardtii. Genes Dev 21:1190–1203

Zheng XF, Kobayashi Y, Takeuchi M (1998) Construction of a low-serine-type-carboxypeptidase-producing mutant of *Aspargillus oryzae* by the expression of antisense RNA and its use a host for heterologous protein secretion. Appl Microbiol Biotechnol 49:39–44

4 Fungal Molecular Response to Heavy Metal Stress

Matthias Gube[1,2]

CONTENTS

Abbreviations

ABC transporters	ATP-binding cassette transporters (Vasiliou et al. 2009; Dassa and Bouige 2001)
CDF transporter	Cation diffusion facilitator transporters (Paulsen and Saier 1997)
CTR transporters	Copper transport protein-like transporters (Dumay et al. 2005)
EPS	Extracellular polymeric substance
GSH	Glutathione
MATE transporters	Multidrug and toxic compound extrusion transporters (Hvorup et al. 2003)
MRP transporters	Multidrug resistant-associated protein-type transporters, members of ABC transporter subfamily C (Vasiliou et al. 2009; Dassa and Bouige 2001)
MT	Metallothioneine
NRAMP transporters	Natural resistance-associated macrophage protein-type transporters (Nevo and Nelson 2006)
PC	Phytochelatin
ROS	Reactive oxygen species
SCF complex	Skp, Cullin, F-box containing complex, promotes ubiquitination (Deshaies 1999)
ZIP transporters	Zrt, Irt-like protein-type transporters (Guerinot 2000)

[1]Soil Science of Temperate Ecosystems, Georg August University Göttingen, Büsgenweg 2, 37077 Göttingen, Germany; e-mail: mgube@uni-goettingen.de
[2]Microbial Communication, Friedrich Schiller University Jena, Neugasse 25, 07743 Jena, Germany; e-mail: mgube@uni-goettingen.de

Biochemistry and Molecular Biology, 3rd Edition
The Mycota III
D. Hoffmeister (Ed.)

I. Introduction

Contamination of soil with heavy metals is a growing problem in the industrialized world. In Europe alone, heavy metals are, together with petrol compounds, the major cause of currently three million polluted sites. These numbers are expected to increase by 50 % by 2025 (EEA 2010). Heavy metal contamination poses a threat not only to ecosystem function but also raises health risks due to accumulation in the food chain. Thus, understanding the role of soil inhabiting organisms in heavy metal mobilization and immobilization, transport and transfer is of crucial importance.

Compared with plants, fungi can generally accumulate far higher amounts of heavy metals in their fruiting bodies (Byrne et al. 1976; Kalač and Svoboda 2000). After the reactor accident of Chernobyl in 1986, it became clear that fungal uptake of radionuclides such as ^{60}Co, ^{90}Sr, and ^{137}Cs may reach harmful levels regarding human consumption. Fungal heavy metal accumulation is so massive that isotope ratios within fungal mycelia are indicative for radioactive contaminations caused by reactor accidents and nuclear weapon tests (Dighton and Horrill 1988; Haselwandter et al. 1988; Rühm et al. 1998). However, metal ion accumulation by fungi is not restricted to radioisotopes. Metatranscriptome analyses at heavy metal-polluted sites suggest a dominant role of fungi in heavy metal response of the affected soil biota (Damon et al. 2012; Lehembre et al. 2013). Since the 1970s, high cadmium contents were detected in species of the genus *Agaricus*, where the Cd ions were found to be bound to phosphoglycoproteins (Meisch et al. 1983; Stijve and Besson 1976). Considering the relevance of *Agaricus* for human consumption, the topic still attracts considerable attention (Sun et al. 2012; Wang et al. 2014). On the other hand, the documented substantial accumulation capacities led to the suggestion of implementing fungi for toxic metal extraction by harvesting metal hyperaccumulating fruiting bodies (García et al. 2005; Gray 1998; Schindler et al. 2012).

Fungi have efficient means to access even mineral-bound elements, to take them up, and, if they are hyphal, translocate them within the mycelium and incorporate them into fruiting bodies and also their symbiotic mycorrhizal partners. This allows for colonization of otherwise unfavorable habitats by both symbionts. However, toxic elements can become mobilized and incorporated by the very same mechanisms. To avoid or at least reduce their detrimental effects, numerous ways of resistance have evolved in fungi. All these mechanisms for uptake and excretion, transport, and deposition have to be coordinated to maintain metal homoeostasis, with respect to all external abiotic and biotic factors. Fungal action to cope with metals also influences the performance of processes involved in environmental metal cycling, such as mineralization, uptake into the food chain, and mobilization and washout into ground and surface water (Büchel et al. 2005; Gadd 2001, 2007; Urban 2011). Also, metal tolerance may confer advantage against competitors in promoting resistance against metal-sensitive antagonists (Naar 2006).

Uptake of metal ions into larger fungal fruiting bodies was quantified for numerous species (e.g., Borovička et al. 2011; Byrne et al. 1976; García et al. 1998; Gast et al. 1988; Haas and Purvis 2006; Kalač and Svoboda 2000; Sarikurkcu et al. 2010; Schindler et al. 2012; Seeger 1978; Svoboda et al. 2000; Urban et al. 2005) and also metal transfer into fungal mycelia and mycorrhizal structures (e.g., Bazała et al. 2008; Brunnert and Zadražil 1983; Colpaert et al. 2005; Frey et al. 2000; Jarosz-Wilkołazka et al. 2006; Lanfranco et al. 2002a; Sácký et al. 2014; Turnau et al. 1994), and fungal weathering mobilizing metal ions was analyzed under field and laboratory conditions (e.g., Bonneville et al. 2011; Bornyasz et al. 2005; Burford et al. 2003; Burgstaller and Schinner 1993; Ehrlich 2006; Fomina et al. 2004, 2007; Hoffland et al. 2004; Hopf et al. 2009; Lee and Parsons 1999; Sterflinger 2000; Wengel et al. 2006; Williams and Rudolph 1974).

The molecular response towards toxic metal exposure is well characterized for ascomycete model organisms, especially for *Saccharomyces cerevisiae* (e.g., Culotta et al. 2005; Hosiner et al. 2014; Jin et al. 2008; Mehra and Winge 1991; Thorsen et al. 2007, 2009; Van Ho et al.

2002; Vido et al. 2001; Wysocki and Tamás 2010) and *Schizosaccharomyces pombe* (e.g., Chen et al. 2003; Clemens and Simm 2003; Kennedy et al. 2008). The transcriptomic or proteomic response upon toxic metal exposure was also explored for the glomeromycete *Rhizophagus* (formerly *Glomus*) *intraradices* (Ouziad et al. 2005), the ascomycetes *Botrytis cinerea* (Cherrad et al. 2012), *Cadophora finlandica* (Gorfer et al. 2009), *Candida albicans* (Enjalbert et al. 2006), *Hansenula polymorpha* (Park et al. 2007b), *Exophiala pisciphila* (Zhao et al. 2015), *Oidiodendron maius* (Chiapello et al. 2015), and *Tuber melanosporum* (Bolchi et al. 2011), and the basidiomycetes *Agaricus blazei* (Wang et al. 2014), *Fibroporia radiculosa* (Tang et al. 2013), *Pisolithus albus* (Majorel et al. 2012), *Paxillus involutus* (Jacob et al. 2004), *Suillus luteus* (Muller et al. 2007; Ruytinx et al. 2011), and *Ganoderma lucidum* (Chuang et al. 2009), mostly under Cd and/or Zn stress, in some cases also after Cu exposure. It became clear from this studies that several response factors function in a similar way in all studied fungi. However, profound diversity in regulation and prevalence of the employed strategies was shown at the same time (Chen et al. 2003; Ehrensberger and Bird 2011). Among other reasons, this is due to diverse fungal ontogenetic features including unicellular yeasts to large mycelial networks and ancient lichen individuals, which led to an evolution of enormous morphological, chemical, and genetic diversity.

Some heavy metals, like Zn, Cu, or Mn, are essential micronutrients but become toxic at higher concentrations. For these metals, toxicity may be exerted at concentrations close to the actual metabolic requirements (Hughes and Poole 1991). On the other hand, many heavy metals such as Cd, Pb, or Hg are toxic even at low concentrations. These usually have no biological function, although exceptions exist (Park et al. 2007a), and are accumulated in competition to essential elements with similar chemical properties or atomic radii (Bellion et al. 2006; Gadd 2001; Urban 2011).

Metal ions can bind to proteins at different moieties, usually forming chelate complexes, but also covalently, as seen in metal ion-triggered DNA protein crosslinking (Kasprzak 2002). Whether they prefer thiols, oxygen, or nitrogen largely depends on their oxidation state. Soft transition metals (Ag, Cd, Hg) prefer sulfur groups, while hard transition metals and metalloids (Cr, Mn, Mo, As, Sb, Se, Te, Bi) prefer oxygen in their higher and sulfur in their lower oxidative state. Metals such as Co, Cu, Fe, Ni, Pb, and Zn may indiscriminately bind to oxygen, sulfur, or nitrogen residues. While this explains specific effects of these metals, fungi utilize these binding preferences also in their detoxification mechanisms (Summers 2009; Wysocki and Tamás 2010). As seen from the above-mentioned binding preferences, sulfur binding plays a dominant role in encountering metal ions. Therefore, additional stress is induced by the depletion of available sulfur (Fauchon et al. 2002).

Metal ions also cause oxidative stress in mediating the formation of **reactive oxygen species** (ROS) including singlet oxygen, the superoxide anion, the hydroxyl radical, and hydrogen peroxide. The mechanisms involved include **Fenton reactions** by redox-active metals such as Fe, Cu, Cr, Co, Ni, and V. In contrast, redox-inactive metals like Cd, Hg, and Pb cause oxidative stress mainly by binding to thiol groups, which may not only lead to depletion of antioxidants such as glutathione and displacement of Zn or the redox-active Fe(II) from metalloproteins but can also directly inactivate enzymes (Enjalbert et al. 2006; Stohs and Bagchi 1995; Valko et al. 2005; Wysocki and Tamás 2010). Zn was described to be oxidatively inert and thus should does not trigger ROS formation (Maret and Li 2009); it may even act as antioxidant (Valko et al. 2005). Especially wood decaying fungi are associated with high production of ROS, for example, due to the deliberate involvement of ROS generation by fungus-triggered Fenton reactions in brown rot. Thus, effective antioxidative mechanisms are crucial (Arantes et al. 2011; Enoki et al. 1997; Jaszek et al. 2006). As relatively weak mutagens, metal ions have been proposed to cause mutagenic effects predominantly by disturbing DNA repair processes, although direct mutagenic effects have been described as well (Beyersmann and Hartwig 2008; Jin et al. 2003). Cd ions are able to replace Zn in metalloproteins, potentially inactivating them, and similar effects can be expected from other metals (Hartwig 2001; Palacios et al. 2011; Schützendübel and Polle 2002). Also, misfolding and aggregation of proteins may be caused by metal and metalloid ions, causing inactivation and potentially subsequent ubiquitination and degradation (Tamás et al. 2014); also, the cytoskeleton organization may be strongly affected (Kennedy et al. 2008; Tuszynska et al. 2006).

To avoid or at least reduce the detrimental effects of metal toxicity, a number of resistance mechanisms have evolved in fungi, which will be reviewed in this chapter. Fe related mechanisms will only be mentioned when other ions are affected, as this topic has been dealt with extensively in several reviews (Haas 2003; Miethke and Marahiel 2007; Schrettl and Haas 2011; Kosman 2003). Also, general stress response mechanisms, which are also triggered by toxic metal exposure, will not be covered in great depth. Detailed studies on this topic are available for several ascomycete model organisms (Causton et al. 2001; Chen et al. 2003; Gasch 2007; Gasch et al. 2000; Rupp 2008).

II. Thou Shall Not Pass!: Extracellular Response upon Metal Toxicity

If toxic areas cannot be avoided by negative chemotaxis (Fomina et al. 2000), extracellular mechanisms are utilized as a first reaction upon toxic metal exposure to reduce the mobile fraction of the heavy metals (Pócsi 2011). Passive sorption to cell wall constituents requires no additional investment of nutrients or energy, while active excretion of chelating compounds requires energy and essential elements, specifically carbon and sulfur. Metal biosorption is rather efficient; fungal cells surpass even clay minerals in their relative sorption values (Morley and Gadd 1995). This is industrially utilized for remediation of metal contaminated water by fungal biomass (Kapoor and Viraraghavan 1995). However, potential sorption sites are prone to be saturated quickly in high-metal conditions; cell wall biosorption may thus play a minor role in natural environments (Jentschke and Godbold 2000).

A. Metal Sorption to the Cell Wall

Exposed carboxy, hydroxy, acetamido, amino, mercapto, or other active functional groups are among the sorption-active molecular components (Baldrian 2003; Say et al. 2001; Volesky and Holan 1995). The major components of fungal cell walls, such as chitin, chitosan, and diverse mucopolysaccharides, thus are all capable of significant metal ion sorption (Gadd 2010; González-Guerrero et al. 2009; Kapoor and Viraraghavan 1995). The composition of the cell wall does not remain constant during the life of an individual hypha, and the metal ion-chelating ability was shown to depend on mycelial age (Yetis et al. 2000). For more information on the fungal cell wall and chitin biosynthesis, see chapters "The Cell Wall Polysaccharides of *Aspergillus fumigatus*" and "Chitin Synthesis and Fungal Cell Morphogenesis" in this volume.

The structure and formation of several chitin–metal complexes has been evaluated in vivo (Bhanoori and Venkateswerlu 2000; Skorik et al. 2010; Tsezos and Volesky 1982). Despite the passive mode of action in terms of metal ion binding, the fungal cell wall may be actively modified upon metal exposure. **Chitosan** differs from chitin in its sorption characteristics (Kapoor and Viraraghavan 1995). Consequently, chitin deacetylase promoting chitosan generation was induced by Cd ions in *Agaricus blazei* (Wang et al. 2014). Treatment of an unidentified ericoid mycorrhizal fungus with millimolar concentrations of Zn led to deviating morphology including hyperbranching, as well as increased amounts of chitin in the affected hyphae (Lanfranco et al. 2002a). Similar morphological changes were observed for numerous basidiomycete hyphae upon metal exposure (Baldrian 2003; Grąz et al. 2009).

Several genes involved in regulation of cell morphology are responsive to treatment with heavy metals (Gorfer et al. 2009; Jin et al. 2008; Kennedy et al. 2008). *Saccharomyces cerevisiae* Skt5, which is involved in chitin biosynthesis regulation (Trilla et al. 1997), was downregulated after Cu exposure (Jin et al. 2008). Yeast deletion mutants of the protein kinases Bck1 and Slt2, which are involved in a MAPK cascade responsible for cell wall integrity (Heinisch et al. 1999), are highly sensitive to Cd (Jin et al. 2008). Differential expression of enzymes directly modifying the cell wall, such as glucoamylases, glycosyltransferases, and exo- and endoglucanases including chitinases, is commonly recorded under high-metal conditions (Cherrad et al. 2012; Chuang et al. 2009; Kennedy et al. 2008; Lanfranco et al. 2002a; Majorel et al. 2012; Michelin et al. 2008; Wang et al. 2014; Zhao et al. 2015).

Sorption of Cd, Cu, Pb, and Zn to **mucopolysaccharides** (extracellular polymeric substances, EPS) has been shown to promote metal resistance in *Curvularia lunata*, *Gloeophyllum trabeum*, *Oidiodendron maius*, and *Trametes versicolor* (Martino et al. 2000; Paraszkiewicz et al. 2007; Vesentini et al. 2007), especially when grown in liquid medium. Complexation of metal ions by EPS presumably represents a major factor in fungal mineral weathering by depleting the mineral substrate of those ions (Barker and Banfield 1996). Sorption to EPS binds metals in the same way as in the chemically similar cell wall components. In creating discrete microzones, EPS further allow for limited control of pH and diffusion of ions (Hoffland et al. 2004). Thus, they may also prevent further ion mobilization (Welch and Vandevivere 1994).

Melanins, polyphenolic compounds present mainly in the cell wall, can also efficiently bind metals, especially copper (Fogarty and Tobin 1996; Gadd and de Rome 1988). Melanized hyphae and structures such as pseudosclerotia and rhizomorphs were consequently shown to contain especially high concentrations of metal ions (Fogarty and Tobin 1996; González-Guerrero et al. 2009; McDougall and Blanchette 1996; Rizzo et al. 1992; Turnau et al. 1994) and melanin production was shown to be induced by metal exposure (Caesar-Tonthat et al. 1995; Martino et al. 2000). In response to Cd exposure, the ectomycorrhiza fungus *Paxillus involutus* produces polyphenolic substances complexing the toxic ions (Jacob et al. 2004). Activity and expression of tyrosinase, which is involved in melanin biosynthesis, is stimulated by metals in various basidiomycete fungi (Gruhn and Miller 1991; Wang et al. 2014). Correspondingly, the well-known induction of basidiomycete laccase gene expression by heavy metals, especially by Cu (Baldrian and Gabriel 2002; Cho et al. 2006; Galhaup et al. 2002; Janusz et al. 2013; Kües and Rühl 2011; Piscitelli et al. 2011), might be partly due to the role of laccases in the generation of metal-chelating polyphenols including melanin (Eisenman et al. 2007; Jacob et al. 2004), although other reasons for this have been described (see below, Baldrian and Gabriel 2002; Fernandez-Larrea and Stahl 1996).

Among fungal pigments, the polyketides **cyathoscavin** A, B, and C of *Cyathus stercoreus* have radical scavenging and metal ion-chelating abilities (Kang et al. 2008). They have been proposed to be functional in protection from Fenton reaction-mediated damage (Zhou and Liu 2010). Further pigments with profound metal-chelating abilities are dibenzoquinones, anthraquinones, and pulvinic acid derivates (Hauck et al. 2009). The best known example is the Cs accumulation in fruiting bodies of *Boletus badius*, which causally depends on the production of the pulvinic acid derivates badion A and norbadion A. These compounds form complexes with either potassium or caesium ions (Kuad et al. 2009; Steffan and Steglich 1984). Whether such compounds also play a role in promoting toxic metal resistance has yet to be evaluated.

Hydrophobins are small, cell-wall-associated proteins rich in cysteine and with low sequence similarity. They self-assemble at hydrophobic–hydrophilic interfaces, including the cell wall, to form an amphipathic monolayer. They are exclusively present in filamentous fungi and absent in yeast, allowing for hyphae to grow into hydrophobic environments including the air space. Due to their high surfactant activity, hydrophobins modify, among other properties, the permeability for ions and water (Linder et al. 2005; Wösten 2001) and have thus been discussed to promote toxic metal resistance (Pawlik-Skowrońska et al. 2002). Hydrophobins have been shown to be upregulated in *Suillus luteus* (Muller et al. 2007) upon exposure to toxic metal ions. However, expression of hydrophobins was decreased in *Paxillus involutus*, *Abortiporus biennis*, and *Cerrena unicolor* after exposure to heavy metals (Jacob et al. 2004; Jarosz-Wilkołazka et al. 2006), which was explained by sulfur-saving due to the increased sulfur demand of other detoxifying mechanisms such as GSH (see below). However, differential expression of several *Wallemia ichthyophaga* hydrophobins in either high or low saline conditions was attributed to their differing functional roles (Zajc et al. 2013). This might also be the case for metal responsive hydrophobins. The fungus-specific cerato-platanins also form self-assembling layers at air–water interfaces. Their known function lies mainly in pathogenicity and pathogen-response, possibly also in hyphal growth in analogy to plant expansins (Baccelli 2014; Frischmann et al. 2013). Like hydrophobins, they might change cell wall permeability and might thus play a role in metal response.

The HSP60 homolog **Glomalin** is a spore and hyphal cell wall glycoprotein of the VAM-forming Glomales (Gadkar and Rillig 2006). Glomalin related soil proteins (GRSP) including Glomalin itself have attracted considerable attention in soil science. Next to being considered major drivers of soil aggregation, they also are able to efficiently sequester various metals, mostly Fe, but also Cr, Cu, Pb, and Cd (Gil-Cardeza et al. 2014; Gonzalez-Chavez et al. 2004; Wright and Upadhyaya 1998). While no detailed studies exist on Glomalin regulation, and no changes in expression were observed in *Rhizophagus* (formerly *Glomus*) *intraradices* upon metal treatment (Ouziad et al. 2005), close homologues are present in most fungi. Correspondingly, HSP60 was specifically induced by Cd in *Candida albicans* (Enjalbert et al. 2006) and by Zn in a Zn-tolerant isolate of *Suillus luteus* (Muller et al. 2007).

Polyamines, such as spermine, spermidine, putrescine, and cadaverine, are known to alleviate oxidative stress (Jamieson 1998), but might also participate in cell wall metal chelation. A substantial increase in spermine production was observed in the ectomycorrhizal *Paxillus involutus* after Zn treatment (Zarb and Walters 1995). In the lichen *Pseudevernia furfuracea*, spermine, spermidine, and putrescine were enriched after treatment with Cd. *Evernia prunastri*, another lichen, produced all those polyamines after Cu exposure, while Pb induced only spermine and putrescine production in this species (Pirintsos et al. 2004). Agmatinase, which is involved in polyamine formation, was induced in *Oidiodendron maius* in response to Zn and Cd treatment (Chiapello et al. 2015).

B. Release of Metal-Binding Substances

Active release of chelating compounds including organic acids, siderophores, and glutathione leads to the formation of relatively immobile metal salts or chelates keeping these metals out of the cell (Jarosz-Wilkolazka and Gadd 2003; Pócsi 2011; Renshaw et al. 2002). However, since the very same substances are used for scavenging essential ions, balancing between these processes is crucial (Gadd 2007; Pócsi 2011).

1. Interaction of Organic Acids with Metals

Many fungi excrete organic acids into the environment (Van Hees et al. 2003), leading to acidification, which in turn may cause protolysis of minerals releasing ions into solution (Dighton 2003; Ehrlich 2006), including nutrients, but also potentially toxic metals. Several of these acids can also chelate cations or form insoluble salts, both of which immobilize the respective ions (Gadd et al. 2014; Manceau and Matynia 2010). This has two main effects on metal availability. On one hand, the solution equilibrium between mineral and its surroundings is changed, potentially causing further ions to become mobilized. This process, together with protolysis, is considered a major driver of fungal weathering (Gadd 2007, 2010; Hoffland et al. 2004). On the other hand, the precipitating chelates and crystallizing salts may be hardly soluble, which effectively prevents their entry into the fungal cell (Gadd 2010; Jarosz-Wilkolazka and Gadd 2003; White et al. 1997). Among the numerous organic acids excreted by fungi, di- and tricarboxylic acids have the strongest chelating abilities (Bellion et al. 2006; Van Hees et al. 2003). **Oxalic** and **citric acid** (Fig. 4.1) are most prominent among these, along with several lichen acids such as

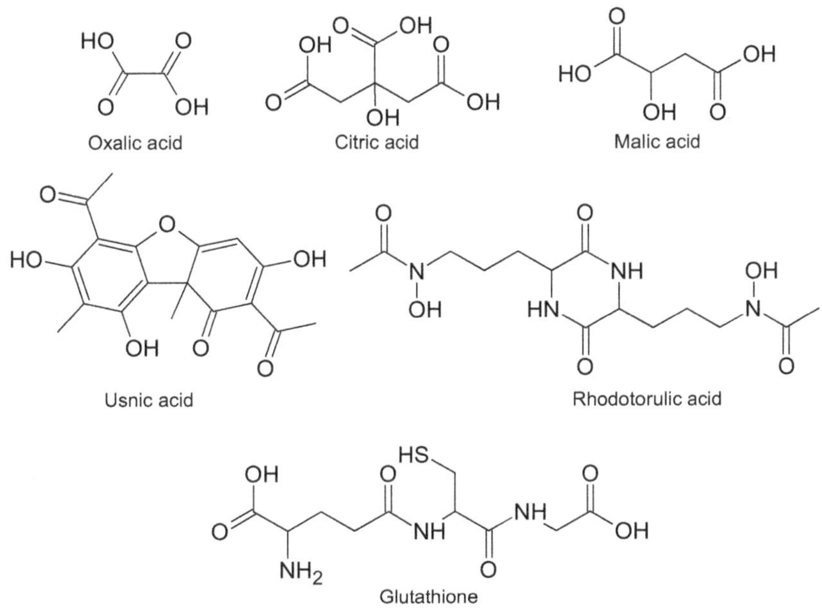

Fig. 4.1 Chemical structures of fungal secreted metal cation chelators

usnic, norstictic, and **psoromic acid** (Dighton 2003; Gadd 2007; Haas and Purvis 2006; Takani et al. 2002).

2. Oxalic Acid

Oxalate anions are very effective metal ion chelators, and the salts of oxalic acid (Fig. 1), particularly those of calcium, belong to the most abundant biominerals. Whether a chelate or salt is formed depends on conditions and the metal ions present (Gadd et al. 2014). Many fungal clades, including representatives of Zygomycota, Ascomycota, and Basidiomycota, secrete oxalic acid, often leading to calcium oxalate crystallization directly on hyphae or in their vicinity (Mäkelä et al. 2010; Tait et al. 1999). Besides this, mycogenic oxalates of numerous heavy metals have been described (Dutton and Evans 1996; Gadd et al. 2014; Haas and Purvis 2006; Jarosz-Wilkolazka and Gadd 2003). A stimulation of oxalate secretion was found after Cu, Cd, Mn, and Zn exposure (Grąz et al. 2009; Green and Clausen 2005; Jarosz-Wilkołazka et al. 2006; Sazanova et al. 2014; Xu et al. 2015). As oxalate production by the citric acid (TCA) and glyoxylate cycles seems mostly constitutive, its amount seems to depend on the activity of **oxalate decarboxylase,** which in turn is regulated by high oxalate levels and low pH (Mäkelä et al. 2010), but at least not directly by metal ions.

3. Further Organic Acids

Other chelating organic acids also play a role in fungal metal response, even though the resulting chelates are usually water soluble. Stimulation of malic and citric acid (Fig. 4.1) exudation after Cu exposure was described for *Aspergillus niger* and *Penicillium citrinum* (Sazanova et al. 2014). Citric acid production of oxalate-nonproducing mutants of *A. niger* is also induced by Mn (Ruijter et al. 1999), which in wild-type strains inhibits citrate production (Kisser et al. 1980). Zinc phosphate induced, next to oxalic acid production, also exudation of acetic, citric, gluconic, succinic, malic, and fumaric acid in several ectomycorrhizal fungi, while pyromophite, a lead phosphate mineral, caused an increase in oxalate, acetate, or formate production (Fomina et al. 2004), although phosphate scavenging might also play a role there. Exudation of malic, formic, and succinic acid was stimulated in *Scleroderma* spp. and *Pisolithus tinctorius* after exposure to metal-rich ashes, but did not seem to promote metal exclusion from the fungal cells (Ray and Adholeya 2009). Lichen acids, which are exclusively produced by lichens, are mostly phenolics such as depsides and depsidones, and are usually derived from **usnic acid** (Fig. 4.1) (Haas and Purvis 2006; Takani et al. 2002). While weakly dissociating in water, they have been shown to be quite mobile in natural environments (Dawson et al. 1984). Due to their metal-chelating abilities, these acids have been postulated to be major drivers of lichen promoted rock weathering (Haas and Purvis 2006; Williams and Rudolph 1974); they might also play a role in resistance mechanisms.

4. Siderophores

Production of siderophores, low molecular weight ligands with high iron affinity, is known from fungi and bacteria. Fungal siderophores, including the well-characterized **ferricrocin** and **rhodotorulic acid** (Fig. 4.1), are mostly of the hydroxymate-type, although fungal carboxylate and phenolate siderophores exist as well (Saha et al. 2013). Only some types of fungal siderophore are excreted for iron scavenging; others serve in intracellular iron storage and targeted transport (Haas 2003; Renshaw et al. 2002). Siderophores can often chelate other metal ions than iron and thus may play a role in toxic metal resistance. This has been widely shown for bacterial siderophores (Kothe et al. 2010), but metal-chelating abilities of fungal hydroxamate siderophores outside of iron have also been demonstrated (Enyedy et al. 2004; Farkas et al. 2008, 2014), especially for actinides (Renshaw et al. 2002; Tircsó et al. 2013). Siderophore production is induced by iron deficiency, and no induction is seen under high metal stress conditions (Finlay and Rosling 2006). For detailed reviews on siderophore function and regulation in fungi, see Haas (2003), Haas et al. (2008), Renshaw et al. (2002), Saha et al.(2013), Varma and Chincholkar (2007).

5. Glutathione

Glutathione (GSH), a tripeptide containing cysteine (Fig. 4.1), is a major antioxidant in most eukaryotes and some prokaryotes (Jamieson 1998), but also has profound metal-chelating properties (Pócsi et al. 2004). Its dominant function in toxic metal response lies in intracellular ion chelation and radical scavenging (Jamieson 1998; Pócsi 2011; Pócsi et al. 2004, see also below). However, secretion of GSH upon As exposure has been described in *S. cerevisiae* with the resulting As-GSH complex unable to enter cells (Thorsen et al. 2012). This mechanism could possibly also function for extracellular chelation of other metalloids or of heavy metals.

III. Leave Someone Holding the Baby: Intracellular Processes and Transport

Intracellular responses to toxic concentrations of metals include transport related processes such as the expression of export systems and the suppression of import systems, and also further chelation processes by substances such as glutathione, phytochelatin, and metallothionein, followed by export or intracellular compartmentalization of the complexes (Fig. 4.2) (Bellion et al. 2006; Pócsi 2011; Urban 2011; Wysocki and Tamás 2010).

A. Regulation of Metal Influx

Toxic metals enter cells either by diffusion or by transporters, which may be otherwise responsible for the uptake of essential nutrients including other cations, but also of anions. Elimination of these transporters might prevent toxic metal uptake, but may also lead to nutrient starvation. Of course, both are only true if no alternative uptake routes exist. Still, deletome analyses provide evidence for a number of transporters conferring metal resistance if they are deleted, suggesting their involvement in metal ion uptake (Pócsi 2011).

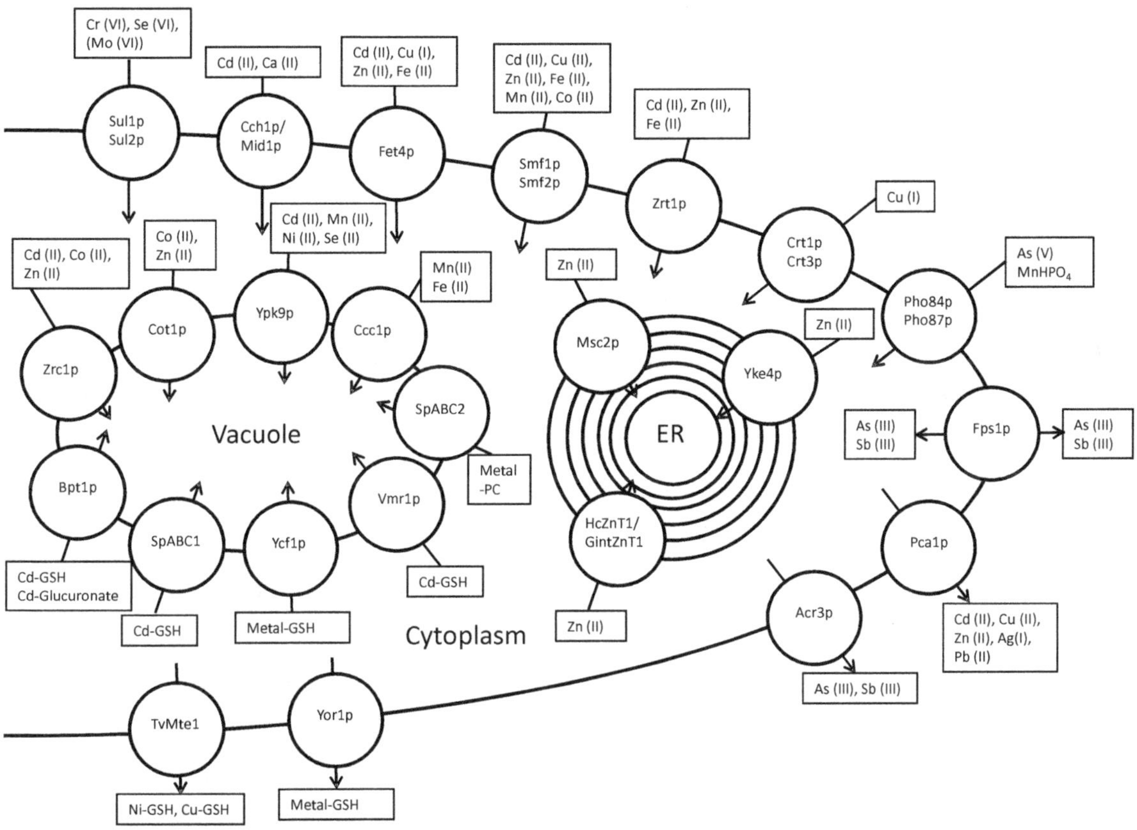

Fig. 4.2 Transporters mediating metal ion and chelate influx, efflux, and compartmentalization. Unless otherwise specified, protein names refer to *S. cerevisiae* proteins. Further transporters are indicated by abbreviating the organism name: Gint = *Rhizophagus* (=*Glomus*) *intraradices*; Hc = *Hebeloma cylindrosporum*; Sp = *Schizosaccharomyces pombe*; Tv = *Tricholoma vaccinum*

Uptake of hexavalent chromium is generally mediated by sulfate uptake systems (Viti et al. 2014). Deletion of the *S. cerevisiae* high affinity **sulfate permeases** Sul1p and Sul2p leads to Cr(VI) and Se(VI) tolerance (Cherest et al. 1997; Pereira et al. 2008). Due to its chemical similarity, a similar function of these sulfate uptake systems is to be expected for hexavalent Mo (Wysocki and Tamás 2010).

Cd tolerance is promoted by deletion of the ZIP (Zrt, Irt-like protein) family Zn transporter Zrt1p and the NRAMP (**natural resistance-associated macrophage protein**) family Mn(II) transporters Smf1p and Smf2p. While Zrt1P may also transport Fe (Guerinot 2000), the latter two transporters may also take Co, Cu, Fe, and Zn as substrates (Chen et al. 1999; Wysocki and Tamás 2010 and references therein). High Zn levels lead to endocytosis and vacuolar degradation of Zrt1p (Gitan et al. 1998); it is however not known whether high Cd levels have the same effect. Further possible uptake paths for Cd are the low affinity iron transporter Fet4p, which also takes up Cu and Zn (Jensen and Culotta 2002; Waters and Eide 2002), and the stretch-activated high-affinity cell membrane Ca transporter Cch1p/Mid1p (Mielniczki-Pereira et al. 2011; Ruta et al. 2014). Homologues of Zrt1p and Fet1p in *S. pombe* were downregulated after Cd exposure (Chen et al. 2003).

Cu uptake in *S. cerevisiae* is generally mediated by the Ctr1 and Crt 3 high-affinity copper transporters and the low affinity Fe/Cu transporter Fet4 (Puig and Thiele 2002). The **copper transport protein** (CTR) family transporters are highly specific for Cu(I). Thus, uptake depends on a ferric/cupric reductase (Hassett and Kosman 1995).

The greater part of fungal Zn uptake is depending on the ZIP transporter family (Eide 2006). Two to eight ZIP transporters are encoded in fungal genomes, and several subgroups are distinguished. For Zn uptake into the cell, only the Zrt1/Zrt2-like group is functional (Tamayo et al. 2014).

The *S. cerevisiae* NRAMP uptake transporter Smf1p is responsible for the majority of Mn taken up by the fungal cell, although other divalent ions including Fe, Cu, and Cd can also be transported (Culotta et al. 2005). Further Mn uptake is mediated by the phosphate transporter Pho84p, which transports inorganic phosphate, but also metal phosphates including $MnHPO_4$ (Fristedt et al. 1999). Excess Mn causes absence of Smf1p and of Smf2p, which is mediating Mn transport from Golgi vesicles, while Pho84p is not downregulated in response to excess Mn levels (Culotta et al. 2005; Tamayo et al. 2014). However, a downregulation of *Candida albicans* pho84 under various stress condition including Cd exposure was observed, which led to its classification as core stress gene for this organism (Enjalbert et al. 2006).

Exposure to the metalloid As(V) was tolerated by deletion mutants of the Pho84p and Pho87p phosphate transporters. Deletion of Pho86p and Gtr1p, both regulating Pho84p functionality, also has this effect. In contrast, tolerance to As(III) and Sb(III) was promoted by deletion of the aquaglyceroporin Fps1p (Wysocki and Tamás 2010, and references therein). Like other aguaglyceroporins, Fps1p works bidirectional and can also export trivalent metalloid ions (Maciaszczyk-Dziubinska et al. 2010). Further, As (III) uptake is promoted by several hexose transporters (Liu et al. 2004).

B. Metal Efflux Systems

Metal efflux systems export either ions or chelates out of the cell. In *S. cerevisiae*, several systems of all types exist, which have been summarized by Pócsi (2011). A Cd and Cu efflux system based on a P_{1B}-type ATPase (Pca1p) is also regulated by Cd, Cu, and Ag exposure (Adle et al. 2007). A homolog of Pca1 has been found in *Tuber melanosporum*, where it is regulated in ectomycorrhizal symbioses without any toxic metal addition (Bolchi et al. 2011). Further P_{1B}-type ATPases outside of Pca1p exist in *S. cerevisiae*, and similar pumps seem to be ubiquitous in Asco- and Basidiomycetes, plants, and bacteria. The fungal variants may be exporting Ag, Cu, Cd, Zn, or Pb ions (Ashrafi et al. 2011; Tamayo et al. 2014). Further yeast efflux systems have been described for the metalloids As and Sb, in the Acr3p plasma membrane transporter, and also the already mentioned bidirectional aquaglyceroporin Fps1p. Transport of Cd and Co into vacuoles of yeast is mediated by the **cation diffusion facilitator** (CDF) family proteins Zrc1p and Cot1p; both can also transport Zn (MacDiarmid et al. 2002). Ypk9p is a vacuole located in **multidrug resistant-associated protein** (MRP, also known as ABCC) transporter in *S. cerevisiae*; its deletion causes sensitivity against Cd, Mn, Ni, and Se (Schmidt et al. 2009). The *Rhizophagus* (=*Glomus*) *intraradices* MRP transporter GintABC1 is strongly upregulated upon Cu and Cd treatment; its cellular localization remains however unclear (González-Guerrero et al. 2010). Ccc1p, a budding yeast mitochondrial carrier family transporter, pumps Fe and Mn into the vacuole, and deletions cause increased sensitivity towards these metals (Li and Kaplan 2004). Deletion of *S. cerevisiae* YKE4, encoding a ZIP family transporter located in the ER membrane, leads to accumulation of toxic zinc levels in the cytosol (Kumánovics et al. 2006). Zn transport from the cytosol is however mainly mediated by CDF transporters.

The already mentioned six-transmembrane domain proteins Zrc1p and Cot1p are responsible for vacuolar transport, while the 12 transmembrane domain transporter Msc2p, which is ubiquitous in fungi, transports Zn into the ER (Montanini et al. 2007; Tamayo et al. 2014). The CDF transporter HcZnT1 of *Hebeloma cylindrosporum* is localized on ER membranes and mediates Zn storage in intracellular vesicles when expressed in yeast (Blaudez and Chalot 2011); a similar transporter, GintZnT1, was found in *Rhizophagus* (formerly *Glomus*) *intraradices* (González-Guerrero et al. 2005). Despite the obvious predominance of compartmentalization for Zn detoxification, metal tolerant isolates of the ectomycorrhizal *Suillus bovinus* were shown to use rather export mechanisms to remove excess Zn from cytoplasm (Ruytinx et al. 2013).

C. Chelation and Chelate Transport

Intracellular chelation is promoted by **glutathione** (GSH), **phytochelatin** (PC), and **metallothioneins** (MT) but also by polyphosphate moieties and even metalloenzymes. Chelation within the cell might lead to either export or compartmentalization in the vacuole (Bellion et al. 2006; Pócsi 2011; Wysocki and Tamás 2010).

GSH, as introduced above (Fig. 4.1), is an atypical tripeptide with high metal-binding capacity (Pócsi et al. 2004). Phytochelatins are polymeric GSH residues and like these represent a major pathway of metal and metalloid detoxification, especially for As and Cd (Clemens 2006; Kennedy et al. 2008). Their biosynthesis is induced by numerous metals including Cd, Cu, Ag, Hg, Zn, and As, and deletion mutants show hypersensitivity against these metals. While metal–PC complexes are generally transported into vacuoles, cell plasma chelation alone already provides metal resistance (Collin-Hansen et al. 2007; Wysocki and Tamás 2010, and references therein). PCs are, however, not common in fungi (Bellion et al. 2006) and even if present may not be differentially expressed upon metal exposure (Zhao et al. 2015).

Metallothioneins are small cytoplasmatic peptides rich in cysteine and lacking aromatic amino acids and histidine. They are ubiquitously distributed in eukaryotes and cyanobacteria. As with GSH and PT, metal binding is promoted by the metal-SH bonds at cysteine residues. MTs are currently classified by their specificity into the extreme Zn- and Cu-metallothioneins, which prefer one and omit the other ligand, and a wide range of intermediates (Palacios et al. 2011).

The considerable metal tolerance of the ectomycorrhizal species *Laccaria laccata*, *Paxillus involutus*, and *Pisolithus tinctorius* was attributed to metallothioneins (Howe et al. 1997; Jacob et al. 2004; Morselt et al. 1986). *S. cerevisiae* contains two MTs, Cup1p and Crs5p (Wysocki and Tamás 2010). Cup1p is induced by high Cu levels, promotes Cu and Cd tolerance, and its deletion leads to increased Cu susceptibility (Ecker et al. 1986; Thiele and Hamer 1986). Crs5p is also upregulated by elevated Cu concentration, but has, apart from Cu and Cd, also a high affinity to Zn, and thus promotes resistance to combined Cu and Zn treatment (Pagani et al. 2007). Induction of MT production might not only depend on metal exposure but also on the life cycle stage of the fungus. A MT of the VAM fungus *Gigaspora margarita* promoting Cu and Cd tolerance was Cu-inducible only in the symbiotic stage and not in germinating spores (Lanfranco et al. 2002b). Metallothioneins are not generally restricted to binding Cu, Zn, and Cd. MTs chelating Ag have been described from the ectomycorrhizal fungi *Amanita strobiliformis* and *Hebeloma mesophaeum*, in the latter case also promoting Ag tolerance to *S. cerevisiae* (Osobová et al. 2011; Sácký et al. 2014).

For catalytic activity, numerous enzymes require metal cofactors, which might excite toxicity, if they were present as free ions in the cytoplasm. Thus, cytoplasmic transport requires vehicles chelating these ions and delivering them to target proteins, which have been termed metallochaperones (O'Halloran and Culotta 2000). In *S. cerevisiae*, several such metallochaperones and their target proteins have been described (e.g., Horng et al. 2004; Sturtz et al. 2001), and orthologs are present in many other fungi (Bellion et al. 2006; Bolchi et al. 2011; Uldschmid et al. 2002). Some metalloenzymes have themselves been suggested to double-act in metal stress response by binding toxic metal ions. Within the fungal kingdom, the best evidence for this phenomenon pertains to the **Cu,Zn superoxide dismutase** (Cu/Zn-SOD), which is otherwise mainly responsible for alleviating oxidative stress (Pócsi 2011, see also below). It was shown that Cu/Zn-SOD can buffer intracellular Cu by incorporating excess Cu (Culotta et al. 1995). Elevated Cu and Zn levels triggered increased expression of two Zn-metalloproteases in *Botrytis cinerea*, for which a similar metal scavenging ability was suggested (Cherrad et al. 2012). Laccases, which contain binding sites for Cu, have also been hypothesized to encounter oxidative stress by chelating excess Cu(II) ions (Fernandez-Larrea and Stahl 1996). ADH1 is required for Cu and As resistance in *S. cerevisiae* and was suggested to scavenge divalent ions in replacement of its Zn cofactors (Jin et al. 2008). Expression of an alcohol dehydrogenase is also induced in *Agaricus blazei* after Cd exposure (Wang et al. 2014).

Intracellular sequestration of metal ions by inorganic polyphosphate represents another way of metal detoxification (Keasling et al. 2000). In fungi, this was shown to be relevant for *Trichoderma harzianum* and *Cryptococcus humicola*, both of which are rich in inorganic polyphosphates, which in turn promoted tolerance to Cd, Co, La, Mn, Ni, and Zn (Andreeva et al. 2014; de Freitas et al. 2011). Deletion of the exopolyphosphatase Ppn1p reduced the lag time of *S. cerevisiae* to adapt to high Mn concentrations as compared to wild-type strains (Andreeva et al. 2013).

For vacuolar transport of GSH complexes of Cd, As, Hg, Pb, and Se, the Ycf1p (yeast cadmium factor) vacuolar GSH S-conjugate transporter of the MRP family is responsible (Lazard et al. 2011; Mielniczki-Pereira et al. 2008; Pócsi 2011, and references therein). Its expression is increased by excess Cd (hence its name) and some other metal and metalloid ions including Hg, Sb, Se, and As and is also induced by the major regulator of yeast oxidative stress response, Yap1p (Wysocki and Tamás 2010). Excess Sb ions block its transport of free and chelated GSH. An ortholog of this transporter is present in *Paxillus involutus*, where it was suggested to promote the high Cd content in the species' vacuoles (Blaudez et al. 2000; Ott et al. 2002). Further yeast MRP transporters mediating Cd resistance are Bpt1p and Vmr1p, which similar to Ycf1p transport glutathione (and glucuronate in Btp1p) conjugates as well as free glutathione into the vacuole. While Vmt1 is inducible by Cd and Zn, Bpt1p is not inducible by Cd stress or by Yap1p (Sharma et al. 2002; Wawrzycka et al. 2010). Yor1p, yet another MRP transporter of *S. cerevisiae*, is situated in the plasma membrane and is responsive for export of metal-GSH chelates (Nagy et al. 2006) from the cell.

While the ABC (ATP-binding cassette) transporter Hmt1 of *S. pombe* preferentially transports metal-GSH complexes, rather than phytochelatin complexes, into the vacuole (Prévéral et al. 2009), a second ABC transporter, Abc2, is required. The gene of the latter transporter has homologs present in many other fungi and plants (Mendoza-Cózatl et al. 2010). **Multidrug and toxic compound extrusion** (MATE) transporters are present in most organisms and allow for export of numerous substances including antibiotics, dyes, and heavy metals (Hvorup et al. 2003). Among MATE proteins of hyphal fungi, only Mte1 of the ectomycorrhizal *Tricholoma vaccinum* was so far characterized. Heterologous expression of the cell membrane localized Mte1 in *S. cerevisiae* promoted resistance to toxic concentrations of various xenobiotics and metal ions, including Ni and Cu, which are transported as GSH chelates. A similar function can be expected for the one to five MATE paralogs present in other fungi (Schlunk et al. 2015). A MATE protein in *S. cerevisiae*, Erc1p, was shown to promote resistance to ethionin (Shiomi et al. 1991), but nothing is known about its metal transport ability.

IV. System Reset: Alleviating Metal-Induced Damage

Several metals, especially Cd, are known to interfere with DNA repair mechanisms. Cd was shown to impair mismatch repair in *S. cerevisiae* (Jin et al. 2003), which was however unaffected in *S. pombe*. Instead, deletion of a MET18 homolog associated with nucleotide excision repair promoted Cd sensitivity in fission yeast (Kennedy et al. 2008). In *Ganoderma lucidum*, DNA polymerase IV, a protein to repair alkylated DNA, an ortholog of the human MRE11 double-strand repair protein, and a FHA domain protein potentially involved in DNA repair were induced after Cd exposure (Chuang et al. 2009).

Selective proteolysis of metal-damaged enzymes and other proteins requires ubiquitylation and transport to proteasomes (Glickman and Ciechanover 2002). Several genes of the ubiquitin-proteasome pathway were induced by Zn treatment in *Suillus luteus* (Muller et al. 2007), and by Cd stress in *Ganoderma lucidum*, in the latter case including Skp1, which is a component of the **SCF ubiquitin ligase** (Chuang et al. 2009). In *S. cerevisiae* and *S. pombe*, SCF (Skp, Cullin, F-box containing) complexes were shown to promote Cd tolerance (Barbey et al. 2005; Harrison et al. 2005). Outside of the ubiquitin-proteasome system, *Botrytis cinerea* expresses numerous proteases upon treatment with several heavy metals, which were postulated to be involved in degradation of metal-inactivated proteins (Cherrad et al. 2012).

As described above, response to many toxic metals depletes sulfur-containing proteins and other substances, especially of GSH. This was seen in the strong induction of several enzymes of the sulfur amino acid and GSH pathways after metal and metalloid exposure in *S. cerevisiae* (Jin et al. 2008; Thorsen et al. 2007; Vido et al. 2001), *Exophiala pisciphila* (Zhao et al. 2015), *Agaricus blazei* (Wang et al. 2014) and *Ganoderma lucidum* (Chuang et al. 2009). Similar effects were seen in *Candida albicans*, where both GSH production and sulfate assimilation were strongly induced (Enjalbert et al. 2006). In *S. pombe*, factors involved in sulfur amino acid production were also induced, while GSH production was not increased. Instead, a GSH transport system homologous to *S. cerevisiae* Hgt1p was strongly induced, suggesting that *S. pombe* invests rather in scavenging of external glutathione than in its production (Chen et al. 2003). In *S. cerevisiae*, it was shown that at least for Cd exposure, the proteomic response was favoring the expression of proteins with low sulfur content, replacing isozymes rich in sulfur. Among these, the aldehyde dehydrogenase Ald4p was replaced by Ald6p, the pyruvate decarboxylase Pdc1p by Pdc6p, and the enolase Eno1p by isozyme Eno2p. In all cases, the genes encoding the substituting enzymes comprised less codons for sulfur-containing amino acids. This was shown to be controlled by Met4p, a factor otherwise controlling the sulfate assimilation pathway, leading to an overall 30 % sulfur saving for the whole proteome (Fauchon et al. 2002; Wysocki and Tamás 2010).

Higher metal concentrations cause considerable oxidative stress by reactive oxygen species, as introduced above. Consequently, antioxidants, mainly GSH, but also catalase, thioredoxin, glutaredoxin, polyamines, superoxide dismutase, ubiquinone, pyridoxal phosphate, or ascorbic acid are produced (Cherrad et al. 2012; Collin-Hansen et al. 2007; Jamieson 1998; Kennedy et al. 2008; Ott et al. 2002; Schützendübel and Polle 2002; Zhao et al. 2015). Balanced expression is crucial here as well, since certain metals can trigger harmful pro-oxidant activities of these substances (Poljšak et al. 2005). The response against oxidative stress also depends on the involved reactive oxygen species. *S. cerevisiae* was shown to exhibit specific response upon treatment with five different ROS (Thorpe et al. 2004).

V. Perspective

This chapter focuses on diverse molecular mechanisms involved in toxic metal response of fungi. In recent years, an increasing number of studies covering a broad variety of fungal species delivered data on the diverse mechanisms involved. From chelation to cell wall constituents and small excreted compounds to regulation of import, export, and compartmentalization processes or repair- and recycling mechanisms, the overall scheme of fungal metal response seems rather uniform. However, the majority of studies is still restricted to ascomycete model organisms. The enormous work done with these organisms allows, in collaboration with bioinformatics, e.g., for impressive large scale de novo whole transcriptome analyses; the number of "proteins of unknown function" in the databases is steadily increasing. While this underlines the enigmatic diversity of kingdom fungi, it also shows the need to further enlarge the knowledge of molecular mechanisms responsible for fungal heavy metal uptake and resistance. This is of crucial importance to understand and possibly modulate their role in geochemical processes such as weathering, speciation changes, and mineral formation. To understand and possibly utilize their abilities for heavy metal resistance in phytoextraction, phytostabilization, or even mycoremediation approaches, a deeper understanding of the resistance mechanisms, their genetic background, and expression profiles seems necessary.

References

Adle DJ, Sinani D, Kim H, Lee J (2007) A cadmium-transporting P1B-type ATPase in yeast *Saccharomyces cerevisiae*. J Biol Chem 282:947–955

Andreeva N, Ryazanova L, Dmitriev V, Kulakovskaya T, Kulaev I (2013) Adaptation of *Saccharomyces cerevisiae* to toxic manganese concentration triggers changes in inorganic polyphosphates. FEMS Yeast Res 13:463–470

Andreeva N, Ryazanova L, Dmitriev V, Kulakovskaya T, Kulaev I (2014) Cytoplasmic inorganic polyphosphate participates in the heavy metal tolerance of *Cryptococcus humicola*. Folia Microbiol 59:381–389

Arantes V, Milagres AM, Filley TR, Goodell B (2011) Lignocellulosic polysaccharides and lignin degradation by wood decay fungi: the relevance of non-enzymatic Fenton-based reactions. J Ind Microbiol Biotechnol 38:541–555

Ashrafi E, Alemzadeh A, Ebrahimi M, Ebrahimie E, Dadkhodaei N (2011) Amino acid features of P1B-ATPase heavy metal transporters enabling small numbers of organisms to cope with heavy metal pollution. Bioinf Biol Insights 5:59

Baccelli I (2014) Cerato-platanin family proteins: one function for multiple biological roles? Front Plant Sci 5:769

Baldrian P (2003) Interactions of heavy metals with white-rot fungi. Enzyme Microb Technol 32:78–91

Baldrian P, Gabriel J (2002) Copper and cadmium increase laccase activity in *Pleurotus ostreatus*. FEMS Microbiol Lett 206:69–74

Barbey R, Baudouin-Cornu P, Lee TA, Rouillon A, Zarzov P, Tyers M, Thomas D (2005) Inducible dissociation of SCFMet30 ubiquitin ligase mediates a rapid transcriptional response to cadmium. EMBO J 24:521–532

Barker W, Banfield J (1996) Biologically versus inorganically mediated weathering reactions: relationships between minerals and extracellular microbial polymers in lithobiontic communities. Chem Geol 132:55–69

Bazała MA, Gołda K, Bystrzejewska-Piotrowska G (2008) Transport of radiocesium in mycelium and its translocation to fruitbodies of a saprophytic macromycete. J Environ Radioact 99:1200–1202

Bellion M, Courbot M, Jacob C, Blaudez D, Chalot M (2006) Extracellular and cellular mechanisms sustaining metal tolerance in ectomycorrhizal fungi. FEMS Microbiol Lett 254:173–181

Beyersmann D, Hartwig A (2008) Carcinogenic metal compounds: recent insight into molecular and cellular mechanisms. Arch Toxicol 82:493–512

Bhanoori M, Venkateswerlu G (2000) In vivo chitin-cadmium complexation in cell wall of *Neurospora crassa*. Biochim Biophys Acta 1523:21–28

Blaudez D, Chalot M (2011) Characterization of the ER-located zinc transporter ZnT1 and identification of a vesicular zinc storage compartment in *Hebeloma cylindrosporum*. Fungal Genet Biol 48:496–503

Blaudez D, Jacob C, Turnau K, Colpaert JV, Ahonen-Jonnarth U, Finlay R, Botton B, Chalot M (2000) Differential responses of ectomycorrhizal fungi to heavy metals in vitro. Mycol Res 104:1366–1371

Bolchi A, Ruotolo R, Marchini G, Vurro E, di Toppi LS, Kohler A, Tisserant E, Martin F, Ottonello S (2011) Genome-wide inventory of metal homeostasis-related gene products including a functional phytochelatin synthase in the hypogeous mycorrhizal fungus *Tuber melanosporum*. Fungal Genet Biol 48:573–584

Bonneville S, Morgan DJ, Schmalenberger A, Bray A, Brown A, Banwart SA, Benning LG (2011) Tree-mycorrhiza symbiosis accelerate mineral weathering: evidences from nanometer-scale elemental fluxes at the hypha–mineral interface. Geochim Cosmochim Acta 75:6988–7005

Bornyasz M, Graham R, Allen M (2005) Ectomycorrhizae in a soil-weathered granitic bedrock regolith: linking matrix resources to plants. Geoderma 126:141–160

Borovička J, Kubrová J, Rohovec J, Randa Z, Dunn CE (2011) Uranium, thorium and rare earth elements in macrofungi: what are the genuine concentrations? Biometals 24:837–845

Brunnert H, Zadražil F (1983) The translocation of mercury and cadmium into the fruiting bodies of 6 higher fungi - a comparative study on species specificity in 5 lignocellulolytic fungi and the cultivated mushroom *Agaricus bisporus*. Eur J Appl Microbiol Biotechnol 17:358–364

Büchel G, Bergmann H, Ebenå G, Kothe E (2005) Geomicrobiology in remediation of mine waste. Chem Erde-Geochem 65:1–5

Burford E, Fomina M, Gadd G (2003) Fungal involvement in bioweathering and biotransformation of rocks and minerals. Mineral Mag 67:1127–1155

Burgstaller W, Schinner F (1993) Leaching of metals with fungi. J Biotechnol 27:91–116

Byrne AR, Ravnik V, Kosta L (1976) Trace element concentrations in higher fungi. Sci Total Environ 6:65–78

Caesar-Tonthat T, Van Ommen KF, Geesey GG, Henson JM (1995) Melanin production by a filamentous soil fungus in response to copper and localization of copper sulfide by sulfide-silver staining. Appl Environ Microbiol 61:1968–1975

Causton HC, Ren B, Koh SS, Harbison CT, Kanin E, Jennings EG, Lee TI, True HL, Lander ES, Young RA (2001) Remodeling of yeast genome expression in response to environmental changes. Mol Biol Cell 12:323–337

Chen X-Z, Peng J-B, Cohen A, Nelson H, Nelson N, Hediger MA (1999) Yeast SMF1 mediates H+-

coupled iron uptake with concomitant uncoupled cation currents. J Biol Chem 274:35089–35094

Chen D, Toone WM, Mata J, Lyne R, Burns G, Kivinen K, Brazma A, Jones N, Bahler J (2003) Global transcriptional responses of fission yeast to environmental stress. Mol Biol Cell 14:214–229

Cherest H, Davidian J-C, Thomas D, Benes V, Ansorge W, Surdin-Kerjan Y (1997) Molecular characterization of two high affinity sulfate transporters in *Saccharomyces cerevisiae*. Genetics 145:627–635

Cherrad S, Girard V, Dieryckx C, Goncalves IR, Dupuy JW, Bonneu M, Rascle C, Job C, Job D, Vacher S, Poussereau N (2012) Proteomic analysis of proteins secreted by *Botrytis cinerea* in response to heavy metal toxicity. Metallomics 4:835–846

Chiapello M, Martino E, Perotto S (2015) Common and metal-specific proteomic responses to cadmium and zinc in the metal tolerant ericoid mycorrhizal fungus *Oidiodendron maius* Zn. Metallomics 7:805–815

Cho N-S, Kim D-H, Cho H-Y, Ohga S, Leonowicz A (2006) Effect of various compounds on the activity of laccases from basidiomycetes and their oxidative and demethoxylating activities. J Fac Agr Kyushu U 51:211–218

Chuang HW, Wang IW, Lin SY, Chang YL (2009) Transcriptome analysis of cadmium response in *Ganoderma lucidum*. FEMS Microbiol Lett 293:205–213

Clemens S (2006) Evolution and function of phytochelatin synthases. J Plant Physiol 163:319–332

Clemens S, Simm C (2003) *Schizosaccharomyces pombe* as a model for metal homeostasis in plant cells: the phytochelatin-dependent pathway is the main cadmium detoxification mechanism. New Phytol 159:323–330

Collin-Hansen C, Pedersen SA, Andersen RA, Steinnes E (2007) First report of phytochelatins in a mushroom: induction of phytochelatins by metal exposure in *Boletus edulis*. Mycologia 99:161–174

Colpaert JV, Adriaensen K, Muller LAH, Lambaerts M, Faes C, Carleer R, Vangronsveld J (2005) Element profiles and growth in Zn-sensitive and Zn-resistant Suilloid fungi. Mycorrhiza 15:628–634

Culotta VC, Joh H-D, Lin S-J, Slekar KH, Strain J (1995) A physiological role for *Saccharomyces cerevisiae* copper/zinc superoxide dismutase in copper buffering. J Biol Chem 270:29991–29997

Culotta VC, Yang M, Hall MD (2005) Manganese transport and trafficking: lessons learned from *Saccharomyces cerevisiae*. Eukaryot Cell 4:1159–1165

Damon C, Lehembre F, Oger-Desfeux C, Luis P, Ranger J, Fraissinet-Tachet L, Marmeisse R (2012) Metatranscriptomics reveals the diversity of genes expressed by eukaryotes in forest soils. PLoS One 7:e28967

Dassa E, Bouige P (2001) The ABC of ABCs: a phylogenetic and functional classification of ABC systems in living organisms. Res Microbiol 152:211–229

Dawson H, Hrutfiord B, Ugolini F (1984) Mobility of lichen compounds from *Cladonia mitis* in arctic soils. Soil Sci 138:40–45

de Freitas LA, Ferreira de Moura G, Barbosa de Lima MA, Mendes de Souza P, Alves da Silva CA, de Campos Takaki GM, Do Nascimento AE (2011) Role of the morphology and polyphosphate in *Trichoderma harzianum* related to cadmium removal. Molecules 16:2486–2500

Deshaies R (1999) SCF and Cullin/Ring H2-based ubiquitin ligases. Annu Rev Cell Dev Biol 15:435–467

Dighton J (2003) Fungi in ecosystem processes. CRC, Boca Raton

Dighton J, Horrill AD (1988) Radiocesium accumulation in the mycorrhizal fungi *Lactarius rufus* and *Inocybe longicystis*, in upland Britain, following the Chernobyl accident. Trans Br Mycol Soc 91:335–337

Dumay QC, Debut AJ, Mansour NM, Saier MH Jr (2005) The copper transporter (Ctr) family of Cu+ uptake systems. J Mol Microbiol Biotechnol 11:10–19

Dutton MV, Evans CS (1996) Oxalate production by fungi: its role in pathogenicity and ecology in the soil environment. Can J Microbiol 42:881–895

Ecker DJ, Butt T, Sternberg E, Neeper M, Debouck C, Gorman J, Crooke S (1986) Yeast metallothionein function in metal ion detoxification. J Biol Chem 261:16895–16900

EEA (2010) The European environment. State and outlook 2010. Soil. European Environment Agency, Copenhagen

Ehrensberger KM, Bird AJ (2011) Hammering out details: regulating metal levels in eukaryotes. Trends Biochem Sci 36:524–531

Ehrlich HL (2006) Geomicrobiology: relative roles of bacteria and fungi as geomicrobial agents. In: Gadd GM (ed) Fungi in biogeochemical cycles. Cambridge University Press, Cambridge, pp 1–27

Eide DJ (2006) Zinc transporters and the cellular trafficking of zinc. Biochim Biophys Acta (BBA)-Mol Cell Res 1763:711–722

Eisenman HC, Mues M, Weber SE, Frases S, Chaskes S, Gerfen G, Casadevall A (2007) *Cryptococcus neoformans* laccase catalyses melanin synthesis from both D-and L-DOPA. Microbiology 153:3954–3962

Enjalbert B, Smith DA, Cornell MJ, Alam I, Nicholls S, Brown AJ, Quinn J (2006) Role of the Hogl stress-activated protein kinase in the global transcriptional response to stress in the fungal pathogen *Candida albicans*. Mol Biol Cell 17:1018–1032

Enoki A, Itakura S, Tanaka H (1997) The involvement of extracelluar substances for reducing molecular oxygen to hydroxyl radical and ferric iron to ferrous iron in wood degradation by wood decay fungi. J Biotechnol 53:265–272

Enyedy ÉA, Pócsi I, Farkas E (2004) Complexation of desferricoprogen with trivalent Fe, Al, Ga, In and divalent Fe, Ni, Cu, Zn metal ions: effects of the linking chain structure on the metal binding ability of hydroxamate based siderophores. J Inorg Biochem 98:1957–1966

Farkas E, Bátka D, Kremper G, Pócsi I (2008) Structure-based differences between the metal ion selectivity of two siderophores desferrioxamine B (DFB) and desferricoprogen (DFC): why DFC is much better Pb (II) sequestering agent than DFB? J Inorg Biochem 102:1654–1659

Farkas E, Szabó O, Parajdi-Losonczi PL, Balla G, Pócsi I (2014) Mn (II)/Mn (III) and Fe (III) binding capability of two *Aspergillus fumigatus* siderophores, desferricrocin and N′, N ″, N‴-triacetylfusarinine C. J Inorg Biochem 139:30–37

Fauchon M, Lagniel G, Aude J-C, Lombardia L, Soularue P, Petat C, Marguerie G, Sentenac A, Werner M, Labarre J (2002) Sulfur sparing in the yeast proteome in response to sulfur demand. Mol Cell 9:713–723

Fernandez-Larrea J, Stahl U (1996) Isolation and characterization of a laccase gene from *Podospora anserina*. Mol Gen Genet 252:539–551

Finlay RD, Rosling A (2006) Integrated nutrient cycles in boreal forest ecosystems-the role of mycorrhizal fungi. In: Gadd GM (ed) Fungi in biogeochemical cycles. Cambridge University Press, Cambridge, pp 28–50

Fogarty RV, Tobin JM (1996) Fungal melanins and their interactions with metals. Enzyme Microb Technol 19:311–317

Fomina M, Ritz K, Gadd GM (2000) Negative fungal chemotropism to toxic metals. FEMS Microbiol Lett 193:207–211

Fomina M, Alexander I, Hillier S, Gadd G (2004) Zinc phosphate and pyromorphite solubilization by soil plant-symbiotic fungi. Geomicrobiol J 21:351–366

Fomina M, Charnock JM, Hillier S, Alvarez R, Gadd GM (2007) Fungal transformations of uranium oxides. Environ Microbiol 9:1696–1710

Frey B, Zierold K, Brunner I (2000) Extracellular complexation of Cd in the Hartig net and cytosolic Zn sequestration in the fungal mantle of *Picea abies-Hebeloma crustuliniforme* ectomycorrhizas. Plant Cell Environ 23:1257–1265

Frischmann A, Neudl S, Gaderer R, Bonazza K, Zach S, Gruber S, Spadiut O, Friedbacher G, Grothe H, Seidl-Seiboth V (2013) Self-assembly at air/water interfaces and carbohydrate binding properties of the small secreted protein EPL1 from the fungus *Trichoderma atroviride*. J Biol Chem 288:4278–4287

Fristedt U, van der Rest M, Poolman B, Konings WN, Persson BL (1999) Studies of cytochrome c oxidase-driven H+-coupled phosphate transport catalyzed by the *Saccharomyces cerevisiae* Pho84 permease in coreconstituted vesicles. Biochemistry 38:16010–16015

Gadd GM (2001) Fungi in bioremediation. Cambridge University Press, Cambridge

Gadd GM (2007) Geomycology: biogeochemical transformations of rocks, minerals, metals and radionuclides by fungi, bioweathering and bioremediation. Mycol Res 111:3–49

Gadd GM (2010) Metals, minerals and microbes: geomicrobiology and bioremediation. Microbiology 156:609–643

Gadd GM, de Rome L (1988) Biosorption of copper by fungal melanin. Appl Microbiol Biotechnol 29:610–617

Gadd GM, Bahri-Esfahani J, Li Q, Rhee YJ, Wei Z, Fomina M, Liang X (2014) Oxalate production by fungi: significance in geomycology, biodeterioration and bioremediation. Fungal Biol Rev 28:36–55

Gadkar V, Rillig MC (2006) The arbuscular mycorrhizal fungal protein glomalin is a putative homolog of heat shock protein 60. FEMS Microbiol Lett 263:93–101

Galhaup C, Goller S, Peterbauer CK, Strauss J, Haltrich D (2002) Characterization of the major laccase isoenzyme from *Trametes pubescens* and regulation of its synthesis by metal ions. Microbiology 148:2159–2169

García MA, Alonso J, Fernandez MI, Melgar MJ (1998) Lead content in edible wild mushrooms in northwest Spain as indicator of environmental contamination. Arch Environ Contam Toxicol 34:330–335

García MA, Alonso J, Melgar MJ (2005) *Agaricus macrosporus* as a potential bioremediation agent for substrates contaminated with heavy metals. J Chem Technol Biotechnol 80:325–330

Gasch AP (2007) Comparative genomics of the environmental stress response in ascomycete fungi. Yeast 24:961–976

Gasch AP, Spellman PT, Kao CM, Carmel-Harel O, Eisen MB, Storz G, Botstein D, Brown PO (2000) Genomic expression programs in the response of yeast cells to environmental changes. Mol Biol Cell 11:4241–4257

Gast CH, Jansen E, Bierling J, Haanstra L (1988) Heavy metals in mushrooms and their relationship with soil characteristics. Chemosphere 17:789–799

Gil-Cardeza ML, Ferri A, Cornejo P, Gomez E (2014) Distribution of chromium species in a Cr-polluted soil: presence of Cr (III) in glomalin related protein fraction. Sci Total Environ 493:828–833

Gitan RS, Luo H, Rodgers J, Broderius M, Eide D (1998) Zinc-induced inactivation of the yeast ZRT1 zinc transporter occurs through endocytosis and vacuolar degradation. J Biol Chem 273:28617–28624

Glickman MH, Ciechanover A (2002) The ubiquitin-proteasome proteolytic pathway: destruction for the sake of construction. Physiol Rev 82:373–428

Gonzalez-Chavez M, Carrillo-Gonzalez R, Wright S, Nichols K (2004) The role of glomalin, a protein produced by arbuscular mycorrhizal fungi, in sequestering potentially toxic elements. Environ Pollut 130:317–323

González-Guerrero M, Azcón-Aguilar C, Mooney M, Valderas A, MacDiarmid CW, Eide DJ, Ferrol N (2005) Characterization of a *Glomus intraradices* gene encoding a putative Zn transporter of the cation diffusion facilitator family. Fungal Genet Biol 42:130–140

González-Guerrero M, Benabdellah K, Ferrol N, Azcón-Aguilar C (2009) Mechanisms underlying heavy metal tolerance in arbuscular mycorrhizas. Mycorrhizas-functional processes and ecological impact. Springer, Berlin, pp 107–122

González-Guerrero M, Benabdellah K, Valderas A, Azcón-Aguilar C, Ferrol N (2010) GintABC1 encodes a putative ABC transporter of the MRP subfamily induced by Cu, Cd, and oxidative stress in *Glomus intraradices*. Mycorrhiza 20:137–146

Gorfer M, Persak H, Berger H, Brynda S, Bandian D, Strauss J (2009) Identification of heavy metal regulated genes from the root associated ascomycete *Cadophora finlandica* using a genomic microarray. Mycol Res 113:1377–1388

Gray SN (1998) Fungi as potential bioremediation agents in soil contaminated with heavy or radioactive metals. Biochem Soc Trans 26:666–670

Grąz M, Jarosz-Wilkołazka A, Pawlikowska-Pawlęga B (2009) *Abortiporus biennis* tolerance to insoluble metal oxides: oxalate secretion, oxalate oxidase activity, and mycelial morphology. Biometals 22:401–410

Green F, Clausen CA (2005) Copper tolerance of brown-rot fungi: oxalic acid production in southern pine treated with arsenic-free preservatives. Int Biodeter Biodegr 56:75–79

Gruhn C, Miller O (1991) Effect of copper on tyrosinase activity and polyamine content of some ectomycorrhizal fungi. Mycol Res 95:268–272

Guerinot ML (2000) The ZIP family of metal transporters. Biochim Biophys Acta (BBA)-Biomembr 1465:190–198

Haas H (2003) Molecular genetics of fungal siderophore biosynthesis and uptake: the role of siderophores in iron uptake and storage. Appl Microbiol Biotechnol 62:316–330

Haas JR, Purvis OW (2006) Lichen biogeochemistry. In: Gadd GM (ed) Fungi in biogeochemical cycles. Cambridge University Press, Cambridge, pp 344–376

Haas H, Eisendle M, Turgeon BG (2008) Siderophores in fungal physiology and virulence. Annu Rev Phytopathol 46:149–187

Harrison C, Katayama S, Dhut S, Chen D, Jones N, Bähler J, Toda T (2005) SCFPof1-ubiquitin and its target Zip1 transcription factor mediate cadmium response in fission yeast. EMBO J 24:599–610

Hartwig A (2001) Zinc finger proteins as potential targets for toxic metal ions: differential effects on structure and function. Antioxid Redox Sign 3:625–634

Haselwandter K, Berreck M, Brunner P (1988) Fungi as bioindicators of radiocesium contamination - Pre-Chernobyl and post-Chernobyl activities. Trans Br Mycol Soc 90:171–174

Hassett R, Kosman DJ (1995) Evidence for Cu (II) reduction as a component of copper uptake by *Saccharomyces cerevisiae*. J Biol Chem 270:128–134

Hauck M, Jürgens S-R, Willenbruch K, Huneck S, Leuschner C (2009) Dissociation and metal-binding characteristics of yellow lichen substances suggest a relationship with site preferences of lichens. Ann Bot 103:13–22

Heinisch JJ, Lorberg A, Schmitz HP, Jacoby JJ (1999) The protein kinase C-mediated MAP kinase pathway involved in the maintenance of cellular integrity in *Saccharomyces cerevisiae*. Mol Microbiol 32:671–680

Hoffland E, Kuyper TW, Wallander H, Plassard C, Gorbushina AA, Haselwandter K, Holmström S, Landeweert R, Lundström US, Rosling A (2004) The role of fungi in weathering. Front Ecol Environ 2:258–264

Hopf J, Langenhorst F, Pollok K, Merten D, Kothe E (2009) Influence of microorganisms on biotite dissolution: an experimental approach. Chem Erde-Geochem 69:45–56

Horng Y-C, Cobine PA, Maxfield AB, Carr HS, Winge DR (2004) Specific copper transfer from the Cox17 metallochaperone to both Sco1 and Cox11 in the assembly of yeast cytochrome C oxidase. J Biol Chem 279:35334–35340

Hosiner D, Gerber S, Lichtenberg-Frate H, Glaser W, Schuller C, Klipp E (2014) Impact of acute metal stress in *Saccharomyces cerevisiae*. PLoS One 9: e83330

Howe R, Evans R, Ketteridge S (1997) Copper-binding proteins in ectomycorrhizal fungi. New Phytol 135:123–131

Hughes MN, Poole RK (1991) Metal speciation and microbial growth—the hard(and soft) facts. Microbiology 137:725–734

Hvorup RN, Winnen B, Chang AB, Jiang Y, Zhou XF, Saier MH (2003) The multidrug/oligosaccharidyl-lipid/polysaccharide (MOP) exporter superfamily. Eur J Biochem 270:799–813

Jacob C, Courbot ML, Martin F, Brun A, Chalot M (2004) Transcriptomic responses to cadmium in the ectomycorrhizal fungus *Paxillus involutus*. FEBS Lett 576:423–427

Jamieson DJ (1998) Oxidative stress responses of the yeast *Saccharomyces cerevisiae*. Yeast 14:1511–1527

Janusz G, Kucharzyk KH, Pawlik A, Staszczak M, Paszczynski AJ (2013) Fungal laccase, manganese peroxidase and lignin peroxidase: gene expression and regulation. Enzyme Microb Technol 52:1–12

Jarosz-Wilkolazka A, Gadd GM (2003) Oxalate production by wood-rotting fungi growing in toxic metal-amended medium. Chemosphere 52:541–547

Jarosz-Wilkołazka A, Grąz M, Braha B, Menge S, Schlosser D, Krauss GJ (2006) Species-specific Cd-stress response in the white rot basidiomycetes *Abortiporus biennis* and *Cerrena unicolor*. Biometals 19:39–49

Jaszek M, Żuchowski J, Dajczak E, Cimek K, Graz M, Grzywnowicz K (2006) Ligninolytic enzymes can participate in a multiple response system to oxidative stress in white-rot basidiomycetes: *Fomes fomentarius* and *Tyromyces pubescens*. Int Biodeter Biodegr 58:168–175

Jensen LT, Culotta VC (2002) Regulation of *Saccharomyces cerevisiae* FET4 by oxygen and iron. J Mol Biol 318:251–260

Jentschke G, Godbold D (2000) Metal toxicity and ectomycorrhizas. Physiol Plant 109:107–116

Jin YH, Clark AB, Slebos RJ, Al-Refai H, Taylor JA, Kunkel TA, Resnick MA, Gordenin DA (2003) Cadmium is a mutagen that acts by inhibiting mismatch repair. Nat Genet 34:326–329

Jin YH, Dunlap PE, McBride SJ, Al-Refai H, Bushel PR, Freedman JH (2008) Global transcriptome and deletome profiles of yeast exposed to transition metals. PLoS Genet 4:e1000053

Kalač P, Svoboda L (2000) A review of trace element concentrations in edible mushrooms. Food Chem 69:273–281

Kang H-S, Kim K-R, Jun E-M, Park S-H, Lee T-S, Suh J-W, Kim J-P (2008) Cyathuscavins A, B, and C, new free radical scavengers with DNA protection activity from the basidiomycete *Cyathus stercoreus*. Bioorg Med Chem Lett 18:4047–4050

Kapoor A, Viraraghavan T (1995) Fungal biosorption—an alternative treatment option for heavy metal bearing wastewaters: a review. Bioresour Technol 53:195–206

Kasprzak KS (2002) Oxidative DNA and protein damage in metal-induced toxicity and carcinogenesis. Free Radic Biol Med 32:958–967

Keasling J, Van Dien S, Trelstad P, Renninger N, McMahon K (2000) Application of polyphosphate metabolism to environmental and biotechnological problems. Biochemistry (Mosc) 65:324–331

Kennedy PJ, Vashisht AA, Hoe KL, Kim DU, Park HO, Hayles J, Russell P (2008) A genome-wide screen of genes involved in cadmium tolerance in *Schizosaccharomyces pombe*. Toxicol Sci 106:124–139

Kisser M, Kubicek C, Röhr M (1980) Influence of manganese on morphology and cell wall composition of *Aspergillus niger* during citric acid fermentation. Arch Microbiol 128:26–33

Kosman DJ (2003) Molecular mechanisms of iron uptake in fungi. Mol Microbiol 47:1185–1197

Kothe E, Dimkpa C, Haferburg G, Schmidt A, Schmidt A, Schütze E (2010) Streptomycete heavy metal resistance: extracellular and intracellular mechanisms. In: Sherameti I, Varma A (eds) Soil heavy metals. Springer, Berlin, pp 225–235

Kuad P, Schurhammer R, Maechling C, Antheaume C, Mioskowski C, Wipff G, Spiess B (2009) Complexation of Cs+, K+ and Na+ by norbadione A triggered by the release of a strong hydrogen bond: nature and stability of the complexes. Phys Chem Chem Phys 11:10299–10310

Kües U, Rühl M (2011) Multiple multi-copper oxidase gene families in basidiomycetes-what for? Curr Genomics 12:72–94

Kumánovics A, Poruk KE, Osborn KA, Ward DM, Kaplan J (2006) YKE4 (YIL023C) encodes a bidirectional zinc transporter in the endoplasmic reticulum of *Saccharomyces cerevisiae*. J Biol Chem 281:22566–22574

Lanfranco L, Balsamo R, Martino E, Perotto S, Bonfante P (2002a) Zinc ions alter morphology and chitin deposition in an ericoid fungus. Eur J Histochem 46:341–350

Lanfranco L, Bolchi A, Ros EC, Ottonello S, Bonfante P (2002b) Differential expression of a metallothionein gene during the presymbiotic versus the symbiotic phase of an arbuscular mycorrhizal fungus. Plant Physiol 130:58–67

Lazard M, Ha-Duong NT, Mounié S, Perrin R, Plateau P, Blanquet S (2011) Selenodiglutathione uptake by the *Saccharomyces cerevisiae* vacuolar ATP-binding cassette transporter Ycf1p. FEBS J 278:4112–4121

Lee M, Parsons I (1999) Biomechanical and biochemical weathering of lichen-encrusted granite: textural controls on organic–mineral interactions and deposition of silica-rich layers. Chem Geol 161:385–397

Lehembre F, Doillon D, David E, Perrotto S, Baude J, Foulon J, Harfouche L, Vallon L, Poulain J, Da Silva C, Wincker P, Oger-Desfeux C, Richaud P, Colpaert JV, Chalot M, Fraissinet-Tachet L, Blaudez D, Marmeisse R (2013) Soil metatranscriptomics for mining eukaryotic heavy metal resistance genes. Environ Microbiol 15:2829–2840

Li L, Kaplan J (2004) A mitochondrial-vacuolar signaling pathway in yeast that affects iron and copper metabolism. J Biol Chem 279:33653–33661

Linder MB, Szilvay GR, Nakari-Setälä T, Penttilä ME (2005) Hydrophobins: the protein-amphiphiles of filamentous fungi. FEMS Microbiol Rev 29:877–896

Liu Z, Boles E, Rosen BP (2004) Arsenic trioxide uptake by hexose permeases in *Saccharomyces cerevisiae*. J Biol Chem 279:17312–17318

MacDiarmid CW, Milanick MA, Eide DJ (2002) Biochemical properties of vacuolar zinc transport systems of *Saccharomyces cerevisiae*. J Biol Chem 277:39187–39194

Maciaszczyk-Dziubinska E, Migdal I, Migocka M, Bocer T, Wysocki R (2010) The yeast aquaglyceroporin Fps1p is a bidirectional arsenite channel. FEBS Lett 584:726–732

Majorel C, Hannibal L, Soupe ME, Carriconde F, Ducousso M, Lebrun M, Jourand P (2012) Tracking nickel-adaptive biomarkers in *Pisolithus albus* from New Caledonia using a transcriptomic approach. Mol Ecol 21:2208–2223

Mäkelä MR, Hildén K, Lundell TK (2010) Oxalate decarboxylase: biotechnological update and prevalence of the enzyme in filamentous fungi. Appl Microbiol Biotechnol 87:801–814

Manceau A, Matynia A (2010) The nature of Cu bonding to natural organic matter. Geochim Cosmochim Acta 74:2556–2580

Maret W, Li Y (2009) Coordination dynamics of zinc in proteins. Chem Rev 109:4682–4707

Martino E, Turnau K, Girlanda M, Bonfante P, Perotto S (2000) Ericoid mycorrhizal fungi from heavy metal polluted soils: their identification and growth in the presence of zinc ions. Mycol Res 104:338–344

McDougall DN, Blanchette RA (1996) Metal ion adsorption by pseudosclerotial plates of *Phellinus weirii*. Mycologia 88:98–103

Mehra RK, Winge DR (1991) Metal ion resistance in fungi: molecular mechanisms and their regulated expression. J Cell Biochem 45:30–40

Meisch HU, Beckmann I, Schmitt JA (1983) A new cadmium-binding phosphoglycoprotein, cadmium-mycophosphatin, from the mushroom, *Agaricus macrosporus*. Biochim Biophys Acta 745:259–266

Mendoza-Cózatl DG, Zhai Z, Jobe TO, Akmakjian GZ, Song W-Y, Limbo O, Russell MR, Kozlovskyy VI, Martinoia E, Vatamaniuk OK (2010) Tonoplast-localized Abc2 transporter mediates phytochelatin accumulation in vacuoles and confers cadmium tolerance. J Biol Chem 285:40416–40426

Michelin M, Ruller R, Ward RJ, Moraes LAB, Jorge JA, Terenzi HF, Maria de Lourdes T (2008) Purification and biochemical characterization of a thermostable extracellular glucoamylase produced by the thermotolerant fungus *Paecilomyces variotii*. J Ind Microbiol Biotechnol 35:17–25

Mielniczki-Pereira AA, Schuch AZ, Bonatto D, Cavalcante CF, Vaitsman DS, Riger CJ, Eleutherio ECA, Henriques JAP (2008) The role of the yeast ATP-binding cassette Ycf1p in glutathione and cadmium ion homeostasis during respiratory metabolism. Toxicol Lett 180:21–27

Mielniczki-Pereira AA, Hahn ABB, Bonatto D, Riger CJ, Eleutherio ECA, Henriques JAP (2011) New insights into the Ca 2+-ATPases that contribute to cadmium tolerance in yeast. Toxicol Lett 207:104–111

Miethke M, Marahiel MA (2007) Siderophore-based iron acquisition and pathogen control. Microbiol Mol Biol Rev 71:413–451

Montanini B, Blaudez D, Jeandroz S, Sanders D, Chalot M (2007) Phylogenetic and functional analysis of the Cation Diffusion Facilitator (CDF) family: improved signature and prediction of substrate specificity. BMC Genomics 8:107

Morley GF, Gadd GM (1995) Sorption of toxic metals by fungi and clay minerals. Mycol Res 99:1429–1438

Morselt AF, Smits WT, Limonard T (1986) Histochemical demonstration of heavy metal tolerance in ectomycorrhizal fungi. Plant Soil 96:417–420

Muller LAH, Craciun AR, Ruytinx J, Lambaerts M, Verbruggen N, Vangronsveld J, Colpaert JV (2007) Gene expression profiling of a Zn-tolerant and a Zn-sensitive *Suillus luteus* isolate exposed to increased external zinc concentrations. Mycorrhiza 17:571–580

Naar Z (2006) Effect of cadmium, nickel and zinc on the antagonistic activity of *Trichoderma* spp. against *Pythium irregulare* Buisman. Acta Phytopathol Entomol Hung 41:193–202

Nagy Z, Montigny C, Leverrier P, Yeh S, Goffeau A, Garrigos M, Falson P (2006) Role of the yeast ABC transporter Yor1p in cadmium detoxification. Biochimie 88:1665–1671

Nevo Y, Nelson N (2006) The NRAMP family of metal-ion transporters. Biochim Biophys Acta (BBA)-Mol Cell Res 1763:609–620

O'Halloran TV, Culotta VC (2000) Metallochaperones, an intracellular shuttle service for metal ions. J Biol Chem 275:25057–25060

Osobová M, Urban V, Jedelský PL, Borovička J, Gryndler M, Ruml T, Kotrba P (2011) Three metallothionein isoforms and sequestration of intracellular silver in the hyperaccumulator *Amanita strobiliformis*. New Phytol 190:916–926

Ott T, Fritz E, Polle A, Schützendübel A (2002) Characterisation of antioxidative systems in the ectomycorrhiza-building basidiomycete *Paxillus involutus* (Bartsch) Fr. and its reaction to cadmium. FEMS Microbiol Ecol 42:359–366

Ouziad F, Hildebrandt U, Schmelzer E, Bothe H (2005) Differential gene expressions in arbuscular mycorrhizal-colonized tomato grown under heavy metal stress. J Plant Physiol 162:634–649

Pagani A, Villarreal L, Capdevila M, Atrian S (2007) The *Saccharomyces cerevisiae* Crs5 metallothionein metal-binding abilities and its role in the response to zinc overload. Mol Microbiol 63:256–269

Palacios Ò, Atrian S, Capdevila M (2011) Zn-and Cu-thioneins: a functional classification for metallothioneins? J Biol Inorg Chem 16:991–1009

Paraszkiewicz K, Frycie A, Słaba M, Długoński J (2007) Enhancement of emulsifier production by *Curvularia lunata* in cadmium, zinc and lead presence. Biometals 20:797–805

Park H, Song B, Morel FM (2007a) Diversity of the cadmium-containing carbonic anhydrase in marine diatoms and natural waters. Environ Microbiol 9:403–413

Park JN, Sohn MJ, Oh DB, Kwon O, Rhee SK, Hur CG, Lee SY, Gellissen G, Kang HA (2007b) Identification of the cadmium-inducible *Hansenula polymorpha* SEO1 gene promoter by transcriptome analysis and its application to whole-cell heavy-metal detection systems. Appl Environ Microbiol 73:5990–6000

Paulsen I, Saier M Jr (1997) A novel family of ubiquitous heavy metal ion transport proteins. J Membr Biol 156:99–103

Pawlik-Skowrońska B, Toppi D, Sanità L, Favali MA, Fossati F, Pirszel J, Skowroński T (2002) Lichens respond to heavy metals by phytochelatin synthesis. New Phytol 156:95–102

Pereira Y, Lagniel G, Godat E, Baudouin-Cornu P, Junot C, Labarre J (2008) Chromate causes sulfur starvation in yeast. Toxicol Sci 106:400–412

Pirintsos S, Kotzabasis K, Loppi S (2004) Polyamine production in lichens under metal pollution stress. J Atmos Chem 49:303–315

Piscitelli A, Giardina P, Lettera V, Pezzella C, Sannia G, Faraco V (2011) Induction and transcriptional regulation of laccases in fungi. Curr Genomics 12:104–112

Pócsi I (2011) Toxic metal/metalloid tolerance in fungi - a biotechnology-oriented approach. In: Banfalvi G (ed) Cellular effects of heavy metals. Springer, Berlin, pp 31–58

Pócsi I, Prade RA, Penninckx MJ (2004) Glutathione, altruistic metabolite in fungi. Adv Microb Physiol 49:1–76

Poljšak B, Gazdag Z, Jenko-Brinovec Š, Fujs Š, Pesti M, Bélagyi J, Plesničar S, Raspor P (2005) Pro-oxidative vs antioxidative properties of ascorbic acid in chromium (VI)-induced damage: an in vivo and in vitro approach. J Appl Toxicol 25:535–548

Préveral S, Gayet L, Moldes C, Hoffmann J, Mounicou S, Gruet A, Reynaud F, Lobinski R, Verbavatz J-M, Vavasseur A (2009) A common highly conserved cadmium detoxification mechanism from bacteria to humans heavy metal tolerance conferred by the ATP-binding cassette (ABC) transporter SpHMT1 requires glutathione but not metal-chelating phytochelatin peptides. J Biol Chem 284:4936–4943

Puig S, Thiele DJ (2002) Molecular mechanisms of copper uptake and distribution. Curr Opin Chem Biol 6:171–180

Ray P, Adholeya A (2009) Correlation between organic acid exudation and metal uptake by ectomycorrhizal fungi grown on pond ash in vitro. Biometals 22:275–281

Renshaw JC, Robson GD, Trinci AP, Wiebe MG, Livens FR, Collison D, Taylor RJ (2002) Fungal siderophores: structures, functions and applications. Mycol Res 106:1123–1142

Rizzo D, Blanchette R, Palmer M (1992) Biosorption of metal ions by *Armillaria* rhizomorphs. Can J Bot 70:1515–1520

Rühm W, Steiner M, Kammerer L, Hiersche L, Wirth E (1998) Estimating future radiocaesium contamination of fungi on the basis of behaviour patterns derived from past instances of contamination. J Environ Radioact 39:129–147

Ruijter GJ, van de Vondervoort PJ, Visser J (1999) Oxalic acid production by *Aspergillus niger*: an oxalate-non-producing mutant produces citric acid at pH 5 and in the presence of manganese. Microbiology 145:2569–2576

Rupp S (2008) Transcriptomics of the fungal pathogens, focusing on *Candida albicans*. In: Brakhage AA, Zipfel PF (eds) The mycota VI, Human and animal relationships. Springer, Berlin, pp 187–222

Ruta LL, Popa VC, Nicolau I, Danet AF, Iordache V, Neagoe AD, Farcasanu IC (2014) Calcium signaling mediates the response to cadmium toxicity in *Saccharomyces cerevisiae* cells. FEBS Lett 588:3202–3212

Ruytinx J, Craciun AR, Verstraelen K, Vangronsveld J, Colpaert JV, Verbruggen N (2011) Transcriptome analysis by cDNA-AFLP of *Suillus luteus* Cd-tolerant and Cd-sensitive isolates. Mycorrhiza 21:145–154

Ruytinx J, Nguyen H, Van Hees M, De Beeck MO, Vangronsveld J, Carleer R, Colpaert JV, Adriaensen K (2013) Zinc export results in adaptive zinc tolerance in the ectomycorrhizal basidiomycete *Suillus bovinus*. Metallomics 5:1225–1233

Sácký J, Leonhardt T, Borovička J, Gryndler M, Briksí A, Kotrba P (2014) Intracellular sequestration of zinc, cadmium and silver in *Hebeloma mesophaeum* and characterization of its metallothionein genes. Fungal Genet Biol 67:3–14

Saha R, Saha N, Donofrio RS, Bestervelt LL (2013) Microbial siderophores: a mini review. J Basic Microbiol 53:303–317

Sarikurkcu C, Tepe B, Semiz DK, Solak MH (2010) Evaluation of metal concentration and antioxidant activity of three edible mushrooms from Mugla, Turkey. Food Chem Toxicol 48:1230–1233

Say R, Denizli A, Arıca MY (2001) Biosorption of cadmium (II), lead (II) and copper (II) with the filamentous fungus *Phanerochaete chrysosporium*. Bioresour Technol 76:67–70

Sazanova K, Osmolovskaya N, Schiparev S, Yakkonen K, Kuchaeva L, Vlasov D (2014) Organic acids induce tolerance to zinc-and copper-exposed fungi under various growth conditions. Curr Microbiol 70:1–8

Schindler F, Gube M, Kothe E (2012) Bioremediation and heavy metal uptake: microbial approaches at field scale. In: Kothe E, Varma A (eds) Bio-geo interactions in metal-contaminated soils. Springer, Berlin, pp 365–383

Schlunk I, Krause K, Wirth S, Kothe E (2015) A transporter for abiotic stress and plant metabolite resistance in the ectomycorrhizal fungus *Tricholoma vaccinum*. Environ Sci Pollut Res. doi:10.1007/s11356-014-4044-8

Schmidt K, Wolfe DM, Stiller B, Pearce DA (2009) Cd 2+, Mn 2+, Ni 2+ and Se 2+ toxicity to *Saccharomyces cerevisiae* lacking YPK9p the orthologue of human ATP13A2. Biochem Biophys Res Commun 383:198–202

Schrettl M, Haas H (2011) Iron homeostasis - Achilles' heel of *Aspergillus fumigatus*? Curr Opin Microbiol 14:400–405

Schützendübel A, Polle A (2002) Plant responses to abiotic stresses: heavy metal-induced oxidative stress and protection by mycorrhization. J Exp Bot 53:1351–1365

Seeger R (1978) Cadmium in mushrooms. Z Lebensm Unters For 166:23–34

Sharma KG, Mason DL, Liu G, Rea PA, Bachhawat AK, Michaelis S (2002) Localization, regulation, and substrate transport properties of Bptlp, a *Saccharomyces cerevisiae* MRP-type ABC transporter. Eukaryot Cell 1:391–400

Shiomi N, Fukuda H, Fukuda Y, Murata K, Kimura A (1991) Nucleotide sequence and characterization of a gene conferring resistance to ethionine in yeast *Saccharomyces cerevisiae*. J Ferment Bioeng 71:211–215

Skorik YA, Pestov AV, Yatluk YG (2010) Evaluation of various chitin-glucan derivatives from *Aspergillus niger* as transition metal adsorbents. Bioresour Technol 101:1769–1775

Steffan B, Steglich W (1984) Fungal pigments 44. Pigments from the cap cuticle of the bay boletus (*Xerocomus badius*). Angew Chem Int Ed 23:445–447

Sterflinger K (2000) Fungi as geologic agents. Geomicrobiol J 17:97–124

Stijve T, Besson R (1976) Mercury, cadmium, lead and selenium content of mushroom species belonging to the genus *Agaricus*. Chemosphere 2:151–158

Stohs S, Bagchi D (1995) Oxidative mechanisms in the toxicity of metal ions. Free Radic Biol Med 18:321–336

Sturtz LA, Diekert K, Jensen LT, Lill R, Culotta VC (2001) A fraction of yeast Cu, Zn-superoxide dismutase and its metallochaperone, CCS, localize to the intermembrane space of mitochondria. A physiological role for SOD1 in guarding against mitochondrial oxidative damage. J Biol Chem 276:38084–38089

Summers AO (2009) Damage control: regulating defenses against toxic metals and metalloids. Curr Opin Microbiol 12:138–144

Sun L, Liu G, Yang M, Zhuang Y (2012) Bioaccessibility of cadmium in fresh and cooked *Agaricus blazei* Murill assessed by in vitro biomimetic digestion system. Food Chem Toxicol 50:1729–1733

Svoboda L, Zimmermannová K, Kalač P (2000) Concentrations of mercury, cadmium, lead and copper in fruiting bodies of edible mushrooms in an emission area of a copper smelter and a mercury smelter. Sci Total Environ 246:61–67

Tait K, Sayer JA, Gharieb MM, Gadd GM (1999) Fungal production of calcium oxalate in leaf litter microcosms. Soil Biol Biochem 31:1189–1192

Takani M, Yajima T, Masuda H, Yamauchi O (2002) Spectroscopic and structural characterization of copper (II) and palladium (II) complexes of a lichen substance usnic acid and its derivatives. Possible forms of environmental metals retained in lichens. J Inorg Biochem 91:139–150

Tamás MJ, Sharma SK, Ibstedt S, Jacobson T, Christen P (2014) Heavy metals and metalloids as a cause for protein misfolding and aggregation. Biomolecules 4:252–267

Tamayo E, Gómez-Gallego T, Azcón-Aguilar C, Ferrol N (2014) Genome-wide analysis of copper, iron and zinc transporters in the arbuscular mycorrhizal fungus *Rhizophagus irregularis*. Front Plant Sci 5:547

Tang JD, Parker LA, Perkins AD, Sonstegard TS, Schroeder SG, Nicholas DD, Diehl SV (2013) Gene expression analysis of copper tolerance and wood decay in the brown rot fungus *Fibroporia radiculosa*. Appl Environ Microbiol 79:1523–1533

Thiele D, Hamer D (1986) Tandemly duplicated upstream control sequences mediate copper-induced transcription of the *Saccharomyces cerevisiae* copper-metallothionein gene. Mol Cell Biol 6:1158–1163

Thorpe GW, Fong CS, Alic N, Higgins VJ, Dawes IW (2004) Cells have distinct mechanisms to maintain protection against different reactive oxygen species: oxidative-stress-response genes. Proc Natl Acad Sci USA 101:6564–6569

Thorsen M, Lagniel G, Kristiansson E, Junot C, Nerman O, Labarre J, Tamás MJ (2007) Quantitative transcriptome, proteome, and sulfur metabolite profiling of the *Saccharomyces cerevisiae* response to arsenite. Physiol Genomics 30:35–43

Thorsen M, Perrone GG, Kristiansson E, Traini M, Ye T, Dawes IW, Nerman O, Tamas MJ (2009) Genetic basis of arsenite and cadmium tolerance in *Saccharomyces cerevisiae*. BMC Genomics 10:105

Thorsen M, Jacobson T, Vooijs R, Navarrete C, Bliek T, Schat H, Tamás MJ (2012) Glutathione serves an extracellular defence function to decrease arsenite accumulation and toxicity in yeast. Mol Microbiol 84:1177–1188

Tircsó G, Garda Z, Kálmán FK, Baranyai Z, Pócsi I, Balla G, Tóth I (2013) Lanthanide (III) complexes of some natural siderophores: a thermodynamic, kinetic and relaxometric study. J Inorg Biochem 127:53–61

Trilla JA, Cos T, Duran A, Roncero C (1997) Characterization of CHS4 (CAL2), a gene of *Saccharomyces cerevisiae* involved in chitin biosynthesis and allelic to SKT5 and CSD4. Yeast 13:795–807

Tsezos M, Volesky B (1982) The mechanism of uranium biosorption by *Rhizopus arrhizus*. Biotechnol Bioeng 24:385–401

Turnau K, Kottke I, Dexheimer J, Botton B (1994) Element distribution in mycelium of *Pisolithus arrhizus* treated with cadmium dust. Ann Bot 74:137–142

Tuszynska S, Davies D, Turnau K, Ashford AE (2006) Changes in vacuolar and mitochondrial motility and tubularity in response to zinc in a *Paxillus involutus* isolate from a zinc-rich soil. Fungal Genet Biol 43:155–163

Uldschmid A, Engel M, Dombi R, Marbach K (2002) Identification and functional expression of tahA, a filamentous fungal gene involved in copper trafficking to the secretory pathway in *Trametes versicolor*. Microbiology 148:4049–4058

Urban A (2011) Metal elements and the diversity and function of ectomycorrhizal communities. In: Rai M, Varma A (eds) Diversity and biotechnology of ectomycorrhizae. Springer, Berlin, pp 231–254

Urban PL, Bazala MA, Asztemborska M, Manjon JL, Kowalska J, Bystrzejewska-Piotrowska G, Pianka D, Steborowski R, Kuthan RT (2005) Preliminary study of platinum accumulation in the fruitbodies of a model fungal species: king oyster mushroom (*Pleurotus eryngii*). Nukleonika 50:S63–S67

Valko M, Morris H, Cronin M (2005) Metals, toxicity and oxidative stress. Curr Med Chem 12:1161–1208

Van Hees P, Godbold D, Jentschke G, Jones D (2003) Impact of ectomycorrhizas on the concentration and biodegradation of simple organic acids in a forest soil. Eur J Soil Sci 54:697–706

Van Ho A, Ward DM, Kaplan J (2002) Transition metal transport in yeast. Annu Rev Microbiol 56:237–261

Varma A, Chincholkar SB (2007) Microbial siderophores. Springer, Berlin

Vasiliou V, Vasiliou K, Nebert DW (2009) Human ATP-binding cassette (ABC) transporter family. Hum Genomics 3:281

Vesentini D, Dickinson DJ, Murphy RJ (2007) The protective role of the extracellular mucilaginous material (ECMM) from two wood-rotting basidiomycetes against copper toxicity. Int Biodeter Biodegr 60:1–7

Vido K, Spector D, Lagniel G, Lopez S, Toledano MB, Labarre J (2001) A proteome analysis of the cadmium response in *Saccharomyces cerevisiae*. J Biol Chem 276:8469–8474

Viti C, Marchi E, Decorosi F, Giovannetti L (2014) Molecular mechanisms of Cr (VI) resistance in bacteria and fungi. FEMS Microbiol Rev 38:633–659

Volesky B, Holan Z (1995) Biosorption of heavy metals. Biotechnol Progr 11:235–250

Wang L, Li H, Wei H, Wu X, Ke L (2014) Identification of cadmium-induced *Agaricus blazei* genes through suppression subtractive hybridization. Food Chem Toxicol 63:84–90

Waters BM, Eide DJ (2002) Combinatorial control of yeast FET4 gene expression by iron, zinc, and oxygen. J Biol Chem 277:33749–33757

Wawrzycka D, Sobczak I, Bartosz G, Bocer T, Ułaszewski S, Goffeau A (2010) Vmr 1p is a novel vacuolar multidrug resistance ABC transporter in *Saccharomyces cerevisiae*. FEMS Yeast Res 10:828–838

Welch SA, Vandevivere P (1994) Effect of microbial and other naturally occurring polymers on mineral dissolution. Geomicrobiol J 12:227–238

Wengel M, Kothe E, Schmidt CM, Heide K, Gleixner G (2006) Degradation of organic matter from black shales and charcoal by the wood-rotting fungus *Schizophyllum commune* and release of DOC and heavy metals in the aqueous phase. Sci Total Environ 367:383–393

White C, Sayer JA, Gadd GM (1997) Microbial solubilization and immobilization of toxic metals: key biogeochemical processes for treatment of contamination. FEMS Microbiol Rev 20:503–516

Williams ME, Rudolph ED (1974) The role of lichens and associated fungi in the chemical weathering of rock. Mycologia 66:648–660

Wösten HA (2001) Hydrophobins: multipurpose proteins. Annu Rev Microbiol 55:625–646

Wright S, Upadhyaya A (1998) A survey of soils for aggregate stability and glomalin, a glycoprotein produced by hyphae of arbuscular mycorrhizal fungi. Plant Soil 198:97–107

Wysocki R, Tamás MJ (2010) How *Saccharomyces cerevisiae* copes with toxic metals and metalloids. FEMS Microbiol Rev 34:925–951

Xu P, Leng Y, Zeng G, Huang D, Lai C, Zhao M, Wei Z, Li N, Huang C, Zhang C (2015) Cadmium induced oxalic acid secretion and its role in metal uptake and detoxification mechanisms in *Phanerochaete chrysosporium*. Appl Microbiol Biotechnol 99:435–443

Yetis U, Dolek A, Dilek FB, Ozcengiz G (2000) The removal of Pb (II) by *Phanerochaete chrysosporium*. Water Res 34:4090–4100

Zajc J, Liu Y, Dai W, Yang Z, Hu J, Gostincar C, Gunde-Cimerman N (2013) Genome and transcriptome sequencing of the halophilic fungus *Wallemia ichthyophaga*: haloadaptations present and absent. BMC Genomics 14:617

Zarb J, Walters D (1995) Polyamine biosynthesis in the ectomycorrhizal fungus *Paxillus involutus* exposed to zinc. Lett Appl Microbiol 21:93–95

Zhao D, Li T, Shen M, Wang J, Zhao Z (2015) Diverse strategies conferring extreme cadmium (Cd) tolerance in the dark septate endophyte (DSE), *Exophiala pisciphila*: evidence from RNA-seq data. Microbiol Res 170:27–35

Zhou Z-Y, Liu J-K (2010) Pigments of fungi (macromycetes). Nat Prod Rep 27:1531–1570

5 Control of Gene Expression in Phytopathogenic Ascomycetes During Early Invasion of Plant Tissue

CANDACE E. ELLIOTT[1]

CONTENTS

I. Introduction

A. Fungal Lifestyles

Plants are host to a wide range of fungi, and the relationships between the two can range from commensal or mutualistic to antagonistic. Traditionally, fungi have been classified as endophytic, **biotrophic, necrotrophic** or **hemibiotrophic**, but many fungi exhibit traits of more than one of these classifications during their life in planta, and it is probably wise to think of these different lifestyles as a continuum. For example, when fungi were isolated from healthy *Brassica napus* plants with the aim to identify endophytes which could be used for biocontrol, three of those identified would normally be considered phytopathogens: *Leptosphaeria biglobosa*, *Botrytis cinerea* and *Rhizoctonia solani* (Zhang et al. 2014). Obviously, as they were isolated from asymptomatic tissue, these pathogens were displaying phases of asymptomatic growth for which they are lesser known. A recent review discusses this dichotomy in lifestyle for *B. cinerea* (Van Kan et al. 2014). Since most fungi do not infect most plants and many do not even attempt to grow as an epiphyte on plants outside their host range, what is it that defines the host range of a fungus and what controls the developmental programme followed by plant-associated ascomycetes? This chapter will highlight recent publications on major ascomyceteous foliar pathogens on which genes are required for getting into the plant.

B. Comparative Genomics

The last 5 years has been a witness to an explosion of information about fungal genes and their expression due to the analysis of fungal genome sequences of many phytopathogenic ascomycetes. The Joint Genome Institute (JGI) project MycoCosm hosts the genome sequence of 234 ascomycetes, 27 of which are plant pathogens, as part of their 1000 fungal genomes project (Grigoriev et al. 2014). The National Centre for Biotechnology Information (NCBI) currently holds 549 genome sequences from the kingdom Ascomycota, 126 of which are from plant pathogens (http://www.ncbi.nlm.nih.gov/genome/browse/). By comparing genome size, gene content, loss, and/or gain between fungi that co-habitate with plants, inferences can be

[1]School of Biosciences, The University of Melbourne, Melbourne, VIC 3010, Australia; e-mail: ce@unimelb.edu.au

Biochemistry and Molecular Biology, 3rd Edition
The Mycota III
D. Hoffmeister (Ed.)

made as to which genes are required or are dispensable for different fungal lifestyles.

Of the genes present in pathogenic fungi, those encoding **secreted proteins** are of particular interest because of the role of those proteins in deciding the outcome of the fungus–plant interaction. Lowe and Howlett (2012) made a comparison of predicted secreted proteins from 47 sequenced genomes of fungal pathogens (including animal and insect pathogens) to see if the proportion of secreted proteins correlated with fungal lifestyle. These secreted proteins included different classes of proteins including carbohydrate-degrading enzymes (CAZys), which are encoded by multigene families, and predicted effector proteins, which are small and usually cysteine rich. Figure 5.1 shows that for some fungi such a relationship exists. As a generalisation, fungi with a biphasic lifestyle, such as *Magnaporthe oryzae*, have the largest proportion of secreted proteins and that animal pathogens have the lowest proportion of secreted proteins and the smallest genomes. Interestingly, saprophytes and necrotrophs have a similar proportion of predicted secreted proteins, which may reflect the fact that necrotrophs spend a considerable proportion of their life cycle feeding as a saprophyte. The observations made by Lowe and Howlett (2012) are echoed in another study that also looked at the effect of lineage and secretome size and content (Krijger et al. 2014). They found that although lifestyle and secretome size were correlated, individual expansion/contraction of gene families correlated more with lineage.

A detailed comparison of genome size and content of approximately 30 plant pathogens, including some basidiomycetes and oomycetes, was undertaken by Raffaele and Kamoun (2012). The ascomycetes they compared ranged in genome sizes from 32 Mb for *Zymoseptoria tritici* that is a pathogen of wheat to 160 Mb for *Golovinomyces orontii*, which is a powdery mildew of the model dicot plant *Arabidopsis thaliana*. Interestingly, genome size does not correlate with numbers of protein-coding genes as the largest genomes of the biotrophic mildews have the least number of protein-coding genes at approximately 6000–7000 (Spanu et al. 2010; Wicker et al. 2013). These genomes have a marked loss of genes encoding secondary metabolites and a very low number of carbohydrate active enzymes (CAZys), which may reflect the mildew biotrophic lifestyle. Also, genes encoding the enzymes responsible for the reduction of inorganic nitrogen and sulphur are entirely absent, possibly reflecting the fact that the mildew acquires nitrogen and sulphur from its host plant through the specialised feeding structure the haustorium (Spanu 2014). Analysis and comparison of the genome sequences of the necrotrophic pathogens *Sclerotinia sclerotiorum* and *B. cinerea* revealed a high level of sequence similarity but with some striking differences (Amselem et al. 2011). For instance, an abundance of CAZys with a specificity for pectin was common to both fungi, which reflects the high pectin content of the cell walls of the leaves of their dicot host plants. However, the amount of transposable elements and types of secondary metabolite gene clusters was quite different. Also of note is that there were no unique features of the genomes of these aggressive pathogens that distinguished them from other pathogens or non-pathogens.

C. Stages of Infection

Valuable information about which of the genes in phytopathogenic fungi are important for establishing disease can be gleaned from the analysis of timing and regulation of gene expression during infection. Figure 5.2 illustrates the stages that will be discussed in this chapter during which major transcriptional changes occur in the fungus to enable it to enter its host plant: surface recognition, germination, entry and establishment of in-planta growth. The first steps, **surface recognition** and **germination** (Fig. 5.2A,B), must be undertaken by all fungi, but after that, there is a diversification of the method used for entry into the plant tissue. Entry can be an active process through invagination of the plant cell or via direct puncture into the cell, or it can be opportunistic through wounds or stomatal apertures (Fig. 5.2C). In-planta growth (Fig. 5.2D) can be established via the invagination of the plant cell and

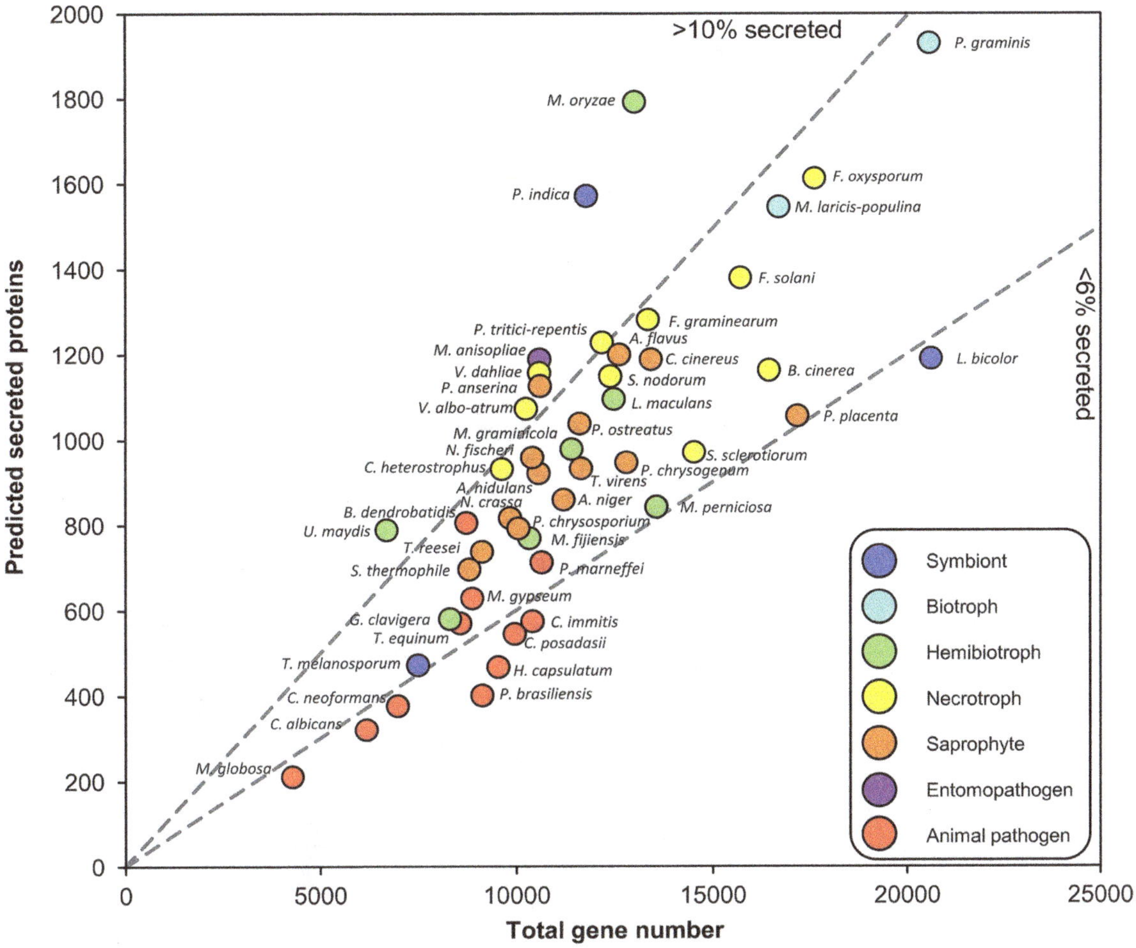

Fig. 5.1 Relationship between predicted secreted protein number and total gene content of fungi. Data are from (Zuccaro et al. 2011) or by applying SignalP to genome releases (indicated by *asterisk*). *Dashed lines* discriminate between fungi with high (>10) or low (<6) % secreted proteins. Animal pathogens: *Batrachochytrium dendrobatidis, Candida albicans, Coccidioides immitis, C. posadasii, Cryptococcus neoformans, Histoplasma capsulatum, Malassezia globosa, Microsporum gypseum, Paracoccidioides brasiliensis, Penicillium marneffei, Trichophyton equinum.* Hemibiotrophs: *Grosmania clavigera*, Leptosphaeria maculans** (Rouxel et al. 2011), *Magnaporthe oryzae*, Mycosphaerella fijiensis, M. graminicola, Moniliophthora perniciosa, Ustilago maydis.* Entomopathogen: *Metarhizium anisopliae** (Gao et al. 2011). Necrotrophs: *Botrytis cinerea, Cochliobolus heterostrophus, Fusarium graminearum, F. oxysporum, F. solani, Pyrenophora tritici-repentis, Stagonospora nodorum*, Sclerotinia sclerotiorum, Verticillium albo-atrum, V. dahliae.* Biotrophs: *Melampsora laricis-populina** (Duplessis et al. 2011), *Puccinia graminis* f. sp. *tritici.* Saprophytes: *Aspergillus flavus, A. nidulans, A. niger, Coprinus cinereus, Neurospora crassa, Neosartorya fischeri, Podospora anserina, Penicillium chrysogenum, Phanerochaete chrysosporium, Pleurotus ostreatus, Postia placenta, Sporotrichum thermophile, Trichoderma reesei, T. virens.* Symbionts: *Laccaria bicolor, Piriformospora indica** (Zuccaro et al. 2011), *Tuber melanosporum** (Martin et al. 2010)

development of a feeding structure such as a haustorium or an invasive hypha. Here the relationship is considered to be one of biotrophy as the plant cell must remain alive for the fungus to acquire nutrients and proliferate. However, in the case of *M. oryzae*, the relationship soon turns sour for the plant as effectors are secreted by the fungus into adjacent cells causing cell death and resulting in a necrotrophic relationship (Khang et al. 2010). Similarly, direct

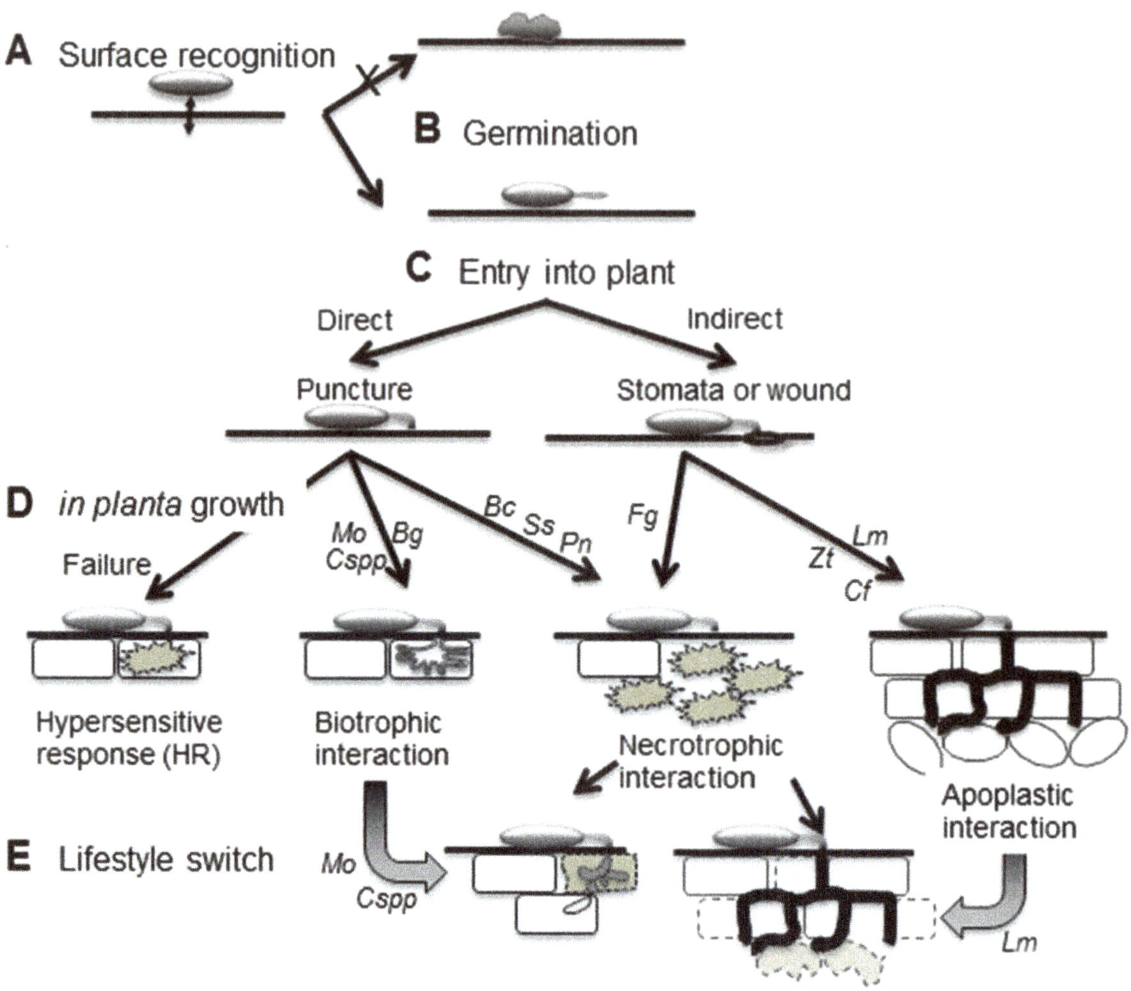

Fig. 5.2 Strategies used by filamentous ascomycetes to invade plant tissue. In order to invade a plant, major transcriptional reprogramming must occur in the fungal spore. (**A**) Upon landing on a host surface, the fungus must recognise an appropriate host and adhere to the surface. (**B**) Germination requires a large transcriptional activation of genes to initiate polar growth and relies on the utilisation of stored lipids for energy. (**C**) Entry into the plant can be an active process via direct puncture of the plant cell wall or opportunistically through wounds or stomatal openings. (**D**) Fungi use diverse strategies to establish in-planta growth. Failure to enter the plant can result in an hypersensitive reaction as seen by localised cell death. Several fungi such as *Blumeria graminis* (*Bg*), *Magnaporthe oryzae* (*Mo*) and *Colletotrichum* species (*Cspp*) establish a biotrophic relationship in the host. In contrast, *Sclerotinia sclerotiorum* (*Ss*), *Botrytis cinerea* (*Bc*), *Parastagonospora nodorum* (*Pn*) and *Fusarium graminearum* (*Fg*) enter the plant directly or opportunistically and establish and necrotrophic relationship. Fungi such as *Leptosphaeria maculans* (*Lm*), *Zymoseptoria tritici* (*Zt*) and *Cladosporium fulvum* (*Cf*) enter through stomata and proliferate in the apoplastic space. (**E**) Some fungi (*Mo*, *Cspp* and *Lm*) ultimately undergo a lifestyle switch to necrotrophy as depicted by the *grey* collapsing cells

puncture of the plant cell can result in cell death and then the relationship is considered necrotrophic. Fungi that enter through stomata often proliferate in the apoplast and experience a latent period of biotrophy before eventually penetrating deeper cell layers, such as the vasculature, to become necrotrophic. As depicted by the arrows in Fig. 5.2, there are many different ways for a fungus to enter and proliferate within the plant, and some fungi use several of these strategies or switch between strategies as they grow through the plant tissue.

These stages serve as a framework for the discussion of recent advances in the understanding of the control of transcription in plant pathogenic ascomycetes in the following sections.

Plants are resistant to infection by most of the hundreds of pathogens to which they are exposed each day. Studies of this type of resistance tend to focus on the attributes and reactions of the plants that make them a non-host for a particular model fungus [reviewed in Fan and Doerner (2012)]. Far fewer studies focus on the fungal side to ask why in general it is that most fungal pathogens are only able to infect a small percentage of the plants they encounter. There are obvious examples of exceptions to this generalisation, such as the broad host range necrotrophic ascomycete *S. sclerotiorum* which has an expansive host range encompassing over 400 plants (Guyon et al. 2014), and the possible reasons for this large host range will be discussed in Sect. III.A.1.

II. Surface Recognition and Breaking Dormancy

A. Surface Recognition

Host surfaces have unique properties and each fungus has a set of preprogrammed tropic responses to the physical and chemical cues presented by its host; hence, the first steps of successful fungal ingress are surface recognition and adhesion (Brand and Gow 2012). Early work on surface perception by ascomycete plant pathogens was conducted on *M. oryzae* and some species of *Colletotrichum* which produce appressoria and have a hemibiotrophic lifestyle (Howard and Valent 1996; Perfect et al. 1999). In order to study the early infection processes without contaminating plant material, researchers sought artificial surfaces upon which they could germinate the fungal spores. By altering surface hardness, hydrophobicity and texture, *M. oryzae* and *C. graminicola* spores could be induced to germinate, thus suggesting that surface recognition by these fungi was mainly determined by physical and chemical cues (Jelitto et al. 1994; Mercure et al. 1994). Although the receptors mediating the perception of these cues were not identified at this time, one of the first indications that a signalling pathway was involved in transducing a surface recognition signal was the finding that the addition of exogenous cyclic AMP to germinating *M. oryzae* spores allowed appressorium formation in the absence of contact with a hydrophobic surface (Lee and Dean 1993).

Cyclic AMP is a well-known second messenger which amplifies a signal received by **G-protein-coupled receptors** after they have bound G-protein subunits, through the activation of protein kinase A cascade (McDonough and Rodriguez 2012). Once researchers were able to induce infection structures in the absence of plant tissue, they were able to begin to identify the genetic components of surface recognition and infection structure differentiation through the study of deletion mutants. Untangling the genetic control of surface recognition by *M. oryzae* from the initiation of polar growth and appressorium formation is extremely difficult as the time between surface recognition and initiation of appressorium (germ tube 'hooking' stage) is merely 4 h (Howard and Valent 1996). Thus, many of the mutants identified in a search for surface recognition determinants have defects in several aspects of early infection. The web of key signalling pathways involved in infection-related morphogenesis has recently been thoroughly reviewed by Li et al. (2012a).

One of the first *M. oryzae* mutants identified as defective in surface sensing and responding to inductive surface cues was the *pth11* mutant, which was defective in appressorium differentiation in that it only made normal infection structures in a very low percentage of cases. The *PTH11* gene encodes a seven transmembrane-spanning putative G-protein-coupled receptor protein which was originally localised via GFP fusions to the cell membrane and vacuoles (DeZwaan et al. 1999). This localisation has recently been re-examined by Ramanujam et al. (2013). Pth11p tagged with the mCherry red fluorescent protein localised to punctate structures in quiescent spores but translocated to tubule-vesicular structures in the germ tube by 2 h and to the plasma membrane by 4 h when the germ tube begins to bend to form the appressorium. Additionally, pth11 co-localises with Rgs1 (regulator of G-protein signalling 1) specifically during early appressorium formation, thus confirming the role of pth11p in G-protein signalling pathways.

Two other receptors, MoSho1, a four transmembrane domain-spanning protein, and MoMsb2, containing a single transmembrane domain, have been identified as sensors for the primary alcohol component of leaf surface waxes and cutin monomers, respectively (Liu et al. 2011). As in *Saccharomyces cerevisiae*, these surface receptors act upstream of the conserved Pmk1 MAP kinase signalling cascade. Interestingly, a synthetic peptide PAF104, a promising lead compound for fungicide development, inhibits appressorium formation by blocking the Pmk1 signalling pathway via the inhibition of expression of *MoMSB2* and *MoMST11*, a mitogen-activated protein/extracellular signal-regulated kinase kinase kinase (Rebollar and Lopez-Garcia 2013). For general information on fungal G proteins, G-protein-coupled receptors and MAP kinase pathways, see also chapters 'Fungal MAP-Kinase-Mediated Regulatory Pathways' and 'Heterotrimeric G Proteins' in this volume.

Similar surface recognition sensors exist in other ascomycetes such as the wheat pathogen *Fusarium graminearum* suggesting that surface recognition is not a process unique to *M. oryzae*. A direct homologue of MgSho1, FgSho1 was shown to directly interact with the FgSte50-Ste11-Ste7 MAP kinase module to regulate pathogenicity (Gu et al. 2014). Whilst this chapter is focused on the ascomycete plant pathogens, it is worth noting that in the basidiomycete smut fungus *Ustilago maydis*, the orthologues of MoSho1 and MoMsb2 have been identified (Lanver et al. 2010) showing that these surface recognition sensors are also found outside the ascomycetes.

The mechanisms of surface recognition by other plant pathogenic ascomycetes are not as well understood because of the difficulty of making targeted mutants in many fungi. For instance, for the obligate biotrophic Erysiphales, the creation of mutants is very difficult, but it is clear that a plant-derived signal is perceived by conidia when they land on an appropriate plant surface. Within minutes of contact with an appropriate surface, an extracellular proteinaceous matrix is released from the conidium, and the conidium is stimulated to uptake low molecular weight anionic compounds (Nielsen et al. 2000). Similarly, the asexual conidia of *Leptosphaeria maculans*, a hemibiotroph, do not germinate on the surface of the non-host crucifer *Arabidopsis thaliana* unless supplemented with a carbon source and are arrested at or before entry into stomatal openings, thus implying some sort of host recognition by the *L. maculans* conidia (Elliott et al. 2008).

One of the first chemical signals a fungus perceives on the host surface is cutin. In addition to being a signal for the fungus, it is also important for some phytopathogenic fungi to be able to attach to and then penetrate this waxy substance. The genomes of plant pathogenic fungi contain multiple copies of cutinases, which are serine esterases that catalyse the hydrolysis of cutin and are proposed to have roles in assisting with the enzymatic degradation of the cuticle (Skamnioti et al. 2008). Interestingly, the genome of *M. oryzae* contains 17 cutin genes in total, but the transcription of only 14 of them has been documented. Four are considered to be constitutively expressed, one is expressed in the first 5 h of germination, whilst six are expressed at the penetration stage, and three are expressed after 48 h in-planta growth. One of those expressed at the penetration stage, CUT2 was shown to be required for surface recognition and penetration peg formation (Skamnioti and Gurr 2007). Similarly, transcripts for a cutinase SsCutA were detected in *S. sclerotiorum* after 1 h growth on leaf tissue (Bashi et al. 2012), and transcripts for two cutinases were detected in *B. graminis* after 6 and 12 h on barley (Hacquard et al. 2013), thus confirming a role for cutinases in early infection.

B. Adhesion

In addition to the recognition of an appropriate surface, the conidium or spore must adhere to the plant surface prior to germination and penetration. This adhesion counteracts the negative environmental forces such as wind, abrasion and rain splash that might dislodge the fungal spore before entry into the plant. It is important especially for fungi that use a direct penetration method to enter plant cells, as the mechanical forces generated could cause the fungal spore to dislodge from the plant surface before successful penetration. The mechanisms by which different fungal spores adhere to the plant surface are varied and include innate surface components such as hydrophobins, as well as contact activated compounds such as esterases, but all require some aspect of hydration (Brand and Gow 2012). The spores of *M. oryzae* initially adhere via a droplet of sticky mucilage produced at the end of the spore tip after hydration (Howard and Valent 1996). Similarly, initial adhesion of spores of *Parastagonospora nodorum*, the necrotrophic pathogen of wheat that causes net blotch, is a rapid but passive process mediated via conidial surface carbohydrates (Zelinger et al. 2006). Subsequent adhesion of *P. nodorum* spores involves active production of extracellular matrix (ECM) that is stimulated by interaction with its wheat host more than with the non-host barley or glass surfaces which suggests that there is surface recognition before penetration (Newey et al. 2007). The conidia of *M. oryzae* also produce ECM composed of fibrous and amorphous material especially under the appressorium (Inoue et al. 2007). The exact composition of the ECM is unknown but is thought to be a matrix of glycoproteins with mannose sugars because of experiments testing the efficaciousness of different enzymes to detach freshly adhered spores. Enzymes such as β-mannosidase, collagenase N-2, collagenase S-1 and gelatinase B were effective in removing germlings from the surface after 1 h of adhesion to glass slides, but they did not inhibit subsequent appressorium formation. However, metalloproteases such as crude collagenase, collagenase S-1 and gelatinase B were effective at removing germinated spores from wheat leaves and from preventing infection and therefore are of interest as agents for infection control (Inoue et al. 2011).

Hydrophobins are small hydrophobic and cysteine-rich proteins secreted by filamentous fungi that have a role in spore adhesion and aerial hyphae production in some fungi. Hydrophobins are secreted as monomers that self-assemble at hydrophobic/hydrophilic interfaces to make a monolayer that coats fungal surfaces, breaking surface tension and aiding in adhesion (Kershaw and Talbot 1998). They can be further characterised into classes I and II based on the distances between the conserved cysteine residues and the strength of the monolayers they produce (Wessels 1994). The Mpg1p protein of *M. oryzae* is a class I hydrophobin required for appressorium formation and lesion formation during infection of rice and hence was reported to be required for surface sensing in *M. oryzae* (Beckerman and Ebbole 1996; Talbot et al. 1996). Similarly, a class I hydrophobin Mhp1p was also shown to be required for fungal development and plant infection (Kim et al. 2005). In total, four hydrophobins have been identified in *M. oryzae* through a global transcriptional analysis of infection structures (Soanes et al. 2012). This study confirmed the upregulation of *MPG1* during appressorium development but did not detect the expression of the other three hydrophobins, suggesting they may have roles other than pathogenicity in *M. oryzae*. Interestingly, the function of hydrophobins has also been studied in other fungi such as *B. cinerea*, a broad host range necrotroph, and *Cladosporium fulvum*, a hemibiotrophic pathogen of tomato. Six hydrophobins were identified in *C. fulvum* with different temporal and physical localisation, and although HCf-6 was shown to have a role in adhesion to glass slides, none of the HCf genes were shown to have a role in adhesion to the plant (Lacroix et al. 2008). Similarly, the three hydrophobins characterised from *B. cinerea* have no role in conferring surface hydrophobicity to conidia or aerial hyphae and no role in virulence, but they have recently been shown to play an important role in apothecia structure (Mosbach et al. 2011; Terhem and van Kan 2014).

Some fungal components are only activated after contact with the plant surface. For instance, the spores of *Blumeria graminis* f. sp. *hordei* and *P. nodorum*, both pathogens of wheat, release a secreted lipase (Lip1) upon contact with the host surface (Feng, 2009, 2011). These lipase homologues have broad affinity for surface waxes, and the application of recombinant Lip1 to the leaf surface releases alkane and primary fatty alcohols and significantly reduces the surface hydrophobicity of wheat leaves, thus aiding spores in adhesion. Intriguingly, long-chain alkanes are potent inducers of appressorium formation in *B. graminis* suggesting the action of BgLip1 may be to release a signal for successful adhesion to allow further fungal differentiation for successful infection (Feng 2009).

C. Germination

Spore germination requires a switch from a presumed quiescent state to active growth and necessitates a change in metabolism and gene expression that must rely on the utilisation of stored nutrients to synthesise energy required for the reorganisation of the actin cytoskeleton and cell wall components. Through the study of diverse types of plant pathogenic ascomycetes, using a variety of techniques, it is apparent that there are general trends that can be seen in the changes in transcription during germination. Transcriptional studies have concentrated on the expression of fungal genes post-penetration but use the pregermination and or pre-penetration stage as a reference (Both et al. 2005a, b; Oh et al. 2008; Mathioni et al. 2011; Soanes et al. 2012; Franck et al. 2013). Studies that have concentrated on documenting the transcriptome and proteome of the very early events in germination have been conducted in vitro and will be discussed first (Seong et al. 2008; Noir et al. 2009; Leroch et al. 2013).

Seong et al. characterised changes in transcription in a microarray experiment during early germination of conidia of *F. graminearum* and found a surprising number of genes expressed in what were presumed to be quiescent spores (Seong et al. 2008). In fact, the number of genes expressed at the zero hour time point was more than at 2 and 8 h after germination and very similar to that of hyphae. Although 76 % of the genes expressed at 0 h were placed into the unclassified protein class at the time, there were genes of several physiologically relevant classes. For instance, genes required for peroxisome biogenesis and transport as well as for the beta-oxidation of fatty acids and the glyoxylate cycle were significantly expressed in fresh spores, which supports the idea that spores utilise stored fatty acids for energy for germination. Also several genes for gluconeogenesis were identified, suggesting that acetyl CoA produced from the glyoxylate cycle could be used to synthesise glucose. Similarly, in the first proteome of ungerminated conidia of *B. graminis*, 75 % of the identified proteins were classed as metabolic proteins including those for lipid catabolism, suggesting spores were primed with proteins in the event of encountering a suitable host (Noir et al. 2009).

The first hours of in vitro germination of *B. cinerea* and *F. graminearum* conidia correspond morphologically to spore swelling and are witness to a massive induction of gene expression (Seong et al. 2008; Leroch et al. 2013). Genes involved in transcription, metabolism, cell cycle and protein synthesis dominate this early stage of activation. The following stage when the germ tube emerges brings about an increase in transcription of genes relating to cell cycle, DNA processing, cell cycle control as well as cell-type differentiation including genes for cell polarity and filament formation. Unique to the *B. cinerea* study, 20.7 % of *B. cinerea* genes induced after 1 h germination were predicted to encode secreted proteins (Leroch et al. 2013).

Whole-scale changes in transcription upon germination have also been documented in experiments conducted on infected plant tissue. Using a cDNA microarray approach to study pre- and post-penetration stages of the obligate biotroph *B. graminis*, Both et al. (2005b) documented the abundant expression of genes for lipid catabolism in spores, which subsequently decreased later in development. Similarly, a microarray analysis comparing gene expression in ungerminated *M. oryzae* spores to ones that had germinated showed that 21 % of the 10,176 genes represented in the array had a significant change in transcription (Oh et al. 2008). The upregulated transcripts were grouped according to gene ontology, and this revealed a high representation of carbohydrate, amino acid and protein metabolism but a downregulation of transcription factors. In a recent study of the *M. oryzae* proteome using nano-liquid chromatography/mass spectrometry, a massive decrease in mitochondrial protein abundance was observed after germination, suggesting that the increased mitochondrial proteins in early germinating spores could be used for rapid energy production (Franck et al. 2013). Additionally, the downregulation of transporters supported the idea that germinating spores rely on stored nutrients for germination.

Germ tube initiation and elongation requires the establishment of a centre of polarity in the spore as well as the mobilisation of cell wall synthesis machinery to that site (Harris and Momany 2004). Cell wall biosynthesis is also covered in chapters 'The Cell Wall Polysaccharides of *Aspergillus fumigatus*' and 'Chitin Synthesis and Fungal Cell Morphogenesis' in this volume. In *S. cerevisiae*, a unicellular model ascomycete, the activation of Cdc42p (cell division cycle 42), a Rho family GTP-binding protein (GTPase), controls cell polarity by conveying positional information through a signalling cascade to the morphogenesis machinery and is essential for cell cycle progression and division (Johnson 1999; Sudbery 2008). For a unicellular organism, cell division and cell cycle progression is a fundamental requirement for viability; however, this is not always the case in the filamentous ascomycetes. For instance, although Veneault-Fourrey et al. showed that cell cycle progression is essential for the germination of *M. oryzae* (Veneault-Fourrey et al. 2006a), this is not the case in other fungi such as *Colletotrichum* sp. In *C. gloeosporioides*, mitosis lags behind germination and appressorial initiation and can occur without nuclear replication (Nesher et al. 2008). Similarly, in the obligate biotroph *B. graminis*, germ tube elongation and swelling does not require the completion of mitosis (Hansjakob et al. 2012).

The signalling cascade governing cell division in the filamentous ascomycetes also differs from that of the yeasts due to the partly overlapping roles of the small GTPases Rho and Rac (Harris 2011). For instance, the knockout *cdc42* of *C. trifolii* proved to be lethal, and by making mutants containing constitutively active and dominant negative versions of *Ccdc42*, Chen et al. (2006) showed that *cdc42* was a positive regulator of germination but a negative regulator of appressorial development (Chen et al. 2006). In contrast, the deletion of the *M. oryzae Cdc42* homologue was not lethal, but the mutant showed abnormal growth and conidiation and was less pathogenic than wild type, suggesting it is not the only component controlling polar growth (Zheng 2009). Recently, the *M. oryzae cdc42* homologue was shown to be required for septin ring assembly at the later penetration peg stage (Dagdas et al. 2012). Differing roles of *B. cinerea* RAC proteins have been reported in promoting cell cycle, as the deletion of one had the same phenotype as the constitutively active version of the other. The deletion of *BcCdc42* resulted in defects in germination and hyphal branching with a twofold increase in the nuclei/cytoplasm ratio (Kokkelink et al. 2011). Similarly, a mutant with constitutively active BcRAC had disrupted polar growth possibly due to mislocation of actin and hyper accumulation of nuclei, thus confirming the *B. cinerea* small GTPase BcRAC was required for cell cycle progression (Dub et al. 2013).

Tetraspanins, small transmembrane-spanning proteins found in animals and fungi, have been implicated in the initiation of polar growth in filamentous fungi (Veneault-Fourrey et al. 2006b). There are three families of fungal tetraspanins, but only two of them are found in ascomycetes. Pls1 was the first tetraspanin identified in *M. oryzae*, and a deletion mutant was unable to make normal infection pegs on rice or barley (Clergeot et al. 2001). A second tetraspanin (*tsp3*) and a tetraspanin-like gene (*tpl1*) more recently identified are expressed in perithecia and mycelia of *M. oryzae*, and mutant strains in those genes are still able to infect barley but make fewer lesions on rice, suggesting they have a different role to *pls1* (Lambou et al. 2008).

III. Entry into the Plant

The next challenge for the fungal germling is entry into the plant tissue. As depicted in Fig. 5.2C, there are different strategies that fungi use to enter plant tissue, and each has its own advantages and pitfalls. The first goal for any strategy is to avoid detection by the plant's basal defences. Microbes shed **MAMPs** or **PAMPs** (**microorganism-** or **pathogen-associated molecular patterns**), for example, fragments of chitin, and the plant detects these via pattern recognition receptors (PRRs) resulting in **PAMP-triggered immunity** (**PTI**) (Jones and Dangl 2006). To subvert this response, the fungus produces effectors. These are often unique small secreted proteins whose targets can be intercellular or intracellular depending on the infection strategy of the fungus [for a review see Giraldo and Valent (2013)]. Some effectors are even expressed before penetration suggesting that not all effectors are made in reaction to PAMP-triggered immunity (Kleemann et al. 2012). For the fungus to be able to fully colonise the plant tissue, the plant must not have receptors or receptor-like proteins that recognise the effectors produced by the fungus, or the fungus must not encode an effector that the plant could detect. The recognition of effectors results in different reactions in the plant depending on the entry strategy used by the fungus. The interaction of intercellular effectors with their receptors is rapid (within hours) and often results in localised cell death (**hypersensitive response** or **HR**) or **effector-triggered immunity** (**ETI**) and death of the fungus. The interaction of apoplastic effectors with their receptors is more gradual, does not necessarily result in cell death and generally limits but does not kill the invading fungus (Stotz et al. 2014). Each of these entry strategies has associated with it massive transcriptional changes that have been recently revealed through large-scale RNA sequencing technologies, and these will be discussed in the following sections.

A. Direct Entry

1. Infection Structure-Initiated Entry

In order for plant pathogenic fungi to force entry into plant tissue, they must modify hyphal tips to create a focal point for close contact with the plant. This might be a simple swelling of the end of the hyphal tip or an elaborate specialised infection structure (Mendgen et al. 1996). For instance, the appressorium made by *M. oryzae* is very elaborate, and its structure and biogenesis has been the subject of intense study. However, not all appressoria are alike. Appressoria made by diverse fungi such as the obligate biotroph *B. graminis*, the hemibiotrophic *Colletotrichum* spp. and the necrotrophs *B. cinerea* and *S. sclerotiorum* amongst others are less elaborate. Compound appressoria or infection cushions are differentiated in response to hard surfaces at the tips of hyphae by fungi such as *S. sclerotiorum* and *B. cinerea* (Backhouse and Willetts 1987; Huang et al. 2008). Infection cushions are formed at the hyphae apex by a concentration of multiple subapical branches in parallel. Other less differentiated structures include hyphopodia which are swellings at the end of hyphae mainly found in mycorrhizal fungi but also identified in *M. oryzae* during root infection as well as occasionally in the initial penetration of wheat with the necrotroph *P. nodorum* (Solomon et al. 2006; Tucker et al. 2010; Bucher et al. 2014). The purpose of these structures is to form an intimate association with the host surface and to gain entry by direct penetration.

a) Developing Appressoria in *Magnaporthe oryzae*

As discussed in Sect. II.A, the ability to form infection structures in vitro and the ability to make random and targeted mutants in *M. oryzae* enabled researchers to identify the genes required for appressorium formation, and many of these have been discussed in this earlier section as the genes are also often required for processes prior to appressorium formation. Appressoria have been carefully studied microscopically and biochemically especially in *M. oryzae* and are found to be the source of enormous turgor pressure. A single *M. oryzae* appressorium is capable of generating over 80 bars of pressure (Howard et al. 1991) that enables the penetration of the plant cuticle by the penetration peg. Chumley and Valent (1990) discovered that mutants unable to produce melanin were unable to infect unwounded rice leaves and that the addition of scytalone, a precursor of melanin not made by the mutant, was able to rescue the non-infective phenotype. A layer of melanin surrounding the appressorium (apart from at the appressorial pore) acts as a barrier to molecules larger than water resulting in the production of this enormous pressure (Howard and Valent 1996). In addition, the production of reactive oxygen species (ROS) was shown to be crucial for appressorium formation in *M. oryzae*. Mutants in either superoxide-generating NADPH oxidase Nox1 or Nox2 do not make functional appressoria and are unable to infect even wounded plants (Egan et al. 2007). More recently characterisation of the fungal transcriptome during the formation of appressoria in vitro has revealed global transcriptional changes at the appressorial stage of infection (Oh et al. 2008; Soanes et al. 2008). Soanes et al. used next-generation sequencing techniques to detect 96 % of the predicted genes in the data set, exceeding the coverage of the previous microarray study by Oh et al. (2008). Consistent observations made are that the beta-oxidation pathway enzymes, melanin biosynthesis and the glyoxylate shunt are upregulated and pivotal for appressorium formation. In addition, through the analysis in parallel of a Pmk1 MAP kinase mutant which does not make appressoria, this MAP kinase pathway was shown to be crucial for the expression of melanin biosynthesis pathway genes, for the multifunctional beta-oxidation enzyme MFP1 and for carnitine acetyl transferases which transport acetyl CoA (Soanes et al. 2012).

b) Gene and Protein Expression in Mature Appressoria

i. Magnaporthe oryzae

Once an appressorium or appressorium-like structure has been elaborated, the fungus is in intimate contact with the plant surface and is preparing for entry. Ideally, one would seek to use infected plant tissue to study the genes expressed in mature appressoria, but the amount of fungal biomass compared to plant biomass at this stage is very small. Therefore, these transcriptional studies of *M. oryzae* at the pre-penetration mature appressorial stage have been conducted on in vitro differentiated appressoria grown on coverslips in the presence of a cutin monomer (Soanes et al. 2012). The types of genes that are upregulated in the mature appressorium reflect the preparation for a hemibiotrophic lifestyle. For *M. oryzae*, which begins its association in planta as a biotroph, evidence of this preparation is in the upregulation of sugar transporters and carbohydrate-degrading enzymes in appressoria as well as expression of genes for quinate transport and utilisation (Soanes et al. 2012). Quinate is not normally present in uninfected rice leaves, but a recent study showed that *M. oryzae*-infected tissue had increase amounts of quinate suggesting that the fungus may divert the plant metabolism for its own benefit (Parker et al. 2009). Quinate can be generated through the diversion of the shikimate pathway of *M. oryzae* possibly diverting intermediates from the phenylpropanoid defence pathway (Soanes et al. 2012). A proteomic study of *M. oryzae* appressoria also identified carbohydrate-modifying proteins involved in remodelling the fungal cell wall or degrading plant cell wall material, such as the cutinase cut2 which was previously shown to be required for host penetration (Skamnioti and Gurr 2007), as well as laccases and peroxidases which may be involved in lignin degradation (Franck et al. 2013).

ii. Colletotrichum *Species*

Four *Colletotrichum* species with different host plants have recently been the focus of detailed genomic and transcriptional analyses—*C. gloeosporioides*, *orbiculare*, *higginsianum* and *graminicola* (Kleemann et al. 2012; O'Connell et al. 2012; Gan et al. 2013). Although all are hemibiotrophs that elaborate appressoria, there are subtle differences to their infection strategy, and hence a comparison of their genomes and transcriptomes aimed to reveal clues to the difference in their biotrophic/necrotrophic switches. An infection time course of *C. higginsianum* after 22 hpi in vitro; 22, 40 and 60 hpi on *Arabidopsis* using RNAseq; and one of *C. orbiculare* on *Nicotiana benthamiana* at 24, 72 and 168 hpi using a custom microarray identified several categories of genes that were upregulated during infection, including transporters, secondary metabolites, candidate secreted effector proteins (CSEPs), CAZymes and transcription factors (O'Connell et al. 2012; Gan et al. 2013). A striking observation made by O'Connell et al. (2012) is that although *C. higginsianum* appressoria made in vitro at 22 hpi were morphologically indistinguishable from those made on *Arabidopsis* leaves at 22 hpi, there was a massive difference in the transcriptome with 1515 genes expressed in planta that were not in vitro. Many of these genes were CSEPs, one of which *ChEC6* was 50,000 times more expressed in appressoria that were differentiated in planta as opposed to the in vitro differentiated ones. This result stresses the importance of assessing transcriptome content on real plant tissue to truly discover the roles of appressoria of different fungi.

One unexpected result for *Colletotrichum* spp. was that secondary metabolite enzymes were most highly expressed during the early infection stages in contrast to the necrotrophic stage (O'Connell et al. 2012; Gan et al. 2013). This suggests that these metabolites might be acting more as effectors secreted to manipulate the host metabolism as opposed to acting as phytotoxins (O'Connell et al. 2012). Also of note, in early infection stages, *C. orbiculare* had notable expression of quinate transporters as was previously described in the *M. oryzae*/ rice interaction, but their expression peaked in necrotrophic stages suggesting that quinate might be also an important carbon source in dead tissue (Gan et al. 2013).

Kleemann et al. (2012) went further to characterise the role of the *C. higginsianum* appressorium in effector delivery by tagging effector proteins with fluorescent tags and including their native promoters. Using this approach, they were able to confirm that the CSEPs were secreted in waves associated with specific developmental stages such as before penetration, during penetration, during biotrophy and during necrotrophy. This coupled with the confirmation of activation of plant defence pathways confirms that the appressorium is more than a physical tool used by the fungus to gain entry to the plant but also a communication conduit between plant and fungus.

iii. Blumeria graminis

Powdery mildew fungi are obligate biotrophs that also use appressoria to enter into their hosts, although few studies exist documenting the development of the appressoria in isolation because they cannot be fully differentiated in vitro (Hueckelhoven and Panstruga 2011). Although powdery mildew appressoria are not melanised, they can produce substantial turgor pressure (Pryce-Jones et al. 1999). Recently, an improved method for studying the germination of *B. graminis* f. sp. *hordei* in culture was developed, and using a combination of DAPI-staining and cell cycle inhibitors, it was shown that the conidium undergoes one round of mitosis in the first 5–6 h and that the application of benomyl prevents the germinating conidium from exiting mitosis and making a fully developed haustorium (Hansjakob et al. 2012).

Several transcriptomic and proteomic characterisations of *B. graminis* have been published as well as a deep RNA sequencing analysis of *G. orontii*, a distantly related powdery mildew that infects *Arabidopsis* (Godfrey et al. 2010; Pedersen et al. 2012; Wessling et al. 2012; Hacquard et al. 2013; Kusch et al. 2014). However, only Hacquard et al. (2013) mention fungal gene expression in the pre-haustorial stage. In one aspect of this elaborate paper, two isolates of *B. graminis* f. sp. *hordei* differing in their recognition of the MLA1 gene of barley (A6 vs K1) were inoculated onto specially adapted immunocompromised *Arabidopsis* plants that expressed the MLA1 gene of barley and had previously been shown to allow full differentiation of powdery mildew in the compatible interaction with A6 or the incompatible one with K1 (Maekawa et al. 2012). By using this model host with a relatively small and well-characterised genome, they were able to map 90 % of the reads to the *Arabidopsis* genome and only 1.79 % to the fungal reference genome and make some observations about gene expression at 6 h postinoculation (prepenetration), 12 hpi (during penetration of host cell wall), 18 hpi (appressorial initials stage) and 24 hpi (mature haustorial stage). At 6 hpi, gene ontology groups for DNA packaging, nucleosome and chromatin structure were highly represented, but this switched at 12 hpi to reflect host entry and pathogenesis. Transcripts for CSEPs were not expressed at 6 hpi but began to accumulate between 12 and 24 hpi, thus supporting a role for these secreted proteins in early pathogenesis before full appressorium development has occurred. Transcripts were grouped according to similarity and were found to be expressed in waves, with members of groups 1 and 3 corresponding to expression at 12 hpi. Interestingly, members of the same structural group differed in their expression profile suggesting sequential delivery of the effectors. This suggests the secretion of these effectors during penetration may aid in the suppression of recognition by the plant. Also notable for their expression at 6 and 12 hpi were two cutinase genes, presumably involved in the penetration process, as well as five genes encoding sugar and amino acid transporters. This was unexpected and suggests that the fungus is already acquiring nutrients from the plant before the formation of the haustorium, which has always been presumed to be the primary nutrient acquisition organ (Bushnell et al. 1967).

iv. Botrytis cinerea

B. cinerea conidia germinate on the host surface and can penetrate the surface directly or enter via a wound (Williamson et al. 2007). In comparison to *M. oryzae*, conidia germinate rapidly forming appressoria after only 4 h and saprotrophic hyphae after 15 h growth on apple wax-coated polystyrene supplemented with fructose (Leroch et al. 2013). In addition, the appressoria produced by *B. cinerea* are much less developed, consisting of a distinct terminal swelling at the end of the germ tube. *B. cinerea* also produces compound appressoria or infection cushions in the presence of a rich carbon source or when growing from a mature mycelium (Backhouse and Willetts 1987). A comparison of genes upregulated in the appressoria of *B. cinerea* to those upregulated in *M. oryzae* showed little overlap which could reflect a difference in lifestyle and appressorial structure. Only one orthologue for acetyl CoA utilisation was upregulated in *B. cinerea* in contrast to all 31 in *M. oryzae*, and none of the genes encoding beta-oxidation pathway enzymes were expressed suggesting that utilisation of fatty acid stores and the glyoxylate cycle do not play an important role in appressorium formation in *B. cinerea*. However, in both fungi, an important role for carbohydrate active enzymes is indicated by the high frequency of genes encoding CAZymes expressed in both fungi.

Interestingly there are no transcriptional studies dedicated to gene expression in the infection cushion in *B. cinerea*. However, recently a homeobox gene *BcHOX8* was identified as highly expressed in infection cushions, and targeted deletion of the gene resulted in multiple defects including malformed infection cushions and reduced pathogenicity on some hosts (Antal et al. 2012). Homeobox genes are transcription factors that normally turn on a cascade of gene expression relating to morphogenesis and were first described in *Drosophila* (Gehring 1993). *BcHOX8* is member of a family of nine putative homeobox genes, two of which might be specific to *B. cinerea* and *S. sclerotiorum*, which also makes infection cushions.

v. Sclerotinia sclerotiorum

S. sclerotiorum is a broad host range necrotroph that has been described by Purdy (1979) as 'among the most nonspecific, omnivorous and successful of plant pathogens' and has been the subject of countless scientific studies. The infection process has been studied microscopically on many different susceptible hosts (Jones 1976; Jamaux et al. 1995). A more recent study examined the development of simple and complex appressoria as they grew from ascospores (Garg et al. 2010). The first signs of germination were visible after only 2 h and germ tube emergence and elongation occurred after only 12 h. By 24 h, the ends of hyphae had already begun to swell, and by 48 h simple appressoria were found. Complex appressoria began to develop by 72 h. By 96 h hyphae had entered the mesophyll layer either via appressorium penetration or through stomata, and the first evidence of necrosis of the plant tissue became visible. Using an alternative inoculation technique, a square of agar containing *S. sclerotiorum* mycelia was placed onto the leaf tissue, and the development of necrosis was much more rapid. After 24 h, a dense mat of mycelia had grown on the host and infection cushions of different sizes developed (Huang et al. 2008). The removal of these infection cushions revealed holes punched through the cuticle by penetration pegs showing that penetration could occur within 24 h.

Despite the detailed knowledge about the infection structure morphology and development, far less is known about the genetic control of the morphogenesis programme in *S. sclerotiorum* because only a handful of researchers around the world have been able to do what is comparatively easy in fungi such as *M. oryzae*, that is, create genetic mutants. A few examples of targeted gene deletion or silencing are as follows. Similar to what was found in *M. oryzae* where the *MAC1* adenylate cyclase mutant was unable to penetrate the rice surface due to lack of appressorium formation (Choi and Dean 1997), a knockout of the *S. sclerotiorum* adenylate cyclase gene *sac1* resulted in an isolate unable to infect unwounded tissue, thus showing that cAMP is required for infection cushion formation (Jurick and Rollins 2007). Similarly, a knockout of gamma-glutamyl transpeptidase, an enzyme that is important for maintaining the balance of cellular levels of glutathione, resulted in an isolate that was unable to make appressoria and other multicellular structures such as sclerotia and apothecia (Li et al. 2012b).

Genome and transcriptome sequencing has revealed much about the genes that are likely to be important for getting into and manipulating the plant and has given some clues as to the basis of the large host range of *S. sclerotiorum*. The genome sequence of *S. sclerotiorum* was published alongside the genome of *B. cinerea* to compare and contrast the genes required for a necrotrophic lifestyle (Amselem et al. 2011). The two genomes were of similar size, and on average the predicted proteins share 83 % amino acid identity between the two species. The complement of CAZymes between the two species is also similar, but the secondary metabolite arsenal of *B. cinerea* was expanded relative to *S. sclerotiorum*. Some analysis of gene expression in planta was conducted for the genome publication; however, the infected tissue was collected after 48 h inoculation on sunflower petals when the infection would have been very advanced and necrotic and therefore sheds little light on which genes are expressed at the pre-penetration stage. However, the genome sequence contained orthologues of genes required for appressoria formation in other fungi such as genes in the MAP kinase and cAMP pathway, the tetraspanin pls1 and members of the *Mas1*-related gene family. This is supported by earlier work by Sexton et al. who characterised genes upregulated in infection cushion and found a high representation of *MAS3* (*Magnaporthe* appressoria specific 3) transcripts (Sexton et al. 2006).

Amselem et al. (2011) identified 603 non-CAZyme non-peptidase transcripts in the genome of *S. sclerotiorum* but observed that more of these were downregulated than upregulated in planta. Therefore, it is surprising that the transcript for a 302 amino acid integrin-like protein, SSITL, was observed after only 1.5–3 h postinoculation. Targeted silencing of this gene resulted in premature activation of the defence genes *PDF1.2* and *PR-1* as well as reduced virulence (Zhu et al. 2013), suggesting that *S. sclerotiorum* is suppressing plant defence at a very early time point using a similar strategy to other fungi discussed in this chapter. Subsequently, the secretome of *S. sclerotiorum* was analysed and predicted to contain 735 secreted proteins with 486 having experimental evidence for expression in planta and 78 considered as effector candidates (Guyon et al. 2014). The expression of a small subset of these effector candidates was tested via quantitative RT-PCR on infected tissue after 6 h, 12 h and 24 h on four different hosts, and they all showed different expression profiles supporting a sequential delivery of effector proteins. This ability to sequentially deliver a variety of effectors to different hosts may explain some of the broad host range of *S. sclerotiorum*.

vi. Parastagonospora nodorum

Much is known about genes required for the penetration of wheat by the necrotrophic pathogen *P. nodorum* due to its amenability to genetic manipulation, and a historical perspective of the advances in this field can be found in Oliver et al. (2012). As mentioned in Sect. II.B, *P. nodorum* adheres to the plant surface using a sticky mucilage and secretes a lipase (Lip1) with affinity for surface waxes to aid in adhesion. A detailed study of the cytology of a susceptible interaction was documented with an isolate transformed with the green fluorescent protein (Solomon et al. 2006). Germination was evident after only 3 h on the plant surface, and penetration attempts began after 8–12 h either through stomata or anticlinal cell walls. Hyphopodia were observed only when penetration was attempted at periclinal cell walls. Once inside the leaf, hyphae proliferated for 4 or 5 days before massive necrosis was observed.

There is a wealth of genetic and genomic resources available to the *P. nodorum* research community. Its genome was the first from a Dothideomycete to be sequenced (Hane et al. 2007), and subsequently more isolates have been sequenced in an effort to identify strain-specific effectors (Syme et al. 2013). This has revealed that each isolate contains a raft of strain-specific genes with no obvious homology to any known genes, and these are likely effector candidates. RNA sequencing analysis of infected wheat leaves at 3, 5, 7 and 10 days postinoculation revealed that the early time points of infection were particularly enriched for differentially expressed genes and a disproportionate number of predicted secreted proteins (Ipcho et al. 2012). Earlier time points before the onset of necrosis were not sampled due to low fungal biomass, and therefore no detailed information about genes required for penetration can be gleaned from this study. However, the authors note that the necrotrophic effectors *SnTox3* and *SnToxA* were upregulated at 3 dpi.

The interaction between *P. nodorum* and wheat has been described as an 'inverse gene for gene' interaction as the fungus produces **necrotrophic effectors**, formerly known as **host-specific** or **host-selective toxins** that correspond to susceptibility genes in wheat conferring susceptibility to the fungus (Oliver et al. 2012). SnTox1, the first identified necrotrophic effector was found by searching the genome for small secreted effector-like proteins and then expressing the candidate genes in yeast to make proteins that were then infiltrated into wheat leaves with the corresponding *Snn1* susceptibility locus (Liu et al. 2012). *SnTox1* is expressed as early as 3 h postinoculation and expression peaks after 3 days. Strains deleted for *SnTox1* were unable to penetrate *Snn1* wheat leaves and continued to grow epiphytically on the wheat leaf surface, thus proving that this necrotrophic effector is essential for penetration. Similarly, SnTox3 is a unique secreted protein that interacts with *Snn3* in wheat to cause infection. SnTox3 expression peaks early in infection coinciding with the appearance of necrosis, and loss of SnTox3 expression results in loss of pathogenicity (Liu et al. 2009).

B. Entry Through Stomata or Wounds

For many fungi, the most efficient method to avoid detection by the plant is to enter with stealth through natural openings such as stomatal apertures or wounds in the epidermis and then proliferate in the apoplast (Fig. 5.2C, D). But how do fungal germlings find the openings into which they enter and are there genes controlling this process?

Fungi such as *Z. tritici* and *F. graminearum*, which infect monocots, and *C. fulvum* and *L. maculans*, which infect dicots, are examples of fungi that can enter their host through stomatal apertures. After germination, hyphae ramify and colonise the host surface until an opening is found, but little is known about the molecular mechanisms controlling the direction of growth and whether there are topographical landmarks recognised by these fungi that aid in finding openings on the leaf surface. Brand and Gow (2012) note that, in general, pathogens of dicots tend to follow the periphery of epidermal cells, whilst pathogens of monocots tend to cross cells at right angles. This makes sense in relation to the probability of the location of stomatal openings as in monocots they are found in staggered rows, whilst in dicots they occur in a mosaic pattern. A histological examination of the hyphae of *L. maculans* growing on the surface of a canola leaf found that the hyphae appeared to ramify and grow across the leaf in a random fashion, seemingly entering stomata at random (Chen and Howlett 1996). Similarly a detailed histological study of *Z. tritici* growing on wheat showed that infection was strictly stomatal, and germinated hyphae entered at random often crossing over stomatal pores (Kema et al. 1996). For *C. fulvum*, spores germinated on tomato leaves forming runner hyphae which did not show any directional growth towards stomata and entered the stoma after 3–4-days growth on the surface (De Wit 1977). When *F. graminearum* was inoculated onto the shoot of wheat seedlings, after 2 days spores had germinated and hyphae had proliferated superficially over the leaf sheaf surface showing no directional growth (Stephens et al. 2008). However, in the case of *F. graminearum*, entry into the plant is complicated as the same fungus causes different diseases, such as *Fusarium* head blight (FHB) and *Fusarium* crown rot (FCR), and entry into the plant can be via stomata, via wounds or directly into the surface of the floret (Zhang et al. 2012). In an elegant study using a green fluorescent protein-tagged isolate, *F. graminearum* was visualised forming foot structures and compound appressorium originating from the epiphytic runner hyphae produced on wheat florets (Boenisch and Schaefer 2011). Interestingly, the mycotoxin trichothecene was also shown to be produced in these compound appressoria. However, it is still not clear which method of entry is the most important for natural infection outside the laboratory, and therefore *F. graminearum* will be included in this section.

Although no evidence for directed growth into stomata has been found in these filamentous ascomycetes, a couple of other studies, one on the ascomycete *Cercospora zeae-maydis* and the other on the basidiomycete wheat stripe rust *Puccinia striiformis* f. sp. *tritici* (*Pst*), indicate that there is a mechanism for **stomatal sensing** in some fungi. The causal agent of grey leaf spot of maize, *Ce. zeae-maydis*, forms appressoria and enters leaves through stomata but only in the presence of a 12 h light/dark cycle. Infections conducted in total darkness failed to develop and *Ce. zeae-maydis* failed to form appressoria in the dark. Furthermore, isolates with loss of the *white collar-1* gene, which is a sensor for blue light, grew randomly over stomata and formed appressoria only at 10 % of the rate of the wild type (Kim et al. 2011). Similarly, when *Pst* urediniospores were inoculated onto non-host broad bean leaves, they germinated, and hyphae proliferated as if they were on their wheat host; however, most failed to enter stomata suggesting that appropriate thigmotropic or chemotropic signals were not being emitted from the non-host broad bean (Cheng et al. 2012).

One could imagine such stomatal-seeking fungi to respond chemotactically to volatile compounds released by the plants via their open stomates; however, there is no direct evidence for this to date. Plants emit volatile compounds to communicate with their environment and as a response to biotic or abiotic stresses (Niinemets et al. 2013). Volatile compounds emitted from green leaf tissue are believed to act as deterrents against herbivory and attract carnivorous animals; however, more recently, the tripartite relationship between plant insects and microbes has come into focus (Ponzio et al. 2013; Lazebnik et al. 2014). These compounds are thought to act to inhibit the ingress of fungal invaders or to induce systemic resistance in noninfected parts of the plant; however, this is not always the case. For instance, in a study of the role of terpenes in transgenic oranges, increased production of the terpene D-limonene in orange fruit correlated with decreased expression of defence genes and increased infection by *Penicillium digitatum* (Rodriguez et al. 2014).

1. Gene Expression During Early Infection

Once these fungi have passed through the stomatal aperture, they enter the apoplastic space of the leaf and begin to proliferate without causing any visible symptoms to the plant. Although the fungi are now inside the leaf, they have not physically penetrated any plant cell walls, and this is normally referenced as the early infection stage in the following gene expression studies.

a) *Zymoseptoria tritici*

It only takes 12 h for the mycelia of *Z. tritici* formed after germination to be found below the stomata of wheat leaves, and by 48 h they are already in the vicinity of the nearest mesophyll cells, but even then there is little cytological evidence of symptoms. It is not until 8 days that extensive proliferation of hyphae is evident, and by 10–12 days necrosis of the plant tissue is evident (Kema et al. 1996). Interestingly, the fungus never penetrates the plant tissue but exists for its entire life cycle in the apoplast. The transcriptome and proteome of fungal and host genes in the *Z. tritici* wheat pathosystem have been extensively studied (Keon et al. 2005; Rudd et al. 2010; Yang et al. 2013b, c). The most recent transcriptional analysis of *Z. tritici* on wheat reported that during the biotrophic phase, at 4 dpi, there was a clear suppression of host defences as assessed by the decreased expression of many wheat defence-related genes which may have been mediated via a large number of fungal transcripts encoding small cysteine-rich secreted unknown proteins (Yang et al. 2013b). Transcripts for two previously characterised LysM domain containing proteins were expressed, one of which was previously predicted to be responsible for the suppression of chitin-mediated defence during the asymptomatic phase of infection (Marshall et al. 2011). The expression of nutrient uptake and transport-related transcripts at the early infection stage contrasted with the expression of primary metabolism genes at the necrotrophic stage. The high levels of transcripts encoding ROS-scavenging genes were notable throughout the infection as the presence of hydrogen peroxide is a notable reaction of wheat to the presence of *Z. tritici* throughout the infection cycle (Yang et al. 2013b).

b) *Fusarium graminearum*

As mentioned above, *F. graminearum* is capable of causing FHB and FCR of wheat and barley, and the epidemiology and cytology of FHB have been well characterised (Sutton 1982; Guenther and Trail 2005). More recently the infection process of FCR was documented microscopically concomitantly with a calculation of fungal biomass and an analysis of fungal gene expression at different infection stages using the *Fusarium* Affymetrix Gene chip (Stephens et al. 2008). Interestingly, although the asymptomatic stage of FCR is much longer than that of FHB, the genes expressed at the early infection stages were very similar. Genes involved in stored nutrient mobilisation and acquisition were upregulated which probably reflects the lack of nutrients available on the surface of the plant. Once the hyphae began to enter the tissue, genes encoding detoxification proteins were upregulated. There have been numerous transcriptomic and proteomic studies of *F. graminearum*-infected tissue that vary in their methodology and focus (Yang et al. 2010, 2013a; Lysoe et al. 2011; Zhang et al. 2012, 2013). The rate of development of disease symptoms in different tissues and cell types can be quite variable, and this complicates the interpretation of fungal gene expression studies from infected tissue. The study by Zhang et al. (2012) addressed this issue by infecting coleoptile tissue with a spore solution and used laser microdissection to remove uninfected plant tissue. This synchronisation of infection was confirmed visually using isolates expressing fluorescent proteins. They found that by 16 hpi hyphae that germinated and entered through the wound continued to grow along the cell periphery with little branching, whereas those that entered through stomata were often stalled by 72 hpi. They preferentially isolated these hyphae avoiding the epiphytic hyphae growing on the surface and those fungal structures stalled under the stomata and profiled the transcription of these growing hyphae using the *Fusarium* Affymetrix chip. They found that there was a set of 344 predicted proteins preferentially expressed in planta and that 41 % of those were predicted to be secreted. Interestingly, the expression of these secreted proteins was 30–40 % higher at the early and mid-infection stages than at the late necrotrophic stages. The expression of cell wall-degrading enzymes targeting the primary chains of pectin was highly upregulated at 16 hpi and those targeting the pectin side chains increased later. Cellulose-degrading enzymes were also upregulated at this early stage. Interestingly, although they preferentially removed the epiphytically growing hyphae during samples at 16 hpi, they still found the fungus expressed preferentially genes involved in beta-oxidation of fatty acids during this early infection stage suggesting that it was relying on stored fatty acids during this early invasion period as opposed to nutrients from in-planta growth.

c) *Leptosphaeria maculans*

L. maculans and *L. biglobosa* are both pathogens of canola and the most well-studied members of the *Leptosphaeria* species complex that has been the subject of recent genomic and transcriptomic studies (Grandaubert et al. 2014; Lowe et al. 2014). Both species infect canola using a similar strategy at the start with fungal inoculum spread by rain splash infecting foliar tissue through stomatal openings or wounds and initially colonising leaf tissue in the apoplastic space. Both species can even be found infecting the same plant (West et al. 2002). However, *L. biglobosa*-infected leaf tissues rapidly turn necrotic, and the fungus rarely induces stem canker infection (van de Wouw et al. 2008). Recently the transcriptomes of these two species were compared by RNAseq with the study tissue from the biotrophic phase at 7 dpi and the necrotrophic phase at 14 dpi. Lowe et al. (2014) showed that after 7 days growth in the cotyledonary tissue, *L. biglobosa* was already causing visible necrosis and expressed more cell wall-degrading genes, such as glycosyl hydrolases and pectinases. In contrast, at 7 dpi, *L. maculans*-infected cotyledons showed no visible necrosis and expressed many genes of the carbohydrate-binding module class, particularly CBM50 genes which are predicted to have a role in chitin binding. This would support the idea that at this stage of infection, *L. maculans* is actively avoiding detection by the basal immune system of the plant. Most of the top 20 most highly expressed genes of *L. maculans* at 7 dpi had no PFAM domain hits but did contain three known avirulence genes, *AvrLm1*, *AvrLm4-7* and *AvrLmJ1* (Gout et al. 2006; Parlange et al. 2009; van de Wouw et al. 2014). In a comparison with the larger group of species in the *Leptosphaeria* complex, 30 of the 100 most highly expressed small secreted proteins were unique to *L. maculans* (Grandaubert et al. 2014). This suggests that *L. maculans* uses a unique and varied arsenal to shield itself from detection by its host plants during early infection. The concerted control of the expression of these in-planta-expressed effector proteins has recently been shown to be related to the state of heterochromatin (Soyer et al. 2014). Silencing of either *LmHP-1* or *LmDIM-5* genes involved in heterochromatin formation and maintenance resulted in the unexpected expression of effector genes in culture. In addition, the transfer of a group of effector genes from their AT-rich locus to a GC-rich region also resulted in expression of the effectors in culture. This shows that *L. maculans* uses chromatin-mediated repression to control effector gene expression to allow for rapid adaptation to environmental changes.

d) *Cladosporium fulvum*

During a compatible interaction, hyphae of *C. fulvum* proliferate in the apoplastic space adjacent to the mesophyll cells, but in an incompatible interaction, hyphae are arrested after 1 or 2 days around the stomata and adjacent cells (De Wit 1977). Compatibility in this pathosystem is specified by the interaction between resistance (R) genes in the host tomato and avirulence (Avr) genes in the fungus (Wulff et al. 2009). To date 12 effectors have been cloned that encode an N-terminal signal peptide for secretion into the tomato apoplast, and for 9 of them the corresponding tomato *Cf* gene has been cloned (Mesarich et al. 2014). Only one global transcriptional study has been done on *C. fulvum*-infected tomato tissue using RNAseq, and no microarray studies have been published. The recent RNAseq publication focussed purely on the cloning of a new avirulence gene *Avr5*, which, until now, had proven impossible to identify using the reverse genetics approaches that had been so successful with all the other Avr genes (Mesarich et al. 2014). Here the authors compared the transcriptome of two isolates differing in their avirulence profile towards Cf5 tomato whilst growing on susceptible tomato plants. The transcriptomes of both isolates were examined for candidate effectors that were less than 300 aa in length, had more than four cysteine residues and had a signal peptide for secretion into the apoplast. Of 44 candidates, only two were identified with polymorphisms in the Avr5-lacking isolate, and the functional copy of only one of these was able to restore avirulence to the Avr5-lacking isolate. Interestingly the authors do not report on any further details of the transcriptome. A little more information about the genes expressed during in-planta growth comes from the *C. fulvum* genome sequencing paper (de Wit et al. 2012). The genome was analysed in comparison with the closely related pathogen *Dothistroma septosporum* which is a hemibiotrophic pathogen of pine, and the expression of genes of interest was confirmed by quantitative PCR analysis of infected tomato tissue. Surprisingly, the genome of *C. fulvum* was found to contain an extensive catalogue of carbohydrate-degrading and pectinolytic enzymes, which is surprising given its biotrophic lifestyle. However, closer analysis showed that these genes were either mutated or not expressed in planta. Also the existence and expression of an alpha-tomatinase enzyme were confirmed which may be related to facilitating the growth of *C. fulvum* in tomato.

IV. Conclusions

At present, a considerable amount of attention has been focused on the role of fungal effectors and their recognition by plants, underlying Flor's now classic 'gene-for-gene' model. However, the importance of the molecules required for the initial ability of fungi to recognise and enter the plant tissue is coming into focus. New highly sensitive technologies such as RNA sequencing have allowed researchers to examine gene expression at the earliest stages of infection when fungal biomass is critically low. The findings described above provide a wealth of knowledge about those genes and increasingly about the timing of their regulation. Whilst there are in some cases commonalities, collectively these studies indicate that most fungi have fairly specialised gene expression profiles on their hosts, between hosts, which are difficult to mimic in vitro.

References

Amselem J, Cuomo CA, van Kan JAL, Viaud M, Benito EP, Couloux A, Coutinho PM, de Vries RP, Dyer PS, Fillinger S, Fournier E, Gout L, Hahn M, Kohn LM, Lapalu N, Plummer KM, Pradier J-M, Quevillon E, Sharon A, Simon A, ten Have A, Tudzynski B, Tudzynski P, Wincker P, Andrew M, Anthouard V, Beever RE, Beffa R, Benoit I, Bouzid O, Brault B, Chen Z, Choquer M, Collemare J, Cotton P, Danchin EG, Da Silva C, Gautier A, Giraud C, Giraud T, Gonzalez C, Grossetete S, Gueldener U, Henrissat B, Howlett BJ, Kodira C, Kretschmer M, Lappartient A, Leroch M, Levis C, Mauceli E, Neuveglise C, Oeser B, Pearson M, Poulain J, Poussereau N, Quesneville H, Rascle C, Schumacher J, Segurens B, Sexton A, Silva E, Sirven C, Soanes DM, Talbot NJ, Templeton M, Yandava C, Yarden O, Zeng Q, Rollins JA, Lebrun M-H, Dickman M (2011) Genomic analysis of the necrotrophic fungal pathogens *Sclerotinia sclerotiorum* and *Botrytis cinerea*. PLoS Genet 7:e1002230. doi:10.1371/journal.pgen.1002230

Antal Z, Rascle C, Cimerman A, Viaud M, Billon-Grand G, Choquer M, Bruel C (2012) The homeobox BcHOX8 gene in *Botrytis cinerea* regulates vegetative growth and morphology. Plos One 7:e48134. doi:10.1371/journal.pone.0048134

Backhouse D, Willetts HJ (1987) Development and structure of infection cushions of *Botrytis cinerea*. Trans Br Mycol Soc 89:89–95

Bashi ZD, Rimmer SR, Khachatourians GG, Hegedus DD (2012) Factors governing the regulation of *Sclerotinia sclerotiorum cutinase A* and *polygalacturonase 1* during different stages of infection. Can J Microbiol 58(5):605–616

Beckerman JL, Ebbole DJ (1996) *MPG1*, a gene encoding a fungal hydrophobin of *Magnaporthe grisea*, is involved in surface recognition. Mol Plant Microbe Interact 9(6):450–456

Boenisch MJ, Schaefer W (2011) *Fusarium graminearum* forms mycotoxin producing infection structures on wheat. BMC Plant Biol 11:110. doi:10.1186/1471-2229-11-110

Both M, Csukai M, Stumpf MPH, Spanu PD (2005a) Gene expression profiles of *Blumeria graminis* indicate dynamic changes to primary metabolism during development of an obligate biotrophic pathogen. Plant Cell 17(7):2107–2122

Both M, Eckert SE, Csukai M, Muller E, Dimopoulos G, Spanu PD (2005b) Transcript profiles of *Blumeria graminis* development during infection reveal a cluster of genes that are potential virulence determinants. Mol Plant Microbe Interact 18(2):125–133

Brand A, Gow NR (2012) Tropic orientation responses of pathogenic fungi. In: Pérez-Martín J, Di Pietro A (eds) Morphogenesis and pathogenicity in fungi, vol 22. Springer, Berlin, pp 21–41

Bucher M, Hause B, Krajinski F, Kuster H (2014) Through the doors of perception to function in arbuscular mycorrhizal symbioses. New Phytol 204(4):833–840

Bushnell WR, Dueck J, Rowell JB (1967) Living haustoria and hyphae of *Erysiphe graminis* f. sp. *hordei* with intact and partly dissected host cells of *Hordeum vulgare*. Can J Bot 45(9):1719–1732

Chen CY, Howlett BJ (1996) Rapid necrosis of guard cells is associated with the arrest of fungal growth in leaves of Indian mustard (*Brassica juncea*) inoculated with avirulent isolates of *Leptosphaeria maculans*. Physiol Mol Plant Pathol 48(2):73–81

Chen CB, Ha YS, Min JY, Memmott SD, Dickman MB (2006) Cdc42 is required for proper growth and development in the fungal pathogen *Colletotrichum trifolii*. Eukaryot Cell 5(1):155–166

Cheng YL, Zhang HC, Yao JN, Wang XJ, Xu JR, Han QM, Wei GR, Huang LL, Kang ZS (2012) Characterization of non-host resistance in broad bean to the wheat stripe rust pathogen. BMC Plant Biol 12:96. doi:10.1186/1471-2229-12-96

Choi W, Dean RA (1997) The adenylate cyclase gene MAC1 of *Magnaporthe grisea* controls appressorium formation and other aspects of growth and development. Plant Cell 9(11):1973–1983

Chumley FG, Valent B (1990) Genetic-analysis of melanin-deficient, nonpathogenic mutants of *Maganporthe grisea*. Mol Plant-Microbe Interact 3(3):135–143. doi:10.1094/mpmi-3-135

Clergeot PH, Gourgues M, Cots J, Laurans F, Latorse MP, Pepin R, Tharreau D, Notteghem JL, Lebrun MH (2001) *PLS1*, a gene encoding a tetraspanin-like protein, is required for penetration of rice leaf by the fungal pathogen *Magnaporthe grisea*. Proc Natl Acad Sci USA 98(12):6963–6968

Dagdas YF, Yoshino K, Dagdas G, Ryder LS, Bielska E, Steinberg G, Talbot NJ (2012) Septin-mediated plant cell invasion by the rice blast fungus, *Magnaporthe oryzae*. Science 336(6088):1590–1595

De Wit PGM (1977) A light and scanning-electron microscopic study of infection of tomato plants by virulent and avirulent races of *Cladosporium fulvum*. Neth J Plant Pathol 83(3):109–122

de Wit PJGM, van der Burgt A, Ökmen B, Stergiopoulos I, Abd-Elsalam KA, Aerts AL, Bahkali AH, Beenen HG, Chettri P, Cox MP, Datema E, de Vries RP, Dhillon B, Ganley AR, Griffiths SA, Guo Y, Hamelin RC, Henrissat B, Kabir MS, Jashni MK, Kema G, Klaubauf S, Lapidus A, Levasseur A, Lindquist E, Mehrabi R, Ohm RA, Owen TJ, Salamov A, Schwelm A, Schijlen E, Sun H, van den Burg HA, van Ham RCHJ, Zhang S, Goodwin SB, Grigoriev IV, Collemare J, Bradshaw RE (2012) The genomes of the fungal plant pathogens *Cladosporium fulvum* and *Dothistroma septosporum* reveal adaptation to different hosts and lifestyles but also signatures of common ancestry. PLoS Genet 8: e1003088. doi:10.1371/journal.pgen.1003088

DeZwaan TM, Carroll AM, Valent B, Sweigard JA (1999) *Magnaporthe grisea* Pth11p is a novel plasma membrane protein that mediates appressorium differentiation in response to inductive substrate cues. Plant Cell 11(10):2013–2030

Dub AM, Kokkelink L, Tudzynski B, Tudzynski P, Sharon A (2013) Involvement of *Botrytis cinerea* small GTPases BcRAS1 and BcRAC in differentiation, virulence, and the cell cycle. Eukaryot Cell 12(12):1609–1618

Duplessis S, Cuomo CA, Lin YC, Aerts A, Tisserant E, Veneault-Fourrey C, Joly DL, Hacquard S, Amselem J, Cantarel BL, Chiu R, Coutinho PM, Feau N, Field M, Frey P, Gelhaye E, Goldberg J, Grabherr MG, Kodira CD, Kohler A, Kues U, Lindquist EA, Lucas SM, Mago R, Mauceli E, Morin E, Murat C, Pangilinan JL, Park R, Pearson M, Quesneville H, Rouhier N, Sakthikumar S, Salamov AA, Schmutz J, Selles B, Shapiro H, Tanguay P, Tuskan GA, Henrissat B, Van de Peer Y, Rouze P, Ellis JG, Dodds PN, Schein JE, Zhong SB, Hamelin RC, Grigoriev IV, Szabo LJ, Martin F (2011) Obligate biotrophy features unraveled by the genomic analysis of rust fungi. Proc Natl Acad Sci USA 108(22):9166–9171

Egan MJ, Wang ZY, Jones MA, Smirnoff N, Talbot NJ (2007) Generation of reactive oxygen species by fungal NADPH oxidases is required for rice blast disease. Proc Natl Acad Sci USA 104(28):11772–11777

Elliott CE, Harjono, Howlett BJ (2008) Mutation of a gene in the fungus *Leptosphaeria maculans* allows increased frequency of penetration of stomatal apertures of *Arabidopsis thaliana*. Mol Plant 1(3):471–481

Fan J, Doerner P (2012) Genetic and molecular basis of nonhost disease resistance: complex, yes; silver bullet, no. Curr Opin Plant Biol 15(4):400–406

Feng J, Hwang R, Hwang SF, Gaudet D, Strelkov SE (2011) Molecular characterization of a *Stagonospora nodorum* lipase gene LIP1. Plant Pathol 60(4):698–708. doi:10.1111/j.1365-3059.2011.02434.x

Feng J, Wang F, Liu G, Greenshields D, Shen W, Kaminskyj S, Hughes GR, Peng Y, Selvaraj G, Zou J, Wei Y (2009) Analysis of a *Blumeria graminis*-secreted lipase reveals the importance of host epicuticular wax components for fungal adhesion and development. Mol Plant-Microbe Interact 22(12):1601–1610. doi:10.1094/mpmi-22-12-1601

Franck WL, Gokce E, Oh Y, Muddiman DC, Dean RA (2013) Temporal analysis of the *Magnaporthe oryzae* proteome during conidial germination and cyclic AMP (cAMP)-mediated appressorium formation. Mol Cell Proteomics 12(8):2249–2265

Gan P, Ikeda K, Irieda H, Narusaka M, O'Connell RJ, Narusaka Y, Takano Y, Kubo Y, Shirasu K (2013) Comparative genomic and transcriptomic analyses reveal the hemibiotrophic stage shift of *Colletotrichum* fungi. New Phytol 197(4):1236–1249

Gao QA, Jin K, Ying SH, Zhang YJ, Xiao GH, Shang YF, Duan ZB, Hu XA, Xie XQ, Zhou G, Peng GX, Luo ZB, Huang W, Wang B, Fang WG, Wang SB, Zhong Y, Ma LJ, St Leger RJ, Zhao GP, Pei Y, Feng MG, Xia YX, Wang CS (2011) Genome sequencing and comparative transcriptomics of the model entomopathogenic fungi *Metarhizium anisopliae* and

M. acridum. PLoS Genet 7:e1001264. doi:10.1371/journal.pgen.1001264

Garg H, Li H, Sivasithamparam K, Kuo J, Barbetti MJ (2010) The infection processes of *Sclerotinia sclerotiorum* in cotyledon tissue of a resistant and a susceptible genotype of *Brassica napus*. Ann Bot 106(6):897–908

Gehring WJ (1993) Exploring the homeobox. Gene 135(1-2):215–221

Giraldo MC, Valent B (2013) Filamentous plant pathogen effectors in action. Nat Rev Microbiol 11(11):800–814

Godfrey D, Bohlenius H, Pedersen C, Zhang Z, Emmersen J, Thordal-Christensen H (2010) Powdery mildew fungal effector candidates share N-terminal Y/F/WxC-motif. BMC Genomics 11:317. doi:10.1186/1471-2164-11-317

Gout L, Fudal I, Kuhn ML, Blaise F, Eckert M, Cattolico L, Balesdent MH, Rouxel T (2006) Lost in the middle of nowhere: the *AvrLm1* avirulence gene of the Dothideomycete *Leptosphaeria maculans*. Mol Microbiol 60(1):67–80

Grandaubert J, Lowe RGT, Soyer JL, Schoch CL, de Wouw APV, Fudal I, Robbertse B, Lapalu N, Links MG, Ollivier B, Linglin J, Barbe V, Mangenot S, Cruaud C, Borhan H, Howlett BJ, Balesdent M-H, Rouxel T (2014) Transposable element-assisted evolution and adaptation to host plant within the *Leptosphaeria maculans-Leptosphaeria biglobosa* species complex of fungal pathogens. BMC Genomics 15:891. doi:10.1186/1471-2164-15-891

Grigoriev IV, Nikitin R, Haridas S, Kuo A, Ohm R, Otillar R, Riley R, Salamov A, Zhao XL, Korzeniewski F, Smirnova T, Nordberg H, Dubchak I, Shabalov I (2014) MycoCosm portal: gearing up for 1000 fungal genomes. Nucleic Acids Res 42(D1):D699–D704

Gu Q, Chen Y, Liu Y, Zhang C, Ma Z (2014) The transmembrane protein FgSho1 regulates fungal development and pathogenicity via the MAPK module Ste50-Ste11-Ste7 in *Fusarium graminearum*. New Phytol 206(1):315–328

Guenther JC, Trail F (2005) The development and differentiation of *Gibberella zeae* (anamorph: *Fusarium graminearum*) during colonization of wheat. Mycologia 97(1):229–237

Guyon K, Balague C, Roby D, Raffaele S (2014) Secretome analysis reveals effector candidates associated with broad host range necrotrophy in the fungal plant pathogen *Sclerotinia sclerotiorum*. BMC Genomics 15:336. doi:10.1186/1471-2164-15-336

Hacquard S, Kracher B, Maekawa T, Vernaldi S, Schulze-Lefert P, van Themaat EVL (2013) Mosaic genome structure of the barley powdery mildew pathogen and conservation of transcriptional programs in divergent hosts. Proc Natl Acad Sci USA 110(24):E2219–E2228

Hane JK, Lowe RGT, Solomon PS, Tan K-C, Schoch CL, Spatafora JW, Crous PW, Kodira C, Birren BW, Galagan JE, Torriani SFF, McDonald BA, Oliver RP (2007) Dothideomycete–plant interactions illuminated by genome sequencing and EST analysis of the wheat pathogen *Stagonospora nodorum*. Plant Cell 19(11):3347–3368. doi:10.1105/tpc.107.052829

Hansjakob A, Riederer M, Hildebrandt U (2012) Appressorium morphogenesis and cell cycle progression are linked in the grass powdery mildew fungus *Blumeria graminis*. Fungal Biol 116(8):890–901

Harris SD (2011) Cdc42/Rho GTPases in fungi: variations on a common theme. Mol Microbiol 79(5):1123–1127

Harris SD, Momany M (2004) Polarity in filamentous fungi: moving beyond the yeast paradigm. Fungal Genet Biol 41(4):391–400

Howard RJ, Valent B (1996) Breaking and entering: host penetration by the fungal rice blast pathogen *Magnaporthe grisea*. Annu Rev Microbiol 50:491–512

Howard RJ, Ferrari MA, Roach DH, Money NP (1991) Penetration of hard substrates by a fungus employing enormous turgor pressures. Proc Natl Acad Sci USA 88(24):11281–11284. doi:10.1073/pnas.88.24.11281

Huang L, Buchenauer H, Han Q, Zhang X, Kang Z (2008) Ultrastructural and cytochemical studies on the infection process of *Sclerotinia sclerotiorum* in oilseed rape. J Plant Dis Protect 115(1):9–16

Hueckelhoven R, Panstruga R (2011) Cell biology of the plant-powdery mildew interaction. Curr Opin Plant Biol 14(6):738–746

Inoue K, Suzuki T, Ikeda K, Jiang S, Hosogi N, Hyon G-S, Hida S, Yamada T, Park P (2007) Extracellular matrix of *Magnaporthe oryzae* may have a role in host adhesion during fungal penetration and is digested by matrix metalloproteinases. J Gen Plant Pathol 73(1):388–398

Inoue K, Onoe T, Park P, Ikeda K (2011) Enzymatic detachment of spore germlings in *Magnaporthe oryzae*. FEMS Microbiol Lett 323(1):13–19

Ipcho SVS, Hane JK, Antoni EA, Ahren D, Henrissat B, Friesen TL, Solomon PS, Oliver RP (2012) Transcriptome analysis of *Stagonospora nodorum*: gene models, effectors, metabolism and pantothenate dispensability. Mol Plant Pathol 13(6):531–545

Jamaux I, Gelie B, Lamarque C (1995) Early stages of infection of rapeseed petals and leaves by *Sclerotinia sclerotiorum* revealed by scanning-electron microscopy. Plant Pathol 44(1):22–30

Jelitto TC, Page HA, Read ND (1994) Role of external signals in regulating the pre-penetration phase of infection by the rice blast fungus, *Magnaporthe grisea*. Planta 194(4):471–477

Johnson DI (1999) Cdc42: an essential Rho-type GTPase controlling eukaryotic cell polarity. Microbiol Mol Biol Rev 63(1):54–105

Jones D (1976) Infection of plant tissue by *Sclerotinia sclerotiorum* - scanning electron microscope study. Micron 7(4):275–279

Jones JDG, Dangl JL (2006) The plant immune system. Nature 444(7117):323–329

Jurick WM II, Rollins JA (2007) Deletion of the adenylate cyclase (*sac1*) gene affects multiple developmental pathways and pathogenicity in *Sclerotinia sclerotiorum*. Fungal Genet Biol 44(6):521–530

Kema GHJ, Yu DZ, Rijkenberg FHJ, Shaw MW, Baayen RP (1996) Histology of the pathogenesis of *Mycosphaerella graminicola* in wheat. Phytopathology 86(7):777–786

Keon J, Rudd JJ, Antoniw J, Skinner W, Hargreaves J, Hammond-Kosack K (2005) Metabolic and stress adaptation by *Mycosphaerella graminicola* during sporulation in its host revealed through microarray transcription profiling. Mol Plant Pathol 6(5):527–540

Kershaw MJ, Talbot NJ (1998) Hydrophobins and repellents: proteins with fundamental roles in fungal morphogenesis. Fungal Genet Biol 23(1):18–33

Khang CH, Berruyer R, Giraldo MC, Kankanala P, Park S-Y, Czymmek K, Kang S, Valent B (2010) Translocation of *Magnaporthe oryzae* effectors into rice cells and their subsequent cell-to-cell movement. Plant Cell 22(4):1388–1403

Kim S, Ahn IP, Rho HS, Lee YH (2005) *MHP1*, a *Magnaporthe grisea* hydrophobin gene, is required for fungal development and plant colonization. Mol Microbiol 57(5):1224–1237

Kim H, Ridenour JB, Dunkle LD, Bluhm BH (2011) Regulation of stomatal tropism and infection by light in *Cercospora zeae-maydis*: evidence for coordinated host/pathogen responses to photoperiod? PLoS Pathog 7:e1002113. doi:10.1371/journal.ppat.1002113

Kleemann J, Rincon-Rivera LJ, Takahara H, Neumann U, van Themaat EVL, van der Does HC, Hacquard S, Stuber K, Will I, Schmalenbach W, Schmelzer E, O'Connell RJ (2012) Sequential delivery of host-induced virulence effectors by appressoria and intracellular hyphae of the phytopathogen *Colletotrichum higginsianum*. PLoS Pathog 8:e1002643. doi:10.1371/journal.ppat.1002643

Kokkelink L, Minz A, Al-Masri M, Giesbert S, Barakat R, Sharon A, Tudzynski P (2011) The small GTPase BcCdc42 affects nuclear division, germination and virulence of the gray mold fungus *Botrytis cinerea*. Fungal Genet Biol 48(11):1012–1019

Krijger J-J, Thon MR, Deising HB, Wirsel SGR (2014) Compositions of fungal secretomes indicate a greater impact of phylogenetic history than lifestyle adaptation. BMC Genomics 15:722. doi:10.1186/1471-2164-15-722

Kusch S, Ahmadinejad N, Panstruga R, Kuhn H (2014) In silico analysis of the core signaling proteome from the barley powdery mildew pathogen (*Blumeria graminis* f.sp. *hordei*). BMC Genomics 15:843. doi:10.1186/1471-2164-15-843

Lacroix H, Whiteford JR, Spanu PD (2008) Localization of *Cladosporium fulvum* hydrophobins reveals a role for HCf-6 in adhesion. FEMS Microbiol Lett 286(1):136–144

Lambou K, Tharreau D, Kohler A, Sirven C, Marguerettaz M, Barbisan C, Sexton AC, Kellner EM, Martin F, Howlett BJ, Orbach MJ, Lebrun M-H (2008) Fungi have three tetraspanin families with distinct functions. BMC Genomics 9:63. doi:10.1186/1471-2164-9-63

Lanver D, Mendoza-Mendoza A, Brachmann A, Kahmann R (2010) Sho1 and Msb2-related proteins regulate appressorium development in the smut fungus *Ustilago maydis*. Plant Cell 22(6):2085–2101

Lazebnik J, Frago E, Dicke M, van Loon JJA (2014) Phytohormone mediation of interactions between herbivores and plant pathogens. J Chem Ecol 40(7):730–741

Lee YH, Dean RA (1993) cAMP regulates infection structure formation in the plant pathogenic fungus *Magnaporthe grisea*. Plant Cell 5(6):693–700

Leroch M, Kleber A, Silva E, Coenen T, Koppenhofer D, Shmaryahu A, Valenzuela PDT, Hahn M (2013) Transcriptome profiling of *Botrytis cinerea* conidial germination reveals upregulation of infection-related genes during the prepenetration stage. Eukaryot Cell 12(4):614–626

Li GT, Zhou XY, Xu JR (2012a) Genetic control of infection-related development in *Magnaporthe oryzae*. Curr Opin Microbiol 15(6):678–684

Li MY, Liang XF, Rollins JA (2012b) *Sclerotinia sclerotiorum* gamma-glutamyl transpeptidase (*Ss-Ggt1*) is required for regulating glutathione accumulation and development of sclerotia and compound appressoria. Mol Plant Microbe Interact 25(3):412–420

Liu Z, Faris JD, Oliver RP, Tan K-C, Solomon PS, McDonald MC, McDonald BA, Nunez A, Lu S, Rasmussen JB, Friesen TL (2009) SnTox3 acts in effector triggered susceptibility to induce disease on wheat carrying the *Snn3* gene. PLoS Pathog 5, e1000581. doi:10.1371/journal.ppat.1000581

Liu W, Zhou X, Li G, Li L, Kong L, Wang C, Zhang H, Xu J-R (2011) Multiple plant surface signals are sensed by different mechanisms in the rice blast fungus for appressorium formation. PLoS Pathog 7(1), e1001261. doi:10.1371/journal.ppat.1001261

Liu ZH, Zhang ZC, Faris JD, Oliver RP, Syme R, McDonald MC, McDonald BA, Solomon PS, Lu SW, Shelver WL, Xu S, Friesen TL (2012) The cysteine rich necrotrophic effector SnTox1 produced by *Stagonospora nodorum* triggers susceptibility of wheat lines harboring *Snn1*. PLoS Pathog 8: e1002467. doi:10.1371/journal.ppat.1002467

Lowe RGT, Howlett BJ (2012) Indifferent, affectionate, or deceitful: lifestyles and secretomes of fungi.

PLoS Pathog 8:e1002515. doi:10.1371/journal.ppat.1002515

Lowe RGT, Cassin A, Grandaubert J, Clark BL, Van de Wouw AP, Rouxel T, Howlett BJ (2014) Genomes and transcriptomes of partners in plant-fungal-interactions between canola (*Brassica napus*) and two *Leptosphaeria* species. Plos One 9:e0103098. doi:10.1371/journal.pone.0103098

Lysoe E, Seong K-Y, Kistler HC (2011) The transcriptome of *Fusarium graminearum* during the infection of wheat. Mol Plant Microbe Interact 24(9):995–1000

Maekawa T, Kracher B, Vernaldi S, van Themaat EVL, Schulze-Lefert P (2012) Conservation of NLR-triggered immunity across plant lineages. Proc Natl Acad Sci USA 109(49):20119–20123

Marshall R, Kombrink A, Motteram J, Loza-Reyes E, Lucas J, Hammond-Kosack KE, Thomma BPHJ, Rudd JJ (2011) Analysis of two *in planta* expressed LysM effector homologs from the fungus *Mycosphaerella graminicola* reveals novel functional properties and varying contributions to virulence on wheat. Plant Physiol 156(2):756–769

Martin F, Kohler A, Murat C, Balestrini R, Coutinho PM, Jaillon O, Montanini B, Morin E, Noel B, Percudani R, Porcel B, Rubini A, Amicucci A, Amselem J, Anthouard V, Arcioni S, Artiguenave F, Aury JM, Ballario P, Bolchi A, Brenna A, Brun A, Buee M, Cantarel B, Chevalier G, Couloux A, Da Silva C, Denoeud F, Duplessis S, Ghignone S, Hilselberger B, Iotti M, Marcais B, Mello A, Miranda M, Pacioni G, Quesneville H, Riccioni C, Ruotolo R, Splivallo R, Stocchi V, Tisserant E, Viscomi AR, Zambonelli A, Zampieri E, Henrissat B, Lebrun MH, Paolocci F, Bonfante P, Ottonello S, Wincker P (2010) Perigord black truffle genome uncovers evolutionary origins and mechanisms of symbiosis. Nature 464(7291):1033–1038

Mathioni SM, Belo A, Rizzo CJ, Dean RA, Donofrio NM (2011) Transcriptome profiling of the rice blast fungus during invasive plant infection and *in vitro* stresses. BMC Genomics 12:49. doi:10.1186/1471-2164-12-49

McDonough KA, Rodriguez A (2012) The myriad roles of cyclic AMP in microbial pathogens: from signal to sword. Nat Rev Microbiol 10(1):27–38

Mendgen K, Hahn M, Deising H (1996) Morphogenesis and mechanisms of penetration by plant pathogenic fungi. Annu Rev Phytopathol 34:367–386

Mercure EW, Leite B, Nicholson RL (1994) Adhesion of ungerminated conidia of *Colletotrichum graminicola* to artificial hydrophobic surfaces. Physiol Mol Plant Pathol 45(6):421–440

Mesarich CH, Griffiths SA, van der Burgt A, Okmen B, Beenen HG, Etalo DW, Joosten MHAJ, de Wit PJGM (2014) Transcriptome sequencing uncovers the *Avr5* avirulence gene of the tomato leaf mold pathogen *Cladosporium fulvum*. Mol Plant Microbe Interact 27(8):846–857

Mosbach A, Leroch M, Mendgen KW, Hahn M (2011) Lack of evidence for a role of hydrophobins in conferring surface hydrophobicity to conidia and hyphae of *Botrytis cinerea*. BMC Microbiol 11:10. doi:10.1186/1471-2180-11-10

Newey LJ, Caten CE, Green JR (2007) Rapid adhesion of *Stagonospora nodorum* spores to a hydrophobic surface requires pre-formed cell surface glycoproteins. Mycol Res 111:1255–1267

Nielsen KA, Nicholson RL, Carver TLW, Kunoh H, Oliver RP (2000) First touch: an immediate response to surface recognition in conidia of *Blumeria graminis*. Physiol Mol Plant Pathol 56(2):63–70

Niinemets U, Kannaste A, Copolovici L (2013) Quantitative patterns between plant volatile emissions induced by biotic stresses and the degree of damage. Front Plant Sci 4:262. doi:10.3389/fpls.2013.00262

Nesher I, Barhoom S, Sharon A (2008) Cell cycle and cell death are not necessary for appressorium formation and plant infection in the fungal plant pathogen *Colletotrichum gloeosporioides*. BMC Biol 6:9. doi:10.1186/1741-7007-6-9

Noir S, Colby T, Harzen A, Schmidt J, Panstruga R (2009) A proteomic analysis of powdery mildew (*Blumeria graminis* f.sp. *hordei*) conidiospores. Mol Plant Pathol 10(2):223–236

O'Connell RJ, Thon MR, Hacquard S, Amyotte SG, Kleemann J, Torres MF, Damm U, Buiate EA, Epstein L, Alkan N, Altmuller J, Alvarado-Balderrama L, Bauser CA, Becker C, Birren BW, Chen Z, Choi J, Crouch JA, Duvick JP, Farman MA, Gan P, Heiman D, Henrissat B, Howard RJ, Kabbage M, Koch C, Kracher B, Kubo Y, Law AD, Lebrun M-H, Lee Y-H, Miyara I, Moore N, Neumann U, Nordstrom K, Panaccione DG, Panstruga R, Place M, Proctor RH, Prusky D, Rech G, Reinhardt R, Rollins JA, Rounsley S, Schardl CL, Schwartz DC, Shenoy N, Shirasu K, Sikhakolli UR, Stuber K, Sukno SA, Sweigard JA, Takano Y, Takahara H, Trail F, van der Does HC, Voll LM, Will I, Young S, Zeng Q, Zhang J, Zhou S, Dickman MB, Schulze-Lefert P, Ver Loren van Themaat E, Ma L-J, Vaillancourt LJ (2012) Lifestyle transitions in plant pathogenic *Colletotrichum* fungi deciphered by genome and transcriptome analyses. Nat Genet 44(9):1060–1065

Oh Y, Donofrio N, Pan H, Coughlan S, Brown DE, Meng S, Mitchell T, Dean RA (2008) Transcriptome analysis reveals new insight into appressorium formation and function in the rice blast fungus *Magnaporthe oryzae*. Genome Biol 9:R85. doi:10.1186/gb-2008-9-5-r85

Oliver RP, Friesen TL, Faris JD, Solomon PS (2012) *Stagonospora nodorum*: from pathology to genomics and host resistance. Annu Rev Phytopathol 50:23–43

Parker D, Beckmann M, Zubair H, Enot DP, Caracuel-Rios Z, Overy DP, Snowdon S, Talbot NJ, Draper J (2009) Metabolomic analysis reveals a common pattern of metabolic re-programming during invasion of three host plant species by *Magnaporthe grisea*. Plant J 59(5):723–737

Parlange F, Daverdin G, Fudal I, Kuhn ML, Balesdent MH, Blaise F, Grezes-Besset B, Rouxel T (2009) *Leptosphaeria maculans* avirulence gene *AvrLm4-7* confers a dual recognition specificity by the *Rlm4* and *Rlm7* resistance genes of oilseed rape, and circumvents Rlm4-mediated recognition through a single amino acid change. Mol Microbiol 71(4):851–863

Pedersen C, van Themaat EVL, McGuffin LJ, Abbott JC, Burgis TA, Barton G, Bindschedler LV, Lu X, Maekawa T, Wessling R, Cramer R, Thordal-Christensen H, Panstruga R, Spanu PD (2012) Structure and evolution of barley powdery mildew effector candidates. BMC Genomics 13:694. doi:10.1186/1471-2164-13-694

Perfect SE, Hughes HB, O'Connell RJ, Green JR (1999) *Colletotrichum* - a model genus for studies on pathology and fungal-plant interactions. Fungal Genet Biol 27(2-3):186–198

Ponzio C, Gols R, Pieterse CMJ, Dicke M (2013) Ecological and phytohormonal aspects of plant volatile emission in response to single and dual infestations with herbivores and phytopathogens. Funct Ecol 27(3):587–598

Pryce-Jones E, Carver T, Gurr SJ (1999) The roles of cellulase enzymes and mechanical force in host penetration by *Erysiphe graminis* f.sp. *hordei*. Physiol Mol Plant Pathol 55(3):175–182

Purdy LH (1979) *Sclerotinia sclerotiorum* - history, disease and symptomatology, host range, geographic-distribution, and impact. Phytopathology 69(8):875–880

Raffaele S, Kamoun S (2012) Genome evolution in filamentous plant pathogens: why bigger can be better. Nat Rev Microbiol 10(6):417–430

Ramanujam R, Calvert ME, Selvaraj P, Naqvi NI (2013) The late endosomal HOPS complex anchors active G-protein signaling essential for pathogenesis in *Magnaporthe oryzae*. PLoS Pathog 9:e1003527. doi:10.1371/journal.ppat.1003527

Rebollar A, Lopez-Garcia B (2013) PAF104, a synthetic peptide to control rice blast disease by blocking appressorium formation in *Magnaporthe oryzae*. Mol Plant Microbe Interact 26(12):1407–1416

Rodriguez A, Shimada T, Cervera M, Alquezar B, Gadea J, Gomez-Cadenas A, De Ollas CJ, Rodrigo MJ, Zacarias L, Pena L (2014) Terpene down-regulation triggers defense responses in transgenic orange leading to resistance against fungal pathogens. Plant Physiol 164(1):321–339

Rouxel T, Grandaubert J, Hane JK, Hoede C, van de Wouw AP, Couloux A, Dominguez V, Anthouard V, Bally P, Bourras S, Cozijnsen AJ, Ciuffetti LM, Degrave A, Dilmaghani A, Duret L, Fudal I, Goodwin SB, Gout L, Glaser N, Linglin J, Kema GHJ, Lapalu N, Lawrence CB, May K, Meyer M, Ollivier B, Poulain J, Schoch CL, Simon A, Spatafora JW, Stachowiak A, Turgeon BG, Tyler BM, Vincent D, Weissenbach J, Amselem J, Quesneville H, Oliver RP, Wincker P, Balesdent MH, Howlett BJ (2011) Effector diversification within compartments of the *Leptosphaeria maculans* genome affected by repeat-induced point mutations. Nat Commun 2:202. doi:10.1038/ncomms1189

Rudd JJ, Antoniw J, Marshall R, Motteram J, Fraaije B, Hammond-Kosack K (2010) Identification and characterisation of *Mycosphaerella graminicola* secreted or surface-associated proteins with variable intragenic coding repeats. Fungal Genet Biol 47(1):19–32

Seong K-Y, Zhao X, Xu J-R, Gueldener U, Kistler HC (2008) Conidial germination in the filamentous fungus *Fusarium graminearum*. Fungal Genet Biol 45(4):389–399

Sexton AC, Cozijnsen AJ, Keniry A, Jewell E, Love CG, Batley J, Edwards D, Howlett BJ (2006) Comparison of transcription of multiple genes at three developmental stages of the plant pathogen *Sclerotinia sclerotiorum*. FEMS Microbiol Lett 258(1):150–160

Skamnioti P, Gurr SJ (2007) *Magnaporthe grisea* cutinase2 mediates appressorium differentiation and host penetration and is required for full virulence. Plant Cell 19(8):2674–2689

Skamnioti P, Furlong RF, Gurr SJ (2008) Evolutionary history of the ancient cutinase family in five filamentous Ascomycetes reveals differential gene duplications and losses and in *Magnaporthe grisea* shows evidence of sub- and neo-functionalization. New Phytol 180(3):711–721

Soanes DM, Alam I, Cornell M, Wong HM, Hedeler C, Paton NW, Rattray M, Hubbard SJ, Oliver SG, Talbot NJ (2008) Comparative genome analysis of filamentous fungi reveals gene family expansions associated with fungal pathogenesis. Plos One 3: e0002300. doi:10.1371/journal.pone.0002300

Soanes DM, Chakrabarti A, Paszkiewicz KH, Dawe AL, Talbot NJ (2012) Genome-wide transcriptional profiling of appressorium development by the rice blast fungus *Magnaporthe oryzae*. PLoS Pathog 8: e1002514. doi:10.1371/journal.ppat.1002514

Solomon PS, Wilson TJG, Rybak K, Parker K, Lowe RGT, Oliver RP (2006) Structural characterisation of the interaction between *Triticum aestivum* and the dothideomycete pathogen *Stagonospora nodorum*. Eur J Plant Pathol 114(3):275–282

Soyer JL, El Ghalid M, Glaser N, Ollivier B, Linglin J, Grandaubert J, Balesdent MH, Connolly LR, Freitag M, Rouxel T, Fudal I (2014) Epigenetic control of effector gene expression in the plant pathogenic fungus *Leptosphaeria maculans*. PLoS Genet 10: e1004227. doi:10.1371/journal.pgen.1004227

Spanu PD (2014) Messages from powdery mildew DNA: how the interplay with a host moulds pathogen genomes. J Integr Agric 13(2):233–236

Spanu PD, Abbott JC, Amselem J, Burgis TA, Soanes DM, Stuber K, van Themaat EVL, Brown JKM, Butcher SA, Gurr SJ, Lebrun MH, Ridout CJ, Schulze-Lefert P, Talbot NJ, Ahmadinejad N, Ametz C, Barton GR, Benjdia M, Bidzinski P, Bindschedler LV, Both M, Brewer MT, Cadle-Davidson L, Cadle-Davidson MM, Collemare J, Cramer R, Frenkel O, Godfrey D, Harriman J, Hoede C, King BC, Klages S, Kleemann J, Knoll D, Koti PS, Kreplak J, Lopez-Ruiz FJ, Lu XL, Maekawa T, Mahanil S, Micali C, Milgroom MG, Montana G, Noir S, O'Connell RJ, Oberhaensli S, Parlange F, Pedersen C, Quesneville H, Reinhardt R, Rott M, Sacristan S, Schmidt SM, Schon M, Skamnioti P, Sommer H, Stephens A, Takahara H, Thordal-Christensen H, Vigouroux M, Wessling R, Wicker T, Panstruga R (2010) Genome expansion and gene loss in powdery mildew fungi reveal tradeoffs in extreme parasitism. Science 330(6010):1543–1546

Stephens AE, Gardiner DM, White RG, Munn AL, Manners JM (2008) Phases of infection and gene expression of *Fusarium graminearum* during Crown Rot disease of wheat. Mol Plant Microbe Interact 21(12):1571–1581

Stotz HU, Mitrousia GK, de Wit PJGM, Fitt BDL (2014) Effector-triggered defence against apoplastic fungal pathogens. Trends Plant Sci 19(8):491–500

Sudbery PE (2008) Regulation of polarised growth in fungi. Fungal Biol Rev 22(2):44–55

Sutton JC (1982) Epidemiology of wheat head blight and maize ear rot caused by *Fusarium graminearum*. Can J Plant Pathol 4(2):195–209

Syme RA, Hane JK, Friesen TL, Oliver RP (2013) Resequencing and comparative genomics of *Stagonospora nodorum*: sectional gene absence and effector discovery. G3 Genes Genom Genet 3(6):959–969

Talbot NJ, Kershaw MJ, Wakley GE, deVries OMH, Wessels JGH, Hamer JE (1996) *MPG1* encodes a fungal hydrophobin involved in surface interactions during infection-related development of *Magnaporthe grisea*. Plant Cell 8(6):985–999

Terhem RB, van Kan JAL (2014) Functional analysis of hydrophobin genes in sexual development of *Botrytis cinerea*. Fungal Genet Biol 71:42–51

Tucker SL, Besi MI, Galhano R, Franceschetti M, Goetz S, Lenhert S, Osbourn A, Sesma A (2010) Common genetic pathways regulate organ-specific infection-related development in the rice blast fungus. Plant Cell 22(3):953–972

van de Wouw AP, Thomas VL, Cozijnsen AJ, Marcroft SJ, Salisbury PA, Howlett BJ (2008) Identification of *Leptosphaeria biglobosa* 'canadensis' on *Brassica juncea* stubble from northern New South Wales, Australia. Australas Plant Dis Notes 3(1):124–128

van de Wouw AP, Lowe RGT, Elliott CE, Dubois DJ, Howlett BJ (2014) An avirulence gene, *AvrLmJ1*, from the blackleg fungus, *Leptosphaeria maculans*, confers avirulence to *Brassica juncea* cultivars. Mol Plant Pathol 15(5):523–530

Van Kan JAL, Shaw MW, Grant-Downton RT (2014) *Botrytis* species: relentless necrotrophic thugs or endophytes gone rogue? Mol Plant Pathol 15(9):957–961

Veneault-Fourrey C, Barooah M, Egan M, Wakley G, Talbot NJ (2006a) Autophagic fungal cell death is necessary for infection by the rice blast fungus. Science 312(5773):580–583

Veneault-Fourrey C, Lambou K, Lebrun MH (2006b) Fungal Pls1 tetraspanins as key factors of penetration into host plants: a role in re-establishing polarized growth in the appressorium? FEMS Microbiol Lett 256(2):179–184

Wessels JGH (1994) Developmental regulation of fungal cell-wall formation. Annu Rev Phytopathol 32:413–437

Wessling R, Schmidt SM, Micali CO, Knaust F, Reinhardt R, Neumann U, van Themaat EVL, Panstruga R (2012) Transcriptome analysis of enriched *Golovinomyces orontii* haustoria by deep 454 pyrosequencing. Fungal Genet Biol 49(6):470–482

West JS, Balesdent MH, Rouxel T, Narcy JP, Huang YJ, Roux J, Steed JM, Fitt BDL, Schmit J (2002) Colonization of winter oilseed rape tissues by A/Tox(+) and B/Tox(0) *Leptosphaeria maculans* (phoma stem canker) in France and England. Plant Pathol 51(3):311–321

Wicker T, Oberhaensli S, Parlange F, Buchmann JP, Shatalina M, Roffler S, Ben-David R, Dolezel J, Simkova H, Schulze-Lefert P, Spanu PD, Bruggmann R, Amselem J, Quesneville H, van Themaat EVL, Paape T, Shimizu KK, Keller B (2013) The wheat powdery mildew genome shows the unique evolution of an obligate biotroph. Nat Genet 45(9):1092–1096

Williamson B, Tudzynsk B, Tudzynski P, van Kan JAL (2007) *Botrytis cinerea*: the cause of grey mould disease. Mol Plant Pathol 8(5):561–580

Wulff BBH, Chakrabarti A, Jones DA (2009) Recognitional specificity and evolution in the tomato-*Cladosporium fulvum* pathosystem. Mol Plant Microbe Interact 22(10):1191–1202

Yang F, Jensen JD, Svensson B, Jorgensen HJL, Collinge DB, Finnie C (2010) Analysis of early events in the interaction between *Fusarium graminearum* and the susceptible barley (*Hordeum vulgare*) cultivar Scarlett. Proteomics 10(21):3748–3755

Yang F, Jacobsen S, Jorgensen HJL, Collinge DB, Svensson B, Finnie C (2013a) *Fusarium graminearum* and its interactions with cereal heads: studies in the proteomics era. Front Plant Sci 4:37. doi:10.3389/fpls.2013.00037

Yang F, Li WS, Jorgensen HJL (2013b) Transcriptional reprogramming of wheat and the hemibiotrophic pathogen *Septoria tritici* during two phases of the compatible interaction. Plos One 8:e0081606. doi:10.1371/journal.pone.0081606

Yang F, Melo-Braga MN, Larsen MR, Jorgensen HJL, Palmisano G (2013c) Battle through signaling between wheat and the fungal pathogen *Septoria tritici* revealed by proteomics and phosphoproteomics. Mol Cell Proteomics 12(9):2497–2508

Zelinger E, Hawes CR, Gurr SJ, Dewey FM (2006) Attachment and adhesion of conidia of *Stagonospora nodorum* to natural and artificial surfaces. Physiol Mol Plant Pathol 68(4–6):209–215

Zhang X-W, Jia L-J, Zhang Y, Jiang G, Li X, Zhang D, Tang W-H (2012) *In Planta* stage-specific fungal gene profiling elucidates the molecular strategies of *Fusarium graminearum* growing inside wheat coleoptiles. Plant Cell 24(12):5159–5176

Zhang X, Fu J, Hiromasa Y, Pan H, Bai G (2013) Differentially expressed proteins associated with *Fusarium* head blight resistance in wheat. Plos One 8: e0082079. doi:10.1371/journal.pone.0082079

Zhang QH, Zhang J, Yang L, Zhang L, Jiang DH, Chen WD, Li GQ (2014) Diversity and biocontrol potential of endophytic fungi in *Brassica napus*. Biol Control 72:98–108

Zheng W, Zhao Z, Chen J, Liu W, Ke H, Zhou J, Lu G, Darvill AG, Albersheim P, Wu S, Wang Z (2009) A Cdc42 ortholog is required for penetration and virulence of *Magnaporthe grisea*. Fungal Genet Biol 46(6–7):450–460. doi:10.1016/j.fgb.2009.03.005

Zhu WJ, Wei W, Fu YP, Cheng JS, Xie JT, Li GQ, Yi XH, Kang ZS, Dickman MB, Jiang DH (2013) A secretory protein of necrotrophic fungus *Sclerotinia sclerotiorum* that suppresses host resistance. Plos One 8:e0053901. doi:10.1371/journal.pone.0053901

Zuccaro A, Lahrmann U, Guldener U, Langen G, Pfiffi S, Biedenkopf D, Wong P, Samans B, Grimm C, Basiewicz M, Murat C, Martin F, Kogel KH (2011) Endophytic life strategies decoded by genome and transcriptome analyses of the mutualistic root symbiont *Piriformospora indica*. PLoS Pathog 7: e1002290. doi:10.1371/journal.ppat.1002290

Signal Transduction

6 Fungal MAP-Kinase-Mediated Regulatory Pathways

STEFAN IRNIGER[1], ÖZLEM SARIKAYA-BAYRAM[2], ÖZGÜR BAYRAM[2]

CONTENTS

I. Introduction

A fundamental feature of living organisms is their capability to sense and adapt to different intra- and extracellular conditions. For this purpose, they use signaling pathways to respond to their environment. This allows them to initiate an appropriate response. Many signals are transmitted by conserved **mitogen-activated protein kinase** (MAPK) modules that are involved in the response to various stimuli, often conveying signals from the cell surface receptors to the nucleus.

A module of three protein kinases represents the central part of the MAPK cascade (Widmann et al. 1999). A MAP kinase kinase kinase (MAPKKK) phosphorylates and activates the MAP kinase kinase (MAPKK). This in turn phosphorylates the downstream MAP kinase (MAPK), resulting in MAPK activation. The MAPK is phosphorylated on threonine and tyrosine residues in a conserved T-loop motif (T–X–Y motif; X represents any amino acid). Signaling through MAPK cascades can be initiated by an upstream activating kinase, such as the p21-activated kinase (PAK), by membrane-bound **G-protein-coupled receptor** (GPCR), or by other membrane receptors but may also be mediated by phosphorelay systems. GPCRs and G proteins are extensively reviewed in the chapter by Kathy Borkovich "Heterotrimeric G Proteins" in this volume. The output of these signaling pathways comprises various responses such as the transcriptional activation or repression of specific genes in the nucleus. MAPK modules often involve specific scaffold proteins that contribute to signaling specificity

[1]Department of Molecular Microbiology and Genetics, Georg August University, Grisebachstr. 8, Göttingen 37077, Germany
[2]Department of Biology, Maynooth University, National University of Ireland Maynooth, Maynooth Co. Kildare, Ireland; e-mail: ozgur.bayram@nuim.ie

Biochemistry and Molecular Biology, 3rd Edition
The Mycota III
D. Hoffmeister (Ed.)

and fidelity (Good et al. 2011). The scaffold proteins act as molecular hubs to assemble relevant molecular components on a protein complex.

Individual MAPK cascades do not act as isolated modules but rather form an interconnected network with other signaling pathways (Saito 2010). To ensure that signals are properly processed, control mechanisms are required which insulate other signaling pathways from the activated signaling pathway. Why do the cells need MAPK modules? Many years ago, it was demonstrated that MAPK cascades/modules serve to amplify signals at the cell surface and to convert the input in a highly specific switch-like response in the nucleus (Peter and Herskowitz 1994; Herskowitz 1995).

Many MAPK components were initially identified in *Saccharomyces cerevisiae*. Indeed, at present time, the most comprehensive knowledge of MAPK signaling is available in this yeast. Five MAPK pathways have been identified in *S. cerevisiae*, which mediate different responses: (1) response to pheromones during mating (pheromone response pathway), (2) filamentous growth under nitrogen starvation conditions, (3) response to osmotic stress and other stress conditions, (4) response to cell wall stress, and (5) spore formation.

This chapter provides an overview of the current knowledge regarding MAPK pathways in fungi, with an emphasis on mating, stress response, and cell integrity pathways. Each section starts with an overview on the state of the art in *S. cerevisiae* and then compares the findings with what is known in the filamentous fungi. Since MAPKs are important for the pathogenicity of fungi (Roman et al. 2007), pathogens are also included in the text. Among pathogenic fungi, more is known about MAPK signaling in *Candida albicans*, but findings with *Aspergillus fumigatus*, *Cryptococcus neoformans*, and plant pathogens are also presented.

II. The Pheromone Response and Filamentation MAPK Pathways

The MAPK cascade regulating the response to pheromones was first discovered in budding yeast. Mutants of the pathway were initially identified in screenings for sterile mutants (*ste*), which were unable to mate with a partner of opposite mating type (Mackay and Manney 1974). Mating of yeast was the first process for which the involvement of MAP kinase cascades was identified. Yeast uses various common components of signaling for mating and filamentation, which include the MAPKKK and MAPKK. Depending on the type of the incoming signal, a different MAPK pathway is activated, leading to an appropriate response. Due to the overlap in the MAPK module in yeast, the two signaling pathways will be discussed in a common chapter.

A. *Saccharomyces cerevisiae*

1. Pheromone Response in *S. cerevisiae*

Haploid yeast cells identify sexual partners through the presence of pheromones (small peptides) secreted by cells of opposite mating types. These pheromones bind to specific receptors and initiate a signal transduction pathway that finally leads to specific morphological changes preparing cells for mating. This involves cell cycle arrest in G1 phase and a typical shmoo-like cell shape (Elion 2000; Bardwell 2005).

Signal transduction in response to pheromones is mediated by the Ste11–Ste7–Fus3 MAPK module (Fig. 6.1). While *ste11* and *ste7* mutants are completely unresponsive to pheromone, yeast strains lacking the *FUS3* gene still have a modest response (Elion et al. 1991). This can be explained by the presence of a second MAPK, Kss1, which can partially replace Fus3, although Kss1 has a primary role on the filamentation pathway. *fus3 kss1* double mutants are entirely defective in pheromone response and mating.

Ste11, Ste7, and Fus3 are assembled on the scaffold protein Ste5, which keeps the three kinases in close proximity and supports signal specificity (Good et al. 2009; Malleshaiah et al. 2010). Furthermore, Ste5 localizes the module to the plasma membrane. Despite the high conservation of MAPK signaling in fungi, Ste5 is specific for budding yeast and is not present even in related species such as *Ashbya gossypii*

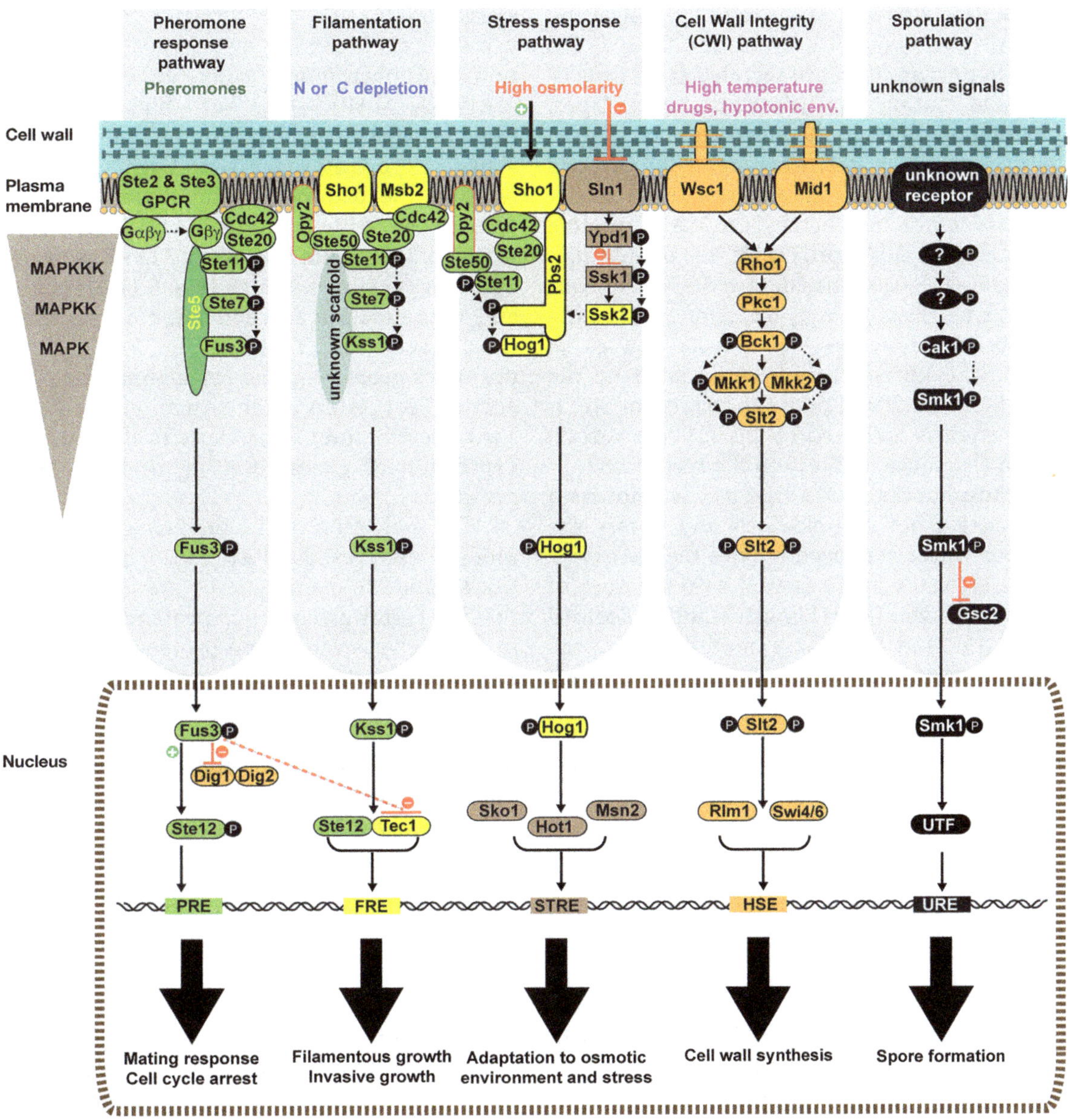

Fig. 6.1 Major MAPK pathways in yeast *S. cerevisiae*. Transduction of environmental stimuli (pheromones, nitrogen or carbon starvation, changes in osmolarity, high temperature, cell wall-disturbing drugs) by regulatory elements of different yeast MAPK modules. In each MAPK pathway, signals are received by the membrane-bound cell surface receptors and transmitted to downstream elements including trimeric G proteins, GTPases, and protein kinases. These signals are further transduced and amplified by the MAPK modules that are often docked by scaffolding proteins that are close to the plasma membrane. Phosphorylation of the final MAPK element results in the migration of the kinase protein into the nucleus, where it interacts and phosphorylates the target regulatory proteins or transcription factors. These transcription factors bind to promoters of the target genes and activate the expression of these genes. Since pheromone pathway components also play roles in filamentous growth pathway as well as osmolarity sensing, they are depicted as green elements. Green, pheromone pathway MAPK module; yellow, filamentation pathway; brown and yellow, high osmolarity pathway; orange, cell wall integrity pathway; black, sporulation pathway. PRE pheromone response element, FRE filamentation and invasion response element, STRE stress response element, HSE heat shock response element, URE unknown response element

or *Candida albicans*. At the top of the signaling pathway act, two GPCRs, Ste2 and Ste3, which bind to pheromones, are secreted from cells of opposite mating type (Bardwell 2005). Upon pheromone binding, the heterotrimeric G protein Gαβγ dissociates, whereupon the dimeric Gβγ subunit binds Ste5 and recruits it to the plasma membrane. Thereby, the MAPK module becomes in close proximity to membrane-bound PAK Ste20. The adaptor Ste50 also contributes to this process by tethering Ste11 to the membrane (Truckses et al. 2006; Wu et al. 2006). The kinase cascade is initiated by the PAK Ste20-mediated phosphorylation of Ste11. Ste20 itself is activated by the GTPase Cdc42, which also localizes Ste20 to the membrane.

Phosphorylated MAPK Fus3 is known to dissociate from scaffold Ste5 and enters the nucleus where it phosphorylates the transcriptional activator Ste12 as well as repressors of Ste12, Dig1, and Dig2 (Elion et al. 1993; Tedford et al. 1997). Upon these modifications, the expression of genes triggering cell cycle arrest and morphological changes, such as shmoo formation, is induced. Cell cycle arrest in G1 phase is mediated by the Far1 protein, an inhibitor of G1-specific cyclin-dependent kinases (CDKs) (Peter et al. 1993).

2. Filamentous Growth of *S. cerevisiae*

Under conditions of nitrogen starvation, diploid yeast cells undergo a morphological transition from budding to pseudohyphal growth (Kron 1997). Yeast cells elongate and bud in a bipolar manner, whereby mother and daughter cells remain connected after cell division. This results in chains of yeast cells displaying a unidirectional growth, which allows the organism to grow in a specific direction to forage nutrients. Haploid yeast also responds to nutrient limitations, particularly to glucose depletion, by a related morphological transition, resulting in a similar growth behavior called invasive growth.

Several components of the signal transduction pathway triggering pheromone response, including the Ste11 and Ste7 kinases, are also involved in signaling in response to nitrogen starvation (Cullen and Sprague 2012). Thus, two different incoming signals, nutrient starvation and pheromone, converge on a single MAPKKK–MAPKK pair, but a highly sophisticated network of mechanisms ensures that each signal is properly processed. A key difference between the pathways is the relevant MAPK component. Although the MAPKK Ste7 activates both Fus3 and Kss1, each MAPK activates different genetic programs (Fig. 6.1). Kss1 triggers accumulation and activation of transcriptional activator Tec1, while Ste12 alone promotes genes required for mating; the presence of Tec1 results in the formation of a Ste12–Tec1 heterodimer that specifically induces expression of genes resulting in filamentous growth.

The activation of the filamentation MAPK module involves the PAK Ste20, the GTPase Cdc42, and the membrane-bound mucin Msb2 that is a highly glycosylated protein. A decrease in Msb2 glycosylation and proteolytic cleavage activates the filamentation pathway (Cullen et al. 2004; Pitoniak et al. 2009; Yang et al. 2009). It was suggested that membrane-inserted Sho1 protein that has a well-defined function in the stress response MAPK pathway (see below) is also involved in initiating signal transduction at the plasma membrane (Cullen and Sprague 2012).

The two different signals converge on common Ste11/Ste7 module, but how do cells ensure signal specificity to receive the proper output signal? It is fundamental that proper activation of either Fus3 or Kss1 is achieved. Several mechanisms contribute to signal specificity (Schwartz and Madhani 2004; Saito 2010). The Ste5 scaffold protein plays an important role for specificity in pheromone response. In contrast, Kss1 is phosphorylated by Ste7 independently of Ste5. Currently, there is no scaffold protein identified which is associated with Kss1.

Further mechanisms for the insulation of the pheromone and filamentation pathways involve cross-pathway regulatory mechanisms. To avoid the expression of genes for filamentation upon pheromone signaling, Fus3 phosphorylates Tec1, and thereby Tec1 is degraded by ubiquitin-mediated proteolysis (Bao et al.

2004; Chou et al. 2004). The instability of Tec1 helps to ensure that genes inducing pseudohyphal growth are not activated when pheromones are present.

B. *Candida albicans*

Candida albicans is a pathogenic yeast that can cause severe systemic infections in immunocompromised patients. Being a dimorphic fungus, this organism is able to alter its morphology depending on environmental conditions. Although the morphological switch from single cell yeast to filamentous form is important for virulence of *C. albicans* (Sudbery 2011), different morphologies are required for host infection. The yeast form seems to initiate infection and more easily disseminates through the bloodstream of the host. Hyphae are more resistant to phagocytosis and hyphae formation within macrophages may kill these defense cells. Hyphae can also penetrate through epithelial layers and thereby invade specific tissues.

Two pathways acting in parallel, the cAMP/PKA and the MAPK pathway consisting of the Ste11–Hst7–Cek1/Cek2 kinase cascade, are involved in switching to filamentous growth. This pathway is involved in mating, morphogenesis, and filamentation (Monge et al. 2006). In response to environmental cues, *C. albicans* can also switch from its normal roundish yeast form (termed *white*) to an elongated form (termed *opaque*). Opaque cells are less virulent than white cells but have the ability to escape phagocytosis, and only opaque cells are mating competent being able to fuse with an opaque cell of opposite mating type. This results in tetraploid cells that can undergo a parasexual cycle.

Deletion of *CPH1* (encoding yeast transcription factor Ste12 homolog) and *HST7* (yeast Ste7 MAPKK homolog) interferes with the ability to mate (Chen et al. 2002). Similarly, a double deletion of *CEK1* and *CEK2* results in inability to mate, while single mutants are only partially defective in mating. Thus, these MAP kinases have moderately overlapping functions in the pheromone response pathway as in *S. cerevisiae*. Downstream elements of the MAPK pathway include the nuclear transcription factors such as Cph1 and Tec1. MAPK signaling is activated by pheromones and involves membrane-bound pheromone receptors and trimeric G proteins. Scaffold proteins corresponding to yeast Ste5 are not involved in signaling in *C. albicans*.

Both white and opaque cells use the Ste11–Hst7–Cek1/Cek2 pheromone response pathway. Both cell types can respond to pheromones, but the outcome of the signaling pathway differs. In opaque cells, the transcription factor Cph1 is the target of the MAP kinase. Activated Cph1 triggers the expression of mating-specific genes. In white cells, a different transcription factor, Tec1, is activated. Tec1 promotes the formation of a biofilm, which is important for the mating of opaque cells in the population (Sahni et al. 2010).

A recent study showed that the activation of MAP kinase signaling by a hyperactive form of Ste11 induces switching from white to opaque form in *Candida* (Ramirez-Zavala et al. 2013). Interestingly, pheromones do not induce this switch. Thus, there seem to be different activation mechanisms for this MAP kinase pathway in white cells, leading to different responses. A pheromone-independent process working through Cph1 induces the switching to opaque form, while pheromone induces biofilm formation via Tec1. In summary, the mating pathway in *Candida* does not only regulate mating but also filamentous growth and morphological changes, which are important for the pathogenicity of this fungus in the host.

C. Filamentous Fungi

1. *Aspergillus nidulans* and *Aspergillus fumigatus*

Pheromone response pathways have been studied in various fungi, and the MAPK module was found to be highly conserved in filamentous fungi. Among other filamentous fungi, much is known in the ascomycete *Aspergillus nidulans*. This model organism grows vegetatively by hyphal growth but has the capability to produce spores by either asexual or sexual modes (Adams et al. 1998; Pöggeler et al. 2006; Bayram and Braus 2012). Asexual spores (conidiospores) are produced in light and the presence of air, while sexual reproduction is favored in darkness and in low O_2 concentrations. In darkness, *A. nidulans* produces large closed fruiting bodies, termed cleistothecia, in which thousands of meiotically derived ascospores are produced (Bayram and Braus 2012; Dyer and O'Gorman 2012). *A. nidulans* is a homothallic fungus but nevertheless contains a pathway related to the pheromone response pathway of yeast (Paoletti et al. 2007).

Studies on this MAPK pathway were initiated with the characterization of *steC* mutants (mutants in MAPKKK AnSte11). *steC* mutants exhibit reduced growth and produce many aberrant conidiophores (Wei et al. 2003). Forced heterokaryon formation resulting in a homozygous diploid for *steC* showed that this diploid fails to form cleistothecia. Therefore, SteC is required for hyphal fusion and sexual fruiting body formation.

The MAPK AnFus3 was initially described as MpkB (Paoletti et al. 2007; Jun et al. 2011). *mpkB* deletion strains are defective in sexual development. They fail to produce cleistothecia and show similar defects as *steC* mutants in hyphal fusion, suggesting that SteC and MpkB act in a common pathway. The missing member of the MAPK module, the MAPKK MkkB (also termed AnSte7), was identified in the genome of *A. nidulans*, and characterization of the deletion phenotype revealed a phenotype similar to *steC* and *mpkB* mutants (Bayram et al. 2012). Phosphorylation of MpkB is dependent on MkkB and SteC, suggesting that these three kinases represent a MAPK module required for the sexual development of this fungus.

A Ste5 homolog-encoding gene is not present in *A. nidulans* genome, but the MAPK module is associated with a SteD (AnSte50) protein displaying high homology to yeast Ste50 (Fig. 6.2). In mutants lacking Ste50, interactions of the module components at the plasma membrane could not be detected. Localization studies using **bimolecular fluorescence complementation** (BiFC) showed that the entire MAPK module is localized to the membrane like in yeast. It was shown that the components of the module interact with each other not only at the membrane, but, in contrast to yeast, the entire module is also detected at the nuclear envelope (Fig. 6.2). Time-lapse imaging supports the model that the entire MAPK module migrates from the plasma membrane to the nuclear envelope, where it probably dissociates, because only MpkB translocates into the nucleus. In the nucleus, MpkB directly interacts with the Ste12 homolog SteA and thereby promotes sexual development. These findings suggest the presence of a docking/scaffolding protein on the nuclear envelope. This MAPK module is also localized to hyphal tips and septa, but the function of these kinases at these locations is not yet known.

Upstream signals regulating MAPK activity are still unknown. Pheromones such as oxylipins have been identified in *A. nidulans*, but whether they induce signaling through the kinase cascade is still unknown. Besides its role in fruiting body formation, the MAPK pathway is also required for secondary metabolism (Bayram et al. 2012). *mpkB* mutants are impaired in the expression of the mycotoxin sterigmatocystin gene cluster. The heterotrimeric velvet complex is a central factor for the coordination of fungal development and secondary metabolism (Bayram et al. 2008; Bayram and Braus 2012). It consists of the velvet domain proteins VeA and VelB and contains also the **LaeA protein**, which is an S-adenosyl-L-methionine (SAM)-dependent methyltransferase (Patananan et al. 2013). Strains lacking any of the genes encoding velvet components fail to produce sterigmatocystin. The LaeA protein is exhaustively reviewed in the chapter "Insight into Fungal Secondary Metabolism from Ten Years of LaeA Research."

In vitro kinase assays show that MpkB directly phosphorylates the VeA subunit of the velvet complex but not the components VelB and LaeA (Bayram et al. 2012). In the absence of MpkB, only low levels of VelB and LaeA proteins are associated with VeA, implying that VeA phosphorylation enhances formation of the velvet complex (Fig. 6.2). From these data, it was concluded that the mating signaling pathway controls secondary metabolism via modulating velvet complex assembly in the nucleus.

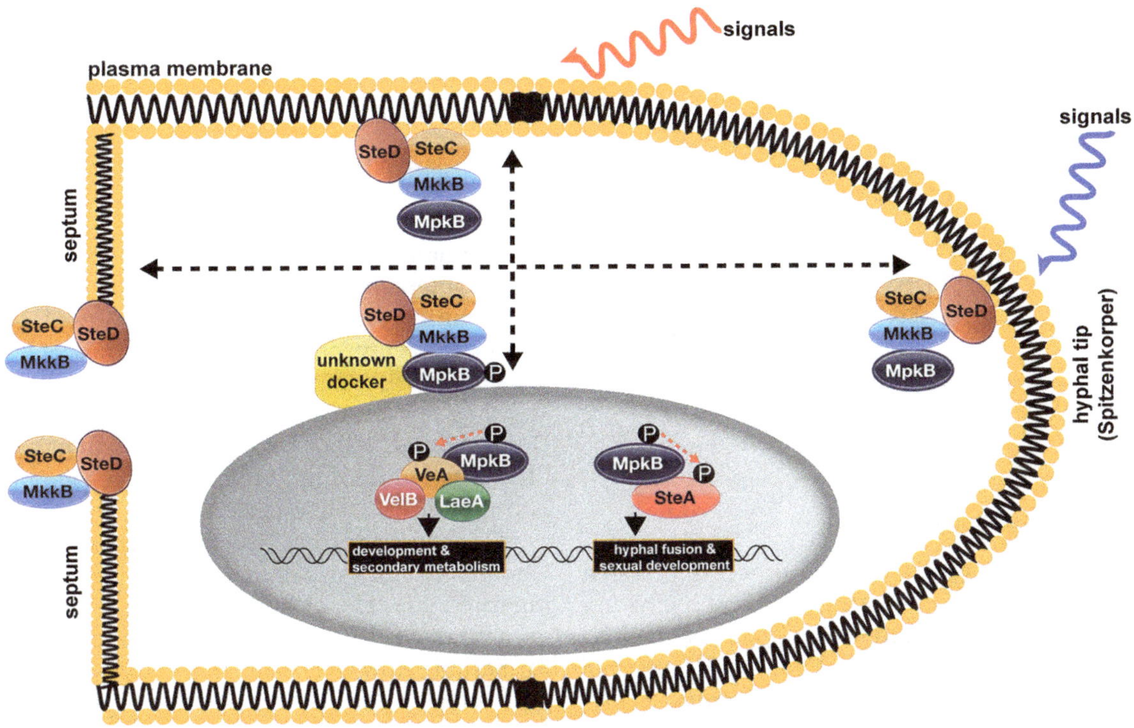

Fig. 6.2 Proposed dynamic interactions of the MAPK pheromone module in filamentous fungi based on *A. nidulans*. The pheromone response MAPK module of the filamentous fungus *A. nidulans* [SteD (AnSte50)–SteC (AnSte11)–MkkB (AnSte7)–MpkB (AnFus3)] shuttles between the plasma membrane and nuclei. In contrast to yeast, the entire MAPK module of *A. nidulans* moves from the plasma membrane to the nuclear envelope where only MpkB is imported into the nucleus. MAPK modules are also located at the hyphal tip (Spitzenkorper), plasma membrane, nuclear envelope, and septum (plural septa). There is no Ste5 scaffold homolog in *A. nidulans*, and the SteD adaptor has partially taken over the role of the Ste5 scaffold. An unknown docking protein proposed at the nuclear envelope to tether the MAPK module to the nuclear envelope for a short time, allowing the delivery of the MpkB into the nucleus. MpkB interacts and phosphorylates the VeA component of the velvet complex (VelB–VeA–LaeA) and the transcription factor AnSte12 (SteA). These molecular interactions control sexual development and secondary metabolite production in this fungus and probably play similar roles in other filamentous fungi. Dashed lines indicate possible transitions of the MAPK module within the cell. Different colored signals indicate various environmental cues such as lipid-derived pheromones or environmental stimuli

Another study showed that *mkpB* mutants also have a reduced expression of genes involved in penicillin and terrequinone A biosynthesis. Since *laeA* mRNA levels are reduced to about 50 % in *mpkB* mutants, it was proposed that the MAP kinase may affect the production of secondary metabolites by regulating *laeA* expression (Atoui et al. 2008; Jun et al. 2011).

In conclusion, the mating signaling pathway regulates secondary metabolite production in *A. nidulans*, expanding the cellular roles of this signal transduction pathway beyond the mating response. Whether this pathway plays similar roles in other *Aspergillus* species producing important secondary metabolites such as the aflatoxin producer *A. flavus* or biotechnologically relevant *A. terreus* is yet unknown. However, in *A. fumigatus*, signaling through the cell integrity pathway affects the biosynthesis of toxic secondary metabolite gliotoxin (see below).

The saprotrophic fungus *Aspergillus fumigatus* is an opportunistic pathogen which can cause severe respiratory and systemic infections in humans and animals.

Mortality of invasive aspergillosis can reach up to 80–95 % in immunocompromised patients. It was thought for long that *A. fumigatus* was an asexual organism exclusively producing conidia. A sexual life cycle of this fungus was discovered only 5 years ago (O'Gorman et al. 2009). *A. fumigatus* can produce cleistothecia that contain ascospores. In contrast to *A. nidulans*, *A. fumigatus* is a heterothallic fungus that needs cells of opposite mating type for sexual reproduction. It is conceivable that signaling pathways are involved in communication between cells in order to prepare for mating. The *A. fumigatus* genome encodes the yeast Ste20, Ste11, Ste7, Fus3, and Ste12 homologous proteins. At present, however, it is not known which MAP kinases are required for sexual development in *A. fumigatus*.

2. *Neurospora crassa*

In *N. crassa*, vegetative fusion occurs between germinated conidia (germlings) and hyphae in a mature colony. Both fusion events are dependent on a MAPK cascade related to *S. cerevisiae* pheromone response pathway. *N. crassa* MAK-2 displays highest similarities to yeast Fus3 and Kss1 MAPKs. *mak-2* disruptants display pleiotropic phenotypes including defects in hyphal fusion, female sterility, as well as morphological abnormalities (Pandey et al. 2004). Mutants of the Ste11-related MAPKKK gene NRC-1 have several similar phenotypes as *mak-2* mutants, in particular defects in hyphal fusion. MAK2 phosphorylation during early colony development is dependent on NRC1 MAPKKK. It is apparent that NRC1 and MAK2 as well as the MAPKK MEK2 represent the *N. crassa* MAP kinase cascade related to yeast pheromone pathway. Thus, a key function of the mating signaling pathway in *N. crassa* is to control cell fusion, in analogy to other fungal organisms. Downstream target of the pathway is the Ste12-related transcription factor PP1 (Leeder et al. 2013). It is currently unknown how this MAP kinase pathway is activated in this fungus.

Recently, it has been shown that factors acting upstream of the kinase cascade involve the COT1 NDR1 kinase complex (Dettmann et al. 2013). This complex contains the HYM1 scaffold protein, which also interacts with the mating kinase module and is required for its activity. Thus, this scaffold protein seems to connect the upstream kinase complex to the MAPK module.

3. *Cryptococcus neoformans*

The human pathogenic basidiomycetous fungus *Cryptococcus neoformans* can cause life-threatening meningoencephalitis, particularly in immunocompromised hosts. This dimorphic organism propagates by yeastlike budding but switches to filamentous growth during mating (Idnurm et al. 2005). In contrast to *C. albicans*, filamentation does not seem to play an important role during pathogenesis.

C. neoformans has a defined sexual cycle. Under nutrient limitation and the presence of pheromones, cells of different mating types (a or alpha) can fuse, which leads to dikaryon formation. However, cells of the alpha mating type can also differentiate by a dimorphic transition from yeastlike budding to hyphal growth and spore production. This differentiation process termed haploid fruiting of alpha cells is important for virulence. Indeed the *MATalpha* locus has been linked to virulence of this organism, because alpha cells are more virulent than a-cells (Lin and Heitman 2006; Kozubowski et al. 2009).

Several components of the MAPK module are encoded by the large *MAT* locus (Chang et al. 2000; Karos et al. 2000; Clarke et al. 2001). Thereby, divergent alleles of these genes are encoded by the *MATa* and *MATalpha* strains, including *STE20*, *STE11*, and *STE12* genes. Other genes are not encoded by the *MAT* locus and therefore are non-cell-type specific, such as Ste7 and the MAPK Cpk1 (Davidson et al. 2003). Cpk1 has about 50 % identity to *S. cerevisiae* Fus3 and Kss1 and *C. albicans* Cek1. The disruption of *CPK1* or *STE7* genes causes severe mating defects and failure to undergo haploid fruiting. A similar phenotype was observed for mating-type-specific *ste11* alpha mutants. Thus, the MAP kinase cascade Ste11–Ste7–Cpk1 is required for both efficient mating and haploid fruiting.

The core kinases of the mating signaling module are not all required for the pathogenicity of *C. neoformans* (Davidson et al. 2003). Disruptants of each of the respective genes exhibit a virulence similar to the wild-type strain in mouse models. Virulence was also tested in serotypes A and D, of which A is more virulent. Ste12 alpha was required for virulence of serotype D but not of serotype A (Yue et al. 1999). Intriguingly, the exact opposite influence on virulence was observed by the deletion of the *STE20* gene, indicating that Ste20 alpha is required for virulence of serotype A but not D (Wang et al. 2002).

To summarize, the *C. neoformans* mating MAP kinase pathway is important for cell fusion and haploid fruiting, being also involved in virulence in a complex fashion that depends on mating-type-specific alleles and serotypes.

4. Plant Pathogenic Fungi

Pheromone response pathways have been studied in various plant pathogenic fungi (Hamel et al. 2012). The basidiomycete *Ustilago maydis* causes maize smut disease, characterized by the formation of swollen tumorlike structures also named galls (Brefort et al. 2009; Dean et al. 2012). This fungus has two different forms, haploid and diploid, respectively. In axenic culture, *U. maydis* grows as a saprobic, haploid fungus that cannot infect the maize plant. Contact of the fungus with the host plant triggers mating of compatible cells and dikaryon formation. Pheromone-mediated cell fusion is required for the formation of the infectious dikaryotic form. Both cAMP and MAPK pathways are involved in the regulation of dikaryon formation. The MAPK module Ubc4–Fuz7–Ubc3 regulates the expression of pheromone-responsive genes (Banuett and Herskowitz 1994; Mayorga and Gold 1999; Muller et al. 1999). The high-mobility group transcription factor Prf1 is a target of this MAPK pathway in this fungus and binds to pheromone response elements (PREs) of the target genes (Hartmann et al. 1996). The pathway is also required for conjugation tube formation, a process that works independently of Prf1. Interestingly, the deletion of genes encoding components of this MAPK module results in defective mating, filamentous growth, and virulence.

In several plant pathogenic species including *Magnaporthe oryzae* (Xu and Hamer 1996), *Cochliobolus heterostrophus* (Lev et al. 1999), and *Colletotrichum orbiculare* (Kojima et al. 2002), disruptants of the Fus3/Kss1 orthologs fail to develop appressoria, that are the infection structures to penetrate the host tissue. Also, other plant pathogens lacking appressoria but using different colonization means, including *Fusarium oxysporum* (Di Pietro et al. 2001), *Fusarium graminearum* (Jenczmionka et al. 2003), and *Botrytis cinerea* (Doehlemann et al. 2006), fail to efficiently colonize plant tissues in the absence of Fus3/Kss1 homologs.

These studies suggest that in plant pathogenic fungi, the pheromone response MAPK module plays major roles during the invasion of the host plant and greatly contributes to pathogenicity.

III. The Stress Response MAPK Pathway

The **high-osmolarity glycerol** (HOG) MAPK pathway was identified in yeast as a pathway required for the adaptation to environments with high osmolarity (Hohmann 2002, 2009; Westfall et al. 2004). Currently, it is known that this conserved pathway modulates multiple cellular functions, in response to many kinds of stress conditions; besides osmotic stress, this pathway is also involved in cellular responses to oxidative, ultraviolet radiation, heavy metal, and heat stress. For pathogenic organisms, it is crucial to rapidly adapt to specific microenvironments in order to ensure survival within the host. Adaptation to changes in the environment such as osmolarity, body temperature, pH, limitations of nutrients, and oxidative stress occurs upon encountering reactive oxygen species (ROS)-producing immune cells. Therefore, stress response MAPK pathway participates in fungal morphogenesis and plays critical roles for virulence in human and plant pathogenic fungi such as *C. albicans*, *A. fumigatus*, *C. neoformans* and *M. oryzae*, and *F. graminearum*, respectively.

A. *Saccharomyces cerevisiae*

The central parts of the HOG pathway in *S. cerevisiae* are the MAPKK Pbs2 and the MAPK Hog1. Signaling through this pathway is activated by two branches. Both branches result in the activation of different MAPKKK, which then converges on Pbs2 that also functions as a scaffold (Posas and Saito 1997). These routes are known as Sln1 and Sho1 branches, because Sln1 and Sho1 are key components acting at the top of the HOG pathway (Fig. 6.1) (Hohmann 2002, 2009; Westfall et al. 2004).

The Sln1 branch involves a **three-component phosphorelay system:** (I) the transmembrane osmosensor Sln1, (II) the phosphotransfer protein Ypd1, and (III) the response regulator Ssk1. Under normal (isotonic) conditions, this system is constitutively active. Sln1 phosphorylates Ypd1 and from this protein, the phosphate is transferred to Ssk1. Phosphorylated Ssk1 is unable to interact with the MAPKKK Ssk2 and Ssk22, and therefore signaling does not occur. Under osmotic stress conditions, the phosphorelay is inactive, and unphosphorylated Ssk1 binds and activates Ssk2/Ssk22 and thereby initiates signaling to Hog1, via Pbs2.

The second branch also involves a transmembrane protein, Sho1, which under osmotic stress acts as sensor activating the PAK Ste20, leading to Ste11 activation. Thus, this branch shares several common components with the pheromone response/filamentation pathway (Fig. 6.1). However, stress conditions result exclusively in activation of Pbs2 but not of Ste7. Phosphorylated Hog1 enters the nucleus where it phosphorylates several osmoresponsive transcription factors such as Msn2/Msn4, Sko1, and Hot1, which mediate the upregulation of nearly 600 genes (O'Rourke and Herskowitz 2004; Westfall et al. 2004).

B. *Candida albicans*

In *C. albicans*, Hog1 phosphorylation is induced in a Pbs2-mediated manner by ionic stress upon sodium chloride addition, which triggers nuclear import of Hog1 (Monge et al. 2006). Similar to yeast, Hog1 activation results in glycerol accumulation within the cell. Upstream branches are involved in regulating MAPK activation. In addition to osmotic stress, the HOG pathway is also important for response to oxidative stress in *C. albicans*. The pathway is activated by oxidants such as hydrogen peroxide, and *pbs2* and *hog1* mutants are susceptible to oxidizing molecules including hydrogen peroxide. Similar to yeast, the upstream Sho1 branch is required for this response (Roman et al. 2005).

C. albicans Hog1 is required for oxidative stress response and for the production of chlamydospores (Alonso-Monge et al. 2003). Therefore, Hog1 has an important role for the survival of *Candida* within phagocytes by conferring resistance to oxidative stress created by these cells. In *hog1* deletion strains, intracellular levels of reactive oxygen species are highly elevated. The *hog1* mutants have reduced virulence in murine models, and mutants are more susceptible to phagocytosis during their encounter with phagocytes. A recent study addressed the question, which of the MAPK kinase pathways is important for the colonization of the mouse gut (Prieto et al. 2014)? Quantification of the fungal population of wild-type and MAPK mutants showed that *hog1* mutants have the most pronounced effect on the establishment of *C. albicans* in the gut flora. Furthermore, *hog1* mutants are defective in the adhesion to gut mucosa.

Besides stress response, the HOG pathway was also shown to be involved in morphogenesis, hyphal growth, and cell wall formation. *hog1* and *pbs2* mutants have an enhanced filamentation under non-inducing conditions. *hog1* mutants are sensitive against cell wall inhibitors such as caspofungin that inhibits beta-glucan synthesis (Staib and Morschhauser 2007; Rauceo et al. 2008).

C. Filamentous Fungi

1. *Aspergillus nidulans* and *Aspergillus fumigatus*

SakA is the *A. nidulans* homolog of Hogl. The *sakA* gene of *A. nidulans* is not essential for survival under hyperosmotic stress, but *sakA* mutants have modest growth defects on high-osmolarity medium (Kawasaki et al. 2002). Mutants are more sensitive to H_2O_2 and exposure to high temperatures (50 °C). SakA is phosphorylated and activated in conidiospores but becomes dephosphorylated during germination (Lara-Rojas et al. 2011). It was also shown that SakA physically and functionally interacts with the transcription factor AtfA. While AtfA is permanently localized in the nucleus, SakA undergoes nuclear import under osmotic and oxidative stress conditions. Furthermore, SakA and AtfA are also involved in sexual development. Deletion of both *sakA* and *atfA* genes results in an increase in the number of sexual fruiting bodies in *A. nidulans*.

The HOG pathway in *A. fumigatus* has been recently reviewed (Ma and Li 2013). In *A. fumigatus*, *sakA* mutants displayed abnormalities in spore germination (Xue et al. 2004). Under nitrogen limitation conditions, spore germination increases in *sakA* mutants compared to wild-type strains. Germination occurs regardless of nitrogen availability in the medium suggesting that the SakA pathway regulates spore germination in response to nitrogen sources. *A. fumigatus* SakA is essential for surviving through osmotic stress, a property that is not shared with *A. nidulans* SakA. In *A. fumigatus*, SakA is also involved in response to oxidative stress because *sakA* mutants are sensitive to oxidants. However, virulence of *sakA* strains has not yet been tested in any infection system. Like in *C. albicans*, SakA of *A. fumigatus* is involved in morphogenesis. *sakA* mutants are retarded in growth and this phenotype was alleviated by H_2O_2 treatment (Du et al. 2006). *A. fumigatus* SakA regulates the transcription of DprA and DprB, factors required for oxidative stress response and protection from killing by phagocytes (Wong Sak Hoi et al. 2011). *sakA* deletion mutants have an increased sensitivity against antifungal agents amphotericin B and itraconazole (Kim et al. 2012). This is in contrast to other pathogens in which *hog1* mutants have an increased resistance to antifungal drugs (Kojima et al. 2004).

A. nidulans has a further MAPK with homology to SakA/Hogl, MpkC, which is conserved in aspergilli (May 2007). Deletion of this gene in *A. nidulans* or *A. fumigatus* does not result in a visible phenotype under normal growth conditions, and mutants are not sensitive to osmotic or oxidative stress (Reyes et al. 2006). It was proposed that this may be due to the overlapping functions of SakA and MpkC. Evidence for the involvement of this MAPK in stress response is provided by showing that MpkC is phosphorylated by PbsB when MpkC is overexpressed (Furukawa et al. 2005).

2. *Neurospora crassa*

In this fungus, OS-2 (osmo-sensitive) MAPK, encoded by the *os-2* gene, is homologous to yeast Hogl protein. OS-2 MAPK can replace the function of Hogl in yeast. Similar to other *hogl* mutants of fungi, *N. crassa os-2* mutants are highly sensitive to high osmolarity and exhibit resistance to fungicides such as phenylpyrrole (Zhang et al. 2002). Moreover, OS-2 also regulates the circadian rhythm by controlling the preparation of fungal cells against hyperosmotic stress and desiccation at sunrise (Vitalini et al. 2007; Lamb et al. 2012). OS-2-responsive transcription factor ASL-1 is required for the rhythmic expression of OS-pathway responsive transcripts, including catalase and glycerol dehydrogenase-encoding genes *cat-1*, *gcy-1*, and *gcy-3*, respectively (Lamb et al. 2012).

3. *Cryptococcus neoformans*

The HOG pathway has an important role for stress response, drug resistance, and virulence of *C. neoformans* (Kozubowski et al. 2009). Similar to *Candida albicans*, *C. neoformans hogl* and *pbs2* mutants are less virulent than wild-type strains, and *hogl* mutants are sensitive to reactive oxygen species. *HOG1* genes were found to be identical in different serotypes, but signaling was found to differ among serotypes A and D (Bahn et al. 2005). In the less virulent serotype D, the HOG pathway seems to function similar as in budding yeast. However, in the more virulent serotype A, which is highly resistant to osmotic shock, the pathway is adapted to regulate virulence. In striking contrast to yeast and most other fungi, Hog1 is dephosphorylated upon osmotic stress in serotype A, and the molecular mechanism for this difference in Hog1 regulation is unknown. It is also still unclear how Hog1 is regulated, since no homologs of Sln1 and Sho1 were identified in *C. neoformans*. Hybrid-sensor kinases Tco1/2, Ypd1, and Ssk1 are thought to be upstream components of the HOG pathway and sense a variety of different signals (Bahn et al. 2006).

4. Plant Pathogenic Fungi

Homologs of Hog1 MAPK have not been as extensively studied in plant pathogenic fungi as the pheromone response MAPK pathway. *M. oryzae* MAPK MoOsm1 is the first Hog1 homolog from plant pathogenic fungi, shown to be required for osmotic stress response and morphological development but not needed for appressorium function or virulence (Dixon et al. 1999).

F. graminearum is one of the most notorious pathogens of cereals and produces mycotoxins zearalenone and deoxynivalenol, which are virulence factors in this pathogen. ΔFgOS-2 strains show reduced virulence in maize and wheat (Van Thuat et al. 2012). Germination of the mutant strains on osmotic media is severely impaired and germ tubes are swollen with multiple nuclei. Furthermore, the ΔFgOS-2 strain cannot produce perithecia and ascospores. Similar to *A. nidulans* and *S. pombe*, FgAtf1 transcription factor interacts with FgOS-2 under osmotic stress conditions, and *Fgatf1* mutants show increased sensitivity to osmotic stress and resistance to oxidative stress (Van Nguyen et al. 2013).

BcSAK1, the Hog1 homolog in *Botrytis cinerea*, is phosphorylated under osmotic and oxidative stress conditions. The mutant strain is unable to produce conidia and impaired in vegetative growth showing increased sclerotia production. Furthermore, the *Bcsak1* mutant cannot penetrate into undamaged plant tissues and therefore has reduced pathogenicity in maize and wheat (Segmuller et al. 2007). These findings and other studies (Kojima et al. 2004; Viaud et al. 2006) suggest that the signaling pathways involving Hog1 homologs do not always contribute to the virulence of the plant pathogenic fungi but play important roles during the development of these fungi.

IV. The Cell Wall Integrity (CWI) Pathway

The fungal cell wall has a crucial role in protecting cells from changes in external osmotic conditions, for morphogenesis and for cell expansion during growth. The cell wall integrity pathway serves to monitor the state of the cell wall. Cell wall stress activates this signaling pathway, which controls cell wall biosynthetic enzymes, cell wall-related genes, and polarization of the actin cytoskeleton. The CWI pathway is not only activated by cell wall damage but also by other types of stresses, such as high temperature and oxidative stress. Oxidative stress therefore signals through both the HOG and CWI pathways.

A. *Saccharomyces cerevisiae*

First components of this pathway were identified in yeast by the analysis of temperature-sensitive mutants. Such conditional mutants lysed at 37 °C, the restrictive temperature, but this phenotype could be suppressed by osmotic stabilization with sorbitol (Torres et al. 1991). Besides their temperature-sensitive phenotype, mutants in the CWI pathway are further characterized by sensitivity to cell antagonists such as calcofluor white and caffeine and to cell wall-degrading enzyme such as Zymolyase (Levin 2005). The central part of the cell integrity pathway is the MAPK module consisting of Bck1 and the pair of redundant MAPKKs Mkk1/2 and Slt2 (also known as Mpk1) (Levin

2005, 2011; Rodicio and Heinisch 2010). Bck1 is activated by protein kinase C (Pkc1) which itself is activated by the small G protein Rho1. CWI signaling is initiated at the plasma membrane through multiple cell surface sensors that recruit the factors such as guanine nucleotide exchange factors for Rho1 to the plasma membrane. Two highly mannosylated membrane receptors Wsc1 and Mid2 are responsible for sensing cell wall stress (Jendretzki et al. 2011).

MAPK signaling influences the nuclear localization of Slt2. It is localized in the nucleus under non-stress conditions but relocates to the cytoplasm upon cell wall stress (van Drogen and Peter 2002). Low levels of heat shock protein (Hsp90) suppress Slt2 kinase activity since Slt2 acts as a substrate for Hsp90 during heat shock (Truman et al. 2007). Nuclear targets of Slt2 include the transcription factors Swi4/Swi6 (Kim et al. 2008) and Rlm1 (Jung et al. 2002). Phosphorylation of Rlm1 by Slt2 occurs in the transcriptional activation domain leading to the enhanced activity of Rlm1, which regulates the expression of many cell wall genes. The MAPK Slt2 is localized to the site of polarized growth by interacting with the protein Spa2. A constitutive shuttling of Slt2 between this site and the nucleus is observed. Also Mkk1 and Mkk2, which are cytoplasmic proteins, are localized to the cell polarity site in a Spa2-dependent manner. Thus, Spa2 is a scaffold protein for at least part of the MAPK module, serving to focus kinase activity at the site of polarized growth. Bck1 is also a cytoplasmic but was not detected at this site.

B. *Candida albicans*

Mkc1 is the homolog of the yeast Slt2. Functional characterization of *mkc1* mutants showed a defect in cell integrity and cell wall formation, manifested by a reduced deposition of surface polymer mannan (Navarro-Garcia et al. 1995). As in *S. cerevisiae*, *C. albicans* mutants in the CWI pathway are sensitive against cell wall-interfering drugs. Mkc1 was found to be activated under different stress conditions, including oxidative and osmotic stress, as well as by antifungal drugs. Activation by osmotic stress is dependent on a functional HOG pathway. Both the HOG and the CWI pathways are activated by osmotic and oxidative stress, but the pheromone MAPK Cek1 is deactivated (Navarro-Garcia et al. 2005). *mkc1* mutants are able to proliferate at 37 °C but not at 42 °C and have a reduced virulence in a mouse model for systemic infections (Diez-Orejas et al. 1997). Reduction in virulence is likely due to reduced response against defense mechanisms of the host, such as reactive oxygen production of immune cells. For *C. albicans*, the reduction in virulence of mutants in the CWI pathway is not as dramatic as found for other signal transduction pathways. Thus, the CWI pathway seems to be required for proliferation inside the host organisms.

C. Filamentous Fungi

1. *Aspergillus nidulans* and *Aspergillus fumigatus*

MpkA is the MAPK related to yeast Slt2 (May 2007). An *mpkA* deletion mutant is viable, but consistent with its role in cell wall integrity, optimal growth of the strain depends on high-osmolarity medium. Furthermore, deletion strains display multiple additional phenotypes that include abnormal colony morphology and hyphal branching, swelling of hyphal tips, and defects in germ tube formation of conidia (Bussink and Osmani 1999). Two homologs of yeast Wsc1, WscA and WscB, have been studied in *A. nidulans*. Both proteins are *N*- and *O*-mannosylated and found in the cell wall and plasma membrane. In addition to swollen hyphae under hypoosmotic conditions, *wscA* mutants have reduced colony size and conidiation (Futagami et al. 2011). *wscA* and *wscB* disruptants show drastic changes in the expression of two α-(1,3)-glucan synthase-encoding genes *agsA* and *agsB*, respectively. As many fungicides interfere with cell wall function, the CWI pathway represents a potential target for novel antifungal drugs.

In *A. fumigatus*, deletion mutants of each of the genes *bck1*, *mkk2*, and *mpkA*, encoding orthologs of the yeast kinase module, are highly sensitive to compounds affecting cell wall function such as calcofluor white or the cell wall-degrading enzyme glucanex (Valiante et al. 2008, 2009). Mutants also have abnormal morphology with respect to hyphal structure and hyphal branching. Cell wall stress induces *mpkA* transcription and phosphorylation of

MpkA, which is dependent on both Bck1 and Mkk2, showing that they act upstream of MpkA. The cellular role of the CWI integrity signal transduction pathway was further investigated by using a transcriptomic analysis of *mpkA* deletion strains (Jain et al. 2011). As expected, genes involved in cell wall remodeling and oxidative stress response were identified. These data further suggest that MpkA functions in adaptation to iron starvation. Furthermore, in *mpkA* mutants, the transcript levels of the gliotoxin gene cluster as well as gliotoxin production are reduced. MpkA-related genes were also found to contribute to virulence in other human pathogenic fungi as well as in plant pathogens (Hamel et al. 2012).

2. *Neurospora crassa*

The deletion of *wsc1* encoding the CWI receptor causes sensitivity against several cell wall-damaging chemicals including caspofungin and calcofluor white, and mutants display strong deformations during asexual sporulation in *N. crassa* (Maddi et al. 2012). As a further downstream element, the *N. crassa* MIK1/MEK1/MAK1 pathway controls colony morphology, hyphal growth, and sexual development (Maerz et al. 2008; Park et al. 2008). MAK1 is the MAPK of the CWI pathway. *mak1* deletion affects vegetative growth and leads to alterations in transcription of more than 500 genes. Interestingly, many of these genes were previously identified as **clock-controlled genes**, revealing a connection between MAPK signaling and the circadian clock (Bennett et al. 2013). Moreover, MAK1 is also involved in cell–cell signaling and hyphal anastomosis because *mak1* mutants are defective in cell–cell fusion and female sterile. In contrast, *wsc1* mutants still undergo sexual development, suggesting a redundant function for receptor WSC1 in signaling through the MAK1 protein.

3. *Cryptococcus neoformans*

MAPK Mpk1 corresponds to Slt2 of yeast in the cell integrity pathway. *mpk1* deletion strains are inviable at 37 °C, but osmotic stabilization of the medium restores growth at this temperature (Kraus et al. 2003). *mpk1* mutants are less virulent in mouse models than the wild-type strains. Phosphorylation of Mpk1 was shown to be induced by antifungal drugs affecting cell wall synthesis. Furthermore, Mpk1 is also phosphorylated in response to thermal stress. Pkc1 acts through Mpk1 to maintain cell integrity but signals independently of Mpk1 in response to oxidative stress (Gerik et al. 2008). *pkc1* mutants have defects in capsule formation and melanin production, both of which are relevant for virulence, but *bck1* mutants do not display this drastic phenotype (Gerik et al. 2005). A recent study showed that two of the three Rho GTPases are required for the CWI pathway in *C. neoformans*. Mpk1 is constitutively phosphorylated in the absence of Rho1 and Rho10. Bck1 and Mkk1, which are two downstream kinases of PKC pathway in this fungus, are further required for the heat shock-induced Mpk1 phosphorylation (Lam et al. 2013).

4. Plant Pathogenic Fungi

The CWI pathway is involved in pathogenicity strategies of plant pathogenic fungi. The deletion of Slt2 homologs in plant pathogenic fungi results in avirulent strains or drastically reduced virulence (Hamel et al. 2012). The studies with the Slt2 homolog of *M. oryzae* suggest that it is required for differentiation and construction of the appressorium wall, because mutants of *slt2* are nonpathogenic and sensitive to cell wall-degrading enzymes (Xu et al. 1998). Furthermore, *slt2* mutants are unable to accumulate the α-(1,3)-glucans that fortify the chitin against plant hydrolases (Fujikawa et al. 2009). Several downstream transcription factors that genetically interact with Slt2 have been identified. These Rlm1 and Swi6 homologs of yeast in *M. oryzae* play similar roles to the Slt2 homolog with defects in virulence.

In *F. graminearum*, the Mkk1 homolog FgMkk1 plays essential roles in the regulation of hyphal growth, pigmentation, conidiation, and secondary metabolite production (Yun et al. 2013). *Fgmkk1* mutants show increased sensitivity to cell wall-damaging agents. FgMkk1 positively regulates MAPK homologs FgSlt2 and FgOS-2, which influences both cell wall integrity and the HOG pathway. Characterization of Rho family GTPases in *F. graminearum* showed that FgRho2 and Rho4 are involved in cell wall integrity (Zhang et al. 2013).

Bck1–Mkk1–Mpk1 pathway of *U. maydis* is required for the exit of the fungal cells from G2 phase, in contrast to yeast where the same pathway is involved in arresting cell cycle in G2 phase in response to cell wall stress (Carbo and Perez-Martin 2010). However, the influence of this pathway in the pathogenicity of *U. maydis* in maize plant remains to be shown.

Overall these and other studies in several plant pathogens indicate that the CWI pathway is an integral part of fungal virulence due to its importance in the protection against cell wall-degrading enzymes of host plants.

V. Spore Morphogenesis in *Saccharomyces cerevisiae*

Sporulation in *S. cerevisiae* is coupled to meiosis resulting in a tetrad containing four haploid ascospores. Spore morphogenesis is controlled by the meiosis-specific MAPK Smk1. This kinase is required for multiple steps in yeast spore morphogenesis, particularly for the formation of the outer spore wall layer (Krisak et al. 1994; Wagner et al. 1999). The *SMK1* gene is not expressed during vegetative growth, but its transcription is induced during meiotic nuclear divisions before ascospore enclosure. Smk1 contains a typical TxY motif in its activation loop but is not phosphorylated by a canonical MAPKK. Instead the protein is phosphorylated by the CDK-activating kinase Cak1 on the threonine residue of the activation loop (Whinston et al. 2013). Furthermore, autophosphorylation on the tyrosine residue occurs and this requires the meiosis-specific protein Ssp2.

β-Glucan synthase Gsc2 was identified as a Smk1 substrate in a screen for Smk1-associated proteins (Huang et al. 2005). Gsc2, which is required for early events in cell wall biosynthesis, is negatively affected by Smk1 activity. It was proposed that Smk1 is required to inhibit this enzyme in order to proceed with late stages of spore wall formation.

Little is known about *SMK1* homologs in other fungi. The MAPK *CSK1* gene of *C. albicans* isolated by PCR amplification from genomic DNA shows high homology to *SMK1* (Chen et al. 2000). A homologous gene from the phytopathogenic fungus *Sclerotinia sclerotiorum* is required for the development of sclerotia, and Smk1 expression and enzymatic activity were enhanced during sclerotia formation (Chen et al. 2004). Smk1 homologs have not been identified in other fungal genomes, including those from *Aspergillus*, *Penicillium*, *Neurospora*, and *Fusarium*.

VI. Conclusions

Since the discovery of the first MAPK cascade in budding yeast, the importance of these signaling pathways has been demonstrated for eukaryotes from yeast to mammals. Initial findings could be applied to study such pathways in mammals where they have fundamental roles in diseases like cancer. The last few years have also provided many new insights into MAPK pathways in filamentous fungi. Of particular significance, recent studies showing that the MAPK modules do not only assemble at the plasma membrane but are able to move to the nuclear envelope challenge the common concept on signal transduction in general. Furthermore, fungal MAPK systems are involved in chemotropic interactions showing oscillations during cell–cell communications (Fleissner et al. 2009). The discovery of novel scaffolds or docks (Fig. 6.2) at the nuclear envelope will open up a new dimension for the understanding of the fungal MAPK complexes at the molecular and functional level.

Filamentous fungi produce numerous secondary metabolites ranging from beneficial antibiotics, antivirals, and antitumor agents to toxic mycotoxins. Besides common roles in sexual development, morphogenesis, and stress response pathways, MAPKs are also involved in regulating secondary metabolism. Understanding the regulation of secondary metabolites by the MAPK pathways will help to use the potential of secondary metabolites in biotechnology and the biomedical field and to develop preventative measures against mycotoxins. MAPK pathways are also important for virulence in diverse pathogenic fungi. A better understanding of MAP kinase signaling should also enable the development of novel antifungal strategies to counteract human and plant pathogenic fungi.

Acknowledgments We would like to thank Prof. Dr. Jesus Aguirre (Universidad Nacional Autónoma de México, Mexico City) for constructive comments and corrections and Dr. Gary Jones and Betim Karahoda (Maynooth University, Ireland) for critical reading of the manuscript. Research of the authors was supported by a grant of the Deutsche Forschungsgemeinschaft (DFG) IR 36/4-1 to SI and a research grant from Science Foundation Ireland (SFI) under its Career Development Award (CDA) program in 2013 with grant number 13/CDA/2142 to OB. Özlem Sarikaya-Bayram has been supported by the Irish Research Council (IRC) Postdoctoral Fellowship (GOIPD/2014/178).

References

Adams TH, Wieser JK, Yu JH (1998) Asexual sporulation in *Aspergillus nidulans*. Microbiol Mol Biol Rev 62:35–54

Alonso-Monge R, Navarro-Garcia F, Roman E, Negredo AI, Eisman B, Nombela C, Pla J (2003) The Hog1 mitogen-activated protein kinase is essential in the oxidative stress response and chlamydospore formation in *Candida albicans*. Eukaryot Cell 2:351–361

Atoui A, Bao D, Kaur N, Grayburn WS, Calvo AM (2008) *Aspergillus nidulans* natural product biosynthesis is regulated by *mpkB*, a putative pheromone response mitogen-activated protein kinase. Appl Environ Microbiol 74:3596–3600

Bahn YS, Kojima K, Cox GM, Heitman J (2005) Specialization of the HOG pathway and its impact on differentiation and virulence of *Cryptococcus neoformans*. Mol Biol Cell 16:2285–2300

Bahn YS, Kojima K, Cox GM, Heitman J (2006) A unique fungal two-component system regulates stress responses, drug sensitivity, sexual development, and virulence of *Cryptococcus neoformans*. Mol Biol Cell 17:3122–3135

Banuett F, Herskowitz I (1994) Identification of fuz7, a *Ustilago maydis* MEK/MAPKK homolog required for a-locus-dependent and -independent steps in the fungal life cycle. Genes Dev 8:1367–1378

Bao MZ, Schwartz MA, Cantin GT, Yates JR 3rd, Madhani HD (2004) Pheromone-dependent destruction of the Tec1 transcription factor is required for MAP kinase signaling specificity in yeast. Cell 119:991–1000

Bardwell L (2005) A walk-through of the yeast mating pheromone response pathway. Peptides 26:339–350

Bayram O, Krappmann S, Ni M, Bok JW, Helmstaedt K, Valerius O, Braus-Stromeyer S, Kwon NJ, Keller NP, Yu JH, Braus GH (2008) VelB/VeA/LaeA complex coordinates light signal with fungal development and secondary metabolism. Science 320:1504–1506

Bayram O, Bayram OS, Ahmed YL, Maruyama J, Valerius O, Rizzoli SO, Ficner R, Irniger S, Braus GH (2012) The *Aspergillus nidulans* MAPK module AnSte11-Ste50-Ste7-Fus3 controls development and secondary metabolism. PLoS Genet 8: e1002816

Bayram O, Braus GH (2012) Coordination of secondary metabolism and development in fungi: the velvet family of regulatory proteins. FEMS Microbiol Rev 36:1–24

Bennett LD, Beremand P, Thomas TL, Bell-Pedersen D (2013) Circadian activation of the mitogen-activated protein kinase MAK-1 facilitates rhythms in clock-controlled genes in *Neurospora crassa*. Eukaryot Cell 12:59–69

Brefort T, Doehlemann G, Mendoza-Mendoza A, Reissmann S, Djamei A, Kahmann R (2009) *Ustilago maydis* as a pathogen. Annu Rev Phytopathol 47:423–445

Bussink HJ, Osmani SA (1999) A mitogen-activated protein kinase (MPKA) is involved in polarized growth in the filamentous fungus, *Aspergillus nidulans*. FEMS Microbiol Lett 173:117–125

Carbo N, Perez-Martin J (2010) Activation of the cell wall integrity pathway promotes escape from G2 in the fungus *Ustilago maydis*. PLoS Genet 6: e1001009

Chang YC, Wickes BL, Miller GF, Penoyer LA, Kwon-Chung KJ (2000) *Cryptococcus neoformans* STE12alpha regulates virulence but is not essential for mating. J Exp Med 191:871–882

Chen C, Harel A, Gorovoits R, Yarden O, Dickman MB (2004) MAPK regulation of sclerotial development in *Sclerotinia sclerotiorum* is linked with pH and cAMP sensing. Mol Plant Microbe Interact 17:404–413

Chen J, Ni J, Chen JY (2000) Molecular cloning of MAPK gene family using degenerate PCR. Sheng Wu Hua Xue Yu Sheng Wu Wu Li Xue Bao (Shanghai) 32:305–311

Chen J, Chen J, Lane S, Liu H (2002) A conserved mitogen-activated protein kinase pathway is required for mating in *Candida albicans*. Mol Microbiol 46:1335–1344

Chou S, Huang L, Liu H (2004) Fus3-regulated Tec1 degradation through SCFCdc4 determines MAPK signaling specificity during mating in yeast. Cell 119:981–990

Clarke DL, Woodlee GL, McClelland CM, Seymour TS, Wickes BL (2001) The *Cryptococcus neoformans* STE11alpha gene is similar to other fungal mitogen-activated protein kinase kinase kinase (MAPKKK) genes but is mating type specific. Mol Microbiol 40:200–213

Cullen PJ, Sabbagh W Jr, Graham E, Irick MM, van Olden EK, Neal C, Delrow J, Bardwell L, Sprague GF Jr (2004) A signaling mucin at the head of the Cdc42- and MAPK-dependent filamentous growth pathway in yeast. Genes Dev 18:1695–1708

Cullen PJ, Sprague GF Jr (2012) The regulation of filamentous growth in yeast. Genetics 190:23–49

Davidson RC, Nichols CB, Cox GM, Perfect JR, Heitman J (2003) A MAP kinase cascade composed of cell type specific and non-specific elements controls mating and differentiation of the fungal pathogen *Cryptococcus neoformans*. Mol Microbiol 49:469–485

Dean R, Van Kan JA, Pretorius ZA, Hammond-Kosack KE, Di Pietro A, Spanu PD, Rudd JJ, Dickman M, Kahmann R, Ellis J, Foster GD (2012) The top 10 fungal pathogens in molecular plant pathology. Mol Plant Pathol 13:414–430

Dettmann A, Illgen J, Marz S, Schurg T, Fleissner A, Seiler S (2013) The NDR kinase scaffold HYM1/MO25 is essential for MAK2 map kinase signaling in *Neurospora crassa*. PLoS Genet 8:e1002950

Di Pietro A, Garcia-MacEira FI, Meglecz E, Roncero MI (2001) A MAP kinase of the vascular wilt fungus *Fusarium oxysporum* is essential for root penetration and pathogenesis. Mol Microbiol 39:1140–1152

Diez-Orejas R, Molero G, Navarro-Garcia F, Pla J, Nombela C, Sanchez-Perez M (1997) Reduced virulence of *Candida albicans* MKC1 mutants: a role for mitogen-activated protein kinase in pathogenesis. Infect Immun 65:833–837

Dixon KP, Xu JR, Smirnoff N, Talbot NJ (1999) Independent signaling pathways regulate cellular turgor during hyperosmotic stress and appressorium-mediated plant infection by *Magnaporthe grisea*. Plant Cell 11:2045–2058

Doehlemann G, Berndt P, Hahn M (2006) Different signalling pathways involving a Galpha protein, cAMP and a MAP kinase control germination of *Botrytis cinerea* conidia. Mol Microbiol 59:821–835

Du C, Sarfati J, Latge JP, Calderone R (2006) The role of the sakA (Hog1) and tcsB (sln1) genes in the oxidant adaptation of *Aspergillus fumigatus*. Med Mycol 44:211–218

Dyer PS, O'Gorman CM (2012) Sexual development and cryptic sexuality in fungi: insights from *Aspergillus* species. FEMS Microbiol Rev 36:165–192

Elion EA, Brill JA, Fink GR (1991) Functional redundancy in the yeast cell cycle: FUS3 and KSS1 have both overlapping and unique functions. Cold Spring Harb Symp Quant Biol 56:41–49

Elion EA, Satterberg B, Kranz JE (1993) FUS3 phosphorylates multiple components of the mating signal transduction cascade: evidence for STE12 and FAR1. Mol Biol Cell 4:495–510

Elion EA (2000) Pheromone response, mating and cell biology. Curr Opin Microbiol 3:573–581

Fleissner A, Leeder AC, Roca MG, Read ND, Glass NL (2009) Oscillatory recruitment of signaling proteins to cell tips promotes coordinated behavior during cell fusion. Proc Natl Acad Sci USA 106:19387–19392

Fujikawa T, Kuga Y, Yano S, Yoshimi A, Tachiki T, Abe K, Nishimura M (2009) Dynamics of cell wall components of *Magnaporthe grisea* during infectious structure development. Mol Microbiol 73:553–570

Furukawa K, Hoshi Y, Maeda T, Nakajima T, Abe K (2005) *Aspergillus nidulans* HOG pathway is activated only by two-component signalling pathway in response to osmotic stress. Mol Microbiol 56:1246–1261

Futagami T, Nakao S, Kido Y, Oka T, Kajiwara Y, Takashita H, Omori T, Furukawa K, Goto M (2011) Putative stress sensors WscA and WscB are involved in hypo-osmotic and acidic pH stress tolerance in *Aspergillus nidulans*. Eukaryot Cell 10:1504–1515

Gerik KJ, Donlin MJ, Soto CE, Banks AM, Banks IR, Maligie MA, Selitrennikoff CP, Lodge JK (2005) Cell wall integrity is dependent on the PKC1 signal transduction pathway in *Cryptococcus neoformans*. Mol Microbiol 58:393–408

Gerik KJ, Bhimireddy SR, Ryerse JS, Specht CA, Lodge JK (2008) PKC1 is essential for protection against both oxidative and nitrosative stresses, cell integrity, and normal manifestation of virulence factors in the pathogenic fungus *Cryptococcus neoformans*. Eukaryot Cell 7:1685–1698

Good M, Tang G, Singleton J, Remenyi A, Lim WA (2009) The Ste5 scaffold directs mating signaling by catalytically unlocking the Fus3 MAP kinase for activation. Cell 136:1085–1097

Good MC, Zalatan JG, Lim WA (2011) Scaffold proteins: hubs for controlling the flow of cellular information. Science 332:680–686

Hamel LP, Nicole MC, Duplessis S, Ellis BE (2012) Mitogen-activated protein kinase signaling in plant-interacting fungi: distinct messages from conserved messengers. Plant Cell 24:1327–1351

Hartmann HA, Kahmann R, Bolker M (1996) The pheromone response factor coordinates filamentous growth and pathogenicity in *Ustilago maydis*. EMBO J 15:1632–1641

Herskowitz I (1995) MAP kinase pathways in yeast: for mating and more. Cell 80:187–197
Hohmann S (2002) Osmotic stress signaling and osmoadaptation in yeasts. Microbiol Mol Biol Rev 66:300–372
Hohmann S (2009) Control of high osmolarity signalling in the yeast *Saccharomyces cerevisiae*. FEBS Lett 583:4025–4029
Huang LS, Doherty HK, Herskowitz I (2005) The Smk1p MAP kinase negatively regulates Gsc2p, a 1,3-beta-glucan synthase, during spore wall morphogenesis in *Saccharomyces cerevisiae*. Proc Natl Acad Sci USA 102:12431–12436
Idnurm A, Bahn YS, Nielsen K, Lin X, Fraser JA, Heitman J (2005) Deciphering the model pathogenic fungus *Cryptococcus neoformans*. Nat Rev Microbiol 3:753–764
Jain R, Valiante V, Remme N, Docimo T, Heinekamp T, Hertweck C, Gershenzon J, Haas H, Brakhage AA (2011) The MAP kinase MpkA controls cell wall integrity, oxidative stress response, gliotoxin production and iron adaptation in *Aspergillus fumigatus*. Mol Microbiol 82:39–53
Jenczmionka NJ, Maier FJ, Losch AP, Schafer W (2003) Mating, conidiation and pathogenicity of *Fusarium graminearum*, the main causal agent of the head-blight disease of wheat, are regulated by the MAP kinase gpmk1. Curr Genet 43:87–95
Jendretzki A, Wittland J, Wilk S, Straede A, Heinisch JJ (2011) How do I begin? Sensing extracellular stress to maintain yeast cell wall integrity. Eur J Cell Biol 90:740–744
Jun SC, Lee SJ, Park HJ, Kang JY, Leem YE, Yang TH, Chang MH, Kim JM, Jang SH, Kim HG, Han DM, Chae KS, Jahng KY (2011) The MpkB MAP kinase plays a role in post-karyogamy processes as well as in hyphal anastomosis during sexual development in *Aspergillus nidulans*. J Microbiol 49:418–430
Jung US, Sobering AK, Romeo MJ, Levin DE (2002) Regulation of the yeast Rlm1 transcription factor by the Mpk1 cell wall integrity MAP kinase. Mol Microbiol 46:781–789
Karos M, Chang YC, McClelland CM, Clarke DL, Fu J, Wickes BL, Kwon-Chung KJ (2000) Mapping of the *Cryptococcus neoformans* MATalpha locus: presence of mating type-specific mitogen-activated protein kinase cascade homologs. J Bacteriol 182:6222–6227
Kawasaki L, Sanchez O, Shiozaki K, Aguirre J (2002) SakA MAP kinase is involved in stress signal transduction, sexual development and spore viability in *Aspergillus nidulans*. Mol Microbiol 45:1153–1163
Kim JH, Chan KL, Faria NC, Martins Mde L, Campbell BC (2012) Targeting the oxidative stress response system of fungi with redox-potent chemosensitizing agents. Front Microbiol 3:88
Kim KY, Truman AW, Levin DE (2008) Yeast Mpk1 mitogen-activated protein kinase activates transcription through Swi4/Swi6 by a noncatalytic mechanism that requires upstream signal. Mol Cell Biol 28:2579–2589
Kojima K, Kikuchi T, Takano Y, Oshiro E, Okuno T (2002) The mitogen-activated protein kinase gene MAF1 is essential for the early differentiation phase of appressorium formation in *Colletotrichum lagenarium*. Mol Plant Microbe Interact 15:1268–1276
Kojima K, Takano Y, Yoshimi A, Tanaka C, Kikuchi T, Okuno T (2004) Fungicide activity through activation of a fungal signalling pathway. Mol Microbiol 53:1785–1796
Kozubowski L, Lee SC, Heitman J (2009) Signalling pathways in the pathogenesis of Cryptococcus. Cell Microbiol 11:370–380
Kraus PR, Fox DS, Cox GM, Heitman J (2003) The *Cryptococcus neoformans* MAP kinase Mpk1 regulates cell integrity in response to antifungal drugs and loss of calcineurin function. Mol Microbiol 48:1377–1387
Krisak L, Strich R, Winters RS, Hall JP, Mallory MJ, Kreitzer D, Tuan RS, Winter E (1994) SMK1, a developmentally regulated MAP kinase, is required for spore wall assembly in *Saccharomyces cerevisiae*. Genes Dev 8:2151–2161
Kron SJ (1997) Filamentous growth in budding yeast. Trends Microbiol 5:450–454
Lam WC, Gerik KJ, Lodge JK (2013) Role of Cryptococcus neoformans Rho1 GTPases in the PKC1 signaling pathway in response to thermal stress. Eukaryot Cell 12:118–131
Lamb TM, Finch KE, Bell-Pedersen D (2012) The Neurospora crassa OS MAPK pathway-activated transcription factor ASL-1 contributes to circadian rhythms in pathway responsive clock-controlled genes. Fungal Genet Biol 49:180–188
Lara-Rojas F, Sanchez O, Kawasaki L, Aguirre J (2011) Aspergillus nidulans transcription factor AtfA interacts with the MAPK SakA to regulate general stress responses, development and spore functions. Mol Microbiol 80:436–454
Leeder AC, Jonkers W, Li J, Glass NL (2013) Early colony establishment in Neurospora crassa requires a MAP kinase regulatory network. Genetics 195:883–898
Lev S, Sharon A, Hadar R, Ma H, Horwitz BA (1999) A mitogen-activated protein kinase of the corn leaf pathogen Cochliobolus heterostrophus is involved in conidiation, appressorium formation, and pathogenicity: diverse roles for mitogen-activated protein kinase homologs in foliar pathogens. Proc Natl Acad Sci USA 96:13542–13547
Levin DE (2005) Cell wall integrity signaling in Saccharomyces cerevisiae. Microbiol Mol Biol Rev 69:262–291
Levin DE (2011) Regulation of cell wall biogenesis in Saccharomyces cerevisiae: the cell wall integrity signaling pathway. Genetics 189:1145–1175
Lin X, Heitman J (2006) The biology of the Cryptococcus neoformans species complex. Annu Rev Microbiol 60:69–105
Ma D, Li R (2013) Current understanding of HOG-MAPK pathway in Aspergillus fumigatus. Mycopathologia 175:13–23

Mackay V, Manney TR (1974) Mutations affecting sexual conjugation and related processes in Saccharomyces cerevisiae. I. Isolation and phenotypic characterization of nonmating mutants. Genetics 76:255–271

Maddi A, Dettman A, Fu C, Seiler S, Free SJ (2012) WSC-1 and HAM-7 are MAK-1 MAP kinase pathway sensors required for cell wall integrity and hyphal fusion in Neurospora crassa. PLoS One 7: e42374

Maerz S, Ziv C, Vogt N, Helmstaedt K, Cohen N, Gorovits R, Yarden O, Seiler S (2008) The nuclear Dbf2-related kinase COT1 and the mitogen-activated protein kinases MAK1 and MAK2 genetically interact to regulate filamentous growth, hyphal fusion and sexual development in Neurospora crassa. Genetics 179:1313–1325

Malleshaiah MK, Shahrezaei V, Swain PS, Michnick SW (2010) The scaffold protein Ste5 directly controls a switch-like mating decision in yeast. Nature 465:101–105

May GS (2007) Mitogen-activated protein kinase pathways in Aspergillus. In: Goldman GH, Osmani SA (eds) The Aspergilli genomics, medical aspects, biotechnology and research methods. CRC, Boca Raton, FL, pp 143–175

Mayorga ME, Gold SE (1999) A MAP kinase encoded by the ubc3 gene of Ustilago maydis is required for filamentous growth and full virulence. Mol Microbiol 34:485–497

Monge RA, Roman E, Nombela C, Pla J (2006) The MAP kinase signal transduction network in Candida albicans. Microbiology 152:905–912

Muller P, Aichinger C, Feldbrugge M, Kahmann R (1999) The MAP kinase kpp2 regulates mating and pathogenic development in Ustilago maydis. Mol Microbiol 34:1007–1017

Navarro-Garcia F, Sanchez M, Pla J, Nombela C (1995) Functional characterization of the MKC1 gene of Candida albicans, which encodes a mitogen-activated protein kinase homolog related to cell integrity. Mol Cell Biol 15:2197–2206

Navarro-Garcia F, Eisman B, Fiuza SM, Nombela C, Pla J (2005) The MAP kinase Mkc1p is activated under different stress conditions in Candida albicans. Microbiology 151:2737–2749

O'Gorman CM, Fuller H, Dyer PS (2009) Discovery of a sexual cycle in the opportunistic fungal pathogen Aspergillus fumigatus. Nature 457:471–474

O'Rourke SM, Herskowitz I (2004) Unique and redundant roles for HOG MAPK pathway components as revealed by whole-genome expression analysis. Mol Biol Cell 15:532–542

Pandey A, Roca MG, Read ND, Glass NL (2004) Role of a mitogen-activated protein kinase pathway during conidial germination and hyphal fusion in Neurospora crassa. Eukaryot Cell 3:348–358

Paoletti M, Seymour FA, Alcocer MJ, Kaur N, Calvo AM, Archer DB, Dyer PS (2007) Mating type and the genetic basis of self-fertility in the model fungus Aspergillus nidulans. Curr Biol 17:1384–1389

Park G, Pan S, Borkovich KA (2008) Mitogen-activated protein kinase cascade required for regulation of development and secondary metabolism in Neurospora crassa. Eukaryot Cell 7:2113–2122

Patananan AN, Palmer JM, Garvey GS, Keller NP, Clarke SG (2013) A novel automethylation reaction in the Aspergillus nidulans LaeA protein generates S-methylmethionine. J Biol Chem 288:14032–14045

Peter M, Gartner A, Horecka J, Ammerer G, Herskowitz I (1993) FAR1 links the signal transduction pathway to the cell cycle machinery in yeast. Cell 73:747–760

Peter M, Herskowitz I (1994) Joining the complex: cyclin-dependent kinase inhibitory proteins and the cell cycle. Cell 79:181–184

Pitoniak A, Birkaya B, Dionne HM, Vadaie N, Cullen PJ (2009) The signaling mucins Msb2 and Hkr1 differentially regulate the filamentation mitogen-activated protein kinase pathway and contribute to a multimodal response. Mol Biol Cell 20:3101–3114

Pöggeler S, Nowrousian M, Kück U (2006) Fruiting-body development in Ascomycetes. In: Fischer K (ed) The Mycota, I. Growth, differentiation and sexuality. Springer, Heidelberg, pp 325–355

Posas F, Saito H (1997) Osmotic activation of the HOG MAPK pathway via Ste11p MAPKKK: scaffold role of Pbs2p MAPKK. Science 276:1702–1705

Prieto D, Roman E, Correia I, Pla J (2014) The HOG pathway is critical for the colonization of the mouse gastrointestinal tract by Candida albicans. PLoS One 9:e87128

Ramirez-Zavala B, Weyler M, Gildor T, Schmauch C, Kornitzer D, Arkowitz R, Morschhauser J (2013) Activation of the Cph1-dependent MAP kinase signaling pathway induces white-opaque switching in Candida albicans. PLoS Pathog 9:e1003696

Rauceo JM, Blankenship JR, Fanning S, Hamaker JJ, Deneault JS, Smith FJ, Nantel A, Mitchell AP (2008) Regulation of the Candida albicans cell wall damage response by transcription factor Sko1 and PAS kinase Psk1. Mol Biol Cell 19:2741–2751

Reyes G, Romans A, Nguyen CK, May GS (2006) Novel mitogen-activated protein kinase MpkC of Aspergillus fumigatus is required for utilization of polyalcohol sugars. Eukaryot Cell 5:1934–1940

Rodicio R, Heinisch JJ (2010) Together we are strong—cell wall integrity sensors in yeasts. Yeast 27:531–540

Roman E, Nombela C, Pla J (2005) The Sho1 adaptor protein links oxidative stress to morphogenesis and cell wall biosynthesis in the fungal pathogen Candida albicans. Mol Cell Biol 25:10611–10627

Roman E, Arana DM, Nombela C, Alonso-Monge R, Pla J (2007) MAP kinase pathways as regulators of fungal virulence. Trends Microbiol 15:181–190

Sahni N, Yi S, Daniels KJ, Huang G, Srikantha T, Soll DR (2010) Tec1 mediates the pheromone response of the white phenotype of Candida albicans:

insights into the evolution of new signal transduction pathways. PLoS Biol 8:e1000363

Saito H (2010) Regulation of cross-talk in yeast MAPK signaling pathways. Curr Opin Microbiol 13:677–683

Schwartz MA, Madhani HD (2004) Principles of MAP kinase signaling specificity in Saccharomyces cerevisiae. Annu Rev Genet 38:725–748

Segmuller N, Ellendorf U, Tudzynski B, Tudzynski P (2007) BcSAK1, a stress-activated mitogen-activated protein kinase, is involved in vegetative differentiation and pathogenicity in Botrytis cinerea. Eukaryot Cell 6:211–221

Staib P, Morschhauser J (2007) Chlamydospore formation in Candida albicans and Candida dubliniensis—an enigmatic developmental programme. Mycoses 50:1–12

Sudbery PE (2011) Growth of Candida albicans hyphae. Nat Rev Microbiol 9:737–748

Tedford K, Kim S, Sa D, Stevens K, Tyers M (1997) Regulation of the mating pheromone and invasive growth responses in yeast by two MAP kinase substrates. Curr Biol 7:228–238

Torres L, Martin H, Garcia-Saez MI, Arroyo J, Molina M, Sanchez M, Nombela C (1991) A protein kinase gene complements the lytic phenotype of Saccharomyces cerevisiae lyt2 mutants. Mol Microbiol 5:2845–2854

Truckses DM, Bloomekatz JE, Thorner J (2006) The RA domain of Ste50 adaptor protein is required for delivery of Ste11 to the plasma membrane in the filamentous growth signaling pathway of the yeast Saccharomyces cerevisiae. Mol Cell Biol 26:912–928

Truman AW, Millson SH, Nuttall JM, Mollapour M, Prodromou C, Piper PW (2007) In the yeast heat shock response, Hsf1-directed induction of Hsp90 facilitates the activation of the Slt2 (Mpk1) mitogen-activated protein kinase required for cell integrity. Eukaryot Cell 6:744–752

Valiante V, Heinekamp T, Jain R, Hartl A, Brakhage AA (2008) The mitogen-activated protein kinase MpkA of Aspergillus fumigatus regulates cell wall signaling and oxidative stress response. Fungal Genet Biol 45:618–627

Valiante V, Jain R, Heinekamp T, Brakhage AA (2009) The MpkA MAP kinase module regulates cell wall integrity signaling and pyomelanin formation in Aspergillus fumigatus. Fungal Genet Biol 46:909–918

van Drogen F, Peter M (2002) Spa2p functions as a scaffold-like protein to recruit the Mpk1p MAP kinase module to sites of polarized growth. Curr Biol 12:1698–1703

Van Nguyen T, Kroger C, Bonnighausen J, Schafer W, Bormann J (2013) The ATF/CREB transcription factor Atf1 is essential for full virulence, deoxynivalenol production, and stress tolerance in the cereal pathogen Fusarium graminearum. Mol Plant Microbe Interact 26:1378–1394

Van Thuat N, Schafer W, Bormann J (2012) The stress-activated protein kinase FgOS-2 is a key regulator in the life cycle of the cereal pathogen Fusarium graminearum. Mol Plant Microbe Interact 25:1142–1156

Viaud M, Fillinger S, Liu W, Polepalli JS, Le Pecheur P, Kunduru AR, Leroux P, Legendre L (2006) A class III histidine kinase acts as a novel virulence factor in Botrytis cinerea. Mol Plant Microbe Interact 19:1042–1050

Vitalini MW, de Paula RM, Goldsmith CS, Jones CA, Borkovich KA, Bell-Pedersen D (2007) Circadian rhythmicity mediated by temporal regulation of the activity of p38 MAPK. Proc Natl Acad Sci USA 104:18223–18228

Wagner M, Briza P, Pierce M, Winter E (1999) Distinct steps in yeast spore morphogenesis require distinct SMK1 MAP kinase thresholds. Genetics 151:1327–1340

Wang P, Nichols CB, Lengeler KB, Cardenas ME, Cox GM, Perfect JR, Heitman J (2002) Mating-type-specific and nonspecific PAK kinases play shared and divergent roles in Cryptococcus neoformans. Eukaryot Cell 1:257–272

Wei H, Requena N, Fischer R (2003) The MAPKK kinase SteC regulates conidiophore morphology and is essential for heterokaryon formation and sexual development in the homothallic fungus Aspergillus nidulans. Mol Microbiol 47:1577–1588

Westfall PJ, Ballon DR, Thorner J (2004) When the stress of your environment makes you go HOG wild. Science 306:1511–1512

Whinston E, Omerza G, Singh A, Tio CW, Winter E (2013) Activation of the Smk1 mitogen-activated protein kinase by developmentally regulated autophosphorylation. Mol Cell Biol 33:688–700

Widmann C, Gibson S, Jarpe MB, Johnson GL (1999) Mitogen-activated protein kinase: conservation of a three-kinase module from yeast to human. Physiol Rev 79:143–180

Wong Sak Hoi J, Lamarre C, Beau R, Meneau I, Berepiki A, Barre A, Mellado E, Read ND, Latge JP (2011) A novel family of dehydrin-like proteins is involved in stress response in the human fungal pathogen Aspergillus fumigatus. Mol Biol Cell 22:1896–1906

Wu C, Jansen G, Zhang J, Thomas DY, Whiteway M (2006) Adaptor protein Ste50p links the Ste11p MEKK to the HOG pathway through plasma membrane association. Genes Dev 20:734–746

Xu JR, Hamer JE (1996) MAP kinase and cAMP signaling regulate infection structure formation and pathogenic growth in the rice blast fungus Magnaporthe grisea. Genes Dev 10:2696–2706

Xu JR, Staiger CJ, Hamer JE (1998) Inactivation of the mitogen-activated protein kinase Mps1 from the rice blast fungus prevents penetration of host cells but allows activation of plant defense responses. Proc Natl Acad Sci USA 95:12713–12718

Xue T, Nguyen CK, Romans A, May GS (2004) A mitogen-activated protein kinase that senses nitrogen regulates conidial germination and growth in Aspergillus fumigatus. Eukaryot Cell 3:557–560

Yang HY, Tatebayashi K, Yamamoto K, Saito H (2009) Glycosylation defects activate filamentous growth Kss1 MAPK and inhibit osmoregulatory Hog1 MAPK. EMBO J 28:1380–1391

Yue C, Cavallo LM, Alspaugh JA, Wang P, Cox GM, Perfect JR, Heitman J (1999) The STE12alpha homolog is required for haploid filamentation but largely dispensable for mating and virulence in Cryptococcus neoformans. Genetics 153:1601–1615

Yun Y, Liu Z, Zhang J, Shim WB, Chen Y, Ma Z (2013) The MAPKK FgMkk1 of Fusarium graminearum regulates vegetative differentiation, multiple stress response, and virulence via the cell wall integrity and high-osmolarity glycerol signaling pathways. Environ Microbiol 7:2023–2037

Zhang C, Wang Y, Wang J, Zhai Z, Zhang L, Zheng W, Zheng W, Yu W, Zhou J, Lu G, Shim WB, Wang Z (2013) Functional characterization of Rho family small GTPases in Fusarium graminearum. Fungal Genet Biol 61:90–99

Zhang Y, Lamm R, Pillonel C, Lam S, Xu JR (2002) Osmoregulation and fungicide resistance: the Neurospora crassa os-2 gene encodes a HOG1 mitogen-activated protein kinase homologue. Appl Environ Microbiol 68:532–538

7 Heterotrimeric G Proteins

Rekha Deka[1], Arit Ghosh[1], Ranjan Tamuli[2], Katherine A. Borkovich[1]

CONTENTS

I. Introduction

All living organisms are exposed to a variety of environmental signals that elicit appropriate cellular responses that dictate cell survival and development. A major system used by eukaryotic organisms to sense and respond to environmental stimuli involves **heterotrimeric G-proteins** (Tesmer 2010) and their cognate membrane-bound receptors, known as **G protein-coupled receptors** or GPCRs (Bock et al. 2014) (Fig. 7.1). Heterotrimeric G proteins are composed of three different subunits, Gα, Gβ, and Gγ (Jastrzebska 2013; Tesmer 2010). The Gα subunit is bound to guanosine diphosphate (GDP) when inactive and exchanges GTP for GDP after activation (Jastrzebska 2013). In the inactive state, the heterotrimer is associated with G-protein-coupled receptors (GPCRs), on the cytoplasmic face of the plasma membrane (Bock et al. 2014; Pierce et al. 2002; Neves et al. 2002). GPCRs function as **guanine nucleotide exchange factors (GEFs)** and are characterized by the presence of seven transmembrane alpha helices (Shukla et al. 2014; Chini et al. 2013). Binding of the extracellular ligand to the GPCR causes exchange of GDP for GTP on the Gα subunit and dissociation of Gα-GTP from the

[1]Department of Plant Pathology and Microbiology, University of California, 900 University Avenue, Riverside, CA 92521, USA; e-mail: katherine.borkovich@ucr.edu
[2]Department of Biosciences and Bioengineering, Indian Institute of Technology Guwahati, Guwahati, Assam 781039, India

Biochemistry and Molecular Biology, 3rd Edition
The Mycota III
D. Hoffmeister (Ed.)

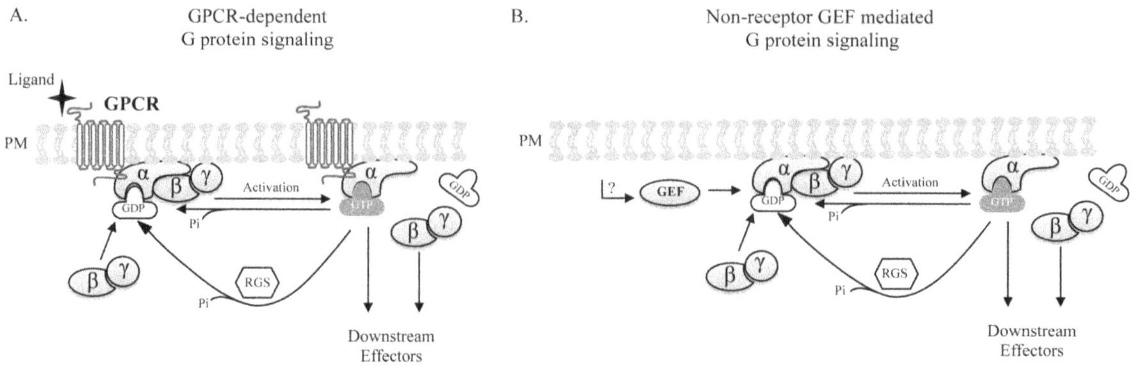

Fig. 7.1 Overview of heterotrimeric G protein pathways. Heterotrimeric G protein signaling mechanisms in fungi are regulated via GPCR-dependent and GPCR-independent, non-receptor GEF pathways. (**a**) Ligand binding to a GPCR stimulates a conformational change in the receptor which leads to GDP–GTP exchange on the Gα protein and subsequent dissociation of the Gα from the Gβγ dimer. The Gα and Gβγ units regulate downstream effectors, such as adenylyl cyclase and MAPK cascades. Hydrolysis of GTP on the Gα subunit occurs via its intrinsic GTPase activity, and this reaction can also be accelerated via RGS proteins. Reassociation of the Gβγ dimer with the GDP-bound Gα returns to the heterotrimeric G protein to the resting state in association with the GPCR. (**b**) Non-receptor GEFs such as RIC8 and ARR4/GET3 can also stimulate GDP-GTP exchange on Gα proteins. In various fungal systems, RIC8 has been shown to interact with Group I and III Gα proteins to regulate cAMP signaling. *Abbreviations*: *GPCR* G protein-coupled receptor, *PM* plasma membrane, *GDP* guanosine diphosphate, *GTP* guanosine triphosphate, *RGS* regulators of G protein signaling, *GEF* guanine nucleotide exchange factor, *Pi* phosphate

Gβγ dimer, allowing both moieties to regulate downstream effectors (Jastrzebska 2013; Chini et al. 2013; Neves et al. 2002; Mende et al. 1998). In addition to GPCRs, GDP–GTP exchange on Gα subunits can also be mediated via non-receptor GEFs, such as RIC8 and GET3 (Miller et al. 2000; Tall et al. 2003; Wilkie and Kinch 2005; Siderovski and Willard 2005; Lee and Dohlman 2008) (Fig. 7.1b). Finally, examples of downstream effectors regulated by G proteins are adenylyl cyclases, ion channels, mitogen-activated protein kinases (MAPK), phosphodiesterases, and phospholipases (Chini et al. 2013; Neves et al. 2002; Mende et al. 1998) (Figs. 7.2 and 7.3).

Termination of the signal is facilitated by hydrolysis of GTP to GDP on the Gα subunit and reassociation of Gα-GDP with the Gβγ dimer and the GPCR at the plasma membrane (Jastrzebska 2013; Chini et al. 2013; Neves et al. 2002; Mende et al. 1998) (Fig. 7.1). The Gα subunit has intrinsic GTPase activity, and this process is accelerated via certain other proteins called **Regulators of G protein Signaling (RGS)** or **GTPase Accelerating Proteins (GAPs)** (Ross and Wilkie 2000; Siderovski et al. 1996; Berman and Gilman 1998).

In this chapter, we will provide a brief summary of heterotrimeric G protein signaling components and effector pathways in fungi. The focus will then shift to select G protein signaling paradigms in several fungal species: *Saccharomyces cerevisiae*, *Schizosaccharomyces pombe*, *Neurospora crassa*, *Aspergillus nidulans*, *Magnaporthe grisea*, *Cryptococcus neoformans*, and *Ustilago maydis*. Due to space limitations, we are unable to reference all of the work in this rapidly expanding field, and we direct the reader to several reviews (Raudaskoski and Kothe 2010; Xue et al. 2008; Yu 2006; Li et al. 2007b; Lengeler et al. 2000).

II. Components of Heterotrimeric G Protein Signaling

A. Gα, Gβ, and Gγ Proteins

Gα subunits are composed of α-helical and a GTPase domains, the latter of which imparts intrinsic GTPase activity to the Gα subunit (Bohm et al. 1997; Birnbaumer 1992). The GTPase domain consists of five α helices sur-

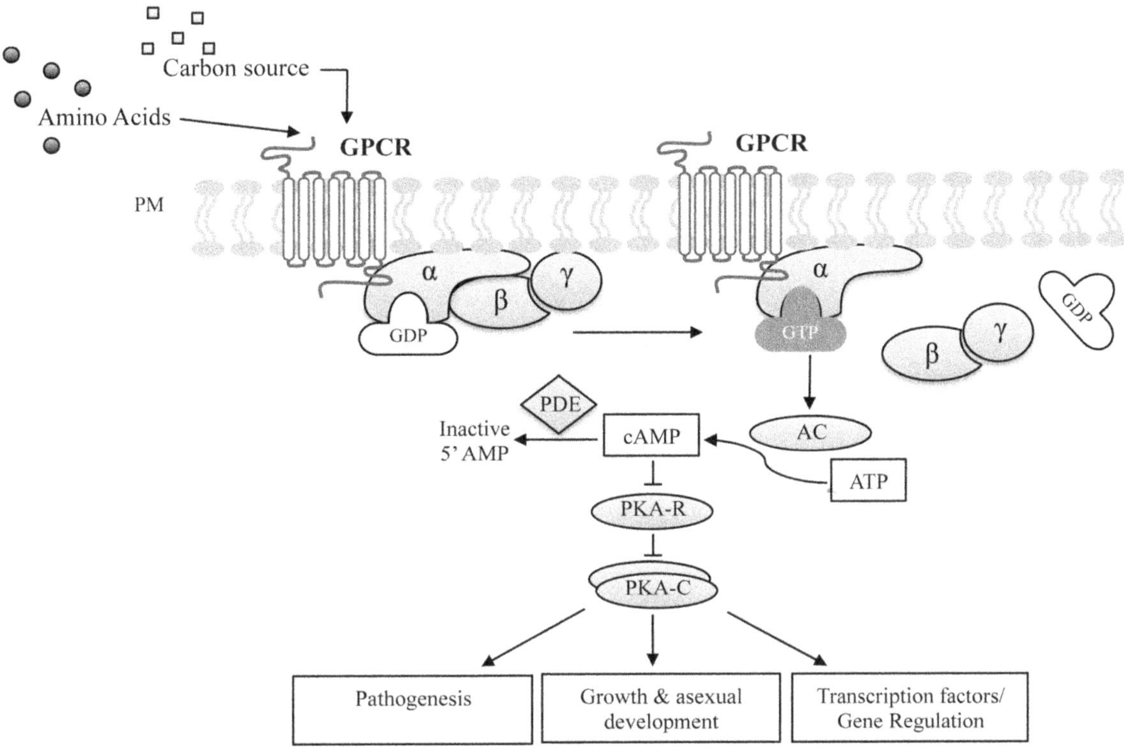

Fig. 7.2 Nutrient sensing. Heterotrimeric G protein-mediated nutrient sensing occurs mainly via the cAMP/PKA signaling pathway. Upstream GPCRs sense carbon or amino acids and couple these responses to downstream G proteins, resulting in activation of adenylyl cyclase and generation of cAMP. The PKA-regulatory subunits binds cAMP, allowing for release of the catalytic subunits that phosphorylate downstream targets, including transcription factors. Additional regulation of cAMP levels takes place via phosphodiesterases, which hydrolyze cAMP to the inactive 5′AMP. GPCRs involved in nutrient sensing in: *N. crassa*—GPR-4 (carbon-sensing), *C. neoformans*—Gpr4 (amino acid sensing), *A. nidulans*—GprC, GprD (putative carbon sensors), *S. cerevisiae*—Gpr1 (carbon-sensing), *S. pombe*—Git3 (carbon sensing). *Abbreviations*: *GPCR* G protein-coupled receptor, *PM* plasma membrane, *GDP* guanosine diphosphate, *GTP* guanosine triphosphate, *cAMP* cyclic adenosine monophosphate, *PDE* phosphodiesterases, *PKA-R* protein kinase A regulatory subunit, *PKA-C* protein kinase A catalytic subunit

rounding a β sheet and connected to the α-helical domain by two linkers. The guanine nucleotide-binding pocket within the GTPase domain contains a consensus sequence GXGXXGKS for GTP binding (Hamm and Gilchrist 1996). Mutations in the GTPase domain of the Gα subunit lead to disruption in its GTPase activity and result in constitutive Gα-GTP signaling (Freissmuth and Gilman 1989; Graziano and Gilman 1989). Finally, many Gα subunits undergo posttranslational lipid modifications, such as palmitoylation and myristoylation, which aid in tethering the protein to the membrane to facilitate interaction with the Gβγ dimer (Resh 1996).

Most filamentous fungi possess three Gα proteins in each of three groups, I–III, but the basidiomycete *Ustilago maydis* also has a fourth Gα (Group IV) (Bolker 1998) (Table 7.1). In contrast, budding and fission yeasts contain only two Gα proteins (Bolker 1998). In general, Group I Gα proteins share high amino acid similarity and regulate sexual and asexual developmental pathways (Li et al. 2007b). The *Neurospora crassa* Group I Gα protein GNA-1 was the first Gα protein identified in filamentous fungi and found to share 55 % identity to members of the mammalian $G\alpha_i$ superfamily (Turner and Borkovich 1993). In addition, the C-terminus of GNA-1 contains the consensus

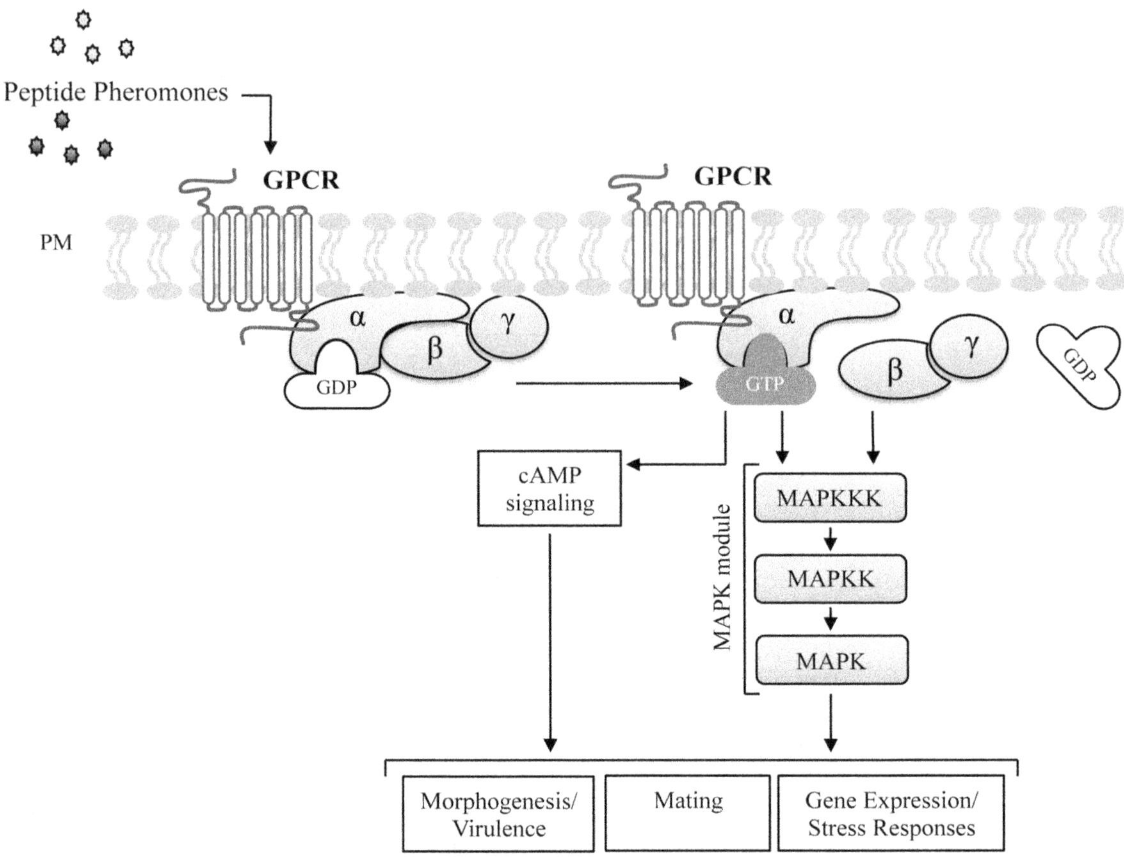

Fig. 7.3 Pheromone sensing and mating. In fungi, typically two different types of peptide pheromones (colored *gray/black*) are responsible for activation of their respective GPCRs (pheromone receptors). The GPCRs respond to the pheromones—α-factor and a-factor in *S. cereviseae*, M-factor and P-factor in *S. pombe*, MFA-1 and CCG-4 in *N. crassa*, and Mfa1-3 and Mfα1-3in *C. neoformans*. In *S. cerevisiae*, it has been demonstrated that activation of the Gα subunit leads to dissociation of the Gβγ heterodimer and subsequent activation of the MAPKKK, leading to sequential phosphorylation of MAPK module components, eventually influencing mating-specific gene expression. In *N. crassa* and other fungi, mating is regulated via both cAMP-dependent and cAMP-independent signaling mechanisms. GPCRs involved in pheromone sensing: PRE-1 and PRE-2 in *N. crassa*, Ste3α and Ste3a in *C. neoformans*, GprA and GprB in *A. nidulans*, Ste2p and Ste3p in *S. cereviseae*, Mam2 and Map3 in *S. pombe*, Pra1 and Pra2 in *U. maydis*. *Abbreviations*: *GPCR* G protein-coupled receptors, *GDP* guanosine diphosphate, *GTP* guanosine triphosphate, *cAMP* cyclic adenosine monophosphate, *MAPKKK* mitogen-activated kinase kinase kinase, *MAPKK* mitogen-activated kinase kinase, *MAPK* mitogen-activated kinase

Table 7.1 Gα, Gβ, and Gγ subunit proteins in various fungi

	Class					
	Gα protein subunits					
Organism	I	II	III	IV	Gβ protein subunit	Gγ protein subunit
Saccharomyces cereviseae	Gpa1	–	Gpa2	–	Ste4	Ste18
Schizosaccharomyces pombe	–	Gpa1	Gpa2	–	Git5, Gnr1	Git11
Neurospora crassa	GNA-1	GNA-2	GNA-3	–	GNB-1	GNG-1
Aspergillus nidulans	FadA	GanA	GanB	–	SfaD	GpgA
Cryptococcus neoformans	Cga1	Gpa3	Gpa1		Gpb1	Gpg1, Gpg2
Magnaporthe oryzae	MAGB	MAGC	MAGA	–	MGB1	MGG-10193.5
Ustilago maydis	Gpa1	Gpa2	Gpa3	Gpa4	Bpp1	UM06109.1

sequence for ADP-ribosylation by pertussis toxin (CAAX) and is modified by pertussis toxin: both properties are a hallmark of the $G\alpha_i$ super family (Turner and Borkovich 1993).

Gβ and Gγ subunits are tightly associated and transduce signals downstream as a heterodimer. In higher eukaryotes, multiple isoforms of Gβ and Gγ subunits are known (Hamm 1998). In fungi, there is one canonical Gβ subunit (Li et al. 2007b) (Table 7.1), but emerging evidence suggests that RACK1 like proteins can also function as Gβ proteins in fungi (Zeller et al. 2007; Wang et al. 2014; see below). Gβ proteins are approximately 340 amino acids and are known to be the most highly conserved of the three subunits. Structurally, the Gβ subunit is composed of an amphipathic α-helix at its N-terminus that extends into seven repeating units of a WD motif of 36–46 amino acid residues with tryptophan (W) and aspartate (D) at the ends (Fong et al. 1986; Neer et al. 1994; Gautam et al. 1998; Downes and Gautam 1999). The seven repeating units form a β-sheet of four antiparallel strands that protrude away from the center such that the overall structure appears to be a seven bladed propeller, called a β-superbarrel. The opposite ends of β-superbarrel interact with Gα and Gγ subunits. In contrast to the Gβ subunit, Gγ subunits are relatively small proteins, consisting of 68–75 amino acid residues. There are numerous Gγ proteins in mammals, but most fungi possess only one canonical Gγ protein (Table 7.1). *C. neoformans* is an exception, with two predicted Gγ genes (Li et al. 2007a). The characteristic feature of Gγ subunits is the presence of a CaaX motif at the C-terminus that provides site for posttranslational lipid modifications such as farnesylation and geranylgeranylation (Lai et al. 1990; Yamane et al. 1990). These modifications allow the Gγ to bind the plasma membrane and aid the interaction with the Gα subunit (Iniguez-Lluhi et al. 1992; Gautam et al. 1998; Zhang and Casey 1996). The Gβ and Gγ proteins in different fungi are summarized in Table 7.1.

In recent years, it has become clear that other proteins may take the place of canonical Gβ and Gγ subunits during signal transduction in fungi. For example, proteins homologous to the RACK1 Gβ-like protein have been found to interact with Gα or/or Gγ subunits and to regulate signaling in various fungi (Zeller et al. 2007; Wang et al. 2014). In yeast, two Kelch repeat proteins can interact with the Gα subunit Gpa2 to activate protein kinase A without intermediate activation of adenylyl cyclase (Peeters et al. 2006).

B. Guanine Nucleotide Exchange Factors (GEFs)

1. G Protein-Coupled Receptors (GPCRs)

The typical GPCR contains a structural signature of conserved **seven-hydrophobic transmembrane (7-TM)** helices that are connected by intracellular and extracellular loops with the N-terminus extending to the outer environment and the C-terminus localized in the cytoplasm (Fig. 7.1a). The most variable regions of GPCRs are the C-terminus, N-terminus, and intracellular loops, with greatest diversity in the N-terminal region (Kobilka 2007; Katritch et al. 2012; Venkatakrishnan et al. 2013).

GPCRs represent the largest and most diverse group of transmembrane receptors, with well over 800 potential GPCRs in the human genome (Hill 2006). Based on sequence similarity in the 7-TM segments, mammalian GPCRs have been clustered into five families: rhodopsin (701 members), adhesion (24 members), frizzled/taste (24 members), glutamate (15 members), and secretin (15 members) (Fredriksson et al. 2003). Several GPCR classification schemes have been proposed for fungal genomes (Galagan et al. 2003; Kulkarni et al. 2005; Lafon et al. 2006; Li et al. 2007b; Zheng et al. 2010; Krishnan et al. 2012; Gruber et al. 2013). However, common to all classification systems, fungal GPCRs include pheromone receptors, cAMP-receptor-like proteins, carbon- and amino acid-sensing receptors, putative nitrogen sensors, and microbial opsin-like proteins.

2. Non-receptor GEFs

Recent studies have revealed the existence of cytosolic non-GPCR proteins that can act as GEFs to positively regulate Gα proteins (Fig. 7.1b). Two examples of such proteins in fungi are RIC8 and GET3/ARR4. RIC8 (resistance to inhibitors of cholinesterase 8; Synembryn) was first identified in the nematode *Caenorhabditis elegans* as a protein that exhibits GEF activity toward Gα proteins (Miller and Rand 2000). RIC8 is found in animals and certain fungi, but absent from plants and the model yeast *Saccharomyces cerevisiae* (Wilkie and Kinch 2005; Wright et al. 2011). In animals, RIC8 has been implicated in cAMP and MAPK signaling and G protein abundance, with downstream regulation of asymmetric cell division, priming of synaptic vesicles, and hormone signaling (Tall 2013; Hampoelz et al. 2005; Schade et al. 2005; Afshar et al. 2004; Miller and Rand 2000).

RIC8 isoforms have been found to be potent GEFs for a subset of Gα proteins, including $G\alpha_q$, $G\alpha_{i1}$, and $G\alpha_o$. Ric8A preferentially interacts with GDP-bound Gα to facilitate the release of GDP and then form a stable nucleotide-free transition state complex with Gα until Gα binds to GTP (Tall et al. 2003; Thomas et al. 2011). The stable transition state complex prevents Gα ubiquitination (Nagai et al. 2010; Chishiki et al. 2013). Further studies on embryonic stem cell lines deleted for Ric8A and Ric8B revealed the functions of RIC8 as a molecular chaperone required for the initial association of nascent Gα subunits with cellular membranes (Gabay et al. 2011; Chan et al. 2013).

The role of RIC8 has been studied in the following four filamentous fungal genera: the rice pathogen *M. oryzae* (Li et al. 2010), the saprobe and model system *N. crassa* (Wright et al. 2011; Eaton et al. 2012), the opportunistic human pathogen *A. fumigatus* and the model system *A. nidulans* (Kwon et al. 2012), and the human pathogen *C. neoformans* (Gong et al. 2014). In *M. oryzae*, *Δmoric8* mutants exhibit defects in formation of infection structures (appressoria) that are necessary for rice blast disease (Li et al. 2010). The appressorial defect is partially rescued by addition of exogenous cAMP, but the induced appressoria are nonfunctional (Li et al. 2010). MoRIC8 is localized in the cytoplasm of vegetative hyphae, conidia, and appressoria, with the highest expression in appressoria, consistent with its role in infection-related morphogenesis (Li et al. 2010). In *N. crassa*, loss of *ric-8* gene leads to extreme phenotypes, including impaired growth, short aerial hyphae, inappropriate conidiation in submerged culture, and female sterility (Wright et al. 2011). These phenotypes are similar to double mutants lacking the *gna-1* and *gna-3* Gα genes (Kays and Borkovich 2004). Many defects of the *Δric8* mutant are suppressed by mutation of the regulatory subunit of PKA, *mcb* (Bruno et al. 1996), which is consistent with RIC8 operating upstream of cAMP signaling in *N. crassa* (Wright et al. 2011). Similar to the situation in *M. oryzae*, *N. crassa* RIC8 is localized to the cytoplasm of vegetative hyphae and mature conidia (Wright et al. 2011). Importantly, like its mammalian homolog, *N. crassa* RIC8 binds and exhibits GEF activity towards the GNA-1 and GNA-3 Gα proteins, with the greatest stimulation of GNA-3 (Wright et al. 2011). In *A. nidulans* and *A. fumigatus*, loss of RIC8 profoundly affects conidial germination, hyphal proliferation, and asexual/sexual fruiting (Kwon et al. 2012). AnRICA physically interacts with the Gα protein GanB in *A. nidulans* (Kwon et al. 2012). A role for RIC8 in capsule formation, melanin production, and virulence has been revealed in the human pathogen *C. neoformans* (Gong et al. 2014). RIC8 interacts with and stimulates GEF activity on the Gpa1 and Gpa2 Gα subunits and impacts cAMP and pheromone signaling in *C. neoformans* (Gong et al. 2014). Taken together, the data from multiple fungal species suggest that RIC8 is a cytoplasmic Gα GEF that interacts with Group I and Group III Gα proteins to modulate cAMP and other signaling pathways (Fig. 7.1).

In contrast to RIC8, GET-3/ARR4 is widely conserved in eukaryotes, including fungi (Lee and Dohlman 2008). ARR4 is so named because of homology to Arsenicals Resistance proteins in bacterial species (Shen et al. 2003). In yeast, Get3p/Arr4p is required for ATP-dependent tail-anchoring of proteins in the endoplasmic reticulum (Denic et al. 2013). Get3p has also been demonstrated to bind to, and exhibit GEF activity towards, the nucleotide-free form of the Gpa1p Ga protein (Lee and Dohlman 2008). Get3p is also required for normal MAPK signaling, gene regulation, and mating in *S. cerevisiae* (Lee and Dohlman 2008).

C. Regulator of G Protein Signaling (RGS) Proteins

RGS proteins are crucial elements in controlling the intensity and duration of G protein signaling. Most RGS proteins are considered as negative regulators of G protein signaling, either by accelerating intrinsic GTPase activity on Gα subunits (GTPase activation) or by competitively blocking the Gα-GTP interaction with downstream effectors (effector antagonism) (Ross and Wilkie 2000; Anger et al. 2004; Wang et al. 2013). All RGS proteins contain a conserved RGS core domain of ~120 to 130 amino acids that is necessary and sufficient for binding to transition state of GTP-Gα subunits for accelerating GTP hydrolysis (Hollinger and Hepler 2002). In addition to the RGS domain, RGS proteins contain additional motifs that allow integration into different G-protein signaling pathways or function as scaffold elements that link G proteins to other related signaling proteins (Wang et al. 2013). These domains include DEP (Dishevelled, Egl-10, Pleckstrin) for membrane targetting, Phosphotyrosine interaction (PTB), Ras binding (RBD), GoLoco for guanine nucleotide exchange, and GGL (Gγ subunit like) for binding Gβ subunits (De Vries et al. 2000; Chasse and Dohlman 2003; Wang et al. 2013). The presence of such diverse domains also makes RGS proteins attractive targets for drugs to modulate cell signaling.

RGS proteins have been discovered in large number of eukaryotes, from yeast to *C. elegans* to humans, and >30 RGS-like proteins have been identified in the human proteome (Chasse and Dohlman 2003; Wang et al. 2013; Hollinger and Hepler 2002). The first RGS protein, Sst2p, was discovered in *S. cerevisiae* through a genetic screen for negative regulators of the pheromone response (Chan and Otte 1982). Sst2p serves as a critical regulator of mating pheromone signaling by directly binding to the C-terminal tail of the pheromone receptor Ste2p via its DEP domain (Chasse et al. 2006). Sst2p also genetically and physically interacts with the Gα protein Gpa1p and accelerates its GTPase activity (Dohlman et al. 1996; Apanovitch et al. 1998). Another *S. cerevisiae* RGS protein, Rgs2p, interacts with the Gα Gpa2p and functions as a multicopy suppressor for Gpa2p-dependent loss of heat shock resistance (Versele et al. 1999).

The functions of RGS and RGS-like proteins have been extensively studied in several filamentous fungi (Liu et al. 2007; Zhang et al. 2011; Shen et al. 2008; Fang et al. 2008; Mukherjee et al. 2011). The first characterized RGS protein in filamentous fungi was *A. nidulans* FlbA (Lee and Adams 1994). FlbA interacts with the FadA Gα protein and is required for regulation of hyphal proliferation, development, and biosynthesis of secondary metabolites (Yu et al. 1996). In addition to FlbA, *A. nidulans* possesses several other RGS proteins, RgsA, RgsB, RgsC, and GprK. RgsA negatively regulates the GanB Gα signaling during stress responses and asexual sporulation (Han et al. 2004). GprK is unique in that it contains both 7-TM and RGS domains, similar to AtRGS1, which negatively regulates the Gpa1 Gα during cellular proliferation in the model plant *A. thaliana* (Lafon et al. 2006; Chen et al. 2003).

D. Effector Pathways

cAMP-dependent protein kinase (PKA) (Fig. 7.2) and MAPK (Fig. 7.3) cascades are two major effector pathways in fungi that act downstream of the activated GTP binding G-proteins to elicit cellular responses such as growth, mating, cell division, cell–cell fusion, morphogenesis, toxicognesis, chemotaxis, and pathogenic development (Bolker 1998; Lengeler et al. 2000).

1. cAMP-Dependent Protein Kinase (PKA)

The key components of the cAMP-PKA pathway include the enzyme adenylyl cyclase and the regulatory and catalytic subunits of PKA (Fig. 7.2). Fungal adenylyl cyclases are cytoplasmic peripheral membrane proteins (Matsumoto et al. 1984; Gold et al. 1994; Adachi and Hamer 1998) and contain multiple domains (from N- to C-terminus), including Gα binding, Ras-association (RA), leucine-rich repeat (LRR), protein phosphatase 2C (PP2C), cyclase catalytic, and cyclase-associated protein 1 (Cap1) binding domains (Wang 2013). The activation of adenylyl cyclase by GTP-bound Gα proteins catalyzes the conversion of adenosine triphospate (ATP) to adenosine-3′5′-cyclic monophosphate (cAMP). The cAMP molecules act as secondary messengers to activate protein kinase A. In the absence of cAMP, the PKA holoenzyme exists as an inactive heterotetramer composed of two regulatory (R) and two catalytic (C) subunits (Robison et al. 1968; Krebs 1989). The R subunits have a conserved domain structure, consisting of an N-terminal dimerization domain, an inhibitory domain, cAMP-binding domain A, and cAMP-binding domain B (Robison et al. 1968; Krebs 1989). R subunits interact with C subunits primarily through the inhibitory site (Francis and Corbin 1999). Binding of four cAMP molecules to the cAMP-binding domains on the two R subunits releases the active catalytic subunits as dimers that can phosphorylate serine and threonine residues on downstream proteins, including transcription factors (Krebs 1989). The intensity of cAMP signaling also depends on the balance between cAMP production and its degradation. 3′5′-cyclic nucleotide phosphodiesterases (PDEs) control the level of cAMP by converting active cAMP to inactive 5′-AMP, thus terminating PKA activation (Fig. 7.2). Nutrient sensing, perhaps the most extensively studied function of the cAMP signaling pathway in yeast and filamentous fungi, is described later in this chapter.

2. Mitogen-Activated Protein Kinase (MAPK)

MAPK signaling pathways are evolutionary-conserved signaling mechanisms that regulate diverse physiological processes, ranging from cell proliferation and differentiation to cell death in eukaryotes (Gustin et al. 1998; Widmann et al. 1999; Qi and Elion 2005; Keshet and Seger 2010). In fungi, MAPKs are involved in regulating mating, morphogenesis, virulence, and stress responses (Kronstad et al. 1998; Lopez-Ilasaca 1998; Mehrabi et al. 2012; Raudaskoski and Kothe 2010). The key components of this pathway are the MAPK kinase kinase (MAPKKK), MAPK kinase (MAPKK), and MAPK (Qi and Elion 2005) (Fig. 7.3). These proteins operate as a three-tiered module to sequentially phosphorylate downstream proteins such as transcription factors, cytoskeletal proteins, kinases, and other enzymes, thereby greatly influencing gene expression, metabolism, cell division, cell morphology, and cell survival (Qi and Elion 2005). The first kinase of the MAPK unit is the MAPKKK. MAPKKKs have an N-terminal autoinhibitory domain, a C-terminal catalytic domain, and a large regulatory motif outside the catalytic domain. The regulatory motifs consists of Pleckstrin homology (PH) and Src homology 3 (SH3) domains, binding sites for GTP-binding proteins, leucine-zipper dimerization sequences, and phosphorylation sites for tyrosine and serine/threonine kinases (Garrington and Johnson 1999). The presence of different kinds of regulatory motifs in MAPKKKs reflects its selective regulation by a variety of upstream inputs, and its flexibility and specificity in responding to a wide range of cellular stimuli through activation of specific MAPKKs and MAPKs of three-tiered modules (Garrington and Johnson 1999). MAPKKs are dually phosphorylated by MAPKKKs at serine and threonine residues within the activation loop of the catalytic domain. They are highly specific for their cognate MAPK and activate it by phosphorylating the MAPK at conserved threonine and tyrosine residues (T–X–Y) within the activation loop (Payne et al. 1991; Ferrell and Bhatt 1997). Depending on the fungal system and environmental stimuli, all or some of the MAPK cascade components may shuttle in and out of the nucleus (Qi and Elion 2005; Nguyen et al. 2002; Bayram et al. 2012). MAPKs generally contain multiple docking domains for binding to substrates, MAPKKs, and scaffolding proteins, allowing high affinity protein–protein and protein–substrate interactions (Tanoue and Nishida 2003). Scaffolding proteins associated with the MAPKs facilitate efficient phosphorylation, preventing cross-talk between different signaling pathways, thus maintaining the integrity of the initiating stimulus (Burack and Shaw 2000). The activated MAPK phosphorylates target proteins, in many cases transcription factors, at serine and threonine residues (Wang et al. 2011; Bardwell 2006). The activated transcription factor is then able to facilitate changes in gene expression in response to the initial stimulus (Fig. 7.3).

There are three conserved MAPKs in fungi (Posas et al. 1998; Hamel et al. 2012). In *S. cerevisiae*, they are required for the pheromone response and mating (Fus3p/Kss1p; Dohlman and Slessareva 2006), cell wall integrity (Mpk1p; Fuchs and Mylonakis 2009), and osmosensing (Hog1p; Saito and Posas 2012). The MAPK cascade required for the pheromone response and mating has been extensively studied in several fungi and will be described later in this chapter. For a detailed review on fungal MAPK pathways, see also chapter "Fungal MAP-Kinase-Mediated Regulatory Pathways" in this volume.

IV. Cellular Functions of G Protein Signaling Components in Yeast and Filamentous Fungi

A. Nutrient Sensing

Nutrients are the source of energy and building blocks for the cell. They act as regulators of metabolism, growth, and development. The key process for the survival of any organism depends on their ability to adapt to changing nutritional conditions. The plasma membrane of the cell is the first boundary that detects the signaling molecules via membrane receptors and transfer nutrients from the external environment to the intracellular environment via its transport system. Certain GPCRs in yeasts and filamentous fungi are known to detect nutrients and activate signal transduction in association with Gα proteins (Holsbeeks et al. 2004). Several examples of such nutrient-sensing pathways are discussed in the following sections, and a generic model diagram is presented in Fig. 7.2.

1. Glucose Sensing in *Saccharomyces cerevisiae* and *Schizoaccharomyces pombe*

In *S. cerevisiae*, Gpr1p is a sugar-sensing GPCR that interacts with the C-terminal end of the Gpa2p Gα protein (Yun et al. 1997). Gpr1p is very effective in sensing glucose and sucrose, while the structurally similar sugars fructose and galactose had no effect and mannose is antagonistic (Lemaire et al. 2004; Rubio-Texeira et al. 2010). Results from cysteine scanning mutagenesis of residues in transmembrane domain VI of *GPR1* support a model in which glucose and sucrose are ligands for Gpr1p (Lemaire et al. 2004). Glucose-mediated Gpr1p signaling in *S. cerevisiae* results in low levels of the storage carbohydrates trehalose and glycogen and reduced expression of stress resistance genes (Tamaki 2007; Rubio-Texeira et al. 2010). The current model is that glucose binding to Gpr1p directs the formation of the GTP-bound active form of Gpa2p, which then stimulates adenylyl cyclase to transiently increase cAMP levels, setting in motion a PKA-mediated protein phosphorylation cascade (Kraakman et al. 1999). In this scenario, the RGS protein Rgs2p negatively controls the activity of Gpa2p to terminate signaling (Versele et al. 1999).

Yeast cells grown in glucose medium are larger in size than those cultured with non-fermentable carbon sources, such as ethanol (Johnston et al. 1979; Busti et al. 2010). *gpr1* and *gpa2* mutant cells are smaller than wild type in glucose medium, leading to the suggestion that Gpr1p–Gpa2p signaling regulates cell size and longevity in *S. cerevisiae* (Tamaki et al. 2005; Tamaki 2007). Signaling through Gpr1p and Gpa2p is also essential for pseudohyphal filamentous growth in *S. cerevisiae* (Lorenz et al. 2000). For instance, diploid *Δgpr1* or *Δgpa2* mutants cannot undergo the yeast to pseudohyphal transition under certain environmental conditions (Lorenz and Heitman 1997; Lorenz et al. 2000).

The fission yeast *S. pombe* primarily responds to glucose via the cAMP-signaling pathway. The components of this pathway include the GPCR Git3, the Gα Gpa2, the Gβ Git5, the Gγ Git11 and the adenylyl cyclase Git2/Cyr1 (Landry and Hoffman 2001; Landry et al. 2000; Hoffman 2005a). Loss of *git3*, *gpa2* or *git2* results in slow cell growth, low intracellular cAMP levels and mating and sporulation even in the absence of nitrogen, supporting the action of these three genes in the same pathway.

Gpa2 binds to and directly activates adenylyl cyclase (Ivey and Hoffman 2005), and mutation of residues important for GTPase activity in *gpa2* leads to a sharp increase in cAMP levels in response to glucose stimulation (Isshiki et al. 1992; Welton and Hoffman 2000). Elevated cAMP levels lead to activation of PKA by binding to the regulatory subunit, Cgs1 (Maeda et al. 1994).

The Git5 Gβ subunit is unique in *S. pombe* as it lacks the N-terminal coiled-coil region which is required for Gγ binding and proper folding of the Gβ subunit (Sondek et al. 1996). However, co-immunoprecipitation studies with Git5 and Git11 support their physical association (Landry and Hoffman 2001). Git11 is required for localization of Git5, and not for its folding, which is supported by the suppression of the *Δgit11* mutation by fusing the Git11 CAAX box to the C-terminus of Git5 (Landry and Hoffman 2001).

The *fbp1* gene encodes the gluconeogenic enzyme fructose-1,6-bisphosphatase (Hoffman and Winston 1990, 1991). Glucose represses transcription of the *fbp1* gene via PKA activation. Strains with the *Δgit3* or *Δgpa2* mutations are derepressed for transcription of *fpb1* on high glucose, while those carrying a constitutively activated *gpa2* allele (*gpa2*^{R176H}) have reduced transcription of *fbp1* (Roux et al. 2009). Simultaneous loss of *gpa2* and *git3* has an additive effect on *fbp1* expression (Welton and Hoffman 2000), suggesting some functions for Git3 independent of Gpa2.

2. Carbon-Sensing Pathways in *Neurospora crassa and Aspergillus nidulans*

Glycerol is one of the significant nutrients utilized by plant pathogenic fungi (Wei et al. 2004), and GPCR-mediated signaling is required for the growth of filamentous fungi in poor carbon sources such as glycerol, arabinose, mannitol, and sorbose (Li and Borkovich 2006). The *N. crassa* carbon-sensing machinery includes a putative GPCR, GPR-4, which is homologous to Gpr1p and Git3 of *S. cerevisiae* and *S. pombe*, respectively (Li and Borkovich 2006). Δ*gpr-4* mutants accumulate much less biomass than wild type in glycerol medium and lack the transient cAMP increase after addition of glucose to glycerol-grown cultures (Li and Borkovich 2006). In addition to glycerol, Δ*gpr-4* mutants exhibit mass defects with mannitol and arabinose as carbon sources (Li and Borkovich 2006). Results from genetic epistasis analysis and the yeast two hybrid assay indicate that the C-terminus of GPR-4 interacts with GNA-1 and that GPR-4 requires GNA-1 to transduce signals downstream (Li and Borkovich 2006).

In *A. nidulans*, the heterotrimeric G protein GanB (Gα subunit)/SfaD(Gβ)/GpgA(Gγ) activates the cAMP-PKA pathway in response to glucose (Seo et al. 2004; Lafon et al. 2005) and controls conidial germination. During the early period of conidial germination, GanB mediates a rapid and transient activation of cAMP synthesis in response to glucose and subsequent stimulation of PKA activity (Lafon et al. 2005). *ΔganB* has altered germination kinetics and defects in trehalose mobilization. Addition to this, germination of *ΔganB* conidia is even impaired in the presence of glycerol, ethanol, galactose, and fructose, indicating the role of GanB in sensing carbon sources other than glucose. Although a carbon-sensing GPCR has not been reported for *A. nidulans*, GprC and GprD are putative GPCRs with homology to *S. cerevisiae* Gpr1p (Affeldt et al. 2014) in the related species *A. flavus. A. flavus* mutants lacking *gprD* have reduced growth on galactose, xylose, and corn oil (Affeldt et al. 2014).

3. Methionine and Glucose Sensing in *Cryptococcus neoformans*

The GPCR Gpr4 identified in *C. neoformans* is known to sense and respond to amino acids in the environment via cAMP signaling pathway (Xue et al. 2006). Gpr4 shares amino acid similarity with the Gpr1 of *S. cerevisiae* and *N. crassa* GPR-4. In spite of its homology to carbon-sensing GPCRs, *C. neoformans* Gpr4 is not required for the carbon-regulated cAMP pathway. In contrast, evidence suggests that Gpr4 senses methionine, as exposure to this amino acid leads to internalization of fluorescently tagged Gpr4 and a transient increase in cAMP levels in wild type, but not *Δgpr4* cells (Xue et al. 2006). Methionine induces mating filamentation in wild-type, but not in *Δgpr4* or *Δgpa1* mutants, and *gpr4* mutants are defective in capsule production and mating, similar to

Δgpa1 mutants, supporting their action in the same pathways. However, Gpa1, but not Gpr4, is required for synthesis of the virulence factor melanin in response to glucose (Alspaugh et al. 1997, 2002). This suggests involvement of other receptors or an alternative mechanism for regulation of Gpa1 functions (Xue et al. 2006). The requirement for Gpr4 and Gpa1 functions in the amino acid-cAMP signaling pathway, as well as Gpa1 for the glucose-cAMP sensing mechanism, presents an example of a nutrient coincidence detection system that remains to be understood in other fungi (Xue et al. 2006, 2008).

B. Mating and Pheromone Response

1. Ste2p/α-Factor and Ste3p/a-Factor in *S. cerevisiae*

S. cerevisiae is a heterothallic yeast, with two mating types, *MATα* and *MATa* (Kurjan 1992). The two known GPCR receptors in yeast, Ste2p and Ste3p, are activated by pheromones α-factor and a-factor, respectively (Kurjan 1992). *MATα* cells produce the 13-residue peptide [WHWLQLKPGQPMY] α-factor, which activates the Ste2p receptor on the *MATa* cell (Burkholder and Hartwell 1985; Nakayama et al. 1985). The *MATa* cell generates the 12-residue lipopeptide [YIIKGVFWDPAC(farnesyl)OCH_3] a-factor that targets the Ste3p receptor (Nakayama et al. 1985; Hagen et al. 1986). Pheromone binding stimulates GDP to GTP exchange on the Gpa1p Gα subunit, allowing release of the Ste4p/Ste18p Gβγ heterodimer to activate downstream signaling (Kurjan 1992); Fig. 7.3). The Ste4p/Ste18p heterodimer recruits a scaffold protein, Ste5p, to the plasma membrane, forming the site of mating projection ("shmoo") formation (Elion 2000, 2001; Merlini et al. 2013). Through its different domains, Ste5p recruits members of a MAPK cascade: the MAPKKK Ste11p, the MAPKK Ste7p, and the MAPK Fus3p (Cairns et al. 1992; Marcus et al. 1994; Whiteway et al. 1995). Ste11p is phosphorylated by the p21-activated kinase (PAK) Ste20p that is in turn activated by the GTP-bound form of Cdc42p (Burbelo et al. 1995; Cabib et al. 1998). Ste11p binds to the adaptor protein Ste50p that also binds to Cdc42p, thereby, bridging Ste11p and active Ste20p (Ramezani-Rad 2003; Wu et al. 1999). Upon phosphorylation by Ste20p, Ste11p catalyzes the sequential phosphorylation and activation of Ste7p and Fus3p (Gartner et al. 1992; Good et al. 2009; Neiman and Herskowitz 1994). Fus3p then phosphorylates substrates in the cytoplasm to induce cell cycle arrest and polarized growth (Elion et al. 1993). Fus3p also phosphorylates two functionally redundant transcription repressors, Dig1p and Dig2p, and the transcription activator, Ste12p, to induce expression of pheromone-responsive genes (Chou et al. 2006; Tedford et al. 1997; Colman-Lerner et al. 2005). Thus, precise quantitative information regarding pheromone bound at the cell surface receptor reaches the nucleus within minutes. A generic model of a fungal pheromone sensing pathway is provided in Fig. 7.3.

2. Map3/M-Factor and Mam2/P-Factor in *S. pombe*

Mating is triggered by nutrient starvation in *S. pombe* (Egel 1971). The mating pheromones are P-factor and M-factor, with P-factor sensed by the Mam2 GPCR expressed on h^- cells and M-factor sensed by the Map3 GPCR expressed on h^+ cells (Kitamura and Shimoda 1991; Tanaka et al. 1993). Mam2 and Map3 are homologous to *S. cerevisiae* Ste3p and Ste2p pheromone receptors, respectively (Tanaka et al. 1993; Kitamura and Shimoda 1991; Hoffman 2005b). In *S. pombe*, pheromone signaling and downstream regulation of the MAPK cascade are activated by the Gα subunit Gpa1, in contrast to *S. cerevisiae*, where the Gβγ heterodimer plays this role (Obara et al. 1991; Hoffman 2005a). The MAPK cascade consists of the MAPKKK Byr2, MAPKK Byr1, and MAPK Spk1, which are homologs of Ste11p, Ste7p, and Fus3p from *S. cerevisiae* (Nadin-Davis and Nasim 1990; Wang et al. 1991; Xu et al. 1994). No homologue of the Ste5p scaffolding protein is found in *S. pombe*, and the mechanism whereby Byr2 is activated by Gpa1 is unknown. However, the mating-specific protein Ste4, which is homologous to *S. cerevisiae* Ste50p, is known to interact with Byr2 and promote its activation (Okazaki et al. 1991; Barr et al. 1996; Ramachander et al. 2002). The small GTPase Ras1, which positively regulates mating and sexual sporulation in *S. pombe*, is located on the plasma membrane and activated by the Ste6 GEF and inactivated by the Gap1 GAP (Hughes et al. 1994). Ras1 activation by Ste6 recruits Byr2 to the plasma membrane and relieves its autoinhibition, thereby causing Byr2 to drive sequential phosphorylation of Byr1 and Spk1. Phosphorylated Spk1 activates Ste11, a homologue of *S. cerevisiae* Ste12p, that acts as a developmental switch from cellular proliferation to sexual differentiation (Sugimoto et al. 1991; Mata and Bahler 2006).

Constitutive expression of Ste11 causes starvation-independent sexual differentiation (Sugimoto et al. 1991), and Ste6 expression is under the direct regulation of Ste11. Once expressed, Ste11 drives its own transcription by binding to the TR box, a ten base motif (5′-TTCTTTGTTY-3′) in its upstream region (Sugimoto et al. 1991; Kunitomo et al. 2000).

3. PRE-1/MFA-2 and PRE-2/CCG-4 in *N. crassa*

N. crassa is a heterothallic fungus with two mating types, *mat A* and *mat a* (Raju 1992). The sexual cycle in *N. crassa* is initiated by nitrogen starvation, light, and low temperature, leading to the formation of the female reproductive structures, protoperithecia (Raju 1992). Specialized hyphae (trichogynes) originating from protoperithecia of one mating type grow towards a conidium of the opposite mating through chemotropic attraction (Bistis 1981, 1983). Trichogyne attraction is regulated by the GPCRs PRE-1 and PRE-2 that recognize their respective pheromone ligands, MFA-1 and CCG-4, respectively (Kim et al. 2002, 2012; Bobrowicz et al. 2002; Kim and Borkovich 2004, 2006). PRE-1 and PRE-2 are homologous to the Ste2p and Ste3p pheromone receptors of *S. cerevisiae* (Galagan et al. 2003; Borkovich et al. 2004). *mat A* strains express high levels of *pre-1* and *ccg-4* mRNA, while expression of *pre-2* and *mfa-1* is elevated in *mat a* strains (Bobrowicz et al. 2002; Kim and Borkovich 2004, 2006). Mating-type-specific gene expression causes *Δpre-1 mat A* trichogynes to be blind and to fail to fuse with *mat a* conidia, while *Δpre-2 mat a* trichogynes are blind to *mat A* conidia (Kim and Borkovich 2004; Kim et al. 2012). Furthermore, mutants lacking the pheromone genes *ccg-1* and *mfa-1* are unable to sense trichogynes are male sterile in the *mat A* or *mat a* background, respectively (Kim and Borkovich 2006). Male fertility in *N. crassa* is due to lack of pheromones, and no other G protein signaling component is required for male fertility (Kim and Borkovich 2006). Finally, as mentioned earlier, the RIC8 cytosolic GEF is required for female fertility in *N. crassa*, but since mutants are blocked in production of female structures, it is not clear whether RIC8 is also essential for the pheromone response pathway (Wright et al. 2011).

The Gα protein GNA-1 is required for chemotropism in both mating types, suggesting a role downstream of both pheromone/ligand systems (Kim and Borkovich 2004). Similar to GNA-1, GNB-1 (Gβ), and GNG-1 (Gγ) are essential for proper trichogyne attraction, since *Δgnb-1* and *Δgng-1* trichogynes are also blind (Krystofova and Borkovich 2005). However, the low levels of GNA-1 protein in *Δgnb-1* and *Δgng-1* mutants may point to an indirect role for GNB-1 and GNG-1 in trichogyne chemotropism (Krystofova and Borkovich 2005). The adenylyl cyclase protein, CR-1, is known to act downstream of GNA-1 (and GNA-3) in *N. crassa* (Ivey et al. 1999; Kays and Borkovich 2004). However, the *cr-1* mutant is female-fertile and only delayed in postfertilization development (Ivey et al. 1996). Exogenous cAMP does not rescue the female-sterile phenotype of the *Δgna-1* mutant, suggesting the existence of CR-1/cAMP-independent signaling pathways during perithecial development (Ivey et al. 2002). Furthermore, the fertility defects of *Δgna-3* and *Δgnb-1* mutants are not corrected by exogenous cAMP (Kays et al. 2000; Yang et al. 2002). Taken together, the data suggest that mating and sexual sporulation requires both cAMP-dependent and cAMP-independent signaling pathways in *N. crassa*.

4. Mating and Cleistothecia Formation in *Aspergillus*

The mating-specific genes have also been identified in *Aspergillus* species, including both sexual and presumed asexual species (Dyer et al.

2003; Paoletti et al. 2005; Pel et al. 2007; Ramirez-Prado et al. 2008). The mating type genes *matA* and *matB* were first described in the homothallic (self-fertile) *A. nidulans*. Deletion of either gene results in production of fewer and smaller sexual fruiting bodies (cleistothecia) and sterile ascospores (Paoletti et al. 2007). Simultaneous overexpression of *matA* and *matB* resulted in vegetative growth arrest and inappropriate development of cleistothecia, further confirming their key roles in sexual development (Paoletti et al. 2007). The predicted GPCRs GprA and GprB are similar to *S. cerevisiae* Ste2p and Ste3p, respectively, and function during self-fertilization to produce the fruiting body, called a cleistothecium (Seo et al. 2004). The *ΔgprA* and *ΔgprB* single mutants produce fewer and smaller cleistothecia, while *ΔgprA ΔgprB* double mutants failed to produce cleistothecia, indicating a role for both receptors in cleistothecia formation (Seo et al. 2004). Genetic evidence supports the action of GprA and GprB upstream of a heterotrimeric G protein consisting of FadA (Gα), SfaD (Gβ), and GpgA (Gγ) (Rosen et al. 1999; Seo et al. 2005). Deletion of any of these genes results in failure of cleistothecia formation. The RGS for FadA, FlbA, is also required for the development of cleistothecia (Yu et al. 1996).

A conserved MAPK cascade acts downstream of G protein signaling during sexual development in *A. nidulans*. The components are Ste20 (PAK), SteC (MAPKKK), Ste7 (MAPKK), and MpkB (MAPK) that activate a final homeodomain protein SteA, which then triggers the further stages of sexual development (Vallim et al. 2000; Wei et al. 2003; Paoletti et al. 2007). Deletion of individual genes in the pathway results in sterility, with failure to form cleistothecia.

5. Mating in *Cryptococcus neoformans*

There are two mating type loci, *MATα* and *MATa*, in the heterothallic basidiomycete fungus *Cryptococcus neoformans* (Kwon-Chung 1975, 1976). The mating type loci are quite large in *C. neoformans*, encoding more than 20 genes (Lengeler et al. 2002; Kozubowski and Heitman 2012). The *MATα* locus contains the peptide pheromone genes *MFα1-3*, which share sequence similarity with other fungal and yeast pheromones and contains a C-terminal CAAX box for lipid modification (Moore and Edman 1993). The *MATa* locus encodes the Mfa1-3 pheromones, that each possess a farnesylation motif (McClelland et al. 2002). Under nutrient starvation, opposite mating type cells of *C. neoformans* form conjugation tubes that fuse to form dikaryotic filaments (Kozubowski and Heitman 2012). *MATα* cells are unique in that they can produce filaments and sporulate (haploid fruiting) in response to nitrogen starvation, even in the absence of *MATa* cells (Kozubowski and Heitman 2012).

The *C. neoformans* mating type loci also encode the pheromone receptors Ste3α and Ste3a (Lengeler et al. 2002) that recognize the Mfa and Mfα type pheromones, respectively (Stanton et al. 2010). The Gpa2 Gα protein is required for mating and has been shown to interact with the pheromone receptors, Gβ subunit Gpb1 and RGS protein Crg1 (Wang et al. 2000; Li et al. 2007a). The two Gγ subunits Gpg1 and Gpg2 interact with Gpb1 and regulate sexual development (Li et al. 2007a). Gpb1 operates upstream of a MAPK cascade homologous to the pheromone response MAPK pathway in *S. cerevisiae* that is required for mating in *C. neoformans* (Wang et al. 2000). Interestingly, in addition to the two Ste3-like GPCRs, the *C. neoformans* genome contains another pheromone receptor-like gene, *CPR2*, that is not encoded by the mating type locus (Hsueh et al. 2009). Evidence suggests that Cpr2 is a constitutively activated GPCR that couples to the same downstream signaling components as the two pheromone receptors, but regulates mating in the absence of pheromone in *C. neoformans* (Hsueh et al. 2009).

C. Pathogenesis and Virulence

1. Appressorium Formation and Pathogenicity in *Magnaporthe oryzae*

M. oryzae causes rice blast disease by forming a specialized infection structure called an appressorium (Li et al. 2012; Liu et al. 2010; Dean et al. 2005). The mature appressorium generates enormous internal turgor pressure that is translated into strong mechanical forces exerted on the penetration peg, allowing the pathogen to penetrate the plant cuticle and cell wall. Following penetration, the pathogen proliferates and fills up the plant cell lumen, resulting in necrotic lesions on the leaf surface.

The first step in the infection-related process is for *M. oryzae* to recognize the plant leaf. Hydrophobicity and the hard surface of the leaf are two major factors that induce appressoria formation (Li et al. 2012; Liu et al. 2010). The predicted GPCR, Pth11, is required for sensing surface hydrophobicity of leaves (DeZwaan et al. 1999). Similar to other filamentous fungi, *M. oryzae* has three Gα (MagA, MagB, and MagC), one Gβ (Mgb1), and one Gγ (Mgg1) proteins (Liu and Dean 1997; Nishimura et al. 2003; Liang et al. 2006). Loss of the *MAGB* Gα gene significantly reduces, but does not completely block, appressorium formation (Liu and Dean 1997). However, exogenous cAMP or 1,16-hexadecanediol restored appressorium formation in Δ*magB* mutants, reflecting a role for intracellular cAMP in initiating appressorium formation on leaf surfaces (Liu and Dean 1997). As mentioned earlier, mutants lacking the *RIC8* GEF gene exhibit a defect in appressorium development that can be partially remediated by exogenous cAMP, but the resulting appressoria are not infectious (Li et al. 2010). Additional support for a role for cAMP during early steps in infection is provided by the demonstration that mutants lacking adenylyl cyclase (*MAC1*) are unable to produce appressoria (Choi and Dean 1997). Furthermore, deletion of the gene encoding the catalytic subunit of PKA (*CPK1*) blocks appressorium formation, even in the presence of exogenous cAMP (Mitchell and Dean 1995; Fernandez and Wilson 2014). Taken together, the findings are consistent with a mechanism in which elevated cAMP activates PKA (Cpk1), initiating appressorial formation. For a generic model of pathogenicity in filamentous fungi, refer to Fig. 7.4.

Deletion of the RGS gene *rgs1* leads to a significant increase in cAMP levels and conidiation, as well as appressoria formation even on the noninducive hydrophilic surfaces (Liu et al. 2007; Zhang et al. 2011). In addition, Rgs1 interacts with all three Gα proteins in *M. oryzae* (Zhang et al. 2011), consistent with a role in regulating GTP occupancy on MagB and downstream appressorium development. In contrast to *MAGB*, *MGB1* and *MGG1* are essential for infection, as deletion of either gene leads to a complete failure to form appressoria (Nishimura et al. 2003; Liang et al. 2006). Exogenous cAMP induces appressorium development in *mgb1* mutants, but these appressoria are defective in shape and cannot initiate infection, suggesting the involvement of other infection-related pathways mediated by Mgb1 and Mgg1 (Nishimura et al. 2003; Liang et al. 2006).

Deletion of genes encoding the MoSom1 and MoCdtf1 transcription factors results in defects in appressorium formation and virulence, and these two proteins interact with CPK1, consistent with roles as downstream targets of CPK1 for initiation of infection (Yan et al. 2011). Taken together, the data suggest that the upstream pathway for initiating infection involves stimulation of the Pth11 GPCR by the leaf surface, which in turn activates MagB, releasing the Mgb1–Mgg1 heterodimer, leading to stimulation of adenylyl cyclase (Mac1), increased cAMP synthesis, and activation of PKA. Of interest, it has recently been shown that MagA, Rgs1, adenylyl cyclase, and the Pth11 GPCR are sequestered on late endosomes during early pathogenic development, presumably to bring these components together to facilitate efficient signal transduction (Ramanujam et al. 2013).

The later stages of appressorial formation and penetration into the leaf surface require two MAPKs, Pmk1 and Mps1, similar to *S. cerevisiae* Fus3p/Kss1p and Slt2p, respectively (Xu and Hamer 1996; Xu et al. 1998). *Δpmk1* mutants recognize leaf surface hydrophobicity, respond to cAMP, and germinate following adhesion, but are unable to form appressoria and invasive hyphae and are thus avirulent (Xu and Hamer 1996). The other components of the Pmk1 pathway are Mst11 (MAPKKK) and Mst7 (MAPKK), forming a Mst11–Mst7–Pmk1 three-tiered MAPK cascade (Zhao et al. 2005). Like the *Δpmk1* mutant, *Δmst11* and *Δmst7* mutants

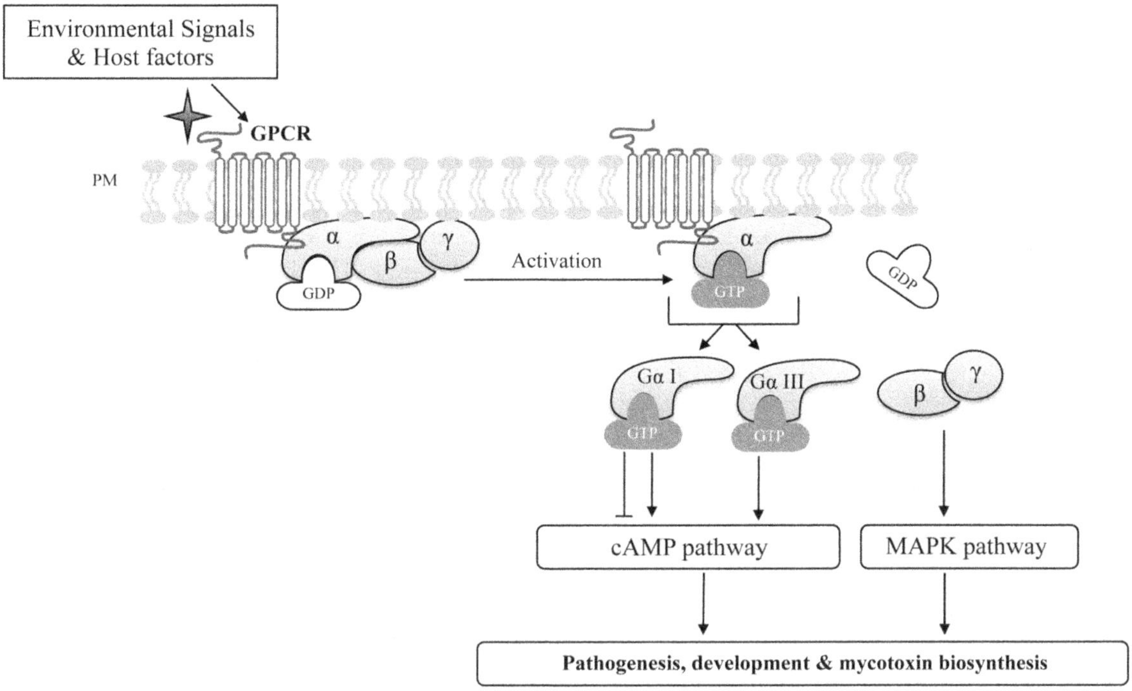

Fig. 7.4 Pathogenicity. Involvement of Group I and III Gα proteins in pathogenicity responses has been demonstrated in a wide variety of filamentous fungi. Appressorium formation in *Magnaporthe oryzae* involves the predicted GPCR Pth11 and the Gα MagB, as well as intracellular cAMP and protein kinase A (Cpk1). In *U. maydis*, the Group III Gα protein Gpa3 is involved in pathogenicity, while in *C. neoformans* Gpa1 is important for virulence and capsule formation. These Gα proteins and the Gβγ dimer impact regulation of PKA and MAPK signaling cascades to control fungal pathogenesis. *Abbreviations*: *GPCR* G protein-coupled receptors, *GDP* guanosine diphosphate, *GTP* guanosine triphosphate, *cAMP* cyclic adenosine monophosphate, *PKA* protein kinase A, *MAPK* mitogen-activated kinase

fail to form appressoria and are nonpathogenic (Zhao et al. 2005). The *M. oryzae* homolog of yeast Ste50p, Mst50, interacts with Mst11 and Mst 7, and *Δmst50* mutants are defective in appressorium formation (Park et al. 2006). Finally, mutants lacking the *MPS1* MAPK trigger early plant-cell defense responses, but are unable to penetrate the plant surface and cause disease (Xu et al. 1998).

2. Pathogenesis in *Ustilago maydis*

U. maydis is a basidiomycete fungus that causes corn smut disease (Brefort et al. 2009). The fungus propagates as haploid budding yeast-like cells and becomes infectious only in the dikaryotic stage that is generated after mating (Brefort et al. 2009; Vollmeister et al. 2012). The dikaryotic stage is characterized by filamentous growth of the fungus, by which it penetrates the meristematic tissue of the plant, culminating in the production of diploid teliospores within tumors on the plant (Brefort et al. 2009). The transition from yeast-like to constitutively filamentous growth requires involvement of G-protein-mediated cAMP signaling and a MAPK cascade (Brefort et al. 2009; Vollmeister et al. 2012). Cell recognition and fusion are regulated by genes at the *a* mating type locus that encodes a pheromone precursor, Mfa1, and pheromone receptors Pra1 and Pra2 (Bakkeren and Kronstad 1996). Following fusion, the dikaryotic stage is maintained by the multiallelic *b* mating type locus which encodes bE and bW homeodomain proteins (Banuett 1995). The pathogenicity of the organism depends on the active formation of bE-bW heterodimers (Banuett 1995; Kahmann and Bolker 1996; Kahmann and Kamper 2004).

Four heterotrimeric G protein Gα subunits (Gpa1, Gpa2, Gpa3 and Gpa4) have been identified in *U. maydis*, with Gpa3 playing an active role in the pheromone response and pathogenic development (Regenfelder et al. 1997). *Δgpa3* mutants are nonpathogenic, fail to respond to pheromone, and are unable to mate. Introduction of a GTPase-deficient allele of *gpa3* ($gpa3^{Q206L}$) causes *U. maydis* cells to mate even in the absence of pheromone signals (Regenfelder et al. 1997). The *Δbpp1* mutant grows filamentously and has reduced expression of pheromone genes, but still causes tumors on corn plants (Muller et al. 2004). Expression of the $gpa3^{Q206L}$ allele suppresses the *Δbpp1* mutant phenotypes (Muller et al. 2004). The nonpathogenic nature of Δ*gpa3* mutants in contrast to *Δbpp1* strains suggests that Gpa3 regulates pathogenic development independently of Bpp1 (Muller et al. 2004). Gpa3 acts upstream of adenylyl cyclase (Uac1) for synthesis of cAMP (Kruger et al. 1998). The finding that mutants lacking either *uac1* or *ubc1* (encoding the regulatory subunit of PKA) are nonpathogenic supports a role for cAMP signaling in pathogenesis (Gold et al. 1994, 1997).

The High Mobility Group (HMG)-domain transcription factor Prf1 (Hartmann et al. 1996) induces the expression of *a* and *b* mating type genes and is phosphorylated by the catalytic subunit of PKA, Adr1 (Kaffarnik et al. 2003). Prf1 is also phosphorylated by the MAPK Kpp2 to trigger expression of *b* mating type genes to undergo dikaryon formation (Kaffarnik et al. 2003). Taken together, the results indicate that cAMP-PKA and pheromone-responsive MAPK pathways regulate mating, filamentous growth, and pathogenesis in *U. maydis*.

3. Melanin and Capsule Formation in *Cryptococcus neoformans*

In the human pathogenic fungus *C. neoformans*, G-protein signaling is an integral part of the virulence pathway. The Gpa1(Gα)/Cac1(adenylyl cyclase)/Pka1(catalytic subunit of PKA) cAMP signaling pathway controls growth, differentiation, and virulence (Alspaugh et al. 1997, 2002). Mutations in the components of this pathway lead to attenuated cAMP signaling, as well as alterations in capsule size and melanin formation. For example, *Δgpa1* mutants fail to produce melanin and capsule and are avirulent (Alspaugh et al. 1997). The RACK1 homolog Gib2 appears to be an essential gene in *C. neoformans*, and overexpression suppresses the melanization and capsule defects of a *gpa1* mutant (Palmer et al. 2006). Gib2 forms heterotrimeric complex with Gpa1 and Gpg1 or Gpg2 (Palmer et al. 2006), further supporting a role in signaling with Gpa1 to control virulence.

4. Quorum Sensing in *Aspergillus*

Quorum sensing is a system of stimulus and response correlated to population density (Garg et al. 2014). Quorum sensing was once thought to exist only in bacteria, but is now well established in fungi (Albuquerque and Casadevall 2012), including *Aspergillus* species. *A. flavus* is a soil-borne pathogen that infects a variety of food and feed crops including maize, peanuts, cottonseed, and tree nuts and also famous for aflatoxin (AF) production, the most potent natural carcinogen known (Squire 1981). In contrast, *A. nidulans* does not produce AF, but terminates the pathway at the penultimate precursor, sterigmatocystin (Alkhayyat and Yu 2014).

One of the primary signals for conidiation and aflatoxin formation in *Aspergillus* species are **oxylipins**, oxygenated polyunsaturated fatty acids (Horowitz Brown et al. 2008; Brown et al. 2009). Oxylipins are synthesized by oxygenases that incorporate oxygen into a fatty acid backbone. At low inoculation density, oxylipin production is low, conidiation is reduced, and there is increased production of aflatoxin in *A. flavus*. Conversely, at high inoculum density, oxylipin levels are high, resulting in increased conidiation with a reduction in aflatoxin production (Herrero-Garcia et al. 2011; Affeldt et al. 2012). Recent work has identified two putative GPCRs (GprC and GprD) in *A. flavus* and GprA and GprD in *A. nidulans*—that are involved in oxylipin-mediated density-dependent shift (Affeldt et al. 2012, 2014). High density cultures of *A. flavus ΔgprC* and *ΔgprD* mutants produce low amounts of conidia and elevated aflatoxin levels that mimic the wild-type culture at low density. Furthermore, the spent medium from wild-type cells cultured at high density failed to induce conidiation in *ΔgprC* and *ΔgprD* strains, locking the fungus in the low density state. Evidence supports *A. nidulans* GprD as receptor that senses oxylipin ligands and initiates cAMP-PKA signaling (Affeldt et al. 2012).

IV. Conclusions

G protein-controlled pathways are used to sense pheromones and nutrients in several fungal species. These two classes of physically diverse signals elicit morphological changes and often synergize to regulate development. Heterotrimeric G proteins are also required for virulence in both animal and plant fungal pathogens.

The pheromone response pathway is perhaps the most conserved G protein pathway in fungi, with similar GPCRs, ligands, G proteins, and MAPK components utilized in diverse species. In many species, the pheromone response/mating and virulence pathways share the same heterotrimeric G protein and MAPK signaling components. In contrast, the GPCR/G protein wiring for nutrient sensing appears to be more variable, although downstream cAMP signaling is a conserved element.

Further analysis will reveal whether the use of alternative binding partners for heterotrimeric G proteins, such as RACK1 proteins, is widespread in fungi and other eukaryotes. It is also anticipated that additional novel upstream regulators, similar to RIC8 and GET3, will continue to be identified in filamentous fungi. Studies on shuttling of signaling proteins between the nucleus and cytoplasm/plasma membrane will undoubtedly reveal mechanistic details underlying both activation and down-regulation of G protein signaling.

It is clear that there are many similarities between G protein signaling pathways found in fungi and higher organisms. Future prospects for determining the mechanisms used in filamentous fungal systems are promising, since many pathway components have been identified in several species. Evolutionary comparisons between mammalian and fungal G protein pathways will continue to be a powerful tool for elucidating the molecular interactions that occur during signal transduction in eukaryotes.

Acknowledgements We acknowledge NIH grant GM086565 to K.A.B.. and an Indo-US Research Fellowship to R.T.

References

Adachi K, Hamer JE (1998) Divergent cAMP signaling pathways regulate growth and pathogenesis in the rice blast fungus *Magnaporthe grisea*. Plant Cell 10(8):1361–1374

Affeldt KJ, Brodhagen M, Keller NP (2012) Aspergillus oxylipin signaling and quorum sensing pathways depend on G protein-coupled receptors. Toxins 4(9):695–717

Affeldt KJ, Carrig J, Amare M, Keller NP (2014) Global survey of canonical *Aspergillus flavus* G protein-coupled receptors. mBio 5(5):e01501–e01514

Afshar K, Willard FS, Colombo K, Johnston CA, McCudden CR, Siderovski DP, Gonczy P (2004) RIC-8 is required for GPR-1/2-dependent Galpha function during asymmetric division of *C. elegans* embryos. Cell 119(2):219–230

Albuquerque P, Casadevall A (2012) Quorum sensing in fungi–a review. Med Mycol 50(4):337–345. doi:10.3109/13693786.2011.652201

Alkhayyat F, Yu JH (2014) Upstream regulation of mycotoxin biosynthesis. Adv Appl Microbiol 86:251–278

Alspaugh JA, Perfect JR, Heitman J (1997) *Cryptococcus neoformans* mating and virulence are regulated by the G-protein alpha subunit GPA1 and cAMP. Genes Dev 11(23):3206–3217

Alspaugh JA, Pukkila-Worley R, Harashima T, Cavallo LM, Funnell D, Cox GM, Perfect JR, Kronstad JW, Heitman J (2002) Adenylyl cyclase functions downstream of the Galpha protein Gpa1 and controls mating and pathogenicity of *Cryptococcus neoformans*. Eukaryot Cell 1(1):75–84

Anger T, Zhang W, Mende U (2004) Differential contribution of GTPase activation and effector antagonism to the inhibitory effect of RGS proteins on Gq-mediated signaling *in vivo*. J Biol Chem 279(6):3906–3915

Apanovitch DM, Slep KC, Sigler PB, Dohlman HG (1998) Sst2 is a GTPase-activating protein for Gpa1: purification and characterization of a cognate RGS-Galpha protein pair in yeast. Biochemistry 37(14):4815–4822

Bakkeren G, Kronstad JW (1996) The pheromone cell signaling components of the Ustilago *a* mating-type loci determine intercompatibility between species. Genetics 143(4):1601–1613

Banuett F (1995) Genetics of *Ustilago maydis*, a fungal pathogen that induces tumors in maize. Annu Rev Genet 29:179–208

Bardwell L (2006) Mechanisms of MAPK signalling specificity. Biochem Soc Trans 34(Pt 5):837–841

Barr MM, Tu H, Van Aelst L, Wigler M (1996) Identification of Ste4 as a potential regulator of Byr2 in the sexual response pathway of *Schizosaccharomyces pombe*. Mol Cell Biol 16(10):5597–5603

Bayram O, Bayram OS, Ahmed YL, Maruyama J, Valerius O, Rizzoli SO, Ficner R, Irniger S, Braus GH (2012) The *Aspergillus nidulans* MAPK module AnSte11-Ste50-Ste7-Fus3 controls development and secondary metabolism. PLoS Genet 8(7): e1002816

Berman DM, Gilman AG (1998) Mammalian RGS proteins: barbarians at the gate. J Biol Chem 273(3):1269–1272

Birnbaumer L (1992) Receptor-to-effector signaling through G proteins: roles for beta gamma dimers as well as alpha subunits. Cell 71(7):1069–1072

Bistis GN (1981) Chemotropic interactions between trichogynes and conidia of opposite mating-type in *Neurospora crassa*. Mycologia 73(5):959–975

Bistis GN (1983) Evidence for diffusible, mating-type-specific trichogyne attractants in *Neurospora crassa*. Exp Mycol 7:292–295

Bobrowicz P, Pawlak R, Correa A, Bell-Pedersen D, Ebbole DJ (2002) The *Neurospora crassa* pheromone precursor genes are regulated by the mating type locus and the circadian clock. Mol Microbiol 45(3):795–804

Bock A, Kostenis E, Trankle C, Lohse MJ, Mohr K (2014) Pilot the pulse: controlling the multiplicity of receptor dynamics. Trends Pharm Sci 35(12):630–638

Bohm A, Gaudet R, Sigler PB (1997) Structural aspects of heterotrimeric G-protein signaling. Curr Opin Biotechnol 8(4):480–487

Bolker M (1998) Sex and crime: heterotrimeric G proteins in fungal mating and pathogenesis. Fungal Genet Biol 25(3):143–156

Borkovich KA, Alex LA, Yarden O, Freitag M, Turner GE, Read ND, Seiler S, Bell-Pedersen D, Paietta J, Plesofsky N, Plamann M, Goodrich-Tanrikulu M, Schulte U, Mannhaupt G, Nargang FE, Radford A, Selitrennikoff C, Galagan JE, Dunlap JC, Loros JJ, Catcheside D, Inoue H, Aramayo R, Polymenis M, Selker EU, Sachs MS, Marzluf GA, Paulsen I, Davis R, Ebbole DJ, Zelter A, Kalkman ER, O'Rourke R, Bowring F, Yeadon J, Ishii C, Suzuki K, Sakai W, Pratt R (2004) Lessons from the genome sequence of *Neurospora crassa*: tracing the path from genomic blueprint to multicellular organism. Microbiol Mol Biol Rev 68(1):1–108

Brefort T, Doehlemann G, Mendoza-Mendoza A, Reissmann S, Djamei A, Kahmann R (2009) *Ustilago maydis* as a pathogen. Annu Rev Phytopathol 47:423–445

Brown SH, Scott JB, Bhaheetharan J, Sharpee WC, Milde L, Wilson RA, Keller NP (2009) Oxygenase coordination is required for morphological transition and the host-fungus interaction of *Aspergillus flavus*. Mol Plant Microbe Interact 22(7):882–894

Bruno KS, Aramayo R, Minke PF, Metzenberg RL, Plamann M (1996) Loss of growth polarity and mislocalization of septa in a Neurospora mutant altered in the regulatory subunit of cAMP-dependent protein kinase. EMBO J 15(21):5772–5782

Burack WR, Shaw AS (2000) Signal transduction: hanging on a scaffold. Curr Opin Cell Biol 12(2):211–216

Burbelo PD, Drechsel D, Hall A (1995) A conserved binding motif defines numerous candidate target proteins for both Cdc42 and Rac GTPases. J Biol Chem 270(49):29071–29074

Burkholder AC, Hartwell LH (1985) The yeast alpha-factor receptor: structural properties deduced from the sequence of the STE2 gene. Nucleic Acids Res 13(23):8463–8475

Busti S, Coccetti P, Alberghina L, Vanoni M (2010) Glucose signaling-mediated coordination of cell growth and cell cycle in *Saccharomyces cerevisiae*. Sensors 10(6):6195–6240

Cabib E, Drgonova J, Drgon T (1998) Role of small G proteins in yeast cell polarization and wall biosynthesis. Annu Rev Biochem 67:307–333

Cairns BR, Ramer SW, Kornberg RD (1992) Order of action of components in the yeast pheromone response pathway revealed with a dominant allele of the STE11 kinase and the multiple phosphorylation of the STE7 kinase. Genes Dev 6(7):1305–1318

Chan RK, Otte CA (1982) Isolation and genetic analysis of *Saccharomyces cerevisiae* mutants supersensitive to G1 arrest by a factor and alpha factor pheromones. Mol Cell Biol 2(1):11–20

Chan P, Thomas CJ, Sprang SR, Tall GG (2013) Molecular chaperoning function of Ric-8 is to fold nascent heterotrimeric G protein alpha subunits. Proc Natl Acad Sci USA 110(10):3794–3799

Chasse SA, Dohlman HG (2003) RGS proteins: G protein-coupled receptors meet their match. Assay Drug Dev Technol 1(2):357–364

Chasse SA, Flanary P, Parnell SC, Hao N, Cha JY, Siderovski DP, Dohlman HG (2006) Genome-scale analysis reveals Sst2 as the principal regulator of mating pheromone signaling in the yeast *Saccharomyces cerevisiae*. Eukaryot Cell 5(2):330–346

Chen JG, Willard FS, Huang J, Liang J, Chasse SA, Jones AM, Siderovski DP (2003) A seven-transmembrane RGS protein that modulates plant cell proliferation. Science 301(5640):1728–1731

Chini B, Parenti M, Poyner DR, Wheatley M (2013) G-protein-coupled receptors: from structural insights to functional mechanisms. Biochem Soc Trans 41(1):135–136

Chishiki K, Kamakura S, Yuzawa S, Hayase J, Sumimoto H (2013) Ubiquitination of the heterotrimeric G protein alpha subunits Galphai2 and Galphaq is prevented by the guanine nucleotide exchange factor Ric-8A. Biochem Biophys Res Commun 435(3):414–419

Choi W, Dean RA (1997) The adenylate cyclase gene *MAC1* of *Magnaporthe grisea* controls appressorium formation and other aspects of growth and development. Plant Cell 9(11):1973–1983

Chou S, Lane S, Liu H (2006) Regulation of mating and filamentation genes by two distinct Ste12 complexes in *Saccharomyces cerevisiae*. Mol Cell Biol 26(13):4794–4805. doi:10.1128/MCB.02053-05

Colman-Lerner A, Gordon A, Serra E, Chin T, Resnekov O, Endy D, Pesce CG, Brent R (2005) Regulated cell-to-cell variation in a cell-fate decision system. Nature 437(7059):699–706

De Vries L, Zheng B, Fischer T, Elenko E, Farquhar MG (2000) The regulator of G protein signaling family. Annu Rev Pharmacol Toxicol 40:235–271

Dean RA, Talbot NJ, Ebbole DJ, Farman ML, Mitchell TK, Orbach MJ, Thon M, Kulkarni R, Xu JR, Pan H, Read ND, Lee YH, Carbone I, Brown D, Oh YY, Donofrio N, Jeong JS, Soanes DM, Djonovic S, Kolomiets E, Rehmeyer C, Li W, Harding M, Kim S, Lebrun MH, Bohnert H, Coughlan S, Butler J, Calvo S, Ma LJ, Nicol R, Purcell S, Nusbaum C, Galagan JE, Birren BW (2005) The genome sequence of the rice blast fungus *Magnaporthe grisea*. Nature 434(7036):980–986

Denic V, Dotsch V, Sinning I (2013) Endoplasmic reticulum targeting and insertion of tail-anchored membrane proteins by the GET pathway. CSL Perspect Biol 5(8):a013334

DeZwaan TM, Carroll AM, Valent B, Sweigard JA (1999) *Magnaporthe grisea* Pth11p is a novel plasma membrane protein that mediates appressorium differentiation in response to inductive substrate cues. Plant Cell 11(10):2013–2030

Dohlman HG, Slessareva JE (2006) Pheromone signaling pathways in yeast. Sci STKE 2006(364):cm6

Dohlman HG, Song J, Ma D, Courchesne WE, Thorner J (1996) Sst2, a negative regulator of pheromone signaling in the yeast *Saccharomyces cerevisiae*: expression, localization, and genetic interaction and physical association with Gpa1 (the G-protein alpha subunit). Mol Cell Biol 16(9):5194–5209

Downes GB, Gautam N (1999) The G protein subunit gene families. Genomics 62(3):544–552

Dyer PS, Paoletti M, Archer DB (2003) Genomics reveals sexual secrets of Aspergillus. Microbiology 149(Pt 9):2301–2303

Eaton CJ, Cabrera IE, Servin JA, Wright SJ, Cox MP, Borkovich KA (2012) The guanine nucleotide exchange factor RIC8 regulates conidial germination through Galpha proteins in *Neurospora crassa*. PLoS One 7(10), e48026

Egel R (1971) Physiological aspects of conjugation in fission yeast. Planta 98(1):89–96

Elion EA (2000) Pheromone response, mating and cell biology. Curr Opin Microbiol 3(6):573–581

Elion EA (2001) The Ste5p scaffold. J Cell Sci 114(Pt 22):3967–3978

Elion EA, Satterberg B, Kranz JE (1993) FUS3 phosphorylates multiple components of the mating signal transduction cascade: evidence for STE12 and FAR1. Mol Biol Cell 4(5):495–510

Fang W, Scully LR, Zhang L, Pei Y, Bidochka MJ (2008) Implication of a regulator of G protein signalling (BbRGS1) in conidiation and conidial thermotolerance of the insect pathogenic fungus *Beauveria bassiana*. FEMS Microbiol Lett 279(2):146–156

Fernandez J, Wilson RA (2014) Cells in cells: morphogenetic and metabolic strategies conditioning rice infection by the blast fungus *Magnaporthe oryzae*. Protoplasma 251(1):37–47

Ferrell JE Jr, Bhatt RR (1997) Mechanistic studies of the dual phosphorylation of mitogen-activated protein kinase. J Biol Chem 272(30):19008–19016

Fong HK, Hurley JB, Hopkins RS, Miake-Lye R, Johnson MS, Doolittle RF, Simon MI (1986) Repetitive segmental structure of the transducin beta subunit: homology with the *CDC4* gene and identification of related mRNAs. Proc Natl Acad Sci USA 83(7):2162–2166

Francis SH, Corbin JD (1999) Cyclic nucleotide-dependent protein kinases: intracellular receptors for cAMP and cGMP action. Crit Rev Clin Lab Sci 36(4):275–328

Fredriksson R, Lagerstrom MC, Lundin LG, Schioth HB (2003) The G-protein-coupled receptors in the human genome form five main families. Phylogenetic analysis, paralogon groups, and fingerprints. Mol Pharm 63(6):1256–1272

Freissmuth M, Gilman AG (1989) Mutations of G_S alpha designed to alter the reactivity of the protein with bacterial toxins. Substitutions at ARG187 result in loss of GTPase activity. J Biol Chem 264(36):21907–21914

Fuchs BB, Mylonakis E (2009) Our paths might cross: the role of the fungal cell wall integrity pathway in stress response and cross talk with other stress response pathways. Eukaryot Cell 8(11):1616–1625

Gabay M, Pinter ME, Wright FA, Chan P, Murphy AJ, Valenzuela DM, Yancopoulos GD, Tall GG (2011) Ric-8 proteins are molecular chaperones that direct nascent G protein alpha subunit membrane association. Sci Signal 4(200):ra79

Galagan JE, Calvo SE, Borkovich KA, Selker EU, Read ND, Jaffe D, FitzHugh W, Ma LJ, Smirnov S, Purcell S, Rehman B, Elkins T, Engels R, Wang S, Nielsen CB, Butler J, Endrizzi M, Qui D, Ianakiev P, Bell-Pedersen D, Nelson MA, Werner-Washburne M, Selitrennikoff CP, Kinsey JA, Braun EL, Zelter A, Schulte U, Kothe GO, Jedd G, Mewes W, Staben C, Marcotte E, Greenberg D, Roy A, Foley K, Naylor J, Stange-Thomann N, Barrett R, Gnerre S, Kamal M, Kamvysselis M, Mauceli E, Bielke C, Rudd S, Frishman D, Krystofova S, Rasmussen C, Metzenberg RL, Perkins DD, Kroken S, Cogoni C, Macino G, Catcheside D, Li W, Pratt RJ, Osmani SA, DeSouza CP, Glass L, Orbach MJ, Berglund JA, Voelker R, Yarden O, Plamann M, Seiler S, Dunlap J, Radford A, Aramayo R, Natvig DO, Alex LA, Mannhaupt G, Ebbole DJ, Freitag M, Paulsen I, Sachs MS, Lander ES, Nusbaum C, Birren B (2003) The genome sequence of the filamentous fungus *Neurospora crassa*. Nature 422(6934):859–868

Garg N, Manchanda G, Kumar A (2014) Bacterial quorum sensing: circuits and applications. Antonie van Leeuwenhoek 105(2):289–305

Garrington TP, Johnson GL (1999) Organization and regulation of mitogen-activated protein kinase signaling pathways. Curr Opin Cell Biol 11(2):211–218

Gartner A, Nasmyth K, Ammerer G (1992) Signal transduction in *Saccharomyces cerevisiae* requires tyrosine and threonine phosphorylation of FUS3 and KSS1. Genes Dev 6(7):1280–1292

Gautam N, Downes GB, Yan K, Kisselev O (1998) The G-protein betagamma complex. Cell Signal 10(7):447–455

Gold S, Duncan G, Barrett K, Kronstad J (1994) cAMP regulates morphogenesis in the fungal pathogen *Ustilago maydis*. Genes Dev 8(23):2805–2816

Gold SE, Brogdon SM, Mayorga ME, Kronstad JW (1997) The *Ustilago maydis* regulatory subunit of a cAMP-dependent protein kinase is required for gall formation in maize. Plant Cell 9(9):1585–1594

Gong J, Grodsky JD, Zhang Z, Wang P (2014) A Ric8/Synembryn homolog promotes Gpa1 and Gpa2 activation to respectively regulate cyclic AMP and pheromone signaling in *Cryptococcus neoformans*. Eukaryot Cell 13(10):1290–1299

Good M, Tang G, Singleton J, Remenyi A, Lim WA (2009) The Ste5 scaffold directs mating signaling by catalytically unlocking the Fus3 MAP kinase for activation. Cell 136(6):1085–1097

Graziano MP, Gilman AG (1989) Synthesis in *Escherichia coli* of GTPase-deficient mutants of Gs alpha. J Biol Chem 264(26):15475–15482

Gruber S, Omann M, Zeilinger S (2013) Comparative analysis of the repertoire of G protein-coupled receptors of three species of the fungal genus Trichoderma. BMC Microbiol 13:108

Gustin MC, Albertyn J, Alexander M, Davenport K (1998) MAP kinase pathways in the yeast *Saccharomyces cerevisiae*. Microbiol Mol Biol Rev 62(4):1264–1300

Hagen DC, McCaffrey G, Sprague GF Jr (1986) Evidence the yeast *STE3* gene encodes a receptor for the peptide pheromone a factor: gene sequence and implications for the structure of the presumed receptor. Proc Natl Acad Sci USA 83(5):1418–1422

Hamel LP, Nicole MC, Duplessis S, Ellis BE (2012) Mitogen-activated protein kinase signaling in plant-interacting fungi: distinct messages from conserved messengers. Plant Cell 24(4):1327–1351

Hamm HE (1998) The many faces of G protein signaling. J Biol Chem 273(2):669–672

Hamm HE, Gilchrist A (1996) Heterotrimeric G proteins. Curr Opin Cell Biol 8(2):189–196

Hampoelz B, Hoeller O, Bowman SK, Dunican D, Knoblich JA (2005) Drosophila Ric-8 is essential for plasma-membrane localization of heterotrimeric G proteins. Nat Cell Biol 7(11):1099–1105

Han KH, Seo JA, Yu JH (2004) Regulators of G-protein signalling in *Aspergillus nidulans*: RgsA downregulates stress response and stimulates asexual sporulation through attenuation of GanB (Galpha) signalling. Mol Microbiol 53(2):529–540

Hartmann HA, Kahmann R, Bolker M (1996) The pheromone response factor coordinates filamentous growth and pathogenicity in *Ustilago maydis*. EMBO J 15(7):1632–1641

Herrero-Garcia E, Garzia A, Cordobes S, Espeso EA, Ugalde U (2011) 8-Carbon oxylipins inhibit germination and growth, and stimulate aerial conidiation in *Aspergillus nidulans*. Fungal Biol 115(4-5):393–400

Hill SJ (2006) G-protein-coupled receptors: past, present and future. Br J Pharm 147(Suppl 1):S27–S37

Hoffman CS (2005a) Except in every detail: comparing and contrasting G-protein signaling in *Saccharomyces cerevisiae* and *Schizosaccharomyces pombe*. Eukaryot Cell 4(3):495–503

Hoffman CS (2005b) Glucose sensing via the protein kinase A pathway in *Schizosaccharomyces pombe*. Biochem Soc Trans 33(Pt 1):257–260

Hoffman CS, Winston F (1990) Isolation and characterization of mutants constitutive for expression of the *fbp1* gene of *Schizosaccharomyces pombe*. Genetics 124(4):807–816

Hoffman CS, Winston F (1991) Glucose repression of transcription of the *Schizosaccharomyces pombe fbp1* gene occurs by a cAMP signaling pathway. Genes Dev 5(4):561–571

Hollinger S, Hepler JR (2002) Cellular regulation of RGS proteins: modulators and integrators of G protein signaling. Pharm Rev 54(3):527–559

Holsbeeks I, Lagatie O, Van Nuland A, Van de Velde S, Thevelein JM (2004) The eukaryotic plasma membrane as a nutrient-sensing device. Trends Biochem Sci 29(10):556–564

Horowitz Brown S, Zarnowski R, Sharpee WC, Keller NP (2008) Morphological transitions governed by density dependence and lipoxygenase activity in *Aspergillus flavus*. Appl Environ Microbiol 74(18):5674–5685

Hsueh YP, Xue C, Heitman J (2009) A constitutively active GPCR governs morphogenic transitions in *Cryptococcus neoformans*. EMBO J 28(9):1220–1233

Hughes DA, Yabana N, Yamamoto M (1994) Transcriptional regulation of a Ras nucleotide-exchange factor gene by extracellular signals in fission yeast. J Cell Sci 107(Pt 12):3635–3642

Iniguez-Lluhi JA, Simon MI, Robishaw JD, Gilman AG (1992) G protein beta gamma subunits synthesized in Sf9 cells. Functional characterization and the significance of prenylation of gamma. J Biol Chem 267(32):23409–23417

Isshiki T, Mochizuki N, Maeda T, Yamamoto M (1992) Characterization of a fission yeast gene, *gpa2*, that encodes a G alpha subunit involved in the monitoring of nutrition. Genes Dev 6(12B):2455–2462

Ivey FD, Hoffman CS (2005) Direct activation of fission yeast adenylate cyclase by the Gpa2 Galpha of the glucose signaling pathway. Proc Natl Acad Sci USA 102(17):6108–6113

Ivey FD, Hodge PN, Turner GE, Borkovich KA (1996) The Galphai homologue *gna-1* controls multiple differentiation pathways in *Neurospora crassa*. Mol Biol Cell 7(8):1283–1297

Ivey FD, Yang Q, Borkovich KA (1999) Positive regulation of adenylyl cyclase activity by a Galphai homolog in *Neurospora crassa*. Fungal Genet Biol 26(1):48–61

Ivey FD, Kays AM, Borkovich KA (2002) Shared and independent roles for a Galpha(i) protein and adenylyl cyclase in regulating development and stress responses in *Neurospora crassa*. Eukaryot Cell 1(4):634–642

Jastrzebska B (2013) GPCR: G protein complexes—the fundamental signaling assembly. Amino Acids 45(6):1303–1314

Johnston GC, Ehrhardt CW, Lorincz A, Carter BL (1979) Regulation of cell size in the yeast *Saccharomyces cerevisiae*. J Bacteriol 137(1):1–5

Kaffarnik F, Muller P, Leibundgut M, Kahmann R, Feldbrugge M (2003) PKA and MAPK phosphorylation of Prf1 allows promoter discrimination in *Ustilago maydis*. EMBO J 22(21):5817–5826

Kahmann R, Bolker M (1996) Self/nonself recognition in fungi: old mysteries and simple solutions. Cell 85(2):145–148

Kahmann R, Kamper J (2004) *Ustilago maydis*: how its biology relates to pathogenic development. New Phytol 164:31–42

Katritch V, Cherezov V, Stevens RC (2012) Diversity and modularity of G protein-coupled receptor structures. Trends Pharm Sci 33(1):17–27

Kays AM, Borkovich KA (2004) Severe impairment of growth and differentiation in a *Neurospora crassa* mutant lacking all heterotrimeric G alpha proteins. Genetics 166(3):1229–1240

Kays AM, Rowley PS, Baasiri RA, Borkovich KA (2000) Regulation of conidiation and adenylyl cyclase levels by the Galpha protein GNA-3 in *Neurospora crassa*. Mol Cell Biol 20(20):7693–7705

Keshet Y, Seger R (2010) The MAP kinase signaling cascades: a system of hundreds of components regulates a diverse array of physiological functions. Methods Mol Biol 661:3–38

Kim H, Borkovich KA (2004) A pheromone receptor gene, *pre-1*, is essential for mating type-specific directional growth and fusion of trichogynes and female fertility in *Neurospora crassa*. Mol Microbiol 52(6):1781–1798

Kim H, Borkovich KA (2006) Pheromones are essential for male fertility and sufficient to direct chemotropic polarized growth of trichogynes during mating in *Neurospora crassa*. Eukaryot Cell 5(3):544–554

Kim H, Metzenberg RL, Nelson MA (2002) Multiple functions of *mfa-1*, a putative pheromone precursor gene of *Neurospora crassa*. Eukaryot Cell 1(6):987–999

Kim H, Wright SJ, Park G, Ouyang S, Krystofova S, Borkovich KA (2012) Roles for receptors, pheromones, G proteins, and mating type genes during sexual reproduction in *Neurospora crassa*. Genetics 190(4):1389–1404

Kitamura K, Shimoda C (1991) The *Schizosaccharomyces pombe mam2* gene encodes a putative pheromone receptor which has a significant homology with the *Saccharomyces cerevisiae* Ste2 protein. EMBO J 10(12):3743–3751

Kobilka BK (2007) G protein coupled receptor structure and activation. Biochim Biophys Acta 1768(4):794–807

Kozubowski L, Heitman J (2012) Profiling a killer, the development of *Cryptococcus neoformans*. FEMS Microbiol Rev 36(1):78–94

Kraakman L, Lemaire K, Ma P, Teunissen AW, Donaton MC, Van Dijck P, Winderickx J, de Winde JH, Thevelein JM (1999) A *Saccharomyces cerevisiae* G-protein coupled receptor, Gpr1, is specifically required for glucose activation of the cAMP pathway during the transition to growth on glucose. Mol Microbiol 32(5):1002–1012

Krebs EG (1989) The Albert Lasker Medical Awards. Role of the cyclic AMP-dependent protein kinase in signal transduction. JAMA 262(13):1815–1818

Krishnan A, Almen MS, Fredriksson R, Schioth HB (2012) The origin of GPCRs: identification of mammalian like rhodopsin, adhesion, glutamate and frizzled GPCRs in fungi. PLoS One 7(1):e29817

Kronstad J, De Maria AD, Funnell D, Laidlaw RD, Lee N, de Sa MM, Ramesh M (1998) Signaling via cAMP in fungi: interconnections with mitogen-activated protein kinase pathways. Arch Microbiol 170(6):395–404

Kruger J, Loubradou G, Regenfelder E, Hartmann A, Kahmann R (1998) Crosstalk between cAMP and pheromone signalling pathways in *Ustilago maydis*. Mol Gen Genet 260(2-3):193–198

Krystofova S, Borkovich KA (2005) The heterotrimeric G-protein subunits GNG-1 and GNB-1 form a Gbetagamma dimer required for normal female fertility, asexual development, and Galpha protein levels in *Neurospora crassa*. Eukaryot Cell 4(2):365–378

Kulkarni RD, Thon MR, Pan H, Dean RA (2005) Novel G-protein-coupled receptor-like proteins in the plant pathogenic fungus *Magnaporthe grisea*. Genome Biol 6(3):R24

Kunitomo H, Higuchi T, Iino Y, Yamamoto M (2000) A zinc-finger protein, Rst2p, regulates transcription of the fission yeast *ste11(+)* gene, which encodes a pivotal transcription factor for sexual development. Mol Biol Cell 11(9):3205–3217

Kurjan J (1992) Pheromone response in yeast. Annu Rev Biochem 61:1097–1129

Kwon NJ, Park HS, Jung S, Kim SC, Yu JH (2012) The putative guanine nucleotide exchange factor RicA mediates upstream signaling for growth and development in Aspergillus. Eukaryot Cell 11(11):1399–1412

Kwon-Chung KJ (1975) A new genus, Filobasidiella, the perfect state of *Cryptococcus neoformans*. Mycologia 67(6):1197–1200

Kwon-Chung KJ (1976) Morphogenesis of *Filobasidiella neoformans*, the sexual state of *Cryptococcus neoformans*. Mycologia 68(4):821–833

Lafon A, Seo JA, Han KH, Yu JH, d'Enfert C (2005) The heterotrimeric G-protein GanB(alpha)-SfaD(-beta)-GpgA(gamma) is a carbon source sensor involved in early cAMP-dependent germination in *Aspergillus nidulans*. Genetics 171(1):71–80

Lafon A, Han KH, Seo JA, Yu JH, d'Enfert C (2006) G-protein and cAMP-mediated signaling in Aspergilli: a genomic perspective. Fungal Genet Biol 43(7):490–502

Lai RK, Perez-Sala D, Canada FJ, Rando RR (1990) The gamma subunit of transducin is farnesylated. Proc Natl Acad Sci USA 87(19):7673–7677

Landry S, Hoffman CS (2001) The git5 Gbeta and git11 Ggamma form an atypical Gbetagamma dimer acting in the fission yeast glucose/cAMP pathway. Genetics 157(3):1159–1168

Landry S, Pettit MT, Apolinario E, Hoffman CS (2000) The fission yeast *git5* gene encodes a Gbeta subunit required for glucose-triggered adenylate cyclase activation. Genetics 154(4):1463–1471

Lee BN, Adams TH (1994) Overexpression of *flbA*, an early regulator of Aspergillus asexual sporulation, leads to activation of *brlA* and premature initiation of development. Mol Microbiol 14(2):323–334

Lee MJ, Dohlman HG (2008) Coactivation of G protein signaling by cell-surface receptors and an intracellular exchange factor. Curr Biol 18(3):211–215

Lemaire K, Van de Velde S, Van Dijck P, Thevelein JM (2004) Glucose and sucrose act as agonist and mannose as antagonist ligands of the G protein-coupled receptor Gpr1 in the yeast *Saccharomyces cerevisiae*. Mol Cell 16(2):293–299

Lengeler KB, Davidson RC, D'Souza C, Harashima T, Shen WC, Wang P, Pan X, Waugh M, Heitman J (2000) Signal transduction cascades regulating fungal development and virulence. Microbiol Mol Biol Rev 64(4):746–785

Lengeler KB, Fox DS, Fraser JA, Allen A, Forrester K, Dietrich FS, Heitman J (2002) Mating-type locus of *Cryptococcus neoformans*: a step in the evolution of sex chromosomes. Eukaryot Cell 1(5):704–718

Li L, Borkovich KA (2006) GPR-4 is a predicted G-protein-coupled receptor required for carbon source-dependent asexual growth and development in *Neurospora crassa*. Eukaryot Cell 5(8):1287–1300

Li L, Shen G, Zhang ZG, Wang YL, Thompson JK, Wang P (2007a) Canonical heterotrimeric G proteins regulating mating and virulence of *Cryptococcus neoformans*. Mol Biol Cell 18(11):4201–4209

Li L, Wright SJ, Krystofova S, Park G, Borkovich KA (2007b) Heterotrimeric G protein signaling in filamentous fungi. Annu Rev Microbiol 61:423–452

Li Y, Yan X, Wang H, Liang S, Ma WB, Fang MY, Talbot NJ, Wang ZY (2010) MoRic8 Is a novel component of G-protein signaling during plant infection by the rice blast fungus *Magnaporthe oryzae*. Mol Plant Microbe Interact 23(3):317–331

Li G, Zhou X, Xu JR (2012) Genetic control of infection-related development in *Magnaporthe oryzae*. Curr Opin Microbiol 15(6):678–684

Liang S, Wang ZY, Liu PJ, Li DB (2006) A Gγ subunit promoter T-DNA insertion mutant—A1-412 of *Magnaporthe grisea* is defective in appressorium formation, penetration and pathogenicity. Chin Sci Bull 51:2214–2218

Liu S, Dean RA (1997) G protein alpha subunit genes control growth, development, and pathogenicity of *Magnaporthe grisea*. Mol Plant Microbe Interact 10(9):1075–1086

Liu H, Suresh A, Willard FS, Siderovski DP, Lu S, Naqvi NI (2007) Rgs1 regulates multiple Galpha subunits in Magnaporthe pathogenesis, asexual growth and thigmotropism. EMBO J 26(3):690–700

Liu J, Wang X, Mitchell T, Hu Y, Liu X, Dai L, Wang GL (2010) Recent progress and understanding of the molecular mechanisms of the rice-*Magnaporthe oryzae* interaction. Mol Plant Pathol 11(3):419–427

Lopez-Ilasaca M (1998) Signaling from G-protein-coupled receptors to mitogen-activated protein (MAP)-kinase cascades. Biochem Pharm 56(3):269–277

Lorenz MC, Heitman J (1997) Yeast pseudohyphal growth is regulated by GPA2, a G protein alpha homolog. EMBO J 16(23):7008–7018

Lorenz MC, Pan X, Harashima T, Cardenas ME, Xue Y, Hirsch JP, Heitman J (2000) The G protein-coupled receptor *gpr1* is a nutrient sensor that regulates pseudohyphal differentiation in *Saccharomyces cerevisiae*. Genetics 154(2):609–622

Maeda T, Watanabe Y, Kunitomo H, Yamamoto M (1994) Cloning of the *pka1* gene encoding the catalytic subunit of the cAMP-dependent protein kinase in *Schizosaccharomyces pombe*. J Biol Chem 269(13):9632–9637

Marcus S, Polverino A, Barr M, Wigler M (1994) Complexes between STE5 and components of the pheromone-responsive mitogen-activated protein kinase module. Proc Natl Acad Sci USA 91(16):7762–7766

Mata J, Bahler J (2006) Global roles of Ste11p, cell type, and pheromone in the control of gene expression during early sexual differentiation in fission yeast. Proc Natl Acad Sci USA 103(42):15517–15522

Matsumoto K, Uno I, Ishikawa T (1984) Identification of the structural gene and nonsense alleles for adenylate cyclase in *Saccharomyces cerevisiae*. J Bacteriol 157(1):277–282

McClelland CM, Fu J, Woodlee GL, Seymour TS, Wickes BL (2002) Isolation and characterization of the *Cryptococcus neoformans MATa* pheromone gene. Genetics 160(3):935–947

Mehrabi R, M'Barek SB, Saidi A, Abrinbana M, de Wit PJ, Kema GH (2012) MAP kinase phosphorylation and cAMP assessment in fungi. Methods Mol Biol 835:571–583

Mende U, Zagrovic B, Cohen A, Li Y, Valenzuela D, Fishman MC, Neer EJ (1998) Effect of deletion of the major brain G-protein alpha subunit (alpha(o)) on coordination of G-protein subunits and on adenylyl cyclase activity. J Neurosci Res 54(2):263–272

Merlini L, Dudin O, Martin SG (2013) Mate and fuse: how yeast cells do it. Open Biol 3(3):130008

Miller KG, Rand JB (2000) A role for RIC-8-(Synembryn) and GOA-1 (G(o)alpha) in regulating a subset of centrosome movements during early embryogenesis in *Caenorhabditis elegans*. Genetics 156(4):1649–1660

Miller KG, Emerson MD, McManus JR, Rand JB (2000) RIC-8 (Synembryn): a novel conserved protein that is required for G(q)alpha signaling in the *C. elegans* nervous system. Neuron 27(2):289–299

Mitchell TK, Dean RA (1995) The cAMP-dependent protein kinase catalytic subunit is required for appressorium formation and pathogenesis by the rice blast pathogen *Magnaporthe grisea*. Plant Cell 7(11):1869–1878

Moore TD, Edman JC (1993) The alpha-mating type locus of *Cryptococcus neoformans* contains a peptide pheromone gene. Mol Cell Biol 13(3):1962–1970

Mukherjee M, Kim JE, Park YS, Kolomiets MV, Shim WB (2011) Regulators of G-protein signalling in *Fusarium verticillioides* mediate differential host-pathogen responses on nonviable versus viable maize kernels. Mol Plant Pathol 12(5):479–491

Muller P, Leibbrandt A, Teunissen H, Cubasch S, Aichinger C, Kahmann R (2004) The Gbeta-subunit-encoding gene bpp1 controls cyclic-AMP signaling in *Ustilago maydis*. Eukaryot Cell 3(3):806–814

Nadin-Davis SA, Nasim A (1990) *Schizosaccharomyces pombe ras1* and *byr1* are functionally related genes of the *ste* family that affect starvation-induced transcription of mating-type genes. Mol Cell Biol 10(2):549–560

Nagai Y, Nishimura A, Tago K, Mizuno N, Itoh H (2010) Ric-8B stabilizes the alpha subunit of stimulatory G protein by inhibiting its ubiquitination. J Biol Chem 285(15):11114–11120

Nakayama N, Miyajima A, Arai K (1985) Nucleotide sequences of STE2 and STE3, cell type-specific sterile genes from *Saccharomyces cerevisiae*. EMBO J 4(10):2643–2648

Neer EJ, Schmidt CJ, Nambudripad R, Smith TF (1994) The ancient regulatory-protein family of WD-repeat proteins. Nature 371(6495):297–300

Neiman AM, Herskowitz I (1994) Reconstitution of a yeast protein kinase cascade in vitro: activation of the yeast MEK homologue STE7 by STE11. Proc Natl Acad Sci USA 91(8):3398–3402

Neves SR, Ram PT, Iyengar R (2002) G protein pathways. Science 296(5573):1636–1639

Nguyen AN, Ikner AD, Shiozaki M, Warren SM, Shiozaki K (2002) Cytoplasmic localization of Wis1 MAPKK by nuclear export signal is important for nuclear targeting of Spc1/Sty1 MAPK in fission yeast. Mol Biol Cell 13(8):2651–2663

Nishimura M, Park G, Xu JR (2003) The G-beta subunit MGB1 is involved in regulating multiple steps of infection-related morphogenesis in *Magnaporthe grisea*. Mol Microbiol 50(1):231–243

Obara T, Nakafuku M, Yamamoto M, Kaziro Y (1991) Isolation and characterization of a gene encoding a G-protein alpha subunit from *Schizosaccharomyces pombe*: involvement in mating and sporulation pathways. Proc Natl Acad Sci USA 88(13):5877–5881

Okazaki N, Okazaki K, Tanaka K, Okayama H (1991) The *ste4⁺* gene, essential for sexual differentiation of *Schizosaccharomyces pombe*, encodes a protein with a leucine zipper motif. Nucleic Acids Res 19(25):7043–7047

Palmer DA, Thompson JK, Li L, Prat A, Wang P (2006) Gib2, a novel Gbeta-like/RACK1 homolog, functions as a Gbeta subunit in cAMP signaling and is essential in *Cryptococcus neoformans*. J Biol Chem 281(43):32596–32605

Paoletti M, Rydholm C, Schwier EU, Anderson MJ, Szakacs G, Lutzoni F, Debeaupuis JP, Latge JP, Denning DW, Dyer PS (2005) Evidence for sexuality in the opportunistic fungal pathogen *Aspergillus fumigatus*. Curr Biol 15(13):1242–1248

Paoletti M, Seymour FA, Alcocer MJ, Kaur N, Calvo AM, Archer DB, Dyer PS (2007) Mating type and the genetic basis of self-fertility in the model fungus *Aspergillus nidulans*. Curr Biol 17(16):1384–1389

Park G, Xue C, Zhao X, Kim Y, Orbach M, Xu JR (2006) Multiple upstream signals converge on the adaptor protein Mst50 in *Magnaporthe grisea*. Plant Cell 18(10):2822–2835

Payne DM, Rossomando AJ, Martino P, Erickson AK, Her JH, Shabanowitz J, Hunt DF, Weber MJ, Sturgill TW (1991) Identification of the regulatory phosphorylation sites in pp 42/mitogen-activated protein kinase (MAP kinase). EMBO J 10(4):885–892

Peeters T, Louwet W, Gelade R, Nauwelaers D, Thevelein JM, Versele M (2006) Kelch-repeat proteins interacting with the Galpha protein Gpa2 bypass adenylate cyclase for direct regulation of protein kinase A in yeast. Proc Natl Acad Sci USA 103(35):13034–13039

Pel HJ, de Winde JH, Archer DB, Dyer PS, Hofmann G, Schaap PJ, Turner G, de Vries RP, Albang R, Albermann K, Andersen MR, Bendtsen JD, Benen JA, van den Berg M, Breestraat S, Caddick MX, Contreras R, Cornell M, Coutinho PM, Danchin EG, Debets AJ, Dekker P, van Dijck PW, van Dijk A, Dijkhuizen L, Driessen AJ, d'Enfert C, Geysens S, Goosen C, Groot GS, de Groot PW, Guillemette T, Henrissat B, Herweijer M, van den Hombergh JP, van den Hondel CA, van der Heijden RT, van der Kaaij RM, Klis FM, Kools HJ, Kubicek CP, van Kuyk PA, Lauber J, Lu X, van der Maarel MJ, Meulenberg R, Menke H, Mortimer MA, Nielsen J, Oliver SG, Olsthoorn M, Pal K, van Peij NN, Ram AF, Rinas U, Roubos JA, Sagt CM, Schmoll M, Sun J, Ussery D, Varga J, Vervecken W, van de Vondervoort PJ, Wedler H, Wosten HA, Zeng AP, van Ooyen AJ, Visser J, Stam H (2007) Genome sequencing and analysis of the versatile cell factory *Aspergillus niger* CBS 513.88. Nat Biotechnol 25(2):221–231

Pierce KL, Premont RT, Lefkowitz RJ (2002) Seven-transmembrane receptors. Nat Rev Mol Cell Biol 3(9):639–650

Posas F, Takekawa M, Saito H (1998) Signal transduction by MAP kinase cascades in budding yeast. Curr Opin Microbiol 1(2):175–182

Qi M, Elion EA (2005) MAP kinase pathways. J Cell Sci 118(Pt 16):3569–3572

Raju NB (1992) Genetic control of the sexual cycle in *Neurospora*. Mycol Res 96:241–262

Ramachander R, Kim CA, Phillips ML, Mackereth CD, Thanos CD, McIntosh LP, Bowie JU (2002) Oligomerization-dependent association of the SAM domains from *Schizosaccharomyces pombe* Byr2 and Ste4. J Biol Chem 277(42):39585–39593

Ramanujam R, Calvert ME, Selvaraj P, Naqvi NI (2013) The late endosomal HOPS complex anchors active G-protein signaling essential for pathogenesis in *Magnaporthe oryzae*. PLoS Pathog 9(8): e1003527

Ramezani-Rad M (2003) The role of adaptor protein Ste50-dependent regulation of the MAPKKK Ste11 in multiple signalling pathways of yeast. Curr Genet 43(3):161–170

Ramirez-Prado JH, Moore GG, Horn BW, Carbone I (2008) Characterization and population analysis of the mating-type genes in *Aspergillus flavus* and *Aspergillus parasiticus*. Fungal Genet Biol 45(9):1292–1299

Raudaskoski M, Kothe E (2010) Basidiomycete mating type genes and pheromone signaling. Eukaryot Cell 9(6):847–859

Regenfelder E, Spellig T, Hartmann A, Lauenstein S, Bolker M, Kahmann R (1997) G proteins in *Ustilago maydis*: transmission of multiple signals? EMBO J 16(8):1934–1942

Resh MD (1996) Regulation of cellular signalling by fatty acid acylation and prenylation of signal transduction proteins. Cell Signal 8(6):403–412

Robison GA, Butcher RW, Sutherland EW (1968) Cyclic AMP. Annu Rev Biochem 37:149–174

Rosen S, Yu JH, Adams TH (1999) The *Aspergillus nidulans sfaD* gene encodes a G protein beta subunit that is required for normal growth and repression of sporulation. EMBO J 18(20):5592–5600

Ross EM, Wilkie TM (2000) GTPase-activating proteins for heterotrimeric G proteins: regulators of G protein signaling (RGS) and RGS-like proteins. Annu Rev Biochem 69:795–827

Roux AE, Leroux A, Alaamery MA, Hoffman CS, Chartrand P, Ferbeyre G, Rokeach LA (2009) Pro-aging effects of glucose signaling through a G protein-coupled glucose receptor in fission yeast. PLoS Genet 5(3):e1000408

Rubio-Texeira M, Van Zeebroeck G, Voordeckers K, Thevelein JM (2010) *Saccharomyces cerevisiae* plasma membrane nutrient sensors and their role in PKA signaling. FEMS Yeast Res 10(2):134–149

Saito H, Posas F (2012) Response to hyperosmotic stress. Genetics 192(2):289–318

Schade MA, Reynolds NK, Dollins CM, Miller KG (2005) Mutations that rescue the paralysis of *Caenorhabditis elegans ric-8* (synembryn) mutants activate the G alpha(s) pathway and define a third major branch of the synaptic signaling network. Genetics 169(2):631–649

Seo JA, Han KH, Yu JH (2004) The gprA and gprB genes encode putative G protein-coupled receptors required for self-fertilization in *Aspergillus nidulans*. Mol Microbiol 53(6):1611–1623

Seo JA, Han KH, Yu JH (2005) Multiple roles of a heterotrimeric G-protein gamma-subunit in governing growth and development of *Aspergillus nidulans*. Genetics 171(1):81–89

Shen J, Hsu CM, Kang BK, Rosen BP, Bhattacharjee H (2003) The *Saccharomyces cerevisiae* Arr4p is involved in metal and heat tolerance. Biometals 16(3):369–378

Shen G, Wang YL, Whittington A, Li L, Wang P (2008) The RGS protein Crg2 regulates pheromone and cyclic AMP signaling in *Cryptococcus neoformans*. Eukaryot Cell 7(9):1540–1548

Shukla AK, Singh G, Ghosh E (2014) Emerging structural insights into biased GPCR signaling. Trends Biochem Sci 39(12):594–602

Siderovski DP, Willard FS (2005) The GAPs, GEFs, and GDIs of heterotrimeric G-protein alpha subunits. Int J Biol Sci 1(2):51–66

Siderovski DP, Hessel A, Chung S, Mak TW, Tyers M (1996) A new family of regulators of G-protein-coupled receptors? Curr Biol 6(2):211–212

Sondek J, Bohm A, Lambright DG, Hamm HE, Sigler PB (1996) Crystal structure of a G-protein beta gamma dimer at 2.1A resolution. Nature 379(6563):369–374

Squire RA (1981) Ranking animal carcinogens: a proposed regulatory approach. Science 214(4523):877–880

Stanton BC, Giles SS, Staudt MW, Kruzel EK, Hull CM (2010) Allelic exchange of pheromones and their receptors reprograms sexual identity in *Cryptococcus neoformans*. PLoS Genet 6(2):e1000860

Sugimoto A, Iino Y, Maeda T, Watanabe Y, Yamamoto M (1991) *Schizosaccharomyces pombe* ste11+ encodes a transcription factor with an HMG motif that is a critical regulator of sexual development. Genes Dev 5(11):1990–1999

Tall GG (2013) Ric-8 regulation of heterotrimeric G proteins. J Recept Signal Transduct Res 33(3):139–143

Tall GG, Krumins AM, Gilman AG (2003) Mammalian Ric-8A (Synembryn) is a heterotrimeric Galpha protein guanine nucleotide exchange factor. J Biol Chem 278(10):8356–8362

Tamaki H (2007) Glucose-stimulated cAMP-protein kinase A pathway in yeast *Saccharomyces cerevisiae*. J Biosci Bioeng 104(4):245–250

Tamaki H, Yun CW, Mizutani T, Tsuzuki T, Takagi Y, Shinozaki M, Kodama Y, Shirahige K, Kumagai H (2005) Glucose-dependent cell size is regulated by a G protein-coupled receptor system in yeast *Saccharomyces cerevisiae*. Genes Cells 10(3):193–206

Tanaka K, Davey J, Imai Y, Yamamoto M (1993) *Schizosaccharomyces pombe map3*$^+$ encodes the putative M-factor receptor. Mol Cell Biol 13(1):80–88

Tanoue T, Nishida E (2003) Molecular recognitions in the MAP kinase cascades. Cell Signal 15(5):455–462

Tedford K, Kim S, Sa D, Stevens K, Tyers M (1997) Regulation of the mating pheromone and invasive growth responses in yeast by two MAP kinase substrates. Curr Biol 7(4):228–238

Tesmer JJ (2010) The quest to understand heterotrimeric G protein signaling. Nat Struct Mol Biol 17(6):650–652

Thomas CJ, Briknarova K, Hilmer JK, Movahed N, Bothner B, Sumida JP, Tall GG, Sprang SR (2011) The nucleotide exchange factor Ric-8A is a chaperone for the conformationally dynamic nucleotide-free state of Galphai1. PLoS One 6(8):e23197

Turner GE, Borkovich KA (1993) Identification of a G protein alpha subunit from *Neurospora crassa* that is a member of the G_i family. J Biol Chem 268(20):14805–14811

Vallim MA, Miller KY, Miller BL (2000) Aspergillus SteA (sterile12-like) is a homeodomain-C2/H2-Zn+2 finger transcription factor required for sexual reproduction. Mol Microbiol 36(2):290–301

Venkatakrishnan AJ, Deupi X, Lebon G, Tate CG, Schertler GF, Babu MM (2013) Molecular signatures of G-protein-coupled receptors. Nature 494(7436):185–194

Versele M, de Winde JH, Thevelein JM (1999) A novel regulator of G protein signalling in yeast, Rgs2, downregulates glucose-activation of the cAMP pathway through direct inhibition of Gpa2. EMBO J 18(20):5577–5591

Vollmeister E, Schipper K, Baumann S, Haag C, Pohlmann T, Stock J, Feldbrugge M (2012) Fungal development of the plant pathogen *Ustilago maydis*. FEMS Microbiol Rev 36(1):59–77

Wang Y (2013) Fungal adenylyl cyclase acts as a signal sensor and integrator and plays a central role in interaction with bacteria. PLoS Pathog 9(10): e1003612

Wang Y, Xu HP, Riggs M, Rodgers L, Wigler M (1991) *byr2*, a *Schizosaccharomyces pombe* gene encoding a protein kinase capable of partial suppression of the *ras1* mutant phenotype. Mol Cell Biol 11(7):3554–3563

Wang P, Perfect JR, Heitman J (2000) The G-protein beta subunit GPB1 is required for mating and haploid fruiting in *Cryptococcus neoformans*. Mol Cell Biol 20(1):352–362

Wang X, Sheff MA, Simpson DM, Elion EA (2011) Ste11p MEKK signals through HOG, mating, calcineurin and PKC pathways to regulate the FKS2 gene. BMC Mol Biol 12:51

Wang Y, Geng Z, Jiang D, Long F, Zhao Y, Su H, Zhang KQ, Yang J (2013) Characterizations and functions of regulator of G protein signaling (RGS) in fungi. Appl Microbiol Biotechnol 97(18):7977–7987

Wang Y, Shen G, Gong J, Shen D, Whittington A, Qing J, Treloar J, Boisvert S, Zhang Z, Yang C, Wang P (2014) Noncanonical Gbeta Gib2 is a scaffolding protein promoting cAMP signaling through functions of Ras1 and Cac1 proteins in *Cryptococcus neoformans*. J Biol Chem 289(18):12202–12216

Wei H, Requena N, Fischer R (2003) The MAPKK kinase SteC regulates conidiophore morphology and is essential for heterokaryon formation and sexual development in the homothallic fungus *Aspergillus nidulans*. Mol Microbiol 47(6):1577–1588

Wei Y, Shen W, Dauk M, Wang F, Selvaraj G, Zou J (2004) Targeted gene disruption of glycerol-3-phosphate dehydrogenase in *Colletotrichum gloeosporioides* reveals evidence that glycerol is a significant transferred nutrient from host plant to fungal pathogen. J Biol Chem 279(1):429–435

Welton RM, Hoffman CS (2000) Glucose monitoring in fission yeast via the Gpa2 Galpha, the Git5 Gbeta and the Git3 putative glucose receptor. Genetics 156(2):513–521

Whiteway MS, Wu C, Leeuw T, Clark K, Fourest-Lieuvin A, Thomas DY, Leberer E (1995) Association of the yeast pheromone response G protein beta gamma subunits with the MAP kinase scaffold Ste5p. Science 269(5230):1572–1575

Widmann C, Gibson S, Jarpe MB, Johnson GL (1999) Mitogen-activated protein kinase: conservation of a three-kinase module from yeast to human. Physiol Rev 79(1):143–180

Wilkie TM, Kinch L (2005) New roles for Galpha and RGS proteins: communication continues despite pulling sisters apart. Curr Biol 15(20): R843–R854

Wright SJ, Inchausti R, Eaton CJ, Krystofova S, Borkovich KA (2011) RIC8 is a guanine-nucleotide exchange factor for Galpha subunits that regulates growth and development in *Neurospora crassa*. Genetics 189:1–12

Wu C, Leberer E, Thomas DY, Whiteway M (1999) Functional characterization of the interaction of Ste50p with Ste11p MAPKKK in *Saccharomyces cerevisiae*. Mol Biol Cell 10(7):2425–2440

Xu JR, Hamer JE (1996) MAP kinase and cAMP signaling regulate infection structure formation and pathogenic growth in the rice blast fungus *Magnaporthe grisea*. Genes Dev 10(21):2696–2706

Xu HP, White M, Marcus S, Wigler M (1994) Concerted action of RAS and G proteins in the sexual response pathways of *Schizosaccharomyces pombe*. Mol Cell Biol 14(1):50–58

Xu JR, Staiger CJ, Hamer JE (1998) Inactivation of the mitogen-activated protein kinase Mps1 from the rice blast fungus prevents penetration of host cells but allows activation of plant defense responses. Proc Natl Acad Sci USA 95(21):12713–12718

Xue C, Bahn YS, Cox GM, Heitman J (2006) G protein-coupled receptor Gpr4 senses amino acids and activates the cAMP-PKA pathway in *Cryptococcus neoformans*. Mol Biol Cell 17(2):667–679

Xue C, Hsueh YP, Heitman J (2008) Magnificent seven: roles of G protein-coupled receptors in extracellular sensing in fungi. FEMS Microbiol Rev 32(6):1010–1032

Yamane HK, Farnsworth CC, Xie HY, Howald W, Fung BK, Clarke S, Gelb MH, Glomset JA (1990) Brain G protein gamma subunits contain an all-trans-geranylgeranylcysteine methyl ester at their carboxyl termini. Proc Natl Acad Sci USA 87(15):5868–5872

Yan X, Li Y, Yue X, Wang C, Que Y, Kong D, Ma Z, Talbot NJ, Wang Z (2011) Two novel transcriptional regulators are essential for infection-related morphogenesis and pathogenicity of the rice blast fungus *Magnaporthe oryzae*. PLoS Pathog 7(12): e1002385

Yang Q, Poole SI, Borkovich KA (2002) A G-protein beta subunit required for sexual and vegetative development and maintenance of normal G alpha protein levels in *Neurospora crassa*. Eukaryot Cell 1(3):378–390

Yu JH (2006) Heterotrimeric G protein signaling and RGSs in *Aspergillus nidulans*. J Microbiol 44(2):145–154

Yu JH, Wieser J, Adams TH (1996) The Aspergillus FlbA RGS domain protein antagonizes G protein signaling to block proliferation and allow development. EMBO J 15(19):5184–5190

Yun CW, Tamaki H, Nakayama R, Yamamoto K, Kumagai H (1997) G-protein coupled receptor from yeast *Saccharomyces cerevisiae*. Biochem Biophys Res Commun 240(2):287–292

Zeller CE, Parnell SC, Dohlman HG (2007) The RACK1 ortholog Asc1 functions as a G-protein beta subunit coupled to glucose responsiveness in yeast. J Biol Chem 282(34):25168–25176

Zhang FL, Casey PJ (1996) Protein prenylation: molecular mechanisms and functional consequences. Annu Rev Biochem 65:241–269

Zhang H, Tang W, Liu K, Huang Q, Zhang X, Yan X, Chen Y, Wang J, Qi Z, Wang Z, Zheng X, Wang P, Zhang Z (2011) Eight RGS and RGS-like proteins orchestrate growth, differentiation, and pathogenicity of *Magnaporthe oryzae*. PLoS Pathog 7(12): e1002450

Zhao X, Kim Y, Park G, Xu JR (2005) A mitogen-activated protein kinase cascade regulating infection-related morphogenesis in *Magnaporthe grisea*. Plant Cell 17(4):1317–1329

Zheng H, Zhou L, Dou T, Han X, Cai Y, Zhan X, Tang C, Huang J, Wu Q (2010) Genome-wide prediction of G protein-coupled receptors in *Verticillium* spp. Fungal Biol 114(4):359–368

Molecular Biology and Biochemistry of Fungal Carbohydrates

8 The Cell Wall Polysaccharides of *Aspergillus fumigatus*

Mark J. Lee[1], Donald C. Sheppard[1,2,3]

CONTENTS

I. Introduction

Cell wall polysaccharides are integral components of the fungal cell wall. Cell wall polysaccharides interact to form structural and amorphous matrices through covalent bonding and non-covalent associations. The resulting cell wall encases the fungus providing structure and protection, mediates adherence to surfaces, and acts as a molecular sieve allowing the selective passage of molecules into and out of the cell. Although all fungi have a cell wall, the polysaccharide structure and composition of the cell wall vary between fungal species. Further, the cell wall can also vary between morphotypes during the life cycle of the fungus, reflecting the need for stage-specific changes to the cell wall. The importance of the cell wall in fungal biology and its absence from human cells have made cell wall polysaccharides a promising target for the development of antifungal therapeutics directed against pathogenic fungi, such as the ubiquitous mold, *Aspergillus fumigatus*.

In immunocompromised patients, *A. fumigatus* can cause invasive pulmonary aspergillosis (IA), a progressive necrotizing pneumonia that often disseminates to other organs, such as the brain. In the absence of a normal immune response, inhaled airborne conidia of *A. fumi-*

[1]Department of Microbiology and Immunology, McGill University, Montréal, QC, Canada; e-mail: donald.sheppard@mcgill.ca
[2]Department of Medicine, McGill University, Montréal, QC, Canada; e-mail: donald.sheppard@mcgill.ca
[3]Department of Microbiology & Immunology, McGill University, 3775 rue University, Room D24, Montréal, Qc, Canada H3A 2B4; e-mail: donald.sheppard@mcgill.ca

Biochemistry and Molecular Biology, 3rd Edition
The Mycota III
D. Hoffmeister (Ed.)

gatus can swell and germinate to grow as hyphae that invade deeper pulmonary tissues (Sheppard 2011). From the time of germination to hyphal invasion of host tissues, the cell wall provides structure and protection of the organism, as well as mediates a variety of host interactions. Cell wall polysaccharides can mediate adherence to host surfaces, modulate inflammation, induce host leukocyte apoptosis, and mask pathogen-associated molecular patterns (PAMPs) produced by *A. fumigatus*. In essence, the cell wall and associated polysaccharides are indispensible in the establishment and progression of fungal infections.

Disrupting the cell wall by interfering with the synthesis of cell wall polysaccharides is, therefore, an attractive strategy in managing and treating IA. However, the effects of inhibiting synthesis of an individual polysaccharide by molecular approaches or chemical agents can be compensated by changes in the production or metabolism of other glycans (Latgé and Calderone 2006). These changes can modulate, and in some cases, even enhance virulence (Lamarre et al. 2009) or host inflammation (Gravelat et al. 2013). These studies highlight that despite decades of research, we are only now beginning to understand the mechanisms underlying the synthesis of the fungal cell wall and the role of its components in the pathogenesis of fungal infection. In this chapter, we review our current understanding of the biosynthesis and function of the polysaccharide components of the *A. fumigatus* hyphal cell wall, as well as the role of the network of cell wall integrity pathways in governing cell wall composition.

II. The Cell Wall of *Aspergillus fumigatus* Hyphae

The cell wall accounts for 20–40 % of the dry weight of *A. fumigatus* mycelia (Gastebois et al. 2009; Mouyna and Fontaine 2009). Electron microscopy has provided valuable insights into the organization of the cell wall. Scanning electron micrographs reveal that the hyphal surface is highly decorated with extracellular matrix material (Fig. 8.1a). Using a high-resolution transmission electron microscopy technique, two distinct cell wall layers with contrasting electron density are visible beneath this looser matrix component (Fig. 8.1b, c). Although, the exact chemical composition of the ECM or cell wall layers remains unclear, cell wall fractionation studies using alkali extraction methods have suggested that the inner cell wall corresponds to alkali-insoluble fractions of cell wall preparations, while the outer cell wall components are found within the alkali-soluble fraction (Fontaine et al. 2000; Latgé 2010).

The alkali-insoluble, inner cell wall fraction contains predominantly β-1,3-glucans (51 %) and chitin (22 %) and minor components of chitosan (7 %), β-1,3;1,4-glucans (6 %), galactosaminogalactan (7 %), and galactomannan (8 %) (Maubon et al. 2006). Together, these polysaccharides form the rigid, structural scaffold of the cell wall. The current model of the inner cell wall suggests that it is composed of a large network of β-1,3-glucan that is highly branched, giving it a three-dimensional shape. While the majority of the branching points are β-1,3 linked, about 4 % are β-1,6 linked (Fontaine et al. 2000). β-1,3-glucans are covalently linked to other inner cell wall polysaccharides, including chitin, which is responsible for the alkali insolubility of the inner cell wall (Fontaine et al. 2000; Hartland et al. 1994).

In contrast to the inner cell wall, the alkali-soluble, outer cell wall fraction contains predominantly α-glucans (92 %), with minor components of galactosaminogalactan (5 %) and galactomannan (3 %) (Maubon et al. 2006). Consistent with the alkali solubility of α-1,3-glucan, immunolabeling with anti-α-1,3-glucan antibody confirmed that α-1,3-glucan is localized in the outer cell wall (Beauvais et al. 2007). Although α-1,3-glucans found in *A. fumigatus* are almost all α-1,3 linked, about 1 % has been reported to have α-1,4 linkages (Latgé 2010; Mouyna and Fontaine 2009).

Interestingly, galactomannan and galactosaminogalactan are present in both the inner and outer layers of the cell wall. While the presence of galactomannan in the inner cell wall can be explained by its conjugation to β-1,3-glucan, the presence of galactosaminogalactan within the inner cell wall is puzzling

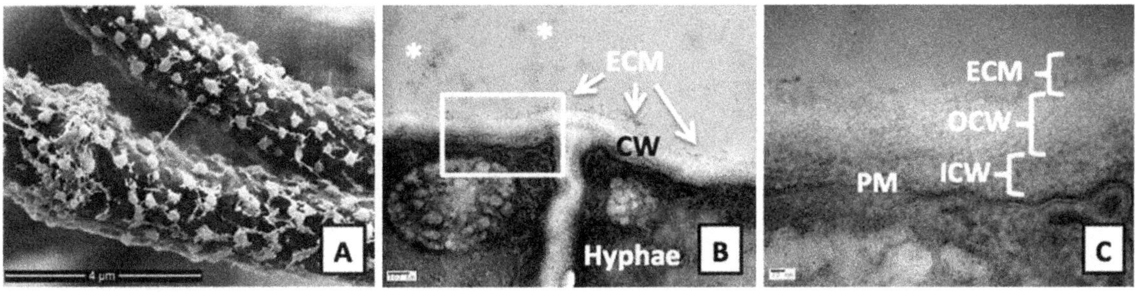

Fig. 8.1 The cell wall of *Aspergillus fumigatus*. Scanning electron micrograph (**a**), and transmission electron micrograph (**b**), of 24 h grown hyphae. Higher magnification of panel (**b**) showing the layers of the cell wall (**c**). Scale bar indicates (**a**) 4 µm, (**b**) 100 nm, and (**c**) 20 nm. *ECM* extracellular matrix, *CW* cell wall, *OCW* outer cell wall, *ICW* inner cell wall, *PM* plasma membrane

since it is not known to be conjugated to other cell wall proteins and polysaccharides. Thus, one would expect that the loosely associated galactosaminogalactan in the inner cell wall would be extracted with those found in the outer cell wall in the alkali-soluble fraction. Extending outward, polysaccharides found in the outer cell wall are also found in the extracellular matrix (ECM) (Beauvais et al. 2007; Loussert et al. 2010), although ratios of the polysaccharides may differ between these two structures (Loussert et al. 2010). Thus, the ECM could be considered an extension of the outer cell wall. The composition of the ECM during infection is not known; however, galactosaminogalactan has been detected surrounding hyphae found in lung tissue samples from both patients with aspergilloma and mice with experimental invasive aspergillosis (Fontaine et al. 2010). Further investigation is needed to better understand the role of this unusual polysaccharide in the cell wall and extracellular matrix.

III. Synthesis and Function of Cell Wall Polysaccharides

A. β-1,3-glucan

β-1,3-glucan is a homopolymer of glucose residues linked by β-1,3-glycosidic bonds (Fig. 8.2). β-1,3-glucan is present in the cell wall of many fungal species. Most importantly, it is a major component of the cell walls of many pathogenic fungi, including *Aspergillus fumigatus*, *Candida albicans*, and *Cryptococcus neoformans* (Douglas 2001). In *A. fumigatus*, β-1,3-glucan is the most abundant cell wall polysaccharide found in the alkali-insoluble fraction of the cell wall, where it can extend up to 1500 residues long (Beauvais et al. 2001). While β-1,3-glucan is essential for maintaining proper cell wall integrity and morphology, it is also a pathogen-associated molecular pattern (PAMP) recognized by the pathogen recognition receptor (PRR), dectin-1. Dectin-1 is a member of the c-type lectin family that is commonly expressed on dendritic cells and macrophages, and can activate various immune responses, including secretion of proinflammatory cytokines like IL-1β and TNF-α (Gravelat et al. 2013; Romani 2011). Thus, minimizing exposure of β-1,3-glucan on its surface is a critical strategy for the fungus to evade immune detection.

β-1,3-glucan is synthesized by a cell membrane-bound synthase complex (Douglas 2001). Two subunits of the **β-1,3-glucan synthase** complex have been identified to date: the catalytic subunit FKS (FK506 sensitive protein) and the regulatory subunit RHO (Beauvais et al. 2001). A single Fks1 (afu6g12400) ortholog encodes the catalytic subunit of the beta-1,3-glucan synthase in *A. fumigatus* (Beauvais et al. 1993). Inhibition of Fks1 expression by RNAi (Mouyna et al. 2004) or by a repressible promoter (Hu et al. 2007) results in defective growth and swelling of the hyphae due to osmotic pressure.

RHO is a small GTP-binding protein that regulates the activity of the β-1,3-glucan synthase complex by switching between GDP-

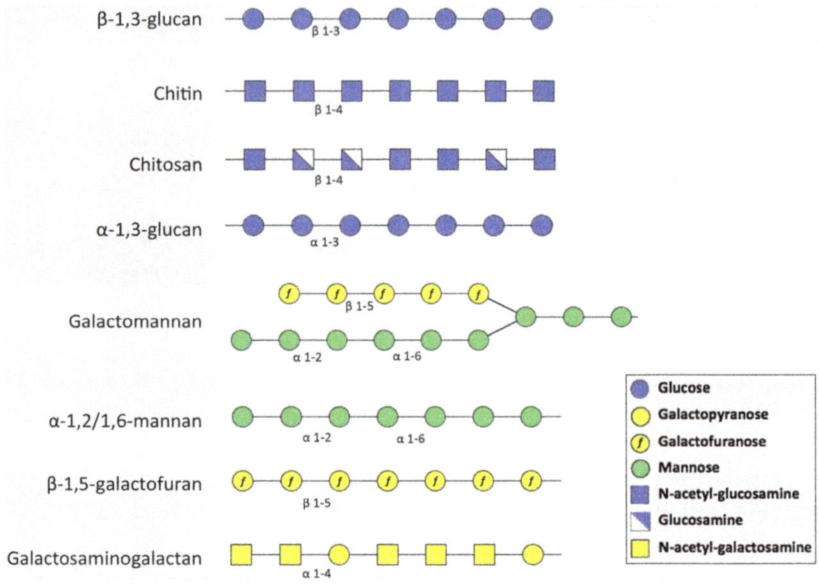

Fig. 8.2 Cell wall polysaccharides. Depiction of cell wall polysaccharides in *A. fumigatus* showing linkages and monosaccharide components

bound active and GTP-bound inactive states, inducing conformational changes of the catalytic subunit (Wei et al. 1997). Unlike *FKS*, the genome of *A. fumigatus* contains 6 putative *RHO* genes. Four of these putative Rho-GTPases have been characterized. Localization studies using GFP-tagged Rho1 or Rho3 revealed that both localize to the hyphal tip, where β-1,3-glucan synthesis is most active (Dichtl et al. 2010). Conditional downregulation of *rho1* (afu6g06900) or deletion of *rho2* (afu3g10340) or *rho4* (afu5g14060) resulted in mutant susceptible to various cell wall-perturbing agents, implicating Rho-GTPases in the cell wall synthesis (Dichtl et al. 2012). Deletion of *rho1* has been reported to have the strongest phenotype, suggesting that it is the primary Rho-GTPase involved in cell wall synthesis (Dichtl et al. 2012). Rho1 also complexes with Fks1, suggesting that Rho1 is the regulatory subunit of the β-1,3-glucan synthase complex (Beauvais et al. 2001).

Rho-GTPases require a guanine exchange factor (GEF) for activation. In *Saccharomyces cerevisiae*, two GEFs, ScRom1 and ScRom2, activate ScRho1 during cell wall stress (Levin 2011). In *A. fumigatus*, co-localization and immunoprecipitation studies have demonstrated that a single GEF ortholog, Rom2 (afu5g08550), interacts with Rho1 (Samantaray et al. 2013). Conditional suppression of *rom2* resulted in a severe growth defect and enhanced susceptibility to cell wall-perturbing agents, suggesting that it is required for cell wall integrity (Samantaray et al. 2013). Specifically, under suppressive conditions, the *rom2* mutant was highly susceptible to the β-1,3-glucan synthase inhibitor caspofungin.

Caspofungin is a member of the echinocandins class of lipopeptides that noncompetitively binds to FKS, thereby inhibiting the synthesis of β-1,3-glucan (Chen et al. 2011). Although echinocandins are highly effective against *Candida albicans*, they show variable activity against other fungi (Abruzzo et al. 1997; Guembe et al. 2007; Maligie and Selitrennikoff 2005). Caspofungin treatment is fungistatic against *A. fumigatus* and induces hyphal swelling, especially at the tip, as well as aberrant hyphal branching (Chen et al. 2011).

Although β-1,3-glucan is a major component of the cell wall, and the only cell wall polysaccharide currently targeted by licensed antifungals, our knowledge of the mechanisms and regulation of β-1,3 glucan synthesis in *A. fumigatus* is only now emerging. A greater understanding of these pathways will be critical for the development of improved next generation inhibitors targeting this important cell wall polysaccharide.

B. Chitin

Chitin is a homopolymer of β-1,4 glycosidically linked N-acetyl-glucosamine (GlcNAc) residues (Fig. 8.2). Chitin is a major component of the fungal cell wall, and it is synthesized through the action of a number of **chitin synthases** (CHS). Although chitin is present in many fungi, the mechanisms of chitin biosynthesis vary among fungal species (Lenardon et al. 2010). In particular, the number of CHS genes varies significantly between species. In *A. fumigatus*, there are eight *CHS* genes categorized into seven different classes based on their amino acid sequence (Gastebois et al. 2009). CHS that belong to classes I–III, grouped as Family I CHS, require trypsin for activation in vitro. Structurally, the catalytic domain of Family I CHS is flanked on either side by transmembrane regions while Family II CHS (class IV–VII) contain only a single C-terminal transmembrane domain (Jimenez-Ortigosa et al. 2012; Mellado et al. 1996a; Roncero 2002).

Among the Family I CHS genes, single deletions of *chsA* (Class I), *chasB* (Class II), or *chsC* (class III) did not result in noticeable defects in chitin synthesis or growth (Mellado et al. 1996a, 1996b). Only deletion of *chsG* (class III) resulted in decreased chitin synthase activity and reduced radial growth (Mellado et al. 1996a). Among the Family II CHS genes, single deletions of *chsD* (class VI), *chsF* (class IV), or *csmB* (class VII) was not associated with any modification in the chitin levels of the hyphal cell wall (Jimenez-Ortigosa et al. 2012; Mellado et al. 1996a, 1996b). Only deletion of *csmA* (class V), previously known as *chsE*, resulted in reduced chitin levels in the hyphal cell wall and the production of swollen hyphae (Aufauvre-Brown et al. 1997). A *chsG* and *csmA* double deletion mutant was still viable, although it displayed severe growth defect and produced only half the amount of chitin in the cell wall compared to the parental strain (Mellado et al. 2003). Interestingly, this decrease in chitin was associated with a compensatory increase in cell wall α-1,3-glucan content. Recently, using an innovative molecular approach (Hartmann et al. 2010), Muszkieta et al. generated quadruple deletion mutants of the Family I (chsACBG) or Family II (csmABchsFG) CHS genes (Muszkieta et al. 2014). The resulting quadruple mutants displayed significant growth and cell wall defects. Although, the Δ*chsACBG* mutant was observed to have reduced chitin synthase activity, it was as virulent as the parental strain.

Chitin synthesis can be inhibited by nikkomycin, a nucleoside-peptide secondary natural product of *Streptomyces tendae* Tü 901 (Dahn et al. 1976; Hector 1993). Nikkomycin contains an UDP-N-acetyl-glucosamine moiety that acts as a competitive inhibitor of chitin synthase, disrupting chitin production, and resulting in osmotic lysis and fungal death (Krainer et al. 1991). Nikkomycin is currently not licensed for clinical use, although animal data suggest that it is effective against *Histoplasma*, *Blastomyces*, and *Coccidioides* infections in vivo (Hector 1993). While nikkomycin is fungicidal against *A. fumigatus* in vitro, these effects are seen at much higher concentrations than those required for other fungi such as *Coccidioides posadasii* (Hector 1993; Hector et al. 1990).

Collectively, these data suggest that consistent with a critical role for chitin in maintaining structure of the organism, *A. fumigatus* has developed multiple redundant chitin synthesis mechanisms. This complexity offers a substantial challenge in dissecting the regulation and mechanisms governing chitin biosynthesis. Conversely, the importance of this polysaccharide in cell wall structural integrity suggests that it may be an attractive antifungal target if these challenges can be overcome. A detailed review on chitin biosynthesis is provided in Chap. 9.

C. Galactomannan

Mannose-containing polysaccharides, termed mannans, are a common component of the fungal cell wall. However, the side chains, branching, and linkages can vary greatly between mannans of different fungal species. In *A. fumigatus*, the main mannose-containing polysaccharide is galactomannan. Unlike the homopolymers chitin and β-1,3-glucan, galactomannan is composed of two glycosyl chains: mannan and galactofuran chains. The backbone of galactomannan is a homopolymer of α-1,2- or 1,6-linked mannose residues (Fig. 8.2)

(Latgé et al. 1991). Along the mannan backbone, side chains of galactofuranose (Gal*f*) oligomers are branched in a β-1,3 and β-1,6 linkage (Fig. 8.2) (Latgé et al. 1994). Galactofuranose is a five-membered cyclic hexose found in many microbial pathogens, which is absent in humans (Latgé 2010). Galactofuranosyl side chains are short β-1,5-linked oligomers of 4–5 residues long (Fig. 8.2) (Latgé et al. 1991).

The synthesis of the galactofuranosyl chain begins in the cytosol, where the precursor of Gal*f*, UDP-galactopyranose (Gal*p*), is produced; UDP-glucose is converted to UDP-Gal*p* through the activity of a UDP-glucose 4-epimerase, Uge5 (Lee et al. 2014b). UDP-Gal*p* is in turn converted to UDP-Gal*f* by the UDP-galactose mutase, UGM (encoded by *ugm1*, also known as *glfA*) (Lamarre et al. 2009; Schmalhorst et al. 2008). Cytosolic UDP-Gal*f* is then transported into the *cis*-Golgi by the activity of the UDP-Gal*f* transporter GlfB (Engel et al. 2009). Deletion of any of these genes results in mutant strains devoid of galactomannan. Interestingly, while Schmalhorst et al., reported attenuated virulence and increased susceptibility to various antifungals in an UGM-deficient mutant, Lamarre et al., reported no significant differences in virulence or antifungal susceptibility in an independently constructed UGM-deficient mutant (Lamarre et al. 2009; Schmalhorst et al. 2008). Differences in strain background or in the design of the phenotypic characterization experiments may have contributed to these conflicting observations.

The synthesis of the mannosyl chain of galactomannan is less well defined. In *S. cerevisiae*, cell wall mannans originate from mannoproteins with extensive N- and O-glycosylations (Levin 2005). Initial protein glycosylation of mannoproteins occurs in the endoplasmic reticulum (Levin 2005). Subsequent extension for cell wall-bound mannoproteins occurs in the *cis*-Golgi, through the activities of multiple mannosyltransferases (Nakanishi-Shindo et al. 1993; Nakayama et al. 1992). In *A. fumigatus*, while these mannosyltransferases have not yet been identified, the synthesis of the mannosyl chain also occurs inside the Golgi (Bruneau et al. 2001) and is dependent on the import of mannose to the Golgi through the action of the GDP-mannose transporter, *gmtA* (Engel et al. 2012).

Although the exact role of galactomannan in *A. fumigatus* is not clearly defined, it is a unique cell wall polysaccharide in several ways. It is the only polysaccharide known to be synthesized in the Golgi and not through the action of membrane-bound synthases. Also, it is the only cell wall polysaccharide that shares transferases involved in protein glycosylation (Jin 2012). Galactomannan detection is also in routine clinical use as a diagnostic marker for *Aspergillus* infection. As a result, there is a great interest in elucidating the complete galactomannan biosynthesis pathway. These include the identification and characterization of the glycosyltransferases involved in the synthesis of galactomannan and the characterization of the mechanisms underlying the extracellular secretion and cell wall trafficking of this important glycan.

D. α-1,3-glucan

α-1,3-glucan is a homopolymer of glucose residues linked in an α 1→3 fashion (Fig. 8.2). Thus, α-1,3-glucan differs from β-1,3-glucan only in the orientation of its linkage. Like β-1,3-glucan, α-1,3-glucan is present in the cell wall of many fungal species, including those of many pathogenic fungi. α-1,3-glucan is essential for correct cell wall morphology and contributes to virulence of several fungal species. In *Schizosaccharomyces pombe*, α-1,3-glucan is an essential cell wall polysaccharide that is required for normal cell wall morphology and cell polarity. In *Histoplasma*, α-1,3-glucan masks β-1,3-glucans from immune recognition by dectin-1 (Rappleye et al. 2007). In *Cryptococcus neoformans*, α-1,3-glucan is required to anchor the capsule, an important virulence factor of this fungus (Reese et al. 2007).

In *A. fumigatus*, α-1,3-glucan is synthesized through the action of three **glucan synthases:** Ags 1, 2, and 3 (Henry et al. 2012). Localization studies have demonstrated that Ags1 and Ags2 are located within the cell membrane of hyphae (Beauvais et al. 2005), and Ags3 is also predicted to be cell membrane-bound. Deletion of either *ags1* or *ags2* resulted in mutants with altered hyphal morphology and impaired conidiation; however, deletion of *ags1* resulted in only partial reduction of α-1,3-glucan levels, while deletion of *ags2* and *ags3* had no effect on levels of this glycan (Beauvais et al. 2005). Increased expression of *ags1* and *ags2* in the *ags3* deletion mutant suggested that compensatory upregulation of *AGS* could contribute to these results (Beauvais et al. 2005). Deletion of all three *AGS* genes resulted in a mutant that is completely devoid of α-1,3-glucan (Beauvais et al. 2013). While the triple mutant displayed normal growth and germination, it was found to have significant alterations in the cell wall organization, with increased exposure of β-1,3-glucans and the presence of extracellular proteins on its conidial cell surface. Importantly, conidia of the triple mutant were more efficiently phagocytosed and showed attenuated virulence in neutropenic mice. Pulmonary histopathology studies from mice infected with this triple mutant revealed an absence of filamentous growth in mice infected, suggesting that the increased survival of these mice was due to enhanced immune recognition of the mutant conidia and/or impaired germination of the mutant strain in vivo.

E. Galactosaminogalactan

Galactosaminogalactan is a heteropolymer composed of α-1,4-linked galactose and N-acetyl-galactosamine (GalNAc) (Fig. 8.2). Unlike other cell wall polysaccharides, the component residues do not seem regularly arranged (Mouyna and Fontaine 2009). Galactosaminogalactan has been described in other *Aspergillus* species, including *A. nidulans*, *A. parasiticus*, *A. niger*, and *A. fumigatus* (Bardalaye and Nordin 1976; Fontaine et al. 2011; Gorin and Eveleigh 1970; Ruperez and Leal 1981), and in a non-*Aspergillus* species, *Bipolaris sorokiniana* (Pringle 1981). The synthesis of galactosaminogalactan begins with the conversion of UDP-galactose from UDP-glucose and UDP-GalNAc from UDP-GlcNAc. Two epimerases, Uge3 and Uge5, are required for the normal synthesis of these sugars. Uge5 is a group 1 epimerase required for UDP-galactose synthesis, while Uge3 is a group 2 bifunctional epimerase that can mediate synthesis of both UDP-galactose and UDP-GalNAc (Lee et al. 2014b). While Uge5 is the main epimerase responsible for mediating the interconversion of UDP-glucose/galactose, in the absence of Uge5, the bifunctional activity of Uge3 is sufficient to produce UDP-galactose for the synthesis of galactosaminogalactan, but not galactomannan. Further studies delineating downstream components would be required to understand this difference in the two synthetic pathways.

Recent studies have found that galactosaminogalactan is required for a number of virulence-associated functions. Blocking galactosaminogalactan synthesis by deletion of *uge3* resulted in a mutant associated with defects in biofilm adherence, increased exposure of β-1,3-glucans, and attenuated virulence in leukopenic mice (Gravelat et al. 2013). In addition, studies using purified fractions of galactosaminogalactan have found that galactosaminogalactan induces natural killer (NK) cell-dependent apoptosis of neutrophils (Robinet et al. 2014), and intranasal administration of these preparations induced anti-inflammatory responses in mice (Fontaine et al. 2011; Gresnigt et al. 2014). Thus, purified galactosaminogalactan seems to have immune suppressive properties and could be potentially developed as a therapeutic agent. The composition and structure of galactosaminogalactan produced in vivo remain unknown. Further, structural and functional characterization of this heteropolymer are required if it is to be exploited as a therapeutic target.

IV. Modifications of Cell Wall Polysaccharides

A. Modification of β-1,3-glucans

β-1,3-glucan undergoes multiple modifications following production of the linear β-1,3-glucan chain by the β-1,3-glucan complex. The most common modification involves a cut/paste action in order to add a new chain of β-1,3-glucan onto an existing chain, form a branching point, or conjugate β-1,3-glucan to other cell wall polysaccharides or glycoproteins. A survey of putative β-1,3-glucan-modifying enzymes in the genome of *A. fumigatus* was recently performed and reported that 29 out of 45 candidate genes are expressed during vegetative growth (Mouyna et al. 2013). These genes were annotated as having glucanase or transferase activity, belonging to eight different GH (glycosyl hydrolase) families as classified on the Carbohydrate-Active enZymes database [CAZy database (http://www.cazy.org)] (Lombard et al. 2014). While functional evidence is lacking to validate the function of most of these 45 putative β-1,3-glucan modifiers, several of these have been studied as detailed below.

1. Endo-β-1,3-glucanase (ENG Proteins)

The endo-β-1,3-glucanase proteins are glycosylphosphatidylinositol (GPI)-anchored proteins that randomly cleave internal residues of soluble β-1,3-glucans (Hartl et al. 2011). To date, two of the genes coding for ENG enzymes, *eng1* and *eng2*, have been deleted and characterized (Hartl et al. 2011; Mouyna et al. 2013). No significant phenotypic differences between the single mutants and the parental strains were observed. Given that there are six other putative ENGs in the genome, it is likely that compensatory activity of other ENG proteins can compensate for the loss of a single family member as is the case for chitin and α-1,3-glucan synthase genes. A triple mutant of *eng1*, *eng2*, and *eng4*, the only three ENG genes expressed in vegetative growth, would likely provide important insights into the role of endo-β-1,3-glucanase in cell wall biosynthesis.

2. β-glucanosyltransferase (GEL Proteins)

Another group of GPI-anchored proteins, the GEL family of β-glucanosyltransferases, has been studied extensively. This class of enzyme promotes the elongation of β-1,3-glucan chains by cleaving an internal β-1,3 linkage of a β-1,3-glucan chain and transferring the newly created reducing end to the nonreducing end of another β-1,3-glucan chain (Hartland et al. 1996). Of the seven putative genes coding for GELs, *gel1*, *gel2*, *gel4*, and *gel7* have been studied. While the deletion of *gel1* did not result in phenotypic differences from the parental strain, deletion of *gel2* or the double deletion of *gel1* and *gel2* was associated with growth defects, abnormal conidiogenesis, and hypovirulence (Mouyna et al. 2005). Deletion of *gel4* was lethal, suggesting that gel4 is essential for viability (Gastebois et al. 2010a). Deletion of *gel7* did not significantly alter germination or susceptibility to cell wall-perturbing agents, but was associated with a conidiation defect (Zhao et al. 2014).

Unlike the ENG enzymes, individual GEL enzymes seem to play important roles in cell wall biogenesis. This observation suggests that while ENG's have redundant functions, each of the GEL proteins may have a distinct role in β-1,3-glucan modification. It is also possible that *A. fumigatus* favors restructuring the cell wall by "recycling" premade β-1,3-glucan by removing these glycans from locations where they are no longer needed and reattaching them as required. This cell wall remodeling strategy using GEL enzymes would offer an energy-saving approach as compared with de novo biosynthesis of β-1,3-glucan chains.

3. Branching Enzymes, BGT Proteins

Branching enzymes are also involved in modification of β-1,3-glucan to generate a complex polysaccharide structure. Of the five putative branching enzymes, all of which are expressed during vegetative growth, Bgt1 and Bgt2 have been studied. Bgt1 transfers the reducing end of one β-1,3-glucan to the hydroxy group at the C6 position of the nonreducing end of another β-1,3-glucan, generating a linear β-1,3;1,6-glucan chain (Gastebois et al. 2010b; Mouyna et al. 1998). Bgt2, on the other hand, transfers the reducing end of a β-1,3-glucan to the hydroxyl on an internal C6, forming a linear β-1,3-glucan with branched β-1,6-linked β-1,3-glucan (Gastebois et al. 2010b). Single or double deletions of *bgt1* and *bgt2* had no effect on *A. fumigatus* growth and resistance to cell wall stressors (Gastebois et al. 2010b). Like the ENG enzymes, it is possible that the other three branching enzymes have redundant function and are able to compensate for the loss of both Bgt1 and Bgt2. A quintuple mutant would ultimately be required to study the role of the branching enzymes, especially since all five are expressed during hyphal growth.

4. Other β-1,3-glucan-Modifying Enzymes (EXO and SUN Proteins)

Other β-1,3-glucan modification enzymes have also been studied. Exo-β-1,3-glucanases, ExoGI and ExoGII, have been purified and found to have specific affinity against β-1,3-glucan and other β-glucans, respectively (Fontaine et al. 1997). To date, no deletion mutants of the corresponding genes have been generated, so their role in cell wall metabolism is unknown.

Another group of β-1,3-glucan modification enzymes belonging to the SUN protein family has been recently studied. Initially characterized in *S. cerevisiae* for their involvement in cell wall biogenesis, this group of enzymes consists of SIM1, UTH1, NCA3, and SUN4 (Camougrand et al. 2000, 2004; Mouassite et al. 2000). In *A. fumigatus*, two SUN protein homologs have been identified. Recombinant Sun1 binds to and hydrolyzes β-1,3-glucan, and deletion of the gene encoding this protein led to defects in growth and cell wall integrity, including swelling and leakage of the hyphal tip (Gastebois et al. 2013). In contrast, deletion of *sun2* did not result in any major phenotypic differences as compared with the parental strain. Thus, Sun1 seems to be directly involved in cell wall biogenesis and required for β-1,3-glucan modification. Further characterization of Sun1, including its role during growth in vivo, will be needed to ascertain its role in pathogenesis.

B. Modification of Chitin

1. Chitinase, CHI Proteins

Modification of chitin is an important process for remodeling of the hyphal cell wall. Similar to hydrolysis of β-1,3-glucan by β-1,3-glucanases, chitin is hydrolyzed by chitinases. The function of chitinases has been well described in yeast, where chitinase activity is important in cell separation during budding (Alcazar-Fuoli et al. 2011; Dunkler et al. 2005; Kuranda and Robbins 1991). In *A. fumigatus*, 14 putative chitinases (CHI) were initially identified and classified into two families based on their sequences and similarities to either plant or bacterial chitinases (Taib et al. 2005). These CHI proteins have more recently been reclassified into three groups: group A, group B, and group C (Alcazar-Fuoli et al. 2011). Single deletion mutants of all *CHI* genes have been constructed and have not displayed unique phenotypes as compared to the parental strain (Alcazar-Fuoli et al. 2011; Jaques et al. 2003). In fact, deletion of all five genes in the group B *CHIs* also was not associated with any changes in cell wall-associated phenotypes (Alcazar-Fuoli et al. 2011). Given that there are 14 putative *CHI* genes in the genome of *A. fumigatus*, it is likely that there are redundancies within this family, and the construction of more multiple deletion mutants will likely be required to elucidate the role of these proteins in cell wall biogenesis.

2. Chitin Deacetylase, CDA Proteins

Chitin can also be modified to produce chitosan through the deacetylation of N-acetyl-glucosamine (GlcNAc) to glucosamine (GlcN) (Fig. 8.2). Chitosan is produced on an industrial scale through chemical deacetylation of chitin found in seashells and other biomass sources (Arbia et al. 2013). In microorganisms, chitosan is produced through the activity of a chitin deacetylase, CDA (Fig. 8.3). Although chitosan is applied to deacetylated chitin, the degree of deacetylation can vary between 50 and 90 % (Arbia et al. 2013). In recent years, chitosan has gained increasing popularity due to its expanding industrial, pharmaceutical, and medical uses (No et al. 2007). Among its many uses, chitosan is becoming a promising option as an antimicrobial coating agent on catheters to fight biofilm-related infections (Cobrado et al. 2013).

Chitosan is synthesized by many fungi and is estimated to account for 4 % of the cell wall of *A. fumigatus* (Lenardon et al. 2010; Mouyna and Fontaine 2009). In *Cryptococcus neoformans*, chitosan is necessary for virulence in mice (Baker et al. 2007). In the pathogenic

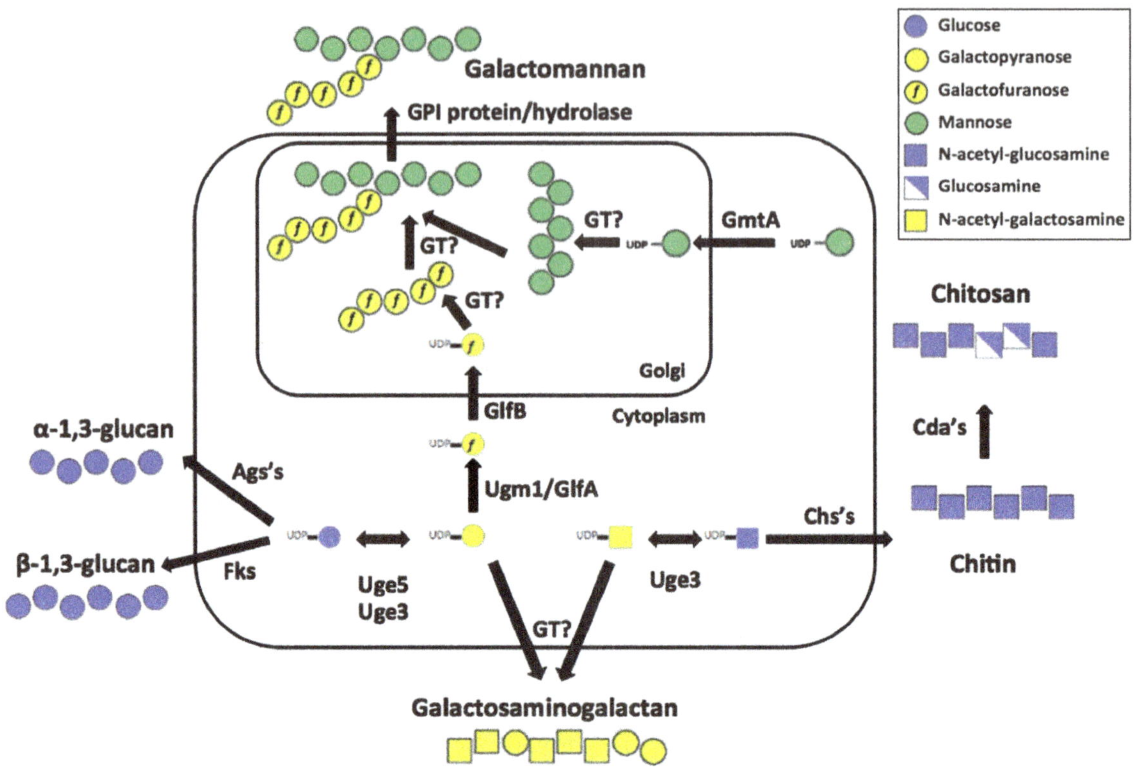

Fig. 8.3 Schematic of cell wall polysaccharide synthesis in *A. fumigatus*

plant fungi, *Puccinia graminis* and *Uromyces fabae*, chitosan is present on the surfaces of invading hyphae, but absent on the surfaces of non-invading hyphae (El Gueddari et al. 2002). In *A. fumigatus*, two putative chitin deacetylase genes, *CDA* (afu4g09940 and afu6g10430), have been identified (Gastebois et al. 2009). However, these genes remain unstudied, and therefore, the contribution of chitin deacetylase to *A. fumigatus* virulence is unknown.

3. Chitosanase, CSN Proteins

Chitosan remodeling requires the action of chitosan hydrolases or chitosanases. In *A. fumigatus*, there are four predicted chitosanase genes (afu3g14980, afu4g01290, afu8g00930, and afu6g00500) in the genome. Secreted chitosanases have been identified in the culture supernatants of *A. fumigatus*, and their expression increases upon exposure to the antifungal voriconazole (da Silva Ferreira et al. 2006). Recently, CsnB was studied in detail (Beck et al. 2014). Deletion of *csnB* did not alter growth or hyphal cell wall morphology. However, the *csnB* deletion mutant was unable to use chitosan as a carbon source, suggesting that the other three putative chitosanases cannot compensate for the extracellular activity of CsnB. Further studies of CsnB and the other three chitosanases will be needed to fully assess the contribution of these enzymes to virulence.

C. Modification of Other Cell Wall Polysaccharides

In contrast to β-1,3-glucan, α-1,3-glucan is not known to be branched or conjugated. Therefore, its post-synthesis modification is predicted to be more limited. One known modification of α-1,3-glucan is hydrolysis of its glycosidic bonds by α-1,3-glucanase, AGN.

In *A. fumigatus*, there are eight putative AGN genes in the genome, all belonging to the GH family 71 which is classified in the CAZy database to have predicted α-1,3-glucanase activity (Lee et al, unpublished). Interestingly, one of the AGN-encoding genes, afu8g06360, is downregulated in *A. fumigatus* in response to voriconazole exposure (da Silva Ferreira et al. 2006). Another AGN gene, afu2g03980, is induced by calcium signaling and has been detected in a secretome analysis from patient sera (Kumar et al. 2011). Given the fact that α-1,3-glucan is the most abundant polysaccharide in the hyphal cell wall, future studies examining the function of the AGN enzymes may provide further insights into the biosynthesis of α-1,3-glucan and its role in the virulence of *A. fumigatus.*

Unlike other cell wall polysaccharides, galactosaminogalactan is a heteropolysaccharide with no distinct repeating units or internal segments. Preliminary studies from our laboratory have identified an N-acetyl galactosamine deacetylase, Agd3, which mediates partial deacetylation of galactosaminogalactan. Deletion of *agd3* is associated with a loss of deacetylation and impairs the ability of galactosaminogalactan to mediate adherence and biofilm formation (Lee et al. 2014a). The role of galactosaminogalactan deacetylation in virulence is currently under study.

V. Towards an Understanding of the Regulation of Cell Wall Biosynthesis

Cell wall biosynthesis in *A. fumigatus* is a tightly controlled process that is required for the survival of the fungus (Fig. 8.3). Active cell wall biogenesis begins when resting conidia sense an appropriate environmental cue and initiate swelling and germination. Upon germination, hyphae must constantly change and remodel the cell wall to accommodate for growth, development, and adaptation to the changing environment. These environmental cues include temperature, moisture, pH, quorum molecules, and nutrients, just to name a few. The need for a constantly adapting cell wall suggests the existence of a sensitive, tightly controlled, and highly rapid set of regulatory pathways that can ensure that enzymes involved in the cell wall biosynthesis are readily available when and where they are needed and efficiently repressed when they are not.

Over the past decade, efforts in deciphering the regulatory pathways and stress-related elements involved in cell wall biosynthesis in *A. fumigatus* have provided us with great insights into this seemingly complicated process. At its core, pathways involved in maintaining the cell wall integrity (CWI) play a prominent role in the regulation of cell wall biosynthesis. These pathways integrate numerous environmental cues and intracellular sensing to activate several signaling cascades that influence the expression of cell wall biosynthetic proteins. Elements and pathways involved in maintaining the CWI include the **mitogen-activated protein kinases** (MAPK), including MpkA and Hogl (Dichtl et al. 2012; Jain et al. 2011; Valiante et al. 2008); calcium signaling through the **calcineurin pathway** (Cramer et al. 2008; Soriani et al. 2008); heat shock protein (Hsp90) (Lamoth et al. 2012); and a number of developmentally regulated transcription factors such as MedA (Gravelat et al. 2010) and StuA (Sheppard et al. 2005). MAP kinases are extensively reviewed in Chap. 6. Understanding these pathways and elements is critical in understanding the complex regulatory web of cell wall biosynthesis and compensatory mechanism, which is emerging as a major challenge in developing cell wall-targeting antifungal agents.

A. Chemogenetic Approaches to Understanding the Regulation of Cell Wall Biosynthesis

Perhaps the best example of the role of the effects of regulatory pathway on cell wall composition comes from studies examining the effect of echinocandins on cell wall composition. As noted previously, echinocandins inhibit the synthesis of β-1,3-glucan by noncompetitive inhibition of the β-1,3-glucan synthase (Chen et al. 2011). Interestingly, echinocandin-mediated inhibition of β-1,3-glucan synthesis is associated with a compensatory increase in cell wall chitin content. A number of studies have therefore examined the effects of mutations in various signaling pathways on the susceptibility of *A. fumigatus* to echinocandins in order to better understand the regulatory pathways governing cell wall biosynthesis.

The best characterized pathway involved in cell wall biosynthesis is the calcineurin pathway. Deletion of key components of this pathway, the transcription factor gene *crz1* or the calcineurin catalytic subunit A gene *cnaA*, increased susceptibility to caspofungin (Fortwendel et al. 2009; Soriani et al. 2008). Importantly, not only did Δ*crz1* or Δ*cnaA* mutants display decreased β-1,3-glucan content, but their chitin levels also remained unchanged compared to the parental wild type (Fortwendel et al. 2009). Further, exposure to echinocandins did not result in an increase in chitin in the cell wall of either mutant strain as was seen in the wild-type parent. In fact, both mutants exhibited severe growth defects compared to the parental strain in the presence of either caspofungin or nikkomycin. Thus, not only does the calcineurin pathway govern β-1,3-glucan synthesis, but it is also a critical element in mediating the compensatory upregulation of chitin in response to β-1,3-glucan deficiency.

Similarly, deletions within the MkpA-MAP kinase pathway are associated with altered susceptibility to echinocandins. Deletion of *wsc1*, a cell surface sensor protein, or the downstream GTPase *rho1* or MAP kinase protein *mkpA* also increased susceptibility to echinocandins as compared to the parental strain (Dichtl et al. 2010, 2012). Whether the MkpA-MAP kinase pathway is required for increasing the synthesis of β-1,3-glucan in the presence of caspofungin or mediates a compensatory increase in the synthesis of chitin or another cell wall polysaccharide is not known.

Signal pathways are not the only integral components of the cell wall integrity network. Repression of an essential heat shock protein, *hsp90*, resulted in a mutant strain with severe cell wall defects, including swollen hyphae with blunted tips, similar to those produced by fungi exposed to echinocandins (Lamoth et al. 2012). Localization studies using fluorescence protein-tagged Hsp90 revealed that while Hsp90 is normally cytoplasmic, exposure to echinocandins results in localization of Hsp90 to the cell wall and septum. Chitin synthesis under repressed conditions were not different in the mutant strain as compared to the parental strain, suggesting that Hsp90 directly affects β-1,3-glucan synthesis rather than compensatory upregulation of chitin synthesis. Further studies will be required to fully understand the role of Hsp90 in the cell wall integrity network.

B. Deciphering the Regulation of Cell Wall Composition Through Direct Molecular Approaches

1. Global Regulators of Cell Wall Biosynthesis

Morphogenesis and development are associated with major changes in the physical structure of fungal cells, and by extension, their cell walls. Studies of development in the model organism *A. nidulans* have identified regulatory pathways governing development whose function seems to be largely conserved in *A. fumigatus*. In both fungi, sequential activation and feedback of transcription factors BrlA, AbaA, and WetA are required for normal development of the fungus, especially in the regulation of conidiation (Marshall and Timberlake 1991; Tao and Yu 2011; Yu 2010). These transcription factors are also covered in Chap. 1. Although detailed cell wall studies of deletion mutants of these transcription factors are lacking, transcriptomic studies revealed that BrlA is required for the production of a number of cell wall-modifying enzymes (Twumasi-Boateng et al. 2009). These include a putative endoglucanse (afu2g14540), endochitinase csn (afu4g01290) (Cheng et al. 2006; Gautam et al. 2011; Schwienbacher et al. 2005), and putative chitosanase (afu8g00930). Given that remodeling of the cell wall is a critical step in later stages of development, the fact that BrlA is required for the regulation of carbohydrate hydrolases fits within this framework.

The core conidiation pathway is also modulated through the action of the temporal and spatial developmental transcription factors,

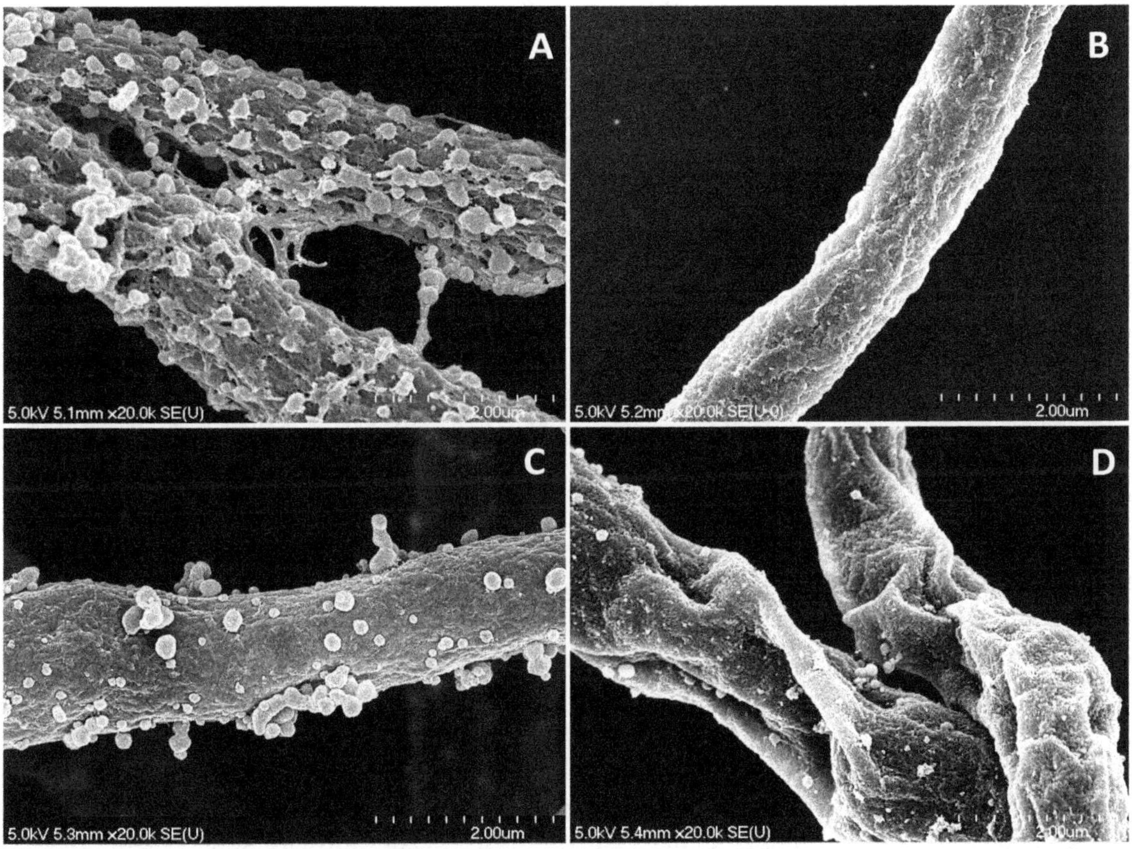

Fig. 8.4 Cell wall morphology of galactosaminogalactan-deficient mutants. Scanning electron micrograph of hyphae of (**a**) wild-type *A. fumigatus* strain Af293, (**b**) Δ*uge3*, (**c**) Δ*stuA*, and (**d**) Δ*medA*

MedA and StuA. These development transcription factors have also been studied in the context of cell wall biosynthesis in *A. fumigatus* (Al Abdallah et al. 2012; Gravelat et al. 2010; Sheppard et al. 2005). Deletion of *medA* or *stuA* resulted in mutants with changes in the cell wall composition, most notably a marked reduction in the production of galactosaminogalactan (Fig. 8.4) (Gravelat et al. 2010, 2013; Sheppard et al. 2005). It is likely that future studies of transcriptional regulators will identify other pathways controlling the synthesis and modification of cell wall polysaccharides in *A. fumigatus.*

2. Other Compensatory Relationships Between Cell Wall Polysaccharides

Deletion analyses of polysaccharide biosynthetic genes have revealed other examples, which altering synthesis of one cell wall polysaccharide leads to increase in the synthesis of another cell wall component. For example, blocking the synthesis of α-1,3-glucan by deletion of the *AGS* genes was associated with an increase in chitin and β-1,3-glucan (Henry et al. 2012). These compensatory relationships between polysaccharides are complex, as inhibition of synthesis of galactomannan by deletion of *ugm1* was associated with an increase in galactosaminogalactan and β-1,3-glucan, while abrogating the synthesis of galactosaminogalactan by deletion of *uge3* had no effect on galactomannan or β-1,3-glucan production (Gravelat et al. 2013; Lamarre et al. 2009).

The molecular mechanism underlying these compensatory changes in cell wall content is unknown. It is possible that, as with chitin and β-1,3-glucan, elements of the cell wall integrity regulatory network play a key role in mediating these compensatory changes, although direct evidence is still lacking. Alternately, alterations in cell wall composition in the face of altered polysaccharide synthesis can reflect substrate shunting whereby the accumulations of sugar substrates from a blocked biosynthetic pathway are utilized for the production of a second cell wall polysaccharide. Substrate shunting is not independent of regulatory effects as it is also likely that the organism detects shifts in intracellular pools of sugars and alters enzyme expression or activity in response to this perturbation in metabolism. One example of this phenomenon can be found in studies examining the biosynthetic pathway of trehalose, a disaccharide that enhances resistance to a variety of stressors and serves as an energy source during germination. Production of trehalose requires dephosphorylation of the intermediate substrate trehalose-6-phosphate (T6P) by a phosphatase OrlA. Deletion of *orlA* leads to the accumulation of T6P, which in turn inhibits hexokinase activity. This inhibition of hexokinase activity in the Δ*orlA* mutant also impairs the synthesis of UDP-GlcNAc, resulting in a mutant deficient in chitin (Puttikamonkul et al. 2010). Since UDP-GlcNAc is also a substrate in the galactosaminogalactan biosynthetic pathway, it is also possible that T6P accumulation could also inhibit the production of this polysaccharide, although this hypothesis has not been tested.

Studying these changes in metabolic activity or intracellular sugar substrate concentration is challenging. Future efforts in developing advanced tools and techniques specific to filamentous fungi will be required to elucidate the role of substrate shunting in *A. fumigatus* cell wall biogenesis.

C. Real World Applications: Targeting Compensatory Changes in Cell Wall Composition to Enhance Antifungal Efficacy

To date, only one class of antifungals targeting the cell wall, the echinocandins, is approved for clinical use. However, echinocandin-mediated inhibition of β-1,3-glucan synthesis results in an increase in the cell wall chitin content (Verwer et al. 2012), which may reduce the efficacy of these agents. Combining echinocandins with inhibitors of chitin synthesis, such as nikkomycin may prove to be an effective strategy to increase the activity of this class of antifungals (Ganesan et al. 2004; Verwer et al. 2012). Further, the development of new agents targeting the synthesis of other cell wall polysaccharides may offer exciting new therapeutic options for the treatment of invasive aspergillosis, alone or in combination with existing antifungals. Indeed, it is possible that agents may be identified that have minimal intrinsic antifungal activity as monotherapy, but which will have important activity when combined with other antifungals that target the cell wall. Thus, the study of cell wall mutants and the corresponding regulatory pathways is likely to be of critical importance in guiding the rational development of combinatorial antifungal strategies that target the cell wall.

VI. Conclusions and Perspectives

The cell wall of *A. fumigatus* is a complex and dynamic structure that must change during fungal growth and development and adapt to rapidly changing environmental conditions. Although recent studies have begun to identify and characterize the polysaccharide components that comprise this structure, our understanding of the dynamic changes that the cell wall undergoes and the regulatory pathways governing these changes are still very limited. Similarly, the important role that cell wall polysaccharides play in virulence is only now beginning to emerge. Antifungals targeting the cell wall biosynthesis have shown great promise in the treatment of invasive aspergillosis. Understanding the regulatory pathways governing the composition of the cell wall during infection and in response to antifungal therapy will be critical in improving our success with existing and novel therapeutics.

References

Abruzzo GK, Flattery AM, Gill CJ, Kong L, Smith JG, Pikounis VB, Balkovec JM, Bouffard AF, Dropinski

JF, Rosen H et al (1997) Evaluation of the echinocandin antifungal MK-0991 (L-743,872): efficacies in mouse models of disseminated aspergillosis, candidiasis, and cryptococcosis. Antimicrob Agents Chemother 41:2333–2338

Al Abdallah Q, Choe SI, Campoli P, Baptista S, Gravelat FN, Lee MJ, Sheppard DC (2012) A conserved C-terminal domain of the *Aspergillus fumigatus* developmental regulator MedA is required for nuclear localization, adhesion and virulence. PLoS One 7:e49959

Alcazar-Fuoli L, Clavaud C, Lamarre C, Aimanianda V, Seidl-Seiboth V, Mellado E, Latge JP (2011) Functional analysis of the fungal/plant class chitinase family in *Aspergillus fumigatus*. Fungal Genet Biol 48:418–429

Arbia W, Arbia L, Adour L, Amrane A (2013) Chitin extraction from crustacean shells using biological methods – a review. Food Technol Biotechnol 51:12–25

Aufauvre-Brown A, Mellado E, Gow NA, Holden DW (1997) Aspergillus fumigatus chsE: a gene related to CHS3 of *Saccharomyces cerevisiae* and important for hyphal growth and conidiophore development but not pathogenicity. Fungal Genet Biol 21:141–152

Baker LG, Specht CA, Donlin MJ, Lodge JK (2007) Chitosan, the deacetylated form of chitin, is necessary for cell wall integrity in *Cryptococcus neoformans*. Eukaryot Cell 6:855–867

Bardalaye PC, Nordin JH (1976) Galactosaminogalactan from cell walls of *Aspergillus niger*. J Bacteriol 125:655–669

Beauvais A, Drake R, Ng K, Diaquin M, Latge JP (1993) Characterization of the 1,3-beta-glucan synthase of *Aspergillus fumigatus*. J Gen Microbiol 139:3071–3078

Beauvais A, Bruneau JM, Mol PC, Buitrago MJ, Legrand R, Latge JP (2001) Glucan synthase complex of *Aspergillus fumigatus*. J Bacteriol 183:2273–2279

Beauvais A, Maubon D, Park S, Morelle W, Tanguy M, Huerre M, Perlin DS, Latge JP (2005) Two alpha(1-3) glucan synthases with different functions in *Aspergillus fumigatus*. Appl Environ Microbiol 71:1531–1538

Beauvais A, Schmidt C, Guadagnini S, Roux P, Perret E, Henry C, Paris S, Mallet A, Prevost MC, Latge JP (2007) An extracellular matrix glues together the aerial-grown hyphae of *Aspergillus fumigatus*. Cell Microbiol 9:1588–1600

Beauvais A, Bozza S, Kniemeyer O, Formosa C, Balloy V, Henry C, Roberson RW, Dague E, Chignard M, Brakhage AA et al (2013) Deletion of the alpha-(1,3)-glucan synthase genes induces a restructuring of the conidial cell wall responsible for the avirulence of *Aspergillus fumigatus*. PLoS Pathog 9:e1003716

Beck J, Broniszewska M, Schwienbacher M, Ebel F (2014) Characterization of the *Aspergillus fumigatus* chitosanase CsnB and evaluation of its potential use in serological diagnostics. Int J Med Microbiol 304:696–702

Bruneau JM, Magnin T, Tagat E, Legrand R, Bernard M, Diaquin M, Fudali C, Latge JP (2001) Proteome analysis of *Aspergillus fumigatus* identifies glycosylphosphatidylinositol-anchored proteins associated to the cell wall biosynthesis. Electrophoresis 22:2812–2823

Camougrand NM, Mouassite M, Velours GM, Guerin MG (2000) The "SUN" family: UTH1, an ageing gene, is also involved in the regulation of mitochondria biogenesis in *Saccharomyces cerevisiae*. Arch Biochem Biophys 375:154–160

Camougrand N, Kissova I, Velours G, Manon S (2004) Uth1p: a yeast mitochondrial protein at the crossroads of stress, degradation and cell death. FEMS Yeast Res 5:133–140

Chen SC, Slavin MA, Sorrell TC (2011) Echinocandin antifungal drugs in fungal infections: a comparison. Drugs 71:11–41

Cheng CY, Chang CH, Wu YJ, Li YK (2006) Exploration of glycosyl hydrolase family 75, a chitosanase from *Aspergillus fumigatus*. J Biol Chem 281:3137–3144

Cobrado L, Silva-Dias A, Azevedo MM, Pina-Vaz C, Rodrigues AG (2013) In vivo antibiofilm effect of cerium, chitosan and hamamelitannin against usual agents of catheter-related bloodstream infections. J Antimicrob Chemother 68:126–130

Cramer RA Jr, Perfect BZ, Pinchai N, Park S, Perlin DS, Asfaw YG, Heitman J, Perfect JR, Steinbach WJ (2008) Calcineurin target CrzA regulates conidial germination, hyphal growth, and pathogenesis of *Aspergillus fumigatus*. Eukaryot Cell 7:1085–1097

da Silva Ferreira ME, Malavazi I, Savoldi M, Brakhage AA, Goldman MH, Kim HS, Nierman WC, Goldman GH (2006) Transcriptome analysis of *Aspergillus fumigatus* exposed to voriconazole. Current Genet 50:32–44

Dahn U, Hagenmaier H, Hohne H, Konig WA, Wolf G, Zahner H (1976) Stoffwechselprodukte von mikroorganismen. 154. Mitteilung. Nikkomycin, ein neuer Hemmstoff der Chitinsynthese bei Pilzen. Arch Microbiol 107:143–160

Dichtl K, Ebel F, Dirr F, Routier FH, Heesemann J, Wagener J (2010) Farnesol misplaces tip-localized Rho proteins and inhibits cell wall integrity signalling in *Aspergillus fumigatus*. Mol Microbiol 76:1191–1204

Dichtl K, Helmschrott C, Dirr F, Wagener J (2012) Deciphering cell wall integrity signalling in *Aspergillus fumigatus*: identification and functional characterization of cell wall stress sensors and relevant Rho GTPases. Mol Microbiol 83:506–519

Douglas CM (2001) Fungal beta(1,3)-D-glucan synthesis. Med Mycol 39(Suppl 1):55–66

Dunkler A, Walther A, Specht CA, Wendland J (2005) Candida albicans CHT3 encodes the functional homolog of the Cts1 chitinase of *Saccharomyces cerevisiae*. Fungal Genet Biol 42:935–947

El Gueddari N, Rauchhaus U, Moerschbacher B, Deising H (2002) Developmentally regulated conversion of surface-exposed chitin to chitosan in cell walls of plant pathogenic fungi. New Phytol 156:103–112

Engel J, Schmalhorst PS, Dörk-Bousset T, Ferrières V, Routier FH (2009) A single UDP-galactofuranose transporter is required for galactofuranosylation in *Aspergillus fumigatus*. J Biol Chem 284:33859–33868

Engel J, Schmalhorst PS, Routier FH (2012) Biosynthesis of the fungal cell wall polysaccharide galactomannan requires intraluminal GDP-mannose. J Biol Chem 287:44418–44424

Fontaine T, Hartland RP, Diaquin M, Simenel C, Latge JP (1997) Differential patterns of activity displayed by two exo-beta-1,3-glucanases associated with the *Aspergillus fumigatus* cell wall. J Bacteriol 179:3154–3163

Fontaine T, Simenel C, Dubreucq G, Adam O, Delepierre M, Lemoine J, Vorgias CE, Diaquin M, Latge JP (2000) Molecular organization of the alkali-insoluble fraction of *Aspergillus fumigatus* cell wall. J Biol Chem 275:27594–27607

Fontaine T, Beauvais A, Loussert C, Thevenard B, Fulgsang CC, Ohno N, Clavaud C, Prevost MC, Latge JP (2010) Cell wall alpha1-3glucans induce the aggregation of germinating conidia of *Aspergillus fumigatus*. Fungal Genet Biol 47:707–712

Fontaine T, Delangle A, Simenel C, Coddeville B, van Vliet SJ, van Kooyk Y, Bozza S, Moretti S, Schwarz F, Trichot C et al (2011) Galactosaminogalactan, a new immunosuppressive polysaccharide of *Aspergillus fumigatus*. PLoS Pathog 7:e1002372

Fortwendel JR, Juvvadi PR, Pinchai N, Perfect BZ, Alspaugh JA, Perfect JR, Steinbach WJ (2009) Differential effects of inhibiting chitin and 1,3-{beta}-D-glucan synthesis in ras and calcineurin mutants of *Aspergillus fumigatus*. Antimicrob Agents Chemother 53:476–482

Ganesan LT, Manavathu EK, Cutright JL, Alangaden GJ, Chandrasekar PH (2004) In-vitro activity of nikkomycin Z alone and in combination with polyenes, triazoles or echinocandins against *Aspergillus fumigatus*. Clin Microbiol Infect 10:961–966

Gastebois A, Clavaud C, Aimanianda V, Latge JP (2009) *Aspergillus fumigatus*: cell wall polysaccharides, their biosynthesis and organization. Future Microbiol 4:583–595

Gastebois A, Fontaine T, Latge JP, Mouyna I (2010a) beta(1-3)Glucanosyltransferase Gel4p is essential for *Aspergillus fumigatus*. Eukaryot Cell 9:1294–1298

Gastebois A, Mouyna I, Simenel C, Clavaud C, Coddeville B, Delepierre M, Latge JP, Fontaine T (2010b) Characterization of a new beta(1-3)-glucan branching activity of *Aspergillus fumigatus*. J Biol Chem 285:2386–2396

Gastebois A, Aimanianda V, Bachellier-Bassi S, Nesseir A, Firon A, Beauvais A, Schmitt C, England P, Beau R, Prevost MC et al (2013) SUN proteins belong to a novel family of beta-(1,3)-glucan-modifying enzymes involved in fungal morphogenesis. J Biol Chem 288:13387–13396

Gautam P, Upadhyay SK, Hassan W, Madan T, Sirdeshmukh R, Sundaram CS, Gade WN, Basir SF, Singh Y, Sarma PU (2011) Transcriptomic and proteomic profile of *Aspergillus fumigatus* on exposure to artemisinin. Mycopathologia 172:331–346

Gorin PAJ, Eveleigh DE (1970) Extracellular 2-acetamido-2-deoxy-D-galacto-D-galactan from *Aspergillus nidulans*. Biochemistry 9:5023–5027

Gravelat FN, Ejzykowicz DE, Chiang LY, Chabot JC, Urb M, Macdonald KD, al-Bader N, Filler SG, Sheppard DC (2010) *Aspergillus fumigatus* MedA governs adherence, host cell interactions and virulence. Cell Microbiol 12:473–488

Gravelat FN, Beauvais A, Liu H, Lee MJ, Snarr BD, Chen D, Xu W, Kravtsov I, Hoareau CMQ, Vanier G et al (2013) *Aspergillus* Galactosaminogalactan Mediates Adherence to Host Constituents and Conceals Hyphal β-Glucan from the Immune System. PLoS Pathog 9(8):e1003575

Gresnigt MS, Bozza S, Becker KL, Joosten LA, Abdollahi-Roodsaz S, van der Berg WB, Dinarello CA, Netea MG, Fontaine T, De Luca A et al (2014) A polysaccharide virulence factor from *Aspergillus fumigatus* elicits anti-inflammatory effects through induction of Interleukin-1 receptor antagonist. PLoS Pathog 10:e1003936

Guembe M, Guinea J, Pelaez T, Torres-Narbona M, Bouza E (2007) Synergistic effect of posaconazole and caspofungin against clinical zygomycetes. Antimicrob Agents Chemother 51:3457–3458

Hartl L, Gastebois A, Aimanianda V, Latge JP (2011) Characterization of the GPI-anchored endo beta-1,3-glucanase Eng2 of *Aspergillus fumigatus*. Fungal Genet Biol 48:185–191

Hartland RP, Vermeulen CA, Klis FM, Sietsma JH, Wessels JG (1994) The linkage of (1-3)-beta-glucan to chitin during cell wall assembly in Saccharomyces *cerevisiae*. Yeast 10:1591–1599

Hartland RP, Fontaine T, Debeaupuis JP, Simenel C, Delepierre M, Latge JP (1996) A novel beta-(1-3)-glucanosyltransferase from the cell wall of *Aspergillus fumigatus*. J Biol Chem 271:26843–26849

Hartmann T, Dumig M, Jaber BM, Szewczyk E, Olbermann P, Morschhauser J, Krappmann S (2010) Validation of a self-excising marker in the human pathogen *Aspergillus fumigatus* by employing the beta-rec/six site-specific recombination system. Appl Environ Microbiol 76:6313–6317

Hector RF (1993) Compounds active against cell walls of medically important fungi. Clin Microbiol Rev 6:1–21

Hector RF, Zimmer BL, Pappagianis D (1990) Evaluation of nikkomycins X and Z in murine models of coccidioidomycosis, histoplasmosis, and blastomycosis. Antimicrob Agents Chemother 34:587–593

Henry C, Latge JP, Beauvais A (2012) alpha1,3 glucans are dispensable in *Aspergillus fumigatus*. Eukaryot Cell 11:26–29

Hu W, Sillaots S, Lemieux S, Davison J, Kauffman S, Breton A, Linteau A, Xin C, Bowman J, Becker J et al (2007) Essential gene identification and drug target prioritization in *Aspergillus fumigatus*. PLoS Pathog 3:e24

Jain R, Valiante V, Remme N, Docimo T, Heinekamp T, Hertweck C, Gershenzon J, Haas H, Brakhage AA (2011) The MAP kinase MpkA controls cell wall integrity, oxidative stress response, gliotoxin production and iron adaptation in *Aspergillus fumigatus*. Mol Microbiol 82:39–53

Jaques AK, Fukamizo T, Hall D, Barton RC, Escott GM, Parkinson T, Hitchcock CA, Adams DJ (2003) Disruption of the gene encoding the ChiB1 chitinase of *Aspergillus fumigatus* and characterization of a recombinant gene product. Microbiology 149:2931–2939

Jimenez-Ortigosa C, Aimanianda V, Muszkieta L, Mouyna I, Alsteens D, Pire S, Beau R, Krappmann S, Beauvais A, Dufrene YF et al (2012) Chitin synthases with a myosin motor-like domain control the resistance of *Aspergillus fumigatus* to echinocandins. Antimicrob Agents Chemother 56:6121–6131

Jin C (2012) Protein glycosylation in Aspergillus fumigatus is essential for cell wall synthesis and serves as a promising model of multicellular eukaryotic development. Int J Microbiol 2012: 654251

Krainer E, Becker JM, Naider F (1991) Synthesis and biological evaluation of dipeptidyl and tripeptidyl polyoxin and nikkomycin analogues as anticandidal prodrugs. J Med Chem 34:174–180

Kumar A, Ahmed R, Singh PK, Shukla PK (2011) Identification of virulence factors and diagnostic markers using immunosecretome of *Aspergillus fumigatus*. J Proteome 74:1104–1112

Kuranda MJ, Robbins PW (1991) Chitinase is required for cell separation during growth of *Saccharomyces cerevisiae*. J Biol Chem 266:19758–19767

Lamarre C, Beau R, Balloy V, Fontaine T, Wong Sak Hoi J, Guadagnini S, Berkova N, Chignard M, Beauvais A, Latgé J-P (2009) Galactofuranose attenuates cellular adhesion of *Aspergillus fumigatus*. Cell Microbiol 11:1612–1623

Lamoth F, Juvvadi PR, Fortwendel JR, Steinbach WJ (2012) Heat shock protein 90 is required for conidiation and cell wall integrity in *Aspergillus fumigatus*. Eukaryot Cell 11:1324–1332

Latgé J-P (2010) Tasting the fungal cell wall. Cell Microbiol 12:863–872

Latgé JP, Calderone R (2006) The fungal cell wall. In: Kues U, Fischer R (eds) The Mycota. Springer, Berlin, pp 73–104

Latgé J-P, Debeaupuis JP, Moutaouakil M, Diaquin M, Sarfati J, Prevost MC, Wieruszeski JM, Leroy Y, Fournet B (1991) Galactomannan and the circulating antigens of *Aspergillus fumigatus*. In: Latgé J-P, Boucias D (eds) Fungal cell wall and immune response. Springer, Heidelberg

Latgé JP, Kobayashi H, Debeaupuis JP, Diaquin M, Sarfati J, Wieruszeski JM, Parra E, Bouchara JP, Fournet B (1994) Chemical and immunological characterization of the extracellular galactomannan of *Aspergillus fumigatus*. Infect Immun 62:5424–5433

Lee MJ, Geller AM, Gravelat FN, Liu H, Snarr BD, Baptista S, Nguyen D, Nitz M, Latge JP, Fontaine T et al (2014a) Deacetylation of *Aspergillus fumigatus* galactosaminogalactan is required for adherence and virulence. Paper presented at: 6th Advances Against Aspergillosis, Madrid, Spain

Lee MJ, Gravelat FN, Cerone RP, Baptista SD, Campoli PV, Choe SI, Kravtsov I, Vinogradov E, Creuzenet C, Liu H et al (2014b) Overlapping and distinct roles of *Aspergillus fumigatus* UDP-glucose 4-epimerases in galactose metabolism and the synthesis of galactose-containing cell wall polysaccharides. J Biol Chem 289:1243–1256

Lenardon MD, Munro CA, Gow NA (2010) Chitin synthesis and fungal pathogenesis. Curr Opin Microbiol 13:416–423

Levin DE (2005) Cell wall integrity signaling in *Saccharomyces cerevisiae*. Microbiol Mol Biol Rev 69:262–291

Levin DE (2011) Regulation of cell wall biogenesis in *Saccharomyces cerevisiae*: the cell wall integrity signaling pathway. Genetics 189:1145–1175

Lombard V, Golaconda Ramulu H, Drula E, Coutinho PM, Henrissat B (2014) The carbohydrate-active enzymes database (CAZy) in 2013. Nucleic Acids Res 42:D490–D495

Loussert C, Schmitt C, Prevost MC, Balloy V, Fadel E, Philippe B, Kauffmann-Lacroix C, Latge JP, Beauvais A (2010) In vivo biofilm composition of *Aspergillus fumigatus*. Cell Microbiol 12:405–410

Maligie MA, Selitrennikoff CP (2005) *Cryptococcus neoformans* resistance to echinocandins: (1,3)beta-glucan synthase activity is sensitive to echinocandins. Antimicrob Agents Chemother 49:2851–2856

Marshall MA, Timberlake WE (1991) *Aspergillus nidulans* wetA activates spore-specific gene expression. Mol Cell Biol 11:55–62

Maubon D, Park S, Tanguy M, Huerre M, Schmitt C, Prevost MC, Perlin DS, Latge JP, Beauvais A (2006) AGS3, an alpha(1-3)glucan synthase gene family member of *Aspergillus fumigatus*, modulates mycelium growth in the lung of experimentally infected mice. Fungal Genet Biol 43:366–375

Mellado E, Aufauvre-Brown A, Gow NA, Holden DW (1996a) The *Aspergillus fumigatus* chsC and chsG genes encode class III chitin synthases with different functions. Mol Microbiol 20:667–679

Mellado E, Specht CA, Robbins PW, Holden DW (1996b) Cloning and characterization of chsD, a chitin synthase-like gene of *Aspergillus fumigatus*. FEMS Microbiol Lett 143:69–76

Mellado E, Dubreucq G, Mol P, Sarfati J, Paris S, Diaquin M, Holden DW, Rodriguez-Tudela JL, Latge JP (2003) Cell wall biogenesis in a double chitin synthase mutant (chsG-/chsE-) of *Aspergillus fumigatus*. Fungal Genet Biol 38:98–109

Mouassite M, Camougrand N, Schwob E, Demaison G, Laclau M, Guerin M (2000) The 'SUN' family: yeast SUN4/SCW3 is involved in cell septation. Yeast 16:905–919

Mouyna I, Fontaine T (2009) Cell wall of *Aspergillus fumigatus*: a dynamic structure. In: Latge JP, Steinbach WJ (eds) *Aspergillus fumigatus* and aspergillosis. ASM Press, Washington, DC, pp 149–158

Mouyna I, Hartland RP, Fontaine T, Diaquin M, Simenel C, Delepierre M, Henrissat B, Latge JP (1998) A 1,3-beta-glucanosyltransferase isolated from the cell wall of *Aspergillus fumigatus* is a homologue of the yeast Bgl2p. Microbiology 144(Pt 11):3171–3180

Mouyna I, Henry C, Doering TL, Latge JP (2004) Gene silencing with RNA interference in the human pathogenic fungus *Aspergillus fumigatus*. FEMS Microbiol Lett 237:317–324

Mouyna I, Morelle W, Vai M, Monod M, Lechenne B, Fontaine T, Beauvais A, Sarfati J, Prevost MC, Henry C et al (2005) Deletion of GEL2 encoding for a beta(1-3)glucanosyltransferase affects morphogenesis and virulence in *Aspergillus fumigatus*. Mol Microbiol 56:1675–1688

Mouyna I, Hartl L, Latge JP (2013) beta-1,3-glucan modifying enzymes in *Aspergillus fumigatus*. Front Microbiol 4:81

Muszkieta L, Aimanianda V, Mellado Encinas P, Beauvais A, Prevost MC, Latge JP (2014) *Aspergillus fumigatus* – deciphering the role of the chitin synthase families I and II by multiple gene targeting deletion. Paper presented at: 6th Advances Against Aspergillosis, Madrid, Spain

Nakanishi-Shindo Y, Nakayama K, Tanaka A, Toda Y, Jigami Y (1993) Structure of the N-linked oligosaccharides that show the complete loss of alpha-1,6-polymannose outer chain from och1, och1 mnn1, and och1 mnn1 alg3 mutants of *Saccharomyces cerevisiae*. J Biol Chem 268:26338–26345

Nakayama K, Nagasu T, Shimma Y, Kuromitsu J, Jigami Y (1992) OCH1 encodes a novel membrane bound mannosyltransferase: outer chain elongation of asparagine-linked oligosaccharides. EMBO J 11:2511–2519

No HK, Meyers SP, Prinyawiwatkul W, Xu Z (2007) Applications of chitosan for improvement of quality and shelf life of foods: a review. J Food Sci 72: R87–R100

Pringle R (1981) Nonspecific adhesion of *Bipolaris sorokiniana* sporelings. Can J Plant 3:9–11

Puttikamonkul S, Willger SD, Grahl N, Perfect JR, Movahed N, Bothner B, Park S, Paderu P, Perlin DS, Cramer RA Jr (2010) Trehalose 6-phosphate phosphatase is required for cell wall integrity and fungal virulence but not trehalose biosynthesis in the human fungal pathogen *Aspergillus fumigatus*. Mol Microbiol 77:891–911

Rappleye CA, Eissenberg LG, Goldman WE (2007) *Histoplasma capsulatum* alpha-(1,3)-glucan blocks innate immune recognition by the beta-glucan receptor. Proc Natl Acad Sci USA 104:1366–1370

Reese AJ, Yoneda A, Breger JA, Beauvais A, Liu H, Griffith CL, Bose I, Kim MJ, Skau C, Yang S et al (2007) Loss of cell wall alpha(1-3) glucan affects *Cryptococcus neoformans* from ultrastructure to virulence. Mol Microbiol 63:1385–1398

Robinet P, Baychelier F, Fontaine T, Picard C, Debre P, Vieillard V, Latge JP, Elbim C (2014) A polysaccharide virulence factor of a human fungal pathogen induces neutrophil apoptosis via NK cells. J Immunol 192:5332–5342

Romani L (2011) Immunity to fungal infections. Nat Rev Immunol 11:275–288

Roncero C (2002) The genetic complexity of chitin synthesis in fungi. Curr Genet 41:367–378

Ruperez P, Leal JA (1981) Extracellular galactosaminogalactan from *Aspergillus parasiticus*. Trans Br Mycol Soc 77:621–625

Samantaray S, Neubauer M, Helmschrott C, Wagener J (2013) Role of the guanine nucleotide exchange factor Rom2 in cell wall integrity maintenance of *Aspergillus fumigatus*. Eukaryot Cell 12:288–298

Schmalhorst PS, Krappmann S, Vervecken W, Rohde M, Müller M, Braus GH, Contreras R, Braun A, Bakker H, Routier FH (2008) Contribution of galactofuranose to the virulence of the opportunistic pathogen *Aspergillus fumigatus*. Eukaryot Cell 7:1268–1277

Schwienbacher M, Weig M, Thies S, Regula JT, Heesemann J, Ebel F (2005) Analysis of the major proteins secreted by the human opportunistic pathogen *Aspergillus fumigatus* under in vitro conditions. Med Mycol 43:623–630

Sheppard DC (2011) Molecular mechanism of *Aspergillus fumigatus* adherence to host constituents. Curr Opin Microbiol 14:375–379

Sheppard DC, Doedt T, Chiang LY, Kim HS, Chen D, Nierman WC, Filler SG (2005) The *Aspergillus fumigatus* StuA protein governs the up-regulation of a discrete transcriptional program during the acquisition of developmental competence. Mol Biol Cell 16:5866–5879

Soriani FM, Malavazi I, da Silva Ferreira ME, Savoldi M, Von Zeska Kress MR, de Souza Goldman MH, Loss O, Bignell E, Goldman GH (2008) Functional characterization of the *Aspergillus fumigatus* CRZ1 homologue, CrzA. Mol Microbiol 67:1274–1291

Taib M, Pinney JW, Westhead DR, McDowall KJ, Adams DJ (2005) Differential expression and extent of fungal/plant and fungal/bacterial chitinases of *Aspergillus fumigatus*. Arch Microbiol 184:78–81

Tao L, Yu JH (2011) AbaA and WetA govern distinct stages of *Aspergillus fumigatus* development. Microbiology 157:313–326

Twumasi-Boateng K, Yu Y, Chen D, Gravelat FN, Nierman WC, Sheppard DC (2009) Transcriptional profiling identifies a role for BrlA in the response to nitrogen depletion and for StuA in the regulation of secondary metabolite clusters in *Aspergillus fumigatus*. Eukaryot Cell 8:104–115

Valiante V, Heinekamp T, Jain R, Hartl A, Brakhage AA (2008) The mitogen-activated protein kinase MpkA of *Aspergillus fumigatus* regulates cell wall signaling and oxidative stress response. Fungal Genet Biol 45:618–627

Verwer PEB, van Duijn ML, Tavakol M, Bakker-Woudenberg IAJM, van de Sande WWJ (2012) Reshuffling of *Aspergillus fumigatus* cell wall components chitin and β-glucan under the influence of caspofungin or nikkomycin Z alone or in combination. Antimicrob Agents Chemother 56:1595–1598

Wei Y, Zhang Y, Derewenda U, Liu X, Minor W, Nakamoto RK, Somlyo AV, Somlyo AP, Derewenda ZS (1997) Crystal structure of RhoA-GDP and its functional implications. Nat Struct Biol 4:699–703

Yu JH (2010) Regulation of Development in *Aspergillus nidulans* and *Aspergillus fumigatus*. Mycrobiology 38:229–237

Zhao W, Li C, Liang J, Sun S (2014) The *Aspergillus fumigatus* beta-1,3-glucanosyltransferase Gel7 plays a compensatory role in maintaining cell wall integrity under stress conditions. Glycobiology 24:418–427

9 Chitin Synthesis and Fungal Cell Morphogenesis

CESAR RONCERO[1], ALBERTO SANCHEZ-DIAZ[2,3], M.-HENAR VALDIVIESO[1]

CONTENTS

[1]Departamento de Microbiologia y Genetica, IBFG, CSIC, Universidad de Salamanca, Calle Zacarias Gonzalez 2, 37007 Salamanca, Spain; e-mail: crm@usal.es; henar@usal.es
[2]Instituto de Biomedicina y Biotecnologia de Cantabria (IBBTEC), Universidad de Cantabria, CSIC, Parque Cientifico y Tecnologico de Cantabria, Calle Albert Einstein 22, 39011 Santander, Spain
[3]Departamento de Biologia Molecular. Facultad de Medicina, Universidad de Cantabria, Cardenal Herrera Oria s/n. 39011 Santander, Spain

I. Introduction

Chitin is an extremely rigid natural polysaccharide and, hence, one of the most important structural biopolymers on earth. It constitutes an essential part of the fungal cell wall, where it is required for the maintenance of cell integrity. Interestingly, its importance does not only derive from its structural role as the major component of the primary septum (PS) but also from its function as a scaffold for the assembly of other components of fungal walls. The interconnections between chitin and other cell wall components determine cell morphogenesis, which is necessarily influenced by environmental changes. The variety observed in cell morphologies and environmental niches across the fungal kingdom would explain the high diversity of chitin synthases (CSs) present there, as well as their different regulatory mechanisms, making chitin synthesis a paradigm for studies addressing fungal morphogenesis. In addition, studies aimed at characterising the regulation of chitin synthesis are the basis for the development of new antifungal therapies.

Biochemistry and Molecular Biology, 3rd Edition
The Mycota III
D. Hoffmeister (Ed.)

II. Chitin Synthases and Deposition of Chitin at the Fungal Cell Wall

After cellulose, chitin is the second most abundant biopolymer. This polymer is present in the fungal cell wall, the exoskeleton of arthropods and some other animal structures. Chitin synthesis is carried out by **chitin synthases (CSs)**, a family of enzymes that use uridine-diphospho-*N*-acetylglucosamine (UDP-GlcNAc) as a substrate and catalyse the $2n$ UDP-GlcNAc→[GlcNAc-β(1,4)-GlcNAc]$_n$+$2n$ UDP reaction. Oligosaccharides are not reaction intermediates. In fungi, chitin synthesis takes place at the plasma membrane (Durán et al. 1975) in a vectorial fashion (Cabib et al. 1983), taking the substrate from the cytoplasm while the nascent fibrils are extruded to the outside of the cell, where the polysaccharide is located. Chapter 8 also covers fungal cell wall polysaccharides. CSs are assumed to be organised in supramolecular complexes at the plasma membrane (PM) in order to facilitate the assembly of chitin fibres. The length of the chitin chains synthesised in vivo and in vitro ranges between 120 and 170 units in *Saccharomyces cerevisiae* (Orlean 1987), the organism in which chitin synthesis has been best characterised. Several CS activities with different in vitro requirements have been described; thus, *Sc*CSI has an optimum pH of 6.5 and is stimulated by the presence of GlcNAc and of Mg^{2+}; Co^{2+} and Ni^{2+} cations inhibit this activity. *Sc*CSII activity levels are approximately 5 % of those of CSI; this enzyme has an optimum pH of 7.5–8.0. Co^{2+} stimulates activity better than Mg^{2+}, while Ni^{2+} inhibits the reaction. Finally, *Sc*CSIII activity is approximately 10 % of the CSI activity level; Mg^{2+} is the best stimulating cation, and neither Co^{2+} nor Ni^{2+} inhibits it. These biochemical differences between the three *S. cerevisiae* CS activities are exploited to determine each of them in the presence of the other two [see Cid et al. (1995) and Roncero (2002) for revision]. CSs from *Candida albicans* also differ in their optimal pH and are stimulated by divalent cations. *Ca*CSII is more active in the presence of Mg^{2+}, while *Ca*CSI and *Ca*CSIII seem to respond slightly better to Co^{2+} (Kim et al. 2002). Recent reports show that the presence of *N*-acetylchito-tetra, -penta and -octaoses in the reaction increases the initial velocity of *Sc*CSI activity, although the nature of this activation is not well understood (Becker et al. 2011). Gyore et al. (2014) have shown that chemically modified analogues of GlcNAc can act as primers in *Sc*CSII in vitro reactions. However, the requirement for a natural primer in the reaction has not been conclusively shown. CSs from other fungi have been found to have similar biochemical characteristics, with requirements for divalent cations and free GlcNAc for optimal activity.

In *S. cerevisiae*, chitin accounts for only 2–3 % of the cell wall dry weight, but it is not uniformly distributed along the cell wall. Most of the chitin accumulates at the septum that will form the bud scar after cytokinesis, whereas the remainder is scattered throughout the cell wall (Molano et al. 1980). Septum synthesis is a complex process that starts early in the cell cycle [see Orlean (2012) for a recent revision]. As soon as a new bud emerges, a chitin ring is synthesised at the base of this bud. As the daughter cell grows, the ring remains at the neck between mother and daughter cells. After mitosis, when nuclei have segregated, more chitin is formed, starting at the ring and continuing centripetally while the plasma membrane invaginates. When the plasma membranes merge, the primary septum (PS) is completed, forming a disc-like structure within a thickened chitin ring. Chitin synthesis at the primary septum is performed by the activity of CSII, while chitin synthesis at the ring and lateral wall is carried out by CSIII. Later, the secondary septum, whose composition is similar to that of the rest of the cell wall, is laid down at both sides of the PS (see Fig. 9.1a for a schematic drawing of the yeast septum). At cytokinesis, the two cells separate through the action of a chitinase (see below) in an asymmetric way; most of the chitin remains in the mother cell, giving rise to the bud scar. A less conspicuous structure—the birth scar—remains on the surface of the daughter cell. Then, the newborn bud starts a process of maturation, which includes growth and some cell wall remodelling. From this moment onwards,

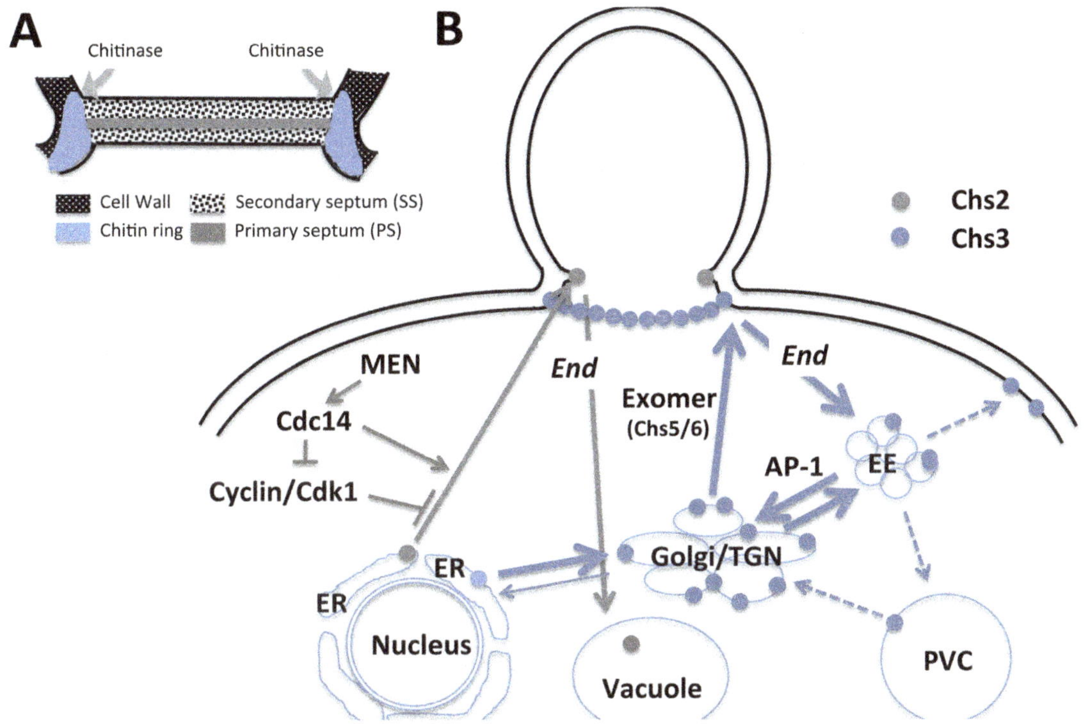

Fig. 9.1 Chitin synthases direct the assembly of yeast septa. (**a**) Structure of the yeast septum showing the two major chitin structures, the chitin ring and the PS. Cell separation occurs after cytokinesis by the action of chitinases from the daughter cell side. (**b**) Traffic and cellular localisation of major yeast CSs, Chs2p and Chs3p. Chs2p travels in a cell cycle controlled form from the ER to the PM at the neck, where it synthesises the PS; later, it is endocytosed and degraded. Chs3p exits the ER and is sorted at the Golgi/TGN boundary towards the PM by the exomer. Chs3p becomes anchored through Chs4p/Bni4p (not shown) to the septin ring to synthesise the chitin ring, being later endocytically recycled back to the Golgi/TGN by means of the AP-1 complex. The alternative routing of Chs3p to the plasma membrane in the absence of AP-1 is indicated by *dashed arrows*. *MEN* mitotic exit network, *ER* endoplasmic reticulum, *TGN* trans-Golgi network, *End* endocytosis, *EE* early endosomes, *PVC* prevacuolar compartment

chitin appears interspersed throughout the lateral wall of the growing bud (Shaw et al. 1991).

In filamentous fungi, chitin is also deposited at the septa, although most of this polymer is distributed uniformly along mycelia after chitin deposition at the tip of the growing hyphae.

III. Chitin Synthases

A. The Diversity of Fungal Chitin Synthases: A Common Catalytic Centre for Multiple Functions

The diversity of fungal CSs was initially uncovered by PCR analysis using degenerated oligonucleotides based on *S. cerevisiae* and *Candida albicans CHS* sequences (Bowen et al. 1992). Since then, many other fungal CS sequences have been identified in more than 200 fungi, and they have been analysed using different comparison algorithms. The number of CS genes identified per species ranges between two and nine, and a detailed study of the multiple CS genes has revealed that three sequence motifs—QXXEY, EDRXL and QXXRW—and seven isolated residues in the core region are conserved in all CS genes [revised in Ruiz-Herrera et al. (2002)]. The QXXRW domain is also present in other glycosyltransferases (GT) (Saxena et al. 1995), and it has been proposed to be the catalytic domain for GT2 enzymes, including CSs. Accordingly, mutations in the

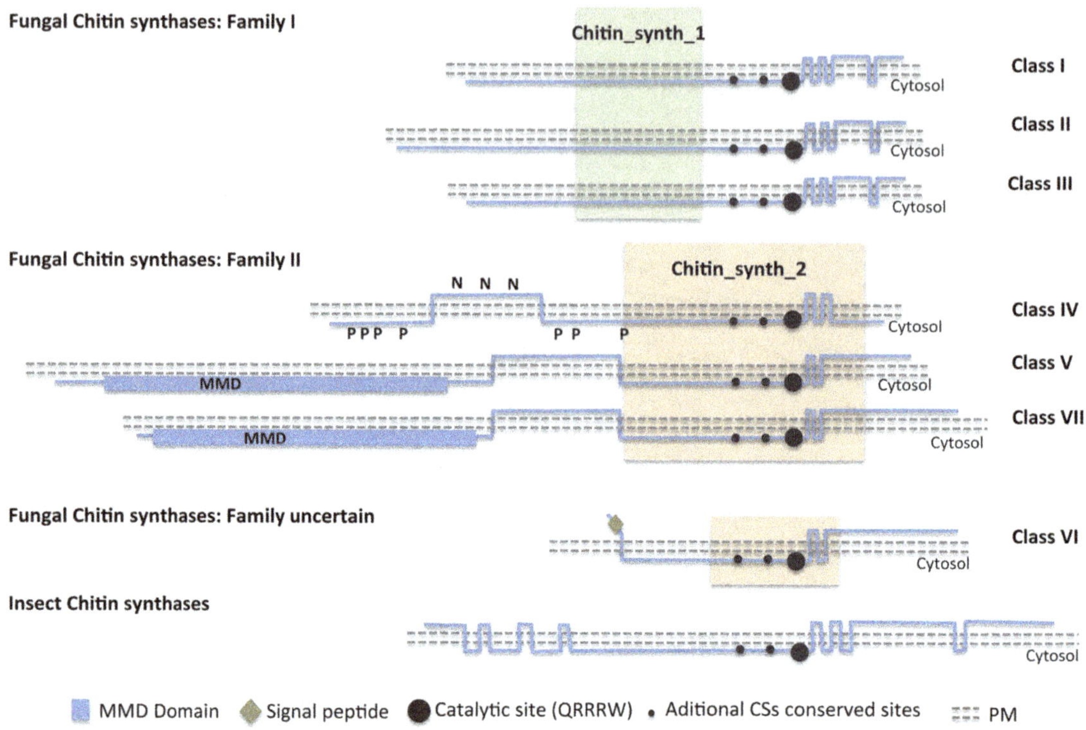

Fig. 9.2 Schematic representation of the different CS classes based on their predicted secondary structure obtained from *S. cerevisiae* and *A. fumigatus* proteins. The images are drawn at approximate scale. *Dashed grey lines* represent the plasma membrane (PM). CS sequences are represented by a *blue line* in which vertical sections represent transmembrane regions. *Shaded boxes* mark the Chitin synthase 1 and 2 domains as described in Pfam. Representative domains are indicated in the figure; note the positional conservation of putative catalytic sites (*large filled circles*). Secondary structure predictions have been only confirmed for *Sc*Chs3p; the N-glycosylation and phosphorylation sites experimentally verified for this protein are included in the scheme

bona fide chitin synthase QRRRW domain result in a loss of both function in vivo and enzymatic activity in vitro in all *S. cerevisiae* CSs (Nagahashi et al. 1995; Cos et al. 1998). Moreover, the *chs2*$^+$ gene from *Schizosaccharomyces pombe*, which lacks these conserved domains, does not code for any CS activity (Martin-Garcia et al. 2003). The consensus sequences identified maintain the same structural features and the same relative location with respect to one another, pointing to their relevance for the proper conformation of the enzymes (Ruiz-Herrera et al. 2002).

The conclusions drawn from multiple analyses have led to the classification of CSs in two large families/divisions, subdivided into seven classes (Roncero 2002; Lenardon et al. 2010). **Family I** genes share the so-called chitin_synth_1 motif (Pfam database accession number: PF01644), and this family includes the *ScCHS1/2*, *CaCHS1/2* and *CaCHS8* genes. All its members show the catalytic domain flanked by a C-terminal region containing multiple transmembrane (TM) domains (Fig. 9.2). The inclusion of additional fungal sequences in multiple comparison analyses allowed the classical subdivision of this family into classes I, II and III. A phylogenetic analysis within the family strongly supported the later appearance of class III enzymes (Ruiz-Herrera and Ortiz-Castellanos 2010); however, a precise separation between Family I enzymes is extremely difficult owing to the high similarity between them and to the lack of functional conservation between the three classes across fungi (see below).

Family II includes the homologues of the *ScCHS3* and *CaCHS3* genes. The genes in this family share the chitin_synth_2 motif (PF03142) and a significant degree of similarity, including a catalytic domain always flanked on both sides by TM regions (Fig. 9.2). Thus, their secondary structure is clearly distinct from the proteins in Family I, a distinction that is also supported by functional assays. While *S. cerevisiae* and *C. albicans* only contain a single member of this family, most fungi contain at least three. Members of this family have been separated into different classes, depending on the author; as a consequence, the subdivision of Family II into classes could be confusing. Class IV, which is present in all fungal groups, is recognised as a unique class in this family, and it usually contains a single member. However, basidiomycetes contain two highly similar members of this class, probably originated by a gene duplication event in the group. Class V CSs are typical of filamentous fungi and are notably absent from the *Saccharomycetales* (yeast) group [see Wang et al. (2009) for a revision on fungal phylogeny]. Based on their catalytic domains, class IV and V CSs are highly related enzymes, although class V CSs contain a myosin motor-like domain (MMD) in their N-terminal region that is absent in class IV members. Further studies separated class V enzymes into two different CS classes that differ in their MMD domains and in their physiological relevance, thus leading to the current separation into class V and VII enzymes. In most cases, the genes encoding these proteins are linked to the chromosome in a head-to-head configuration, sharing their promoter regions; this suggests a common evolutionary origin. The MMD domain is required for the actin-based localisation of these enzymes to sites of active chitin synthesis during hyphal growth. The distinct absence of class V/VII CSs in the *Saccharomycetales* group is probably due to a gene loss effect, in all probability associated with the fact that they only grow as yeasts.

Additional CSs genes have been described in filamentous fungi. Based on sequence comparisons and secondary structures, these CSs seem to define the new class VI of CSs with an uncertain position between Families I and II. However, the experimental evidence about this class is limited, and hence the suggestion of a new family, as recently proposed, should await further studies (Gandia et al. 2014).

B. The Ancient Evolutionary Origin of Chitin Synthases

Although chitin synthesis is best characterised in fungi, CS genes have been also found in Diatomea, Choanoflagellates and several Metazoa (Ruiz-Herrera and Ortiz-Castellanos 2010; Merzendorfer 2011; Zakrzewski et al. 2014). In a recent systematic analysis of *CHS* genes, the authors found that enzymes from Choanoflagellates and Metazoa can be grouped into two clades [I and II; Zakrzewski et al. (2014)]. Clade I is more homogeneous and includes enzymes from all the groups analysed; chitin synthases grouped in the ancient class A from insects belong to this clade. Clade II is a paraphyletic group with a more complex evolution that includes the *CHS2* genes from the insect class B and from nematodes. Interestingly, *CHS* genes from Metazoa exhibit a core region with the chitin_synth_2 motif (Fig. 9.2), suggesting a closer relationship with Family II of CSs (Roncero 2002; Merzendorfer 2011; Zakrzewski et al. 2014). Finally, CS-like genes have also been identified in bacteria, where chitin oligomers participate in nodulation (Peters 1997). The *NodC* gene from *Sinorhizobium meliloti* also shows significant similarity with Family II, suggesting a very ancient origin for the genes that cluster in this family. Based on these data, it is possible to propose an evolutionary trend in chitin synthases. Their origin would be very old, starting with a gene already present in bacteria and probably related to the current Family II of *CHS* genes. Ancient gene duplication at the base of the fungal kingdom would have originated the two fungal *CHS* families reported to date, which now have somewhat different biological functions. The only gene of Family I would have also duplicated very early, and hence all fungi contain class I and II genes. More recent duplication in filamentous fungi would have led to the third class currently observed in Family I. In specific cases, such as in *Aspergillus fumigatus*, more recent duplications would have produced isoenzymes within the same class. A similar situation can be found in Family II, in which duplication from an initial ancestor present in all fungi (class IV) would have led to the class V genes in filamentous fungi after the acquisition, by genetic recombination, of the MMD region; a later duplication would have originated classes V and VII. Interestingly, some metazoan chitin synthases also contain a MMD domain, apparently originated in a fully independent evolutionary event (Zakrzewski et al. 2014). This suggests that the incorporation of MMD domains to CSs would be a common evolutionary trend in the regulation of chitin synthases. An interesting example of evolutionary trends in fungal CSs can be found in *S. pombe*, a relatively modern yeast in which there are two *CHS*-like genes but only one of them (*chs1*$^{+}$) is functional (Arellano et al. 2000; Martin-Garcia et al. 2003). *chs1*$^{+}$ belongs to class I and only synthesises minute amounts of chitin during sporulation. Since vegetative cells do not contain appreciable amounts of chitin, it is not surprising that other *CHS* genes have been lost along the evolution of this fission yeast.

It should be noted, however, that the current classification of CSs is mostly based on sequence comparison and therefore does not take into account the different secondary structure shown by Families I and II of fungal CSs and by metazoan CSs. Moreover, the function of most of the CSs is not yet fully understood (see below); accordingly, the complete evolutionary story of chitin synthases is still unwritten.

IV. Chitin Synthesis and Fungal Cell Wall Assembly

A. Chitin Synthesis and Cell Wall Assembly

Chitin localises to the inner part of the cell wall, close to the PM [see Lesage and Bussey (2006), Orlean (2012), and Cabib and Arroyo (2013) for reviews]. Considerable efforts have been made to characterise the connections between the different polymers of the yeast cell wall. Initially, the existence of a β(1,4)-linkage between the reducing GlcNAC terminus of a chitin chain and the nonreducing terminal glucose of β(1,3)-glucan was described (Kollar et al. 1997). Later, it was reported that all the cell wall components were associated, forming a complex in which β(1,6)-glucan occupied a central position; chitin and β(1,3)-glucan were directly attached by glycosidic linkages, whereas mannoproteins were joined to the polysaccharide through the lipidless remains of a glycosylphosphatidylinositol anchor. Despite this, a complex was described that lacked the mannoproteins (Kollar et al. 1997), and Pir proteins (Proteins with internal repeats) were found to be linked to β(1,3)-glucan through an alkali-sensitive linkage (Ecker et al. 2006). More recently, it has been found that one fraction of chitin is linked to β(1,3)-glucan, another to β(1,6)-glucan, while a third one remains free. These linkages depend on the localisation of chitin (Cabib and Duran 2005); chitin in the PS is mostly free, with only a small fraction linked to β(1,3)-glucan. Chitin dispersed throughout the lateral walls is mostly attached to β(1,6)-glucan. Finally, chitin in the ring of early budding cells is mostly attached to β(1,3)-glucan chains. The presence of alkali-insoluble high-molecular weight β(1,3)-glucan attached to the chitin ring controls morphogenesis at the mother-bud neck by preventing local cell wall remodelling (Blanco et al. 2012). This static fraction of β(1,3)-glucan would keep the width of the neck constant during cell division, allowing the proper segregation of organelles and cell division (Cabib and Arroyo 2013). All these results suggest the existence of different modules that join to form a lattice whose characteristics vary with location and would be influenced by internal and external signals. Several studies have identified the transglycosidase activities involved in the formation of cell wall interlinkages. ScCrh1p and ScCrh2p are Glycosylphosphatidylinositol (GPI)-anchored proteins, with homology to the glycosidase hydrolase family 16. They localise to the cell wall sites where chitin is present, acting as in vivo transglycosylases that form the linkages between chitin and both β(1–3)- and β(1–6)-glucans (Cabib and Arroyo 2013). *CHR1/2*-dependent activities are also required for maintaining the Gas1p transglycosylase at the neck, where it contributes to cell wall assembly by forming additional inter-glucan linkages (Rolli et al. 2009).

Structural studies in *C. albicans* cell wall composition indicate that there are no significant differences between the hyphal and yeast forms, except for a fourfold increase in chitin content in hyphae [reviewed in Klis et al. (2001) and Chaffin (2008)]. The cell wall structure of *C. albicans* contains similar polymers and linkages to those described for *S. cerevisiae*, hinting that the overall cell wall organisation is similar in both organisms. However, there is evidence that in *C. albicans*, β(1,6)-glucan can bind chitin through the C6 position of the GlcNAc residues, a linkage that has not been found in *S. cerevisiae*. Antiparallel chitin chains can bind through hydrogen bonds, resulting in highly insoluble structures. These linkages produce a meshwork of structural fibrillar polysaccharides to which GPI-anchored and Pir cell wall proteins bind covalently. Additionally, there are proteins embedded in the cell wall that are non-covalently bound to the polysaccharides [see Klis et al. (2001) and Chaffin (2008) for reviews].

All these assemblies lead to the formation of the fungal cell wall in which septum synthesis is considered to be a paradigm of fungal morphogenesis (Cid et al. 1995; Cabib et al. 2001; Cabib and Arroyo 2013). In yeast, the chitin ring contributes to septum assembly, while the synthesis of PS is the critical point in the cytokinesis process. Although there are significant differences among organisms, in yeast (and probably in all fungi) the synthesis of the chitin forming the PS is concomitant with the contraction of an actomyosin ring [see Roncero and Sanchez (2010) and Bi and Park (2012) for reviews], both processes being highly interdependent. Mechanistically, the process seems to have been well conserved across fungi, even in organisms lacking chitin such as *S. pombe*, where the role of chitin in PS synthesis has been replaced by the linear β(1–3)-glucan made by the *bgs1*$^{+}$ glucan synthase (Cortes et al. 2007).

Much less is known about cell wall assembly in other forms of fungal growth. Hyphae exhibit a tubular morphology because growth takes place at their tips, where the cell wall is constantly being remodelled. Chitin is usually more abundant in hyphae than in yeasts and, together with other polymers, is continuously incorporated to the cell wall while new bonds are being formed. Studies by Wessels et al. (1983) suggest that the newly-formed chitin is not fully organised, which renders it more susceptible to enzymatic remodelling. As the hypha grows, chitin is crosslinked to glucans at the subapical zone, increasing cell wall rigidity. In *A. fumigatus* chitin, galactomannan and the linear β(1,3) β(1,4)-glucans are covalently linked to the nonreducing end of β(1,3)-glucan side chains while β(1,6)-glucan is absent in this microorganism, as well as in most fungi (Free 2013). Chitin is linked via a β(1,4)-linkage to β(1,3)-glucan [see Latge et al. (2005) for a review on the *A. fumigatus* cell wall]. Chapter 8 also covers *Aspergillus* cell wall polysaccharides. Recent data suggest that chitin synthesis is critical as a scaffold for cell wall assembly during hyphal extension [see below and Jimenez-Ortigosa et al. (2012)]. Septum synthesis during mycelial growth is a poorly known process, although data on *C. albicans* indicate that septum closure in hyphae also depends on chitin, suggesting a mechanism similar to that reported for yeast. The potential cell wall changes during fungal conidiogenesis have not yet been extensively studied, but chitin synthesis seems to be an essential part of this process since many *chs* mutants (see below) show significant defects in sporulation. Moreover, chitin has been shown to be an important component of conidial cell walls. In *S. cerevisiae* spores, chitin is deacetylated into chitosan (Christodoulidou et al. 1999), which serves as a scaffold for the dityrosine outermost layer that confers the characteristic strength of spores. Similarly, chitin is an essential component of *A. fumigatus* conidial cell walls, where it also performs scaffold functions necessary for its characteristic rodlet coverage (Jimenez-Ortigosa et al. 2012).

B. Chitin Degradation and Cell Wall Assembly

Chitinases hydrolyse the β(1,4)-glycosidic bonds of chitin, contributing actively to fungal cell wall remodelling. In *S. cerevisiae*, the *Cts1p* chitinase is required for cell separation while *Sc*Cts2p participates in sporulation [see Adams (2004) for a review on fungal chitinases]. The daughter cell directs cell separation through the asymmetric expression of some genes (including *CTS1*) from the Ace2p transcription factor [reviewed in Lesage and Bussey (2006)]. The potential excess in chitinase activity is normally buffered through cell wall assembly mechanisms; however, deregulation of *CTS1* becomes critical after some cytokinesis defects have emerged, including the absence of the otherwise marginal CSI activity (Cabib et al. 1992; Gomez et al. 2009). A similar function has been described for the *CHT3* chitinase in *C. albicans* (Dunkler et al. 2005). Moreover, inhibition of this activity seems to occur during hyphal growth to prevent mycelium fragmentation (Gonzalez-Novo et al. 2008). An additional family of chitinases similar to *ScCTS2* has been tentatively implicated in sporulation in different organisms (Dunkler et al. 2008). Filamentous fungi have several potential chitinases in their genomes, although no morphogenetic role has been assigned to any of them and most of them probably have nutritional functions (Alcazar-Fuoli et al. 2011).

C. Chitin Synthesis in Response to Cell Stress

As an essential component of the fungal cell wall, chitin synthesis is regulated by stress. In yeasts, cell wall damage triggers a compensatory mechanism to guarantee cell survival; the most apparent effect of this mechanism is the synthesis of chitin-rich amorphous salvage septa after failures in PS synthesis (Shaw et al. 1991). Moreover, cell wall damage caused either by external agents, such as caspofungin, zymolyase or calcofluor, or by mutations in cell wall-related genes triggers a compensatory response mediated by the Cell Wall Integrity (CWI) signalling pathway [reviewed in Levin (2011)]. According to the data accumulated in *S. cerevisiae* and *C. albicans*, this response is known to be complex and interconnected with other signalling routes. Full coverage of this response is out of the scope of this manuscript, but it is worthwhile noting that an essential part of this response is the increase in chitin synthesis mediated by alterations in the intracellular traffic of CSs (see below) and also by the upregulation of *ScGFA1*, which code for a glutamine:fructose-6-phosphate amidotransferase involved in the synthesis of metabolic precursors for chitin. The latter effect is believed to act after cell damage by funnelling metabolic fluxes to the synthesis of the substrate for CSs, increasing chitin synthesis and cell wall strength (Lagorce et al. 2002). Accordingly, *ScGFA1* overexpression can alleviate the otherwise lethal effect of some cell wall-related mutations (Gomez et al. 2009). These compensatory mechanisms not only challenge the efficiency of some antifungal drugs directed against the cell wall (Munro 2013; Walker et al. 2013) but also offer new opportunities for exploring synergistic treatments combining chitin synthase inhibitors with other antifungal agents.

V. A Single Polymer but Distinct Functions for Chitin Synthase Enzymes

A. The Biological Function of Family I CSs

In eukaryotic cells, a contractile ring of actin, type II myosin, and many other proteins assemble underneath the plasma membrane before the end of mitosis. This assembly defines the site at which cytokinesis will occur. Once cells have segregated their nuclei, the actomyosin ring contracts; this contraction is coupled to plasma membrane ingression (Wloka and Bi 2012). Yeast cells need to coordinate actomyosin ring contraction with the deposition of the primary septum that forms centripetally between the dividing cells. Cells then synthesise secondary septa on both sides of the primary septum, before digesting partially the primary septum to allow cell separation (Roncero and Sanchez 2010).

1. Chitin Synthase I

It is after cell separation when the class I CS CHS1 performs its function in *S. cerevisiae* (Cabib et al. 1989). *chs1*Δ mutants are viable, exhibiting very mild phenotypes under non-stressed conditions. ScChs1p acts as a repair enzyme that synthesises a minor fraction of the cell wall chitin in the bud scar immediately after cytokinesis to overcome the possible damage caused by excessive chitinase activity. In fact, disruption of *ScCTS1*, the chitinase required for cell separation, suppresses the lysis associated with *chs1*Δ cells (Cabib et al. 1989, 1992). *C. albicans* has two class I CS enzymes: *CaCHS2* and *CaCHS8*. Disruption of *CHS8* alone or in conjunction with a disruption in *CHS2* generates mutant cells with normal morphology and growth rates in both the yeast and hyphal forms. However, both enzymes have been shown to contribute to chitin synthesis in vivo and to the synthesis of salvage septa in the absence of other CSs (Lenardon et al. 2007; Walker et al. 2013).

2. Chitin Synthase II

In *S. cerevisiae*, formation of the primary septum is carried out by CSII, whose catalytic subunit is Chs2p. At the end of the cell cycle, Chs2p synthesises the primary septum, which grows centripetally, contributing (together with actomyosin ring contraction) to plasma membrane ingression and separation of the two cells (Schmidt et al. 2002). Thus, cells have developed regulatory mechanisms to ensure that primary septum formation will be coordinated with cell cycle events and will only occur once cells have segregated their chromosomes and have assembled their actomyosin ring (Wloka and Bi 2012). Deletion of *CHS2* is either lethal or causes extremely severe defects during cytokinesis, depending on the genetic backgrounds, because the primary septum is not synthesised in *chs2Δ* mutants (Sburlati and Cabib 1986; Shaw et al. 1991). The viability of *chs2Δ* mutants depends on the chitin produced by CS III, which promotes the assembly of a highly abnormal salvage septum (Schmidt et al. 2002; Cabib and Schmidt 2003). Assembly and contraction of the actomyosin ring promote efficient septum deposition at the correct location. Conversely, septum deposition modulates the dynamics and stability of the contractile ring during cytokinesis. Therefore, the assembly and contraction of the actomyosin ring and septum deposition are interdependent processes, since defects in one of them clearly affect the other (Bi 2001; Schmidt et al. 2002; VerPlank and Li 2005). The function of *ScCHS2* orthologue in *C. albicans* (*CaCHS1*) is clearly conserved, since lack of *CaCHS1* has deleterious effects on cell viability due to alterations of essential processes such as primary septum formation (Munro et al. 2001; Lenardon et al. 2010). Similarly to *S. cerevisiae chs2Δ* mutants, *C. albicans chs1Δ* cells induce alternative forms of septation that rescue the lethal cell division defect produced by the absence of *CaCHS1* (Walker et al. 2013).

3. Other Class I, II and III Fungal CSs

While the function of Family I CS in yeast seems straightforward, as described above, the function of these enzymes in filamentous fungi is much more uncertain. Single and double mutants in class I, II and III enzymes have been constructed in several fungi. Their characterisation has led to the conclusion that Family I enzymes always account for most of the CS activity in vitro, as has been described for yeast. However, the morphogenetic phenotypes reported for these mutants vary, depending on the species, making it extremely difficult to obtain an overall view of the biological function of these enzymes. Individual class I and II mutants have marginal or null phenotypes in several fungi, but the characterisation of some double mutants points to a potentially redundant role for both classes in cell wall assembly

(Ichinomiya et al. 2005). In any case, the defects observed are fairly mild, even in the double mutants. The only notable exception seems to be *Yarrowia lipolytica*, whose class II enzymes seem to perform functions in septum assembly (Sheng et al. 2013) similar to those reported in *S. cerevisiae* and *C. albicans*. This exception is probably linked to the intermediate evolutionary position of this dimorphic fungus (see above). By contrast, deletion of class III enzymes in several filamentous ascomycetous fungi produces dramatic morphological alterations. These effects were initially described in *Neurospora crassa* (Yarden and Yanofsky 1991) and later in multiple systems, such as several *Aspergillus* species (Borgia et al. 1996; Mellado et al. 1996; Muller et al. 2002), *Penicillium* (Liu et al. 2013), *Magnaporthe* (Kong et al. 2012), or *Botrytis* (Soulie et al. 2006). However, there is no conclusive experimental evidence pinpointing the precise role of this class of CS in cell wall assembly. Surprisingly, to date no phenotype for class III enzymes has been described in basidiomycetous fungi.

In light of all these results, it seems that, with the unexplained exception of basidiomycetes, Family I CSs perform an essential function in fungal cell morphogenesis. This function has apparently been displaced from class II enzymes in yeast to class III in filamentous fungi, a process probably associated with the evolutionary diversification between Family I chitin synthases.

B. The Biological Function of Family II CSs

1. Chitin Synthase III (Class IV)

The function of class IV chitin synthases became clear after the initial characterisation of the *Scchs3Δ* mutant, which contained as little as 10 % of the chitin content of a wild-type strain. This result was a clear indication that in *S. cerevisiae* class IV, CSIII activity was the major in vivo activity in budding yeast. Most of this chitin forms a ring that surrounds the neck constriction at the mother side. This ring is not essential for cell survival but performs a homeostatic function during cytokinesis, and it becomes an essential structure when other neck assembly functions are compromised. In extreme cases, *Sc*Chs3p is able to synthesise salvage septa that functionally replace the absence of PS (Cabib 2004). A very similar function for *CaChs3p* has been described in the dimorphic yeast *C. albicans* (Walker et al. 2013), where *CaChs3p* also proved to be responsible for most of chitin synthesis during hyphal growth. Class IV enzymes have also been described as the major in vivo CSs in other yeast-like organisms such as *Ustilago maydis* (Weber et al. 2006), *Y. lipolytica* (Sheng et al. 2013), *Wangiella dermatitidis* (Wang et al. 1999) or *Cryptococcus neoformans* (Banks et al. 2005), although in the latter this chitin is mostly deacetylated into chitosan, which forms part of the capsular polysaccharide. Interestingly, some of these organisms are dimorphic, exhibiting both the yeast and hyphal forms. This makes it possible to test the importance of class IV enzymes in hyphal development. Unfortunately, all the class IV mutants characterised in these systems lacked apparent phenotypes during mycelial growth and hence no relationship between class IV CSs and hyphal development has been found other than that demonstrated for *C. albicans*, where Chs3 is involved in that process {Sanz et al. 2005 #1405}. Additional characterisation of class IV mutants in true filamentous fungi, such as *N. crassa*, *A. nidulans*, *A. fumigatus* or *M. oryzae*, did not provide additional information because none of the mutants exhibited an altered phenotype, even after extensive characterisation. Surprisingly, some of these "nonfunctional" class IV enzymes have been shown to have functional catalytic sites based on heterologous expression (Jimenez et al. 2010). Accordingly, their role in chitin synthesis currently remains a mystery.

2. Chitin Synthases with Myosin Motor-Like Domain

To a certain extent it could be expected that the other Family II CS members (the class V/VII enzymes, only present in filamentous fungi) could perform a major role in chitin synthesis during mycelial growth. Extensive characterisation of multiple class V/VII mutants failed to provide conclusive evidence of a direct involvement of these enzymes in bulk chitin synthesis. However, all class V/VII mutants characterised to date show morphogenetic defects, probably associated with cell wall alterations. Moreover, most class V/VII mutants characterised are hypersensitive to cell wall-related drugs, such as calcofluor white or caspofungin, and most class V mutants are hypovirulent in their correspondent hosts [see Madrid et al. (2003), Weber et al. (2006), and Kong et al. (2012), for specific references]. In addition, it has recently been shown that AfcsmA and AfcsmB mutants have reduced in vitro chitin synthase activity (Jimenez-Ortigosa et al. 2012). The picture that emerges from these data is that class V/VII CSs would participate in chitin synthesis, although their contribution to the total amount of chitin is low. The phenotypes observed in these mutants suggest that this chitin might become a critical anchor for other cell wall polymers, such that its absence would contribute to the significant cell wall alterations reported. Most class V/VII mutants also show sporulation defects, suggesting a specific role for these enzymes in cellular differentiation. Moreover, the class V *AfCsmA* mutant has recently been shown to have significantly reduced chitin levels in its conidial cell wall, which triggers alterations in the conidial surface (Jimenez-Ortigosa et al. 2012). This observation seems to be the first conclusive evidence of a direct role of class V enzymes in chitin synthesis.

The biological relevance of the genetic duplication that led to class V and VII homologues in fungi is uncertain. CSs from both classes have been described to have redundant functions in *A. nidulans* due to the synthetic lethality of the double V/VII mutant (Takeshita et al. 2006); however, this result has not been reproduced in any other fungi studied, including the close relative *A. fumigatus* (Jimenez-Ortigosa et al. 2012). In all other fungi studied, class V mutants show the strongest phenotypes, although additional deletion of the corresponding class VII gene exacerbates the defects. Therefore, from the studies undertaken to date it may be concluded that class V and VII enzymes perform partially redundant functions, class V enzymes being more important in biological terms. In this scenario, it is tempting to speculate that a direct relationship between the Family II CSs and chitin synthesis might be masked by compensatory effects between the different activities. However, this possibility seems unlikely since no clear alteration in the transcription patterns of these genes has been found in the mutants in which this hypothesis has been analysed.

In sum, although further analyses are required, the studies undertaken to date have revealed the morphogenetic importance of Family II CS across the fungal kingdom. Class IV enzymes perform essential functions during yeast growth that are important both quantitatively and qualitatively for cell wall assembly. Their biological roles have apparently been replaced by class V/VII enzymes during hyphal growth and sexual differentiation, with a relevance that is more qualitative than quantitative. Both roles highlight the importance of chitin as a scaffold for other more abundant cell wall polymers and the critical relevance of chitin–glucan linkages in fungal cell wall assembly and fungal morphogenesis, as recently proposed (Cabib and Arroyo 2013).

C. The Elusive Function of Class VI CSs

The biological function of fungal class VI CSs remains as controversial as its evolutionary relationship with Families I and II (see above). Mutants in this class do not show any apparent morphological or physiological phenotype, although they do have reduced virulence in some fungal pathogens, such as *B. cinerea* and *M. oryzae*. In *M. oryzae*, this trait has been linked to a specific defect in appressorium formation during the initial step of infection (Kong et al. 2012). Thus, despite the scarce information, class VI enzymes could possibly be associated with specialised functions in fungal plant pathogens.

VI. Regulation of Chitin Synthases

An Overview CSs have been recognised for decades as integral membrane proteins. In view of this, a critical question in the field has been to determine where and how chitin synthases are activated, assuming that chitin synthesis along the secretory pathway could be catastrophic. According to the results of in vivo experiments, the early view of CSs as zymogenic enzymes activated by proteolytic processing no longer holds. The work carried out in *S. cerevisiae* has led to the understanding that the regulation of chitin synthesis is essentially a question of the intracellular transport of the CS enzymes to the plasma membrane (PM) and their potential interaction with other proteins at the cell surface. Interestingly, Family I and II CSs differ considerably in their secondary structures (see Fig. 9.2), even though they are able to synthesise a similar polymer. This indicates that the most conserved region of these enzymes (the catalytic domain followed by a complex transmembrane region) should suffice for directing chitin polymerisation at the PM, while the additional regions of these proteins would be involved in their regulation and/or traffic, processes that are essentially different in both families (see below). In any case, the existence of specific activators for the different CSs at the PM cannot be fully excluded, although their hypothetical biochemical roles remain unknown.

An additional and very important question regarding the regulation of fungal chitin synthesis is the relevance of the coupling between polymerisation and crystallisation. This coupling was reported many years ago (Roncero and Durán 1985) and should be dependent on homo- or hetero-oligomerisation of CSs into supramolecular complexes, as has been demonstrated for cellulose synthases (Atanassov et al. 2009). Unfortunately, the experimental evidence for fungal CS oligomerisation is scarce and very recent (Sacristan et al. 2013). Thus, no clear conclusions can yet be inferred regarding its importance in the regulation of chitin synthesis.

A. Regulation of Family I CSs

1. Regulation of Chitin Synthase I

A cell cycle regulation analysis using synchronised *S. cerevisiae* cultures indicated that the expression of *ScCHS1* mRNA peaks at the M/G1 transition, although the *Sc*Chs1 protein is fairly stable and its level does not change significantly during the cell cycle. It has been suggested that Chs1p would be transported in specialised vesicles, the so-called chitosomes, from where it would be activated and located at the PM (Chuang and Schekman 1996; Spellman et al. 1998). In *C. albicans*, a transcription analysis during the cell cycle revealed that the expression of *CaCHS8* peaked in G2, whereas the expression of *CaCHS2* was non-periodic (Cote et al. 2009). The expression of *CaCHS2* increased shortly after the induction of hypha formation, and both enzymes seem to be required to synthesise chitin during hyphal growth (Gow et al. 1994; Munro et al. 2001, 2003).

2. Regulation of Chitin Synthase II

The activity of *Sc*Chs2p is essential for the synthesis of the primary septum that is laid down behind the contractile ring during cytokinesis (Shaw et al. 1991). The expression, localisation and enzymatic activity of *Sc*Chs2p are temporally and spatially regulated to ensure that septum formation proceeds only after all chromosomes have been pulled apart (Choi et al. 1994; Chuang and Schekman 1996; Spellman et al. 1998; VerPlank and Li 2005). CSII is deposited at the division site through the secretory pathway a few minutes before mitotic spindle breakdown. The delivery of *Sc*Chs2p occurs just before and during actomyosin ring contraction (Chuang and Schekman 1996; VerPlank and Li 2005; Fang et al. 2010). Unlike *ScCHS2* regulation, there is no deep insight into the regulation of its orthologue in *Candida*. The expression of *CaCHS1* peaks in G2, which corresponds to its function in the synthesis of the primary septum during cell division (Munro et al. 2001; Cote et al. 2009).

Cyclin-Dependent Kinases (CDKs) govern progression through the cell cycle. Cdks bind to specific regulators called cyclins, and kinase activity associated with different cyclin/Cdk complexes promotes phase-specific events. At the end of the cell cycle, mitotic forms of Cdk must be down-regulated to allow cells to exit from mitosis, perform cytokinesis and start a new cell cycle (Sanchez-Diaz et al. 2012). The kinase Cdk1 (encoded by *ScCDC28*) is the master regulator of the cell cycle in the budding yeast *S. cerevisiae* and controls *Sc*Chs2p dynamics. Indeed, *Sc*Chs2p is a substrate of Cdk1 in vivo, and *Sc*Chs2p is phosphorylated at the N-terminus of the protein (Chin et al. 2012). During mitosis, when Cdk activity is still high, *Sc*Chs2p is synthesised and held at the endoplasmic reticulum (ER) precisely through its phosphorylation by Cdk1 at four consensus Cdk1-phosphorylation sites in its N-terminal tail (Zhang et al. 2006; Teh et al. 2009).

Once chromosomes have been segregated at the end of mitosis, budding yeast cells activate a signalling pathway, the so-called Mitotic Exit Network (MEN), which consists of a small GTPase, Tem1p and protein kinases Cdc15p, Dbf2p and Dbf20p, together with a few regulatory proteins (Meitinger et al. 2012). The main goal of MEN is the release of the conserved phosphatase Cdc14p from the nucleolus to the entire cell to dephosphorylate key substrates, which then drives cells out of mitosis through an efficient and rapid cytokinesis and the start of a new cell cycle (Bouchoux and Uhlmann 2011; Meitinger et al. 2012; Palani et al. 2012; Sanchez-Diaz et al. 2012). In order to promote exit from mitosis, the phosphatase Cdc14p induces the inactivation of mitotic Cdk activity by dephosphorylation, and hence activation, of the Cdk inhibitor Sic1p, its transcription factor Swi5p, and Cdh1p, an activator of the Anaphase Promoting Complex (APC) that induces the degradation of mitotic cyclin. In parallel, ScChs2p can be dephosphorylated by Cdc14p, which determines *Sc*Chs2 export from the ER (Chin et al. 2012). In fact, it has been shown that Cdc14p promotes the capture of ScChs2p into COPII transport vesicles at the ER (Zhang et al. 2006; Jakobsen et al. 2013). In addition, the dephosphorylation of *Sc*Chs2p stimulates interaction with the COPII component Sec24p, which is considered to be responsible for binding to membrane cargo proteins at the ER, concentrating them in the forming vesicles (Jakobsen et al. 2013). Subsequently, *Sc*Chs2p is delivered to the division site via the secretory pathway using actin cables and type V myosin, together with the exocytic machinery (Chuang and Schekman 1996; VerPlank and Li 2005).

The assembly of the cytokinetic machinery in budding yeast is a sequential and orchestrated process that starts with the assembly of a septin ring early in the cell cycle (Wloka and Bi 2012). Soon after the formation of a new bud, the type II myosin Myo1p forms a ring at the bud neck that later on during mitosis associates with other factors. Among them are actin-nucleating and bundling factors such as formins and the Iqg1p protein, which promote the assembly of the contractile actomyosin ring at the end of anaphase (Wloka and Bi 2012). Components of the actomyosin ring interact with *Sc*Chs2p and play a critical role in regulating PS formation. At the onset of cytokinesis, the septin ring is split into two cortical rings, which act as diffusion barriers at the division site during cytokinesis (Dobbelaere and Barral 2004), while *Sc*Chs2p and actomyosin ring components are sandwiched between the hourglass cortical rings. By FRAP (Fluorescence Recovery After Photobleaching) analysis, it has been shown that *Sc*Chs2 initial localisation is fairly dynamic and its delivery is completed in a few minutes after its initial arrival (Wloka et al. 2013). Soon after, *Sc*Chs2p becomes immobile during cytokinesis, and its immobility depends on Myo1p (Wloka et al. 2013).

Localisation at the cleavage furrow seems to be insufficient for *Sc*Chs2p to synthesise the PS (Nishihama et al. 2009), since *Sc*Chs2p must be activated in vivo by a poorly understood mechanism that can be mimicked in vitro by protease treatment. Indeed, the proteolysis of cell membranes containing *Sc*Chs2p by trypsin stimulates the chitin activity associated with *Sc*Chs2p in vitro (Sburlati and Cabib 1986; Martinez-Rucobo et al. 2009), suggesting that *Sc*Chs2p requires further activation once it has been deposited at the division site. This activation defines an additional level of control to ensure the appropriate coordination of actomyosin ring contraction, PM ingression and PS formation (Schmidt et al. 2002).

Recent findings suggest that the actomyosin ring components Hof1p, Inn1p and Cyk3p coordinate these events during cytokinesis

(Sanchez-Diaz et al. 2008; Jendretzki et al. 2009; Nishihama et al. 2009). *Sc*Chs2p interacts directly with Hof1p and stabilises it at the cleavage site. In addition, Hof1p binds to Myo1p, which could contribute to coupling actomyosin ring contraction to primary septum formation (Oh et al. 2013). It also appears that Cyk3p could regulate CSII activity, since an increased dosage of Cyk3p stimulates *Sc*Chs2p-dependent chitin synthesis and the formation of PS-like structures at the bud neck (Meitinger et al. 2010; Oh et al. 2012). However, Cyk3p plays a dual role during cytokinesis since it inhibits secondary septum formation while simultaneously promoting primary septum deposition (Onishi et al. 2013).

Moreover, another key factor in the regulation of *Sc*Chs2p activity is Inn1p, a protein originally described to be essential for the coordination of actomyosin ring contraction, plasma membrane ingression and primary septum deposition (Sanchez-Diaz et al. 2008; Nishihama et al. 2009). Inn1p associates with the actomyosin ring and co-purifies with Hof1p, Iqg1p and Cyk3p (Sanchez-Diaz et al. 2008; Nishihama et al. 2009; Palani et al. 2012; Nkosi et al. 2013). Regarding the dynamics of *Sc*Chs2p during cell division, the localisation and protein–protein interactions of these factors are controlled by CDK phosphorylation and subsequent dephosphorylation by Cdc14p, which highlights the key role of Cdc14p during cytokinesis (Meitinger et al. 2010; Sanchez-Diaz et al. 2012). Indeed, Cdc14p-dependent dephosphorylation of Inn1p contributes to Inn1p–Cyk3p complex formation (Palani et al. 2012). In Inn1p-depleted cells, the contracting actomyosin ring appears to be unstable and often collapses, similar to the phenotype of cells lacking *Sc*Chs2p (VerPlank and Li 2005; Sanchez-Diaz et al. 2008). Inn1 protein has a C2 domain at the amino terminus of the protein that is essential for ingression of the plasma membrane, whereas the remainder of the protein is required for the timely localisation of Inn1p at the bud neck (Sanchez-Diaz et al. 2008). It has recently been described that hypermorphic versions of *Sc*Chs2p suppress the defects in PM ingression produced by an inactive form of the C2-domain of Inn1p (Devrekanli et al. 2012). These dominant suppressor mutations in ScChs2p are located at conserved sites in the catalytic domain and exhibit enhanced CS activity, suggesting that the ScChs2p mutations suppress the loss of Inn1 function through its enhanced CS activity. Moreover, Inn1p associates with *Sc*Chs2p in yeast cell extracts. Together, these results suggest that Inn1p might directly regulate the CS activity associated with *Sc*Chs2p at the division site (Devrekanli et al. 2012).

After its role during exit from mitosis, MEN kinases regulate cytokinesis by acting on components of the contractile actomyosin ring (Meitinger et al. 2012). *Sc*Chs2 protein is unable to efficiently localise to the division site in the absence of MEN activity, and chitin deposition is therefore reduced (Meitinger et al. 2010). It has been suggested that such a defect could be mediated via Cyk3p, which is unable to localise to the bud neck in the absence of MEN function (Meitinger et al. 2010). In addition, Dbf2 kinase directly phosphorylates *Sc*Chs2p and triggers its dissociation from the actomyosin ring during the late stage of cytokinesis (Oh et al. 2012). Soon after actomyosin ring contraction has ended, *Sc*Chs2p undergoes internalisation in endosome-like vesicles and is degraded in the vacuole, depending on the major vacuolar protease Pep4p (Chuang and Schekman 1996). Direct phosphorylation of *Sc*Chs2p would facilitate its separation from the ring and subsequent endocytic removal from the division site (Oh et al. 2012).

3. Regulation of Other Class I, II and III Fungal CSs

Very little work has been carried out to decipher the in vivo CS regulation of these enzymes in other fungi, and the results reported are mostly descriptive. *Neurospora crassa* proteins *Nc*Chs1, *Nc*Chs3 and *Nc*Chs6 have been shown to be transported to the sites of new cell wall synthesis in a type of specialised vesicles called chitosomes. Microscopy evidence suggests the existence of distinct populations of vesicles, whose transport depends on actin but not on microtubules, as inferred from the use of drugs interfering with actin or microtubule assembly (Riquelme et al. 2007; Sanchez-Leon et al. 2011). The peculiarities of this transport have led some authors to propose that CSs might be transported by a route other than the classical pathway from the ER via the Golgi apparatus to the plasma membrane (Riquelme et al. 2007).

B. Regulation of Family II CSs

1. Regulation of Chitin Synthase III

The characterisation of calcofluor resistance in *S. cerevisiae* has provided an extensive list of proteins involved in CSIII (class IV) activity. While *Sc*Chs3p is the catalytic subunit, the other proteins are markers of an intracellular transport route that delivers ScChs3p to the PM in a polarised way, allowing chitin synthesis at the neck constriction. The characterisation of this route has made *Sc*Chs3p a paradigm for the study of the intracellular traffic of proteins. While a detailed description of this transport is out of the scope of this chapter, its essential rules are described succinctly below.

The exit of *Sc*Chs3p from the ER depends on a dedicated ER membrane chaperone named Chs7p (Trilla et al. 1999; Kota and Ljungdahl 2005), also being modulated through *Sc*Chs3p oligomerisation (Sacristan et al. 2013) and palmitoylation (Lam et al. 2006). Together, these factors allow the progression of *Sc*Chs3p through the Golgi, preventing its recycling back to the ER (Sacristan et al. 2013). After ER exit, Chs3p populates the designated Golgi/Trans Golgi Network (TGN) boundary, where proteins are classified for their delivery to the PM. *Sc*Chs3p exit from Golgi/TGN depends on the Chs5p and Chs6p proteins, which form a specialised transport complex named exomer and required for polarised transport of Chs3p (see Fig. 9.1b, full lines) (Santos and Snyder 1997; Trautwein et al. 2006; Wang et al. 2006). Chs6p is one of the four Chs5p- and Arf1p-binding proteins (**ChAPs**), a set of homologous proteins that associate with the Chs5p scaffold and might assemble in alternative exomer complexes involved in the transport of different proteins to the PM (Trautwein et al. 2006; Rockenbauch et al. 2012). During the last part of its transit to the PM, Chs3p interacts with Chs4p, which leads to CSIII activation at the PM through an unknown mechanism (Reyes et al. 2007). In addition, Chs4p serves as a Chs3p anchor to the neck through its interaction with Bni4p and the septin ring (DeMarini et al. 1997; Kozubowski et al. 2003; Sanz et al. 2004). This anchor regulates the onset of Chs3p endocytosis, regulating chitin synthesis (Sacristan et al. 2012). Upon endocytosis, Chs3p is recycled to the Golgi/TGN by a complex mechanisms not yet fully understood that mostly relies on the AP-1 complex (Valdivia et al. 2002). This recycling prevents Chs3p degradation in the vacuole and allows the maintenance of a massive reservoir of Chs3p at the Golgi/TGN, from where Chs3 can be mobilised upon cellular request. Deletion of the AP-1 complex in an exomer-defective background activates an alternative route (see Fig. 9.1b, dashed arrows) for *Sc*Chs3p delivery to the PM (Valdivia et al. 2002; Starr et al. 2012). Alterations in this transport allow cells to regulate the amount of Chs3p present at the PM to increase chitin synthesis after environmental stress; this is well documented, at least after cell wall (García-Rodríguez et al. 2000) or temperature (Valdivia and Schekman 2003) stress. Not surprisingly, post-translational modification of Chs3p contributes to the regulation of its traffic and to the regulation of chitin synthesis [Lenardon et al. (2010) and our unpublished results].

The rules governing the transport of Family II CSs in other fungi have not been studied extensively, but the fact that all these CSs are localised to the growing tips and septa suggests the existence of mechanisms for polarised transport. Interestingly, the proteins involved in the transport route described above for the class IV *CHS3* enzyme are well conserved across the fungal kingdom, with the notable exception of *S. pombe*, which lacks the otherwise omnipresent and well-conserved Chs7 protein (Fig. 9.3). This absence probably reflects the proposed coordinated evolution of class IV CSs and Chs7 proteins (Jimenez et al.

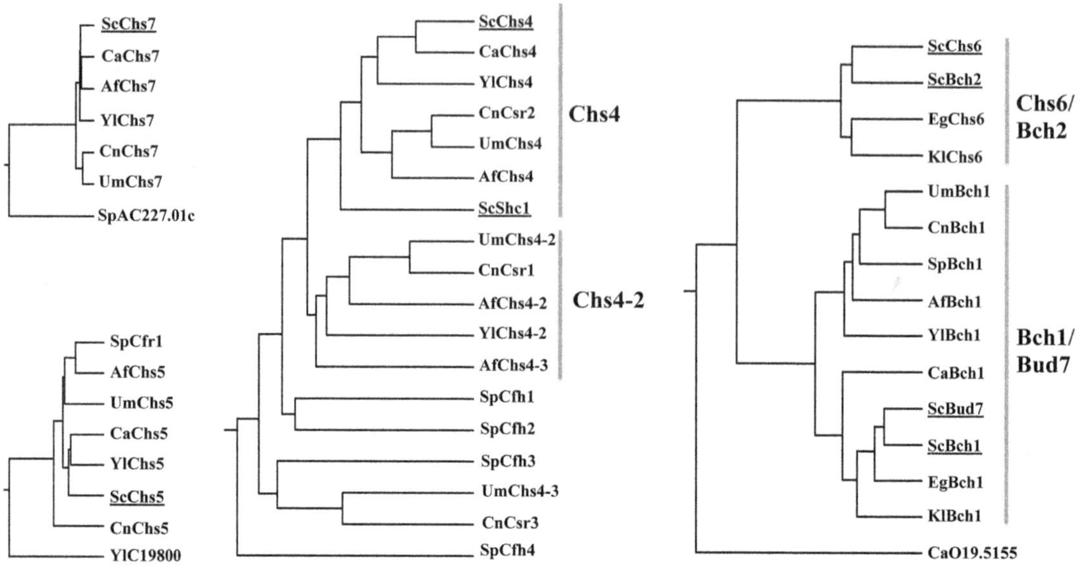

Fig. 9.3 NJ rooted phylogenetic trees after clustal W analysis of proteins homologous to *Sc*Chs4, *Sc*Chs5, *Sc*Chs6 and *Sc*Chs7. Analysis was performed with sequences from *A. fumigatus, C. neoformans, U. maydis, Y. lipolytica, S. cerevisiae, C. albicans, K. lactis, E. gosypii* and *S. pombe*. All sequences with significant similarity after BLAST searches were included in the analysis, acting the most divergent sequence as outgroup for the rooting. Note the high degree of conservation between the Chs5 and Chs7 proteins, but the divergence between Chs4 and Chs6 proteins that led to the separation in well-defined subgroups. For comparison purposes, all the trees are represented at the same scale. See text for additional explanations

2010), both of which are absent in *S. pombe*. *ScCHS7* is well conserved evolutionarily and performs similar functions in *S. cerevisiae* and *C. albicans*, allowing Chs3p export from the ER (Trilla et al. 1999; Sanz et al. 2005). Strikingly, *CHS7* does not seem to be required for the transport of the class V/VII enzymes from the ER (Martin-Urdiroz et al. 2008; Jimenez et al. 2010).

The **exomer** is a well-conserved complex in fungi, as suggested by the observation that *CHS5* and *CHS6* orthologues are present in all of them [Trautwein et al. (2006), Fig. 9.3]. Chs5p is the essential component of the exomer, its N-terminal region being sufficient and necessary for its function (Martin-Garcia et al. 2011; Paczkowski et al. 2012). Not surprisingly, this region has been extremely well conserved along evolution (Fig. 9.3). Comparison of the ChAPs (Chs5- and Arf1-binding proteins) hints at a fascinating story about exomer evolution. Most fungal ChAPs group together with the *Sc*Bch1/*Sc*Bud7 paralogues (Fig. 9.3), suggesting the very early origin of these proteins and probably also a conserved function within the exomer. However, analysis of sequences from the fungal post-WGD (whole-genome duplication) clade and related genera among the Saccharomycotina (Wang et al. 2009) supports an early duplication event between the ChAPs that led to the appearance of the *ScCHS6* homologues which, according to the work undertaken in *S. cerevisiae*, would be specialised in the transport of CSs. Later on, *ScBCH1* and *ScCHS6* would have duplicated into *ScBUD7* and *ScBCH2*, respectively, because of the well-documented whole-genome duplication that occurred in the bona fide WGD clade. This duplication would have in turn led to a further specialisation of the different ChAPs in *S. cerevisiae*, the exomer always acting as a kind of specialised complex in the polarised transport of different cargos (Trautwein et al. 2006; Barfield et al. 2009; Ritz et al. 2014). These studies raise several questions regarding the exomer, such as whether it is involved in chitin synthesis in all fungi, whether it is always involved in polarised transport and how many cargos are really dependent on the exomer.

Unfortunately, the exomer has been studied in detail only in *Saccharomyces*. There is some, but not much, information from the distantly related yeast *S. pombe* that suggests that the exomer might be conserved mechanistically (Martin-Garcia et al. 2011). However, the data from this fission yeast do not link it to polarised transport [Cartagena-Lirola et al. (2006) and our unpublished results]; instead, the *S. pombe* exomer might play some minor role in intracellular traffic. Thus, the answers to these questions will require the characterisation of exomer components in other systems.

Meanwhile, the extensive work that is currently ongoing in *S. cerevisiae* will probably shed light on the reasons for the extreme ChAPs specialisation observed in this organism.

The Chs4 protein is required for chitin synthesis in *S. cerevisiae*, *C. albicans* and *C. neoformans* (Trilla et al. 1997; Ono et al. 2000; Banks et al. 2005), a clear indication of the conserved function of this protein in the regulation of class IV CSs. Moreover, *S. cerevisiae* contains a paralogue (*SHC1*) that is regulated differentially during sporulation (Sanz et al. 2002). Fungal CHS4 orthologues can be unambiguously separated into two distinct groups, one of which has been lost along evolution in the Saccharomycotina group. Unfortunately, there is no information about the physiological function of the second group, although recently the truncation of *A. nidulans* AN3445, the true orthologue of *AfCHS4-3*, has been shown to confer calcofluor resistance (He et al. 2014), raising the possibility of a broad function of these proteins in cell wall assembly. The demonstration of a direct role for them in chitin synthesis will require additional characterisation.

2. Regulation of Fungal Chitin Synthases During Mycelial Growth: Chitin Synthases with a Myosin Motor-Like Domain

As described above, the function of class IV CSs in filamentous fungi is minor or null. Therefore, the most relevant issue has been to determine how the transport of class V/VII enzymes (both having a myosin motor domain in their N-terminal region) is achieved during hyphal growth. For class V enzymes, it has been well documented that both the MMD and CS domains are required for its function; it was therefore assumed that the MMD domain serves for the polarised transport of these enzymes along the hyphal axes (Takeshita et al. 2005). However, the detailed work carried out by Steinberg's group in *U. maydis* suggested that the mechanism was more complicated than anticipated (Treitschke et al. 2010; Schuster et al. 2012). *Um*Mcs1p moves back and forward to the hyphal tip through the cytoskeleton; it seems that this movement is independent of the MMD domain, which would be specifically required for the final actin-dependent translocation of Mcs1p into the PM, a process that affects only 15 % of the Mcs1p molecules arriving at the tip (Schuster et al. 2012). Interestingly, the MMD domain and actin have been also proposed to facilitate a minor lateral delivery of Mcs1 protein, which could be functionally equivalent to the alternative delivery route described in *S. cerevisiae* (Valdivia et al. 2002). It remains to be tested whether this non-polarised delivery of Mcs1p is altered under stress conditions. However, this model does not address how some filamentous fungi could support chitin synthesis at the hyphal tip in the absence of class V/VII CSs. Information about this issue has been gained from studies in yeasts by comparing budding and hyperpolarised growth. During budding, Chs3p localisation at the neck relies on a delicate balance between anterograde delivery and endocytosis (Reyes et al. 2007). However, during hyperpolarised growth, the endocytic recycling of Chs3p is reduced, allowing Chs3p accumulation along the mating projections in *S. cerevisiae* or at the hyphal tip in *C. albicans* (Sacristan et al. 2012). Thus, the modulation of the endocytic recycling of class IV enzymes could suffice to switch the chitin synthesis programme from bud to hyphal growth. It is tempting to speculate about the possibility of a role for endocytosis in Mcs1p traffic, since endocytic recycling could replenish, at least partially, the Mcs1p-containing pool of vesicles (see above). However, this hypothesis has not yet been tested.

VII. Chitin Synthesis and Antifungal Therapies

Fungal cell walls have traditionally been considered as an attractive target for antifungal therapies. Among the various cellular components, chitin emerged as an ideal target owing to its absence in plants and animals, both of which are hosts for fungal infections. This promise was somehow fulfilled after the initial discovery of Polyoxin D, a molecule with in vivo antifungal activity through the inhibition of chitin synthesis (Endo et al. 1970). Further work led to the identification of new polyoxins and the related family of nikkomycins, all of them acting as antifungal agents through competitive inhibition of chitin synthases [reviewed in Debono and Gordee (1994)]. Moreover, genetic analyses performed in *S. cerevisiae* revealed the synthetic lethality of the *chs2Δ chs3Δ* double mutant (Shaw et al. 1991), reinforcing the view of chitin as an essential component of the fungal cell wall and of chitin synthesis inhibition as an ideal antifungal target. Unfortunately, after decades of research in the field none of these compounds has attained therapeutic value owing to their poor efficacy in vivo. The causes of this failure are not fully understood, but two major reasons are apparent: (1) the poor bioavailability of the drugs due to inefficient transport systems (Debono and Gordee 1994), and (2) the presence of several CS isoenzymes in the cell with dramatic differences in the sensitivity to these compounds. This was demonstrated for Nikkomycin Z, which proved to be a specific inhibitor of *Sc*CSIII with null effects on *Sc*CSII, explaining the poor in vivo effect of this drug on the growth of *S. cerevisiae* (Gaughran et al. 1994). Similarly, RO-09-3143 has been identified as a specific inhibitor of CaCSII activity, showing only a fungistatic effect on *C. albicans* growth (Sudoh et al. 2000). In view of the diversity of the fungal CSs determined to date and the potential biological redundancy between them, the poor antifungal efficacy of polyosins/nikkomycins in vivo is not surprising. We must therefore await the identification of better chitin synthase inhibitors with broader specificities against the different classes of chitin synthases. Success in this will probably be linked to a better knowledge of the distinct regulation of different chitin synthases.

Interestingly, chitin synthesis not only depends on CS activity, but also on the metabolic synthesis of its substrate (UDP-NAcGln), a metabolic bottle neck that depends on the activity of Gfa1p (see above). This makes it an additional target for antifungal therapies (Bulik et al. 2003). In this line of work, inhibitors of *Af*Agm1p (*A. fumigatus N*-acetylphosphoglucosamine mutase, another enzyme involved in the synthesis of the substrate for CSs) have recently been identified (Fang et al. 2013), opening new perspectives for the future development of antifungal agents.

Recent evidence indicates that chitin synthesis has not only failed to hold its promise as an antifungal target, but indeed has become a problem in alternative antifungal therapies using β-glucan synthase inhibitors. The clinical use of caspofungin revealed the so-called paradoxical effect of this drug, which is less active against *C. albicans* in vivo at higher concentrations. This effect proved to be dependent on an increase in chitin synthesis that affects cell wall architecture, making cells more resistant to the drug (Munro 2013) and less accessible to immune recognition by the host (Marakalala et al. 2013).

After the initial perspectives and decades of work on the identification of inhibitors of chitin synthesis, it is now becoming accepted that no "magic bullet" is waiting in the wings. Instead, the solution for the treatment of fungal infections can be achieved by exploiting the synergistic effect of chitin synthase inhibitors and some alternative antifungal therapies. This strategy is now being implemented in some animal studies with good results (Clemons and Stevens 2006), making the future introduction of some of these chitin synthesis inhibitors into clinical therapeutics possible.

VIII. Concluding Remarks

The genomic era has provided an extraordinary thrust in the characterisation of the diversity of chitin synthases in fungi, but this progress has not been translated to the same extent in our understanding of the functions of the different chitin synthases. Accordingly, the promise of chitin as ideal antifungal target has not been fulfilled.

Nevertheless, the extensive molecular work carried out in *S. cerevisiae* has made chitin synthases a paradigm in studies on the intracellular traffic of proteins. Chs3p traffic has proved to be extremely complex, highlighting some poorly known aspects about the intracellular traffic of polytopic proteins, such as the relevance of the exomer and AP-1 in their recycling. Our knowledge about this transport is underexplored in fungi and it is expected that more exhaustive work in filamentous fungi could provide additional information about CS regulation in the near future. In addition, studies on *Sc*Chs2p have provided evidence on how yeast cells coordinate actomyosin ring contraction with primary septum formation, whose relevance would be important to understand extracellular matrix remodelling during eukaryotic cytokinesis. Thus, CS studies still remain an exciting field of research.

Acknowledgments We acknowledge N. Skinner for language revision. We apologise to those authors we were unable to cite due to space limitations, their contributions are highly appreciated. We are indebted to E. Cabib for his seminal and continued work on chitin synthesis and to A. Duran for transferring the enthusiasm for this field of research to Salamanca, where some of us are still puzzled by chitin biosynthesis and fungal morphogenesis. ASD was funded from the Cantabria International Campus and via grant BBU2014-58081-P from the CICYT/FEDER programme. CR and HV were supported by grants BFU2013-48582-C2-1P/2P, and grant SA073U14 from Junta de Castilla to HV and CRM.

References

Adams DJ (2004) Fungal cell wall chitinases and glucanases. Microbiology 150:2029–2035

Alcazar-Fuoli L, Clavaud C, Lamarre C, Aimanianda V, Seidl-Seiboth V, Mellado E, Latge JP (2011) Functional analysis of the fungal/plant class chitinase family in Aspergillus fumigatus. Fungal Genet Biol 48:418–429

Arellano M, Cartagena-Lirola H, Nasser Hajibagheri MA, Durán A, Valdivieso MH (2000) Proper ascospore maturation requires the *chs1*$^{+}$ chitin synthase gene in *Schizosaccharomyces pombe.* Mol Microbiol 35:79–89

Atanassov II, Pittman JK, Turner SR (2009) Elucidating the mechanisms of assembly and subunit interaction of the cellulose synthase complex of *Arabidopsis* secondary cell walls. J Biol Chem 284:3833–3841

Banks IR, Specht CA, Donlin MJ, Gerik KJ, Levitz SM, Lodge JK (2005) A chitin synthase and its regulator protein are critical for chitosan production and growth of the fungal pathogen *Cryptococcus neoformans.* Eukaryot Cell 4:1902–1912

Barfield RM, Fromme JC, Schekman R (2009) The exomer coat complex transports Fus1p to the plasma membrane via a novel plasma membrane sorting signal in yeast. Mol Biol Cell 20:4985–4996

Becker HF, Piffeteau A, Thellend A (2011) *Saccharomyces cerevisiae* citin biosynthesis activation by N-acetylchitooses depends on size and structure of chitooligosaccharides. BMC Res Notes 4:454–459

Bi E (2001) Cytokinesis in budding yeast: the relationship between actomyosin ring function and septum formation. Cell Struct Funct 26:529–537

Bi E, Park HO (2012) Cell polarization and cytokinesis in budding yeast. Genetics 191:347–387

Blanco N, Reidy M, Arroyo J, Cabib E (2012) Crosslinks in the cell wall of budding yeast control morphogenesis at the mother-bud neck. J Cell Sci 125:5781–5789

Borgia PT, Iartchouk N, Riggle PJ, Winter KR, Koltin Y, Bulawa CE (1996) The chsB gene of *Aspergillus nidulans* is necessary for normal hyphal growth and development. Fungal Genet Biol 20:193–203

Bouchoux C, Uhlmann F (2011) A quantitative model for ordered Cdk substrate dephosphorylation during mitotic exit. Cell 147:803–814

Bowen AR, Chen-Wu JL, Momany M, Young R, Szaniszlo JP, Robbins PW (1992) Classification of fungal chitin synthases. Proc Natl Acad Sci USA 89:519–523

Bulik DA, Olczak M, Lucero HA, Osmond BC, Robbins PW, Specht CA (2003) Chitin synthesis in *Saccharomyces cerevisiae* in response to supplementation of growth medium with glucosamine and cell wall stress. Eukaryot Cell 2:886–900

Cabib E (2004) The septation apparatus, a chitin-requiring machine in budding yeast. Arch Biochem Biophys 426:201–207

Cabib E, Arroyo J (2013) How carbohydrates sculpt cells: chemical control of morphogenesis in the yeast cell wall. Nat Rev Microbiol 11:648–655

Cabib E, Duran A (2005) Synthase III-dependent chitin is bound to different acceptors depending on location on the cell wall of budding yeast. J Biol Chem 280:9170–9179

Cabib E, Schmidt M (2003) Chitin synthase III activity, but not the chitin ring, is required for remedial septa formation in budding yeast. FEMS Microbiol Lett 224:299–305

Cabib E, Bowers B, Roberts R (1983) Vectorial synthesis of a polysaccharide by isolated plasma membranes. Proc Natl Acad Sci USA 80:3318–3321

Cabib E, Sburlati A, Bowers B, Silverman SJ (1989) Chitin synthase 1, an auxiliary enzyme for chitin synthesis in *Saccharomyces cerevisiae*. J Cell Biol 108:1665–1672

Cabib E, Silverman S, Shaw J (1992) Chitinase and chitin synthase 1: counterbalancing activities in cell separation of *Saccharomyces cerevisiae*. J Gen Microbiol 138:97–102

Cabib E, Roh D, Schmidt M, Crotti L, Varma A (2001) The yeast cell wall and septum as paradigms of cell growth and morphogenesis. J Biol Chem 276:19679–19682

Cartagena-Lirola H, Duran A, Valdivieso MH (2006) The *Schizosaccharomyces pombe cfr1*$^+$ gene participates in mating through a new pathway that is independent of *fus1*$^+$. Yeast 23:375–388

Chaffin WL (2008) *Candida albicans* cell wall proteins. Microbiol Mol Biol Rev 72:495–544

Chin CF, Bennett AM, Ma WK, Hall MC, Yeong FM (2012) Dependence of Chs2 ER export on dephosphorylation by cytoplasmic Cdc14 ensures that septum formation follows mitosis. Mol Biol Cell 23:45–58

Choi WJ, Sburlati A, Cabib E (1994) Chitin synthase 3 from yeast has zymogenic properties that depend on both the *CAL1* and the *CAL3* genes. Proc Natl Acad Sci USA 91:4727–4730

Christodoulidou A, Briza P, Ellinger A, Bouriotis V (1999) Yeast ascospore wall assembly requires two chitin deacetylase isozymes. FEBS Lett 460:275–279

Chuang JS, Schekman RW (1996) Differential trafficking and timed localization of two chitin synthase proteins, Chs2p and Chs3p. J Cell Biol 135:597–610

Cid VJ, Durán A, del Rey F, Snyder MP, Nombela C, Sánchez M (1995) Molecular basis of cell integrity and morphogenesis in *Saccharomyces cerevisiae*. Microbiol Rev 59:345–386

Clemons KV, Stevens DA (2006) Animal models testing monotherapy versus combination antifungal therapy: lessons learned and future directions. Curr Opin Infect Dis 19:360–364

Cortes JC, Konomi M, Martins IM, Munoz J, Moreno MB, Osumi M, Duran A, Ribas JC (2007) The (1,3)beta-D-glucan synthase subunit Bgs1p is responsible for the fission yeast primary septum formation. Mol Microbiol 65:201–217

Cos T, Ford RA, Trilla JA, Duran A, Cabib E, Roncero C (1998) Molecular analysis of Chs3p participation in chitin synthase III activity. Eur J Biochem 256:419–426

Cote P, Hogues H, Whiteway M (2009) Transcriptional analysis of the Candida albicans cell cycle. Mol Biol Cell 20:3363–3373

Debono M, Gordee RS (1994) Antibiotics that inhibit fungal cell wall development. Annu Rev Microbiol 48:471–497

DeMarini DJ, Adams AE, Fares H, De Virgilio C, Valle G, Chuang JS, Pringle JR (1997) A septin-based hierarchy of proteins required for localized deposition of chitin in the *Saccharomyces cerevisiae* cell wall. J Cell Biol 139:75–93

Devrekanli A, Foltman M, Roncero C, Sanchez-Diaz A, Labib K (2012) Inn1 and Cyk3 regulate chitin synthase during cytokinesis in budding yeasts. J Cell Sci 125:5453–5466

Dobbelaere J, Barral Y (2004) Spatial coordination of cytokinetic events by compartmentalization of the cell cortex. Science 305:393–396

Dunkler A, Walther A, Specht CA, Wendland J (2005) Candida albicans CHT3 encodes the functional homolog of the Cts1 chitinase of Saccharomyces cerevisiae. Fungal Genet Biol 42:935–947

Dunkler A, Jorde S, Wendland J (2008) An Ashbya gossypii cts2 mutant deficient in a sporulation-specific chitinase can be complemented by *Candida albicans CHT4*. Microbiol Res 163:701–710

Durán A, Bowers B, Cabib E (1975) Chitin synthetase zymogen is attached to the yeast plasma membrane. Proc Natl Acad Sci USA 72:3952–3955

Ecker M, Deutzmann R, Lehle L, Mrsa V, Tanner W (2006) Pir proteins of *Saccharomyces cerevisiae* are attached to beta-1,3-glucan by a new protein-carbohydrate linkage. J Biol Chem 281:11523–11529

Endo A, Kakiki K, Misato T (1970) Mechanism of action of the antifugal agent polyoxin D. J Bacteriol 104:189–196

Fang X, Luo J, Nishihama R, Wloka C, Dravis C, Travaglia M, Iwase M, Vallen EA, Bi E (2010) Biphasic targeting and cleavage furrow ingression directed by the tail of a myosin II. J Cell Biol 191:1333–1350

Fang W, Du T, Raimi OG, Hurtado-Guerrero R, Marino K, Ibrahim AF, Albarbarawi O, Ferguson MA, Jin C, Van Aalten DM (2013) Genetic and structural validation of *Aspergillus fumigatus* N-acetylphosphoglucosamine mutase as an antifungal target. Biosci Rep 33

Free SJ (2013) Fungal cell wall organization and biosynthesis. Adv Genet 81:33–82

Gandia M, Harries E, Marcos JF (2014) The myosin motor domain-containing chitin synthase PdChsVII is required for development, cell wall integrity and virulence in the citrus postharvest pathogen *Penicillium digitatum*. Fungal Genet Biol 67:58–70

García-Rodríguez LJ, Trilla JA, Castro C, Valdivieso MH, Durán A, Roncero C (2000) Characterization of the chitin biosynthesis process as a compensatory mechanism in the *fks1* mutant of *Saccharomyces cerevisiae*. FEBS Lett 478:84–88

Gaughran JP, Lai MH, Kirsch DR, Silverman SJ (1994) Nikkomycin Z is a specific inhibitor of *Saccharomyces cerevisiae* chitin synthase isozyme Chs3 *in vitro* and *in vivo*. J Bacteriol 176:5857–5860

Gomez A, Perez J, Reyes A, Duran A, Roncero C (2009) Slt2 and Rim101 contribute independently to the correct assembly of the chitin ring at the budding yeast neck in *Saccharomyces cerevisiae*. Eukaryot Cell 8:1449–1459

Gonzalez-Novo A, Correa-Bordes J, Labrador L, Sanchez M, Vazquez de Aldana CR, Jimenez J (2008) Sep7 is essential to modify septin ring dynamics and inhibit cell separation during *Candida albicans* hyphal growth. Mol Biol Cell 19:1509–1518

Gow N, Robbins PW, Lester JW, Brown AJ, Fonzi WA, Chapman T, Kinsman OS (1994) A hyphal-specific chitin synthase gene (*CHS2*) is not essential for growth, dimorphism, or virulence of *Candida albicans*. Proc Natl Acad Sci USA 91:6216–6220

Gyore J, Parameswar AR, Hebbard CFF, Oh Y, Bi E, Demchenko AV, Price NP, Orlean P (2014) 2-Acylamido analogues of N-acetylglucosamine prime formation of chitin oligosaccharides by yeast chitin synthase 2. J Biol Chem 289:12835–12841

He X, Li S, Kaminskyj SG (2014) Using *Aspergillus nidulans* to identify antifungal drug resistance mutations. Eukaryot Cell 13:288–294

Ichinomiya M, Yamada E, Yamashita S, Ohta A, Horiuchi H (2005) Class I and class II chitin synthases are involved in septum formation in the filamentous fungus *Aspergillus nidulans*. Eukaryot Cell 4:1125–1136

Jakobsen MK, Cheng Z, Lam SK, Roth-Johnson E, Barfield RM, Schekman R (2013) Phosphorylation of Chs2p regulates interaction with COPII. J Cell Sci 126:2151–2156

Jendretzki A, Ciklic I, Rodicio R, Schmitz HP, Heinisch JJ (2009) Cyk3 acts in actomyosin ring independent cytokinesis by recruiting Inn1 to the yeast bud neck. Mol Genet Genomics 282:437–451

Jimenez C, Sacristan C, Roncero MI, Roncero C (2010) Amino acid divergence between the CHS domain contributes to the different intracellular behaviour of Family II fungal chitin synthases in *Saccharomyces cerevisiae*. Fungal Genet Biol 47:1034–1043

Jimenez-Ortigosa C, Aimanianda V, Muszkieta L, Mouyna I, Alsteens D, Pire S, Beau R, Krappmann S, Beauvais A, Dufrene YF, Roncero C, Latge JP (2012) Chitin synthases with a myosin motor-like domain control the resistance of *Aspergillus fumigatus* to echinocandins. Antimicrob Agents Chemother 56:6121–6131

Kim M, Park H, Kim C, Park H, Choi W (2002) Inhibitory effect of nikkomycin Z on chitin synthases in Candida albicans. Yeast 19:341–349

Klis F, De Groot P, Hellingwerf K (2001) Molecular organization of the cell wall of *Candida albicans*. Med Mycol 39:1–8

Kollar R, Reinhold BB, Petrakova E, Yeh HJ, Ashwell G, Drgonova J, Kapteyn JC, Klis FM, Cabib E (1997) Architecture of the yeast cell wall. β(1,6)-glucan interconnects mannoprotein, β(1,3)-glucan, and chitin. J Biol Chem 272:17762–17775

Kong LA, Yang J, Li GT, Qi LL, Zhang YJ, Wang CF, Zhao WS, Xu JR, Peng YL (2012) Different chitin synthase genes are required for various developmental and plant infection processes in the rice blast fungus Magnaporthe oryzae. PLoS Pathog 8, e1002526

Kota J, Ljungdahl PO (2005) Specialized membrane-localized chaperones prevent aggregation of polytopic proteins in the ER. J Cell Biol 168:79–88

Kozubowski L, Panek H, Rosenthal A, Bloecher A, DeMarini DJ, Tatchell K (2003) A Bni4-Glc7 phosphatase complex that recruits chitin synthase to the site of bud emergence. Mol Biol Cell 14:26–39

Lagorce A, Le Berre-Anton V, Aguilar-Uscanga B, Martin-Yken H, Dagkessamanskaia A, Francois J (2002) Involvement of *GFA1*, which encodes glutamine-fructose-6-phosphate amidotransferase, in the activation of the chitin synthesis pathway in response to cell-wall defects in *Saccharomyces cerevisiae*. Eur J Biochem 269:1697–1707

Lam KK, Davey M, Sun B, Roth AF, Davis NG, Conibear E (2006) Palmitoylation by the DHHC protein Pfa4 regulates the ER exit of Chs3. J Cell Biol 174:19–25

Latge JP, Mouyna I, Tekaia F, Beauvais A, Debeaupuis JP, Nierman W (2005) Specific molecular features in the organization and biosynthesis of the cell wall of *Aspergillus fumigatus*. Med Mycol 43(Suppl 1):S15–S22

Lenardon MD, Whitton RK, Munro CA, Marshall D, Gow NA (2007) Individual chitin synthase enzymes synthesize microfibrils of differing structure at specific locations in the *Candida albicans* cell wall. Mol Microbiol 66:1164–1173

Lenardon MD, Munro CA, Gow NA (2010) Chitin synthesis and fungal pathogenesis. Curr Opin Microbiol 13:416–423

Lesage G, Bussey H (2006) Cell wall assembly in *Saccharomyces cerevisiae*. Microbiol Mol Biol Rev 70:317–343

Levin DE (2011) Regulation of cell wall biogenesis in *Saccharomyces cerevisiae*: the cell wall integrity signaling pathway. Genetics 189:1145–1175

Liu H, Zheng Z, Wang P, Gong G, Wang L, Zhao G (2013) Morphological changes induced by class III chitin synthase gene silencing could enhance penicillin production of *Penicillium chrysogenum*. Appl Microbiol Biotechnol 97:3363–3372

Madrid MP, Di Pietro A, Roncero MI (2003) Class V chitin synthase determines pathogenesis in the vascular wilt fungus *Fusarium oxysporum* and mediates resistance to plant defence compounds. Mol Microbiol 47:257–266

Marakalala MJ, Vautier S, Potrykus J, Walker LA, Shepardson KM, Hopke A, Mora-Montes HM, Kerrigan A, Netea MG, Murray GI, Maccallum DM, Wheeler R, Munro CA, Gow NA, Cramer RA, Brown AJ, Brown GD (2013) Differential adaptation of *Candida albicans* in vivo modulates immune recognition by dectin-1. PLoS Pathog 9, e1003315

Martinez-Rucobo FW, Eckhardt-Strelau L, Terwisscha van Scheltinga AC (2009) Yeast chitin synthase 2 activity is modulated by proteolysis and phosphorylation. Biochem J 417:547–554

Martin-Garcia R, Duran A, Valdivieso MH (2003) In *Schizosaccharomyces pombe* chs2p has no chitin synthase activity but is related to septum formation. FEBS Lett 549:176–180

Martin-Garcia R, de Leon N, Sharifmoghadam MR, Curto MA, Hoya M, Bustos-Sanmamed P, Valdivieso MH (2011) The FN3 and BRCT motifs in the exomer component Chs5p define a conserved module that is necessary and sufficient for its function. Cell Mol Life Sci 68:2907–2917

Martin-Urdiroz M, Roncero MI, Gonzalez-Reyes JA, Ruiz-Roldan C (2008) ChsVb, a class VII chitin synthase involved in septation, is critical for pathogenicity in *Fusarium oxysporum*. Eukaryot Cell 7:112–121

Meitinger F, Petrova B, Lombardi IM, Bertazzi DT, Hub B, Zentgraf H, Pereira G (2010) Targeted localization of Inn1, Cyk3 and Chs2 by the mitotic-exit network regulates cytokinesis in budding yeast. J Cell Sci 123:1851–1861

Meitinger F, Palani S, Pereira G (2012) The power of MEN in cytokinesis. Cell Cycle 11:219–228

Mellado E, Specht C, Robbins P, Holden D (1996) Cloning and characterization of chsD, a chitin synthase-like gene of *Aspergillus fumigatus*. FEMS Microbiol Lett 143:69–76

Merzendorfer H (2011) The cellular basis of chitin synthesis in fungi and insects: common principles and differences. Eur J Cell Biol 90:759–769

Molano J, Bowers B, Cabib E (1980) Distribution of chitin in the yeast cell wall. An ultrastructural and chemical study. J Cell Biol 85:199–212

Muller C, Hjort CM, Hansen K, Nielsen J (2002) Altering the expression of two chitin synthase genes differentially affects the growth and morphology of *Aspergillus oryzae*. Microbiology 148:4025–4033

Munro CA (2013) Chitin and glucan, the yin and yang of the fungal cell wall, implications for antifungal drug discovery and therapy. Adv Appl Microbiol 83:145–172

Munro CA, Winter K, Buchan A, Henry K, Becker JM, Brown AJ, Bulawa CE, Gow NA (2001) Chs1 of *Candida albicans* is an essential chitin synthase required for synthesis of the septum and for cell integrity. Mol Microbiol 39:1414–1426

Munro CA, Whitton RK, Hughes HB, Rella M, Selvaggini S, Gow NA (2003) CHS8-a fourth chitin synthase gene of *Candida albicans* contributes to in vitro chitin synthase activity, but is dispensable for growth. Fungal Genet Biol 40:146–158

Nagahashi S, Sudoh M, Ono N, Sawada R, Yamaguchi E, Uchida Y, Mio T, Takagi M, Arisawa M, Yamada-Okabe H (1995) Characterization of Chitin synthase 2 of *Saccharomyces cerevisiae*. J Biol Chem 270:13961–13967

Nishihama R, Schreiter JH, Onishi M, Vallen EA, Hanna J, Moravcevic K, Lippincott MF, Han H, Lemmon MA, Pringle JR, Bi E (2009) Role of Inn1 and its interactions with Hof1 and Cyk3 in promoting cleavage furrow and septum formation in *S. cerevisiae*. J Cell Biol 185:995–1012

Nkosi PJ, Targosz BS, Labib K, Sanchez-Diaz A (2013) Hof1 and Rvs167 have redundant roles in actomyosin ring function during cytokinesis in budding yeast. PLoS One 8, e57846

Oh Y, Chang KJ, Orlean P, Wloka C, Deshaies R, Bi E (2012) Mitotic exit kinase Dbf2 directly phosphorylates chitin synthase Chs2 to regulate cytokinesis in budding yeast. Mol Biol Cell 23:2445–2456

Oh Y, Schreiter J, Nishihama R, Wloka C, Bi E (2013) Targeting and functional mechanisms of the cytokinesis-related F-BAR protein Hof1 during the cell cycle. Mol Biol Cell 24:1305–1320

Onishi M, Ko N, Nishihama R, Pringle JR (2013) Distinct roles of Rho1, Cdc42, and Cyk3 in septum formation and abscission during yeast cytokinesis. J Cell Biol 202:311–329

Ono N, Yabe T, Sudoh M, Nakajima T, Yamada-Okabe T, Arisawa M, Yamada-Okabe H (2000) The yeast Chs4p protein stimulates the trypsin-sensitive activity of chitin synthase 3 through an apparent protein-protein interaction. Microbiology 146:385–391

Orlean P (1987) Two chitin synthases in *Saccharomyces cerevisiae*. J Biol Chem 262:5732–5739

Orlean P (2012) Architecture and biosynthesis of the *Saccharomyces cerevisiae* cell wall. Genetics 192:775–818

Paczkowski JE, Richardson BC, Strassner AM, Fromme JC (2012) The exomer cargo adaptor structure reveals a novel GTPase-binding domain. EMBO J 31:4191–4203

Palani S, Meitinger F, Boehm ME, Lehmann WD, Pereira G (2012) Cdc14-dependent dephosphorylation of Inn1 contributes to Inn1-Cyk3 complex formation. J Cell Sci 125:3091–3096

Peters NK (1997) Nodulation: finding the lost common denominator. Curr Biol 7:R223–R226

Reyes A, Sanz M, Duran A, Roncero C (2007) Chitin synthase III requires Chs4p-dependent translocation of Chs3p into the plasma membrane. J Cell Sci 120:1998–2009

Riquelme M, Bartnicki-Garcia S, Gonzalez-Prieto JM, Sanchez-Leon E, Verdin-Ramos JA, Beltran-Aguilar A, Freitag M (2007) Spitzenkorper localization and intracellular traffic of green fluorescent protein-labeled CHS-3 and CHS-6 chitin synthases in living hyphae of *Neurospora crassa*. Eukaryot Cell 6:1853–1864

Ritz AM, Trautwein M, Grassinger F, Spang A (2014) The prion-like domain in the exomer-dependent cargo Pin2 serves as a trans-Golgi retention motif. Cell Rep 7:249–260

Rockenbauch U, Ritz AM, Sacristan C, Roncero C, Spang A (2012) The complex interactions of Chs5p, the ChAPs, and the cargo Chs3p. Mol Biol Cell 23:4402–4415

Rolli E, Ragni E, Calderon J, Porello S, Fascio U, Popolo L (2009) Immobilization of the glycosylphosphatidylinositol-anchored Gas1 protein into the chitin ring and septum is required for proper morphogenesis in yeast. Mol Biol Cell 20:4856–4870

Roncero C (2002) The genetic complexity of chitin synthesis in fungi. Curr Genet 41:367–378

Roncero C, Durán A (1985) Effect of Calcofluor White and Congo red on fungal cell wall morphogenesis: *in vivo* activation of chitin polymerization. J Bacteriol 163:1180–1185

Roncero C, Sanchez Y (2010) Cell separation and the maintenance of cell integrity during cytokinesis in yeast: the assembly of a septum. Yeast 27:521–530

Ruiz-Herrera J, Ortiz-Castellanos L (2010) Analysis of the phylogenetic relationships and evolution of the cell walls from yeasts and fungi. FEMS Yeast Res 10:225–243

Ruiz-Herrera J, Gonzalez Prieto JM, Ruiz-Medrano R (2002) Evolution and phylogenetic relationships of chitin synthases from yeast and fungi. FEMS Yeast Res 1:247–256

Sacristan C, Reyes A, Roncero C (2012) Neck compartmentalization as the molecular basis for the different endocytic behaviour of Chs3 during budding or hyperpolarized growth in yeast cells. Mol Microbiol 83:1124–1135

Sacristan C, Manzano-Lopez J, Reyes A, Spang A, Muniz M, Roncero C (2013) Oligomerization of the chitin synthase Chs3 is monitored at the Golgi and affects its endocytic recycling. Mol Microbiol 90:252–266

Sanchez-Diaz A, Marchesi V, Murray S, Jones R, Pereira G, Edmondson R, Allen T, Labib K (2008) Inn1 couples contraction of the actomyosin ring to membrane ingression during cytokinesis in budding yeast. Nat Cell Biol 10:395–406

Sanchez-Diaz A, Nkosi PJ, Murray S, Labib K (2012) The Mitotic Exit Network and Cdc14 phosphatase initiate cytokinesis by counteracting CDK phosphorylations and blocking polarised growth. EMBO J 31:3620–3634

Sanchez-Leon E, Verdin J, Freitag M, Roberson RW, Bartnicki-Garcia S, Riquelme M (2011) Traffic of chitin synthase 1 (CHS-1) to the Spitzenkorper and developing septa in hyphae of *Neurospora crassa*: actin dependence and evidence of distinct microvesicle populations. Eukaryot Cell 10:683–695

Santos B, Snyder M (1997) Targeting of chitin synthase 3 to polarized growth sites in yeast requires Chs5p and Myo2p. J Cell Biol 136:95–110

Sanz M, Trilla JA, Durán A, Roncero C (2002) Control of chitin synthesis through Shc1p, a functional homologue of Chs4p specifically induced during sporulation. Mol Microbiol 43:1183–1195

Sanz M, Castrejon F, Duran A, Roncero C (2004) *Saccharomyces cerevisiae* Bni4p directs the formation of the chitin ring and also participates in the correct assembly of the septum structure. Microbiology 150:3229–3241

Sanz M, Carrano L, Jimenez C, Candiani G, Trilla JA, Duran A, Roncero C (2005) *Candida albicans* strains deficient in CHS7, a key regulator of chitin synthase III, exhibit morphogenetic alterations and attenuated virulence. Microbiology 151:2623–2636

Saxena IM, Brown R Jr, Fevre M, Geremia RA, Henrissat B (1995) Multidomain architecture of β-Glycosyl transferases: implications for mechanism of action. J Bacteriol 177:1419–1424

Sburlati A, Cabib E (1986) Chitin synthetase 2, a presumptive participant in septum formation in *Saccharomyces cerevisiae*. J Biol Chem 261:15147–15152

Schmidt M, Bowers B, Varma A, Roh DH, Cabib E (2002) In budding yeast, contraction of the actomyosin ring and formation of the primary septum at cytokinesis depend on each other. J Cell Sci 115:293–302

Schuster M, Treitschke S, Kilaru S, Molloy J, Harmer NJ, Steinberg G (2012) Myosin-5, kinesin-1 and myosin-17 cooperate in secretion of fungal chitin synthase. EMBO J 31:214–227

Shaw JA, Mol PC, Bowers B, Silverman SJ, Valdivieso MH, Durán A, Cabib E (1991) The function of chitin synthases 2 and 3 in the *Saccharomyces cerevisiae* cell cycle. J Cell Biol 114:111–123

Sheng W, Yamashita S, Ohta A, Horiuchi H (2013) Functional differentiation of chitin synthases in Yarrowia lipolytica. Biosci Biotechnol Biochem 77:1275–1281

Soulie MC, Perino C, Piffeteau A, Choquer M, Malfatti P, Cimerman A, Kunz C, Boccara M, Vidal-Cros A (2006) Botrytis cinerea virulence is drastically reduced after disruption of chitin synthase class III gene (Bcchs3a). Cell Microbiol 8:1310–1321

Spellman P, Sherlock G, Zhang MQ, Iyer VR, Anders K, Eisen MB, Brown PO, Botstein D, Futcher B (1998) Comprehensive identification of cell cycle-regulated genes of the yeast *Saccharomyces cerevisiae* by microarray hybridization. Mol Biol Cell 9:3273–3297

Starr TL, Pagant S, Wang CW, Schekman R (2012) Sorting signals that mediate traffic of chitin synthase III between the TGN/endosomes and to the plasma membrane in yeast. PLoS One 7, e46386

Sudoh M, Yamazaki T, Masubuchi K, Taniguchi M, Shimma N, Arisawa M, Yamada-Okabe H (2000)

Identification of a novel inhibitor specific to the fungal chitin synthase. Inhibition of chitin synthase 1 arrests the cell growth, but inhibition of chitin synthase 1 and 2 is lethal in the pathogenic fungus *Candida albicans*. J Biol Chem 275:32901–32905

Takeshita N, Ohta A, Horiuchi H (2005) CsmA, a class V chitin synthase with a myosin motor-like domain, is localized through direct interaction with the actin cytoskeleton in *Aspergillus nidulans*. Mol Biol Cell 16:1961–1970

Takeshita N, Yamashita S, Ohta A, Horiuchi H (2006) *Aspergillus nidulans* class V and VI chitin synthases CsmA and CsmB, each with a myosin motor-like domain, perform compensatory functions that are essential for hyphal tip growth. Mol Microbiol 59:1380–1394

Teh EM, Chai CC, Yeong FM (2009) Retention of Chs2p in the ER requires N-terminal CDK1-phosphorylation sites. Cell Cycle 8:2964–2974

Trautwein M, Schindler C, Gauss R, Dengjel J, Hartmann E, Spang A (2006) Arf1p, Chs5p and the ChAPs are required for export of specialized cargo from the Golgi. EMBO J 25:943–954

Treitschke S, Doehlemann G, Schuster M, Steinberg G (2010) The myosin motor domain of fungal chitin synthase V is dispensable for vesicle motility but required for virulence of the maize pathogen *Ustilago maydis*. Plant Cell 22:2476–2494

Trilla JA, Cos T, Durán A, Roncero C (1997) Characterization of *CHS4* (*CAL2*), a gene of *Saccharomyces cerevisiae* involved in chitin biosynthesis and allelic to *SKT5* and *CSD4*. Yeast 13:795–807

Trilla JA, Durán A, Roncero C (1999) Chs7p, a new protein involved in the control of protein export from the endoplasmic reticulum that is specifically engaged in the regulation of chitin synthesis in *Saccharomyces cerevisiae*. J Cell Biol 145:1153–1163

Valdivia RH, Schekman R (2003) The yeasts Rho1p and Pkc1p regulate the transport of chitin synthase III (Chs3p) from internal stores to the plasma membrane. Proc Natl Acad Sci USA 100:10287–10292

Valdivia RH, Baggott D, Chuang JS, Schekman RW (2002) The yeast clathrin adaptor protein complex 1 is required for the efficient retention of a subset of late Golgi membrane proteins. Dev Cell 2:283–294

VerPlank L, Li R (2005) Cell cycle-regulated trafficking of Chs2 controls actomyosin ring stability during cytokinesis. Mol Biol Cell 16:2529–2543

Walker LA, Lenardon MD, Preechasuth K, Munro CA, Gow NA (2013) Cell wall stress induces alternative fungal cytokinesis and septation strategies. J Cell Sci 126:2668–2677

Wang Z, Zheng L, Hauser M, Becker JM, Szaniszlo PJ (1999) WdChs4p, a homolog of chitin synthase 3 in *Saccharomyces cerevisiae*, alone cannot support growth of *Wangiella* (*Exophiala*) *dermatitidis* at the temperature of infection. Infect Immun 67:6619–6630

Wang CW, Hamamoto S, Orci L, Schekman R (2006) Exomer: a coat complex for transport of select membrane proteins from the trans-Golgi network to the plasma membrane in yeast. J Cell Biol 174:973–983

Wang H, Xu Z, Gao L, Hao B (2009) A fungal phylogeny based on 82 complete genomes using the composition vector method. BMC Evol Biol 9:195

Weber I, Assmann D, Thines E, Steinberg G (2006) Polar localizing class V myosin chitin synthases are essential during early plant infection in the plant pathogenic fungus *Ustilago maydis*. Plant Cell 18:225–242

Wessels J, Sietsma J, Sonnenberg A (1983) Wall synthesis and assembly during hyphal morphogenesis in *Schizophyllum comune*. J Gen Microbiol 129:1607–1616

Wloka C, Bi E (2012) Mechanisms of cytokinesis in budding yeast. Cytoskeleton (Hoboken) 69:710–726

Wloka C, Vallen EA, The L, Fang X, Oh Y, Bi E (2013) Immobile myosin-II plays a scaffolding role during cytokinesis in budding yeast. J Cell Biol 200:271–286

Yarden O, Yanofsky C (1991) Chitin synthase 1 of *Neurospora crassa*. Genes Dev 5:2420–2430

Zakrzewski AC, Weigert A, Helm C, Adamski M, Adamska M, Bleidorn C, Raible F, Hausen H (2014) Early divergence, broad distribution and high diversity of animal chitin synthases. Genome Biol Evol 6:316–325

Zhang G, Kashimshetty R, Ng KE, Tan HB, Yeong FM (2006) Exit from mitosis triggers Chs2p transport from the endoplasmic reticulum to mother-daughter neck via the secretory pathway in budding yeast. J Cell Biol 174:207–220

10 Trehalose Metabolism: Enzymatic Pathways and Physiological Functions

Marta Rubio-Texeira[1,2], Griet Van Zeebroeck[1,2], Johan M. Thevelein[1,2]

CONTENTS

Abbreviations

TPS	Trehalose-6-phosphate synthase
Tre6P	Trehalose-6-phosphate
TPP	Tre6P phosphatase
TS	Trehalose synthase
TreY	Maltooligosyl trehalose synthase
TreZ	Maltooligosyl trehalose trehalohydrolase
TreP	Trehalose phosphorylase
TreT	Trehalose glycosyl-transferring synthase
UDPG	UDP-glucose
G6P	Glucose-6-phosphate
ADPG	ADP-glucose
GDPG	GDP-glucose
TDPG	Thymidine-DPG
PKA	Protein kinase A: cAMP-dependent protein kinase
Hsp	Heat shock protein

[1]Laboratory of Molecular Cell Biology, Institute of Botany and Microbiology, KU Leuven, Leuven, Belgium; e-mail: johan.thevelein@mmbio.vib-kuleuven.be
[2]Department of Molecular Microbiology, VIB, Kasteelpark Arenberg 31, 3001 Leuven-Heverlee, Flanders, Belgium; e-mail: johan.thevelein@mmbio.vib-kuleuven.be

Biochemistry and Molecular Biology, 3rd Edition
The Mycota III
D. Hoffmeister (Ed.)

HSE	Heat shock element
STRE	Stress-responsive element
TORC1	Target of rapamycin complex 1
ROS	Reactive oxygen species
GPCR	G-protein-coupled receptor
AC	Adenylate cyclase
FGM	Fermentable growth medium
Gly3P	Glycerol-3-phosphate
AdOx	Oxidized adenosine
SAM	*S*-adenosyl-L-methionine
PP2A	Protein phosphatase 2A
PDS	Postdiauxic shift
PP1	Protein phosphatase 1
NLS	Nuclear localization sequence
ORF	Open reading frame
PFK	Phosphofructokinase
AZC	Azetidine 2-carboxylic acid
GAPDH	Glyceraldehyde-3-phosphate dehydrogenase
AMP	Adenosine monophosphate
CCR	Carbon catabolite repression

I. Introduction

Trehalose is a nonreducing disaccharide (1,1-α-D-glucopyranosyl α-D-glucopyranoside) (Fig. 10.1a) commonly found in many fungi and present in particularly high concentrations in quiescent cells and survival forms. For a long time trehalose was only considered as a **storage carbohydrate**, accumulated under conditions of imminent carbon shortage and mobilized under prolonged starvation conditions. However, in comparison to structurally related sugars, trehalose displays unique characteristics, which provide it with functions going beyond that of a mere storage compound. Trehalose levels are often closely correlated with stress tolerance and numerous in vivo and in vitro experiments have shown a remarkable capacity of high trehalose levels in conferring protection to denaturation of proteins, membranes, cells, and organisms under stress conditions. As a result, the role of trehalose as a **stress protectant** is now well established. An important outcome of these findings has been the evaluation of enzymes of trehalose metabolism as possible targets for drugs against infectious diseases. For pathogenic yeasts and fungi, trehalose metabolism can be important to resist the stressful conditions caused by the host defense mechanisms. Moreover, in the case of human infections, the use of these targets provides the advantage that trehalose genes are absent in mammals.

Trehalose metabolism has been particularly useful as a target of nutrient-induced signaling pathways since under appropriate conditions of growth stimulation, rapid changes in trehalose levels and trehalase activity can be triggered by multiple essential nutrients. This has led to the identification of novel nutrient-sensing mechanisms in yeast, some of which have already been shown to work similarly in other fungi and eukaryotic species. The molecular biology of trehalose metabolism has seen a dramatic expansion. After their initial identification in yeast, numerous genes for trehalases and trehalose biosynthesis enzymes have been identified in other eukaryotic organisms including fungi, insects, nematodes, crustaceans, and plants. New enzymatic activities involved in trehalose metabolism have also been discovered which are apparently absent in yeast. The systematic sequencing projects are continuously adding new genes to the available repertoire. The most surprising output with respect to trehalose metabolism undoubtedly was the identification of trehalose metabolism genes in green plants. Yeast mutants in trehalose metabolism have proved to be particularly useful for the functional characterization of these plant genes. Although trehalose is barely detectable in green plants, all data obtained up to now are consistent with an important pleiotropic regulatory role of trehalose metabolism in plant development and carbohydrate metabolism. In addition, an important role of trehalose metabolism in the control of glycolysis has been discovered in yeast and other fungi. An active **trehalose-6-phosphate synthase (TPS)** is essential for proper homeostasis of the initial metabolites of glycolysis. Several studies made in different organisms all point at the trehalose precursor, trehalose-6-phosphate (Tre6P), as a signaling molecule with an important role in the regulation of carbon and energy metabolism.

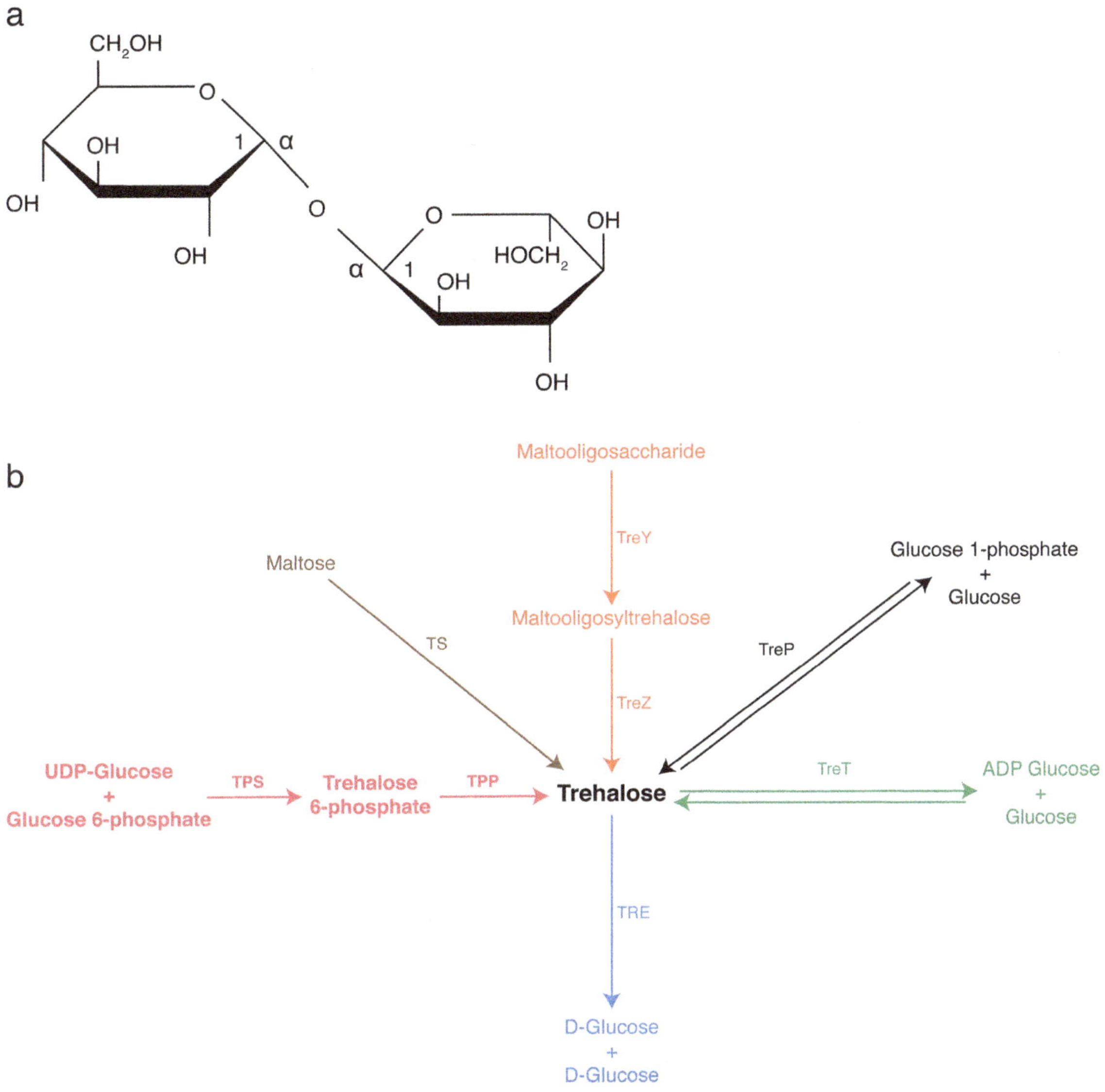

Fig. 10.1 Structure, biosynthesis, and degradation of trehalose. (**a**) Chemical structure of trehalose. Trehalose is a nonreducing disaccharide composed of two α-D-glucose units linked through the C1 carbon atoms (1,1-α-D-glucopyranosyl α-D-glucopyranoside). (**b**) The trehalose biosynthetic and degradation pathways. The TPS-TPP pathway for trehalose biosynthesis is the most widespread found in eubacteria, archaea, fungi, insects, and plants. The TreT, TreY-TreZ, and TS pathways have been reported only in prokaryotes, while the TreP pathway is found in prokaryotes and fungi. The universal trehalose degradation pathway involves trehalase activity. TPS, trehalose-6-P synthase; TPP, trehalose-6-P phosphatase; TS, trehalose synthase; TreY, maltooligosyltrehalose synthase; TreZ, maltooligosyltrehalose trehalohydrolase; TreP, trehalose phosphorylase; TreT, glycosyl-transferring trehalose synthase; TRE, trehalase

II. Occurrence of Trehalose

Trehalose biosynthesis occurs in a wide variety of organisms, such as bacteria, fungi, insects, invertebrates, and plants (reviewed in Avonce et al. 2006; Elbein et al. 2003). Trehalose has gained particular attention because of its multifunctional properties in the different organisms where it occurs. Trehalose is a crucial molecule for the survival of many different species, from a source of carbon and energy to a molecular chaperone involved in protection against a broad spectrum of environmental stresses. As a source of carbon and energy and a regulator of carbon metabolism, it is involved in processes as diverse as yeast/fungal spore germination, exit from quiescence, insect flight, virulence of bacterial, yeast and fungal pathogens, control of sucrose levels, chitin synthesis, as well as cellular differentiation (Kramer et al. 1985; Liu et al. 2013a; O'Hara et al. 2013; Sakamoto et al. 2013; Shi et al. 2010; Thevelein 1984c). As a stress protectant it preserves the integrity of cells against heat, cold, desiccation, dehydration, and oxidation, mostly by preventing protein denaturation. Multiple studies, started by the pioneering work of Singer and Lindquist (1998a), have demonstrated that trehalose interacts with proteins and acts as a chemical chaperone, reducing protein aggregation and maintaining proteins in a partially folded state, facilitating their further refolding by protein chaperones. For example, although it is not naturally produced in mammals, trehalose can still inhibit amyloid formation of insulin in vitro (Arora et al. 2004) and prevent aggregation of β-amyloid associated with Alzheimer disease (Liu et al. 2005). Trehalose can inhibit polyglutamine (polyQ)3-mediated protein aggregation in vitro, reduce mutant huntingtin aggregates and toxicity in cell models, and alleviate polyQ-induced pathology in the R6/2 mouse model of Huntington disease (Tanaka et al. 2004). Moreover, trehalose has been shown to act as an mTOR-independent autophagy activator in mammalian cell lines. Trehalose-induced autophagy enhanced the clearance of autophagy substrates like mutant huntingtin and the A30P and A53T mutants of α-synuclein, associated with Huntington and Parkinson disease, respectively (Sarkar et al. 2007). In addition, trehalose has been shown to preserve the integrity of membranes during dehydration in vitro (Crowe et al. 1992, 1998).

The dual role of trehalose as nutrient source and metabolic regulator on the one hand and chemical chaperone on the other hand explains its occurrence in particular organisms. For example, trehalose is present in organisms subjected to anhydrobiosis, including nematodes, embryonic cysts of crustaceans, and yeast. Anhydrobiosis is a highly stable state of suspended animation in an organism due to its desiccation, which is followed by recovery after rehydration (Goyal et al. 2005). Trehalose concentrations ranging from 10 to 18 % of the total dry weight have been reported for anhydrobiotic cysts of brine shrimps from *Artemia* genus (Yang et al. 2013), the nematode *Aphelenchus avenae* (Goyal et al. 2005), and the Arctic collembolan *Onychiurus arcticus* during partial desiccation at sub-zero temperatures (Clark et al. 2007) and in the larvae of the African anhydrobiotic chironomid insect *Polypedilum vanderplanki* (Cornette and Kikawada 2011; Watanabe et al. 2002). In the latter, high temperatures and high moisture levels have reduced the survival rate of desiccated larvae due to the change from the glassy to the rubbery state of trehalose (Sakurai et al. 2008). Westh and Ramløv (1991) reported a 20-fold accumulation of trehalose during transition to anhydrobiosis in the eutardigrade *Adorybiotus coronifer*. Accumulation of trehalose has also been found in tardigrade embryos and loss of resurrection ability in *Paramacrobiotus richtersi* has been linked to loss of trehalose protection due to exposition to high temperature and humidity levels (Hengherr et al. 2008; Rebecchi et al. 2009). Trehalose also accumulates in *Drosophila* larvae subjected to dessication stress and has been proposed as a biochemical marker of drought to monitor the effects of global warming in regional climate change (Thorat et al. 2012).

In insects, trehalose is the main sugar in hemolymph (80–90 %) and in thorax muscles, where it is consumed during flight (Becker et al. 1996; Behm 1997). Trehalose also fuels other processes such as embryo/larval development (Xu et al. 2013; Yang et al. 2013) and chitin synthesis (Chen et al. 2010). In the red flour beetle *Tribolium castaneum*, trehalose seems to play a role in starvation resistance not only as energy source but also as chemical chaperone (Xu et al. 2013). Insects have also developed mechanisms such as tissue-specific membrane transporters and even taste receptors tuned for trehalose, to adsorb this sugar from yeast and fungi present in their ecological niche and gut microbiota (Chyb et al. 2003; Kanamori et al. 2010).

With the exception of certain drought-tolerant vascular plants (also known as "resurrection plants") (Goddijn and Smeekens 1998; Marquez-Escalante et al. 2006), trehalose accumulation has been difficult to detect in higher plants, in which sucrose acts as the major reserve carbohydrate. However, the discovery of a wide repertoire of genes for trehalose synthesis and degradation in *Arabidopsis* and other plant species (Leyman et al. 2001) highlights the possible importance of trehalose also in this kingdom (Eastmond and Graham 2003; O'Hara et al. 2013). Trehalose, and especially its precursor, Tre6P, is currently considered as a signaling molecule acting in the regulation of sucrose and starch metabolism in plants (Ponnu et al. 2011). *TPS1* loss-of-function mutants are embryonic lethal (Eastmond et al. 2002). This lethality has also been observed in *Drosophila* (Chen and Haddad 2004) and in *C. elegans* embryos (Kormish and McGhee 2005). This underscores the importance of trehalose and Tre6P in the developmental programs of many different organisms.

When considering the situation in fungi, reproductive stages, such as spores, and other survival forms, such as sclerotia, are usually associated with accumulation of trehalose, which can largely exceed that of other carbohydrates such as glycogen or sugar alcohols, like mannitol. In many instances, such as in the ascospores of *Saccharomyces cerevisiae*, trehalose is virtually the only sugar present (Thevelein 1984c). Intensive trehalose accumulation in fungi is observed not only during sporulation and differentiation processes but in general during periods of reduced growth rate or quiescence, for example, during starvation in vegetative cells (Thevelein 1984c). *S. cerevisiae* cells can continue to accumulate trehalose for many hours when fed with low sugar concentrations (Grba et al. 1975). This property is being used by commercial baker's yeast producers to enhance the trehalose content of the cells to up to 20 % of the dry weight (Gélinas et al. 1989; Trivedi and Jacobson 1986). In the yeast *S. cerevisiae* reduced growth rate correlates well with higher trehalose content (Kuenzi and Fiechter 1972) and cells starved for nitrogen, phosphate, or sulfate in the presence of glucose accumulate large amounts of trehalose (Lillie and Pringle 1980). Also sublethal stress treatments often trigger trehalose accumulation (Wiemken 1990). This appears to be most pronounced in the case of heat stress. Incubation of yeast cells at sublethal temperatures induces strong trehalose accumulation (Grba et al. 1975, 1979; Hottiger et al. 1987b).

In general, resumption and stimulation of growth are associated with trehalose mobilization. This is particularly prominent during the induction of growth in spores and sclerotia, where the high trehalose content is usually rapidly mobilized during the initial stages of germination. A similar mobilization of trehalose is observed upon addition of nutrients to stationary-phase yeast cells (Thevelein 1984c). Prolonged starvation of trehalose-containing yeast cells also causes very slow mobilization of trehalose (Lillie and Pringle 1980; Panek 1963), a process which is also known to occur during storage of baker's yeast (Stewart et al. 1950; Suomalainen and Pfäffli 1961). Rapid trehalose mobilization has also been observed in glucose-limited self-synchronized chemostat cultures of *S. cerevisiae* during initiation of the budding phase of the cell cycle (Küenzi and Fiechter 1969; von Meyenburg 1969). The correct interpretation of this phenomenon appears to be that the cells temporarily stay in the stationary G0 phase because of the glucose limitation and then suddenly, triggered by an unknown factor, enter the G1 phase of the cell cycle in a synchronous way. Hence, mobilization of trehalose in this system is also associated with resumption of growth in stationary-phase cells, rather than associated with a specific phase of the cell cycle as was concluded previously (Küenzi and Fiechter 1969).

III. Biosynthetic Pathways for Trehalose

In bacteria there are five different biosynthetic pathways for trehalose, whereas in fungi, plants, and animals, there is only one. The most widely distributed pathway, present in all the abovementioned kingdoms, involves two enzymatic steps catalyzed by TPS and

Tre6P phosphatase (TPP). TPS catalyzes the transfer of glucose from UDP-glucose (UDPG) to glucose-6-phosphate (G6P) to yield Tre6P and UDP, while TPP dephosphorylates Tre6P to trehalose and inorganic phosphate (Fig. 10.1b) (Cabib and Leloir 1958). The second biosynthetic pathway, found in eubacteria, consists of the enzyme trehalose synthase (TS), which isomerizes the α1–α4 bond of maltose to an α1–α1 bond, to produce trehalose (Fig. 10.1b) (Elbein 1974; Elbein et al. 2003; Nishimoto et al. 1996). A third pathway, found in thermophilic archaea of the genus *Sulfolobus*, involves a maltooligosyl trehalose synthase (TreY), which promotes transglycosylation of the last glucose moiety at the reduced end of maltodextrins (maltooligosaccharides, glycogen, and starch) from an α1–α4 to an α1–α1 bond leading to maltooligosyltrehalose, containing a trehalose moiety at the end of the polymer. In a next step, a maltooligosyl trehalose trehalohydrolase (TreZ) catalyzes the hydrolytic release of trehalose (Fig. 10.1b) (Fang et al. 2004; Gueguen et al. 2001; Streeter and Bhagwat 1999). In a fourth pathway, trehalose phosphorylase (TreP), present in some fungi, catalyzes the reversible hydrolysis of trehalose in the presence of inorganic phosphate, by transfer of phosphate to one of the glucose molecules, generating glucose 1-phosphate, and release of the other glucose residue (Fig. 10.1b) (Schiraldi et al. 2002; Wannet et al. 1998). More recently an additional biosynthetic pathway has been found in the hyperthermophilic archaeon *Thermococcus litoralis*, involving the trehalose glycosyl-transferring synthase (TreT), which catalyzes the reversible formation of trehalose from ADP-glucose (ADPG) and glucose (Qu et al. 2004; Ryu et al. 2005). TreT can also use UDPG and GDP-glucose (GDPG), although less efficiently. The TreT enzyme transfers the glucose moiety from ADPG and joins it at position 1 of another glucose molecule to form trehalose (Fig. 10.1b). In all organisms containing one or more of these pathways, a trehalose-degrading enzyme is present, namely, **trehalase**, which hydrolyzes trehalose into two glucose molecules (Fig. 10.1b).

A. Trehalose-6-Phosphate Synthase and Phosphatase

The enzymes responsible for trehalose synthesis have been studied in many fungi. Figure 10.2a shows a phylogenetic tree with the trehalose-6-phosphate synthase (Tps1) proteins from different yeast species and filamentous fungi. In *S. cerevisiae* the activities of Tre6P synthase and Tre6P phosphatase copurify (Bell et al. 1992; Cabib and Leloir 1958; Londesborough and Vuorio 1991; Vandercammen et al. 1989). Both activities reside in one complex suggesting that the Tre6P might be channeled, at least in part, inside the complex rather than released into the cytosol. Although in *Escherichia coli* the two activities reside in separate enzymes (Kaasen et al. 1992, 1994), an artificially constructed chimeric fusion enzyme containing the two activities displayed an improved catalytic efficiency (Seo et al. 2000). This suggested that the presence of the two activities in a complex in yeast might provide a benefit for rapid trehalose accumulation capacity.

In support for this theory, an actual prokaryotic trehalose synthase complex has recently been identified in the hyperthermophilic crenarchaeon *Thermoproteus tenax* (Zaparty et al. 2013). Interestingly, in this organism Tre6P synthase and phosphatase are expressed as a single bifunctional protein (TPSP). The N-terminal part contains the TPS and the C-terminal part the TPP domain.

The possible channeling of Tre6P inside the yeast trehalose synthase complex has implications for the discovery that Tre6P acts as an inhibitor of hexokinase (Blazquez et al. 1993).

The purified complex, displaying the two activities, has a total MW of 600–800 kD and appears to consist of three types of subunits with a MW of 130 kD, 100 kD, and 56 kD, respectively. The precise ratio between the different subunits is unknown. In the absence of strong protease inhibitors, subunits with lower MW are obtained. Partial proteolysis also abolishes the strong activation of Tre6P synthase by fructose-6-phosphate. Phosphate

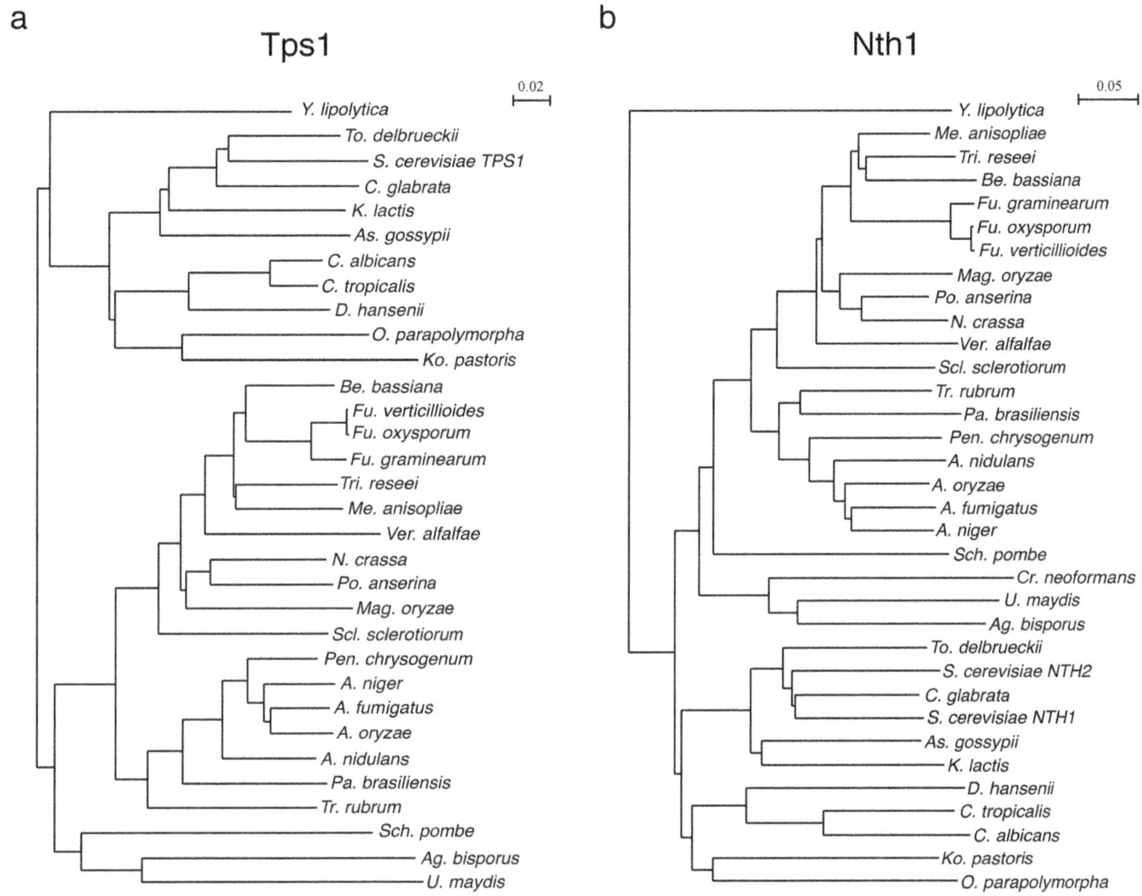

Fig. 10.2 Evolutionary trees for the Tps1 and Nth1 proteins. (**a**) Tps1. Two groups of related Tps1 proteins can be discerned. The first group contains Tps1 from *S. cerevisiae* and most other yeast species, while the other group contains Tps1 proteins from filamentous fungi. The *Y. lipolytica* Tps1 protein displays a special intermediate position between the yeast and the filamentous fungi proteins. (**b**) Nth1. A similar division between yeast and filamentous fungi proteins is observed for Nth1. The *Y. lipolytica* protein again takes a special position between the two groups

acts as an inhibitor of Tre6P synthase activity but stimulates Tre6P phosphatase activity (Londesborough and Vuorio 1991; Vandercammen et al. 1989). Partial proteolysis has been reported to activate both the *S. cerevisiae* (Londesborough and Vuorio 1991) and the *Candida utilis* (Vicente-Soler et al. 1989) enzyme. The group of Ghosh has made significant progress in the purification of active trehalose synthase complex from *S. cerevisiae*, which mainly has to do with their methodology involving the use of polyanions such as heparin in the purification steps (Chaudhuri et al. 2008, 2009). According to their studies, the Tre6P synthase Tps1 is a highly hydrophobic protein, prone to aggregation, whose activity is directly proportional to protein concentration and inversely proportional to ionic strength (Chaudhuri et al. 2008). In terms of substrate specificity, Tps1 showed the greatest affinity for UDPG, as opposed to GDPG or thymidine-DPG (TDPG). Among the glucosyl acceptors, Tps1 showed maximum activity with glucose-6-P, followed by mannose-6-P and fructose-6-P. Metal cofactors, specifically $MnCl_2$ and $ZnCl_2$, acted as stimulators. Enzyme inhibitors such as 5′-sulfosalicylic acid, sodium fluoride, sodium azide, iodoacetic acid, iodoacetamide, or *N*-ethylmaleimide had very little effect on purified Tps1 activity. Temperature and pH optima of the purified enzyme were determined to be 40 °C and pH 8.5, respectively. The enzyme activity was stable between 0 and 40 °C and at alkaline pH (Chaudhuri et al. 2009).

The genes encoding the three subunits have been cloned and sequenced. The *TPS1* gene encodes the smallest subunit of 56 kD which is responsible for Tre6P synthase activity (Bell et al. 1992; Vuorio et al. 1993). The *TPS2* gene encodes the 100 kD subunit which is responsible for Tre6P phosphatase activity (De Virgilio et al. 1993). The *TSL1* and *TPS3* genes encode the largest subunit of 130 kD which appears to confine regulatory properties to the trehalose synthase complex (Bell et al. 1998; Vuorio et al. 1993). Deletion of *TPS1* abolishes Tre6P synthase activity (Bell et al. 1992) and expression of *TPS1* in *E. coli otsA* mutants which are defective in Tre6P synthase activity restores trehalose accumulation (McDougall et al. 1993). Unexpectedly, the *TPS1* (*TSS1*) gene turned out to be identical to the *GGS1*=*FDP1* =*BYP1*=*CIF1* gene. The same gene was cloned independently by complementation of the *fdp1*, *byp1*, and *cif1* mutants which show a specific growth defect on rapidly fermented sugars, like glucose, fructose, mannose, and sucrose (Gonzalez et al. 1992; Hohmann et al. 1992; Van Aelst et al. 1993). This has revealed a regulatory role of trehalose metabolism in the control of glycolysis (see further). The *TPS1* gene displays significant sequence similarity with part of the *TSL1* gene (37 % identity at the amino acid level). Truncation of the *TSL1* gene, resulting in a MW decrease of the protein from 130 kD to 90 kD, causes insensitivity to the allosteric regulators fructose-6-phosphate and phosphate, an effect similar to that of partial proteolysis (Vuorio et al. 1993). Deletion of the *TPS2* gene causes accumulation of Tre6P rather than trehalose and under extreme conditions such accumulation is apparently toxic to the cells. A temperature-sensitive mutant probably defective in Tre6P phosphatase was previously isolated (Piper and Lockheart 1988), but it is unclear whether it is affected in the 100 kD subunit. The *TPS2* gene is considerably longer than the *otsB* gene of *E. coli*, which also encodes Tre6P phosphatase (Kaasen et al. 1992). The extra part, which is not present in *E. coli otsB*, shows sequence similarity to the yeast *TPS1* gene. This difference might be related to the fact that the *E. coli* phosphatase is a separate enzyme while the yeast phosphatase forms part of the trehalose synthase complex. The *TPS3* gene has been identified by systematic sequencing of the yeast genome. The predicted amino acid sequence shows 63 % identity to the amino acid sequence of the *TSL1* gene product over a large stretch of 798 aa. *TPS3* represents a second gene encoding the large 130 kD subunit (Bell et al. 1998). Also the *TPS3* gene has a part with sequence similarity to *TPS1*. Hence, all subunits of the yeast trehalose synthase complex have a part that shows sequence similarity with Tps1. The most obvious explanation is that the homologous parts aid in the formation of the **trehalose synthase complex**. The Tps1 protein is apparently also found in considerable amounts in a form not bound to the trehalose synthase complex (Bell et al. 1998).

Given their sequence similarity, the function of the 130 kD regulatory subunits encoded by *TSL1* and *TPS3* was initially thought to be more or less equivalent in providing stability and regulation of the complex. However, recent research has shown that these two proteins are far from equivalent in their contribution to the activity of the TPS complex. Early investigations already indicated that deletion of these two genes reduces enzymatic activity. The *tsl1*Δ *tps3*Δ double deletion mutant was unable to enhance trehalose synthase activity during diauxic growth while only a modest enhancement was detected in a *tsl1*Δ mutant, as opposed to wild type and *tps3Δ* strains (Ferreira et al. 1997). On the other hand, different reports indicated that although Tps1 activity was found to be reduced in the *tps3*Δ strain, the synthase activity was only weakly affected in *tsl1*Δ cells submitted to heat stress, thus concluding that neither trehalose accumulation nor ability to grow on glucose was significantly affected by deletions of the *TSL1* and *TPS3* genes (Bell et al. 1998; Reinders et al. 1997). This conclusion has now been challenged by the recent results of Trevisol and colleagues (Trevisol et al. 2013). This group has found that after a heat treatment at 40 °C, both *tps3*Δ and *tsl1*Δ mutants show reduced Tps1 activity.

Under their working conditions, Tps1 activity in the *tps3*Δ mutant was significantly lower than in the wild type, while its activity was not induced in the *tsl1*Δ mutant, in which no significant levels of Tre6P nor of trehalose accumulated. These results indicate that, contrary to previous observations, Tps3 and Tsl1 subunits are essential for the normal functioning of the TPS complex under heat shock. One possible reason for the discrepancy of these newer data with the previous work is that Trevisol and coworkers have used glucose as carbon source, whereas galactose was used in the earlier studies. Since the level of catabolic repression determines the state of activation of the TPS complex (Livas et al. 2011; Lutfiyya et al. 1998) and galactose is a less effective repressor than glucose, this might explain the differences observed between the levels of Tps1 activity in cells grown on glucose versus galactose. Moreover, although other groups using stationary-phase cells instead of heat-shocked cells could not detect any alteration in a *tps3*Δ mutant with respect to Tps1 activity, they still reported none or only a weak activation of Tps1 in *tsl1*Δ mutants exposed to these conditions (Bell et al. 1998; Ferreira et al. 1996). This highlights the differential effect and requirement for Tps3 and Tsl1 depending on the stress condition (see further sections). Although in the *tps3*Δ strain, Tps1 activity was strongly reduced upon heat shock, Trevisol and coworkers (2013) could also detect higher relative accumulation of Tre6P in the *tps3*Δ mutant compared to the wild type, suggesting reduction of Tps2 activity also in the *tps3*Δ mutant. The accumulation of Tre6P in the *tps3*Δ mutant is thought to be the cause for the reduced levels of Tps1 in the *tps3*Δ strain because of feedback inhibition of Tps1 by Tre6P. This result also fits with the previous observation that interrupting *TPS2* in a *tps3*Δ mutant results in significant reduction of synthase activity (Reinders et al. 1997; Bell et al. 1998). In this regard, an important role for the Tps3 subunit involving phosphorylation is discussed in a further section of this chapter. Supporting the importance of the regulatory subunits of TPS complex for cell survival, a recent study has provided evidence of a unique role for Tsl1 in "bet-hedging" (Levy et al. 2012). The "bet-hedging" theory is based on the fact that populations of genetically identical unicellular organisms include both fast- and slow-growing cells. The former thrive in optimal conditions, while the latter hedge against future bad times, being capable of surviving stresses where the rest of the population is annihilated. Levy and coworkers have shown that *S. cerevisiae* aged, slow-growing cells resisted stress better than fast-growing cells, due at least in part to their higher content of Tsl1. This places Tsl1 and preservation of the ability to synthesize trehalose as a crucial factor in evolutionary survival within rapidly changing environments.

In *C. utilis*, two UDPG-utilizing Tre6P synthase enzymes have been reported (Vicente-Soler et al. 1989). One of the two enzymes was suggested to be regulated by phosphorylation (see further). In *Aspergillus niger* two genes encoding Tre6P synthase have also been cloned. Disruption of *tpsA* did not influence the capacity of *A. niger* to grow on glucose and had no effect on the germination of conidia. Whereas *AnTPSA* is expressed constitutively during vegetative growth, *AnTPSB* is strongly induced under heat shock conditions (Wolschek and Kubicek 1997). In *Candida albicans* there is only one, or at least one major, Tre6P synthase implicated in the synthesis of trehalose. Disruption of both chromosomal *CaTPS1* copies in this diploid organism caused lack of growth on glucose at certain temperatures, impairment of hypha formation, and decrease of infectivity (Zaragoza et al. 1998). In *Schizosaccharomyces pombe*, one Tre6P encoding gene has been cloned. Disruption of this *tps1*$^+$ gene had no noticeable effect on growth on glucose, but it prevented germination of spores carrying the disruption. Overexpression caused enhanced trehalose levels (Blazquez et al. 1994). Disruption of *TPS1* in the necrotrophic fungal pathogen *Stagonospora nodorum* strongly reduced trehalose accumulation, affecting fitness of the fungus in vitro and in planta as well as its tolerance to different stresses (Lowe et al. 2009). The pycnidiospores also exhibited a reduced germination rate under stress. The *TPS1* gene from the thermophilic yeast *Hansenula polymorpha* (*Pichia angusta*) has been cloned. Although the deletion of this gene did not result in any obvious growth defect on a glucose-containing medium, it still compromised the ability of cells to acquire thermotolerance (Reinders et al. 1999). Similarly, studies of expression of the recently identified and cloned *TPS1* gene of the psychrotolerant yeast *Guehomyces pullulans* indicated progressive upregulation upon temperature increases from 10 to 25 °C, 15 °C being the optimal growth temperature for this yeast (Zhang et al. 2013). The *TPS1* homologue from the distantly related yeast *Kluyveromyces lactis* has also been cloned. The *KlTPS*1 gene is 74 % and 79 % identical at the nucleotide and amino acid level, respectively, to the *S. cerevisiae* counterpart. As in *S. cerevisiae*, inactivation of *KlTPS1* not only causes inability to accumulate trehalose but also a very similar growth defect on rapidly fermented sugars, apparently caused by deregulation of glycolysis, and it also causes defects in certain glucose-induced regulatory events (Luyten et al. 1993). Hence, both *S. cerevisiae* and *K. lactis* require *TPS1* for growth on rapidly fermented sugars. Since then additional examples have been found of species in which *TPS1* is required for growth on fermentable carbon sources. The *tps1Δ* mutant of the pathogenic fungus *Magnaporthe grisea* was unable to grow on glucose or other simple sugars including fructose, sucrose, maltose, and trehalose (Foster et al. 2003). This mutant also failed to synthesize trehalose, sporulated poorly, and was greatly attenuated in pathogenicity. The *Yarrowia lipolytica TPS1* gene has been cloned by complementation of the glucose growth defect of the *S. cerevisiae tps1* mutant (Flores et al. 2011). The *Yltps1* null mutant grew in glucose although the *Y. lipolytica* hexokinase is extremely sensitive to inhibition by Tr6P. The simultaneous presence of a glucokinase, insensitive to Tre6P, which constitutes about 80 % of the glucose-phosphorylating capacity during growth in glucose, may explain the absence of the growth defect on glucose.

Shortly after the identification of the *S. cerevisiae TPS2* gene, the Tre6P phosphatase-encoding *TPS2* genes of *C. albicans* (Van Dijck et al. 2002; Zaragoza et al. 2002) and *S. pombe* (Franco et al. 2000) were also identified and cloned. Their deletion abolishes Tre6P phosphatase activity and causes hyperaccumulation of Tre6P. Other *TPS2* homologous genes have been identified in *Zygosaccharomyces rouxii* (AccNr AAK69413 and AAF80562) and in some other fungi.

B. Trehalose Phosphorylase

The enzyme **trehalose phosphorylase** (EC 2.4.1.64), originally discovered in *Euglena gracilis* (Belocopitow and Marechal 1970) has now also been detected in many fungi. The enzyme functions in the mobilization as well as in the biosynthesis of trehalose and has therefore also been called trehalose synthase. In the basidiomycete *Grifola frondosa*, it catalyzes the synthesis of trehalose from D-glucose and α-D-glucose-1-phosphate with high substrate specificity (Saito et al. 1998). Only trehalose is phosphorylated among various disaccharides and only α-D-glucose-1-phosphate serves as a donor substrate to D-glucose as the acceptor in trehalose synthesis. This is also the case for the *Agaricus bisporus* enzyme (Wannet et al. 1998), while in *Catellatospora ferruginea* other sugars could apparently also serve as sugar acceptors in the biosynthetic reaction (Aisaka et al. 1998). The trehalose phosphorylase from *A. bisporus* has been purified to homogeneity (Wannet et al. 1998). The enzyme is apparently a tetramer of 240 kDa consisting of four identical subunits of 61 kDa. The optimum pH ranges for trehalose degradation and synthesis were 6.0–7.5 and 6.0–7.0, respectively. Trehalose degradation was inhibited by ATP, whereas the synthetic activity was inhibited by P_i. The purified enzyme from *Schizophyllum commune*, on the other hand, was a monomer of 61 kDa; it contains a tightly bound Mg^{2+} ion and did not show evidence of allosteric regulation (Eis et al. 1998, 2001; Eis and Nidetzky 1999). The presence of trehalose phosphorylase activity was also detected in *Pichia fermentans* (Schick et al. 1995), in *Neurospora crassa*, and in three other basidiomycetes: *Pleurotus sajor-caju*, *Flammulina velutipes* (GenBank accession number AAF22230), and *Pleurotus ostreatus* (Kitamoto et al. 1988). The trehalose phosphorylase gene in *N. crassa* is induced during conidiation and is also photoinducible. The gene has also been isolated as a clock-controlled gene (Bell-Pedersen et al. 1996). In *P. fermentans* trehalose phosphorylase expression was maximal at the transition from exponential to stationary phase and was also induced by osmotic stress (Schick et al. 1995), fitting with a function of trehalose phosphorylase in trehalose accumulation upon entry into stationary phase and under osmotic stress. On the other hand, trehalose was rapidly degraded when the carbon source in the medium was exhausted indicating a function of trehalose phosphorylase also in trehalose mobilization for carbon and energy provision. Hence, the trehalose phosphorylase in *P. fermentans* appears to play a crucial role both in trehalose accumulation and degradation. Also, in *A. bisporus* developmental regulation of trehalose levels through regulation of trehalose phosphorylase has been reported. The enzyme apparently plays a crucial role both in trehalose accumulation and mobilization since no trehalase and trehalose synthase activities could be detected in this fungus (Wannet et al. 1998, 1999).

IV. Enzymes of Trehalose Hydrolysis: Trehalases

Most of the **fungal trehalases** are classically divided in two types according to their optimal pH for enzymatic activity and their subcellular localization: a **first type**, mainly cell surface localized, characterized by an **acidic pH optimum** and a high heat stability, and a **second type**, cytosolic, characterized by a **neutral pH optimum** and a low heat stability (Thevelein 1984c). The original acid/neutral classification has, in more recent times, been challenged by the successive isolation of new trehalases from diverse fungi, displaying mixed properties (Lucio-Eterovic et al. 2005). For example, the rice fungus *M. grisea* possesses only one trehalase (encoded by *NTH1*) that has characteristics of both neutral and acid trehalases (Foster et al. 2003). Therefore, it has recently been proposed to classify **fungal trehalases** as **either extracellular** or **cytosolic** (Parrou et al. 2005). As discussed below, a certain level of controversy still exists over the site of action of certain extracellular trehalases, such as Ath1 from *S. cerevisiae*, the reason why the classification acid versus neutral is still being applied.

Acid trehalases show widespread occurrence in fungi (Thevelein 1984c). In most cases their extracellular localization is well established.

The localization of *S. cerevisiae* acid trehalase, however, has for some time remained controversial, although the most recent results point to its main action taking place at the periplasmic space (He et al. 2009). This protein was first identified by Wiemken and coworkers in 1982, as associated to vacuole-enriched fractions obtained by density-gradient centrifugation (Keller et al. 1982). Subsequent work led to the isolation of the *ATH1* gene as encoding vacuolar acid trehalase (Destruelle et al. 1995). Later studies, however, supported a role for Ath1 in the extracellular degradation of trehalose, which was difficult to explain by a strict vacuolar localization of the protein. A vacuolar localization also failed to explain studies where Ath1 was found to copurify with extracellular invertase (Mittenbuhler and Holzer 1988) and to be required for growth on extracellular trehalose as only carbon source (Nwaka et al. 1996). The authors of the latter work tried to integrate their results with the previously described vacuolar localization of Ath1 by hypothesizing the entry of extracellular trehalose in the vacuole either by a mechanism involving endocytosis or by transport to the cytosol followed by vacuolar delivery via autophagy (Nwaka and Holzer 1998). In the meantime another group identified the high-affinity H^+-symporter Agt1 as being responsible for trehalose uptake in *S. cerevisiae* and pointed at the additional existence of a yet unidentified low-affinity mechanism for transport of this disaccharide (Malluta et al. 2000). Further work, however, still placed most of the Ath1 activity at the periplasmic space and uncovered a dual system in yeast for the utilization of extracellular trehalose, the first one dependent on the concerted action of the H^+-trehalose symporter, Agt1, with the cytosolic neutral trehalase Nth1, and the second one dependent on the extracellular localization of Ath1 (Jules et al. 2004). *MAL*-deficient laboratory strains (e.g., BY) exclusively rely on the second mechanism since they fail to express the *MAL*-dependent *AGT1* gene. Although this seemed to settle the discussion on the localization of Ath1 activity, a more recent study provided further evidence that Ath1 is targeted to the vacuole via the multivesicular body pathway (Huang et al. 2007). In this study, an N-terminally GFP-tagged Ath1 was analyzed and described as an N-glycosylated, type II transmembrane protein, whose activity mostly copurified with vacuoles. This study convincingly demonstrated that GFP-Ath1 is strictly vacuolar even in an endocytosis-deficient background. To try to understand the divergence in their own results, the group of Parrou developed C-terminally GFP or Cherry-tagged Ath1 constructs, demonstrating that, although eventually trafficked to the vacuole, Ath1 could also reach the cell surface (He et al. 2009). Moreover, they showed that fusing the N-terminal first 131 amino acids of Ath1 to a truncated version of invertase allowed its periplasmic localization and also the growth of a *suc2*Δ strain on sucrose. Conversely, fusion of the signal peptide of invertase to an N-terminally truncated Ath1 allowed the *ath1*Δ mutant to grow on trehalose, whereas the signal sequence of the vacuolar-targeted Pep4 constrained Ath1 in the vacuole and prevented growth of this mutant on trehalose. Given these last results, it appears legitimate to conclude that Ath1 most likely has a dual vacuole-cell surface localization and that its activity toward extracellular trehalose mostly depends on its localization at the cell periphery.

Studies in other organisms have confirmed the role of acid trehalases in utilization of extracellular trehalose. For example, the cell wall-linked acid trehalase Atc1 from *C. albicans* is required for the ability to grow on trehalose. Its importance for dimorphism and infectivity has also been shown, as its deletion decreases the ability to form pseudohyphae and hyphae (Pedreno et al. 2004, 2007). In contrast to *S. cerevisiae* Ath1, *C. albicans* Atc1 is not as sensitive to pH, since its activity can still be detected in exponentially growing cells at neutral pH (Sanchez-Fresneda et al. 2009). A role in regulating intracellular trehalose accumulation is also indicated by the fact that *atc1* mutants display increased resistance to multiple stresses, including heat shock, salt, and oxidative stress. The dual role of acid trehalases in regulating intracellular trehalose levels is also underscored by the studies of *S. pombe* acid trehalase. Interestingly, although no homologous sequence to acid trehalase genes from other yeasts and fungi could be found in the *S. pombe* genome (Parrou et al. 2005), this enzymatic activity is present at the spore stage of the life cycle of fission yeasts and could be isolated as a soluble protein with kinetic parameters, optimum pH and temperature, thermal denaturation, and salt responses which are closely similar to that of other conventional acid trehalases (Vicente-Soler et al. 2009). This enzyme seems to act as an auxiliary activity complementing that of neutral trehalase in the mobilization of intracellular trehalose required for spore germination (Beltran et al. 2000).

The second category, cytosolic **neutral trehalases**, has been found in many different organisms including *S. cerevisiae* and other fungal species, as the main activity responsible for the rapid changes in trehalose content caused by stimulation of the **protein kinase A (PKA)** pathway upon repletion of a given nutrient to cells previously starved for it (Thevelein 1984c; van der Plaat 1974). The neutral trehalase, Nth1, in *S. cerevisiae* is active in its phosphorylated form and appears to form dimers (Dellamora-Ortiz et al. 1986; Londesborough and Varimo 1984). *S. pombe* neutral trehalase, Ntp1, has similar features, being dimeric and/or trimeric in its active form (Franco et al. 2003; Soto et al. 2002). As described in more detail in a further section, regulation of neutral trehalases by phosphorylation depends on cAMP-PKA (Schepers et al. 2012; Veisova et al. 2012; Wera et al. 1999). In addition, the enzymatic activity of yeast neutral trehalases is also regulated in a Ca^{2+}/Mg^{2+}-dependent manner (App and Holzer 1989; Dellamora-Ortiz et al. 1986; Ortiz et al. 1983; Uno et al. 1983). After identification and sequence comparison of neutral trehalases from multiple organisms, common sequence elements have emerged. Neutral trehalases typically contain a **common C-terminal domain** that likely contains the catalytic core of the enzyme and a more **variable N-terminal regulatory domain** with at least one perfect and several imperfect consensus sites for cAMP-PKA-mediated phosphorylation. In addition, a **Ca^{2+}-binding** site is found in close proximity to the conserved PKA-recognition site (Amaral et al. 1997; d'Enfert et al. 1999; Franco et al. 2003; Kopp et al. 1993; Thevelein 1984c). The existence of this putative EF-like Ca^{2+}-binding motif was first observed in neutral trehalases from *S. cerevisiae* and *K. lactis*, and later this was confirmed in neutral trehalases from other yeasts and filamentous fungi, including *S. pombe*, *C. albicans*, *Aspergillus nidulans*, *N. crassa*, the insect pathogenic fungus *Metarhizium anisopliae*, and the rice blast fungus *M. grisea* (Amaral et al. 1997; d'Enfert et al. 1999; Eck et al. 1997; Foster et al. 2003; Franco et al. 2003; Xia et al. 2002). The presence of aspartic acid residues in specific positions of this motif seems to be essential for the binding of Ca^{2+}, as verified in *S. pombe* neutral trehalase, Ntp1 (Franco et al. 2003).

S. cerevisiae has a paralogous gene, *NTH2*, encoding an additional neutral trehalase with 77 % sequence similarity to Nth1. Given that the major function of Nth1 is to hydrolyze intracellular trehalose both during the normal life cycle and under different stress conditions, the question was raised why yeast expresses a second enzymatic activity (San Miguel and Arguelles 1994). Although its transcriptional regulation is similar to that of *NTH1*, being upregulated upon entry in stationary phase, no effect on trehalose levels or neutral trehalase activity was initially found in *nth2Δ* mutants (Nwaka et al. 1995a, b). Nevertheless, *NTH2* was shown to be upregulated upon heat stress and to have an effect on the recovery from heat shock (Nwaka and Holzer 1998). More recent investigations have highlighted specific situations in which the Nth2 function is particularly apparent, matching changes in its expression level. For example, Nth2 has been shown to play a substantial role in trehalose mobilization in *tps1Δ nth1Δ* mutants growing on trehalose (Jules et al. 2008). Its enzymatic level was also nearly fivefold higher in *nth1Δ* cells cultivated on trehalose compared to the same cells in stationary phase after growth on glucose. In addition, the role of Nth2 in trehalose mobilization during recovery from saline stress has been made apparent in backgrounds where Nth1 and Ath1 are abolished (Garre and Matallana 2009; Hirasawa et al. 2006; Mahmud et al. 2009). More puzzling is the fact that a certain degree of trehalose mobilization can still take place in *nth1Δ nth2Δ ath1Δ* triple mutant cells (Garre and Matallana 2009; Jules et al. 2008). This implies the existence of yet unidentified activities which can, to certain extent, replace the function of these three main trehalases under particular situations. Another interesting fact is that mobilization of intracellular cytosolic trehalose in a *tps1Δ* strain cultivated on extracellular trehalose has also been shown not to be dependent on Nth1 and Nth2 but to require instead extracellular Ath1 (Jules et al. 2008). This involves a mechanism for extracellular export of intracellularly accumulated trehalose.

The enzymatic properties of neutral trehalase from *Candida* are very similar to those of the *S. cerevisiae* trehalase except that the enzyme appears to be a tetramer with a MW of 280,000 (Arguelles and Gacto 1986; Argüelles et al. 1986). In spite of this similarity, researchers have long struggled to find a comparable function for the Ca*NTC1* gene, encoding neutral trehalase, during morphological changes and pathogenicity in *C. albicans* (Eck et al. 1997). The absence of a specific role of Ca*NTC1* in the cellular physiology of *C. albicans* is particularly intriguing. While *Ca*Ntc1 displays high activity in exponential phase regardless of the carbon source (glucose, trehalose, or glycerol), its activity is similar under neutral (pH 7.1) or acid (pH 4.5) conditions, strongly inhibited by ATP, weakly stimulated by divalent cations (Ca^{2+} or Mn^{2+}) and unaffected by the presence of cAMP. Its expression is similarly upregulated by stress-activated signaling pathways. In contrast to its homologue in *S. cerevisiae*, *Ca*Ntc1 activity decreases in stationary phase, except in glycerol-grown cultures (Sanchez-Fresneda et al. 2009). The regulation of neutral trehalases in other yeasts, such as *S. pombe* Ntp1, shows a higher resemblance to that of *Sc*Nth1. For example, activation of *Sp*Ntp1 induced by salt stress or oxidative conditions in glucose-repressed cultures of *S. pombe* is also dependent upon the activity of protein kinases *Sp*Pka1 (homologue of *Sc*PKA) or *Sp*Sck1 (homologue of *Sc*Sch9) (Fernandez et al. 1997b, 1998). Activation of *Sp*Ntp1 by addition of glucose or a nitrogen source to derepressed cultures of *S. pombe* additionally depends on the protein kinase *Sp*Sck1, whose overexpression has been shown to suppress loss of *Sp*Pka1 function (Jin et al. 1995; Soto et al. 1997). A single change in the conserved consensus site for Pka1-dependent phosphorylation (Ser71), as well as point mutations in two other putative phosphorylation sites (Ser6, Ser51), produced inactive *Sp*Ntp1, unresponsive to stress (Franco et al. 2005). Based on the atypical results with *C. albicans* neutral trehalase, identification of trehalase activities in new model organisms, with a different mode of regulation and/or enzymatic characteristics, should not be surprising. Certain organisms, such as *C. utilis*, express an extracellular enzyme with combined invertase and neutral trehalase activities (Lahiri et al. 2012).

Many trehalase-encoding genes have now been cloned and sequenced. Remarkably, there is no significant sequence similarity between neutral and acid trehalases. Figure 10.2b shows a phylogenetic tree with Nth1 proteins from different yeast species and filamentous fungi. The gene encoding the *S. cerevisiae* neutral trehalase shows also sequence similarity to the periplasmic trehalase from *E. coli* (Gutierrez et al. 1989) and to the rabbit small intestinal trehalase (Ruf et al. 1990).

V. Functions of Trehalose

A. Trehalose as Storage Carbohydrate

The accumulation of trehalose during sporulation, its presence in large amounts in spores and resting cells, its slow utilization during the resting stage, and its rapid mobilization during the resumption of growth have always been taken as evidence for a role of trehalose as storage carbohydrate (Elbein 1974; Thevelein 1984c). Subsequently, however, extensive evidence has been obtained that trehalose also functions as a stress protectant. This has led to speculations that trehalose might not act at all as a reserve carbohydrate and only have a function in stress resistance (Van Laere 1989; Wiemken 1990). However, in their argumentation, both authors overlooked a number of important data. One argument in support of their view was that in yeast accumulation and mobilization of trehalose do not match the behavior of a typical reserve carbohydrate as observed in higher eukaryotes. Rather than being accumulated under conditions of excess sugar and mobilized under conditions of shortage, trehalose apparently behaved in the opposite way. However, in yeast glycogen levels generally follow the same pattern as trehalose levels although the fluctuations are less pronounced (François et al. 1987, 1988; Lillie and Pringle 1980).

Apparently, the behavior of a reserve carbohydrate in microorganisms cannot be compared to that in higher multicellular organisms. In the former reserves are accumulated in anticipation of periods of shortage rather than during periods of abundance. Under plentiful conditions, microorganisms prefer to divert all available resources maximally to cell proliferation rather than to accumulation of reserves. Only a surplus of a specific nutrient that cannot be used for proliferation is stored as reserve. This is probably best explained by the fact that the survival of a microbial species is best supported by a maximal rate of proliferation under nutrient-rich conditions rather than maximizing storage compound accumulation, which requires expenditure of energy and nutrients which would reduce proliferation rate. Indeed, accumulation of storage compounds in microorganisms generally increases strongly at the end of the exponential growth phase, apparently in anticipation of the subsequent nutrient starvation period, which can last for a very long time.

Another argument put forward by Van Laere (1989) and Wiemken (1990) was that during germination of *S. cerevisiae* ascospores and spores of other fungi, and also during growth induction in stationary-phase yeast cells resuspended in a glucose-containing nutrient-medium, the supply of glucose from trehalose breakdown appears insignificant compared to glucose uptake from the medium (Barton et al. 1982; Van Laere 1986a). However, these studies were performed under laboratory conditions with spores germinating rapidly in liquid media under optimal conditions, i.e., with a large supply of external sugar. Under natural conditions, i.e., slow germination with limited external nutrients and on solid media, trehalose mobilization might easily turn out to be more important for the supply of energy and carbon to the germinating spore. This has actually been observed in *A. nidulans* where trehalose-deficient spores germinated considerably slower than wild type spores under conditions of carbon limitation (d'Enfert et al. 1999). Also in *S. pombe* spores deletion of neutral trehalase resulted both in slower trehalose mobilization and in a reduced germination rate. Moreover, additional inhibition of the acid trehalase completely blocked germination while this inhibition was without effect in wild type spores (Beltran et al. 2000). These results support that trehalose mobilization is important for fungal spore germination and that the degree varies with the species and with the germination conditions.

Another important aspect not considered by Van Laere (1989) and Wiemken (1990) is that many fungal spores do not need exogenous nutrients for germination and some are even able to germinate in distilled water (Gottlieb 1978). Typical examples are *Neurospora* ascospores (Sussman 1954), *Dictyostelium* spores (Cotter 1975), and spores of *Cunninghamella* (Tereshina et al. 1988). They all consume trehalose, the only storage carbohydrate they contain, during the germination in distilled water. For *Cunninghamella* spores it was reported that they use their trehalose stock at a reduced rate when provided with exogenous glucose and that mutants with a lower trehalose content and reduced trehalase activity showed a slower rate of spore germination (Tereshina et al. 1988). Other examples of spores that probably use their trehalose reserve as energy and carbon source during germination are sporangiospores of *Mucor* and *Pilobolus*. Their germination can be induced by non-metabolizable glucose analogues in media lacking a carbon source (Bourret 1986; Tripp and Paznokas 1982).

It remains intriguing why stationary-phase yeast cells and yeast ascospores mobilize their large trehalose reserve when supplied with plenty exogenous glucose. This behavior is probably best understood by comparison with a completely different biological system displaying a similar behavior, i.e., mobilization of reserves upon supply of exogenous nutrients. Newly hatched chickens contain a yolk sac of which the contents are rapidly used during initial development. However, when the newly hatched chickens are starved for food, they do not use the contents of the yolk sac and it remains present as such (Noy and Sklan 2001; Noy et al. 1996). Why do the chickens use this reserve when they have food and not when they are starved? The answer in this case appears to be that when the chickens are fed they start to develop and the resorption of the yolk sac is part of their developmental program. When they are starved, initiation of the developmental program is postponed and the reserve is not used for the time being. Hence, the presence of food does not trigger directly the use of the reserve but rather the start of a developmental program of which depletion of the reserve is a normal part. A similar conclusion might well be true in the case of germinating fungal spores and during resumption of growth in starved yeast cells. The presence of nutrients does not directly trigger the mobilization of the trehalose but triggers a developmental program, i.e., germination or growth induction, of which the complete usage of the trehalose is only a part. When, for instance, this program is arrested, trehalose mobilization is also arrested. When yeast ascospores or stationary-phase yeast cells are given only glucose and no other nutrients, germination or growth induction is initiated but subsequently quickly arrested and the mobilization of the trehalose reserve is only transient (Thevelein 1984c). After some time the cells rapidly resynthesize trehalose from the glucose in the medium. In this case, the argument put forward by Van Laere (1989) and Wiemken (1990) that the cells only synthesize trehalose when starved for external carbohydrate clearly does not apply. Yeast cells starved for nitrogen or another essential nutrient in the presence of glucose accumulate large amounts of trehalose (Lillie and Pringle 1980).

Yeast cells and other fungi do not only accumulate trehalose in resting cells. Growing cells of many fungal species (Elbein 1974) and also Streptomycetes (Braña et al. 1986) contain trehalose in quantities which appear too low to act as effective stress protectant. In conclusion, it appears very likely that trehalose in addition to its role as stress protectant also serves as a reserve carbohydrate both during growth, for survival during starvation periods and for the stimulation of germination as during the induction of growth in resting cells.

B. Trehalose as Stress Protectant

1. Trehalose as Stress Protectant In Vitro

After the first clear-cut report by Crowe et al. (1984), many papers dealing with in vitro protective effects of trehalose on isolated proteins and membranes have been published. The role of trehalose as stress protectant has been extensively reviewed (Argüelles 2000; Crowe et al. 1992, 2001; Newman et al. 1993; Van Laere 1989; Wiemken 1990). The protective effects of trehalose have been demonstrated during desiccation of biological structures by drying at ambient temperatures (Colaço et al. 1992; Crowe et al. 1990; Roser 1991a).

Because of this remarkable property, trehalose has been used increasingly as cryoprotectant for freeze storage of a variety of intact cells and organisms (e.g., Anchordoguy et al. 1988; Bhandal et al. 1985; Coutinho et al. 1988; De Antoni et al. 1989; Honadel and Killian 1988; Momose et al. 2010) and also for the active preservation in dried form of restriction enzymes (Colaço et al. 1992), antibodies (Blakeley et al. 1990), vaccines (Arya 2000), mammalian blood cells (Wolkers et al. 2001), a wide variety of food products (Kopjar et al. 2013; Roser 1991b), as well as platelet storage and organ preservation (Guibert et al. 2011; Wang et al. 2014b). More recently it was reported that trehalose could be used to inhibit protein aggregation associated with Huntington disease (Tanaka et al. 2004), Alzheimer disease (Liu et al. 2005), and oculopharyngeal muscular dystrophy (Davies et al. 2006). Due to the recent advances in tissue engineering, cell transplantation, and genetic engineering, and the increasing need for efficient long-term storage of mammalian cells, trehalose is often included as xeroprotectant (Chen et al. 2001) and to increase desiccation tolerance (Chakraborty et al. 2012; Kikawada et al. 2007; Nakahara et al. 2010), reviewed by Guibert et al. (2011).

A combination of in vitro and in vivo studies suggests two not mutually exclusive models to explain the protective role of trehalose, most recently reviewed in Wyatt et al. (2013). First, trehalose plays a role in the protection of cells and biomolecules by **replacing water** that is normally bonded to these molecules through hydrogen bonds (Clegg et al. 1982). Second, it is involved in the formation of a **glassy matrix** in the cytoplasm, which protects proteins and membranes (Crowe et al. 1998; Crowe 2002; Wolkers et al. 2002).

Though the exact mechanism is not known, some of the physical properties of trehalose are believed to play an important role in its bioprotective role. One important property is the existence of a number of **polymorphs** in **crystalline as well as amorphous state**, the most common one being trehalose dihydrate. Dihydration of this form is reversible without any loss of integrity of the crystalline structure (Sussich et al. 2001). Together with the high glass transition temperature of trehalose, the highest of all disaccharides, this explanation is proposed as a **first** possible **hypothesis** (see below) (Kilburn et al. 2006). Trehalose solutions have the tendency to undergo **glass** rather than crystal **formation** upon drying, resulting in the establishment of a physical state particularly protective for embedded macromolecules. In nature, sugar-protein glasses have been found to occur in plant seeds (Buitink and Leprince 2004). The glass capsule around the macromolecules would freeze their native shape and in this way also prevent any distortion of their structure during dehydration (Burke 1985; Colaço et al. 1992; Franks et al. 1991; Levine and Slade 1992; Roser 1991a; Slade and Levine

1988). This theory, however, fails to explain the protective effects of trehalose at temperatures much higher than the glass transition temperature (reviewed by Chen et al. (2000). The **second hypothesis** (the "water-replacement hypothesis") proposes that trehalose replaces water molecules that are hydrogen-bonded to the surface of biological macromolecules and that are essential for the maintenance of tertiary structure. Hydrogen bonds of the polar head groups of phospholipids with the multiple hydroxyl groups of trehalose would result in a better stabilizing effect under adverse physical conditions, such as desiccation, heat, and freezing, compared to hydrogen bonds with water molecules (Clegg 1985; Otting et al. 1991; Saenger 1989; Singer and Lindquist 1998a, b). The crucial importance of direct interaction between trehalose and the macromolecules has been emphasized by Crowe et al. (1990) for desiccation resistance as opposed to freezing resistance. Later research, however, showed that sugars are excluded from the dry bilayer, questioning the validity of this theory (Koster et al. 2003; Lenné et al. 2006). Research by Golovina et al. (2009), using molecular-dynamics simulations, on the other hand, supported the theory. Similar conclusions were also obtained by other groups (Hengherr et al. 2011; Sakurai et al. 2008). The main problem with both theories appears to be the superiority of trehalose in conferring stress protection compared to molecules with a very similar structure, such as glucose, maltose, and sucrose (Colaço et al. 1992; Crowe et al. 1990; Green and Angell 1989). However, as pointed out by Roser (1991a, b), trehalose possesses a number of physical properties which together might give it a unique functionality, different from other sugars. These properties include a very high hydrophilicity, nonhygroscopic glass formation, very high chemical stability, nonreducing character, and absence of internal hydrogen bond formation resulting in unusual flexibility of the disaccharide bond. In a review by Crowe and collaborators, various pro's and contra's for the mechanism by which trehalose exerts its protective function have been discussed (Crowe et al. 2001).

2. Trehalose as Stress Protectant In Vivo

One of the main arguments used for a physiological role of trehalose as stress protectant in living organisms is its natural presence in very high amounts in a wide variety of biological structures able to withstand extreme stress conditions, such as desiccation and heat (Crowe et al. 1992; Van Laere 1989; Wiemken 1990). Well-known examples in fungi are survival forms like spores and sclerotia, of which many contain high trehalose levels. A well-known commercial application of the correlation between trehalose content and stress resistance is the production of commercial baker's yeast, where the extremely high trehalose content (15–20 % of the dry weight) is a crucial parameter for both the activity and the stability. This is even more true for "Instant Active Dry Yeast," commercial baker's yeast that not only remains active in dried form but also does not need rehydration prior to mixing with flour (Trivedi and Jacobson 1986). The baker's yeast-producing companies were well aware of the importance of trehalose for the stress resistance of their yeast many years before academia appreciated the importance of trehalose as stress protectant. An early publication by Pollock and Holmstrom (1951) already suggested a possible relationship between trehalose content and resistance against drying of baker's yeast. Suomalainen and Pfäffli (1961) still reported a relatively low trehalose content of 8.5 % of the dry weight, while at present values between 15 and 20 % are common (Cerrutti et al. 2000; Gélinas et al. 1989; Trivedi and Jacobson 1986). These values amount to 1–2 M of trehalose in the cytoplasm. The culture conditions of commercial baker's yeast have been continuously optimized over the years in order to obtain such high trehalose content (Jorgensen et al. 2002).

A strong correlation between trehalose content and stress resistance has been observed in many systems and conditions. It has been known for a long time that stationary-phase yeast cells are more stress-resistant than exponentially growing cells and this correlates with their much higher trehalose content (Elliott and Futcher 1993; Gadd et al. 1987; Iida and Yahara

1984; Parry et al. 1976; Plesset et al. 1987; Schenberg-Frascino and Moustacchi 1972; Walton et al. 1979). Yeast cells growing at high temperature are also known to contain a higher trehalose level (Grba et al. 1979). The group of Wiemken has made use of sublethal heat treatment and subsequent cooling to increase and to reduce again the trehalose content of *S. cerevisiae* cells. They demonstrated a close correlation between trehalose content and resistance against heat shock (a few min at 50 °C) (Hottiger et al. 1987a). Similar results were reported for *S. pombe* (De Virgilio et al. 1990), *N. crassa* (Neves et al. 1991), *C. albicans* (Argüelles 1997), and *Y. lipolytica* (Flores et al. 2011), using the same approach.

Yeast strains with mutations causing directly or indirectly reduced activity of PKA contain more trehalose and are more heat resistant than wild type strains while the opposite is true for strains with mutations causing enhanced activity of PKA (Cameron et al. 1988; Hottiger et al. 1989; Iida 1988; Martegani et al. 1986; Panek et al. 1989; Shin et al. 1987). Attfield et al. (1992) demonstrated a correlation between the trehalose level and resistance against heat and freeze-thaw stress in a series of related yeast strains. Contrary to their claim however, there is no reason to assume that the strains were only different in the trehalose level and not in other properties. The reason for the difference in the trehalose level was not known. It is important to point out that the trehalose levels present in laboratory yeast strains, e.g., in stationary-phase cells or in cells given a sublethal heat treatment, are still only about 3–4 % of the dry weight and therefore much lower than the 15–20 % which is present in commercial baker's yeast. Whereas it is difficult to see how a trehalose concentration of about 1–2 M in the cytoplasm of commercial baker's yeast cells would *not* have an effect on stress resistance, this is not so obvious for the concentrations present in yeast cells grown under normal conditions.

McBride and Ensign (1987) demonstrated a positive correlation between trehalose content and heat resistance in spores of *Streptomyces griseus*. Ascospores of *Talaromyces macrosporus*, a fungus involved in food spoilage, are highly resistant to a variety of stress treatments, including pasteurization. These ascospores contain up to 9–17 % wet weight of trehalose and this sugar is thought to play an important role in their resistance to stress (Dijksterhuis et al. 2002). In *E. coli* the *otsA* and *otsB* genes encoding Tre6P synthase and phosphatase, respectively, are induced during transition into stationary phase and are involved in stationary-phase thermotolerance (Hengge-Aronis et al. 1991). Conidia of *A. nidulans* accumulated trehalose in response to a heat shock or to an oxidative shock. Deletion of the *A. nidulans* Tre6P synthase (*tpsA* gene) results in an increased sensitivity to moderate stress conditions consistent with a role of trehalose in the acquisition of stress tolerance (Fillinger et al. 2001). In *A. niger*, disruption of *tpsA* and the accompanying reduction of the conidial trehalose concentration led to a reduction in the heat stability of the conidia. In contrast to *AnTPSA*, the *AnTPSB* transcript was hardly detectable during vegetative growth of *A. niger* but accumulated strongly during heat shock (Wolschek and Kubicek 1997). The *ScTPS1* (Van Aelst et al. 1993) and *SpTPS1* (Blazquez et al. 1994) genes are also induced by heat shock. In *S. cerevisiae* the enzymes required for trehalose synthesis as well as hydrolysis behave as general stress-responsive proteins. Like heat shock proteins (Hsps), Tps protein expression is induced upon heat shock (Winderickx et al. 1996). Hsp's in yeast are induced by interaction of the Hsf or the Msn2 and Msn4 transcription factors with the heat shock element (HSE) or stress-responsive element (STRE), respectively, in their promoter (Martinez-Pastor et al. 1996). Also in the promoter of the genes encoding the subunits of the trehalose synthase complex, STRE elements have been found. However, the expression of *TPS1* and *TPS2* differs from that of a typical STRE-driven gene. In particular, these genes are less sensitive to downregulation of transcription during growth on glucose (Winderickx et al. 1996). Overexpression of *TPS1* and deletion of trehalase genes have been used to enhance trehalose content in yeast and this was reported to result in an increase in tolerance to several stress conditions (An et al. 2011; Mahmud et al. 2009; Soto et al. 1999; Tan et al. 2014).

3. Trehalose and Thermotolerance

The levels of trehalose increase after heat shock along with the accumulation of Hsps. This increase in trehalose is mainly due to increase in translation of the genes involved in the synthesis of trehalose as well as the substrate required for the synthesis. A recent report using a strain overexpressing *TSL1* and thus having increased TPS activity as well as trehalose content resulted in enhanced carbon flux at increased temperatures (Ge et al. 2013). These results were consistent with the findings that *TSL1* is not only involved in trehalose synthesis but also in glucose signaling (Apweiler et al. 2012). A correlation between heat tolerance and trehalose content has also been reported in bacteria. Recent experiments with the microsymbiont of *Phaseolus vulgaris*, *Rhizobium etli*, showed the relevance of trehalose for tolerance to high temperature (Reina-Bueno et al. 2012). Growth of the trehalose-deficient *otsAch* strain was impaired at high temperatures. Similarly in *E. coli*, thermotolerance has been shown to depend upon the *otsAB* genes for trehalose synthesis (Hengge-Aronis et al. 1991).

The correlation between the level of trehalose and heat stress resistance, however, is not always present. Yeast strains in which the neutral trehalase gene *NTH1* was deleted contained very high levels of trehalose in stationary phase. Upon addition of glucose to such cells, the trehalose level remained high, in contrast to the situation in a wild type strain where trehalose is rapidly mobilized. Heat stress experiments with cells harvested during the start of the fermentation clearly demonstrated that despite the high trehalose content, the cells of the *nth1Δ* mutant became as sensitive to heat stress as wild type cells (Van Dijck et al. 1995). Therefore additional factors seem to be required for conferring stress resistance.

A major problem with nearly all studies showing a correlation between trehalose level and thermotolerance in yeast is that virtually all conditions and mutations used are known to also affect Hsp synthesis (reviewed by Piper 1998; Singer and Lindquist 1998b). Disruption of *TPS1* prevents cells from synthesizing trehalose and severely compromises their thermotolerance (De Virgilio et al. 1994), but it was found that this mutant is also impaired in Hsp synthesis (Hazell et al. 1995). The same decrease in thermotolerance was also observed in the *tps1Δ* mutant of the gray mold fungus *Botrytis cinerea* (Doehlemann et al. 2006b), as well as in the *tpsA* mutants of *A. nidulans* (Fillinger et al. 2001) and *tps1* mutants of the yeast *Y. lipolytica* (Flores et al. 2011). Three hypotheses have been proposed to explain the diminished thermotolerance of *tps1* mutants (Singer and Lindquist 1998b): (1) the Tps1 protein is needed for expression of *HSP* genes (Hazell et al. 1995); (2) trehalose only protects cells against mildly elevated temperatures, allowing the synthesis of Hsps, which enables cells to survive severe heat stress (Nwaka et al. 1994); and (3) trehalose acts directly to prevent protein denaturation at high temperatures, with those proteins that do denature being bound by Hsps and prevented from aggregating (De Virgilio et al. 1994; Hottiger et al. 1994). From the studies with the *tps1* mutant of the yeast *S. pombe*, it was concluded that the importance of Hsp and trehalose synthesis in the acquisition of thermotolerance is strongly dependent on the temperature during the conditioning heat shock. The two adaptative responses have different temperature optima for maximal induction (Ribeiro et al. 1997). This correlation of trehalose involvement and temperature range of heat shock is also discussed in the section on trehalose accumulation during sublethal heat treatment (Sect. VI.B.4). Unexpectedly, addition of canavanine induces Hsps without causing an increase in the trehalose level or thermotolerance (Hottiger et al. 1989). Subsequent more detailed studies, however, revealed that canavanine also increased both the trehalose level and thermotolerance in *S. pombe* and *S. cerevisiae* cells but only after a long lag phase. The increases in trehalose and thermotolerance were closely correlated (Hottiger et al. 1992). Induction of heat resistance without any effect on the trehalose level was observed upon treatment of *S. cerevisiae* cells with a plant cytokinin, confirming that other factors exist, which are able to confer stress resistance (Coote et al. 1992). Recent research indicated that stepwise adaptation enables *S.*

cerevisiae to acquire thermotolerance by induction of both stress-response genes and enhanced intracellular trehalose levels (Satomura et al. 2013).

It was found that trehalose also reduces aggregation of proteins that have already denatured, a function previously thought to be carried out exclusively by Hsps. However, high concentrations of trehalose inhibit the reactivation of denatured proteins in vitro by molecular chaperones (Singer and Lindquist 1998a). In vivo evidence that trehalose breakdown is important for recovery from heat shock has been obtained with yeast strains expressing a series of trehalase alleles that were mutated in evolutionarily conserved putative phosphorylation sites resulting in different levels of activity. Reduction of trehalase activity below a certain threshold level impaired recovery from a sublethal heat shock. This suggests that trehalose breakdown is required for efficient recovery from heat shock (Wera et al. 1999).

In *S. pombe* heat-induced accumulation of trehalose and thermotolerance are insensitive to the presence of cycloheximide, supporting the idea that for acquired thermotolerance during sublethal heat treatment, trehalose accumulation is more important than Hsp synthesis (De Virgilio et al. 1990). However, in an *S. pombe pka1*-disrupted cell, the inhibitory effect of cycloheximide on the accumulation of trehalose was less evident, which implies the existence of **posttranslational activation of** the **Tre6P synthase** complex in addition to de novo synthesis. This mutant exhibited a considerable heat shock resistance as compared to control cells (Fernandez et al. 1997a). In *S. cerevisiae*, on the other hand, inhibition of protein synthesis reduces acquired thermotolerance, but since it also reduces trehalose accumulation, it remains unclear whether the absence of Hsp synthesis or the reduction in trehalose accumulation is responsible (Coote et al. 1992; De Virgilio et al. 1991b; Hall 1983). *De novo* protein synthesis is essential for thermotolerance acquisition in an *S. cerevisiae tps1* mutant. Inhibition of protein synthesis during heat shock totally abolishes acquisition of thermotolerance in the *tps1* and wild type strain, but not trehalose accumulation in the control strain, supporting the conclusion that Hsps rather than (only) trehalose are key factors in heat-induced acquisition of thermotolerance (Gross and Watson 1998). Some of the discrepancies between the conclusions on the protective effect of trehalose and Hsps might be due to the requirement of a minimal concentration of trehalose for its protective effect to be observed and second to the fact that trehalose only prevents stress-induced damage while Hsps might also be able to correct stress-induced damage. Gadd et al. (1987) provided evidence that a certain minimal intracellular trehalose concentration would be required to confer dehydration resistance in yeast. They also showed that it was impossible to increase the desiccation resistance of exponential-phase cells above 10 % survival by addition of high external concentrations of trehalose, whereas in stationary-phase cells, 100 % survival was easily obtained. This supports the idea that trehalose is only effective as a stress protectant in the presence of one or more other factors found in stationary-phase cells and not in exponential-phase cells. This conclusion is in agreement with our data indicating that prevention of glucose-induced trehalose mobilization in stationary-phase cells does not prevent glucose-induced loss of stress resistance (Van Dijck et al. 1995). Other data indicated an approximate rather than a precise correlation between trehalose level and freeze-thaw tolerance (Lewis et al. 1993).

4. Trehalose and Freeze Tolerance

The freeze-thaw tolerance of yeast is dependent on numerous factors, such as physiological conditions, genetic background of the strain, as well as the freezing conditions. Several studies have shown a positive correlation between trehalose content and freeze tolerance of yeast (Gélinas et al. 1989; Hino et al. 1990; Oda et al. 1986). Yeast cells rapidly deplete their trehalose level upon contact with rich media containing fermentable sugars, and this is considered to be the major reason why baker's yeast rapidly loses its freeze tolerance during dough preparation (Oda et al. 1986). Special dough preparation conditions minimizing fermentation activity prior to freezing (e.g., rapid mixing at low temperature) have to be used in order to preserve the vitality and gassing power of baker's yeast for freeze storage of the dough (Kline and Sugihara 1968; Merritt 1960; Sugihara and Kline 1968). The gassing power of frozen doughs made with cells of a diploid homozygous neutral trehalase mutant, an acid trehalase mutant, and the double mutant was greater than the gassing power of doughs made with the baker's yeast parent strain. The mutants deficient in trehalase accumulate higher levels of trehalose and this accumulation correlates with high freeze tolerance, also in frozen-dough baking (Kim et al. 1996; Shima et al. 1999). In a similar way, induction of Tre6P synthase, resulting in elevated intracellular trehalose levels, increases the resistance to freezing in *S. pombe* (Soto et al. 1999). A method has been reported to improve freeze tolerance of bakers' yeast by loading the cells with exogenous trehalose. This enhanced the trehalose content and resulted in higher freeze tolerance in dough (Hirasawa et al. 2001). Recent research indicated that overexpression of the transcription activator Msn2 increases the freeze-thaw tolerance, probably due to activation of the Msn2-regulated response resulting in increased intracellular trehalose levels (Sasano et al. 2012a). Further research by this group has reported that simultaneous accumulation of both proline and trehalose increased the freeze tolerance as well as the fermentation ability after freezing (Sasano et al. 2012b). Also in *Phycomyces* and in *Neurospora* spores, a positive correlation between freeze resistance and trehalose content has been demonstrated (Van Laere 1989). In the psychrotolerant yeast *G. pullulans*, trehalose seems to play a more important role in the adaptation to high temperature rather than to low temperature (Zhang et al. 2013). A genome-wide screening of the *S. cerevisiae* deletion strain collection for genes important for freeze-thaw tolerance revealed overlap with those involved in oxidative stress and cell wall stress (Ando et al. 2007).

5. Trehalose and Osmo- and Dehydration Tolerance

A positive correlation between trehalose content and resistance against osmotic stress (Hounsa et al. 1998; Mackenzie et al. 1988) as well as desiccation resistance has been demonstrated in *S. cerevisiae* (Gadd et al. 1987; Hottiger et al. 1987a; Zikmanis et al. 1985, 1988) and in spores of *Phycomyces* (Van Laere 1986b), *Mucor* (Van Laere and Slegers 1987), *Streptomyces* (Martin et al. 1986; McBride and Ensign 1987), and *Myxococcus* (Kimura et al. 2014). Also in other bacteria, a role of trehalose in the osmostress response has been widely demonstrated, including *Sinorhizobium meliloti*, *Bradyrhizobium japonicum*, and *R. etli* (Dominguez-Ferreras et al. 2009; Suarez et al. 2008; Sugawara et al. 2010). In *S. pombe*, on the other hand, the activity of the trehalase enzyme is upregulated upon osmostress. Though this is difficult to rationalize, it could be important to fine-tune the levels of intracellular trehalose or could provide glucose units for the synthesis of glycerol, a compatible solute during stress (Fernandez et al. 1997b). In *C. albicans*, the trehalose content is increased as a response to oxidative stress, while mainly glycerol is produced in response to osmotic stress (Sanchez-Fresneda et al. 2013), supporting the existence of selective stress-responsive mechanisms controlling the intracellular storage of osmoprotectants.

S. cerevisiae mutants unable to produce trehalose (*tps1*Δ *tps2*Δ and *tps1*Δ *hxk2*Δ) were more sensitive to severe osmotic stress than the isogenic wild type strain. Although initial experiments indicated that hyperaccumulation of trehalose in the *nth1* mutant did not improve survival rates under osmostress compared to the wild type (Hounsa et al. 1998), recent results showed that increased trehalose accumulation, due to deletion of the three trehalase-encoding genes, improved the growth under saline stress (Mahmud et al. 2009). Moreover, a whole-genome expression analysis of *S. cerevisiae* under saline stress revealed upregulation of genes encoding enzymes involved in both glycerol and trehalose synthesis (Garre et al. 2009; Hirasawa et al. 2006). On the other hand, the presence of trehalose apparently allows yeast cells to produce more glycerol upon osmotic shock (Hazell et al. 1997). Significant in this respect is also that many microorganisms synthesize large amounts of trehalose upon osmotic shock (Breedveld et al. 1991; Mackay et al. 1984; Reed et al. 1986; Strøm et al. 1986; Zevenhuizen 1992). In *E. coli* TPS and TPP, encoded by the *otsA* and *otsB* genes, and also periplasmic trehalase, encoded by the *treA* gene, are strongly induced upon osmotic shock and mutants deficient in trehalose synthesis display an osmotically sensitive phenotype (Boos et al. 1987; Giæver et al. 1988; Gutierrez et al. 1989; Kaasen et al. 1992; McDougall et al. 1993; Strom 1998). In more recent research, *E. coli* strains overproducing trehalose have been constructed and this resulted in improved growth in the presence of a variety of osmotic stress agents (Purvis et al. 2005).

Desiccation of *S. cerevisiae* cells is known to trigger mobilization of glycogen and concomitant accumulation of trehalose (Marino et al. 1989; Payen 1949; Pollock and Holmstrom 1951). It was shown that when the trehalose content is greater than 2–3 % of the cell dry weight, the amount of bound water decreases and the viability of the dried cells is enhanced. It was also suggested that a minimal level of trehalose is required to protect yeast cells and that this value corresponds to the amount of sugar needed to replace bound water and to fully cover the proteins and intracellular structures of the yeast cell (Sano et al. 1999). By analyzing the desiccation tolerance of mutants in the trehalose biosynthetic pathway at various growth stages, it was found that, although high intracellular trehalose concentrations can improve the desiccation tolerance, the correlation between the tolerance and the trehalose level is poor, indicating that other, possibly more important, factors besides trehalose must also be involved (Ratnakumar and Tunnacliffe 2006). The desiccation tolerance of exponential cells can be increased by inhibition of Ras-cAMP as well as **target of rapamycin complex 1 (TORC1)** activity (Welch et al. 2013). This regulation is likely occurring via the transcription factors Msn2 and Msn4 as well as Sfp1; the latter is a ribosome biogenesis transcription factor (see also further). It has been found that Hsp12 acts in an analogous manner to trehalose and protects liposomal membrane integrity against desiccation damage. It was suggested that Hsp12 interacts electrostatically with charged groups present on membrane surfaces. Serine and threonine groups on Hsp12 would mimic the action of water by shielding the charge repulsion of any free phosphate groups. It is not clear whether trehalose and Hsp12 have separate or synergistic roles (Sales et al. 2000).

6. Trehalose and Other Stress Conditions

Changes in the level of trehalose also correlate with changes in the **barotolerance** of yeast cells. Using mutant strains lacking the ability to accumulate trehalose and/or Hsp104, both factors were shown to be important for barotolerance, with trehalose playing a major role (Fujii et al. 1996; Iwahashi et al. 1997a, b). In further research this group has also demonstrated a positive correlation between *NTH1* expression and barotolerance (Iwahashi et al. 2000). Apparently, trehalase activity is needed for recovery after high-pressure treatment. Therefore, although the importance of both trehalose and trehalase seems at first sight contradictory, increasing evidence is pointing to a role of trehalose in intrinsic barotolerance and neutral trehalase in recovery from high-pressure stress. One possibility is that trehalase activity indirectly promotes refolding of proteins by disrupting their interaction with trehalose by causing trehalose breakdown. Other stress conditions, such as exposure to **toxic chemicals** (ethanol, copper sulfate, hydrogen peroxide), also induce trehalose accumulation indicating that it may be part of a general stress response in yeast (Attfield 1987). The precise role of trehalose in ethanol tolerance of *S. cerevisiae*, however, remains controversial, mainly due to discrepancies in the results, likely caused in part by the diversity of genetic backgrounds and experimental conditions employed. Some research groups have reported a positive effect (Jung and Park 2005; Kim et al. 1996), while others have reported no impact (Alexandre et al. 1998; Gomes et al. 2002). Recent results using BY4742 wild type and the deletion strains, *tsl1*Δ and *nth1*Δ, showed that trehalose plays a role in ethanol tolerance at lethal, but not at sublethal ethanol concentrations (Bandara et al. 2009). Other studies using genome-wide screens reached similar conclusions. In these cases trehalose metabolism-associated genes were shown to be important for ethanol tolerance at high ethanol concentrations (Kubota et al. 2004), but not during sublethal ethanol stress (Fujita et al. 2006; Kubota et al. 2004). Ethanol concentrations of 10 % (v/v) and higher are known to cause inhibition of cytosolic protein function (Sebollela et al. 2004; Walker 1998). Through its role in protein folding, trehalose could protect enzyme integrity, facilitating cell survival at high ethanol concentrations (Sebollela et al. 2004). On the other hand, results of this group have again highlighted that trehalase activity is necessary for recovery after ethanol-induced stress. Their results indicate that trehalose, although protecting proteins from misfolding during the stress treatment, can also inhibit their function during recovery. This may be the reason why its degradation seems to be important for survival after a stress treatment. Sebollela and colleagues demonstrated that trehalose can inhibit important enzymatic activities such as glutathione reductase. Other groups showed the inhibitory effect of trehalose on other enzyme activities, such as cytosolic pyrophosphatase (Lopes et al. 1999) and glucose 6-phosphate dehydrogenase (Sola-Penna et al. 1997). A putative effect of trehalose in inhibiting glutathione reductase has also been discussed in experiments with the strict aerobic filamentous fungus, *Phycomyces blakesleeanus* (Rua et al. 2014).

Recent research using industrial yeast strains, engineered to accumulate high levels of intracellular trehalose, clearly demonstrated the protective effect of trehalose also during very high gravity fermentations (Wang et al. 2014a). Cells pregrown under increasing NaCl concentrations were more ethanol tolerant than controls, suggesting an overlap between osmotolerance and ethanol endurance in *S. cerevisiae* (Sharma 1997). Ethanol also causes enhanced acetic acid tolerance in *S. cerevisiae*, correlated with the extent of trehalose accumulation (Arneborg et al. 1997). In the brewing process, a strong correlation between ethanol and oxidative stress has been reported. Only in recent years, the possible antioxidant activity of trehalose under these conditions has been evaluated. It has been reported that trehalose protects proteins against oxidative damage caused by exposure to H_2O_2 (Benaroudj et al. 2001). This group proposed this protection to be the result of trehalose acting as a **free radical scavenger**. Oxygen radicals are known to cause damage to the lipids in membranes and in some circumstances, to a greater extent than they do to proteins. Herdeiro et al. (2006) observed that treatments leading to enhanced trehalose accumulation increased both cell survival and cellular antioxidant capacity, depending on the type of oxidative stress. Their results indicated an ability of trehalose to reduce the concentration of intracellular **reactive oxygen species (ROS)**, consistent with its role as a ROS scavenger. Apart from this effect, accumulation of trehalose at both sides of the lipid bilayer of the plasma membrane has also been proposed as being important for the protective mechanism involving trehalose. The importance of trehalose binding to the outside of the plasma membrane was already demonstrated by the increased protection of strains deleted in the trehalose transporter Agt1 (da Costa Morato Nery et al. 2008). Thus, Herdeiro et al. (2006) proposed that trehalose is synthesized internally and then transported to the exterior where it can protect membranes by interaction with the lipid components. In support of this hypothesis, Oku et al. (2003) had already demonstrated that in vitro one trehalose moiety can interact specifically with one cis-double bond of an unsaturated fatty acid, through hydrogen bond formation.

The role of trehalose in stress resistance has also been demonstrated in other yeast species and fungi. For example, $tps1^+$ overexpression in *S. pombe* increases resistance to temperature stress, freezing, dehydration, and the presence of high levels of salt or ethanol. The elevated levels of intracellular trehalose found in the overexpression strain correlated with increased tolerance to multiple stress factors, indicating that trehalose is an important stress tolerance factor in *S. pombe* (Soto et al. 1999). *C. albicans* has been shown to accumulate trehalose in response to oxidative and osmotic stress (Sanchez-Fresneda et al. 2013). *P. blakesleeanus* accumulates glycogen and trehalose in the mycelium and the high levels of trehalose accumulated seem to be part of the stress response under glucose starvation or in the presence of acetate (Rua et al. 2014).

7. Other Stress Protectants

Although there is a correlation between the level of trehalose and stress resistance, this correlation is not always present. Additional factors, depending on the species and the condition, seem to be required for conferring stress resistance.

The highly conserved Hsps, whose expression is induced in response to a wide variety of stresses, appear to have a general role in stress response and in particular the acquisition of thermotolerance (Garrido et al. 2001; Mager and Moradas-Ferreira 1993). Although the term "heat shock protein" is often used as equivalent with "chaperone," a distinction must be made since not all Hsps are chaperones and vice versa (Gong et al. 2009). In yeast, increased expression of Hsps is mediated by a single heat shock transcription factor (Hsf1), which has a DNA-binding domain, three leucine zipper repeats responsible for trimerization of the factor, a carboxy-terminal transactivation domain, and a yeast specific transcriptional activation domain at the amino terminus (Nieto-Sotelo et al. 1999). How exactly these Hsps function in response to stress is generally unclear. Exceptions in this respect appear to be the yeast heat shock protein Hsp104 and the heat shock-induced catalase T, encoded by the *CTT1* gene. Deletion of *CTT1* causes a reduction in thermotolerance possibly because at higher temperatures damage due to oxidative stress is more pronounced than at regular temperatures (Wieser et al. 1991).

More recent research has confirmed the initial finding that Hsp104, a member of the Hsp100/ClpB family of hexameric AAA$^+$-ATPases, has the ability to rescue denatured proteins from aggregates, in concert with Hsp40 and Hsp70 (Glover and Lindquist 1998; Lo Bianco et al. 2008; Perrin et al. 2007). Although Hsp104 is not required for yeast viability, it is well known to play a role in prion propagation and distribution and in inheritance of damaged proteins (reviewed by Grimminger-Marcquardt and Lashuel 2009). An *hsp104Δ* strain is very sensitive to a wide variety of stresses such as high ethanol, heat, freezing, or desiccation and acquires thermotolerance more slowly than the corresponding wild type strain (Sanchez and Lindquist 1990; Sanchez et al. 1992). The same *hsp104Δ* strain has been used by Winkler et al. (1991), who reported that deletion of *HSP104* did not prevent trehalose accumulation during sublethal heat treatment (40 min at 39 °C), while, on the other hand, the cells did not acquire thermotolerance (heat shock at 50 °C). However, they measured trehalose content and thermotolerance only at one time point. In yeast, it was shown that Hsp104 enhances survival by only tenfold when cells are exposed to moderately high temperatures (44 °C) (Sanchez et al. 1992), while at extreme temperatures (50 °C), Hsp104 enhances survival from 100- to 10,000-fold (Parsell et al. 1994), suggesting that the effect of Hsp104 on thermotolerance induction is more important at elevated temperatures. De Virgilio et al. (1991b) measured trehalose levels and resistance against heat stress (a few min at 50 °C) of *hsp104Δ* cells incubated for longer times at a sublethal temperature of 40 °C. They observed both an increase in trehalose and in heat resistance confirming a possible role for trehalose as stress protectant. However, the *hsp104Δ* strain acquired thermotolerance more slowly than the corresponding wild type strain in agreement with the data of the Holzer group (Winkler et al. 1991) and the original report by Sanchez and Lindquist (1990). The *tps1* and *hsp104* null mutants show moderate heat shock sensitivity in stationary-phase cells.

However, the two proteins apparently act synergistically since the double mutant was extremely heat shock sensitive, suggesting that trehalose and Hsp104 somehow cooperate to produce heat shock resistance (Elliott et al. 1996). In addition, the *hsp104* mutant showed lower Tre6P synthase and phosphatase activities after heat shock treatment than the wild type, suggesting that Hsp104 also contributes to trehalose biosynthesis under heat shock conditions (Iwahashi et al. 1998). The importance of Hsps might be related more to repair of stress-induced damage (Yost and Lindquist 1991) rather than physical protection against stress, as in the case of trehalose and other small molecules acting as compatible solutes like glycerol, mannitol, sucrose, etc.

As mentioned above, trehalose is not the only protection metabolite. Depending on the species, compatible solutes, such as the polyols glycerol and mannitol, provide protection to different types of stress. The effectiveness of the solute appears to be species and stress specific. In yeast, for example, the introduction of genes coding for enzymes supporting mannitol and sorbitol production in a glycerol-deficient mutant indicated that neither mannitol nor sorbitol could properly substitute for the protective function of glycerol (Shen et al. 1999).

C. Trehalose as Carbon Source

For many yeast species, extracellular trehalose is a potential carbon source (Barnett 1976). In *S. cerevisiae*, its assimilation can occur by two distinct pathways. In the first one, acid trehalase is secreted as a free protein into the periplasmic space where it cleaves trehalose into glucose, which can then enter the cells via hexose transporters (Jules et al. 2004). This assimilation mechanism could explain the results of Nwaka and coworkers (1996) with *S. cerevisiae* that acid trehalase is required for growth on trehalose, as well as results obtained with other fungi such as *C. albicans* (Ram et al. 1984), *Mucor rouxii* (de Almeida et al. 1997), and *A. nidulans* (d'Enfert and Fontaine 1997). Basu et al. (2006) also reported the utilization of extracellular trehalose by secreted trehalase both in *S. cerevisiae* and *C. utilis*. In the second one, trehalose is transported into the cells by Agt1 after which it is hydrolyzed by the intracellular neutral trehalase Nth1. Recent work has shown that the acid trehalase Atc1 is also required for growth on exogenous trehalose in *Candida parapsilosis* (Sanchez-Fresneda et al. 2014). Also in some higher eukaryotes, trehalose is used as an energy source, for instance, in the hemolymph and thorax muscles of insects, like the cockroach, *Periplaneta americana*, and the long-distance fliers like locusts (Chino et al. 1992; Thompson 2003).

1. Transport of Trehalose

Early work by the group of Panek suggested the possibility of *S. cerevisiae* expressing two types of Tre6P synthases, the first type using UDPG and the second being maltose inducible and using ADPG (Paschoalin et al. 1989). However, the yeast genome does not contain any other *TPS1* homologue besides the *TPS2*, *TSL1*, and *TPS3* genes. This claim originated from the observations that the *fdp1* and *sst1* mutants, later found to be allelic with *TPS1*, were unable to synthesize trehalose under normal conditions and could only accumulate trehalose when incubated with maltose or when a constitutive *MAL* gene (*MAL*^{c}) was present in the strain (Ferreira et al. 1997; Operti et al. 1982). Later on, the group of François demonstrated that accumulation of trehalose in these mutants was not due to a second Tre6P synthase gene but that instead resulted from its uptake by the maltose-inducible disaccharide transporter Agt1, such that *tps1* mutants could use this transporter to accumulate trehalose in rich media containing yeast extract (which contains trehalose) as opposed to minimal media which do not contain trehalose (Plourde-Owobi et al. 1999).

Although a trehalose transport system was early on described in *Trichosporon cutaneum* (Mortberg and Neujahr 1986), few data are available on transport of trehalose in fungi. Also in *S. cerevisiae* evidence for a trehalose/H^+ symport mechanism had been reported earlier (Kotyk and Michaljanicova 1979). It was also shown later that in *S. cerevisiae* trehalose transport is derepressed upon glucose exhaustion while addition of glucose to stationary-phase cells results in loss of trehalose transport, and this requires protein synthesis for reversibility (Crowe et al. 1991). They also investigated trehalose transport in isolated membrane vesicles and obtained in this way new evidence for trehalose/H^+ symport (De Araujo et al. 1991). The kinetic properties of trehalose transport in intact yeast cells were investigated and it was shown that two different transport activities can be detected, a high-affinity H^+-trehalose symport ($K_m = 4$ mM) and a low-affinity transport activity ($K_m >$ 100 mM). The high-affinity H^+-trehalose symport system was repressed by glucose, whereas the low-affinity uptake system was constitutively expressed (Stambuk et al. 1996). The expression of high-affinity H^+-trehalose symport requires *MAL*-activator genes. It is highly expressed under growth conditions in which intracellular trehalose accumulates (Stambuk et al. 1998). It was demonstrated that trehalose transport is essential during the germination process. Germination is a response to the abrupt addition of nutrients, which can cause osmotic stress against which cells generally protect themselves by accumulating trehalose. Most probably, a trehalose permease carries part of the endogenously accumulated trehalose to the outer side of the bilayer, thus protecting the membrane against the stress caused by germination (Cuber et al. 1997). Characterization of a new disaccharide permease, Agt1, showed that this is the carrier responsible for the transport of trehalose in addition to other disaccharides. Trehalose appears to be the best substrate for the Agt1 carrier followed by sucrose (Han et al. 1995; Stambuk et al. 1998, 1999). This carrier is also responsible for the uptake of trehalose during growth in rich medium containing yeast extract in which trehalose is present (Plourde-Owobi et al. 2000). The Agt1 permease can transport trehalose and sucrose ($K_m = \sim 8$ mM) as well as maltose, maltotriose, and α-methylglucoside ($K_m = 20$–35 mM) (Stambuk and de Araujo 2001).

Most recent studies on the Agt1 transporter focused on its ability to transport maltose and maltotriose. This is an important issue for the brewing industry, since incomplete fermentation of maltotriose (which accounts for ca. 15–20 % by mass of the fermentable sugars in wort) is a common problem in brewing fermentations (Alves et al. 2008). Most brewing strains screened harbor the *AGT1* gene (Jespersen et al. 1999; Vidgren et al. 2005). The *AGT1* gene has been found both in ale (*S. cerevisiae*) and lager (*Saccharomyces pastorianus*, *S. carlsbergensis*, and *S. uvarum*) strains (Hammond 1993). However, in most lager strains, it encodes a truncated nonfunctional protein (Nakao et al. 2009; Vidgren et al. 2005, 2010). This difference apparently has to do with the environmental temperature used for the fermentations in which the two types of yeast have evolved. Lager beers are traditionally made at a lower temperature (6–14 °C) than ales (15–25 °C). When their maltose transport at different temperatures was compared, lager and ale strains had similar transport activities at 20 °C, but at 0 °C the lager strains had fivefold greater activity (Vidgren et al. 2010). Besides Agt1, other maltose transporters, such as Mtt1 (see below), are co-expressed under these conditions. It seems that the loss of Agt1 transporters during the evolution of lager strains may have provided plasma membrane space for the Mtt1 transporters that perform better at a low temperature. It has also been shown that on maltose the expression level of the *AGT1* promoter (a bidirectional promoter that co-regulates expression of other transporter genes) is much higher in ale strains than in lager strains and that glucose represses the expression particularly in ale strains (Vidgren et al. 2005). To analyze how *AGT1* expression is regulated, promoter regions up to 1.9 kbp upstream of the *AGT1* gene have been sequenced from three brewer's yeast strains and the laboratory yeast strain CEN.PK-1D. The promoter sequence of the laboratory strain was identical to the *AGT1* promoter sequence of the S288c strain in the *Saccharomyces* Genome Database, whereas the promoter sequences of the industrial strains diverged markedly from that of S288c. Apparently, the promoter of the ale strain was identical to the promoter of the lager strains with exception of one 22 bp deletion and two 94 and 95 bp insertions in the ale strain promoter. Additional Migl and *MAL*-activator-binding sites in the ale promoter may account for differences in expression. The different distribution of Migl-binding sites renders the ale promoter less responsive to derepression (Vidgren et al. 2011). The amino acids Thr505 and Ser557, respectively, located in the transmembrane segment 11 and in the intracellular loop after transmembrane segment 12 in Agt1, are critical for efficient transport of maltotriose in *S. cerevisiae* (Smit et al. 2008). The Ser557 or Thr557 residues are known to be involved in protein phosphorylation that leads to catabolite inactivation of maltose transporters (Brondijk et al. 1998). It would be interesting to gain more insight into the modulation of trehalose/maltose transport of Agt1 and other *MAL* transporters by regulatory kinases such as PKA.

Another gene from *S. carlsbergensis* (PYCC 4457) designated as *MTT1* [mty-like transporter (Dietvorst et al. 2005) or *MTY1* (maltotriose transport in yeast) (Salema-Oom et al. 2005)] encodes a transporter that shares 90 % and 54 % identity with *MAL31* and *AGT1*, respectively. Mty1 has a higher affinity for maltotriose (K_m = ~20 mM) than for maltose (K_m = ~70 mM) and can to some extent also transport trehalose (Salema-Oom et al. 2005). Nakao et al. (2009) have recently sequenced the genome of the lager strain, Weihenstephan 34/70. They found a gene, *LBYG13187*, which they believed to be the *Saccharomyces bayanus* counterpart of the *S. cerevisiae AGT1* gene, because it was the closest homologue, with 79 % identity, to the *AGT1* sequence in the *Saccharomyces* genome database (SGDB, where *AGT1* is referred to as *MAL11*).

Transport of trehalose has also been shown in *C. utilis* (Rolim et al. 2003), where trehalose is transported into the cell by an inducible trehalose transporter (K_m = 8 mM). The activity of this transporter is high in cells growing in media containing trehalose or maltose and very low or absent during growth in glucose or glycerol. Aside from transporting trehalose, acidic and neutral trehalase activities are also present in this organism and increased during growth on trehalose, with neutral trehalase contributing about 70 % of the total activity.

D. Other Functions of Trehalose

The budding yeast is a valuable model for unraveling mechanisms involved in cellular aging, which strongly depends on oxidative stress resistance. Yeast aging can be measured in two ways, replicative, as the number of daughter cells one single mother cell can produce, or chronological, as the time one cell remains viable in a nondividing state or the stationary-phase survival. Both of them can be slowed down by calorie restriction. Research by Goldberg et al. (2009) revealed that yeast cells define long-term viability by designing a diet-specific pattern of metabolism and organelle dynamics. Calorie restriction among others promotes the accumulation of trehalose by stimulating the synthesis of its biosynthetic enzymes and as such shifts carbohydrate metabolism toward glucose formation via gluconeogenesis. The elevation of trehalose during postdiauxic growth protects aggregation of the proteins that were denatured due to ROS exposure. This most likely contributes to the enhanced survival in the following stationary phase. In accordance with this, deletion of *ATH1* or *NTH1* resulted in increased life span, while mutants unable to accumulate trehalose had a shorter life span (Favre et al. 2008; Trevisol et al. 2011). These results indicate that trehalose may be important for maintenance of cell longevity, most likely by preventing oxidative damage to proteins. Moreover, calorie restriction acts by altering the pattern of age-related changes in trehalose concentration (Kyryakov et al. 2012). The beneficial effect of trehalose on longevity is not restricted to yeast. Also in *C. elegans* it extends life span even when administered late in life (Honda et al. 2010). It is therefore possible that the accumulation of carbon sources, a conserved response to starvation stress, maximizes long-term survival and allows their utilization later, during growth resumption.

The function of trehalose appears to be somewhat species dependent. Although it is found in many bacteria, its function in bacteria has not been well studied. In *Mycobacterium* species, trehalose serves as a structural component when incorporated into glycolipids. Trehalose has also been suggested to function as a long-distance translocation carbohydrate in fungi (Cochrane 1958). Evidence for such a role has been reported for hyphae of *Serpula lacrymans* (Brownlee and Jennings 1981) and *A. bisporus* (Hammond and Nichols 1976). Translocation of trehalose from the hyphae of mycorrhizal fungi into the cells of orchid seedlings has also been demonstrated (Smith 1967).

VI. Regulation of Trehalose Metabolism

A. Cellular Signaling Pathways Controlling Trehalose Metabolism

The molecular mechanism of nutrient sensing and signaling has been a very active research field in yeast and other fungi. Because of its conspicuous sensitivity to the presence of all essential nutrients, trehalose metabolism and in particular short-term nutrient-induced activation of trehalase have been favorite readouts for studies on nutrient sensing and signaling in yeast and other fungi (Conrad et al. 2014; Thevelein 1984c, 1994).

1. Glucose-Induced Trehalose Mobilization: cAMP as Second Messenger

Glucose-induced activation of trehalase in *S. cerevisiae* is preceded by a rapid transient increase in the cAMP level both in derepressed vegetative cells (van der Plaat 1974) and in ascospores (Thevelein 1984b). Although the response of the cAMP level to glucose addition varies in different organisms and tissues, a comparable glucose-induced rise in the cAMP level preceding the activation of trehalase was also observed in cells of *S. pombe* (Carrillo et al. 1994a, b), *Pachysolen tannophilus* (Soto et al. 1996), and *C. utilis* (Carrillo et al. 1995) as well as during the initiation of spore germination in *Phycomyces* (Van Mulders and Van Laere 1984), *Mucor* (Dewerchin and Van Laere 1984), and *Pilobolus* (Bourret 1986). Also in *N. crassa,* cAMP is involved in the activation of neutral trehalase at the onset of germination (de Pinho et al. 2001).

In *S. cerevisiae,* adenylate cyclase (AC) is the key enzyme in activation of the cAMP-PKA pathway upon addition of glucose to glucose-derepressed cells or cells grown on a nonfermentable carbon source. The activity of AC, encoded by *CYR1*, is controlled in two branches both under control of a G protein: the G-protein-coupled receptor system (GPCR) composed of Gpr1 and its associated Gα protein, Gpa2, is responsible for extracellular glucose sensing while an intracellular sensing system that is dependent on glucose uptake and hexokinase-mediated phosphorylation results in activation of the Ras proteins in a way that is not well understood (Rolland et al. 2000) (Fig. 10.3). The fact that the extracellular glucose-sensing GPCR system can only activate adenylate cyclase when the latter is made responsive by activation of the Ras proteins made the elucidation of the involved mechanisms of glucose-induced cAMP signaling difficult (Rolland et al. 2000). For details on G proteins and the GPCR system, see Chap. 7. Moreover, the activation of the Ras proteins requires glucose transport and phosphorylation (Colombo et al. 2004), making the extracellular sensing system dependent on intracellular conversion of glucose in metabolism. A possible mechanism of activation is that a glycolytic intermediate functions as allosteric activator of the Ras protein system. The precise mechanism of Ras activation by glucose catabolism remains unknown.

The first indication for a possible role of the Gα protein Gpa2 came from the sequence similarity with its mammalian counterpart, the Gα subunit of the heterotrimeric G protein. Although overexpression of *GPA2* enhanced cAMP levels, a *gpa2Δ* strain still showed a glucose-induced cAMP signal, making its precise function unclear (Nakafuku et al. 1988; Papasavvas et al. 1992). Gpr1 was originally isolated as an interaction partner of Gpa2 in a two-hybrid screen with Gpa2 as bait (Xue et al. 1998) and in a screen for mutants showing delayed glucose-induced loss of heat tolerance (Kraakman et al. 1999). Subsequently, Gpr1 was shown to be a highly specific sensor for glucose and sucrose (Lemaire et al. 2004). The intrinsic GTPase activity of Gpa2 is stimulated by the

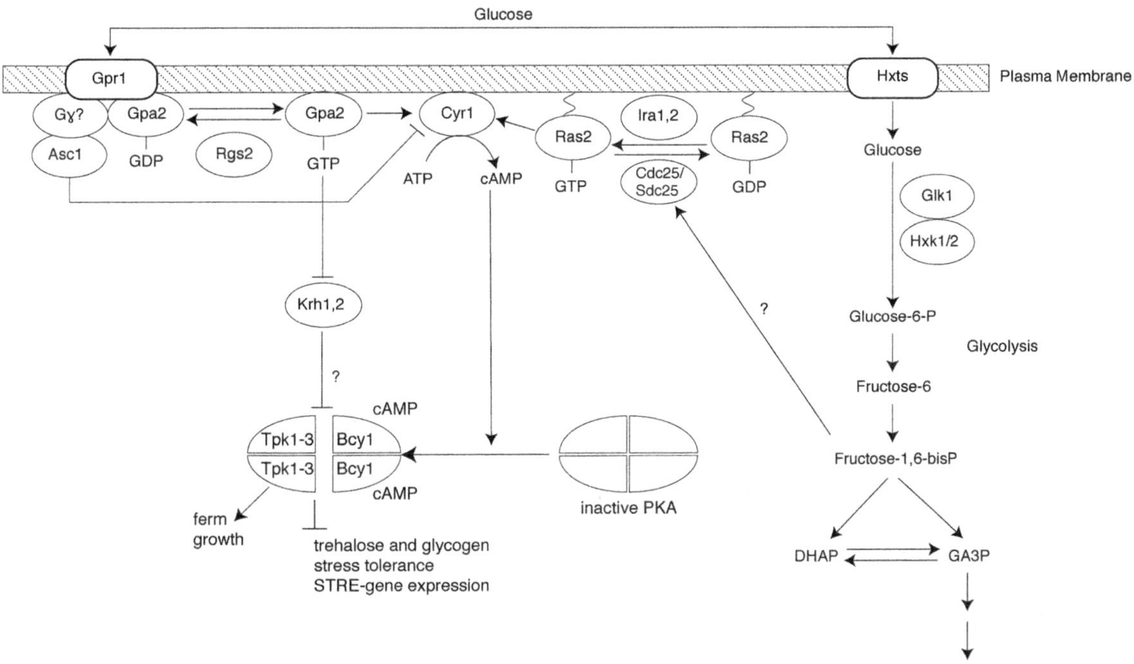

Fig. 10.3 Overview of the glucose signaling pathway controlling trehalose metabolism. Glucose causes activation of adenylate cyclase (Cyrl) via two distinct G proteins, one involved in sensing extracellular glucose and one involved in sensing intracellular glucose. Extracellular glucose is sensed via a GPCR system, consisting of Gprl, the Gα protein Gpa2 bound with Ascl and its intrinsic GTPase-activating protein Rgs2. Activation of adenylate cyclase through the GPCR system is dependent on the intracellular sensing of glucose by a system that depends on active glucose catabolism in glycolysis through the small G protein Ras2. Activation of Gpa2 also triggers an adenylate cyclase bypass pathway that inhibits the kelch repeat proteins Krh1,2 that are in turn negative regulators of PKA activity. Hence, this pathway leads to activation of PKA in a cAMP-independent way. Upon binding of cAMP to the regulatory subunits of PKA, the catalytic subunits are released and affect many downstream targets by phosphorylation

Rgs2 protein, which thus acts as an inhibitor (Versele et al. 1999). Remarkably, Gpa2 is not part of a classical heterotrimeric G-protein complex, but instead associates with the kelch repeat-containing proteins, Krhl and Krh2, initially called Gpb2 and Gpbl (Batlle et al. 2003; Harashima and Heitman 2002; Hoffman 2005a, b; Peeters et al. 2006). Later work, however, showed that these proteins function independently of AC acting directly on PKA as negative regulators (Lu and Hirsch 2005; Peeters et al. 2006). The kelch repeat proteins directly bind to the catalytic subunits of PKA and thereby stimulate association of the catalytic and regulatory subunits, lowering PKA activity (Peeters et al. 2006). Further work has shown that Krhl/2 affect both the abundance and phosphorylation state of Bcyl, such that its levels increase upon glucose limitation in a Krh-dependent manner. PKA establishes a negative feedback loop by phosphorylation of Bcyl in Ser145, which targets Bcyl for degradation unless it is protected by Krh proteins (Budhwar et al. 2010, 2011). Several proposals for alternative subunits have been made, such as Ascl that preferably interacts with Gpa2-GDP, but a clear role for this protein still needs to be defined (Zeller et al. 2007).

The Ras proteins have been implicated in maintaining basal levels of AC activity (Toda et al. 1985). They are capable of binding both GDP and GTP and have intrinsic GTPase activity, which can be stimulated by the GTPase-activating proteins Iral and Ira2 (Tanaka et al. 1989, 1990a, b). GTP loading on the other hand is stimulated by the guanine nucleotide exchange factors Sdc25 and Cdc25 (Boy-Marcotte et al. 1996; Broek et al. 1987; Camonis et al. 1986). Although the presence of C-terminal features that determine tethering to membranes was shown in the Ras proteins many years ago (Kato et al. 1992), only recently localization studies with the Ras proteins as well as the Ira proteins, Cdc25, and some of their downstream effectors, e.g., Cyr1, revealed that they are not only localized at the plasma membrane but are also associated with internal membranes, in compartments such as the endoplasmatic reticulum, mitochondria, and nucleus (Belotti et al. 2011, 2012; Broggi et al. 2013; Dong and Bai 2011). The carbon source and the activity and localization of other components of the cAMP-PKA pathway determine the relative distribution between plasma and internal membranes. Glycolytic enzymes such as Hxk2 also seem to play an important role in the localization of active Ras (Broggi et al. 2013). An additional level of regulation of AC activity is carried out by the molecular chaperone Hsp90 and its cochaperone Sgt1 (Flom et al. 2012). Also in *C. albicans,* Hsp90 represses signaling by PKA and as such affects the yeast to hyphae transition (Shapiro et al. 2009). Recent results indicate a role for the deubiquitinating enzyme, Ubp3, as negative regulator of cAMP-PKA signaling through control of Ira2 ubiquitination (Li and Wang 2013).

The transient character of the glucose-induced cAMP increase in derepressed *S. cerevisiae* cells is accomplished at different levels. A first level is the inhibition of cAMP synthesis by downregulation of AC or its regulatory proteins; a second level is stimulation of cAMP breakdown by the phosphodiesterases, Pde1 and Pde2. Also the regulation of adenylate cyclase as well as the phosphodiesterases occurs at multiple levels: transcription, mRNA stability, and posttranslational regulation and localization. PKA establishes a negative feedback loop by regulating the activity of the Pde proteins. For example, Pde1 is activated by PKA-dependent phosphorylation (Ma et al. 1999) and PKA regulates the localization and protein concentration of Pde2 (Hu et al. 2010). Other targets for the PKA feedback-inhibition mechanism have been proposed, but up to now it has not been possible to mimic the very high cAMP levels observed in yeast strains with attenuated PKA activity by inactivation of one or more phosphorylation sites in such putative target proteins (reviewed by Gancedo 2013; Vandamme et al. 2012).

In addition to activation by glucose, intracellular acidification also stimulates Ras-dependent AC activity (Thevelein 1991). Under starvation conditions, the ATP/ADP ratio in the cell drops, resulting in lower activity of the plasma membrane H^+-ATPase, less proton export, and thus a lower intracellular pH. The rise in cAMP activates PKA and the resulting stimulation of storage carbohydrates catabolism restores ATP levels, leading in turn to a rise in intracellular pH and downregulation of the cAMP-PKA pathway. In this way, the cAMP-PKA pathway may serve to maintain internal energy homeostasis under starvation conditions.

Deletion of *TPS1*, which abolishes production of Tre6P, causes a defect in glucose-induced cAMP synthesis and other glucose-induced regulatory effects (Van Aelst et al. 1993). Later it was shown that additional deletion of *HXK2* in a *tps1Δ* mutant restored growth on glucose as well as glucose-induced signaling but not the trehalose level (Hohmann et al. 1993). This indicated that the *TPS1* gene product might not be directly involved in glucose-induced signaling but that the observed defects were a side effect of the metabolic deregulation of the *tps1* mutants. Although the precise cause of this metabolic deregulation is not yet clear (see further), no new evidence has been obtained implicating the Tps1 protein directly in the glucose-sensing process.

The glucose-induced cAMP signal is only observed in glucose-derepressed wild type cells and in exponential-phase glucose-grown cells of glucose-repression mutants, suggesting the possible occurrence of a glucose-repressible protein in the pathway. An alternative explanation might be that AC itself is glucose repressible to some extent and therefore able to react much faster to glucose addition in derepressed cells compared to repressed cells. Such a difference between repressed and derepressed cells was also observed in *S. pombe* (Carrillo et al. 1992, 1994a) and *C. utilis* (Carrillo et al. 1995) for glucose-induced activation of trehalase. In

S. pombe it was apparently also due to a different responsiveness of adenylate cyclase, since exogenous cAMP was able to trigger trehalase activation in both cell types.

In *S. pombe,* a similar glucose-sensing system is responsible for activation of the AC pathway. It consists of the Git3 G-protein-coupled receptor and the Gpa2 Gα protein (Nocero et al. 1994; Welton and Hoffman 2000). As opposed to *Sc*Gpa2, *Sp*Gpa2 was shown to interact with classical Gβ (Git5) and Gγ (Git11) proteins (Landry and Hoffman 2001). The Ras protein of *S. pombe* does not act on AC (Engelberg et al. 1990; Fukui et al. 1986) and is not required for glucose activation of cAMP synthesis or trehalase (Soto et al. 1995a). As in *S. cerevisiae* (Griffioen et al. 2000), the cytoplasm-nucleus redistribution of the catalytic and regulatory subunits of PKA is regulated by cAMP and triggered by growth on glucose (Matsuo et al. 2008). Recent results indicate a direct role for Hsp90 in the *S. pombe* glucose-induced cAMP pathway (Alaamery and Hoffman 2008). Resting cells, but not derepressed cells, of the *S. pombe pka1*-deficient mutant were deficient in glucose-induced activation of trehalase indicating the existence of an alternative PKA-independent pathway for trehalase activation (Soto et al. 1995a). The alternative protein kinase was later identified as Sck1, the homologue of the yeast Sch9 protein kinase (Soto et al. 1997). Both overexpression of *SCH9* in *S. cerevisiae* (Toda et al. 1988) and *SCK1* in *S. pombe* (Jin et al. 1995) can suppress the phenotypes of a PKA-deficient mutant. The *Sp*Sck1-dependent, cAMP-independent pathway is the main signaling pathway controlling trehalase activation under derepression conditions. Derepressed cells of a *Sp*Sck1-deficient strain did not activate trehalase in the presence of glucose. However, in repressed cells, *Sp*Pka1 was able to activate trehalase in the absence of *Sp*Sck1 function. These results suggested that *Sp*Pka1 is required for trehalase activation during repression, whereas *Sp*Sck1 is needed for trehalase activation under derepression conditions (Soto et al. 1997). Recent results indicate a dual role for *Sp*Sck1 as a negative regulator of Gpa2 not only in glucose but in general nutrient sensing, acting in parallel with PKA (Mudge et al. 2014). Recent research with *Neurospora crassa* has described that the *Nc*GPR-4 protein (a GPCR) interacts with Gα GNA-1 to regulate carbon-source-dependent growth and development (Freitas et al. 2010). Just as the cAMP-PKA pathway is a key regulator of filamentation in *S. cerevisiae*, PKA is an important contributor to cell shape transitions in diverse plant and human fungal pathogens. In *C. albicans* it was recently shown that the receptor, *Ca*Gpr1, is a regulator of trehalose metabolism and as such of Hsp90-mediated cell elongation. This phenomenon is glucose dependent although *Ca*Gpr1 has no role in rapid glucose signaling to adenylate cyclase for cAMP production (Maidan et al. 2005; Serneels et al. 2012).

2. Fermentable-Growth-Medium-Induced Trehalose Mobilization: Nutrient Sensing by Transceptors

Many years ago, Van der Plaat (1974) demonstrated that glucose-induced mobilization of trehalose was transient and that the presence of nitrogen in the medium sustained trehalose mobilization for a much longer time. Moreover, during diauxic shift, the cells start to accumulate trehalose when the glucose levels drop (Panek and Mattoon 1977; Polakis and Bartley 1966; Quain and Haslam 1979; Suomalainen and Pfäffli 1961) and starvation of yeast cells for any essential nutrient, in the presence of glucose, results in growth arrest and trehalose accumulation (Lillie and Pringle 1980). These early results indicated that the combined presence of a fermentable carbon source and a complete growth medium in some way controlled trehalose levels. However, how and why was not clear for many years. Much later, in 1992, it was shown that addition of nitrogen, phosphate, or sulfate to cells starved for that specific nutrient each triggered rapid activation of trehalase (Hirimburegama et al. 1992). This pointed to the existence of specific nutrient-sensing systems for activation of trehalase and let to the concept of a "fermentable-growth-medium (FGM)-induced pathway" that controls trehalase and apparently also other PKA targets during the growth of yeast cells on rap-

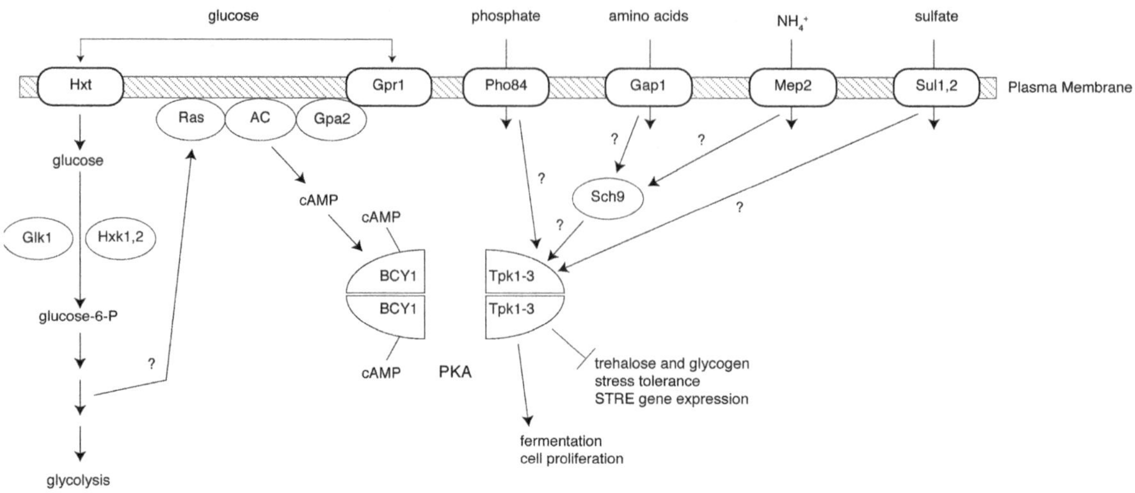

Fig. 10.4 Overview of the fermentable-growth-medium-induced signaling pathway controlling trehalose metabolism. Addition of nitrogen, phosphate, or sulfate to cells starved for nitrogen, phosphate, or sulfate, respectively, triggers rapid activation of PKA targets. This response requires the detection of glucose by either the GPCR system (Gpr1–Gpa2) or intracellular sensing via the Ras proteins. Their action leads to activation of adenylate cyclase (AC), increase in cAMP, and (partial) liberation of Bcy1 regulatory subunits from Tpk catalytic subunits. The sensors for activation of the pathway by the other nutrients are active transporters and acting as receptors or "transceptors" for the specific nutrients: Gap1 for amino acids, Pho84 for phosphate, Mep2 for ammonium, and Sul1,2 for sulfate. The protein kinase Sch9 is only required for nitrogen activation (amino acids and ammonium). Transceptor signaling is supposed to further activate catalytic subunits of PKA that have been liberated from Bcy1 regulatory subunits after binding of cAMP to the latter. Active Tpk1–3 catalytic subunits downregulate, e.g., trehalose and glycogen levels, stress tolerance, and STRE-controlled gene expression, and stimulate, e.g., fermentation and cell proliferation

idly fermentable sugars like glucose (Thevelein 1994), most recently reviewed in Rubio-Texeira et al. (2010) and Schothorst et al. (2013).

In contrast to glucose activation in derepressed yeast cells, addition of nitrogen, phosphate, or sulfate to cells deprived for that specific nutrient does not cause a cAMP spike although it requires the presence of glucose (Hirimburegama et al. 1992). This nutrient activation is still observed in the absence of the regulatory subunit of PKA, providing evidence for additional regulatory mechanisms acting on the free catalytic subunits of PKA (Durnez et al. 1994; Giots et al. 2003) (Fig. 10.4). Although the presence of glucose is detected by the same dual glucose-sensing system activating the cAMP-PKA pathway, in this case both systems can act independently of each other (Donaton et al. 2003; Giots et al. 2003; Rolland et al. 2000). Investigation of the nutrient-sensing mechanism involved in this rapid "FGM signaling" phenomenon has led to the identification of transporters that display a receptor function for activation of the PKA pathway by their nutrient substrate in addition to their established transport function. We have proposed the name of "transceptors" for proteins with a double function as transporter and receptor (Holsbeeks et al. 2004; Thevelein and Voordeckers 2009).

a) Nitrogen-Induced Trehalose Mobilization

Investigation of the nitrogen- induced activation of PKA targets, such as trehalase in *S. cerevisiae* nitrogen-starved cells, revealed that the general amino acid permease Gap1 acts as an amino acid receptor for activation of the FGM pathway (Donaton et al. 2003). Similar work indicated that Mep2 and Mep1 function as transceptors for ammonium activation of the FGM pathway in nitrogen-starved cells (Van Nuland et al. 2006). In both cases, it has been shown that neither transport nor metabolism of the substrate is required for the signaling function. Moreover, the identification of a nontransported agonist of the signaling function has provided a strong new argument for the signaling function of the transceptors and has also shown that transport through the transceptor is not required to trigger signaling (Van Zeebroeck et al. 2009). More recent work has shown that Gap1-mediated transport of particular dipeptides leads to persistent activation of the PKA target trehalase and that at least part of this persistent signal is transduced by Gap1 after it has been internalized in endosomes (Rubio-Texeira et al. 2012). So far Gap1 is the first nutrient transceptor shown to maintain signaling capacity after endocytosis. In a separate work, evidence has been gathered of transported substrates that elicit Gap1 endocytosis without signaling (Van Zeebroeck et al. 2014). These amino acid substrates (L-histidine and L-lysine) were also unable to sustain growth as only nitrogen source, so in principle it could be argued that signaling is coupled to the quality of the nitrogen source. However, in the same work it was also shown that different transported but non-metabolizable amino acid analogues can elicit signaling. Thus, it seems that different substrates have different abilities, inherent in their molecular structure, to trigger endocytosis and signaling and that the two phenomena are not necessarily interconnected. We also discovered that a nontransported dipeptide which previously showed competitive inhibition of L-citrulline transport, L-Asp-γ-Phe (Van Zeebroeck et al. 2009), can trigger ubiquitination of Gap1 without triggering its endocytosis. Since this dipeptide cannot enter the cells, this result shows that interaction with certain substrate analogues can trigger ubiquitination of the transceptor without triggering its endocytosis. In summary, we found conditions in which we could separate ubiquitination from transport, endocytosis, and signaling. This suggests that the process of signaling is dependent on the conformational change that a particular substrate can elicit in the transceptor, regardless of its transport and its quality as nutrient source.

Although little is known on how Gap1 or Mep2 connects to downstream signaling mediators, requirement of the Sch9 protein kinase for amino acid and ammonium activation in nitrogen-starved cells has been demonstrated (Crauwels et al. 1997). Stimulation of trehalase activity by nitrogen sources was also observed during yeast ascospore germination (Thevelein et al. 1982). Remarkably, in this case activation of trehalase by a nitrogen source is associated with a clear spike in the cAMP level (Thevelein 1984b). The mechanism involved has not been investigated further. Activation of trehalase by nitrogen sources was also reported for other yeasts. In *C. utilis* nitrogen activation of trehalase is not associated with an increase in cAMP and evidence was obtained that cAMP is not needed for this response, as opposed to glucose activation (Carrillo et al. 1995). Addition of a nitrogen source to *S. pombe* or *P. tannophilus* cells deprived of nitrogen triggers a distinct cAMP signal preceding the activation of trehalase (Soto et al. 1995b). For nitrogen-induced activation of trehalase, a similar situation was found as for glucose activation. In derepressed cells, the *Sp*Sck1 protein kinase was the main mediator of trehalase activation, while in repressed cells, PKA was responsible for trehalase activation (Soto et al. 1997). Hence, it appears that in different yeast species slightly different mechanisms have evolved for glucose and nitrogen regulation of trehalase.

b) Phosphate-Induced Trehalose Mobilization

As in the case of nitrogen addition to nitrogen-starved cells, addition of phosphate to phosphate-starved cells triggers rapid activation of trehalase. The presence of glucose is required and the glucose can be detected by any one of the two glucose-sensing mechanisms involved in stimulation of cAMP synthesis by glucose. Phosphate signaling was not affected by deletion of *GPR1* (encoding the GPCR glucose sensor) or *HXK1*, *HXK2*, and/or *GLK1* (encoding the three glucose-phosphorylating glycolytic enzymes). However, deletion of both systems together abolishes phosphate signaling (Giots et al. 2003). As in the case of nitrogen sources, phosphate addition does not trigger a distinct cAMP signal as is observed after glucose addition to glucose-deprived cells. In contradiction with the nitrogen addition, however, absence of Sch9 does not prevent phosphate-induced activation, indicating that this is not a general requirement for transceptor signaling (Giots et al. 2003). *S. cerevisiae* cells contain five genes encoding phosphate carriers: Pho84, Pho87, Pho89, Pho90, and Pho91. Deletion of all five genes abolishes phosphate transport and strains with strong overexpression of only one phosphate carrier have been constructed (Wykoff and O'Shea 2001). In phosphate-starved cells of such strains, phosphate transport and consumption from the medium are highest with Pho90 and Pho91, but phosphate-induced trehalase activation is apparently not sustained by these carriers. On the other hand, phosphate transport and consumption with Pho84 and Pho87 are much slower but both carriers are able to sustain rapid activation of trehalase upon addition of phosphate (Giots et al. 2003). Detailed research on phosphate-induced activation of trehalase revealed that Pho84 and Pho87 function as specific phosphate transceptors for activation of trehalase in phosphate-starved cells. Discovery of the nontransported agonist glycerol-3-phosphate (Gly3P) has demonstrated that binding of this compound into the phosphate-binding site of Pho84 was enough to trigger signaling (Popova et al. 2010). Although it is transported by two other carriers, Git1 and Pho91, Gly3P is a competitive inhibitor of transport through Pho84, indicating direct interaction with its phosphate-binding site. Another compound, phosphonoacetic acid, was identified as a competitive inhibitor of transport without agonist function for signaling. This indicates that mere binding to the substrate-binding site is apparently not enough for signaling and that only certain molecules are able to elicit the additional conformational change needed to trigger this function.

Further evidence that the signaling function can be separated from the transport function for the transceptor Pho84 was recently obtained by Samyn et al. (2012). Using a three-dimensional model of Pho84 created by analogy to the GlpT permease, single mutations were introduced in residues potentially involved in either phosphate or proton binding, respectively, during transport. In this way it was found that Asp358 and Lys492 (in transmembrane domains 7 and 11) are critical for the transport function and may be part of the putative substrate-binding pocket of Pho84. Moreover, alleles mutated in the putative proton-binding site Asp358 were still capable of strongly activating PKA pathway targets, despite their severely reduced transport activity. This evidence indicates again that signaling does not require transport and suggests that mutagenesis of amino acid residues involved in binding of the co-transported ion may constitute a promising general approach to separate the transport and signaling functions in transceptors. This idea has recently been discussed in detail using this and other examples (Schothorst et al. 2013). Additional work has also shown that Pho84 endocytosis is delayed in strains with reduced PKA activity, suggesting that Pho84-mediated activation of the PKA pathway may play a role in its downregulation by phosphorylation and ubiquitination followed by internalization and vacuolar degradation (Lundh et al. 2009).

c) Sulfate-Induced Trehalose Mobilization

The first indication for sulfate-induced activation of PKA upon re-addition to sulfate-starved cells was obtained many years ago (Hirimburegama et al. 1992). As was the case for nitrogen- and phosphate-induced activation, sulfate-induced activation also occurs independent of a cAMP increase and is dependent on the presence of glucose in the medium. More recent work in our laboratory has uncovered the existence of two sulfate transceptors in *S. cerevisiae*, the sulfate transporters, Sul1 and Sul2 (Kankipati et al., manuscript in submission).

d) Induction of Trehalose Mobilization by Other Nutrients

The maintenance of a **"high-PKA phenotype"** in *S. cerevisiae*, which includes high trehalase activity and low trehalose levels, depends on the combined presence of a sufficiently high level of a rapidly fermented sugar in the medium as well as all other nutrients required for growth. When the concentration of the fermentable sugar drops or when an essential nutrient runs out, the cells will switch to the **"low-PKA phenotype."** Trehalase activity will drop and the cells start to accumulate trehalose (Lillie and

Pringle 1980; San Miguel and Arguelles 1994). Apparently, the cells integrate information from different nutrient-sensing systems to activate a pathway that controls the targets of PKA, which we have called the **FGM pathway** (Thevelein 1994). However, for rapid activation of the PKA targets in cells deprived of a fermentable sugar or another essential nutrient, the requirements are less strict. For glucose activation of the cAMP-PKA pathway in glucose-deprived cells, the presence of glucose is enough to trigger the cAMP signal and the effects on the PKA targets. In nitrogen-, phosphate-, or sulfate-starved cells, the presence of glucose is essential for the rapid activation of trehalase by re-addition of a nitrogen, phosphate, or sulfur source, respectively. In both cases the glucose can be detected by the Gpr1 GPCR or by the glucose-phosphorylating enzymes (Donaton et al. 2003; Giots et al. 2003). In all cases the effects of these nutrients on the PKA targets are transient if the cells are not in a medium that sustains cell growth. A similar situation pertains for yeast ascospore germination where a fermentable sugar and a complete medium sustaining germination are required for maintenance of the "**high-PKA phenotype**" (Thevelein and Jones 1983). Addition of only glucose to stationary-phase yeast cells or to yeast ascospores causes only a transient mobilization of trehalose followed by resynthesis. The trehalose pattern closely correlates with the proliferation behavior since stationary-phase cells rapidly arrest again in G0 in the absence of essential nutrients like nitrogen, phosphate, or sulfate. Ascospores also need a full medium for complete germination. The course of trehalase activity closely follows the changes in the trehalose level. When only glucose is given, both trehalase activation and trehalose mobilization are transient. When a full medium with a fermentable carbon source is provided, trehalase remains activated for a much longer time and trehalose is completely degraded (Thevelein 1984c). A further argument for the close connection between trehalase activity and growth rate is the observation that growth stimulation by addition of a good nitrogen source to cells growing on a poor nitrogen source also triggers rapid activation of trehalase (Hirimburegama et al. 1992).

B. Posttranslational and Transcriptional Control of Trehalose Biosynthesis and Degradation

In *S. cerevisiae* there is a strong correlation between growth rate and trehalose content. Yeast cells growing on fermentable sugars like glucose and fructose have the highest growth rate and the lowest trehalose level, stationary-phase cells have the highest trehalose level, while cells growing on nonfermentable carbon sources have a low growth rate and an intermediate trehalose level (Thevelein 1984c). Yeast cells incubated with a fermentable carbon source such as glucose and starved for another essential nutrient, such as nitrogen accumulate large amounts of trehalose (Kuenzi and Fiechter 1972; Lillie and Pringle 1980). In the absence of an external sugar, conversion of glycogen into trehalose can be induced under specific conditions: desiccation (Marino et al. 1989; Payen 1949) and formation of ascospores (Roth 1970).

1. Posttranslational Regulation of Trehalose Biosynthesis

One possible explanation for the differences in the trehalose level observed during the growth cycle is control of the enzymes of trehalose metabolism at the posttranslational level: activation of trehalase and inactivation of Tre6P synthase by cAMP-dependent protein phosphorylation. Whereas control of trehalase by cAMP-dependent protein phosphorylation is well established, there has been controversy whether Tre6P synthase in *S. cerevisiae* is regulated by phosphorylation. Initial claims that the synthase is reversibly inactivated by phosphorylation (Panek et al. 1987) have been contradicted afterward (Vandercammen et al. 1989). On the other hand, the initial observation that yeast mutants with increased or decreased activity of cAMP-PKA had lower and higher activity of Tre6P synthase, respectively (Panek et al. 1987), has been confirmed (Francois et al. 1991). Moreover, Panek et al. (1987) also reported glucose-induced inactivation of Tre6P synthase. This effect was studied in more detail by Francois et al. (1991)) who showed that glucose-induced inactivation was dependent on the presence of a nitrogen source in a similar way as previously reported for activation of trehalase (Thevelein and Beullens 1985). Probably, glucose-induced inactivation of Tre6P synthase is triggered by the same FGM-induced pathway, which controls trehalase activity (Thevelein 1991). Based on reversibility studies and experiments with cycloheximide, Francois et al. (1991) concluded that glucose-induced inactivation of Tre6P synthase was probably due to proteolytic inactivation and that the effect of high protein kinase activity was probably exerted at the transcriptional level rather than by posttranslational modification as suggested by Panek et al. (1987). The latter conclusion fits with data showing that expression of the *TPS1* gene is repressed in mutants with enhanced activity of PKA and derepressed in mutants with reduced activity of PKA (Winderickx et al. 1996). The group of Panek has re-investigated Tre6P synthase activity using different assay methods and confirmed the validity of the assay, the differences in activity in the cAMP-PKA pathway mutants, the increase in activity upon entry into stationary phase, and the subsequent inactivation by re-addition of glucose to the cells (Panek et al. 1990). On the other hand, the previously reported results on in vitro activation of Tre6P synthase by dephosphorylation and in vitro inhibition of the enzyme by phosphorylation have never been confirmed. Because of the sensitivity of the complex to proteolysis and the activation of synthase activity by partial proteolysis (Londesborough and Vuorio 1991), results obtained in vitro with this enzyme have to be regarded with much caution. It appears safe to conclude at present that most of the differences observed in Tre6P synthase activity are due to regulation at the transcriptional level. Although the evidence is still preliminary, the rapid drop in activity observed after addition of glucose is probably due to proteolysis but a mechanism involving phosphorylation followed by proteolysis cannot be excluded. In general, involvement of phosphorylation in regulation of the enzyme remains uncertain. Another candidate for direct control of TPS activity was the Rim15 protein kinase, which interacts with Tps1 in a yeast two-hybrid screen (Reinders et al. 1998). However, no role for Rim15 in direct control of TPS activity could be demonstrated.

More recently, in silico and proteomic analyses, however, have indicated the existence of putative phosphorylation sites in both Tps3 and Tsl1, in particular four cAMP-PKA motifs in Tps3 and two in Tsl1, while Tps1 and Tps2 do not contain PKA motifs (Sadowski et al. 2013); PhosphoGRID database http://www.phosphogrid.org/). These observations strongly suggest that phosphorylation events may occur on the regulatory subunits of the TPS complex and not on the catalytic subunits. Based on this fact and on their results (see comments in Sect. III.B), Trevisol et al. (2013) propose that PKA-dependent phosphorylation of the Tps3 subunit happens during stress recovery. This phosphorylation would result in inhibition of Tps2 activity, resulting in Tre6P accumulation, which would in turn lead to Tps1 inhibition. This way, Tps3 would play a major role in maintaining Tps2 active during stress and in downregulating the activity of the TPS complex via Tps2, upon return to unstressed conditions. Tsl1 would, on the other hand, be required to maintain the complex properly activated during stress. Its function is proposed to be structural, being necessary for proper formation and integrity of the TPS complex under stressful conditions. Further work on the putative phosphorylation of Tps3 and Tsl1 is thus expected to improve our understanding of the regulation of trehalose synthesis in response to stress.

Additional layers of posttranslational regulation have emerged from recent studies by the group of Ghosh and coworkers. They have provided evidence that trehalose metabolism is affected by methylation, as inferred from their studies using the potent methylation inhibitor oxidized adenosine (AdOx) and the universal methyl group donor S-adenosyl-L-methionine (SAM) (Bhattacharyya et al. 2005). In addition, they have also shown that purified TPS was stimulated in the presence of thiol modifiers like iodoacetic acid and iodoacetamide but not with N-ethylmaleimide, suggesting carboxymethylation of cysteine residues in TPS (Chaudhuri et al. 2009). In a further work, they have purified the TPS complex from SAM-

or AdOx-treated versus non-treated control cells (Sengupta et al. 2011). Differences in mobility of methylated, methylation-inhibited, and control TPS in acidic native gel electrophoresis as well as MALDI-TOF analysis confirmed the occurrence of induced methylation. The AdOx treatment led to reduced levels of trehalose in the cells whereas AdoMet increased it, indicating that trehalose production was enhanced due to methylation of TPS arising from carboxymethylation of cysteine residues (Bhattacharyya et al. 2005; Sengupta et al. 2011).

Studies of posttranslational regulation of TPS complexes from other yeast species have indicated the existence of similar regulatory mechanisms as those described for *S. cerevisiae*. In *C. utilis*, two UDPG-utilizing Tre6P synthase enzymes have been reported. Evidence was presented for in vitro activation of the first enzyme by partial proteolysis and in vitro regulation of the second enzyme by phosphorylation/dephosphorylation (Vicente-Soler et al. 1989, 1991). As in *S. cerevisiae*, evidence for the latter has to be considered preliminary until more proof is shown on whether catalytic or instead regulatory subunits may be the direct targets for this modification. Addition of glucose to stationary-phase cells of *C. utilis* nevertheless also caused a rapid partial loss of total Tre6P synthase activity. Later studies by the group of Ghosh in parallel to the above mentioned studies of methylation in *S. cerevisiae* have confirmed that methylation must also act as a positive regulatory mechanism for activation of the TPS complex in *C. utilis* (Sengupta et al. 2012).

2. Posttranslational Regulation of Trehalose Degradation

Before cAMP-PKA dependent phosphorylation during activation of neutral trehalase could be demonstrated in vivo (Schepers et al. 2012), a large number of early in vivo and in vitro results were already consistent with this mode of regulation (Thevelein 1984c). Neutral trehalase was activated in vitro by cAMP-dependent protein kinase and this activation was associated with incorporation of phosphate into the enzyme (App and Holzer 1989; Uno et al. 1983). Its in vivo activation upon nutrient repletion to starved cells (Coutinho et al. 1992; Durnez et al. 1994), upon heat shock (De Virgilio et al. 1991a), and during diauxic shift (Coutinho et al. 1992; François et al. 1987) seemed to depend on the presence of active cAMP-PKA. Mutations that reduced PKA activity lowered activation of trehalase by glucose (Mbonyi et al. 1990) and mutations that increased PKA activity lead to enhanced trehalase activity (Durnez et al. 1994). However, the lability of the neutral trehalase during purification had so far hampered more detailed studies of its activation by phosphorylation.

The *Nth1* enzyme contains several consensus sites for cAMP-dependent protein phosphorylation. Eight potential PKA phosphorylation sites were identified in the predicted protein sequence from *S. cerevisiae*, which are also conserved in the corresponding sequences of *K. lactis* and *C. albicans*: Ser20, Ser21, Ser60, Ser83, Ser475, Thr58, Thr135, and Thr149 (Wera et al. 1999). Two of these sites, Ser20 and Ser83, are in fact "perfect" in terms of their similarity to the consensus PKA phosphorylation sequence. Although site-directed mutagenesis of these eight consensus sequences was carried out for *S. cerevisiae* Nth1, none of these mutations alone abolished activation by PKA of *Sc*Nth1 in vitro or in vivo. However, for several of these mutants, the basal activity of trehalase was significantly lowered. Moreover, simultaneous mutagenesis of all eight sites resulted in a totally inactive *Sc*Nth1 enzyme. Only the combination of three mutations, the two perfect consensus sites and an additional third "perfect" consensus site (Thr260), resulted in an inactive trehalase unable to be activated by PKA or glucose. Hence, it was suggested that activation of trehalase is mediated by phosphorylation of more than one site (Wera et al. 1999).

Mass spectrometry studies have again pointed to the ability of PKA to phosphorylate purified Nth1 in vitro, unveiling Ser20, Ser21, Ser60, and Ser83 as target phosphorylation residues (Veisova et al. 2012). At the same time, after developing site- and phospho-specific antibodies, direct evidence of this mechanism taking place in vivo was provided by Schepers et al. (2012). Glucose and nitrogen activation of trehalase in vivo is associated with phosphorylation of Ser21 and Ser83 of Nth1. Activation of Nth1, however, is not only dependent on phosphorylation by PKA but also on the concomitant dephosphorylating activities. In the same work it has been shown that mutants with reduced PKA activity show constitutive Nth1 phosphorylation despite showing reduced levels of trehalase activation. The same phenotype was observed upon deletion of the catalytic subunits of yeast protein phosphatase 2A (PP2A), suggesting that lower PKA activity causes reduced trehalase dephosphorylation. In this regard, it has been shown in another recent report that in *S. cerevisiae* the two main protein phosphatases, PP2A and PP1, are rapidly activated by addition of glucose to cells growing on a nonfermentable carbon source and that this activation is dependent on glucose activation of the cAMP-PKA pathway, indicating PKA-dependent activation of the phosphatases (Castermans et al. 2012).

Two additional phenomena influencing Nth1 activity must be mentioned to gain a better understanding of this regulation. The first one is the interesting effect mediated by the Dcs1,2 mRNA decapping enzymes (Liu et al. 2002). As glucose becomes scarce and the diauxic shift starts, STRE genes, including *NTH1*, *DCS1*, and *DCS2*, are upregulated. Even though the transcription of *NTH1* increases, its activity paradoxically becomes reduced (DeRisi et al. 1997). De Mesquita et al. (1997) reported that, under this condition, most Nth1 remains in a non-phosphorylated inactive state and suggested that this was due to its interaction with a negative regulatory protein, which, by two-hybrid analysis, was identified as Dcs1. Further work has supported a negative regulatory effect of Dcs1 on *NTH1* function, both at the transcriptional and enzymatic activity levels (De Mesquita et al. 2003). One possibility suggested in this work is that interaction of Nth1 with Dcs1 could control Nth1 phosphorylation and consequently its activation. According to previous work of this group, the trehalase inhibitory protein was a Ca^{2+}/calmodulin ligand and possibly a substrate for the Ca^{2+}/calmodulin protein kinase (CaM kinase II) isozymes encoded by the *CMK1* and *CMK2* genes (Souza et al. 2002). They proposed that the inhibitory protein acts as a mediator between the Ca^{2+} signal and trehalase activation by PKA. In agreement with this proposal, the presence of Ca^{2+} ions was not sufficient to activate trehalase in crude extracts from the *dcs1* mutant, in contrast to the twofold activation observed in the control strain (De Mesquita et al. 2003). However, this possibility has been questioned by results obtained with *S. pombe* Ntp1 by Franco et al. (2003), which indicated Ca^{2+}-binding through the Ca^{2+}-binding motif as being an essential step because it allows proper oligomerization. In the meantime, Malys et al. (2004) have demonstrated that at least part of the effects exerted by Dcs1 on Nth1 activity may occur in a more indirect manner. Dcs1,2 are required to maintain low levels of mRNA turnover products, which may otherwise hamper initiation of translation. When these activities are absent, accumulation of such products and inhibition of translation trigger a stress response, which includes induction of STRE genes such as *NTH1*. Although they hypothesize that further effects observed in *dcs1* mutants are caused by accumulation of such molecules, how they affect Nth1 at the protein level remains to be elucidated.

A second level of regulation taking place at the level of protein activity has been shown to happen through interaction of Nth1 with the Bmh proteins. Yeast Bmh1,2 are members of the 14-3-3 family of proteins, which basically serve as molecular chaperones, binding their clients in a phosphorylation-dependent manner and in this way modulating the activity, subcellular localization, structure, and/or stability of hundreds of proteins (reviewed by van Heusden 2009). Panni et al. (2008) first identified Nth1 as a partner, coprecipitating with Bmh1,2. They subsequently reconstituted this interaction in vitro, showing interaction of Bmh1,2 with the phosphorylated N-terminal, Ca^{2+}-binding domain, and the active site-containing region of Nth1. Moreover, the presence of one of the Bmh proteins is necessary for complete activation of Nth1. Based on their results, they hypothesized that during the diauxic shift, non-phosphorylated Nth1 accumulates and is partially phosphorylated by PKA on Ser21. Phosphorylated Ser 21 serves then as docking place for Bmh proteins which then bind to and further activate Nth1. Recent results have shown that phosphorylation on Ser60 and Ser81 of trehalase is important for Bmh interaction and mediates activation of Nth1 and that interaction with the Bmh proteins most likely causes changes in the tertiary structure, which increase accessibility of the active site (Macakova et al. 2013; Veisova et al. 2012). Schepers et al. (2012) have demonstrated that the presence of the Bmh proteins is required for in vivo phosphorylation on Ser21 and Ser83 as well as for Nth1 activation. Most importantly, this work has also provided further support for a role of Dcs1 in the direct regulation of Nth1 activity, involving the Bmh proteins. Deletion of *DCS1* caused constitutive activation and phosphorylation of trehalase, which in turn resulted in stronger binding of the Bmh proteins to trehalase, suggesting that Dcs1 may inhibit Nth1 by preventing Bmh binding.

3. Transcriptional Regulation of Trehalose Biosynthesis and Degradation

Expression of the Tre6P synthase/phosphatase complex in *S. cerevisiae*, encoded by the *TPS1*, *TSL1*, *TPS2*, and *TPS3* genes, is co-induced by stress conditions and nutrient starvation. *NTH1* and *2* are similarly coregulated at the transcriptional level. The cAMP-PKA pathway negatively regulates the *TPS* genes and, therefore, trehalose accumulation (Winderickx et al. 1996). *TPS* genes contain one or multiple repeats of 5′-AGGGG-3′ or 5′-CCCCT-3′ **stress response elements (STRE)** in their promoter (Ruis and Schuller 1995) (Fig. 10.5). Activation through these sequences occurs in response to a broad range of stress factors ranging from oxidative and hyperosmotic stress to nitrogen starvation (Boy-Marcotte et al. 1998; Moskvina et al. 1998; Treger et al. 1998). The STRE system is considered a general, nonspecific stress response (Ruis and Schuller 1995). STREs serve as a binding site for two Cys_2His_2 zinc finger-containing transcription factors, Msn2 and its paralog, Msn4 (Martinez-Pastor et al. 1996; Schmitt and McEntee 1996). In *msn2Δ msn4Δ* mutants, the induction of the STRE-controlled genes is reduced but not abolished, indicating involvement of additional transcription factors (Gorner et al. 1998). The presence of a STRE element in the TPS gene promoters is also not sufficient for coregulated transcriptional control with other STRE-driven genes under different stress conditions. Ethanol stress, which is known to enhance expression of *CTT1*, did not significantly alter the activity of the *TPS1* promoter (Winderickx et al. 1996). These observations suggest that additional transcription factors are involved in conferring differential responses depending on the type of stress. For example, Hot1 and Msn1 have also been found to activate some STRE-driven genes upon osmotic stress (Rep et al. 1999). Expression of *TPS1* is also influenced by Gts1 (Xu et al. 2004). Gts1 increases heat tolerance by mainly activating the Snf1 kinase-dependent derepression of *HSP104* and *TPS1* under glucose-derepressed conditions, thus acting as transcriptional modulator for the complex between Snf1 kinase and RNA polymerase II complex. The *Saccharomyces* Genome Database describes up to 133 regulators of *TPS1* at the transcriptional level, most of them identified through chromatin immunoprecipitation experiments (Venters et al. 2011).

Apart from the PKA-dependent signaling network, additional components are also controlling trehalose accumulation at the transcriptional level. Increasing evidence has shown that overall activity of the transcription factors involved in entry into quiescence and/or setting up stress responses, Hsf1, which binds to **heat shock elements (HSE)**; Gis1, which binds to the **postdiauxic shift (PDS)** elements (Pedruzzi et al. 2000); and Msn2,4, which bind to STRE elements, is coordinately regulated according to an interplay between multiple global regulatory networks, including Ras-cAMP-PKA, TOR, SNF1, and high-osmolarity glycerol pathways (Petrenko et al. 2013; Sadeh et al. 2011; Zaman et al. 2008). To discern the contributions of other regulators, Sadeh et al. (2011) examined the effects of single deletions

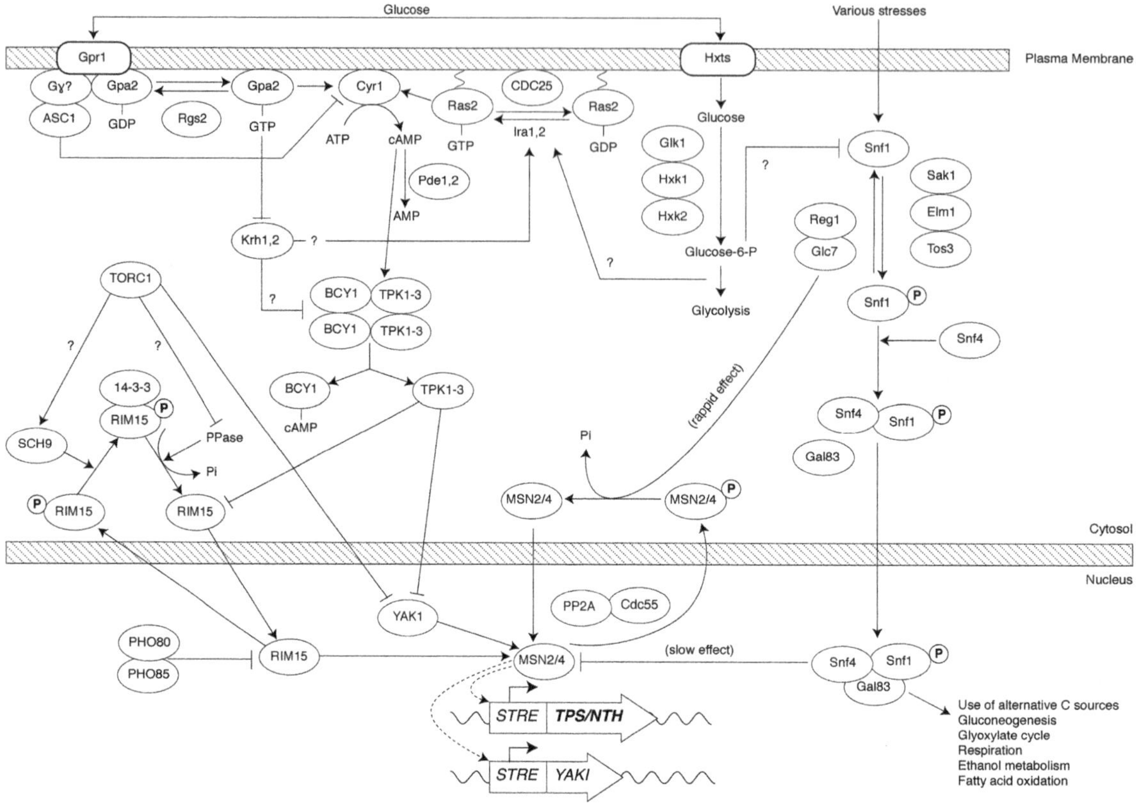

Fig. 10.5 Overview of transcriptional regulation of trehalose biosynthesis. The *TPS1,2* and *NTH1* genes are controlled by STRE elements in their promoter through the transcription factors Msn2,4. Nuclear localization and activity of the latter are regulated by several signaling pathways including the PKA, Snf1, and TORC1 pathways (for further details see text)

of 35 Msn2,4 interacting partners on Msn2,4-mediated gene expression following exposure, or absence to, of three different stress conditions. They found that the majority of Msn2,4 activators include genes encoding for protein kinases (*MCK1*, *RIM15*, *YAK1*, *SNF1*, and *HOG1*) and phosphatases (*PSR1*, *PSR2*, and *GLC7*). They also found that the deletion of *WHI2*, encoding a scaffold protein known to bind Msn2 and Psr1 (Kaida et al. 2002), results in a dramatic decrease in Msn2,4 activity. The *RPD3* and *HDA3* genes, encoding chromatin remodelers, were also strictly required for Msn2,4 activity. Deletion of either the *TOR1* or *TOR2* gene, belonging to the TOR pathway, and *RAS2*, *CDC25*, *CYR1*, *TPK1*, or *TPK3*, belonging to the cAMP-PKA pathway, significantly elevated Msn2,4 activity. In addition, the deletion of *UMP1*, encoding a proteasome maturation factor, significantly increased Msn2,4 activity. Interestingly, Sadeh and colleagues also found that *TPK1* and *TPK3* exert a repressive effect on Msn2,4-directed expression, whereas *TPK2* instead had a positive effect.

Msn2,4 are known to be activated by inhibition of the PKA or TOR pathway, which allows them to induce expression of genes containing STRE elements in their promoter (Gorner et al. 1998; Martinez-Pastor et al. 1996; Santhanam et al. 2004; Smith et al. 1998). So far, nuclear/cytosolic shuffling of Msn2,4 has been considered as the primary regulatory step for Msn2,4 activity (Gorner et al. 1998). In glucose-grown cells, PKA-dependent phosphorylation of sites within the nuclear import domain of Msn2 blocks its migration into the nucleus, while PKA-dependent phosphorylation of a site within the nuclear export domain promotes its exit from the nucleus (Gorner et al. 2002). Glucose starvation and stress conditions, such as heat shock or osmotic shock, cause dephosphorylation of these sites by PP1 and nuclear accumulation of active Msn2 (De Wever et al. 2005; Santhanam et al. 2004). Apart from PKA, the protein kinase Snf1 can also directly modify one of the Msn2 phosphorylation sites (S582) and thereby repress Msn2 function. It has recently been proposed that PP1 promotes glucose-dependent Msn2 entry into the nucleus on a fast time scale while Snf1 would stimulate exit at a slower rate (Petrenko et al. 2013). This effect could be better discerned in *tpk*-wimp allele mutants, since higher PKA levels yield lower Snf1 activity. This group has also provided evidence that, while inhibition of TORC1 may sensitize Msn2 localization to other stresses, the immediate perception of stress from nitrogen downshift proceeds mainly through PKA rather than through TORC1. An important regulatory effect occurring both upstream and downstream of Msn2,4 is that exerted by Yak1 kinase. In conditions of active cell growth, PKA-dependent phosphorylation of Yak1 on Ser295 and two other minor sites inhibits Yak1 nuclear localization (Lee et al. 2011). In low-PKA/TOR conditions, for example, under glucose depletion or rapamycin treatment, Yak1 is translocated to the nucleus (Martin et al. 2004; Moriya et al. 2001; Schmelzle et al. 2004). This allows Yak1 kinase to activate the transcription factors Hsf1, Msn2,4, and Gis1 by phosphorylation (Lee et al. 2008). In the absence of PKA activity, the PAS kinase Rim15 activates Msn2,4 by phosphorylation in the nucleus, which activates the transcription of *YAK1*, causing Yak1 to further block cellular growth (Cameroni et al. 2004; Lee et al. 2008). In the presence of PKA, the protein kinase Rim15 and the transcription factors Msn2,4 can be phosphorylated and exported to the cytoplasm, which inactivates this regulatory loop (Gorner et al. 1998; Reinders et al. 1998). Rim15-dependent phosphorylation of the endosulfines Igo1,2 is also essential for mRNAs, which are transcriptionally controlled by the stress response (STRE) and postdiauxic shift (PDS) transcription factors Msn2,4 and Gis1, respectively, to be sheltered from degradation via the 50–30 mRNA decay pathway (Bontron et al. 2013; Cameroni et al. 2004; Luo et al. 2011; Pedruzzi et al. 2000; Talarek et al. 2010).

Expression of STRE genes in other yeast species seems to depend on different transcription factors. For example, in *S. pombe* it is dependent on Atf1, whereas the *S. cerevisiae* homolog of *Sp*Atf1, Sko1, is only activated in response to osmotic shock (Gasch 2007). Msn2 and Msn4-like transcription factors have been found in *C. albicans*, *Ca*Msn4 and Mnl1, respectively. However further work has shown functions unrelated to stress response for these proteins (Nicholls et al. 2004). In fact, further research has shown that both *C. albicans* and *S. pombe* use different strategies compared to *S. cerevisiae* to regulate their core stress response genes and this is partly due to differential regulation of **stress-activated protein kinase (SAPK)** pathways. In *S. pombe*, the response to environmental stress is controlled predominantly by the Sty1 SAPK (Chen et al. 2003), whereas, as we have mentioned previously, in *S. cerevisiae* this response is governed by multiple regulatory pathways. In fact, in *S. cerevisiae* the homolog of the fission yeast Sty1 SAPK, Hog1 SAPK, mainly responds to changes in osmolarity (Brewster et al. 1993). In contrast, *C. albicans* shows a regulatory network more similar to that of *S. pombe* with the system also depending on a **multiple-stress-responsive Hog1 SAPK homolog** (Smith et al. 2004). Interestingly, in other filamentous fungi, such as the entomopathogenic fungi *Beauveria bassiana* and *Metarhizium anisopliae*, Msn2 orthologs have been found which are vital for conidiation, virulence, and multi-stress response (Liu et al. 2013b).

4. Trehalose Accumulation During Sublethal Heat Treatment

The mechanism of trehalose accumulation by sublethal heat treatment appears to be complex. The activity of Tre6P synthase increases during heat treatment in *S. pombe* (De Virgilio et al. 1990) and in *S. cerevisiae* (Hottiger et al. 1987b; Neves and Francois 1992). The latter might be due to the known heat shock induction of the

TPS1 gene (Bell et al. 1992; Winderickx et al. 1996). Heat shock also induces the *TPS2* gene and a concomitant increase in Tre6P phosphatase activity (De Virgilio et al. 1993). The increase in Tre6P synthase activity during sublethal heat treatment, however, is relatively slow (Hottiger et al. 1987b; Neves and Francois 1992). When measured after a heat treatment of only 20 min, no or only a small change in activity is observed in spite of a large increase in trehalose content (Winkler et al. 1991). Interestingly, in *S. cerevisiae* (De Virgilio et al. 1991b; Neves and Francois 1992; Winkler et al. 1991), as well as in *S. pombe* (De Virgilio et al. 1990) and *N. crassa* (Neves et al. 1991), heat-induced accumulation of trehalose is largely insensitive to cycloheximide. On the other hand, a strong inhibition was reported with the RNA synthesis inhibitors, acridine orange and ethidium bromide (Attfield 1987). These compounds, however, are known to affect other processes at similar concentrations (e.g., cAMP synthesis: (Thevelein and Beullens 1985). This indicates that a substantial part of heat shock-induced accumulation of trehalose occurs by regulation of trehalose synthesis and degradation at the posttranslational level. Although Tre6P synthase activity still increases in the presence of cycloheximide in *S. pombe* (De Virgilio et al. 1990) and to some extent in *S. cerevisiae* (De Virgilio et al. 1991b), apparently due to posttranslational regulation, the rapid kinetics of trehalose accumulation compared to the changes in enzyme activity may also indicate stimulation at the level of increased substrate availability. According to Winkler et al. (1991), the major reason for trehalose accumulation during sublethal heat treatment is the enhanced level of the substrates of Tre6P synthase, G6P and UDPG (Winkler et al. 1991). Their increase might be due to a differential effect of temperature on glucose influx and glycolytic capacity causing metabolites upstream in glycolysis to accumulate. This explanation has been challenged by Neves and Francois (1992), who reported a decrease in G6P levels during sublethal heat treatment rather than an increase (Neves and Francois 1992). The level of UDPG, however, also showed a heat-induced increase. These authors claimed that temperature-induced changes in the kinetic characteristics of trehalase and Tre6P synthase/phosphatase strongly favor trehalose synthesis and proposed this as the main cause for heat-induced trehalose accumulation. In support of the increased substrate-level hypothesis, recent work has shown that about 85 % of the increase in glycolytic flux upon shift of *S. cerevisiae* cultures to higher temperature is due to changes in the metabolic environment of each glycolytic enzyme, rather than to changes in their catalytic capacity (Postmus et al. 2008).

To study the effect of temperature on growth rate, the strain CEN.PK113-7D was grown in aerated, pH- and temperature-controlled glucose-excess batch fermentors at temperatures in the range of 27–41 °C. The concentrations of different glycolytic intermediates were substantially changed within this temperature interval. In particular, the concentration of Tre6P decreased which could account for the observed increase in flux through hexokinase. The concentration of the known activator of phosphofructokinase (PFK), fructose 2,6-bisphosphate, was also increased, which could contribute to PFK converting fructose 6-phosphate into fructose 1,6-bisphosphate, even when the concentration of the latter would be high. A higher level of fructose 1,6-bisphosphate, a potent activator of pyruvate kinase, could help maintain a high flux through pyruvate kinase at 38 °C, even when the concentration of phosphoenolpyruvate was much lower. As for the pyruvate branch point, intersection of glycolysis with the tricarboxylic acid cycle and C2 metabolism, a much higher intracellular concentration of pyruvate at 38 °C was detected, in agreement with the occurrence of alcoholic fermentation, because high intracellular concentrations of pyruvate are thought to favor the pyruvate decarboxylase reaction. The higher glycolytic flux led to a switch from respiratory to respiro-fermentative metabolism, which led to lower ATP yield. The authors explain this change as probably related to mitochondrial function being more sensitive to high temperature, an observation already reported in *S. cerevisiae* and also in other fungi such as *A. niger* (Abrashev et al. 2013; Kawai et al. 2001), or to the need to limit oxidative metabolism under increasing temperature, which could otherwise harm the cells if the temperature reaches lethal levels (Davidson and Schiestl 2001).

In addition to the changes in substrate availability and the posttranslational effects on the activity of trehalose metabolism enzymes, gene induction still explains an important part of heat shock-induced accumulation of trehalose. Three major transcription factors are responsible for the bulk of the transcriptional response to increased temperatures, the **heat shock response**. The transcription factor Hsf1 is responsible for activation of heat shock-specific genes, most of them encoding chaperones, whereas Msn2,4 are responsible for activation of genes that form part of a more generalized stress response, including trehalose metabolism genes (Gasch et al. 2000; Grably et al. 2002; Yamamoto et al. 2008), most recently reviewed by Morano et al. (2012). Hsf1 and Msn2,4 share certain target genes, such as *HSP104* (Grably et al. 2002). It has been found that Hsf1 also regulates the transcription of genes encoding trehalose-metabolizing enzymes (Yamamoto et al. 2005) and that the transcriptional activity of Hsf1 is positively regulated by trehalose (Conlin and Nelson 2007). How Msn2,4 are activated in response to heat shock is still poorly understood. What is now clear is that mere accumulation of misfolded proteins is not the signal triggering Msn2,4-dependent transcription upon heat shock. Treatment of cells with the toxic proline analogue azetidine 2-carboxylic acid (AZC) results in G1 arrest in yeast cells in a manner similar to heat shock leading to transcriptional activation of Hsf1-dependent but not Msn2,4-dependent genes (Rowley et al. 1993; Trotter et al. 2001). Treatment with sublethal concentrations of ethanol (6–8 %) also induced Hsf1 but not Msn2,4 (Takemori et al. 2006). Lastly, inhibition of proteasomal degradation with the specific inhibitor MG132 was also an effective activator of only Hsf1 (Lee and Goldberg 1998).

Like Hsf1, Msn2,4 are hyperphosphorylated in response to heat shock, but this modification is inhibited by cAMP, suggesting that it is not directly mediated by PKA (Garreau et al. 2000). This effect, however, could still be explained in terms of cAMP-mediated activation of PKA leading to negative regulation of Yak1, as expected, since activation of the cAMP-PKA pathway opposes activation of Msn2,4. The PKA-negatively regulated kinases, Yak1 and Rim15, are two of the few regulatory proteins known to influence both Hsf1 and Msn2,4 by phosphorylation (Griffioen et al. 2001; Lallet et al. 2004; Lee et al. 2008, 2013). Heat shock and oxidative and ethanol stress were shown to cause reduced levels of the Ras2 activator Cdc25. This reduction was accompanied by a reduced ability of cells to activate accumulation of cAMP via Ras (Wang et al. 2004). The wild type cells still accumulated cAMP because of adenylate cyclase feedback activation by PKA so this reduction in Ras-mediated cAMP accumulation could only be detected in tpk^{w} strains (Nikawa et al. 1987). This indicates that at least part of the heat shock response starts by downregulation of upstream signaling components.

The involvement of trehalose synthesis upregulation is strongly linked to the temperature range at which the heat shock is performed (Parrou et al. 1997). In general, it has been shown that during sublethal heat treatment (shift from 28 °C to 37 ° C) both Hsf1 and Msn2,4 are activated to produce elements that protect against heat inactivation (Yamamoto et al. 2008). This group showed that more severe heat shock (48 °C) inactivated Hsf1-dependent expression because of RNA polymerase II inactivation. Accumulation of misfolded proteins reactivated Hsf1 during the recovery from heat shock (return to 28 °C). In this study, the activity of Msn2,4 was crucial to prepare cells for subsequent severe heat shock but was dispensable during the recovery period. The results for *TPS* expression are more or less in agreement with a major part of the upregulation happening during sublethal heat treatment. For temperature upshifts to between 33 and 37 °C, the transcriptional induction of *TPS1* was required for accumulation of trehalose (indicating that transcriptional activation contributes to this activity). Between 37 and 40 °C, small changes in the absolute temperature resulted in a strong reduction of *TPS1* expression, but in pronounced changes in the kinetic properties of the TPS complex and trehalase (Parrou et al. 1997). When Trevisol and coworkers (2013) measured transcription of *TPS1*, *TPS2*, *TPS3*, and *TSL1* upon shift from 28 to 40 °C, they observed induction mostly for *TSL1* while the other genes were either not induced or only showed a slight increase. This nearly matches with the previous results from Gasch et al. (2000), for which a more sensitive approach by DNA microarray analysis of cells shifted from 25 to 37 °C had revealed upregulation of *TPS1* expression by 2-fold, no changes in *TPS3* expression, and strong upregulation of *TPS2* and *TSL1* expression, which increased almost 4- and 12-fold, respectively. Thus, according to these newer data, the increase in trehalose level during heat shock seems to be due, at least in part, to the induction of *TPS2* expression and, mainly of *TSL1*, which subsequently shows a strong influence on Tps1 activity. All these data are consistent with no significant levels of trehalose accumulating at 28 °C because of the TPS complex not being completely/properly formed in that condition. Since *TSL1* is the most strongly expressed gene upon heat shock, as mentioned above, synthesis of Tsl1 seems to be a crucial factor required for activation of the TPS complex.

In accordance with this idea, it was shown that overexpression of *TSL1* at 38 °C increased the glycolytic flux and stress tolerance of *S. cerevisiae* by increasing the intracellular levels of trehalose (Ge et al. 2013). Trevisol and colleagues (2013) also analyzed protein extracts from wild type and *tps3Δ* cells shifted from 28 to 40 °C and then back to 28 °C in the presence of a phosphorylation cocktail and cAMP. This experiment indicated that upon return from heat stress to nonstressed conditions, cAMP-PKA phosphorylation seems to negatively regulate activity of the TPS complex in a Tps3-dependent manner. On the other hand, the purified trehalose synthase enzyme from *S. cerevisiae* was shown to have optimum activity at 40 °C, suggesting that temperature itself may also lead to fluctuating changes in kinetics (Chaudhuri et al. 2009). Therefore, all these different mechanisms (enhanced level of substrates and subunits of the TPS complex, changes in kinetic properties, and posttranslational effects on the catalytic and/or regulatory subunits) could together explain the overall changes in trehalose accumulation during and after heat shock.

Paradoxically, sublethal heat treatment not only causes increased trehalose biosynthesis but also activates trehalase. Trehalose accumulates to a much higher level in yeast strains lacking neutral trehalase, indicating that heat shock might induce a strong futile cycle of trehalose synthesis and breakdown (Parrou et al. 1997). The relevance of such a futile cycle for metabolism or survival at high temperature remains unclear. Studies of trehalose accumulation during quiescence may provide an answer to the existence of this futile cycle (Shi et al. 2010). This group has noticed that trehalose is important not only to endure quiescence but also in fueling metabolic demands of a cell upon rapid exit from quiescence. Cells that do not accumulate trehalose initiate growth more slowly and display poor survival. The group hypothesizes that in nature, yeast cells are more likely to undergo periodic bursts of rapid growth that alternate with long periods of quiescence, rather than continuous rapid growth that typically occurs under laboratory growth conditions. Rapid accessibility to a pathway in which cycles of trehalose storage and breakdown can rapidly occur can help wild populations to quickly adapt to the environmental changes, including fast recovery from heat shock. Work from other groups also supports a main role of neutral trehalase in recovery after stress by rapid induction of trehalose mobilization. Since early on it has been observed that trehalase activation happened during the rapid cooling of cells after a heat treatment (De Virgilio et al. 1991a). The latter effect had previously been shown for heat-induced activation of trehalase in yeast ascospores (Thevelein 1984a). Recent studies in the arbuscular mycorrhizal fungus *Glomus intraradices* indicated that sublethal heat shock, i.e., incubation at 10 °C above its normal growth temperature, induced trehalose accumulation by *TPS/NTH* genes similar to that of *S. cerevisiae* (Ocon et al. 2007). In contrast to the latter, heat shock did not transcriptionally induce

GiNTH1 but its activity was strongly increased, indicating posttranslational modification. *GiTPS2* was also upregulated but the activity of both enzymes is reduced to accommodate them to an overall increase in trehalose. The authors postulate that neutral trehalase activity is initially enhanced to cope with a potential excess in Tre6P, since trehalose hydrolysis would enhance the flux of conversion of Tre6P to trehalose. In yeast and other fungi, disruption of the Tre6P phosphatase gene leads to different levels of thermosensitivity (Borgia et al. 1996; De Virgilio et al. 1993; Franco et al. 2000; Piper and Lockheart 1988; Van Dijck et al. 2002; Zaragoza et al. 2002). In *S. cerevisiae*, the thermosensitive phenotype of *tps2Δ* seems to correlate with accumulation of the intermediate Tre6P rather than with lack of trehalose accumulation (Elliott et al. 1996). Experiments with Tre6P phosphatase mutants in *A. nidulans* and *C. albicans* indicate a possible correlation between the accumulation of Tre6P and defects in the cell wall biosynthesis/assembly (Borgia et al. 1996; Zaragoza et al. 2002). Therefore, conversion of Tre6P to trehalose and then to glucose would avoid a toxic accumulation of Tre6P. Later in the heat shock, these activities would be reduced in favor of trehalose accumulation and finally after heat shock the trehalase activity would prevail during the recovery stage. Evidence has been presented that the *TPS2* gene and other STRE-controlled genes might be induced during sublethal heat treatment by sphingolipid signaling (Dickson et al. 1997). The level of several sphingolipids shows transient increases upon heat stress. Dihydrosphingosine is able to induce trehalose accumulation as well as transcription of *TPS2* and of an STRE-LacZ reporter gene. Although more research into this direction remains to be done, it is also known that the Ypk-Pkh signaling cascade activates the transient sphingolipid induction upon heat stress (Sun et al. 2012). Interestingly, proteins such as PKA catalytic subunits and Sch9 are phosphorylation targets for the Pkh kinases (Haesendonckx et al. 2012; Voordeckers et al. 2011), so at least part of this regulation could take place via regulation of cAMP-PKA and/or TOR-mediated effects in the activation of Msn2,4.

VII. Regulatory Functions of Trehalose Metabolism

A. Control of Growth, Cell Cycle Progression, and Sporulation

When conditions are unfavorable for proliferation, cells have the capacity to enter a nondividing state while retaining their ability to reenter the proliferative cell cycle. In this state they stop investing metabolic effort into cell growth and division and focus instead on the activation of pathways that allow them to resist long periods of famine and other stressful conditions. Yeast and other fungi can enter a quiescent, nondividing state in two ways. First, exhaustion of one or more nutrients in the environment will cause cells to stop the cell cycle and enter G0, or stationary phase (De Virgilio 2012; Gray et al. 2004). Second, diploid cells, in response to similar cues, will initiate sporulation (Neiman 2011). Acquisition of both states is generally associated with, among other phenomena, accumulation of reserve carbohydrates, glycogen and trehalose. Therefore, a strong correlation exists between the trehalose level and growth rate in yeast and in many other fungi. In *S. cerevisiae*, exponential-phase cells growing on glucose have the highest growth rate and a very low trehalose level. G0 cells which are in stationary phase have the highest trehalose level, while cells growing exponentially on glycerol display intermediate values both for growth rate and trehalose level. Cells in the G1 phase already accumulate trehalose and glycogen, and this accumulation, particularly in the case of trehalose, is correlated with the duration of the G1 phase (Futcher 2006; Paalman et al. 2003; Sillje et al. 1997). This accumulation is transient since the carbohydrates were metabolized again before bud emergence, suggesting that a transient increase in ATP flux obtained from their consumption may be required to fuel progression from this step into further stages of the cell cycle.

S. cerevisiae cells transiently enter a stationary-phase period when depletion of fermentable nutrients forces them to switch from fermentative to respiratory metabolism. This moment is collectively known as **diauxic shift** and is associated with an extensive reprogramming of the transcription machinery, resulting from activation of the transcription factors Msn2,4 and Gis1 in response to changes in the cAMP-PKA, TOR, Snf1, and Rim15 pathways (see previous Sect. VI.B.2 and Galdieri et al. 2010). During the diauxic shift, cells form two distinct populations, with a different density (Shi et al. 2010). The heavier population is unbudded, indicating cell cycle arrest, replicatively younger, and exhibits properties typical of quiescence such as higher long-term viability and reproductive capacity, greater thermotolerance, and decreased aging and oxidative stress signs. Their higher density is also correlated with higher accumulation of the storage carbohydrates, glycogen and, more particularly, trehalose. While non-quiescent cells seem unable to turn off pro-growth signaling pathways such as cAMP-PKA, a requirement for survival in stationary phase, the heavier quiescent cells have a transcriptome consistent not only with increased stress resistance but also with these cells being poised for return to favorable stimuli, such that they can synchronously reenter the cell cycle when provided with nutrients (Allen et al. 2006). For example, they show increased levels of mRNAs not only for *TPK1* but also for *BCY1*, which helps to set activation of PKA ready to happen but on hold (Aragon et al. 2008). Shi et al. (2010) have observed that cells lacking trehalose exit quiescence more slowly, suggesting that trehalose is the carbohydrate of choice for cell cycle reentry from quiescence in *S. cerevisiae*. Trehalose may be a carbohydrate of choice for cell cycle reentry compared to glycogen simply because its cleavage can supply more molecules of glucose per hydrolyzed bond.

Successive studies have led to an emergent model to explain the physiological characteristics of quiescent and non-quiescent cells in stationary phase (Allen et al. 2006; Davidson et al. 2011). According to this model, prior to the diauxic shift a cell division takes place that already distinguishes quiescent and non-quiescent cells. Each of them still possesses the ability to divide, but the non-quiescent cells accumulate ROS species and eventually become unable to reproduce and ultimately apoptotic and necrotic. These cells do not accumulate glycogen and trehalose because apparently they not only activate autophagy but also consume their intracellular reserves of glycogen and trehalose. This way they do not compete for resources with the quiescent cells while still producing nonfermentable carbon sources, e.g., ethanol and glycerol, that the quiescent population, via its activated respiratory metabolism, can consume. In other words, the non-quiescent cells remain alive as a long-term storage to allow further survival of the quiescent population (Davidson et al. 2011). From that point of view, glycogen and trehalose also help for this non-quiescent sector of the population to contribute to the survival of the more physiologically fortunate quiescent cells. This population heterogeneity has also been described during high-level production of bioethanol in aerated fed-batch processes in a fermentation phase where high levels of ethanol impair further growth. During this phase, cells continue producing ethanol and this is dependent on the high respiratory capacity of the fraction of denser, glycogen- and trehalose-enriched quiescent-like cells (Benbadis et al. 2009). In another study, the retention of metabolic capacity after starvation for nitrogen, carbon, or both was strongly dependent on the ability to accumulate glycogen and trehalose when cells were analyzed in anaerobic conditions (Thomsson et al. 2005). However, this was not so crucial if cells were starved in aerobiosis, which again highlights the importance of respiratory capacity of the cells in the quiescent population to sustain metabolism (Albers et al. 2007).

Diploid homozygous *tps1Δ* and *fdp1* strains are unable to sporulate, while homozygous *byp1* diploids show reduced sporulation (Van Aelst et al. 1993). Homozygous diploids of *sst1*, another allele of *TPS1* (Charlab et al. 1985), show both strongly reduced sporulation and spore viability (Panek and Bernardes 1983). During yeast sporulation intensive trehalose synthesis occurs and mature ascospores contain exclusively trehalose as reserve carbohydrate (Barton et al. 1982; Roth 1970; Thevelein et al. 1982). It remains unclear whether the synthesis of trehalose itself is in some way required for the sporulation process. Sporulation occurs in the absence of glucose and deletion of *HXK2*, which suppresses the glycolytic deregulation of *tps1* mutants, does not prevent the sporulation defect (Neves et al. 1995). Hence, the sporulation defect of *tps1* mutants appears to be unrelated to their defect in the control of glucose influx into glycolysis. On the other hand, the *glc6* point mutation, which is allelic to *TPS1* and causes enhanced TPS activity, also reduces sporulation capacity, but in this case the defect is suppressed by deletion of the hexokinase 2 gene (De Silva-Udawatta and Cannon 2001). Activation of the sporulation process is dependent on induction of the central regulator, Ime1. Apparently, one important effect mediated by Tps1 or its metabolic product consists of activating Mck1, which in turn activates expression of Ime1 (De Silva-Udawatta and Cannon 2001).

ScMCK1 encodes a protein kinase homologous to metazoan glycogen synthase kinase-3. This kinase has been found to exert a negative regulatory effect on the activity of the glycolytic step catalyzed by pyruvate kinase. This effect occurs because Mck1 binds to PKA catalytic subunit Tpk1, which in turn inhibits Tpk1-dependent phosphorylation and activation of pyruvate kinase (Rayner et al. 2002). Mck1 has also been shown to phosphorylate Bcy1 (Griffioen et al. 2003). Mck1, along with two other glycogen synthase kinase-3 homologues, Rim11 and Mrk1, and the protein kinase Rim15 reinforce the binding of Ime1 to its partner, Ume6, to allow expression of early sporulation genes (Xiao and Mitchell 2000). The *IME1* promoter undergoes a variety or regulatory mechanisms and several transcription factors bind to it, Msn2,4 among them (Sagee et al. 1998). In this respect it may be interesting to note that recent work has found that deletion of *MCK1* almost abolishes Msn2,4 activity, revealing Mck1 as a main kinase activator for these transcription factors (Sadeh et al. 2011). In addition, it has been shown that Mck1 and the other glycogen synthase kinase-3 homologues regulate binding of Msn2 to the STRE promoters (Hirata et al. 2003). It is thus plausible that gene products originating from Msn2,4 controlled expression, such as the enzymes involved in trehalose synthesis, may also have positive effects on the pathways that are positively regulated by Mck1.

Trehalose has also an important role in **spore survival** and **germination**, not only in *S. cerevisiae* but also in other fungi. The levels of trehalose are high in fungal conidia, and activation of trehalase has been shown to accompany sexual and asexual spore (conidia) germination in numerous fungi like *N. crassa*, *A. nidulans*, and *Aspergillus oryzae* (d'Enfert et al. 1999; d'Enfert and Fontaine 1997; Horikoshi and Ikeda 1966). Previous studies have demonstrated that ungerminated conidia contain the highest levels of trehalase detected during *N. crassa* development (Hill and Sussman 1964), effectively priming conidia for rapid germination and colonization. Trehalose content increases during the developmental life cycle of *Aspergillus fumigatus*, throughout which two putative trehalose synthase genes, *tpsA* and *tpsB*, are significantly expressed. Deletion of both genes stops trehalose accumulation during development resulting in a strong reduction of their viability after heat shock and increased susceptiblility to oxidative stress (Al-Bader et al. 2010). Trehalose is also accumulated in spores of *A. nidulans* in which the effects of a putative transcription factor designated as VosA have recently been described (Ni and Yu 2007). The *vosA* gene is specifically expressed during the formation of both sexual and asexual spores. Its deletion results in absence of trehalose in spores, reducing their resistance to heat and oxidative stress as well as their viability.

B. Control of Glycolysis by Tre6P

Several *S. cerevisiae* mutants with specific growth defects on glucose and other rapidly fermented sugars, such as *fdp1*, *cif1*, *byp1*, and *sst1*, turned out to be allelic with *TPS1* (Bell et al. 1992; Charlab et al. 1985; Gonzalez et al. 1992; Neves et al. 1995; Van Aelst et al. 1993). Deletion of *TPS1* causes the same phenotype, indicating that lack of Tps1 function is responsible for the observed defects. The characteristic growth defect in these mutants is accompanied by rapid ATP depletion within less than 1 min upon exposure to glucose and hyperaccumulation of phosphorylated sugars upstream of glyceraldehyde-3-phosphate dehydrogenase (GAPDH), depletion of intracellular phosphate, absence of cAMP signaling, and lack of repression of gluconeogenic enzymes (Gonzalez et al. 1992; Hohmann et al. 1996; Van Aelst et al. 1993). This indicates an imbalance between energy-consuming and energy-recovering steps at the level of glycolysis. Apparently, the influx of glucose into glycolysis is so rapid that the second part of glycolysis cannot keep up with it. The very low ATP level is apparently mainly used by hexokinase for the further accumulation of sugar phosphates causing all cellular phosphate after some time to end up mainly in fructose-1,6-bisphosphate. Another interesting mutant identified based on its glycogen accumulation deficiency is *glc6-1* (now known as tps1-H223Y). The *glc6-1* mutant was also allelic to *TPS1* although it could grow on glucose (Cannon et al. 1994). Later it was found that, in contrast to the wild type Tps1, this mutant protein was not inhibited by phosphate and therefore behaved as a dominant overactive form of Tps1 (De Silva-Udawatta and Cannon 2001). The increase in Tre6P synthase activity in the mutant *glc6-1* could explain the diminished glycogen levels through reduced availability of UDPG and G6P for glycogen synthesis and lowered glycogen synthase activity due to a decrease in the concentration of its allosteric activator, G6P.

Extragenic suppressors of the phenotypes characteristic of loss of *TPS1* function can be classified into two main categories: functions that reduce the glycolytic flux and functions that reduce the blockage apparently taking place at the GAPDH step, responsible for the accumulation of hexose phosphates. An example of genes within the first category is the overexpression of regulators that reduce transcription of hexose transporters, such as *MIG1*, *MTH1*, or *SNF4* (Blazquez and Gancedo 1994, 1995; Gamo et al. 1994; Hohmann et al. 1992; Lafuente et al. 2000). Another important suppressor of *tps1* within this category, which restored growth on glucose and reduced the accumulation of glycolytic intermediates, was the deletion of *HXK2*, which encodes hexokinase II (Blazquez and Gancedo 1994; Hohmann et al. 1993). Overexpression of the glycerol channel *FPS1* and the glycerol-3-phosphate dehydrogenase gene, *GPD1*, both diverting the glycolytic flux toward glycerol production were examples of the second category. Thus, either reducing the rate of the initial steps of glycolysis or enhancing the flux toward glycerol production, which reduces sugar phosphate accumulation and allows regeneration of P_i, can suppress to different extents the growth and metabolic defects of the *tps1* mutant on glucose.

Interestingly, the *tps1* mutant can grow on galactose, which enters glycolysis below the hexokinase step. This is consistent with a role for either Tps1 itself or its metabolic products, Tre6P or trehalose, in regulating glycolytic flux at the level of hexokinase. In accordance with this idea, Tre6P has been found to inhibit the hexokinases of *S. cerevisiae* and also those of other fungi to different extents. Tre6P inhibits *S. cerevisiae* hexokinases I and II, but not glucokinase, in vitro (Blazquez et al. 1993) (Fig. 10.6). The inhibition constant in vitro was about 40 μM, which fitted approximately with the concentration of Tre6P in vivo. *Y. lipolytica* hexokinase is also extremely sensitive to inhibition by Tre6P, but the simultaneous presence of a glucokinase insensitive to Tre6P, which constitutes around 80 % of the glucose-phosphorylating capacity during growth on glucose, may account for growth of the *YlTPS1* deletion mutant on glucose (Flores et al. 2011). Although Tre6P-dependent inhibition of hexokinase II as a way of regulating glycolytic flux is an attractive model, several results have been obtained which indicate that this cannot fully explain the glycolytic defect in the *tps1Δ* mutant.

In an attempt to explain the *tps1* phenotype, initially two alternative models were proposed (Thevelein and Hohmann 1995). The first model suggested that the Tre6P synthase protein itself could be a regulator of hexokinase, rather than only Tre6P. Support for this possibility came from studies in which the *ScTPS1* gene was replaced by the *otsA* homologue of *E. coli*. Although Tre6P levels were restored in this strain, it only showed partial restoration of growth on glucose and regulation of glycolysis (Bonini et al. 2000). On the other hand, so far

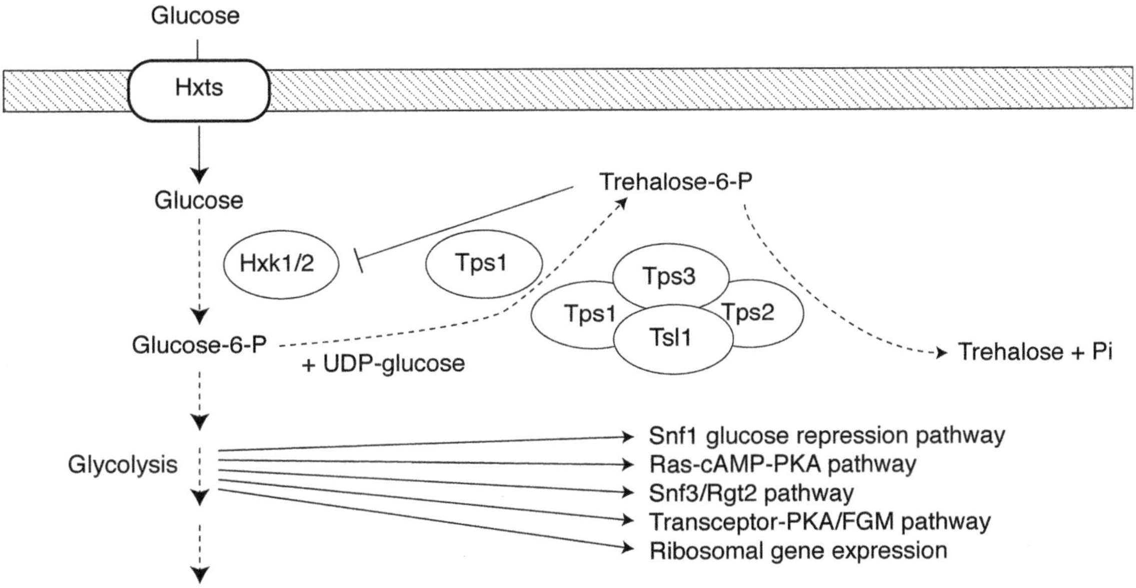

Fig. 10.6 Feedback inhibition of hexokinase in glycolysis by Tre6P. Trehalose-6-phosphate, the intermediate of trehalose biosynthesis, is accumulated in the cytosol through release from the trehalose synthase complex and/or synthesis by free Tps1 enzyme. Tre6P (and possibly Tps1 itself) is supposed to cause feedback inhibition on Hxk1,2 and to restrict in this way influx of glucose into glycolysis. Glucose catabolism in glycolysis affects multiple glucose signaling pathways

no evidence has been found for a physical interaction between Tps1 and hexokinase. The second hypothesis proposed that the block in glycolysis in the *tps1Δ* mutant could be due to the rapid drop in the level of phosphate, which is a substrate in the reaction catalyzed by GAPDH. As a result, the intermediates in the second part of glycolysis would become depleted and ATP not properly regenerated after its initial use in the first steps of glycolysis. The low ATP level would only be used for further fueling the first glycolytic steps, thereby leading to further accumulation of sugar phosphates up to the point where most of the cellular phosphate eventually becomes sequestrated in the form of fructose-1,6-bisphosphate. The inevitable outcome of this process would be the inability of the cells to grow on glucose, ultimately followed by death. To distinguish between the different models, we made use of the *byp1* mutant. This allele contains a nonsense mutation in the *TPS1* gene, which is to a low extent read through by a tRNA suppressor, resulting in a lowered amount of Tre6P synthase (Hohmann et al. 1994). Deletion of *TPS2* in this strain was expected to enhance the level of Tre6P while reducing phosphate recovery because of the absence of conversion of Tre6P into trehalose. Unexpectedly, the *byp1 tps2Δ* strain showed better growth on glucose than the *byp1* strain, contradicting the phosphate-recovery model (Hohmann et al. 1996). Also, as pointed out by Francois and Parrou (2001), the rate of glycolytic flux is most likely exceeding the rate of trehalose formation in yeast, making it difficult for the latter pathway to supply the necessary phosphate for the sustained functioning of glycolysis. All these results seem to make a putative phosphate-recovery role of the trehalose biosynthetic pathway in the regulation of glycolysis rather unlikely.

Even though growth on glucose could be reestablished, the initial hyperaccumulation of sugar phosphates after addition of glucose observed in *byp1* was still present in *byp1 tps2Δ* supporting the first hypothesis that the Tre6P synthase enzyme might play a role in controlling hexokinase activity (Hohmann et al. 1996). Restoration of growth on glucose without apparent restoration of the proper control on glucose influx into glycolysis was also observed in *tps1Δ* strains expressing the *otsA* gene from *E. coli* (Bonini et al. 2000) or the plant *TPS1* homologues from *S. lepidophylla* and *A. thaliana* (Zentella et al. 1999). When these genes were expressed in a *tps1Δ tps2Δ* strain, they led to an even better restoration of growth on glucose although the initial hyper-accumulation of sugar phosphates and depletion of ATP and phosphate were not different from those observed in the *tps1Δ* mutant. Hence, the evidence for control of hexokinase activity in vivo remained meager. Even though accumulation of Tre6P in these backgrounds could rescue growth via control of hexokinase activity, the presence of accumulated hexose phosphates and their inability to be fermented has led to proposals that Tre6P, trehalose, or Tps1 itself could have an additional role in controlling the second part of the glycolysis (Gancedo and Flores 2004). For example, a phosphoglucose isomerase mutant was isolated which is unable to ferment glucose. The authors hypothesized that it should still be able to ferment fructose, which enters the pathway after the blocked step (Boles et al. 1993). However, the mutant was also unable to ferment fructose and accumulated glycolytic intermediates in a way closely similar to that of *tps1Δ*. Addition of small amounts of glucose along with fructose restored normal fermentative metabolism. Based on this result, it has been suggested that Tre6P or a related metabolite derived from glucose could be responsible for this suppressive effect most likely by influencing one or more steps in the second part of glycolysis. Consistent with this conclusion is the observation that addition of glucose to yeast cells causes a transient increase in the level of Tre6P in the first few hours (Hohmann et al. 1996; Walther et al. 2013), which is easier to reconcile with a stimulatory effect of Tre6P downstream in glycolysis rather than with inhibition of hexokinase at a moment when the cells are keen to increase their fermentation rate in order to rapidly produce ethanol.

If the inhibition of hexokinases was solely dependent on Tre6P, one would expect strong overexpression of hexokinase to result in a similar growth defect on glucose. However, strong overexpression of hexokinase activity in wild type *S. cerevisiae* cells did not cause a glucose growth defect. Only the transient increase in sugar phosphates and drop in ATP observed just after addition of glucose are more pronounced than in a wild type strain (Ernandes et al. 1998). Later on, it was shown that overexpression of Tre6P- (as well as G6P-) insensitive hexokinase from *S. pombe* in a wild type or in a mutant strain devoid of the three endogenous glucose-phosphorylating enzymes did not affect growth on glucose (Bonini et al. 2003). Even though the hexokinase expressed was Tre6P insensitive in in vitro assays, deletion of *TPS1* in the triple mutant strain still caused a pronounced glucose growth defect, but this was associated with reduced sugar phosphate accumulation rather than hyperaccumulation as observed in a regular *tps1Δ* strain (Bonini et al. 2003). These results argue against the possibility that the glucose-sensitive phenotype of *tps1Δ* is merely due to lack of Tre6P or especially of Tre6P-dependent inhibition of hexokinase. The accumulation of sugar phosphates also does not seem to be the cause for the glucose growth defect. This accumulation, in particular that of fructose-1,6-bisphosphate, causes inhibition of respiration in Crabtree-positive yeasts (Diaz-Ruiz et al. 2008), recently shown to be due to inhibition of the mitochondrial unspecific channel (Rosas-Lemus et al. 2014). This prompted investigations to address the question whether the trehalose synthesis pathway is also involved in the regulation of oxidative phosphorylation activity in yeast mitochondria (Noubhani et al. 2009). It was found that the quantity of respiratory chain components and respiratory rate decreased or increased in the absence of Tps1 or Tps2, respectively. Previously, the group of Rigoulet had shown that overactivation of the cAMP-PKA pathway has a positive effect on the content of cytochromes and maximal respiratory capacity (Dejean et al. 2002). In the subsequent work (Noubhani et al. 2009), it was shown that cAMP levels positively correlated with respiratory capacity and that somehow this effect is dependent on Tps1- and Tps2-dependent regulation of hexokinase. Therefore, even though not all effects in *tps1Δ* can be explained solely on the basis of hexokinase deregulation, the latter observations do suggest that an important part of the role of the TPS complex in control of glycolysis and regulation of respiratory versus fermentative growth occurs at the hexokinase step.

More recently, the phenotypic defects of the *tps1Δ* mutant have been readdressed by Walther et al. (2013). This group has found that *tps1Δ* causes irreversible conversion of the entire **intracellular adenosine nucleotide pool** into inosine upon exposure to glucose. Abolishing the accumulation of inosine in *tps1Δ* cells by deleting the adenosine monophosphate (AMP) deaminase-encoding gene, *AMD1*, caused irreversible accumulation of AMP without restoration of growth on fermentable carbon sources. The authors therefore hypothesized that depletion of cytosolic phosphate is the major cause for inosine accumulation in *tps1Δ* cells since phosphate is required to catalyze the conversion of inosine into hypoxanthine. This also highlights the possible inability of *tps1Δ* mutants to maintain sufficiently high concentrations of free cytosolic phosphate in order to sustain sufficient GAPDH activity, as a possible cause of the *tps1Δ* growth defect on glucose. However, since the capacity of the trehalose pathway is too small to regenerate sufficient amounts of phosphate, Walther and colleagues directed their attention to the previously reported observation that the *tps1Δ* mutant failed to activate plasma membrane H^+-ATPase in response to glucose (Van Aelst et al. 1993). In fact, regulation of the plasma membrane H^+-ATPase could serve as a more potent phosphate-regenerating system than trehalose synthesis itself. Walther et al. showed that in a *tps1Δ* strain cytosolic acidification takes place due to failure of the plasma membrane H^+-ATPase. Conversely, hyperaccumulation of Tre6P in a *tps2* mutant caused alkalinization of the cytosol and reduced growth on nonfermentable carbon sources. Expression of *YlTPS1*, which results in 90 % reduced levels of Tre6P in *S. cerevisiae*, was sufficient to restore growth and fermentation of the *tps1Δ* strain on fermentable carbon sources. This result was reminiscent of what was earlier observed with the low accumulation of Tre6P observed in the *byp1 tps2Δ* mutant (Hohmann et al. 1996). On the other hand, these results differ from the previous observations made with strains expressing the bacterial Tre6P synthase *otsA* (Bonini et al. 2000) or the leaky *tps1* allele, *byp1-3*, (Hohmann et al. 1992), in which low levels of Tre6P were not sufficient to cause proper suppression of the *tps1Δ* phenotypes.

One possible explanation for the differences is the use of other genetic backgrounds in some of these studies. For example, deletion of *HXK2* in CEN.PK (the strain used by Walther et al. (2013)) causes a collapse in cell growth rates and derepression in the presence of glucose, whereas it barely affects the same phenotypes in the S288C background. Apparently CEN.PK has lower Snf1 kinase activity and higher PKA activity compared to S288C, and this influences glucose signaling and regulation (Kummel et al. 2010). Although growth, ethanol production, plasma membrane H^+-ATPase activation, and cytosolic pH were largely restored to wild type under the conditions used by Walther et al. (2013), the low Tre6P level still caused strong accumulation of phosphorylated sugars and a pronounced perturbation of the ATP level in response to glucose. The use of alternative conditions leading to hypo- or hyperaccumulation of Tre6P has allowed this group to conclude that yeast cells can tolerate an up to tenfold decrease and a fourfold increase in Tre6P without any noticeable impact on glycolytic flux or on the ability to grow on fermentable carbon sources. Above or below these thresholds, defects in growth and fermentation as well as abnormal accumulation of glycolytic intermediates take place. Overall the results of this group point to the possibility of reduced plasma membrane H^+-ATPase activity as a possible cause for the lack of growth on glucose of the *tps1Δ* mutant. However, since the ATP levels are not totally restored in the *YlTps1* condition even when growth on glucose is, it is still not possible to make a clear correlation between defective plasma membrane H^+-ATPase activity and Tre6P levels. Although their results provide additional evidence that accumulation of sugar phosphates and drop in ATP are not *in se* causatives for the absence of growth on glucose, how Tre6P or alternatively Tps1 itself impinges on this plasma membrane H^+-ATPase function remains unclear. In any case, this group has presented additional phenotypes and new evidence for the important role of Tre6P/Tre6P synthase in controlling regular glycolytic flux.

The putative function of Tre6P synthase, and/or its products, in the control of glycolysis raises many interesting questions. For example, since the Tre6P synthase and phosphatase enzymes are present in a large multienzyme complex, one would expect Tre6P to be channeled inside the complex rather than released into the cytosol. Since the hexokinase enzymes are present in the cytosol, an effect of Tre6P on their activities would imply Tre6P release from the TPS complex in a highly regulated manner. On the other hand, part of the Tps1 protein is not located in the trehalose synthase complex but appears to be free in the cytosol (Bell et al. 1998), suggesting that regulation of trehalose levels and cytosolic Tre6P might happen through different mechanisms. The cytosolic Tre6P concentration should be controlled precisely in order to sustain proper glycolytic flux. Association of enzymes in complexes is not limited to the TPS complex but is also known to occur, for instance, among glycolytic enzymes (Campanella et al. 2005). In a recent report, four yeast glycolytic enzymes previously shown to associate with each other (Ashmarina et al. 1984; Ovadi and Keleti 1978; Tompa et al. 1986) were isolated from cell extracts and analyzed for their susceptibility to inhibition by trehalose (Araiza-Olivera et al. 2010). This study indicated that GAPDH and hexokinase II were inhibited by trehalose when tested as separate enzymes. Interestingly, both enzymes, when mixed with either aldolase or phosphoglycerate kinase, were protected against inhibition. Only combination of specific enzymes led to protection from trehalose inhibition suggesting that association into specific complexes could regulate to a certain extent the access of trehalose and possibly also Tre6P to these glycolytic enzymes. The same group has further tested association of glycolytic enzymes in the presence of actin and has shown that polymerized F-actin stabilizes these complexes, supporting the hypothesis that glycolytic enzymes could work at least to some extent as a **cytoskeleton-associated metabolon** (Araiza-Olivera et al. 2013). The regulated formation of a glycolytic metabolon is a very attractive idea that still needs further confirmation in vivo. In this respect it would be interesting to find out whether the TPS complex or the free Tps1 enzyme can conditionally associate to parts of this metabolon, which in turn could help to understand how the TPS enzymes and/or their products can regulate glycolysis. Association of plasma membrane and cytoskeleton could facilitate access of such structures to plasma membrane functions such as glucose sensors, hexose transporters, and/or H^+-ATPase, which could in turn be responsible for additional levels of regulation of the activities of these membrane proteins in response to the glycolytic flux and vice versa. In this sense it is worth mentioning that recent studies have shown that Tps1 and sugar phosphates activate potassium transport via Trk transporters, which in turn reduces sensitivity of cells to toxic cations (Mulet et al. 2004). This effect does not seem to happen indirectly through activation of H^+-ATPase nor does it seem to depend on accumulation of Tre6P or trehalose. Deletions in *TPS1*, phosphoglucomutase, or hexokinase all caused a similar effect in reducing Trk activity and increasing sensitivity to toxic cations. This has led to the suggestion that Tps1, phosphoglucomutase, and hexokinase have an effect on potassium transport through their modulation of sugar phosphate levels.

As mentioned above, deletion of *TPS1* in certain yeast species, such as *K. lactis*, causes a similar growth defect on glucose as in *S. cerevisiae* (Luyten et al. 1993). However, in other fungi deletion of *TPS1* does not cause the same phenotype. In *S. pombe* only the germination of spores on glucose medium is prevented while growth of vegetative cells on glucose is not affected (Blazquez et al. 1994). Moreover, *Sp*Hxk2 is not inhibited in vitro by Tre6P (Blazquez et al. 1993). A related phenotype has been shown in *C. albicans*. Growth of a *tps1/tps1* homozygous deletion strain at 30 °C is indistinguishable from that of the wild type. However, at 42 °C the mutant did not grow on glucose or fructose but grew normally on galactose or glycerol (Zaragoza et al. 1998). In *A. niger* deletion of *TpsA* or *TpsB* also did not prevent growth on glucose (Wolschek and Kubicek 1997). In *A. niger* deletion of *TpsA* resulted in faster initiation of citrate production from high sucrose levels, whereas overexpression produced the opposite effect. Since *A. niger* hexokinase is only weakly inhibited by Tre6P in vitro this could also point to the existence of an additional control by Tre6P in glycolysis (Arisan-Atac et al. 1996). In a later report, however, *A. niger* hexokinase was shown to be inhibited strongly by low levels of Tre6P (0.1–0.2 mM) but in a competitive way with the substrate (Panneman et al. 1998). Hence, the physiological importance of Tre6P for regulation of glycolysis in *A. niger* remains unclear. In the meantime, studies in *Magnaporthe grisea* have shown that loss of Tps1 also causes inability to grow on glucose, but growth could be restored by addition of free amino acids to the medium, suggesting a cross talk between sugar signaling and nitrogen metabolism in this fungus (Foster et al. 2003).

C. Yeast as a Model for Plant Trehalose Metabolism

Yeast trehalose metabolism has served as an important model and tool for the discovery of the presence and important regulatory role of trehalose metabolism in plants. With the exception of resurrection plants living in extreme habitats subject to drought and other harsh conditions, most plants do not accumulate trehalose to high enough levels for this molecule being able to act as a major stress protectant (Gaff 1971; Salerno and Curatti 2003). In plants, the functions of sugar storage and protection are provided by sucrose rather than trehalose. However, it is paradoxical that higher plants possess up to 11 *TPS*-like genes and up to 10 *TPP*-like genes, such as in *A. thaliana*, in spite of the fact that higher plants do not accumulate significant amounts of trehalose. The *TPS*-like genes are divided in classes I and II according to their sequence similarity either to the *ScTPS1* or to the *ScTPS2* gene, respectively (Eastmond and Graham 2003; Leyman et al. 2001; Thaller et al. 1998).

Heterologous expression of each of these genes in *S. cerevisiae* has helped to understand their function. *TPS*-like genes have been expressed in *S. cerevisiae* and tested for complementation of the *tps1Δ* growth defect on glucose (Vandesteene et al. 2010). Interestingly, this study has shown that out of the class I genes present in *A. thaliana*, only *AtTPS1* encodes an active Tre6P synthase. Similarly, heterologous expression of the class II genes in *S. cerevisiae* has shown that none of them apparently codes for a catalytically active Tre6P phosphatase based on this assay (Ramon et al. 2009; Vogel et al. 1998). Although apparently lacking catalytic activity, the tissue-specific and developmental pattern of expression of the Sc*TPS2*-like genes in the plant suggests that their products may exert important regulatory functions (Ramon et al. 2009; Vandesteene et al. 2010; Zimmermann et al. 2004). The plant *TPP* genes, on the other hand, show no sequence similarity to the yeast trehalose biosynthesis genes except for the presence of the TPP catalytic phosphatase box domain. However, they were all catalytically active when expressed in yeast (Vandesteene et al. 2012). Their abundance and tissue-specific expression pattern in plants again suggest that these enzymes may be responsible for maintaining a tightly regulated Tre6P/sucrose ratio at the tissue, cellular, and even subcellular levels (Satoh-Nagasawa et al. 2006; Yadav et al. 2014). In fact, whereas *TPS1* is expressed in most plant cells, the expression of different *TPP* genes seems to be restricted at the cell-type level (Delorge et al. 2014). Moreover their expression is also influenced by environmental conditions (Van Houtte et al. 2013). Trehalase (*TRE1*), on the other hand, is expressed in plants from a single gene and its expression also seems to be cell-type specific (Lopez et al. 2008; Muller et al. 2001; van Dijken et al. 2004). Expression of *AtTRE1* in *S. cerevisiae* has indicated plasma membrane localization of this protein and ability to functionally substitute Sc*ATH1* (Frison et al. 2007). In the plant it is localized in the apoplast membrane and its catalytic domain oriented toward the interior of this compartment, in a way also reminiscent of *Sc*Ath1 (Muller et al. 2001).

The use of yeast has been useful not only to ascertain the localization and catalytic versus regulatory activity of many different plant genes but has also helped to understand their interactions through application of the yeast two-hybrid technology. For example, individual overexpression of different class II *TPS* genes from rice caused enhanced tolerance of this plant to cold and salinity (Li et al. 2011). Two-hybrid analysis in yeast suggests that this effect most likely results from interaction between each of these proteins and the catalytically active *Oryza sativa TPS1* (Zang et al. 2011). These results support the hypothesis that complex formation between several of these catalytically inactive class II proteins and catalytically active class I and/or TPP proteins is crucial to regulate the levels of trehalose, and particularly of Tre6P (Geelen et al. 2007). As it has recently been hypothesized for yeast (Araiza-Olivera et al. 2013; Trevisol et al. 2013), the interaction of these different variants in complex with the catalytically active enzymes may channel Tre6P intramolecularly to conversion into trehalose, thus controlling its subcellular concentration and in turn its effect on signaling pathways and potential toxicity. In support for this hypothesis, the expression of fusion constructs linking TPS and TPP domains has helped to avoid aberrant phenotypes resulting from heterologous gene expression (see also below) (Garg et al. 2002; Goddijn et al. 1997; Jang et al. 2003; Karim et al. 2007).

Heterologous *TPS* and *TPP* genes, including those of *S. cerevisiae*, have been introduced in plants in an effort to confer trehalose accumulation capacity and higher stress tolerance. However, in some cases this has led to opposite results from what was initially expected. In other cases, improvement in stress tolerance has been reached but at the expense of unexpected and undesirable side effects on other phenotypes. The first attempts to obtain transgenic plants accumulating trehalose were carried out in tobacco, employing the *E. coli otsA* gene or Sc*TPS1* (Holmström et al. 1996; Pilon-Smits et al. 1998; Romero et al. 1997). In each case, the resulting plants displayed improved drought tolerance. Later on, this approach was also applied to tomato and potato plants with similar success in terms of enhanced stress tolerance (Cortina and Culiáñez-Macià 2005; Yeo et al. 2000). However, in all these cases, the plants also exhibited unexpected morphological and growth abnormalities. After discovery of endogenous plant trehalose metabolism, it was realized that these side effects were likely due to interference with a role of trehalose metabolism, and in particular Tre6P, in the regulation of plant carbohydrate metabolism and development. Introduction of constitutively expressed heterologous *TPS* or *TPP* genes may be causing an imbalance in the Tre6P/sucrose ratio at the tissue- and cell-type levels, which in turn affects specific signaling and developmental pathways (Delorge et al. 2014; Yadav et al. 2014). Introduction of new heterologous *TPS* and *TPP* genes, for example, newly identified genes from fungi, still appears to be a promising approach for enhancing stress tolerance in plants as long as the appropriate expression system is first selected. Tissue- or cell-specific endogenous promoters may be a more suitable approach compared to constitutive overexpression. However, targeting to the appropriate subcompartment and cell type seems to be crucial for the success of this approach.

VIII. Trehalose Metabolism as a Target for Antifungal Compounds

Invasion of a host organism by a fungal pathogen elicits multiple stress responses in the fungus. The hostile environment in which the fungal pathogen has to survive and reproduce represents many challenges including the defense mechanisms of the host and other stress-inducing conditions such as inadequate nutrient composition, high temperature, oxidative stress, and high osmolarity. At this level, trehalose may play an important role as stress protectant. Moreover, evidence has been accumulated indicating that trehalose metabolism is important not only for survival within the host but also for the development of virulence factors. More detailed aspects on gene regulation in plant pathogenic fungi are covered in Chap. 5.

Plant pathogenic fungi. The importance of trehalose synthesis/degradation enzymes in fungal pathogenicity and virulence is illustrated in studies of the **rice blast fungus *Magnaporthe oryzae*** and the related species, *M. grisea*. This fungus infects rice after a three-celled conidium lands on the surface of the leaf and germinates (Wilson and Talbot 2009). In a nutrient-free and hydrophobic environment such as the leaf surface, the germ tube swells and forms a dome-shaped infectious cell called the appressorium. High turgor in the appressorium, caused by the accumulation of glycerol, provokes the emergence of a thin penetration peg from the base of the cell, which ruptures the leaf cuticle, allowing entry into the plant. It has recently been found that the rapid changes in metabolism during appressorium maturation largely depend on Tre6P synthase (*Mg*Tps1) (Wilson et al. 2010). Invasion of the rice leaf exposes the fungus to sugar-rich conditions, resulting in increased levels of G6P. *Mg*Tps1 directly senses G6P and activates glucose-6-phosphate dehydrogenase in the oxidative pentose phosphate pathway, leading to increased production of NADPH. As NADPH levels increase at the expense of $NADP^+$, three transcriptional repressors, *Mg*Nmr1-3, are inactivated resulting in the activation of at least three GATA factors (including *Mg*Nut1) as well as the expression of genes required for pathogenicity. Through its G6P-sensing role and its effects on NADPH, *Mg*Tps1 simultaneously regulates nitrogen metabolite and carbon catabolite-repressible (CCR) genes. CCR, which represses genes for metabolism of alternative carbon sources, is activated when virulence genes are induced. The use of nitrate as a nitrogen source is only possible via G6P sensing by *Mg*Tps1 and increase in NADPH, necessary for reduction of nitrate to ammonia (Wilson et al. 2007). Growth on nitrate further activates hexokinase I which in turn makes more G6P. Excess glucose metabolism leads to increased levels of citrate that need to be extruded via the MATE-family efflux pump, *Mg*Mtd1, to avoid activation of *Mg*Nmr1-3 repressors and CCR inactivation which would negatively affect the early stages of infectivity (Fernandez et al. 2012). Glycogen and the cAMP-PKA pathway also play an important role in the early stages of infection by *Magnaporthe*. Spore germination at the leaf surface involves glycogen mobilization in a cAMP-dependent process. The *MgGPH1* glycogen phosphorylase and *MgAGL1* amyloglucosidase genes, encoding enzymes required for cytosolic glycogen breakdown, are important virulence factors not merely because of their role in glycogen mobilization but rather because of their requirement for activation of *MgTPS1* expression (Badaruddin et al. 2013). Ras2-dependent activation of cAMP synthesis by *Mg*Mac1 (adenylate cyclase) is important for surface recognition and appressorium development. The Srv2-homologue Cap1 plays a role not only by activating *Mg*Mac1 but also through its actin-binding function, by which it is also thought to be involved in cytoskeleton reorganization during appressorium morphogenesis (Zhou et al. 2012). Finally, deletion of the neutral trehalase gene, *MgNTH1*, still allowed the fungus to infect plants normally, but it caused attenuated pathogenicity due to a decreased ability to colonize plant tissue (Foster et al. 2003).

Fusarium graminearum is an ascomycete that causes ***Fusarium* head blight** in wheat, maize, and other small grain cereal crops worldwide. In this fungus, *TPS1* regulates nitrogen utilization, as was similarly reported in *M. grisea* (Fernandez and Wilson 2011; Fernandez et al. 2012). *F. graminearum tps1Δ* mutants do not have defects in development and virulence (Song et al. 2014). Nevertheless, a *tps1Δ tps2Δ* strain shows a stronger growth defect than the single *TPS1* or *TPS2* deletion mutants in stress conditions, such as in the presence of H_2O_2. Loss of *FgTPS2* causes a 99 % reduction in virulence accompanied by multiple phenotypes, including abolishment of sporulation and sexual development, alteration of cell polarity and cell wall structure, and reduction of chitin biosynthesis. The effects on cell wall composition alone can account only for up to 60 % reduced virulence (Xu et al. 2010). Apparently, the reduction in virulence has to do with Tre6P toxicity since the effect is counteracted by additional deletion of *FgTPS1*. Such toxicity has already been observed in *C. neoformans* (see below) (Petzold et al. 2006). However, this effect does not seem to be correlated with changes in hexokinase activity (Song et al. 2014). Accumulation of high Tre6P was associated with a decrease in free phosphate, such as shown in a *TPS2*-deficient strain of the human pathogen *A. fumigatus* (Puttikamonkul et al. 2010). An important additional effect of *AfTPS2* deletion is the strong reduction of deoxynivalenol mycotoxin biosynthesis (Song et al. 2014). In this sense, recent work in *Fusarium verticillioides* has shown that *TPS1* mutants also produce significantly less fumonisin B1 than wild type and are less pathogenic on maize (Boudreau et al. 2013). Interestingly, in this fungus deletion of *HXK1* had a similar effect on fungal metabolites produced during colonization of maize kernels. The *Fvhxk1Δ* mutant produced approximately 50 % less trehalose and 80 % less fumosine B1 than the wild type and showed increased sensitivity to stress (Kim et al. 2011).

Botrytis cinerea is a necrotrophic fungus that **causes** severe **pre- and postharvest diseases** in more than 200 plant species. The interaction of this fungus with its host begins with conidia attachment to the host surface and continues with conidia germination and penetration into the host (Epton and Richmond 1980). Germination is dependent on the presence of nutrients, mainly sugars, and can be stimulated by exudates from mature fruits (Doehlemann et al. 2005). It can also be stimulated by contact with hydrophobic surfaces (such as the fruit surfaces). Germination induced by a chemical (sugar) signal is dependent on the Ga3 subunit of the heterotrimeric G protein, cAMP, and the mitogen-activated protein kinase BMP1, whereas germination by contact to a hydrophobic surface is absolutely dependent on BMP1 (Doehlemann et al. 2006a). Trehalose metabolism plays a significant role during carbon source-induced germination, but not during germination induced on hydrophobic surfaces in the absence of external nutrients (Doehlemann et al. 2006b). Although the latter takes place more slowly and does not induce trehalose mobilization, addition of cAMP can cause trehalose breakdown in these conditions. This is thought to happen through cAMP-PKA dependent phosphorylation of the neutral trehalase, Tre1. It is therefore likely that in *B. cinerea*, hexoses are sensed and activate trehalose mobilization during conidia germination via activation of G-protein-dependent cAMP signaling. Trehalose metabolism plays a role in conidial germination, first, by providing an initial source of energy for the germinating spore when external nutrient sources are limiting and second through Tre6P regulation of the glycolytic flux (via regulation of hexokinase activity), which is also important for induction of germination. Knockout mutant studies have indeed shown that in *B. cinerea*, Hxk1 is required for fungal development and virulence in the presence of hexoses (Rui and Hahn 2007).

A particular case in which fungus-plant interaction can be symbiotic or at minimum mutualistic is in **ectomycorrhizas**, soil fungi that interact with trees of boreal and temperate forests. When fungal hyphae recognize an emerging fine root of a compatible plant partner, they grow toward it and colonize the root surface, by forming a sheath or mantle of hyphae, which encloses the root and isolates it from the surrounding soil (Blasius et al. 1986; Martin et al. 2001). After or at the same time as sheath formation, fungal hyphae grow inside the infected root, forming highly branched structures in the apoplast of the rhizodermis (angiosperms) or in the root cortex (gymnosperms). This so-called Hartig net generates a large surface area between the two partners (Kottke and Oberwinkler 1987). The "Hartig net" becomes an interface between plant and fungus, adapted to the exchange of plant-derived carbohydrates and fungus-derived nutrients. Experiments performed between the ectomycorrhiza *Amanita muscaria* and the poplar plant host *Populus tremula x tremuloides* have indicated major expression of *TPS* genes as well as trehalose synthesis taking place in the hyphae present in the "Hartig net" interface, rather than at the sheath (Lopez et al. 2007). The resulting immediate transformation of glucose derived from the plant into trehalose helps to maintain constant low concentrations of glucose at the roots, which the plant tries to compensate by increasing sugar transport toward the roots. At the same time, the fungus benefits from continuously piling the sugar up in the form of storage trehalose.

Human pathogenic fungi. Multiple studies in human pathogenic fungi have also revealed the importance of the trehalose biosynthesis pathway in pathogenicity and virulence. ***Aspergillus fumigatus*** is an opportunistic mold that causes acute and often fatal **pulmonary disease** in immunocompromised patients. Up to four Tre6P synthase genes have been identified in this organism. Deletion of two of them, *tpsA* and *tpsB*, causes hypervirulence associated with alterations in the cell wall resulting in a significant reduction in macrophage adherence and phagocytosis (Al-Bader et al. 2010). On the other hand, deletion of the only putative Tre6P phosphatase (*TPS2* orthologue) gene *orlA* caused avirulence, also due to cell wall defects (Puttikamonkul et al. 2010). In this case trehalose synthesis was not blocked because of activation of two trehalose phosphorylase genes. However, accumulation of Tre6P led to strongly reduced levels of hexokinase I activity, which in turn affects synthesis of cell wall components such as *N*-acetyl-D-glucosamine. Thus, trehalose synthesis enzymes have both positive and negative effects on virulence suggesting that they could be successful antifungal targets. The cAMP-PKA pathway plays an important role in *A. fumigatus* virulence. Apart from affecting growth and sporulation, elevated PKA activity led to increased expression of the polyketide synthase that is essential for production of dihydroxynaphthalene melanin in *A. fumigatus,* contributing to virulence (Grosse et al. 2008).

Candida albicans is a major pathogenic fungus in humans. It exists as a commensal in the urogenital and gastrointestinal tracts and on the skin. It can cause **mucosal infections** in healthy individuals and potentially fatal **infections in the bloodstream and internal organs** of immunocompromised patients. Subjecting the fungus to oxidative and cationic stress forms part of the defense mechanism displayed by macrophages and granulocytes in the blood. These immune system cells engulf yeast cells and destroy them by using high concentrations of enzymes and reactive oxygen species. *C. albicans* has developed a variety of strategies to avoid being destroyed by macrophages, including the formation of hyphae (which rupture the macrophage allowing the yeast to escape), the synthesis of molecules (such as trehalose), and/or the activation of protective enzymes and induction of apoptosis in macrophages through surface glycolipids (Ibata-Ombetta et al. 2003) or secreted proteins (Alvarez-Peral et al. 2002; Fernandez-Arenas et al. 2009; Martinez-Esparza et al. 2009).

The *CaTPS1* gene is induced in response to oxidative stress and mutants unable to synthesize trehalose are very sensitive to oxidative stress in vitro (Alvarez-Peral et al. 2002). Moreover, when inoculated in mice, the Ca*TPS1*-deficient strain displays a lower infection rate and lower resistance to killing by macrophages (Martinez-Esparza et al. 2007; Zaragoza et al. 1998). Expression of trehalose synthesis genes as well as accumulation of trehalose during stationary phase or upon oxidative stress is dependent on Cap1, a basic region-leucine zipper (bZip) transcription factor, homologue of *Sc*Yap1, participating in oxidative stress tolerance in *C. albicans* (Cao et al. 2008). *C. albicans* has adapted to detect levels of glucose below the regular concentrations in blood, and, in contrast to *S. cerevisiae*, it responds to this sugar by activating genes involved in protection against a wide variety of stresses, including upregulation of trehalose synthesis genes (Rodaki et al. 2009). Thus, *Candida* interprets glucose as a signal for being in the blood and thus in potential danger of attack by macrophages. Although trehalose has not yet been accumulated, the mRNA for the *TPS* genes accumulates, preparing the cells to eventual trehalose synthesis for use as stress protectant. Lack of *CaTPS1* alters the composition of the outer cell wall layer (Martinez-Esparza et al. 2011). In liquid culture, the *tps1Δ/tps1Δ* strain induced lower levels of ERK1,2 phosphorylation and TNF-α production in macrophages, increasing the resistance to killing by these cells. However, on solid medium, in which over-glycosylation was less evident, the *tps1Δ/tps1Δ* strain showed similar macrophage interaction as the wild type but was less resistant to killing, underlining the protective role of trehalose. Lack of Tre6P phosphatase in the *tps2Δ/tps2Δ* strain also seems to have a virulence attenuating effect, likely due to accumulation of Tre6P interfering with the assembly of a normal cell wall (Van Dijck et al. 2002; Zaragoza et al. 2002). This highlights the importance of trehalose and carbohydrate metabolism in cell wall organization, which in turn provides different levels of protection against macrophage killing. Human body temperature induces filamentous growth in *C. albicans*, and this process is triggered by the release from repression by the essential molecular chaperone Hsp90 on its target proteins (Shapiro et al. 2009). Heat-induced filamentation is severely reduced in the absence of *Ca*Gpr1 (Serneels et al. 2012). Whereas a *gpr1Δ/gpr1Δ* mutant shows only decreased virulence, additional deletion of *TPS2* causing accumulation of high levels of Tre6P was sensitive to high temperatures and resulted in complete avirulence in a mouse infection model (Maidan et al. 2008). *Ca*Gpr1 is required for control of basal trehalose levels, in particular when cells are metabolically active and growing in favorable conditions. Absence of the receptor seems to mimic a nutrient-poor environment for the cells, leading to trehalose accumulation. This is consistent with absence of Gpr1 leading to reduced cAMP-PKA signaling, which in turn reduces repression of TPS genes. Supporting this conclusion, mutants in proteins that activate the pathway, such as in Cdc25, display higher trehalose content, whereas mutants in repressors of the pathway such as Ira2 have reduced levels of trehalose. Reactivation of the PKA pathway by serum restores filamentation in a *gpr1Δ/gpr1Δ* strain (Maidan et al. 2005). Use of the **trehalase inhibitor validamycin** and of the **Hsp90 inhibitor geldanamycin** have confirmed that the *gpr1Δ/gpr1Δ* defect in filamentation is due to increased levels of trehalose caused by downregulation of the cAMP-PKA pathway. This accentuates the repressive chaperoning function of *Ca*Hsp90 on some of its clients supporting filamentation (Serneels et al. 2012). How accumulation of trehalose affects *Ca*Hsp90 activity is still not known. Three possibilities have been proposed: trehalose could be acting directly on Hsp90 as a chemical cochaperone that stabilizes its interaction with the client. A second possibility is that levels of trehalose indirectly affect expression of *HSP90* by inhibiting heat shock-triggered hyperphosphorylation of the transcription factor Hsf1 (Conlin and Nelson 2007; Nicholls et al. 2009). Third, accumulation of trehalose could affect Hsp90 nuclear/cytosolic distribution (Tapia and Morano 2010). As for the effects of trehalose mobilization in *C. albicans*, deletion of the cell wall-associated acid trehalase, but not of the neutral trehalase, has a negative impact on morphogenesis and virulence in mice (Pedreno et al. 2007). Trehalose levels increase dramatically in the first 6 h of biofilm formation, but decline in mature biofilms (Zhu et al. 2013). One possibility is that trehalose exerts an antioxidant function in the early stages of biofilm formation and is later used as energy source as the biofilm matures. It is also possible that its presence in early biofilm formation prevents filament formation. Recent work has shown that in the yeast *C. parapsilosis* deletion of the acid trehalase gene, *ATC1*, results in reduced pathogenicity, again indicating the importance of proper trehalose metabolism for virulence (Sanchez-Fresneda et al. 2014).

Recent studies in the yeast ***Cryptococcus neoformans*** have also demonstrated the importance of the trehalose biosynthesis pathway for infectivity and survival in the host. *C. neoformans* is an **opportunistic heterobasidiomycetous fungus**, with a propensity for producing **meningoencephalitis** in immunocompromised patients. While elimination of the orthologue of *S. cerevisiae NTH1* had no striking effect on virulence, deletions of trehalose-synthesizing enzymes resulted in loss of virulence (Petzold et al. 2006). This was in part due to a reduced capacity to withstand the mammalian body temperature (37 °C). The loss of virulence was more accentuated in the *tps2*Δ than in the *tps1*Δ strain, suggesting Tre6P accumulation rather than lack of trehalose buildup to be the cause. *C. gattii* is a related species for which certain strains were found to be virulent also in non-immunocompromised patients. Deletion of *CgTPS1* and *CgTPS2* caused, respectively, inability to grow and death at 37 °C in laboratory media and decreased survival in the invertebrate and vertebrate hosts (Ngamskulrungroj et al. 2009). Although one important role of *CgTPS1* is to control hexokinases I and II and thus glycolysis and growth at 37 °C, presence of the TPS enzymes has also consequences beyond glycolysis, affecting the development of other virulence factors such as melanin synthesis, capsule production, mating, cell wall integrity, and protein secretion. *C. neoformans* capsule synthesis is also transcriptionally regulated by elements of the cAMP-PKA pathway (Cramer et al. 2006; D'Souza et al. 2001; Pukkila-Worley et al. 2005). Strains with mutations in core cAMP signaling elements display defective expression of capsule and are attenuated for virulence in animal models. *C. neoformans* also uses PKA and the Rim pathway to regulate the localization, activation, and processing of the *Cn*Rim101 transcription factor, which is needed for development of virulence factors such as capsule formation (O'Meara et al. 2014).

The large number of reports documenting the importance of trehalose metabolism for fungal pathogenicity highlights its potential as a target for new antifungals. This is particularly true in mammals where trehalose metabolism is completely absent.

IX. Conclusions and Perspectives

Trehalose was originally discovered as a storage disaccharide in fungi. Since then, the biological role of trehalose and trehalose metabolism has seen a remarkable expansion. First, it was found to serve also as a stress protectant, leading to attempts to engineer trehalose accumulation in higher plants in order to enhance their stress tolerance. This resulted in highly unexpected side effects on the physiology and development of the plants. A similar unexpected event happened in yeast research when the cloning of genes involved in trehalose metabolism and those affected in several mutants with serious growth defects on glucose and related fermentable sugars turned out to be the same indicating that trehalose metabolism played a crucial role in the control of glycolysis. Although the role of Tre6P as feedback inhibitor of hexokinase is well established in vitro, numerous results obtained in vivo suggest that this is not the only crucial target of Tre6P and may not even be the most important target. On the other hand, the role of Tre6P in regulation of carbon development in other fungi remains more unclear. Research on trehalose metabolism in plants has seen remarkable expansion. It has revealed that, similarly to yeast, Tre6P plays a role in the regulation of carbon metabolism, although this does not seem to involve control of hexokinase activity. It is clear that both in yeast and in plants, the regulatory role of trehalose metabolism will continue to attract much attention. Also the protective role of trehalose against stress-induced damage will continue to attract great interest because of the fundamental importance of stress protection mechanisms in general and the obvious importance in many commercial applications. A possible role of trehalose metabolism as target for novel antifungal compounds also appears to be promising. The conspicuous effects in yeast of nutrient withdrawal and re-addition on trehalose accumulation and mobilization, respectively, and its connection with growth rate and cell cycle progression will continue to serve as a convenient readout for studies on nutrient sensing and signaling. They may finally lead to a better understanding of the molecular mechanisms involved in nutrient control of cell growth and proliferation.

References

Abrashev R, Stoitsova S, Krumova E, Pashova S, Paunova-Krasteva T, Vassilev S, Dolashka-Angelova P, Angelova M (2013) Temperature-stress tolerance of the fungal strain *Aspergillus niger* 26: physiological and ultrastructural changes. World J Microbiol Biotechnol. doi:10.1007/s11274-013-1586-8

Aisaka K, Masuda T, Chikamune T, Kamitori K (1998) Purification and characterization of trehalose phosphorylase from *Catellatospora ferruginea*. Biosci Biotechnol Biochem 62(4):782–787

Alaamery MA, Hoffman CS (2008) *Schizosaccharomyces pombe* Hsp90/Git10 is required for glucose/cAMP signaling. Genetics 178(4):1927–1936. doi:10.1534/genetics.107.086165

Al-Bader N, Vanier G, Liu H, Gravelat FN, Urb M, Hoareau CM, Campoli P, Chabot J, Filler SG, Sheppard DC (2010) Role of trehalose biosynthesis in *Aspergillus fumigatus* development, stress response, and virulence. Infect Immun 78(7):3007–3018. doi:10.1128/iai.00813-09

Albers E, Larsson C, Andlid T, Walsh MC, Gustafsson L (2007) Effect of nutrient starvation on the cellular composition and metabolic capacity of *Saccharomyces cerevisiae.* Appl Environ Microbiol 73(15):4839–4848. doi:10.1128/aem.00425-07

Alexandre H, Plourde L, Charpentier C, Francois J (1998) Lack of correlation between trehalose accumulation, cell viability and intracellular acidification as induced by various stresses in *Saccharomyces cerevisiae.* Microbiology 144(Pt 4):1103–1111

Allen C, Buttner S, Aragon AD, Thomas JA, Meirelles O, Jaetao JE, Benn D, Ruby SW, Veenhuis M, Madeo F, Werner-Washburne M (2006) Isolation of quiescent and nonquiescent cells from yeast stationary-phase cultures. J Cell Biol 174(1):89–100. doi:10.1083/jcb.200604072

Alvarez-Peral FJ, Zaragoza O, Pedreno Y, Arguelles JC (2002) Protective role of trehalose during severe oxidative stress caused by hydrogen peroxide and the adaptive oxidative stress response in *Candida albicans.* Microbiology 148(Pt 8):2599–2606

Alves SL Jr, Herberts RA, Hollatz C, Trichez D, Miletti LC, de Araujo PS, Stambuk BU (2008) Molecular analysis of maltotriose active transport and fermentation by *Saccharomyces cerevisiae* reveals a determinant role for the AGT1 permease. Appl Environ Microbiol 74(5):1494–1501. doi:10.1128/aem.02570-07

Amaral FC, Van Dijck P, Nicoli JR, Thevelein JM (1997) Molecular cloning of the neutral trehalase gene from *Kluyveromyces lactis* and the distinction between neutral and acid trehalases. Arch Microbiol 167(4):202–208

An MZ, Tang YQ, Mitsumasu K, Liu ZS, Shigeru M, Kenji K (2011) Enhanced thermotolerance for ethanol fermentation of *Saccharomyces cerevisiae* strain by overexpression of the gene coding for trehalose-6-phosphate synthase. Biotechnol Lett 33(7):1367–1374. doi:10.1007/s10529-011-0576-x

Anchordoguy TJ, Crowe JH, Griffin FJ, Clark WH (1988) Cryopreservation of sperm from the marine shrimp *Sicyonia ingentis.* Cryobiology 25:238–243

Ando A, Nakamura T, Murata Y, Takagi H, Shima J (2007) Identification and classification of genes required for tolerance to freeze-thaw stress revealed by genome-wide screening of *Saccharomyces cerevisiae* deletion strains. FEMS Yeast Res 7(2):244–253. doi:10.1111/j.1567-1364.2006.00162.x

App H, Holzer H (1989) Purification and characterization of neutral trehalase from the yeast ABYS1 mutant. J Biol Chem 264(29):17583–17588

Apweiler E, Sameith K, Margaritis T, Brabers N, van de Pasch L, Bakker LV, van Leenen D, Holstege FC, Kemmeren P (2012) Yeast glucose pathways converge on the transcriptional regulation of trehalose biosynthesis. BMC Genomics 13:239. doi:10.1186/1471-2164-13-239

Aragon AD, Rodriguez AL, Meirelles O, Roy S, Davidson GS, Tapia PH, Allen C, Joe R, Benn D, Werner-Washburne M (2008) Characterization of differentiated quiescent and nonquiescent cells in yeast stationary-phase cultures. Mol Biol Cell 19(3):1271–1280. doi:10.1091/mbc.E07-07-0666

Araiza-Olivera D, Sampedro JG, Mujica A, Pena A, Uribe-Carvajal S (2010) The association of glycolytic enzymes from yeast confers resistance against inhibition by trehalose. FEMS Yeast Res 10(3):282–289. doi:10.1111/j.1567-1364.2010.00605.x

Araiza-Olivera D, Chiquete-Felix N, Rosas-Lemus M, Sampedro JG, Pena A, Mujica A, Uribe-Carvajal S (2013) A glycolytic metabolon in *Saccharomyces cerevisiae* is stabilized by F-actin. FEBS J 280(16):3887–3905. doi:10.1111/febs.12387

Argüelles JC (1997) Thermotolerance and trehalose accumulation induced by heat shock in yeast cells of *Candida albicans.* FEMS Microbiol Lett 146:65–71

Argüelles JC (2000) Physiological roles of trehalose in bacteria and yeasts: a comparative analysis. Arch Microbiol 174:217–224

Arguelles JC, Gacto M (1986) Comparative study of two trehalases from *Candida utilis.* Microbiologia 2(2):105–114

Argüelles JC, Vicente-Soler J, Gacto M (1986) Protein phosphorylation and trehalase activation in *Candida utilis.* FEMS Microbiol Lett 34:361–365

Arisan-Atac I, Wolschek MF, Kubicek CP (1996) Trehalose-6-phosphate synthase A affects citrate accumulation by *Aspergillus niger* under conditions of high glycolytic flux. FEMS Microbiol Lett 140(1):77–83

Arneborg N, Moos MK, Jakobsen M (1997) Induction of acetic acid tolerance and trehalose accumulation by added and produced ethanol in *Saccharomyces cerevisiae.* Biotechnol Lett 19:931–933

Arora A, Ha C, Park CB (2004) Inhibition of insulin amyloid formation by small stress molecules. FEBS Lett 564(1–2):121–125. doi:10.1016/s0014-5793(04)00326-6

Arya SC (2000) Stabilization of vaccines: to be or not to be. Vaccine 19:595–597

Ashmarina LI, Muronetz VI, Nagradova NK (1984) Immobilized glyceraldehyde-3-phosphate dehydrogenase forms a complex with phosphoglycerate kinase. Biochem Int 9(4):511–521

Attfield PV (1987) Trehalose accumulates in *Saccharomyces cerevisiae* during exposure to agents that induce heat shock response. FEBS Lett 225(1–2):259–263

Attfield PV, Raman A, Northcott CJ (1992) Construction of *Saccharomyces cerevisiae* strains that accumulate relatively low concentrations of trehalose, and their application in testing the contribution of the disaccharide to stress tolerance. FEMS Microbiol Lett 94:271–276

Avonce N, Mendoza-Vargas A, Morett E, Iturriaga G (2006) Insights on the evolution of trehalose biosynthesis. BMC Evol Biol 6:109. doi:10.1186/1471-2148-6-109

Badaruddin M, Holcombe LJ, Wilson RA, Wang ZY, Kershaw MJ, Talbot NJ (2013) Glycogen metabolic genes are involved in trehalose-6-phosphate synthase-mediated regulation of pathogenicity by the rice blast fungus *Magnaporthe oryzae*. PLoS Pathog 9(10):e1003604. doi:10.1371/journal.ppat.1003604

Bandara A, Fraser S, Chambers PJ, Stanley GA (2009) Trehalose promotes the survival of *Saccharomyces cerevisiae* during lethal ethanol stress, but does not influence growth under sublethal ethanol stress. FEMS Yeast Res 9(8):1208–1216. doi:10.1111/j.1567-1364.2009.00569.x

Barnett JA (1976) The utilization of sugars by yeasts. Adv Carbohydr Chem Biochem 32:125–234

Barton JK, den Hollander JA, Hopfield JJ, Shulman RG (1982) 13C Nuclear magnetic resonance study of trehalose mobilization in yeast spores. J Bacteriol 151:177–185

Basu A, Bhattacharyya S, Chaudhuri P, Sengupta S, Ghosh AK (2006) Extracellular trehalose utilization by *Saccharomyces cerevisiae*. Biochim Biophys Acta 1760(2):134–140. doi:10.1016/j.bbagen.2005.11.016

Batlle M, Lu A, Green DA, Xue Y, Hirsch JP (2003) Krh1p and Krh2p act downstream of the Gpa2p G(alpha) subunit to negatively regulate haploid invasive growth. J Cell Sci 116(Pt 4):701–710

Becker A, Schloder P, Steele JE, Wegener G (1996) The regulation of trehalose metabolism in insects. Experientia 52(5):433–439

Behm CA (1997) The role of trehalose in the physiology of nematodes. Int J Parasitol 27(2):215–229

Bell W, Klaassen P, Ohnacker M, Boller T, Herweijer M, Schoppink P, Van der Zee P, Wiemken A (1992) Characterization of the 56-kDa subunit of yeast trehalose-6-phosphate synthase and cloning of its gene reveal its identity with the product of *CIF1*, a regulator of carbon catabolite inactivation. Eur J Biochem 209(3):951–959

Bell W, Sun W, Hohmann S, Wera S, Reinders A, De Virgilio C, Wiemken A, Thevelein JM (1998) Composition and functional analysis of the *Saccharomyces cerevisiae* trehalose synthase complex. J Biol Chem 273(50):33311–33319

Bell-Pedersen D, Shinohara ML, Loros JJ, Dunlap JC (1996) Circadian clock-controlled genes isolated from *Neurospora crassa* are late night- to early morning-specific. Proc Natl Acad Sci USA 93(23):13096–13101

Belocopitow E, Marechal LR (1970) Trehalose phosphorylase from *Euglena gracilis*. Biochim Biophys Acta 198(1):151–154

Belotti F, Tisi R, Paiardi C, Groppi S, Martegani E (2011) PKA-dependent regulation of Cdc25 Ras-GEF localization in budding yeast. FEBS Lett 585(24):3914–3920. doi:10.1016/j.febslet.2011.10.032

Belotti F, Tisi R, Paiardi C, Rigamonti M, Groppi S, Martegani E (2012) Localization of Ras signaling complex in budding yeast. Biochim Biophys Acta 1823(7):1208–1216. doi:10.1016/j.bbamcr.2012.04.016

Beltran FF, Castillo R, Vicente-Soler J, Cansado J, Gacto M (2000) Role for trehalase during germination of spores in the fission yeast *Schizosaccharomyces pombe*. FEMS Microbiol Lett 193(1):117–121

Benaroudj N, Lee DH, Goldberg AL (2001) Trehalose accumulation during cellular stress protects cells and cellular proteins from damage by oxygen radicals. J Biol Chem 276(26):24261–24267. doi:10.1074/jbc.M101487200

Benbadis L, Cot M, Rigoulet M, Francois J (2009) Isolation of two cell populations from yeast during high-level alcoholic fermentation that resemble quiescent and nonquiescent cells from the stationary phase on glucose. FEMS Yeast Res 9(8):1172–1186. doi:10.1111/j.1567-1364.2009.00553.x

Bhandal IS, Hauptmann RM, Widholm JM (1985) Trehalose as cryoprotectant for the freeze preservation of carrot and tobacco cells. Plant Physiol 78:430–432

Bhattacharyya S, Chaudhuri P, Basu A, Sengupta S, Ghosh AK (2005) Regulation of trehalose metabolism by Adox and AdoMet in *Saccharomyces cerevisiae*. Indian J Exp Biol 43(4):360–368

Blakeley D, Tolliday B, Colaço C, Roser B (1990) Dry instant blood typing plate for bedside use. Lancet 336:854–855

Blasius D, Feil W, Kottke I, Oberwinkler F (1986) Hartig net structure and formation of fully ensheated ectomycorrhizas. Nord J Bot 6:837–842

Blazquez MA, Gancedo C (1994) Identification of extragenic suppressors of the *cif1* mutation in *Saccharomyces cerevisiae*. Curr Genet 25(2):89–94

Blazquez MA, Gancedo C (1995) Mode of action of the *qcr9* and *cat3* mutations in restoring the ability of *Saccharomyces cerevisiae tps1* mutants to grow on glucose. Mol Gen Genet 249(6):655–664

Blazquez MA, Lagunas R, Gancedo C, Gancedo JM (1993) Trehalose-6-phosphate, a new regulator of yeast glycolysis that inhibits hexokinases. FEBS Lett 329:51–54

Blazquez MA, Stucka R, Feldmann H, Gancedo C (1994) Trehalose-6-P synthase is dispensable for growth on glucose but not for spore germination in *Schizosaccharomyces pombe*. J Bacteriol 176:3895–3902

Boles E, Heinisch J, Zimmermann FK (1993) Different signals control the activation of glycolysis in the yeast *Saccharomyces cerevisiae*. Yeast 9(7):761–770. doi:10.1002/yea.320090710

Bonini BM, Van Vaeck C, Larsson C, Gustafsson L, Ma P, Winderickx J, Van Dijck P, Thevelein JM (2000) Expression of *Escherichia coli otsA* in a *Saccharomyces cerevisiae tps1* mutant restores trehalose 6-phosphate levels and partly restores growth and fermentation with glucose and control of glucose influx into glycolysis. Biochem J 350(Pt 1):261–268

Bonini BM, Van Dijck P, Thevelein JM (2003) Uncoupling of the glucose growth defect and the deregulation of glycolysis in *Saccharomyces cerevisiae* Tps1 mutants expressing trehalose-6-phosphate-insensitive hexokinase from *Schizosaccharomyces pombe*. Biochim Biophys Acta 1606(1–3):83–93

Bontron S, Jaquenoud M, Vaga S, Talarek N, Bodenmiller B, Aebersold R, De Virgilio C (2013) Yeast endosulfines control entry into quiescence and chronological life span by inhibiting protein phosphatase 2A. Cell Rep 3(1):16–22. doi:10.1016/j.celrep.2012.11.025

Boos W, Ehmann U, Bremer E, Middendorf A, Postma P (1987) Trehalase of *Escherichia coli*. Mapping and cloning of its structural gene and identification of the enzyme as a periplasmic protein induced under high osmolarity growth conditions. J Biol Chem 262:13212–13218

Borgia PT, Miao Y, Dodge CL (1996) The *orlA* gene from *Aspergillus nidulans* encodes a trehalose-6-phosphate phosphatase necessary for normal growth and chitin synthesis at elevated temperatures. Mol Microbiol 20(6):1287–1296

Boudreau BA, Larson TM, Brown DW, Busman M, Roberts ES, Kendra DF, McQuade KL (2013) Impact of temperature stress and validamycin A on compatible solutes and fumonisin production in *F. verticillioides*: role of trehalose-6-phosphate synthase. Fungal Genet Biol 57:1–10. doi:10.1016/j.fgb.2013.06.001

Bourret JA (1986) Evidence that a glucose-mediated rise in cyclic AMP triggers germination of *Pilobolus longipes* spores. Exp Mycol 10:60–66

Boy-Marcotte E, Ikonomi P, Jacquet M (1996) SDC25, a dispensable Ras guanine nucleotide exchange factor of *Saccharomyces cerevisiae* differs from CDC25 by its regulation. Mol Biol Cell 7(4):529–539

Boy-Marcotte E, Perrot M, Bussereau F, Boucherie H, Jacquet M (1998) Msn2p and Msn4p control a large number of genes induced at the diauxic transition which are repressed by cyclic AMP in *Saccharomyces cerevisiae*. J Bacteriol 180(5):1044–1052

Braña AF, Mendez C, Diaz LA, Manzanal MB, Hardisson C (1986) Glycogen and trehalose accumulation during colony development in *Streptomyces antibioticus*. J Gen Microbiol 132(Pt 5):1319–1326

Breedveld MW, Zevenhuizen LPTM, Zehnder AJB (1991) Osmotically-regulated trehalose accumulation and cyclic beta-(1,2)-glucan excretion by *Rhizobium leguminosarum* biovar trifolii TA-1. Arch Microbiol 156:501–506

Brewster JL, de Valoir T, Dwyer ND, Winter E, Gustin MC (1993) An osmosensing signal transduction pathway in yeast. Science 259(5102):1760–1763

Broek D, Toda T, Michaeli T, Levin L, Birchmeier C, Zoller M, Powers S, Wigler M (1987) The *S. cerevisiae CDC25* gene product regulates the RAS/adenylate cyclase pathway. Cell 48(5):789–799

Broggi S, Martegani E, Colombo S (2013) Live-cell imaging of endogenous Ras-GTP shows predominant Ras activation at the plasma membrane and in the nucleus in *Saccharomyces cerevisiae*. Int J Biochem Cell Biol 45(2):384–394. doi:10.1016/j.biocel.2012.10.013

Brondijk TH, van der Rest ME, Pluim D, de Vries Y, Stingl K, Poolman B, Konings WN (1998) Catabolite inactivation of wild-type and mutant maltose transport proteins in *Saccharomyces cerevisiae*. J Biol Chem 273(25):15352–15357

Brownlee C, Jennings DH (1981) The content of soluble carbohydrates and their translocation in mycelium of *Serpula lacrymans*. Trans Br Mycol Soc 77:615–619

Budhwar R, Lu A, Hirsch JP (2010) Nutrient control of yeast PKA activity involves opposing effects on phosphorylation of the Bcy1 regulatory subunit. Mol Biol Cell 21(21):3749–3758. doi:10.1091/mbc.E10-05-0388

Budhwar R, Fang G, Hirsch JP (2011) Kelch repeat proteins control yeast PKA activity in response to nutrient availability. Cell Cycle 10(5):767–770

Buitink J, Leprince O (2004) Glass formation in plant anhydrobiotes: survival in the dry state. Cryobiology 48(3):215–228. doi:10.1016/j.cryobiol.2004.02.011

Burke MJ (1985) The glassy state and survival of anhydrous biological systems. In: Leopold AC (ed) Membranes, metabolism and dry organisms. Cornell University Press, Ithaca, NY, pp 358–363

Cabib E, Leloir LF (1958) The biosynthesis of trehalose phosphate. J Biol Chem 231(1):259–275

Cameron S, Levin L, Zoller M, Wigler M (1988) cAMP-independent control of sporulation, glycogen metabolism and heat shock resistance in *S. cerevisiae*. Cell 53:555–566

Cameroni E, Hulo N, Roosen J, Winderickx J, De Virgilio C (2004) The novel yeast PAS kinase Rim 15 orchestrates G0-associated antioxidant defense mechanisms. Cell Cycle 3(4):462–468

Camonis JH, Kalekine M, Gondre B, Garreau H, Boy-Marcotte E, Jacquet M (1986) Characterization, cloning and sequence analysis of the *CDC25* gene which controls the cyclic AMP level of *Saccharomyces cerevisiae*. EMBO J 5(2):375–380

Campanella ME, Chu H, Low PS (2005) Assembly and regulation of a glycolytic enzyme complex on the human erythrocyte membrane. Proc Natl Acad Sci USA 102(7):2402–2407. doi:10.1073/pnas.0409741102

Cannon JF, Pringle JR, Fiechter A, Khalil M (1994) Characterization of glycogen-deficient glc mutants of *Saccharomyces cerevisiae*. Genetics 136(2):485–503

Cao Y, Wang Y, Dai B, Wang B, Zhang H, Zhu Z, Xu Y, Cao Y, Jiang Y, Zhang G (2008) Trehalose is an important mediator of Cap1p oxidative stress response in *Candida albicans*. Biol Pharm Bull 31(3):421–425

Carrillo D, Vicente-Soler J, Gacto M (1992) Activation of neutral trehalase by fermentable sugars and cAMP in the fission yeast *Schizosaccharomyces pombe*. FEMS Microbiol Lett 98:61–66

Carrillo D, Vicente-Soler J, Gacto M (1994a) Cyclic AMP signalling pathway and trehalase activation in the fission yeast *Schizosaccharomyces pombe*. Microbiology 140:1467–1472

Carrillo D, Vicente-Soler J, Gacto M (1994b) Sensitivity of fructose-1,6-bisphosphatase to glucose and cyclic AMP in the fission yeast *Schizosaccharomyces pombe*. Microbios 79:73–79

Carrillo D, Vicente-Soler J, Fernandez J, Soto T, Cansado J, Gacto M (1995) Activation of cytoplasmic trehalase by cyclic-AMP-dependent and cyclic-AMP-independent signalling pathways in the yeast *Candida utilis*. Microbiology 141(Pt 3):679–686

Castermans D, Somers I, Kriel J, Louwet W, Wera S, Versele M, Janssens V, Thevelein JM (2012) Glucose-induced posttranslational activation of protein phosphatases PP2A and PP1 in yeast. Cell Res 22(6):1058–1077. doi:10.1038/cr.2012.20

Cerrutti P, Segovia de Huergo M, Galvagno M, Schebor C, del Pilar Buera M (2000) Commercial baker's yeast stability as affected by intracellular content of trehalose, dehydration procedure and the physical properties of external matrices. Appl Microbiol Biotechnol 54(4):575–580

Chakraborty N, Menze MA, Elmoazzen H, Vu H, Yarmush ML, Hand SC, Toner M (2012) Trehalose transporter from African chironomid larvae improves desiccation tolerance of Chinese hamster ovary cells. Cryobiology 64(2):91–96. doi:10.1016/j.cryobiol.2011.11.007

Charlab R, Oliveira DE, Panek AD (1985) Investigation of the relationship between *sst1* and *fdp* mutations in yeast and their effect on trehalose synthesis. Braz J Med Biol Res 18:447–454

Chaudhuri P, Basu A, Ghosh AK (2008) Aggregation dependent enhancement of trehalose-6-phosphate synthase activity in *Saccharomyces cerevisiae*. Biochim Biophys Acta 1780(2):289–297. doi:10.1016/j.bbagen.2007.12.002

Chaudhuri P, Basu A, Sengupta S, Lahiri S, Dutta T, Ghosh AK (2009) Studies on substrate specificity and activity regulating factors of trehalose-6-phosphate synthase of *Saccharomyces cerevisiae*. Biochim Biophys Acta 1790(5):368–374. doi:10.1016/j.bbagen.2009.03.008

Chen Q, Haddad GG (2004) Role of trehalose phosphate synthase and trehalose during hypoxia: from flies to mammals. J Exp Biol 207(Pt 18):3125–3129. doi:10.1242/jeb.01133

Chen T, Fowler A, Toner M (2000) Literature review: supplemented phase diagram of the trehalose-water binary mixture. Cryobiology 40:277–282

Chen T, Acker JP, Eroglu A, Cheley S, Bayley H, Fowler A, Toner M (2001) Beneficial effect of intracellular trehalose on the membrane integrity of dried mammalian cells. Cryobiology 43(2):168–181. doi:10.1006/cryo.2001.2360

Chen D, Toone WM, Mata J, Lyne R, Burns G, Kivinen K, Brazma A, Jones N, Bahler J (2003) Global transcriptional responses of fission yeast to environmental stress. Mol Biol Cell 14(1):214–229. doi:10.1091/mbc.E02-08-0499

Chen J, Tang B, Chen H, Yao Q, Huang X, Chen J, Zhang D, Zhang W (2010) Different functions of the insect soluble and membrane-bound trehalase genes in chitin biosynthesis revealed by RNA interference. PLoS One 5(4):e10133. doi:10.1371/journal.pone.0010133

Chino H, Lum PY, Nagao E, Hiraoka T (1992) The molecular and metabolic essentials for long-distance flight in insects. J Comp Phys B Biochem Syst Environ Phys 162:101–106

Chyb S, Dahanukar A, Wickens A, Carlson JR (2003) *Drosophila* Gr5a encodes a taste receptor tuned to trehalose. Proc Natl Acad Sci USA 100(Suppl 2):14526–14530. doi:10.1073/pnas.2135339100

Clark MS, Thorne MA, Purac J, Grubor-Lajsic G, Kube M, Reinhardt R, Worland MR (2007) Surviving extreme polar winters by desiccation: clues from Arctic springtail (*Onychiurus arcticus*) EST libraries. BMC Genomics 8:475. doi:10.1186/1471-2164-8-475

Clegg JS (1985) The physical properties and metallic status of *Artemia* cysts at low water contents: The 'water replacement hypothesis'. In: Leopold AC (ed) Membranes, metabolism and dry organisms. Cornell University Press, Ithaca, NY, pp 169–187

Clegg JS, Seitz P, Seitz W, Hazlewood CF (1982) Cellular responses to extreme water loss: the water-replacement hypothesis. Cryobiology 19(3):306–316

Cochrane VW (1958) The physiology of fungi. Wiley, New York

Colaço C, Sen S, Thangavelu M, Pinder S, Roser B (1992) Extraordinary stability of enzymes dried in trehalose: simplified molecular biology. Biotechnology 10:1007–1011

Colombo S, Ronchetti D, Thevelein JM, Winderickx J, Martegani E (2004) Activation state of the Ras2 protein and glucose-induced signaling in *Saccharomyces cerevisiae*. J Biol Chem 279(45):46715–46722. doi:10.1074/jbc.M405136200

Conlin LK, Nelson HC (2007) The natural osmolyte trehalose is a positive regulator of the heat-induced activity of yeast heat shock transcription factor. Mol Cell Biol 27(4):1505–1515. doi:10.1128/mcb.01158-06

Conrad M, Schothorst J, Kankipati HN, Van Zeebroeck G, Rubio-Texeira M, Thevelein JM (2014) Nutrient sensing and signaling in the yeast *Saccharomyces cerevisiae*. FEMS Microbiol Rev 38(2):254–299. doi:10.1111/1574-6976.12065

Coote PJ, Jones MV, Edgar K, Cole MB (1992) *TPK* gene products mediate cAMP-independent thermotolerance in *Saccharomyces cerevisiae*. J Gen Microbiol 138:2551–2557

Cornette R, Kikawada T (2011) The induction of anhydrobiosis in the sleeping chironomid: current status of our knowledge. IUBMB Life 63(6):419–429. doi:10.1002/iub.463

Cortina C, Culiáñez-Macià FA (2005) Tomato abiotic stress enhanced tolerance by trehalose biosynthesis. Plant Sci 169:75–82

Cotter DA (1975) Spores of the cellular slime mold *Dictyostelium discoideum*. In: Gerhardt P, Costilow RN, Sadoff HL (eds) Spores VI. American Society for Microbiology, Washington, pp 61–72

Coutinho C, Bernardes E, Felix D, Panek A (1988) Trehalose as cryoprotectant for preservation of yeast strains. J Biotechnol 7:23–32

Coutinho CC, Silva JT, Panek AD (1992) Trehalase activity and its regulation during growth of *Saccharomyces cerevisiae*. Biochem Int 26(3):521–530

Cramer KL, Gerrald QD, Nichols CB, Price MS, Alspaugh JA (2006) Transcription factor Nrgl mediates capsule formation, stress response, and pathogenesis in *Cryptococcus neoformans*. Eukaryot Cell 5(7):1147–1156. doi:10.1128/ec.00145-06

Crauwels M, Donaton MC, Pernambuco MB, Winderickx J, de Winde JH, Thevelein JM (1997) The Sch9 protein kinase in the yeast *Saccharomyces cerevisiae* controls cAPK activity and is required for nitrogen activation of the fermentable-growth-medium-induced (FGM) pathway. Microbiology 143:2627–2637

Crowe LM (2002) Lessons from nature: the role of sugars in anhydrobiosis. Comp Biochem Physiol A Mol Integr Physiol 131(3):505–513

Crowe JH, Crowe LM, Chapman D (1984) Preservation of membranes in anhydrobiotic organisms: the role of trehalose. Science 223:701–703

Crowe JH, Carpenter JF, Crowe LM, Anchordoguy TJ (1990) Are freezing and dehydration similar stress vectors? A comparison of modes of interaction of stabilizing solutes with biomolecules. Cryobiology 27:219–231

Crowe JH, Panek AD, Crowe LM, Panek AC, De Araujo PS (1991) Trehalose transport in yeast cells. Biochem Int 24(4):721–730

Crowe JH, Hoekstra FA, Crowe LM (1992) Anhydrobiosis. Annu Rev Physiol 54:579–599. doi:10.1146/annurev.ph.54.030192.003051

Crowe JH, Carpenter JF, Crowe LM (1998) The role of vitrification in anhydrobiosis. Annu Rev Physiol 60:73–103. doi:10.1146/annurev.physiol.60.1.73

Crowe JH, Crowe LM, Oliver AE, Tsvetkova N, Wolkers W, Tablin F (2001) The trehalose myth revisited: introduction to a symposium on stabilization of cells in the dry state. Cryobiology 43:89–105

Cuber R, Eleutherio EC, Pereira MD, Panek AD (1997) The role of the trehalose transporter during germination. Biochim Biophys Acta 1330(2):165–171

d'Enfert C, Fontaine T (1997) Molecular characterization of the *Aspergillus nidulans treA* gene encoding an acid trehalase required for growth on trehalose. Mol Microbiol 24(1):203–216

d'Enfert C, Bonini BM, Zapella PD, Fontaine T, da Silva AM, Terenzi HF (1999) Neutral trehalases catalyse intracellular trehalose breakdown in the filamentous fungi *Aspergillus nidulans* and *Neurospora crassa*. Mol Microbiol 32(3):471–483

D'Souza CA, Alspaugh JA, Yue C, Harashima T, Cox GM, Perfect JR, Heitman J (2001) Cyclic AMP-dependent protein kinase controls virulence of the fungal pathogen *Cryptococcus neoformans*. Mol Cell Biol 21(9):3179–3191. doi:10.1128/mcb.21.9.3179-3191.2001

da Costa Morato Nery D, da Silva CG, Mariani D, Fernandes PN, Pereira MD, Panek AD, Eleutherio EC (2008) The role of trehalose and its transporter in protection against reactive oxygen species. Biochim Biophys Acta 1780(12):1408–1411. doi:10.1016/j.bbagen.2008.05.011

Davidson JF, Schiestl RH (2001) Cytotoxic and genotoxic consequences of heat stress are dependent on the presence of oxygen in *Saccharomyces cerevisiae*. J Bacteriol 183(15):4580–4587. doi:10.1128/jb.183.15.4580-4587.2001

Davidson GS, Joe RM, Roy S, Meirelles O, Allen CP, Wilson MR, Tapia PH, Manzanilla EE, Dodson AE, Chakraborty S, Carter M, Young S, Edwards B, Sklar L, Werner-Washburne M (2011) The proteomics of quiescent and nonquiescent cell differentiation in yeast stationary-phase cultures. Mol Biol Cell 22(7):988–998. doi:10.1091/mbc.E10-06-0499

Davies JE, Sarkar S, Rubinsztein DC (2006) Trehalose reduces aggregate formation and delays pathology in a transgenic mouse model of oculopharyngeal muscular dystrophy. Hum Mol Genet 15(1):23–31. doi:10.1093/hmg/ddi422

de Almeida FM, Lucio AKB, Polizeli ML, Jorge JA, Terenzi HF (1997) Function and regulation of the acid and neutral trehalases of *Mucor rouxii*. FEMS Microbiol Lett 155:73–77

De Antoni GL, Perez P, Abraham A, Anon MC (1989) Trehalose, a cryoprotectant for *Lactobacillus bulgaricus*. Cryobiology 26:149–153

De Araujo PS, Panek AC, Crowe JH, Crowe LM, Panek AD (1991) Trehalose-transporting membrane vesicles from yeasts. Biochem Int 24(4):731–737

de Mesquita JF, Paschoalin VM, Panek AD (1997) Modulation of trehalase activity in *Saccharomyces cerevisiae* by an intrinsic protein. Biochim Biophys Acta 1334(2–3):233–239

De Mesquita JF, Panek AD, de Araujo PS (2003) *In silico* and *in vivo* analysis reveal a novel gene in *Saccharomyces cerevisiae* trehalose metabolism. BMC Genomics 4(1):45. doi:10.1186/1471-2164-4-45

de Pinho CA, de Lourdes M, Polizeli TM, Jorge JA, Terenzi HF (2001) Mobilisation of trehalose in mutants of the cyclic AMP signalling pathway, *cr-1* (CRISP-1) and *mcb* (microcycle conidiation), of *Neurospora crassa*. FEMS Microbiol Lett 199:85–89

De Silva-Udawatta MN, Cannon JF (2001) Roles of trehalose phosphate synthase in yeast glycogen metabolism and sporulation. Mol Microbiol 40(6):1345–1356

De Virgilio C (2012) The essence of yeast quiescence. FEMS Microbiol Rev 36(2):306–339. doi:10.1111/j.1574-6976.2011.00287.x

De Virgilio C, Simmen U, Hottiger T, Boller T, Wiemken A (1990) Heat shock induces enzymes of trehalose metabolism, trehalose accumulation, and thermotolerance in *Schizosaccharomyces pombe*, even in the presence of cycloheximide. FEBS Lett 273(1–2):107–110

De Virgilio C, Burckert N, Boller T, Wiemken A (1991a) A method to study the rapid phosphorylation-related modulation of neutral trehalase activity by temperature shifts in yeast. FEBS Lett 291(2):355–358

De Virgilio C, Piper P, Boller T, Wiemken A (1991b) Acquisition of thermotolerance in *Saccharomyces cerevisiae* without heat shock protein hsp 104 and in the absence of protein synthesis. FEBS Lett 288(1–2):86–90

De Virgilio C, Burckert N, Bell W, Jeno P, Boller T, Wiemken A (1993) Disruption of *TPS2*, the gene encoding the 100-kDa subunit of the trehalose-6-phosphate synthase/phosphatase complex in *Saccharomyces cerevisiae*, causes accumulation of trehalose-6-phosphate and loss of trehalose-6-phosphate phosphatase activity. Eur J Biochem 212(2):315–323

De Virgilio C, Hottiger T, Dominguez J, Boller T, Wiemken A (1994) The role of trehalose synthesis for the acquisition of thermotolerance in yeast. I. Genetic evidence that trehalose is a thermoprotectant. Eur J Biochem 219(1–2):179–186

De Wever V, Reiter W, Ballarini A, Ammerer G, Brocard C (2005) A dual role for PP1 in shaping the Msn2-dependent transcriptional response to glucose starvation. EMBO J 24(23):4115–4123. doi:10.1038/sj.emboj.7600871

Dejean L, Beauvoit B, Bunoust O, Guerin B, Rigoulet M (2002) Activation of Ras cascade increases the mitochondrial enzyme content of respiratory competent yeast. Biochem Biophys Res Commun 293(5):1383–1388. doi:10.1016/s0006-291x(02)00391-1

Dellamora-Ortiz GM, Ortiz CH, Maia JC, Panek AD (1986) Partial purification and characterization of the interconvertible forms of trehalase from *Saccharomyces cerevisiae*. Arch Biochem Biophys 251(1):205–214

Delorge I, Janiak M, Carpentier S, Van Dijck P (2014) Fine tuning of trehalose biosynthesis and hydrolysis as novel tools for the generation of abiotic stress tolerant plants. Front Plant Sci 5:147. doi:10.3389/fpls.2014.00147

DeRisi JL, Iyer VR, Brown PO (1997) Exploring the metabolic and genetic control of gene expression on a genomic scale. Science 278(5338):680–686

Destruelle M, Holzer H, Klionsky DJ (1995) Isolation and characterization of a novel yeast gene, ATH1, that is required for vacuolar acid trehalase activity. Yeast 11(11):1015–1025. doi:10.1002/yea.320111103

Dewerchin MA, Van Laere AJ (1984) Trehalase activity and cyclic AMP content during early development of *Mucor rouXii* spores. J Bacteriol 158:575–579

Diaz-Ruiz R, Averet N, Araiza D, Pinson B, Uribe-Carvajal S, Devin A, Rigoulet M (2008) Mitochondrial oxidative phosphorylation is regulated by fructose 1,6-bisphosphate. A possible role in Crabtree effect induction? J Biol Chem 283(40):26948–26955. doi:10.1074/jbc.M800408200

Dickson RC, Nagiec EE, Skrzypek M, Tillman P, Wells GB, Lester RL (1997) Sphingolipids are potential heat stress signals in *Saccharomyces*. J Biol Chem 272(48):30196–30200

Dietvorst J, Londesborough J, Steensma HY (2005) Maltotriose utilization in lager yeast strains: *MTT1* encodes a maltotriose transporter. Yeast 22(10):775–788. doi:10.1002/yea.1279

Dijksterhuis J, van Driel KG, Sanders MG, Molenaar D, Houbraken JA, Samson RA, Kets EP (2002) Trehalose degradation and glucose efflux precede cell ejection during germination of heat-resistant ascospores of *Talaromyces macrosporus*. Arch Microbiol 178(1):1–7. doi:10.1007/s00203-002-0410-x

Doehlemann G, Molitor F, Hahn M (2005) Molecular and functional characterization of a fructose specific transporter from the gray mold fungus *Botrytis cinerea*. Fungal Genet Biol 42(7):601–610. doi:10.1016/j.fgb.2005.03.001

Doehlemann G, Berndt P, Hahn M (2006a) Different signalling pathways involving a Galpha protein, cAMP and a MAP kinase control germination of

Botrytis cinerea conidia. Mol Microbiol 59(3):821–835. doi:10.1111/j.1365-2958.2005.04991.x

Doehlemann G, Berndt P, Hahn M (2006b) Trehalose metabolism is important for heat stress tolerance and spore germination of *Botrytis cinerea*. Microbiology 152(Pt 9):2625–2634. doi:10.1099/mic.0.29044-0

Dominguez-Ferreras A, Soto MJ, Perez-Arnedo R, Olivares J, Sanjuan J (2009) Importance of trehalose biosynthesis for *Sinorhizobium meliloti* osmotolerance and nodulation of Alfalfa roots. J Bacteriol 191(24):7490–7499. doi:10.1128/jb.00725-09

Donaton MC, Holsbeeks I, Lagatie O, Van Zeebroeck G, Crauwels M, Winderickx J, Thevelein JM (2003) The Gap1 general amino acid permease acts as an amino acid sensor for activation of protein kinase A targets in the yeast *Saccharomyces cerevisiae*. Mol Microbiol 50(3):911–929

Dong J, Bai X (2011) The membrane localization of Ras2p and the association between Cdc25p and Ras2-GTP are regulated by protein kinase A (PKA) in the yeast *Saccharomyces cerevisiae*. FEBS Lett 585(8):1127–1134. doi:10.1016/j.febslet.2011.03.057

Durnez P, Pernambuco MB, Oris E, Arguelles JC, Mergelsberg H, Thevelein JM (1994) Activation of trehalase during growth induction by nitrogen sources in the yeast *Saccharomyces cerevisiae* depends on the free catalytic subunits of cAMP-dependent protein kinase, but not on functional Ras proteins. Yeast 10(8):1049–1064. doi:10.1002/yea.320100807

Eastmond PJ, Graham IA (2003) Trehalose metabolism: a regulatory role for trehalose-6-phosphate? Curr Opin Plant Biol 6(3):231–235

Eastmond PJ, van Dijken AJ, Spielman M, Kerr A, Tissier AF, Dickinson HG, Jones JD, Smeekens SC, Graham IA (2002) Trehalose-6-phosphate synthase 1, which catalyses the first step in trehalose synthesis, is essential for *Arabidopsis* embryo maturation. Plant J 29(2):225–235

Eck R, Bergmann C, Ziegelbauer K, Schonfeld W, Kunkel W (1997) A neutral trehalase gene from *Candida albicans*: molecular cloning, characterization and disruption. Microbiology 143(Pt 12):3747–3756

Eis C, Nidetzky B (1999) Characterization of trehalose phosphorylase from *Schizophyllum commune*. Biochem J 341(Pt 2):385–393

Eis C, Albert M, Dax K, Nidetzky B (1998) The stereochemical course of the reaction mechanism of trehalose phosphorylase from *Schizophyllum commune*. FEBS Lett 440(3):440–443

Eis C, Watkins M, Prohaska T, Nidetzky B (2001) Fungal trehalose phosphorylase: kinetic mechanism, pH-dependence of the reaction and some structural properties of the enzyme from *Schizophyllum commune*. Biochem J 356(Pt 3):757–767

Elbein AD (1974) The metabolism of a, a-trehalose. Adv Carbohydr Chem Biochem 30:227–256

Elbein AD, Pan YT, Pastuszak I, Carroll D (2003) New insights on trehalose: a multifunctional molecule. Glycobiology 13(4):17R–27R. doi:10.1093/glycob/cwg047

Elliott B, Futcher B (1993) Stress resistance of yeast cells is largely independent of cell cycle phase. Yeast 9(1):33–42

Elliott B, Haltiwanger RS, Futcher B (1996) Synergy between trehalose and Hsp104 for thermotolerance in *Saccharomyces cerevisiae*. Genetics 144(3):923–933

Engelberg D, Poradosu E, Simchen G, Levitzki A (1990) Adenylyl cyclase activity of the fission yeast *Schizosaccharomyces pombe* is not regulated by guanyl nucleotides. FEBS Lett 261:413–418

Epton H, Richmond DV (1980) Formation, structure and germination of conidia. In: Coley-Smith JR, Verhoeff K, Jarvis WR (eds) The biology of *Botrytis*. Academic, New York, pp 41–81

Ernandes JR, De Meirsman C, Rolland F, Winderickx J, de Winde J, Brandao RL, Thevelein JM (1998) During the initiation of fermentation overexpression of hexokinase PII in yeast transiently causes a similar deregulation of glycolysis as deletion of Tps1. Yeast 14(3):255–269. doi:10.1002/(sici)1097-0061(199802)14:3<255::aid-yea228>3.0.co;2-n

Fang TY, Hung XG, Shih TY, Tseng WC (2004) Characterization of the trehalosyl dextrin-forming enzyme from the thermophilic archaeon *Sulfolobus solfataricus* ATCC 35092. Extremophiles 8(4):335–343. doi:10.1007/s00792-004-0393-4

Favre C, Aguilar PS, Carrillo MC (2008) Oxidative stress and chronological aging in glycogen-phosphorylase-deleted yeast. Free Radic Biol Med 45(10):1446–1456. doi:10.1016/j.freeradbiomed.2008.08.021

Fernandez J, Wilson RA (2011) The sugar sensor, trehalose-6-phosphate synthase (Tps1), regulates primary and secondary metabolism during infection by the rice blast fungus: will *Magnaporthe oryzae*'s "sweet tooth" become its "Achilles' heel"? Mycology 2:46–53

Fernandez J, Soto T, Vicente-Soler J, Cansado J, Gacto M (1997a) Heat-shock response in *Schizosaccharomyces pombe* cells lacking cyclic AMP-dependent phosphorylation. Curr Genet 31:112–118

Fernandez J, Soto T, Vicente-Soler J, Cansado J, Gacto M (1997b) Osmo-stress-induced changes in neutral trehalase activity of the fission yeast *Schizosaccharomyces pombe*. Biochim Biophys Acta 1357(1):41–48

Fernandez J, Soto T, Franco A, Vicente-Soler J, Cansado J, Gacto M (1998) Enhancement of neutral trehalase activity by oxidative stress in the fission yeast *Schizosaccharomyces pombe*. Fungal Genet Biol 25(2):79–86. doi:10.1006/fgbi.1998.1084

Fernandez J, Wright JD, Hartline D, Quispe CF, Madayiputhiya N, Wilson RA (2012) Principles of carbon catabolite repression in the rice blast fungus: Tps1, Nmr1-3, and a MATE-family pump regulate glucose metabolism during infection. PLoS Genet 8(5):e1002673. doi:10.1371/journal.pgen.1002673

Fernandez-Arenas E, Bleck CK, Nombela C, Gil C, Griffiths G, Diez-Orejas R (2009) *Candida albicans* actively modulates intracellular membrane trafficking in mouse macrophage phagosomes. Cell Microbiol 11(4):560–589. doi:10.1111/j.1462-5822.2008.01274.x

Ferreira JC, Silva JT, Panek AD (1996) A regulatory role for TSL1 on trehalose synthase activity. Biochem Mol Biol Int 38(2):259–265

Ferreira JC, Thevelein JM, Hohmann S, Paschoalin VM, Trugo LC, Panek AD (1997) Trehalose accumulation in mutants of *Saccharomyces cerevisiae* deleted in the UDPG-dependent trehalose synthase-phosphatase complex. Biochim Biophys Acta 1335(1-2):40–50

Fillinger S, Chaveroche MK, van Dijck P, de Vries R, Ruijter G, Thevelein J, d'Enfert C (2001) Trehalose is required for the acquisition of tolerance to a variety of stresses in the filamentous fungus *Aspergillus nidulans*. Microbiology 147(Pt 7):1851–1862

Flom GA, Langner E, Johnson JL (2012) Identification of an Hsp90 mutation that selectively disrupts cAMP/PKA signaling in *Saccharomyces cerevisiae*. Curr Genet 58(3):149–163. doi:10.1007/s00294-012-0373-7

Flores CL, Gancedo C, Petit T (2011) Disruption of *Yarrowia lipolytica TPS1* gene encoding trehalose-6-P synthase does not affect growth in glucose but impairs growth at high temperature. PLoS One 6(9):e23695. doi:10.1371/journal.pone.0023695

Foster AJ, Jenkinson JM, Talbot NJ (2003) Trehalose synthesis and metabolism are required at different stages of plant infection by *Magnaporthe grisea*. EMBO J 22(2):225–235. doi:10.1093/emboj/cdg018

Franco A, Soto T, Vicente-Soler J, Guillen PV, Cansado J, Gacto M (2000) Characterization of *tpp1(+)* as encoding a main trehalose-6P phosphatase in the fission yeast *Schizosaccharomyces pombe*. J Bacteriol 182(20):5880–5884

Franco A, Soto T, Vicente-Soler J, Paredes V, Madrid M, Gacto M, Cansado J (2003) A role for calcium in the regulation of neutral trehalase activity in the fission yeast *Schizosaccharomyces pombe*. Biochem J 376(Pt 1):209–217. doi:10.1042/bj20030825

Franco A, Soto T, Madrid M, Vicente-Soler J, Gacto M, Cansado J (2005) Functional characterization of *Schizosaccharomyces pombe* neutral trehalase altered in phosphorylatable serine residues. Arch Microbiol 183(6):394–400. doi:10.1007/s00203-005-0005-4

Francois J, Parrou JL (2001) Reserve carbohydrates metabolism in the yeast *Saccharomyces cerevisiae*. FEMS Microbiol Rev 25(1):125–145

François JM, Eraso P, Gancedo C (1987) Changes in the concentration of cAMP, fructose-2,6-bisphosphate and related metabolites and enzymes in *Saccharomyces cerevisiae* during growth on glucose. Eur J Biochem 164:369–373

François J, Villanueva ME, Hers HG (1988) The control of glycogen metabolism in yeast. 1. Interconversion *in vivo* of glycogen synthase and glycogen phosphorylase induced by glucose, a nitrogen source or uncouplers. Eur J Biochem 174:551–559

Francois J, Neves MJ, Hers HG (1991) The control of trehalose biosynthesis in *Saccharomyces cerevisiae*: evidence for a catabolite inactivation and repression of trehalose-6-phosphate synthase and trehalose-6-phosphate phosphatase. Yeast 7(6):575–587. doi:10.1002/yea.320070605

Franks F, Hatley RHM, Mathias SF (1991) Materials science and the production of shelf-stable biologicals. BioPharm 4:38–42

Freitas FZ, de Paula RM, Barbosa LC, Terenzi HF, Bertolini MC (2010) cAMP signaling pathway controls glycogen metabolism in *Neurospora crassa* by regulating the glycogen synthase gene expression and phosphorylation. Fungal Genet Biol 47(1):43–52. doi:10.1016/j.fgb.2009.10.011

Frison M, Parrou JL, Guillaumot D, Masquelier D, Francois J, Chaumont F, Batoko H (2007) The *Arabidopsis thaliana* trehalase is a plasma membrane-bound enzyme with extracellular activity. FEBS Lett 581(21):4010–4016. doi:10.1016/j.febslet.2007.07.036

Fujii S, Iwahashi H, Obuchi K, Fujii T, Komatsu Y (1996) Characterization of a barotolerant mutant of the yeast *Saccharomyces cerevisiae*: importance of trehalose content and membrane fluidity. FEMS Microbiol Lett 141:97–101

Fujita K, Matsuyama A, Kobayashi Y, Iwahashi H (2006) The genome-wide screening of yeast deletion mutants to identify the genes required for tolerance to ethanol and other alcohols. FEMS Yeast Res 6(5):744–750. doi:10.1111/j.1567-1364.2006.00040.x

Fukui Y, Kozasa T, Kaziro Y, Takeda T, Yamamoto M (1986) Role of a ras homolog in the life cycle of *Schizosaccharomyces pombe*. Cell 44:329–336

Futcher B (2006) Metabolic cycle, cell cycle, and the finishing kick to Start. Genome Biol 7(4):107. doi:10.1186/gb-2006-7-4-107

Gadd GM, Chalmers K, Reed RH (1987) The role of trehalose in dehydration resistance of *Saccharomyces cerevisiae*. FEMS Microbiol Lett 48:249–254

Gaff DF (1971) Desiccation-tolerant flowering plants in southern Africa. Science 174(4013):1033–1034. doi:10.1126/science.174.4013.1033

Galdieri L, Mehrotra S, Yu S, Vancura A (2010) Transcriptional regulation in yeast during diauxic shift and stationary phase. OMICS 14(6):629–638. doi:10.1089/omi.2010.0069

Gamo FJ, Lafuente MJ, Gancedo C (1994) The mutation *DGT1-1* decreases glucose transport and alleviates carbon catabolite repression in *Saccharomyces cerevisiae*. J Bacteriol 176(24):7423–7429

Gancedo JM (2013) Biological roles of cAMP: variations on a theme in the different kingdoms of life. Biol Rev Camb Philos Soc 88(3):645–668. doi:10.1111/brv.12020

Gancedo C, Flores CL (2004) The importance of a functional trehalose biosynthetic pathway for the life of yeasts and fungi. FEMS Yeast Res 4(4-5):351–359

Garg AK, Kim JK, Owens TG, Ranwala AP, Choi YD, Kochian LV, Wu RJ (2002) Trehalose accumulation in rice plants confers high tolerance levels to different abiotic stresses. Proc Natl Acad Sci USA 99(25):15898–15903. doi:10.1073/pnas.252637799

Garre E, Matallana E (2009) The three trehalases Nth1p, Nth2p and Ath1p participate in the mobilization of intracellular trehalose required for recovery from saline stress in *Saccharomyces cerevisiae*. Microbiology 155(Pt 9):3092–3099. doi:10.1099/mic.0.024992-0

Garre E, Perez-Torrado R, Gimeno-Alcaniz JV, Matallana E (2009) Acid trehalase is involved in intracellular trehalose mobilization during postdiauxic growth and severe saline stress in *Saccharomyces cerevisiae*. FEMS Yeast Res 9(1):52–62. doi:10.1111/j.1567-1364.2008.00453.x

Garreau H, Hasan RN, Renault G, Estruch F, Boy-Marcotte E, Jacquet M (2000) Hyperphosphorylation of Msn2p and Msn4p in response to heat shock and the diauxic shift is inhibited by cAMP in *Saccharomyces cerevisiae*. Microbiology 146(Pt 9):2113–2120

Garrido C, Gurbuxani S, Ravagnan L, Kroemer G (2001) Heat shock proteins: endogenous modulators of apoptotic cell death. Biochem Biophys Res Commun 286(3):433–442. doi:10.1006/bbrc.2001.5427

Gasch AP (2007) Comparative genomics of the environmental stress response in ascomycete fungi. Yeast 24(11):961–976. doi:10.1002/yea.1512

Gasch AP, Spellman PT, Kao CM, Carmel-Harel O, Eisen MB, Storz G, Botstein D, Brown PO (2000) Genomic expression programs in the response of yeast cells to environmental changes. Mol Biol Cell 11(12):4241–4257

Ge XY, Xu Y, Chen X (2013) Improved carbon metabolic flux in *Saccharomyces cerevisiae* at high temperature by overexpressed *TSL1* gene. J Ind Microbiol Biotechnol 40(3-4):345–352. doi:10.1007/s10295-013-1233-2

Geelen D, Royackers K, Vanstraelen M, De Bus M, Inzé D, Van Dijck P, Thevelein JM, Leyman B (2007) Trehalose-6-P synthase *At*TPS1 high molecular weight complexes in yeast and *Arabidopsis*. Plant Sci 173:426–437

Gélinas P, Fiset G, LeDuy A, Goulet J (1989) Effect of growth conditions and trehalose content on cryotolerance of bakers' yeast in frozen doughs. Appl Environ Microbiol 55:2453–2459

Giæver HM, Styrvold OB, Kaasen J, Strom AR (1988) Biochemical and genetic characterization of osmoregulatory trehalose synthesis in *Escherichia coli*. J Bacteriol 170:2841–2849

Giots F, Donaton MC, Thevelein JM (2003) Inorganic phosphate is sensed by specific phosphate carriers and acts in concert with glucose as a nutrient signal for activation of the protein kinase A pathway in the yeast *Saccharomyces cerevisiae*. Mol Microbiol 47(4):1163–1181

Glover JR, Lindquist S (1998) Hsp104, Hsp70, and Hsp40: a novel chaperone system that rescues previously aggregated proteins. Cell 94:73–82

Goddijn O, Smeekens S (1998) Sensing trehalose biosynthesis in plants. Plant J 14(2):143–146

Goddijn OJ, Verwoerd TC, Voogd E, Krutwagen RW, de Graaf PT, van Dun K, Poels J, Ponstein AS, Damm B, Pen J (1997) Inhibition of trehalase activity enhances trehalose accumulation in transgenic plants. Plant Physiol 113(1):181–190

Goldberg AA, Bourque SD, Kyryakov P, Gregg C, Boukh-Viner T, Beach A, Burstein MT, Machkalyan G, Richard V, Rampersad S, Cyr D, Milijevic S, Titorenko VI (2009) Effect of calorie restriction on the metabolic history of chronologically aging yeast. Exp Gerontol 44(9):555–571. doi:10.1016/j.exger.2009.06.001

Golovina EA, Golovin AV, Hoekstra FA, Faller R (2009) Water replacement hypothesis in atomic detail—factors determining the structure of dehydrated bilayer stacks. Biophys J 97:490–499

Gomes FC, Pataro C, Guerra JB, Neves MJ, Correa SR, Moreira ES, Rosa CA (2002) Physiological diversity and trehalose accumulation in *Schizosaccharomyces pombe* strains isolated from spontaneous fermentations during the production of the artisanal Brazilian cachaca. Can J Microbiol 48(5):399–406

Gong Y, Kakihara Y, Krogan N, Greenblatt J, Emili A, Zhang Z, Houry WA (2009) An atlas of chaperone-protein interactions in *Saccharomyces cerevisiae*: implications to protein folding pathways in the cell. Mol Syst Biol 5:275. doi:10.1038/msb.2009.26

Gonzalez MI, Stucka R, Blazquez MA, Feldmann H, Gancedo C (1992) Molecular cloning of *CIF1*, a yeast gene necessary for growth on glucose. Yeast 8(3):183–192. doi:10.1002/yea.320080304

Gorner W, Durchschlag E, Martinez-Pastor MT, Estruch F, Ammerer G, Hamilton B, Ruis H, Schuller C (1998) Nuclear localization of the C2H2 zinc finger protein Msn2p is regulated by stress and protein kinase A activity. Genes Dev 12(4):586–597

Gorner W, Durchschlag E, Wolf J, Brown EL, Ammerer G, Ruis H, Schuller C (2002) Acute glucose starvation activates the nuclear localization signal of a stress-specific yeast transcription factor. EMBO J 21(1–2):135–144. doi:10.1093/emboj/21.1.135

Gottlieb D (1978) The germination of fungus spores. Meadowfield Press, Shildon

Goyal K, Walton LJ, Browne JA, Burnell AM, Tunnacliffe A (2005) Molecular anhydrobiology: identifying molecules implicated in invertebrate anhydrobiosis. Integr Comp Biol 45(5):702–709. doi:10.1093/icb/45.5.702

Grably MR, Stanhill A, Tell O, Engelberg D (2002) HSF and Msn2/4p can exclusively or cooperatively activate the yeast HSP104 gene. Mol Microbiol 44(1):21–35

Gray JV, Petsko GA, Johnston GC, Ringe D, Singer RA, Werner-Washburne M (2004) "Sleeping beauty": quiescence in *Saccharomyces cerevisiae*. Microbiol Mol Biol Rev 68(2):187–206. doi:10.1128/mmbr.68.2.187-206.2004

Grba S, Oura E, Suomalainen H (1975) On the formation of glycogen and trehalose in baker's yeast. Eur J Appl Microbiol 2:29–37

Grba S, Oura E, Suomalainen H (1979) Formation of trehalose and glycogen in growing baker's yeast. Finn Chem Lett 1979:61–64

Green JL, Angell CA (1989) Phase relations and vitrification in saccharide-water solutions and the trehalose anomaly. J Phys Chem 93:2880–2882

Griffioen G, Anghileri P, Imre E, Baroni MD, Ruis H (2000) Nutritional control of nucleocytoplasmic localization of cAMP-dependent protein kinase catalytic and regulatory subunits in *Saccharomyces cerevisiae*. J Biol Chem 275(2):1449–1456

Griffioen G, Branduardi P, Ballarini A, Anghileri P, Norbeck J, Baroni MD, Ruis H (2001) Nucleocytoplasmic distribution of budding yeast protein kinase A regulatory subunit Bcy1 requires Zds1 and is regulated by Yak1-dependent phosphorylation of its targeting domain. Mol Cell Biol 21(2):511–523. doi:10.1128/mcb.21.2.511-523.2001

Griffioen G, Swinnen S, Thevelein JM (2003) Feedback inhibition on cell wall integrity signaling by Zds1 involves Gsk3 phosphorylation of a cAMP-dependent protein kinase regulatory subunit. J Biol Chem 278(26):23460–23471. doi:10.1074/jbc.M210691200

Grimminger-Marcquardt V, Lashuel HA (2009) Structure and function of the molecular chaperone Hsp104 from yeast. Biopolymers 93:252–276

Gross C, Watson K (1998) De novo protein synthesis is essential for thermotolerance acquisition in a *Saccharomyces cerevisiae* trehalose synthase mutant. Biochem Mol Biol Int 45:663–671

Grosse C, Heinekamp T, Kniemeyer O, Gehrke A, Brakhage AA (2008) Protein kinase A regulates growth, sporulation, and pigment formation in *Aspergillus fumigatus*. Appl Environ Microbiol 74(15):4923–4933. doi:10.1128/aem.00470-08

Gueguen Y, Rolland JL, Schroeck S, Flament D, Defretin S, Saniez MH, Dietrich J (2001) Characterization of the maltooligosyl trehalose synthase from the thermophilic archaeon *Sulfolobus acidocaldarius*. FEMS Microbiol Lett 194(2):201–206

Guibert EE, Petrenko AY, Balaban CL, Somov AY, Rodriguez JV, Fuller BJ (2011) Organ preservation: current concepts and new strategies for the next decade. Transfus Med Hemother 38(2):125–142. doi:10.1159/000327033

Gutierrez C, Ardourel M, Bremer E, Middendorf A, Boos W, Ehmann U (1989) Analysis and DNA sequence of the osmoregulated treA gene encoding the periplasmic trehalase of *Escherichia coli* K12. Mol Gen Genet 217:347–354

Haesendonckx S, Tudisca V, Voordeckers K, Moreno S, Thevelein JM, Portela P (2012) The activation loop of PKA catalytic isoforms is differentially phosphorylated by Pkh protein kinases in *Saccharomyces cerevisiae*. Biochem J 448(3):307–320. doi:10.1042/bj20121061

Hall BG (1983) Yeast thermotolerance does not require protein synthesis. J Bacteriol 156:1363–1365

Hammond JRM (1993) Brewer's yeasts. In: Rose AH, Harrison JS (eds) The yeasts, vol 5. Academic, London, pp 7–67

Hammond JBW, Nichols R (1976) Carbohydrate metabolism in *Agaricus bisporus* (Lange) Sing.: changes in soluble carbohydrate during growth of mycelium and sporphore. J Gen Microbiol 95:309–320

Han EK, Cotty F, Sottas C, Jiang H, Michels CA (1995) Characterization of *AGT1* encoding a general alpha-glucoside transporter from *Saccharomyces*. Mol Microbiol 17(6):1093–1107

Harashima T, Heitman J (2002) The Galpha protein Gpa2 controls yeast differentiation by interacting with kelch repeat proteins that mimic Gbeta subunits. Mol Cell 10(1):163–173

Hazell BW, Nevalainen H, Attfield PV (1995) Evidence that the *Saccharomyces cerevisiae CIF1 (GGS1/TPS1)* gene modulates heat shock response positively. FEBS Lett 377:457–460

Hazell BW, Kletsas S, Nevalainen H, Attfield PV (1997) Involvement of *CIF1 (GGS1/TPS1)* in osmotic stress response in *Saccharomyces cerevisiae*. FEBS Lett 414:353–358

He S, Bystricky K, Leon S, Francois JM, Parrou JL (2009) The *Saccharomyces cerevisiae* vacuolar acid trehalase is targeted at the cell surface for its physiological function. FEBS J 276(19):5432–5446. doi:10.1111/j.1742-4658.2009.07227.x

Hengge-Aronis R, Klein W, Lange R, Rimmele M, Boos W (1991) Trehalose synthesis genes are controlled by the putative sigma factor encoded by *rpoS* and are involved in stationary-phase thermotolerance in *Escherichia coli*. J Bacteriol 173(24):7918–7924

Hengherr S, Heyer AG, Kohler HR, Schill RO (2008) Trehalose and anhydrobiosis in tardigrades–evidence for divergence in responses to dehydration. FEBS J 275(2):281–288. doi:10.1111/j.1742-4658.2007.06198.x

Hengherr S, Heyer AG, Brummer F, Schill RO (2011) Trehalose and vitreous states: desiccation tolerance of dormant stages of the crustaceans *Triops* and *Daphnia*. Physiol Biochem Zool 84(2):147–153. doi:10.1086/658499

Herdeiro RS, Pereira MD, Panek AD, Eleutherio EC (2006) Trehalose protects *Saccharomyces cerevisiae* from lipid peroxidation during oxidative stress. Biochim Biophys Acta 1760(3):340–346. doi:10.1016/j.bbagen.2006.01.010

Hill EP, Sussman AS (1964) Development of trehalase and invertase activity in *Neurospora*. J Bacteriol 88:1556–1566

Hino A, Mihara K, Nakashima K, Takano H (1990) Trehalose levels and survival ratio of freeze-tolerant versus freeze-sensitive yeasts. Appl Environ Microbiol 56:1386–1391

Hirasawa R, Yokoigawa K, Isobe Y, Kawai H (2001) Improving the freeze tolerance of bakers' yeast by loading with trehalose. Biosci Biotechnol Biochem 65:522–526

Hirasawa T, Nakakura Y, Yoshikawa K, Ashitani K, Nagahisa K, Furusawa C, Katakura Y, Shimizu H, Shioya S (2006) Comparative analysis of transcriptional responses to saline stress in the laboratory and brewing strains of *Saccharomyces cerevisiae* with DNA microarray. Appl Microbiol Biotechnol 70(3):346–357. doi:10.1007/s00253-005-0192-6

Hirata Y, Andoh T, Asahara T, Kikuchi A (2003) Yeast glycogen synthase kinase-3 activates Msn2p-dependent transcription of stress responsive genes. Mol Biol Cell 14(1):302–312. doi:10.1091/mbc.E02-05-0247

Hirimburegama K, Durnez P, Keleman J, Oris E, Vergauwen R, Mergelsberg H, Thevelein JM (1992) Nutrient-induced activation of trehalase in nutrient-starved cells of the yeast *Saccharomyces cerevisiae*: cAMP is not involved as second messenger. J Gen Microbiol 138(10):2035–2043

Hoffman CS (2005a) Except in every detail: comparing and contrasting G-protein signaling in *Saccharomyces cerevisiae* and *Schizosaccharomyces pombe*. Eukaryot Cell 4(3):495–503. doi:10.1128/ec.4.3.495-503.2005

Hoffman CS (2005b) Glucose sensing via the protein kinase A pathway in *Schizosaccharomyces pombe*. Biochem Soc Trans 33(Pt 1):257–260. doi:10.1042/bst0330257

Hohmann S, Huse K, Valentin E, Mbonyi K, Thevelein JM, Zimmermann FK (1992) Glucose-induced regulatory defects in the *Saccharomyces cerevisiae* *byp1* growth initiation mutant and identification of *MIG1* as a partial suppressor. J Bacteriol 174(12):4183–4188

Hohmann S, Neves MJ, de Koning W, Alijo R, Ramos J, Thevelein JM (1993) The growth and signalling defects of the *ggs1 (fdp1/byp1)* deletion mutant on glucose are suppressed by a deletion of the gene encoding hexokinase PII. Curr Genet 23(4):281–289

Hohmann S, Van Dijck P, Luyten K, Thevelein JM (1994) The *byp1-3* allele of the *Saccharomyces cerevisiae GGS1/TPS1* gene and its multi-copy suppressor tRNA(GLN) (CAG): Ggs1/Tps1 protein levels restraining growth on fermentable sugars and trehalose accumulation. Curr Genet 26:295–301

Hohmann S, Bell W, Neves MJ, Valckx D, Thevelein JM (1996) Evidence for trehalose-6-phosphate-dependent and -independent mechanisms in the control of sugar influx into yeast glycolysis. Mol Microbiol 20(5):981–991

Holmström KO, Welin B, Mandal A, Kristiansdottir I, Teeri TH, Lamark T, Strom AR, Palva ET (1996) Production of the *Escherichia coli* betaine-aldehyde dehydrogenase, an enzyme required for synthesis of the osmoprotectant glycine betaine, in transgenic plants. Plant J 5:749–758

Holsbeeks I, Lagatie O, Van Nuland A, Van de Velde S, Thevelein JM (2004) The eukaryotic plasma membrane as a nutrient-sensing device. Trends Biochem Sci 29(10):556–564. doi:10.1016/j.tibs.2004.08.010

Honadel TE, Killian GJ (1988) Cryopreservation of murine embryos with trehalose and glycerol. Cryobiology 25:331–337

Honda Y, Tanaka M, Honda S (2010) Trehalose extends longevity in the nematode *Caenorhabditis elegans*. Aging Cell 9(4):558–569. doi:10.1111/j.1474-9726.2010.00582.x

Horikoshi K, Ikeda Y (1966) Trehalase in conidia of *Aspergillus oryzae*. J Bacteriol 91(5):1883–1887

Hottiger T, Boller T, Wiemken A (1987a) Rapid changes of heat and desiccation tolerance correlated with changes of trehalose content in *Saccharomyces cerevisiae* cells subjected to temperature shifts. FEBS Lett 220:113–115

Hottiger T, Schmutz P, Wiemken A (1987b) Heat-induced accumulation and futile cycling of trehalose in *Saccharomyces cerevisiae*. J Bacteriol 169(12):5518–5522

Hottiger T, Boller T, Wiemken A (1989) Correlation of trehalose content and heat resistance in yeast mutants altered in the RAS/adenylate cyclase pathway: Is trehalose a thermoprotectant? FEBS Lett 255:431–434

Hottiger T, De Virgilio C, Bell W, Boller T, Wiemken A (1992) Canavanine treatment of yeast induces thermotolerance. Yeast 8 (Special Issue):S91

Hottiger T, De Virgilio C, Hall MN, Boller T, Wiemken A (1994) The role of trehalose synthesis for the acquisition of thermotolerance in yeast. II. Physiological concentrations of trehalose increase the thermal stability of proteins in vitro. Eur J Biochem 219:187–193

Hounsa CG, Brandt EV, Thevelein J, Hohmann S, Prior BA (1998) Role of trehalose in survival of *Saccharomyces cerevisiae* under osmotic stress. Microbiology 144:671–680

Hu Y, Liu E, Bai X, Zhang A (2010) The localization and concentration of the *PDE2*-encoded high-affinity cAMP phosphodiesterase is regulated by cAMP-dependent protein kinase A in the yeast *Saccharomyces cerevisiae*. FEMS Yeast Res 10(2):177–187. doi:10.1111/j.1567-1364.2009.00598.x

Huang J, Reggiori F, Klionsky DJ (2007) The transmembrane domain of acid trehalase mediates ubiquitin-independent multivesicular body pathway sorting. Mol Biol Cell 18(7):2511–2524. doi:10.1091/mbc.E06-11-0995

Ibata-Ombetta S, Idziorek T, Trinel PA, Poulain D, Jouault T (2003) *Candida albicans* phospholipomannan promotes survival of phagocytosed yeasts through modulation of bad phosphorylation and macrophage apoptosis. J Biol Chem 278(15):13086–13093. doi:10.1074/jbc.M210680200

Iida H (1988) Multistress resistance of *Saccharomyces cerevisiae* is generated by insertion of retrotransposon Ty into the 5' coding region of the adenylate cyclase gene. Mol Cell Biol 8:5555–5560

Iida H, Yahara I (1984) Specific early-G1 blocks accompanied with stringent response in *Saccharomyces cerevisiae* lead to growth arrest in resting state similar to the Go of higher eukaryotes. J Cell Biol 98:1185–1193

Iwahashi H, Obuchi K, Fujii S, Komatsu Y (1997a) Barotolerance is dependent on both trehalose and heat shock protein 104 but is essentially different from thermotolerance in *Saccharomyces cerevisiae*. Lett Appl Microbiol 25:43–47

Iwahashi H, Obuchi K, Fujii S, Komatsu Y (1997b) Effect of temperature on the role of Hsp104 and trehalose in barotolerance of *Saccharomyces cerevisiae*. FEBS Lett 416:1–5

Iwahashi H, Nwaka S, Obuchi K, Komatsu Y (1998) Evidence for the interplay between trehalose metabolism and Hsp104 in yeast. Appl Environ Microbiol 64:4614–4617

Iwahashi H, Nwaka S, Obuchi K (2000) Evidence for contribution of neutral trehalase in barotolerance of *Saccharomyces cerevisiae*. Appl Environ Microbiol 66(12):5182–5185

Jang IC, Oh SJ, Seo JS, Choi WB, Song SI, Kim CH, Kim YS, Seo HS, Choi YD, Nahm BH, Kim JK (2003) Expression of a bifunctional fusion of the *Escherichia coli* genes for trehalose-6-phosphate synthase and trehalose-6-phosphate phosphatase in transgenic rice plants increases trehalose accumulation and abiotic stress tolerance without stunting growth. Plant Physiol 131(2):516–524. doi:10.1104/pp.007237

Jespersen L, Cesar LB, Meaden PG, Jakobsen M (1999) Multiple alpha-glucoside transporter genes in brewer's yeast. Appl Environ Microbiol 65(2):450–456

Jin M, Fujita M, Culley BM, Apolinario E, Yamamoto M, Maundrell K, Hoffman CS (1995) *sck1*, a high copy number suppressor of defects in the cAMP-dependent protein kinase pathway in fission yeast, encodes a protein homologous to the *Saccharomyces cerevisiae SCH9* kinase. Genetics 140(2):457–467

Jorgensen H, Olsson L, Ronnow B, Palmqvist EA (2002) Fed-batch cultivation of baker's yeast followed by nitrogen or carbon starvation: effects on fermentative capacity and content of trehalose and glycogen. Appl Microbiol Biotechnol 59(2–3):310–317. doi:10.1007/s00253-002-1017-5

Jules M, Guillou V, Francois J, Parrou JL (2004) Two distinct pathways for trehalose assimilation in the yeast *Saccharomyces cerevisiae*. Appl Environ Microbiol 70(5):2771–2778

Jules M, Beltran G, Francois J, Parrou JL (2008) New insights into trehalose metabolism by *Saccharomyces cerevisiae*: *NTH2* encodes a functional cytosolic trehalase, and deletion of *TPS1* reveals Ath1p-dependent trehalose mobilization. Appl Environ Microbiol 74(3):605–614. doi:10.1128/aem.00557-07

Jung YJ, Park HD (2005) Antisense-mediated inhibition of acid trehalase (*ATH1*) gene expression promotes ethanol fermentation and tolerance in *Saccharomyces cerevisiae*. Biotechnol Lett 27(23–24):1855–1859. doi:10.1007/s10529-005-3910-3

Kaasen I, Falkenberg P, Styrvold OB, Strom AR (1992) Molecular cloning and physical mapping of the *otsBA* genes, which encode the osmoregulatory trehalose pathway of *Escherichia coli* - Evidence that transcription is activated by KatF (AppR). J Bacteriol 174:889–898

Kaasen I, Mcdougall J, Strom AR (1994) Analysis of the *otsBA* operon for osmoregulatory trehalose synthesis in *Escherichia coli* and homology of the OtsA and OtsB proteins to the yeast trehalose-6-phosphate synthase/phosphatase complex. Gene 145(1):9–15

Kaida D, Yashiroda H, Toh-e A, Kikuchi Y (2002) Yeast Whi2 and Psr1-phosphatase form a complex and regulate STRE-mediated gene expression. Genes Cells 7(6):543–552

Kanamori Y, Saito A, Hagiwara-Komoda Y, Tanaka D, Mitsumasu K, Kikuta S, Watanabe M, Cornette R, Kikawada T, Okuda T (2010) The trehalose transporter 1 gene sequence is conserved in insects and encodes proteins with different kinetic properties involved in trehalose import into peripheral tissues. Insect Biochem Mol Biol 40(1):30–37. doi:10.1016/j.ibmb.2009.12.006

Karim S, Aronsson H, Ericson H, Pirhonen M, Leyman B, Welin B, Mantyla E, Palva ET, Van Dijck P,

Holmstrom KO (2007) Improved drought tolerance without undesired side effects in transgenic plants producing trehalose. Plant Mol Biol 64(4):371–386. doi:10.1007/s11103-007-9159-6

Kato K, Cox AD, Hisaka MM, Graham SM, Buss JE, Der CJ (1992) Isoprenoid addition to Ras protein is the critical modification for its membrane association and transforming activity. Proc Natl Acad Sci USA 89(14):6403–6407

Kawai A, Nishikawa S, Hirata A, Endo T (2001) Loss of the mitochondrial Hsp70 functions causes aggregation of mitochondria in yeast cells. J Cell Sci 114(Pt 19):3565–3574

Keller F, Schellenberg M, Wiemken A (1982) Localization of trehalase in vacuoles and of trehalose in the cytosol of yeast (*Saccharomyces cerevisiae*). Arch Microbiol 131(4):298–301

Kikawada T, Saito A, Kanamori Y, Nakahara Y, Iwata K, Tanaka D, Watanabe M, Okuda T (2007) Trehalose transporter 1, a facilitated and high-capacity trehalose transporter, allows exogenous trehalose uptake into cells. Proc Natl Acad Sci USA 104(28):11585–11590. doi:10.1073/pnas.0702538104

Kilburn D, Townrow S, Meunier V, Richardson R, Alam A, Ubbink J (2006) Organization and mobility of water in amorphous and crystalline trehalose. Nat Mater 5(8):632–635. doi:10.1038/nmat1681

Kim J, Alizadeh P, Harding T, Hefner-Gravink A, Klionsky DJ (1996) Disruption of the yeast *ATH1* gene confers better survival after dehydration, freezing, and ethanol shock: potential commercial applications. Appl Environ Microbiol 62(5):1563–1569

Kim H, Smith JE, Ridenour JB, Woloshuk CP, Bluhm BH (2011) HXK1 regulates carbon catabolism, sporulation, fumonisin B(1) production and pathogenesis in *Fusarium verticillioides*. Microbiology 157(Pt 9):2658–2669. doi:10.1099/mic.0.052506-0

Kimura Y, Kawasaki S, Tuchimoto R, Tanaka N (2014) Trehalose biosynthesis in *Myxococcus xanthus* under osmotic stress and during spore formation. J Biochem 155(1):17–24. doi:10.1093/jb/mvt091

Kitamoto Y, Akashi H, Tanaka H, Mori N (1988) a-Glucose-1-phosphate formation by a novel trehalose phosphorylase from *Flammulina velutipes*. FEMS Microbiol Lett 55:147–150

Kline L, Sugihara TF (1968) Factors affecting the stability of frozen bread doughs. I. Prepared by the straight dough method. Bakers Dig 42:44–69

Kopjar M, Hribar J, Simcic M, Zlatic E, Pozrl T, Pilizota V (2013) Effect of trehalose addition on volatiles responsible for strawberry aroma. Nat Prod Commun 8(12):1767–1770

Kopp M, Muller H, Holzer H (1993) Molecular analysis of the neutral trehalase gene from *Saccharomyces cerevisiae*. J Biol Chem 268(7):4766–4774

Kormish JD, McGhee JD (2005) The *C. elegans* lethal gut-obstructed *gob-1* gene is trehalose-6-phosphate phosphatase. Dev Biol 287(1):35–47. doi:10.1016/j.ydbio.2005.08.027

Koster KL, Maddocks KJ, Bryant G (2003) Exclusion of maltodextrins from phosphatidylcholine multilayers during dehydration: effects on membrane phase behaviour. Eur Biophys J 32(2):96–105. doi:10.1007/s00249-003-0277-z

Kottke I, Oberwinkler F (1987) The cellular structure of the Hartig net: coenocytic and transfer cell-like organization. Nord J Bot 7:85–95

Kotyk A, Michaljanicova D (1979) Uptake of trehalose by *Saccharomyces cerevisiae*. J Gen Microbiol 110(2):323–332

Kraakman L, Lemaire K, Ma P, Teunissen AW, Donaton MC, Van Dijck P, Winderickx J, de Winde JH, Thevelein JM (1999) A *Saccharomyces cerevisiae* G-protein coupled receptor, Gpr1, is specifically required for glucose activation of the cAMP pathway during the transition to growth on glucose. Mol Microbiol 32(5):1002–1012

Kramer KJ, Dzladik-Turner C, Koga D (1985) Chitin metabolism in insects. Comprehensive Insect Physiology, Biochemistry and Pharmacology 3:75–115

Kubota S, Takeo I, Kume K, Kanai M, Shitamukai A, Mizunuma M, Miyakawa T, Shimoi H, Iefuji H, Hirata D (2004) Effect of ethanol on cell growth of budding yeast: genes that are important for cell growth in the presence of ethanol. Biosci Biotechnol Biochem 68(4):968–972

Küenzi MT, Fiechter A (1969) Changes in carbohydrate composition and trehalase activity during the budding cycle of *Saccharomyces cerevisiae*. Arch Mikrobiol 64:396–407

Kuenzi MT, Fiechter A (1972) Regulation of carbohydrate composition of *Saccharomyces cerevisiae* under growth limitation. Arch Mikrobiol 84(3):254–265

Kummel A, Ewald JC, Fendt SM, Jol SJ, Picotti P, Aebersold R, Sauer U, Zamboni N, Heinemann M (2010) Differential glucose repression in common yeast strains in response to *HXK2* deletion. FEMS Yeast Res 10(3):322–332. doi:10.1111/j.1567-1364.2010.00609.x

Kyryakov P, Beach A, Richard VR, Burstein MT, Leonov A, Levy S, Titorenko VI (2012) Caloric restriction extends yeast chronological lifespan by altering a pattern of age-related changes in trehalose concentration. Front Physiol 3:256. doi:10.3389/fphys.2012.00256

Lafuente MJ, Gancedo C, Jauniaux JC, Gancedo JM (2000) Mth1 receives the signal given by the glucose sensors Snf3 and Rgt2 in *Saccharomyces cerevisiae*. Mol Microbiol 35(1):161–172

Lahiri S, Basu A, Sengupta S, Banerjee S, Dutta T, Soren D, Chattopadhyay K, Ghosh AK (2012) Purification and characterization of a trehalase-invertase enzyme with dual activity from *Candida utilis*.

Arch Biochem Biophys 522(2):90–99. doi:10.1016/j.abb.2012.03.026

Lallet S, Garreau H, Poisier C, Boy-Marcotte E, Jacquet M (2004) Heat shock-induced degradation of Msn2p, a *Saccharomyces cerevisiae* transcription factor, occurs in the nucleus. Mol Genet Genomics 272(3):353–362. doi:10.1007/s00438-004-1063-z

Landry S, Hoffman CS (2001) The git5 Gbeta and git11 Ggamma form an atypical Gbetagamma dimer acting in the fission yeast glucose/cAMP pathway. Genetics 157:1159–1168

Lee DH, Goldberg AL (1998) Proteasome inhibitors cause induction of heat shock proteins and trehalose, which together confer thermotolerance in *Saccharomyces cerevisiae*. Mol Cell Biol 18(1):30–38

Lee P, Cho BR, Joo HS, Hahn JS (2008) Yeast Yak1 kinase, a bridge between PKA and stress-responsive transcription factors, Hsf1 and Msn2/Msn4. Mol Microbiol 70(4):882–895. doi:10.1111/j.1365-2958.2008.06450.x

Lee P, Paik SM, Shin CS, Huh WK, Hahn JS (2011) Regulation of yeast Yak1 kinase by PKA and autophosphorylation-dependent 14-3-3 binding. Mol Microbiol 79(3):633–646. doi:10.1111/j.1365-2958.2010.07471.x

Lee P, Kim MS, Paik SM, Choi SH, Cho BR, Hahn JS (2013) Rim15-dependent activation of Hsf1 and Msn2/4 transcription factors by direct phosphorylation in *Saccharomyces cerevisiae*. FEBS Lett 587(22):3648–3655. doi:10.1016/j.febslet.2013.10.004

Lemaire K, Van de Velde S, Van Dijck P, Thevelein JM (2004) Glucose and sucrose act as agonist and mannose as antagonist ligands of the G protein-coupled receptor Gpr1 in the yeast *Saccharomyces cerevisiae*. Mol Cell 16(2):293–299. doi:10.1016/j.molcel.2004.10.004

Lenné T, Bryant G, Garvey CJ, Keiderling U, Koster KL (2006) Location of sugars in multilamellar membranes at low hydration. Phys B Condens Matter 385–386:862–864

Levine H, Slade L (1992) Another view of trehalose for drying and stabilizing biological materials. BioPharm 5:36–40

Levy SF, Ziv N, Siegal ML (2012) Bet hedging in yeast by heterogeneous, age-correlated expression of a stress protectant. PLoS Biol 10(5), e1001325. doi:10.1371/journal.pbio.1001325

Lewis JG, Learmonth RP, Watson K (1993) Role of growth phase and ethanol in freeze-thaw stress resistance of *Saccharomyces cerevisiae*. Appl Environ Microbiol 59:1065–1071

Leyman B, Van Dijck P, Thevelein JM (2001) An unexpected plethora of trehalose biosynthesis genes in *Arabidopsis thaliana*. Trends Plant Sci 6(11):510–513

Li Y, Wang Y (2013) Ras protein/cAMP-dependent protein kinase signaling is negatively regulated by a deubiquitinating enzyme, Ubp3, in yeast. J Biol Chem 288(16):11358–11365. doi:10.1074/jbc.M112.449751

Li HW, Zang BS, Deng XW, Wang XP (2011) Overexpression of the trehalose-6-phosphate synthase gene *OsTPS1* enhances abiotic stress tolerance in rice. Planta 234(5):1007–1018. doi:10.1007/s00425-011-1458-0

Lillie SH, Pringle JR (1980) Reserve carbohydrate metabolism in *Saccharomyces cerevisiae*: responses to nutrient limitation. J Bacteriol 143(3):1384–1394

Liu H, Rodgers ND, Jiao X, Kiledjian M (2002) The scavenger mRNA decapping enzyme DcpS is a member of the HIT family of pyrophosphatases. EMBO J 21(17):4699–4708

Liu R, Barkhordarian H, Emadi S, Park CB, Sierks MR (2005) Trehalose differentially inhibits aggregation and neurotoxicity of beta-amyloid 40 and 42. Neurobiol Dis 20(1):74–81. doi:10.1016/j.nbd.2005.02.003

Liu K, Dong Y, Huang Y, Rasgon JL, Agre P (2013a) Impact of trehalose transporter knockdown on *Anopheles gambiae* stress adaptation and susceptibility to *Plasmodium falciparum* infection. Proc Natl Acad Sci USA 110(43):17504–17509. doi:10.1073/pnas.1316709110

Liu Q, Ying SH, Li JG, Tian CG, Feng MG (2013b) Insight into the transcriptional regulation of Msn2 required for conidiation, multi-stress responses and virulence of two entomopathogenic fungi. Fungal Genet Biol 54:42–51. doi:10.1016/j.fgb.2013.02.008

Livas D, Almering MJ, Daran JM, Pronk JT, Gancedo JM (2011) Transcriptional responses to glucose in *Saccharomyces cerevisiae* strains lacking a functional protein kinase A. BMC Genomics 12:405. doi:10.1186/1471-2164-12-405

Lo Bianco C, Shorter J, Regulier E, Lashuel H, Iwatsubo T, Lindquist S, Aebischer P (2008) Hsp104 antagonizes alpha-synuclein aggregation and reduces dopaminergic degeneration in a rat model of Parkinson disease. J Clin Invest 118(9):3087–3097. doi:10.1172/JCI35781

Londesborough J, Varimo K (1984) Characterization of two trehalases in baker's yeast. Biochem J 219(2):511–518

Londesborough J, Vuorio O (1991) Trehalose-6-phosphate synthase/phosphatase complex from bakers' yeast: purification of a proteolytically activated form. J Gen Microbiol 137:323–330

Lopes DH, Meyer-Fernandes JR, Sola-Penna M (1999) Effects of trehalose and ethanol on yeast cytosolic pyrophosphatase. Z Naturforsch C 54(3–4):186–190

Lopez MF, Manner P, Willmann A, Hampp R, Nehls U (2007) Increased trehalose biosynthesis in Hartig net hyphae of ectomycorrhizas. New Phytol

174(2):389–398. doi:10.1111/j.1469-8137.2007.01983.x

Lopez M, Tejera NA, Iribarne C, Lluch C, Herrera-Cervera JA (2008) Trehalose and trehalase in root nodules of *Medicago truncatula* and *Phaseolus vulgaris* in response to salt stress. Physiol Plant 134(4):575–582. doi:10.1111/j.1399-3054.2008.01162.x

Lowe RG, Lord M, Rybak K, Trengove RD, Oliver RP, Solomon PS (2009) Trehalose biosynthesis is involved in sporulation of *Stagonospora nodorum*. Fungal Genet Biol 46(5):381–389. doi:10.1016/j.fgb.2009.02.002

Lu A, Hirsch JP (2005) Cyclic AMP-independent regulation of protein kinase A substrate phosphorylation by Kelch repeat proteins. Eukaryot Cell 4(11):1794–1800. doi:10.1128/ec.4.11.1794-1800.2005

Lucio-Eterovic AK, Jorge JA, Polizeli Mde L, Terenzi HF (2005) Biochemical characterisation of the trehalase of thermophilic fungi: an enzyme with mixed properties of neutral and acid trehalase. Biochim Biophys Acta 1723(1–3):201–207. doi:10.1016/j.bbagen.2005.02.011

Lundh F, Mouillon JM, Samyn D, Stadler K, Popova Y, Lagerstedt JO, Thevelein JM, Persson BL (2009) Molecular mechanisms controlling phosphate-induced downregulation of the yeast Pho84 phosphate transporter. Biochemistry 48(21):4497–4505. doi:10.1021/bi9001198

Luo X, Talarek N, De Virgilio C (2011) Initiation of the yeast G0 program requires Igo1 and Igo2, which antagonize activation of decapping of specific nutrient-regulated mRNAs. RNA Biol 8(1):14–17

Lutfiyya LL, Iyer VR, DeRisi J, DeVit MJ, Brown PO, Johnston M (1998) Characterization of three related glucose repressors and genes they regulate in *Saccharomyces cerevisiae*. Genetics 150(4):1377–1391

Luyten K, de Koning W, Tesseur I, Ruiz MC, Ramos J, Cobbaert P, Thevelein JM, Hohmann S (1993) Disruption of the *Kluyveromyces lactis GGS1* gene causes inability to grow on glucose and fructose and is suppressed by mutations that reduce sugar uptake. Eur J Biochem 217:701–713

Ma P, Wera S, Van Dijck P, Thevelein JM (1999) The *PDE1*-encoded low-affinity phosphodiesterase in the yeast *Saccharomyces cerevisiae* has a specific function in controlling agonist-induced cAMP signaling. Mol Biol Cell 10:91–104

Macakova E, Kopecka M, Kukacka Z, Veisova D, Novak P, Man P, Obsil T, Obsilova V (2013) Structural basis of the 14-3-3 protein-dependent activation of yeast neutral trehalase Nth1. Biochim Biophys Acta 1830(10):4491–4499. doi:10.1016/j.bbagen.2013.05.025

Mackay MA, Norton RS, Borowitzka LJ (1984) Organic osmoregulatory solutes in cyanobacteria. J Gen Microbiol 130:2177–2191

Mackenzie KF, Singh KK, Brown AD (1988) Water stress plating hypersensitivity of yeasts: protective role of trehalose in *Saccharomyces cerevisiae*. J Gen Microbiol 134:1661–1666

Mager WH, Moradas-Ferreira P (1993) Stress response of yeast. Biochem J 290(Part 1):1–13

Mahmud SA, Nagahisa K, Hirasawa T, Yoshikawa K, Ashitani K, Shimizu H (2009) Effect of trehalose accumulation on response to saline stress in *Saccharomyces cerevisiae*. Yeast 26(1):17–30. doi:10.1002/yea.1646

Maidan MM, De Rop L, Serneels J, Exler S, Rupp S, Tournu H, Thevelein JM, Van Dijck P (2005) The G protein-coupled receptor Gpr1 and the Galpha protein Gpa2 act through the cAMP-protein kinase A pathway to induce morphogenesis in *Candida albicans*. Mol Biol Cell 16(4):1971–1986. doi:10.1091/mbc.E04-09-0780

Maidan MM, De Rop L, Relloso M, Diez-Orejas R, Thevelein JM, Van Dijck P (2008) Combined inactivation of the *Candida albicans GPR1* and *TPS2* genes results in avirulence in a mouse model for systemic infection. Infect Immun 76(4):1686–1694. doi:10.1128/iai.01497-07

Malluta EF, Decker P, Stambuk BU (2000) The Kluyver effect for trehalose in *Saccharomyces cerevisiae*. J Basic Microbiol 40(3):199–205. doi:10.1002/1521-4028(200007)40:3<199::aid-jobm199>3.0.co;2-e

Malys N, Carroll K, Miyan J, Tollervey D, McCarthy JE (2004) The 'scavenger' m7GpppX pyrophosphatase activity of Dcs1 modulates nutrient-induced responses in yeast. Nucleic Acids Res 32(12):3590–3600. doi:10.1093/nar/gkh687

Marino C, Curto M, Bruno R, Rinaudo MT (1989) Trehalose synthase and trehalose behaviour in yeast cells in anhydrobiosis and hydrobiosis. Int J Biochem 21:1369–1375

Marquez-Escalante JA, Figueroa-Soto CG, Valenzuela-Soto EM (2006) Isolation and partial characterization of trehalose 6-phosphate synthase aggregates from *Selaginella lepidophylla* plants. Biochimie 88(10):1505–1510. doi:10.1016/j.biochi.2006.06.004

Martegani E, Baroni M, Vanoni M (1986) Interaction of cAMP with the CDC25-mediated step in the cell cycle of budding yeast. Exp Cell Res 162:544–548

Martin MC, Diaz LA, Manzanal MB, Hardisson C (1986) Role of trehalose in the spores of *Streptomyces*. FEMS Microbiol Lett 35:49–54

Martin F, Duplessis S, Ditengou F, Lagrange H, Voiblet C, Lapeyrie F (2001) Developmental cross talking in the ectomycorrhizal symbiosis: signals and communication genes. New Phytol 151:145–154

Martin DE, Soulard A, Hall MN (2004) TOR regulates ribosomal protein gene expression via PKA and the Forkhead transcription factor FHL1. Cell 119(7):969–979. doi:10.1016/j.cell.2004.11.047

Martinez-Esparza M, Aguinaga A, Gonzalez-Parraga P, Garcia-Penarrubia P, Jouault T, Arguelles JC

(2007) Role of trehalose in resistance to macrophage killing: study with a *tpsl/tpsl* trehalose-deficient mutant of *Candida albicans*. Clin Microbiol Infect 13(4):384–394. doi:10.1111/j.1469-0691.2007.01663.x

Martinez-Esparza M, Martinez-Vicente E, Gonzalez-Parraga P, Ros JM, Garcia-Penarrubia P, Arguelles JC (2009) Role of trehalose-6P phosphatase (TPS2) in stress tolerance and resistance to macrophage killing in *Candida albicans*. Int J Med Microbiol 299(6):453–464. doi:10.1016/j.ijmm.2008.12.001

Martinez-Esparza M, Tapia-Abellan A, Vitse-Standaert A, Garcia-Penarrubia P, Arguelles JC, Poulain D, Jouault T (2011) Glycoconjugate expression on the cell wall of *tpsl/tpsl* trehalose-deficient *Candida albicans* strain and implications for its interaction with macrophages. Glycobiology 21(6):796–805. doi:10.1093/glycob/cwr007

Martinez-Pastor MT, Marchler G, Schuller C, Marchler-Bauer A, Ruis H, Estruch F (1996) The *Saccharomyces cerevisiae* zinc finger proteins Msn2p and Msn4p are required for transcriptional induction through the stress response element (STRE). EMBO J 15(9):2227–2235

Matsuo Y, McInnis B, Marcus S (2008) Regulation of the subcellular localization of cyclic AMP-dependent protein kinase in response to physiological stresses and sexual differentiation in the fission yeast *Schizosaccharomyces pombe*. Eukaryot Cell 7(9):1450–1459. doi:10.1128/ec.00168-08

Mbonyi K, van Aelst L, Arguelles JC, Jans AW, Thevelein JM (1990) Glucose-induced hyperaccumulation of cyclic AMP and defective glucose repression in yeast strains with reduced activity of cyclic AMP-dependent protein kinase. Mol Cell Biol 10(9):4518–4523

McBride MJ, Ensign JC (1987) Effects of intracellular trehalose content on *Streptomyces griseus* spores. J Bacteriol 169:4995–5001

McDougall J, Kaasen I, Strøm AR (1993) A yeast gene for trehalose-6-phosphate synthase and its complementation of an *Escherichia coli otsA* mutant. FEMS Microbiol Lett 107:25–30

Merritt PP (1960) The effect of preparation on the stability and performance of frozen, unbaked, yeast-leavened doughs. Bakers Digest 34:57–58

Mittenbuhler K, Holzer H (1988) Purification and characterization of acid trehalase from the yeast suc2 mutant. J Biol Chem 263(17):8537–8543

Momose Y, Matsumoto R, Maruyama A, Yamaoka M (2010) Comparative analysis of transcriptional responses to the cryoprotectants, dimethyl sulfoxide and trehalose, which confer tolerance to freeze-thaw stress in *Saccharomyces cerevisiae*. Cryobiology 60(3):245–261. doi:10.1016/j.cryobiol.2010.01.001

Morano KA, Grant CM, Moye-Rowley WS (2012) The response to heat shock and oxidative stress in *Saccharomyces cerevisiae*. Genetics 190(4):1157–1195. doi:10.1534/genetics.111.128033

Moriya H, Shimizu-Yoshida Y, Omori A, Iwashita S, Katoh M, Sakai A (2001) Yak1p, a DYRK family kinase, translocates to the nucleus and phosphorylates yeast Pop2p in response to a glucose signal. Genes Dev 15(10):1217–1228. doi:10.1101/gad.884001

Mortberg M, Neujahr HY (1986) Transport and hydrolysis of disaccharides by *Trichosporon cutaneum*. J Bacteriol 168(2):734–738

Moskvina E, Schuller C, Maurer CT, Mager WH, Ruis H (1998) A search in the genome of *Saccharomyces cerevisiae* for genes regulated via stress response elements. Yeast 14(11):1041–1050. doi:10.1002/(sici)1097-0061(199808)14:11<1041::aid-yea296>3.0.co;2-4

Mudge DK, Yang F, Currie BM, Kim JM, Yeda K, Bashyakarla VK, Ivey FD, Hoffman CS (2014) Sck1 negatively regulates Gpa2-mediated glucose signaling in *Schizosaccharomyces pombe*. Eukaryot Cell 13(2):202–208. doi:10.1128/ec.00277-13

Mulet JM, Alejandro S, Romero C, Serrano R (2004) The trehalose pathway and intracellular glucose phosphates as modulators of potassium transport and general cation homeostasis in yeast. Yeast 21(7):569–582. doi:10.1002/yea.1126

Muller J, Aeschbacher RA, Wingler A, Boller T, Wiemken A (2001) Trehalose and trehalase in *Arabidopsis*. Plant Physiol 125(2):1086–1093

Nakafuku M, Obara T, Kaibuchi K, Miyajima I, Miyajima A, Itoh H, Nakamura S, Arai K, Matsumoto K, Kaziro Y (1988) Isolation of a second yeast *Saccharomyces cerevisiae* gene (*GPA2*) coding for guanine nucleotide-binding regulatory protein: studies on its structure and possible functions. Proc Natl Acad Sci USA 85(5):1374–1378

Nakahara Y, Imanishi S, Mitsumasu K, Kanamori Y, Iwata K, Watanabe M, Kikawada T, Okuda T (2010) Cells from an anhydrobiotic chironomid survive almost complete desiccation. Cryobiology 60(2):138–146. doi:10.1016/j.cryobiol.2009.10.004

Nakao Y, Kanamori T, Itoh T, Kodama Y, Rainieri S, Nakamura N, Shimonaga T, Hattori M, Ashikari T (2009) Genome sequence of the lager brewing yeast, an interspecies hybrid. DNA Res 16(2):115–129. doi:10.1093/dnares/dsp003

Neiman AM (2011) Sporulation in the budding yeast *Saccharomyces cerevisiae*. Genetics 189(3):737–765. doi:10.1534/genetics.111.127126

Neves MJ, Francois J (1992) On the mechanism by which a heat shock induces trehalose accumulation in *Saccharomyces cerevisiae*. Biochem J 288(Pt 3):859–864

Neves MJ, Jorge JA, Francois JM, Terenzi HF (1991) Effects of heat shock on the level of trehalose and glycogen, and on the induction of thermotolerance in *Neurospora crassa*. FEBS Lett 283(1):19–22

Neves MJ, Hohmann S, Bell W, Dumortier F, Luyten K, Ramos J, Cobbaert P, de Koning W, Kaneva Z, Thevelein JM (1995) Control of glucose influx into glycolysis and pleiotropic effects studied in different isogenic sets of *Saccharomyces cerevisiae* mutants in trehalose biosynthesis. Curr Genet 27:110–122

Newman YM, Ring SG, Colaço C (1993) The role of trehalose and other carbohydrates in biopreservation. Biotechnol Genet Eng Rev 11:263–294

Ngamskulrungroj P, Himmelreich U, Breger JA, Wilson C, Chayakulkeeree M, Krockenberger MB, Malik R, Daniel HM, Toffaletti D, Djordjevic JT, Mylonakis E, Meyer W, Perfect JR (2009) The trehalose synthesis pathway is an integral part of the virulence composite for *Cryptococcus gattii*. Infect Immun 77(10):4584–4596. doi:10.1128/iai.00565-09

Ni M, Yu JH (2007) A novel regulator couples sporogenesis and trehalose biogenesis in *Aspergillus nidulans*. PLoS One 2(10), e970. doi:10.1371/journal.pone.0000970

Nicholls S, Straffon M, Enjalbert B, Nantel A, Macaskill S, Whiteway M, Brown AJ (2004) Msn2- and Msn4-like transcription factors play no obvious roles in the stress responses of the fungal pathogen *Candida albicans*. Eukaryot Cell 3(5):1111–1123. doi:10.1128/ec.3.5.1111-1123.2004

Nicholls S, Leach MD, Priest CL, Brown AJ (2009) Role of the heat shock transcription factor, Hsf1, in a major fungal pathogen that is obligately associated with warm-blooded animals. Mol Microbiol 74(4):844–861. doi:10.1111/j.1365-2958.2009.06883.x

Nieto-Sotelo J, Kannan KB, Martinez LM, Segal C (1999) Characterization of a maize heat-shock protein 101 gene, *HSP101*, encoding a ClpB/Hsp100 protein homologue. Gene 230(2):187–195

Nikawa J, Cameron S, Toda T, Ferguson KM, Wigler M (1987) Rigorous feedback control of cAMP levels in *Saccharomyces cerevisiae*. Genes Dev 1(9):931–937

Nishimoto T, Nakano M, Nakada T, Chaen H, Fukuda S, Sugimoto T, Kurimoto M, Tsujisaka Y (1996) Purification and properties of a novel enzyme, trehalose synthase, from Pimelobacter sp. R48. Biosci Biotechnol Biochem 60(4):640–644

Nocero M, Isshiki T, Yamamoto M, Hoffman CS (1994) Glucose repression of fbp1 transcription of *Schizosaccharomyces pombe* is partially regulated by adenylate cyclase activation by a G protein alpha subunit encoded by gpa2 (git8). Genetics 138:39–45

Noubhani A, Bunoust O, Bonini BM, Thevelein JM, Devin A, Rigoulet M (2009) The trehalose pathway regulates mitochondrial respiratory chain content through hexokinase 2 and cAMP in *Saccharomyces cerevisiae*. J Biol Chem 284(40):27229–27234. doi:10.1074/jbc.M109.029454

Noy Y, Sklan D (2001) Yolk and exogenous feed utilization in the posthatch chick. Poult Sci 80(10):1490–1495

Noy Y, Uni Z, Sklan D (1996) Routes of yolk utilisation in the newly-hatched chick. Br Poult Sci 37(5):987–995. doi:10.1080/00071669608417929

Nwaka S, Holzer H (1998) Molecular biology of trehalose and the trehalases in the yeast *Saccharomyces cerevisiae*. Prog Nucleic Acid Res Mol Biol 58:197–237

Nwaka S, Kopp M, Burgert M, Deuchler I, Kienle I, Holzer H (1994) Is thermotolerance of yeast dependent on trehalose accumulation? FEBS Lett 344:225–228

Nwaka S, Kopp M, Holzer H (1995a) Expression and function of the trehalase genes *NTH1* and *YBR0106* in *Saccharomyces cerevisiae*. J Biol Chem 270(17):10193–10198

Nwaka S, Mechler B, Destruelle M, Holzer H (1995b) Phenotypic features of trehalase mutants in *Saccharomyces cerevisiae*. FEBS Lett 360(3):286–290

Nwaka S, Mechler B, Holzer H (1996) Deletion of the *ATH1* gene in *Saccharomyces cerevisiae* prevents growth on trehalose. FEBS Lett 386(2–3):235–238

O'Hara LE, Paul MJ, Wingler A (2013) How do sugars regulate plant growth and development? New insight into the role of trehalose-6-phosphate. Mol Plant 6(2):261–274. doi:10.1093/mp/sss120

O'Meara TR, Xu W, Selvig KM, O'Meara MJ, Mitchell AP, Alspaugh JA (2014) The *Cryptococcus neoformans* Rim101 transcription factor directly regulates genes required for adaptation to the host. Mol Cell Biol 34(4):673–684. doi:10.1128/mcb.01359-13

Ocon A, Hampp R, Requena N (2007) Trehalose turnover during abiotic stress in arbuscular mycorrhizal fungi. New Phytol 174(4):879–891. doi:10.1111/j.1469-8137.2007.02048.x

Oda Y, Uno K, Ohta S (1986) Selection of yeasts for breadmaking by the frozen dough method. Appl Environ Microbiol 52:941–943

Oku K, Watanabe H, Kubota M, Fukuda S, Kurimoto M, Tsujisaka Y, Komori M, Inoue Y, Sakurai M (2003) NMR and quantum chemical study on the OH. . .pi and CH. . .O interactions between trehalose and unsaturated fatty acids: implication for the mechanism of antioxidant function of trehalose. J Am Chem Soc 125(42):12739–12748. doi:10.1021/ja034777e

Operti MS, Oliveira DE, Freitas-Valle AB, Oestreicher EG, Mattoon JR, Panek AD (1982) Relationships between trehalose metabolism and maltose utilization in *Saccharomyces cerevisiae*: III. Evidence for alternative pathways of trehalose synthesis. Curr Genet 5(1):69–76. doi:10.1007/bf00445744

Ortiz CH, Maia JC, Tenan MN, Braz-Padrao GR, Mattoon JR, Panek AD (1983) Regulation of yeast trehalase by a monocyclic, cyclic AMP-dependent phosphorylation-dephosphorylation cascade system. J Bacteriol 153(2):644–651

Otting G, Liepinsh E, Wuthrich K (1991) Protein hydration in aqueous solution. Science 254:974–980

Ovadi J, Keleti T (1978) Kinetic evidence for interaction between aldolase and D-glyceraldehyde-3-phosphate dehydrogenase. Eur J Biochem 85(1):157–161

Paalman JW, Verwaal R, Slofstra SH, Verkleij AJ, Boonstra J, Verrips CT (2003) Trehalose and glycogen accumulation is related to the duration of the G1 phase of *Saccharomyces cerevisiae*. FEMS Yeast Res 3(3):261–268

Panek AD (1963) Function of trehalose in baker's yeast (*Saccharomyces cerevisiae*). Arch Biochem Biophys 100:422–425

Panek AD, Bernardes EJ (1983) Trehalose: its role in germination of *Saccharomyces cerevisiae*. Curr Genet 7:393–397

Panek AD, Mattoon JR (1977) Regulation of energy metabolism in *Saccharomyces cerevisiae*. Relationships between catabolite repression, trehalose synthesis, and mitochondrial development. Arch Biochem Biophys 183:306–316

Panek AC, de Araujo PS, Moura Neto V, Panek AD (1987) Regulation of the trehalose-6-phosphate synthase complex in *Saccharomyces*. I. Interconversion of forms by phosphorylation. Curr Genet 11(6–7):459–465

Panek AD, Ferreira R, Panek AC (1989) Comparative studies between the glucose-induced phosphorylation signal and the heat shock response in mutants of *Saccharomyces cerevisiae*. Biochimie 71:313–318

Panek AC, Araujo PS, Poppe SC, Panek AD (1990) On the determination of trehalose-6-phosphate synthase in *Saccharomyces*. Biochem Int 21(4):695–704

Panneman H, Ruijter GJ, van den Broeck HC, Visser J (1998) Cloning and biochemical characterisation of *Aspergillus niger* hexokinase–the enzyme is strongly inhibited by physiological concentrations of trehalose 6-phosphate. Eur J Biochem 258(1):223–232

Panni S, Landgraf C, Volkmer-Engert R, Cesareni G, Castagnoli L (2008) Role of 14-3-3 proteins in the regulation of neutral trehalase in the yeast *Saccharomyces cerevisiae*. FEMS Yeast Res 8(1):53–63. doi:10.1111/j.1567-1364.2007.00312.x

Papasavvas S, Arkinstall S, Reid J, Payton M (1992) Yeast alpha-mating factor receptor and G-protein-linked adenylyl cyclase inhibition requires RAS2 and GPA2 activities. Biochem Biophys Res Commun 184(3):1378–1385

Parrou JL, Teste MA, Francois J (1997) Effects of various types of stress on the metabolism of reserve carbohydrates in *Saccharomyces cerevisiae*: genetic evidence for a stress-induced recycling of glycogen and trehalose. Microbiology 143(Pt 6):1891–1900

Parrou JL, Jules M, Beltran G, Francois J (2005) Acid trehalase in yeasts and filamentous fungi: localization, regulation and physiological function. FEMS Yeast Res 5(6–7):503–511. doi:10.1016/j.femsyr.2005.01.002

Parry JM, Davies PJ, Evans WE (1976) The effects of 'cell age' upon the lethal effects of physical and chemical mutagens in the yeast *Saccharomyces cerevisiae*. Mol Gen Genet 146:27–35

Parsell DA, Kowal AS, Singer MA, Lindquist S (1994) Protein disaggregation mediated by heat-shock protein Hsp104. Nature 372:475–478

Paschoalin VMF, Silva JT, Panek AD (1989) Identification of an ADPG-dependent trehalose synthase in *Saccharomyces*. Curr Genet 16:81–87

Payen R (1949) Variation des teneurs en glycogène et en tréhalose pendant le séchage de la levure. Can J Res 27B:749–756

Pedreno Y, Maicas S, Arguelles JC, Sentandreu R, Valentin E (2004) The *ATC1* gene encodes a cell wall-linked acid trehalase required for growth on trehalose in *Candida albicans*. J Biol Chem 279(39):40852–40860. doi:10.1074/jbc.M400216200

Pedreno Y, Gonzalez-Parraga P, Martinez-Esparza M, Sentandreu R, Valentin E, Arguelles JC (2007) Disruption of the *Candida albicans ATC1* gene encoding a cell-linked acid trehalase decreases hypha formation and infectivity without affecting resistance to oxidative stress. Microbiology 153(Pt 5):1372–1381. doi:10.1099/mic.0.2006/003921-0

Pedruzzi I, Burckert N, Egger P, De Virgilio C (2000) *Saccharomyces cerevisiae* Ras/cAMP pathway controls post-diauxic shift element-dependent transcription through the zinc finger protein Gis1. EMBO J 19(11):2569–2579. doi:10.1093/emboj/19.11.2569

Peeters T, Louwet W, Gelade R, Nauwelaers D, Thevelein JM, Versele M (2006) Kelch-repeat proteins interacting with the Galpha protein Gpa2 bypass adenylate cyclase for direct regulation of protein kinase A in yeast. Proc Natl Acad Sci USA 103(35):13034–13039. doi:10.1073/pnas.0509644103

Perrin V, Regulier E, Abbas-Terki T, Hassig R, Brouillet E, Aebischer P, Luthi-Carter R, Deglon N (2007) Neuroprotection by Hsp104 and Hsp27 in lentiviral-based rat models of Huntington's disease. Mol Ther 15(5):903–911. doi:10.1038/mt.sj.6300141

Petrenko N, Chereji RV, McClean MN, Morozov AV, Broach JR (2013) Noise and interlocking signaling pathways promote distinct transcription factor dynamics in response to different stresses. Mol Biol Cell 24(12):2045–2057. doi:10.1091/mbc.E12-12-0870

Petzold EW, Himmelreich U, Mylonakis E, Rude T, Toffaletti D, Cox GM, Miller JL, Perfect JR (2006) Characterization and regulation of the trehalose synthesis pathway and its importance in the pathogenicity of *Cryptococcus neoformans*. Infect Immun 74(10):5877–5887. doi:10.1128/iai.00624-06

Pilon-Smits EAH, Terry N, Sears T, Kim H, Zayed A, Hwang S, van Dun K, Voogd E, Verwoerd TC, Krutwagen RWHH, Goddijn OJM (1998) Trehalose-producing transgenic tobacco plants show improved growth performance under drought stress. J Plant Physiol 152:525–532

Piper P (1998) Differential role of Hsps and trehalose in stress tolerance. Trends Microbiol 6:43–44

Piper PW, Lockheart A (1988) A temperature-sensitive mutant of *Saccharomyces cerevisiae* defective in the specific phosphatase of trehalose biosynthesis. FEMS Microbiol Lett 49(2):245–250. doi:10.1111/j.1574-6968.1988.tb02724.x

Plesset J, Ludwig J, Cox B, McLaughlin C (1987) Effect of cell cycle position on thermotolerance in *Saccharomyces cerevisiae*. J Bacteriol 169:779–784

Plourde-Owobi L, Durner S, Parrou JL, Wieczorke R, Goma G, Francois J (1999) *AGT1*, encoding an alpha-glucoside transporter involved in uptake and intracellular accumulation of trehalose in *Saccharomyces cerevisiae*. J Bacteriol 181(12):3830–3832

Plourde-Owobi L, Durner S, Goma G, Francois J (2000) Trehalose reserve in *Saccharomyces cerevisiae*: phenomenon of transport, accumulation and role in cell viability. Int J Food Microbiol 55:33–40

Polakis ES, Bartley W (1966) Changes in dry weight, protein, deoxyribonucleic acid, ribonucleic acid and reserve and structural carbohydrate during the aerobic growth cycle of yeast. Biochem J 98:883–887

Pollock GE, Holmstrom CD (1951) The trehalose content and the quality of active dry yeast. Cereal Chem 28:498–505

Ponnu J, Wahl V, Schmid M (2011) Trehalose-6-phosphate: connecting plant metabolism and development. Front Plant Sci 2:70. doi:10.3389/fpls.2011.00070

Popova Y, Thayumanavan P, Lonati E, Agrochao M, Thevelein JM (2010) Transport and signaling through the phosphate-binding site of the yeast Pho84 phosphate transceptor. Proc Natl Acad Sci USA 107(7):2890–2895. doi:10.1073/pnas.0906546107

Postmus J, Canelas AB, Bouwman J, Bakker BM, van Gulik W, de Mattos MJ, Brul S, Smits GJ (2008) Quantitative analysis of the high temperature-induced glycolytic flux increase in *Saccharomyces cerevisiae* reveals dominant metabolic regulation. J Biol Chem 283(35):23524–23532. doi:10.1074/jbc.M802908200

Pukkila-Worley R, Gerrald QD, Kraus PR, Boily MJ, Davis MJ, Giles SS, Cox GM, Heitman J, Alspaugh JA (2005) Transcriptional network of multiple capsule and melanin genes governed by the *Cryptococcus neoformans* cyclic AMP cascade. Eukaryot Cell 4(1):190–201. doi:10.1128/ec.4.1.190-201.2005

Purvis JE, Yomano LP, Ingram LO (2005) Enhanced trehalose production improves growth of *Escherichia coli* under osmotic stress. Appl Environ Microbiol 71(7):3761–3769. doi:10.1128/aem.71.7.3761-3769.2005

Puttikamonkul S, Willger SD, Grahl N, Perfect JR, Movahed N, Bothner B, Park S, Paderu P, Perlin DS, Cramer RA Jr (2010) Trehalose 6-phosphate phosphatase is required for cell wall integrity and fungal virulence but not trehalose biosynthesis in the human fungal pathogen *Aspergillus fumigatus*. Mol Microbiol. doi:10.1111/j.1365-2958.2010.07254.x

Qu Q, Lee SJ, Boos W (2004) TreT, a novel trehalose glycosyltransferring synthase of the hyperthermophilic archaeon *Thermococcus litoralis*. J Biol Chem 279(46):47890–47897. doi:10.1074/jbc.M404955200

Quain DE, Haslam JM (1979) Changes in glucose 6-phosphate and storage carbohydrates during catabolite derepression in *Saccharomyces cerevisiae*. J Gen Microbiol 113:195–198

Ram SP, Romana LK, Shepherd MG, Sullivan PA (1984) Exo-(1——3)-beta-glucanase, autolysin and trehalase activities during yeast growth and germ-tube formation in *Candida albicans*. J Gen Microbiol 130(5):1227–1236

Ramon M, De Smet I, Vandesteene L, Naudts M, Leyman B, Van Dijck P, Rolland F, Beeckman T, Thevelein JM (2009) Extensive expression regulation and lack of heterologous enzymatic activity of the Class II trehalose metabolism proteins from *Arabidopsis thaliana*. Plant Cell Environ 32(8):1015–1032. doi:10.1111/j.1365-3040.2009.01985.x

Ratnakumar S, Tunnacliffe A (2006) Intracellular trehalose is neither necessary nor sufficient for desiccation tolerance in yeast. FEMS Yeast Res 6(6):902–913. doi:10.1111/j.1567-1364.2006.00066.x

Rayner TF, Gray JV, Thorner JW (2002) Direct and novel regulation of cAMP-dependent protein kinase by Mck1p, a yeast glycogen synthase kinase-3. J Biol Chem 277(19):16814–16822. doi:10.1074/jbc.M112349200

Rebecchi L, Cesari M, Altiero T, Frigieri A, Guidetti R (2009) Survival and DNA degradation in anhydrobiotic tardigrades. J Exp Biol 212(Pt 24):4033–4039. doi:10.1242/jeb.033266

Reed RH, Borowitzka LJ, Mackay MA, Chudek JA, Forster R, Warr RSC, Moore DJ, Stewart WDP (1986) Organic solute accumulation in osmotically stressed cyanobacteria. FEMS Microbiol Rev 39:51–56

Reina-Bueno M, Argandona M, Nieto JJ, Hidalgo-Garcia A, Iglesias-Guerra F, Delgado MJ, Vargas C (2012) Role of trehalose in heat and desiccation tolerance in the soil bacterium *Rhizobium etli*.

BMC Microbiol 12:207. doi:10.1186/1471-2180-12-207

Reinders A, Burckert N, Hohmann S, Thevelein JM, Boller T, Wiemken A, De Virgilio C (1997) Structural analysis of the subunits of the trehalose-6-phosphate synthase/phosphatase complex in *Saccharomyces cerevisiae* and their function during heat shock. Mol Microbiol 24(4):687–695

Reinders A, Burckert N, Boller T, Wiemken A, De Virgilio C (1998) *Saccharomyces cerevisiae* cAMP-dependent protein kinase controls entry into stationary phase through the Rim15p protein kinase. Genes Dev 12(18):2943–2955

Reinders A, Romano I, Wiemken A, De Virgilio C (1999) The thermophilic yeast *Hansenula polymorpha* does not require trehalose synthesis for growth at high temperatures but does for normal acquisition of thermotolerance. J Bacteriol 181:4665–4668

Rep M, Reiser V, Gartner U, Thevelein JM, Hohmann S, Ammerer G, Ruis H (1999) Osmotic stress-induced gene expression in *Saccharomyces cerevisiae* requires Msn1p and the novel nuclear factor Hot1p. Mol Cell Biol 19(8):5474–5485

Ribeiro MJ, Reinders A, Boller T, Wiemken A, De Virgilio C (1997) Trehalose synthesis is important for the acquisition of thermotolerance in *Schizosaccharomyces pombe*. Mol Microbiol 25:571–581

Rodaki A, Bohovych IM, Enjalbert B, Young T, Odds FC, Gow NA, Brown AJ (2009) Glucose promotes stress resistance in the fungal pathogen *Candida albicans*. Mol Biol Cell 20(22):4845–4855. doi:10.1091/mbc.E09-01-0002

Rolim MF, de Araujo PS, Panek AD, Paschoalin VM, Silva JT (2003) Shared control of maltose and trehalose utilization in *Candida utilis*. Braz J Med Biol Res 36(7):829–837

Rolland F, De Winde JH, Lemaire K, Boles E, Thevelein JM, Winderickx J (2000) Glucose-induced cAMP signalling in yeast requires both a G-protein coupled receptor system for extracellular glucose detection and a separable hexose kinase-dependent sensing process. Mol Microbiol 38(2):348–358

Romero C, Belles JM, Vaya JL, Serrano R, Culianez-Macia FA (1997) Expression of the yeast trehalose-6-phosphate synthase gene in transgenic tobacco plants: pleiotropic phenotypes include drought tolerance. Planta 201(3):293–297. doi:10.1007/s004250050069

Rosas-Lemus M, Uribe-Alvarez C, Chiquete-Felix N, Uribe-Carvajal S (2014) In *Saccharomyces cerevisiae* fructose-1,6-bisphosphate contributes to the Crabtree effect through closure of the mitochondrial unspecific channel. Arch Biochem Biophys 555–556:66–70. doi:10.1016/j.abb.2014.05.027

Roser BJ (1991a) Trehalose drying: a novel replacement for freeze-drying. BioPharm 5:44–53

Roser BJ (1991b) Trehalose, a new approach to premium dried foods. Trends Food Sci Technol 2:166–169

Roth R (1970) Carbohydrate accumulation during the sporulation of yeast. J Bacteriol 101(1):53–57

Rowley A, Johnston GC, Butler B, Werner-Washburne M, Singer RA (1993) Heat shock-mediated cell cycle blockage and G1 cyclin expression in the yeast *Saccharomyces cerevisiae*. Mol Cell Biol 13(2):1034–1041

Rua J, de Castro C, de Arriaga D, Garcia-Armesto MR, Busto F, Del Valle P (2014) Stress in *Phycomyces blakesleeanus* by glucose starvation and acetate growth: Response of the antioxidant system and reserve carbohydrates. Microbiol Res. doi:10.1016/j.micres.2013.12.007

Rubio-Texeira M, Van Zeebroeck G, Voordeckers K, Thevelein JM (2010) *Saccharomyces cerevisiae* plasma membrane nutrient sensors and their role in PKA signaling. FEMS Yeast Res 10(2):134–149. doi:10.1111/j.1567-1364.2009.00587.x

Rubio-Texeira M, Van Zeebroeck G, Thevelein JM (2012) Peptides induce persistent signaling from endosomes by a nutrient transceptor. Nat Chem Biol 8(4):400–408. doi:10.1038/nchembio.910

Ruf J, Wacker H, James P, Maffia M, Seiler P, Galand G, von Kieckebusch A, Semenza G, Mantei N (1990) Rabbit small intestinal trehalase. Purification, cDNA cloning, expression, and verification of glycosylphosphatidylinositol anchoring. J Biol Chem 265:15034–15039

Rui O, Hahn M (2007) The *Botrytis cinerea* hexokinase, Hxk1, but not the glucokinase, Glk1, is required for normal growth and sugar metabolism, and for pathogenicity on fruits. Microbiology 153(Pt 8):2791–2802. doi:10.1099/mic.0.2007/006338-0

Ruis H, Schuller C (1995) Stress signaling in yeast. Bioessays 17(11):959–965. doi:10.1002/bies.950171109

Ryu SI, Park CS, Cha J, Woo EJ, Lee SB (2005) A novel trehalose-synthesizing glycosyltransferase from *Pyrococcus horikoshii*: molecular cloning and characterization. Biochem Biophys Res Commun 329(2):429–436. doi:10.1016/j.bbrc.2005.01.149

Sadeh A, Movshovich N, Volokh M, Gheber L, Aharoni A (2011) Fine-tuning of the Msn2/4-mediated yeast stress responses as revealed by systematic deletion of Msn2/4 partners. Mol Biol Cell 22(17):3127–3138. doi:10.1091/mbc.E10-12-1007

Sadowski I, BJ Breitkreutz, C Stark, TC Su, M Dahabieh, S Raithatha, W Bernhard, R Oughtred, K Dolinski, K Barreto, M Tyers (2013) The PhosphoGRID *Saccharomyces cerevisiae* protein phosphorylation site database: version 2.0 update. Database (Oxford) 2013:bat026. doi:10.1093/database/bat026

Saenger W (1989) Structure and dynamics of water surrounding biomolecules. Annu Rev Biophys Biophys Chem 16:93–114

Sagee S, Sherman A, Shenhar G, Robzyk K, Ben-Doy N, Simchen G, Kassir Y (1998) Multiple and distinct activation and repression sequences mediate the regulated transcription of IME1, a transcriptional activator of meiosis-specific genes in *Saccharomyces cerevisiae*. Mol Cell Biol 18(4):1985–1995

Saito K, Kase T, Takahashi E, Horinouchi S (1998) Purification and characterization of a trehalose synthase from the basidiomycete *Grifola frondosa*. Appl Environ Microbiol 64(11):4340–4345

Sakamoto K, Kim MJ, Rhoades ER, Allavena RE, Ehrt S, Wainwright HC, Russell DG, Rohde KH (2013) Mycobacterial trehalose dimycolate reprograms macrophage global gene expression and activates matrix metalloproteinases. Infect Immun 81(3):764–776. doi:10.1128/iai.00906-12

Sakurai M, Furuki T, Akao K, Tanaka D, Nakahara Y, Kikawada T, Watanabe M, Okuda T (2008) Vitrification is essential for anhydrobiosis in an African chironomid, *Polypedilum vanderplanki*. Proc Natl Acad Sci USA 105(13):5093–5098. doi:10.1073/pnas.0706197105

Salema-Oom M, Valadao Pinto V, Goncalves P, Spencer-Martins I (2005) Maltotriose utilization by industrial *Saccharomyces* strains: characterization of a new member of the alpha-glucoside transporter family. Appl Environ Microbiol 71(9):5044–5049. doi:10.1128/aem.71.9.5044-5049.2005

Salerno GL, Curatti L (2003) Origin of sucrose metabolism in higher plants: when, how and why? Trends Plant Sci 8(2):63–69. doi:10.1016/S1360-1385(02)00029-8

Sales K, Brandt W, Rumbak E, Lindsey G (2000) The LEA-like protein HSP12 in *Saccharomyces cerevisiae* has a plasma membrane location and protects membranes against desiccation and ethanol-induced stress. Biochim Biophys Acta 1463:267–278

Samyn DR, Ruiz-Pavon L, Andersson MR, Popova Y, Thevelein JM, Persson BL (2012) Mutational analysis of putative phosphate- and proton-binding sites in the *Saccharomyces cerevisiae* Pho84 phosphate:H(+) transceptor and its effect on signalling to the PKA and PHO pathways. Biochem J 445(3):413–422. doi:10.1042/bj20112086

San Miguel PF, Arguelles JC (1994) Differential changes in the activity of cytosolic and vacuolar trehalases along the growth cycle of *Saccharomyces cerevisiae*. Biochim Biophys Acta 1200(2):155–160

Sanchez Y, Lindquist S (1990) HSP104 required for induced thermotolerance. Science 248:1112–1115

Sanchez Y, Taulien J, Borkovich KA, Lindquist S (1992) Hsp104 is required for tolerance to many forms of stress. EMBO J 11:2357–2364

Sanchez-Fresneda R, Gonzalez-Parraga P, Esteban O, Laforet L, Valentin E, Arguelles JC (2009) On the biochemical classification of yeast trehalases: *Candida albicans* contains two enzymes with mixed features of neutral and acid trehalase activities. Biochem Biophys Res Commun 383(1):98–102. doi:10.1016/j.bbrc.2009.03.134

Sanchez-Fresneda R, Guirao-Abad JP, Arguelles A, Gonzalez-Parraga P, Valentin E, Arguelles JC (2013) Specific stress-induced storage of trehalose, glycerol and D-arabitol in response to oxidative and osmotic stress in *Candida albicans*. Biochem Biophys Res Commun 430(4):1334–1339. doi:10.1016/j.bbrc.2012.10.118

Sanchez-Fresneda R, Martinez-Esparza M, Maicas S, Arguelles JC, Valentin E (2014) In *Candida parapsilosis* the *ATC1* gene encodes for an acid trehalase involved in trehalose hydrolysis, stress resistance and virulence. PLoS One 9(6):e99113. doi:10.1371/journal.pone.0099113

Sano F, Asakawa N, Inoue Y, Sakurai M (1999) A dual role for intracellular trehalose in the resistance of yeast cells to water stress. Cryobiology 39:80–87

Santhanam A, Hartley A, Duvel K, Broach JR, Garrett S (2004) PP2A phosphatase activity is required for stress and Tor kinase regulation of yeast stress response factor Msn2p. Eukaryot Cell 3(5):1261–1271. doi:10.1128/ec.3.5.1261-1271.2004

Sarkar S, Davies JE, Huang Z, Tunnacliffe A, Rubinsztein DC (2007) Trehalose, a novel mTOR-independent autophagy enhancer, accelerates the clearance of mutant huntingtin and alpha-synuclein. J Biol Chem 282(8):5641–5652. doi:10.1074/jbc.M609532200

Sasano Y, Haitani Y, Hashida K, Ohtsu I, Shima J, Takagi H (2012a) Overexpression of the transcription activator Msn2 enhances the fermentation ability of industrial baker's yeast in frozen dough. Biosci Biotechnol Biochem 76(3):624–627

Sasano Y, Haitani Y, Hashida K, Ohtsu I, Shima J, Takagi H (2012b) Simultaneous accumulation of proline and trehalose in industrial baker's yeast enhances fermentation ability in frozen dough. J Biosci Bioeng 113(5):592–595. doi:10.1016/j.jbiosc.2011.12.018

Satoh-Nagasawa N, Nagasawa N, Malcomber S, Sakai H, Jackson D (2006) A trehalose metabolic enzyme controls inflorescence architecture in maize. Nature 441(7090):227–230. doi:10.1038/nature04725

Satomura A, Katsuyama Y, Miura N, Kuroda K, Tomio A, Bamba T, Fukusaki E, Ueda M (2013) Acquisition of thermotolerant yeast *Saccharomyces cerevisiae* by breeding via stepwise adaptation. Biotechnol Prog 29(5):1116–1123. doi:10.1002/btpr.1754

Schenberg-Frascino A, Moustacchi E (1972) Lethal and mutagenic effects of elevated temperature on haploid yeast. Mol Gen Genet 115:243–257

Schepers W, Van Zeebroeck G, Pinkse M, Verhaert P, Thevelein JM (2012) In vivo phosphorylation of Ser21 and Ser83 during nutrient-induced activation of the yeast protein kinase A (PKA) target trehalase. J Biol Chem 287(53):44130–44142. doi:10.1074/jbc.M112.421503

Schick I, Haltrich D, Kulbe KD (1995) Trehalose phosphorylase from *Pichia fermentans* and its role in the metabolism of trehalose. Appl Microbiol Biotechnol 43(6):1088–1095

Schiraldi C, Di Lernia I, De Rosa M (2002) Trehalose production: exploiting novel approaches. Trends Biotechnol 20(10):420–425

Schmelzle T, Beck T, Martin DE, Hall MN (2004) Activation of the RAS/cyclic AMP pathway suppresses a TOR deficiency in yeast. Mol Cell Biol 24(1):338–351

Schmitt AP, McEntee K (1996) Msn2p, a zinc finger DNA-binding protein, is the transcriptional activator of the multistress response in *Saccharomyces cerevisiae*. Proc Natl Acad Sci USA 93(12):5777–5782

Schothorst J, Kankipati HN, Conrad M, Samyn DR, Van Zeebroeck G, Popova Y, Rubio-Texeira M, Persson BL, Thevelein JM (2013) Yeast nutrient transceptors provide novel insight in the functionality of membrane transporters. Curr Genet 59(4):197–206. doi:10.1007/s00294-013-0413-y

Sebollela A, Louzada PR, Sola-Penna M, Sarone-Williams V, Coelho-Sampaio T, Ferreira ST (2004) Inhibition of yeast glutathione reductase by trehalose: possible implications in yeast survival and recovery from stress. Int J Biochem Cell Biol 36(5):900–908. doi:10.1016/j.biocel.2003.10.006

Sengupta S, Chaudhuri P, Lahiri S, Dutta T, Banerjee S, Majhi R, Ghosh AK (2011) Possible regulation of trehalose metabolism by methylation in *Saccharomyces cerevisiae*. J Cell Physiol 226(1):158–164. doi:10.1002/jcp.22317

Sengupta S, Lahiri S, Banerjee S, Dutta T, Dhar TK, Ghosh AK (2012) Methylation dependent enhancement of trehalose production in *Candida utilis*. Carbohydr Res 361:175–181. doi:10.1016/j.carres.2012.09.002

Seo HS, Koo YJ, Lim JY, Song JT, Kim CH, Kim JK, Lee JS, Choi YD (2000) Characterization of a bifunctional enzyme fusion of trehalose-6-phosphate synthetase and trehalose-6-phosphate phosphatase of *Escherichia coli*. Appl Environ Microbiol 66:2484–2490

Serneels J, Tournu H, Van Dijck P (2012) Tight control of trehalose content is required for efficient heat-induced cell elongation in *Candida albicans*. J Biol Chem 287(44):36873–36882. doi:10.1074/jbc.M112.402651

Shapiro RS, Uppuluri P, Zaas AK, Collins C, Senn H, Perfect JR, Heitman J, Cowen LE (2009) Hsp90 orchestrates temperature-dependent *Candida albicans* morphogenesis via Ras1-PKA signaling. Curr Biol 19(8):621–629. doi:10.1016/j.cub.2009.03.017

Sharma SC (1997) A possible role of trehalose in osmotolerance and ethanol tolerance in *Saccharomyces cerevisiae*. FEMS Microbiol Lett 152:11–15

Shen B, Hohmann S, Jensen RG, Bohnert aH (1999) Roles of sugar alcohols in osmotic stress adaptation. Replacement of glycerol by mannitol and sorbitol in yeast. Plant Physiol 121(1):45–52

Shi L, Sutter BM, Ye X, Tu BP (2010) Trehalose is a key determinant of the quiescent metabolic state that fuels cell cycle progression upon return to growth. Mol Biol Cell 21(12):1982–1990. doi:10.1091/mbc.E10-01-0056

Shima J, Hino A, Yamada-Iyo C, Suzuki Y, Nakajima R, Watanabe H, Mori K, Takano H (1999) Stress tolerance in doughs of *Saccharomyces cerevisiae* trehalase mutants derived from commercial Baker's yeast. Appl Environ Microbiol 65:2841–2846

Shin D-Y, Matsumoto K, Iida H, Uno I, Ishikawa T (1987) Heat shock response of *Saccharomyces cerevisiae* mutants altered in cyclic AMP-dependent protein phosphorylation. Mol Cell Biol 7:244–250

Sillje HH, ter Schure EG, Rommens AJ, Huls PG, Woldringh CL, Verkleij AJ, Boonstra J, Verrips CT (1997) Effects of different carbon fluxes on G1 phase duration, cyclin expression, and reserve carbohydrate metabolism in *Saccharomyces cerevisiae*. J Bacteriol 179(21):6560–6565

Singer MA, Lindquist S (1998a) Multiple effects of trehalose on protein folding *in vitro* and *in vivo*. Mol Cell 1(5):639–648

Singer MA, Lindquist S (1998b) Thermotolerance in *Saccharomyces cerevisiae*: the Yin and Yang of trehalose. Trends Biotechnol 16:460–468

Slade L, Levine H (1988) Non-equilibrium behavior of small carbohydrate-water systems. Pure Appl Chem 60:1841–1864

Smit A, Moses SG, Pretorius IS, Cordero Otero RR (2008) The Thr505 and Ser557 residues of the *AGT1*-encoded alpha-glucoside transporter are critical for maltotriose transport in *Saccharomyces cerevisiae*. J Appl Microbiol 104(4):1103–1111. doi:10.1111/j.1365-2672.2007.03671.x

Smith SE (1967) Carbohydrate translocation in orchid mycorrhizas. New Phytol 66:371–378

Smith A, Ward MP, Garrett S (1998) Yeast PKA represses Msn2p/Msn4p-dependent gene expression to regulate growth, stress response and glycogen accumulation. EMBO J 17(13):3556–3564. doi:10.1093/emboj/17.13.3556

Smith DA, Nicholls S, Morgan BA, Brown AJ, Quinn J (2004) A conserved stress-activated protein kinase regulates a core stress response in the human pathogen *Candida albicans*. Mol Biol Cell 15(9):4179–4190. doi:10.1091/mbc.E04-03-0181

Sola-Penna M, Ferreira-Pereira A, Lemos AP, Meyer-Fernandes JR (1997) Carbohydrate protection of enzyme structure and function against guanidinium chloride treatment depends on the nature of carbohydrate and enzyme. Eur J Biochem 248(1):24–29

Song XS, Li HP, Zhang JB, Song B, Huang T, Du XM, Gong AD, Liu YK, Feng YN, Agboola RS, Liao YC (2014) Trehalose 6-phosphate phosphatase is required for development, virulence and mycotoxin biosynthesis apart from trehalose biosynthesis in *Fusarium graminearum*. Fungal Genet Biol 63:24–41. doi:10.1016/j.fgb.2013.11.005

Soto T, Fernandez J, Vicente-Soler J, Cansado J, Gacto M (1995a) Glucose-induced, cyclic-AMP-independent signalling pathway for activation of neutral trehalase in the fission yeast *Schizosaccharomyces pombe*. Microbiology 141:2665–2671

Soto T, Fernandez J, Vicente-Soler J, Cansado J, Gacto M (1995b) Nitrogen-source-induced activation of neutral trehalase in *Schizosaccharomyces pombe* and *Pachysolen tannophilus*: role of cAMP as second messenger. FEMS Microbiol Lett 132:229–232

Soto T, Fernandez J, Vicente-Soler J, Cansado J, Gacto M (1996) Posttranslational regulatory control of trehalase induced by nutrients, metabolic inhibitors, and physical agents in *Pachysolen tannophilus*. Fungal Genet Biol 20:143–151

Soto T, Fernandez J, Cansado J, Vicente-Soler J, Gacto M (1997) Protein kinase Sck1 is involved in trehalase activation by glucose and nitrogen source in the fission yeast *Schizosaccharomyces pombe*. Microbiology 143(Pt 7):2457–2463

Soto T, Fernandez J, Vicente-Soler J, Cansado J, Gacto M (1999) Accumulation of trehalose by overexpression of *tps1*, coding for trehalose-6-phosphate synthase, causes increased resistance to multiple stresses in the fission yeast *Schizosaccharomyces pombe*. Appl Environ Microbiol 65(5):2020–2024

Soto T, Franco A, Padmanabhan S, Vicente-Soler J, Cansado J, Gacto M (2002) Molecular interaction of neutral trehalase with other enzymes of trehalose metabolism in the fission yeast *Schizosaccharomyces pombe*. Eur J Biochem 269(15):3847–3855

Souza AC, De Mesquita JF, Panek AD, Silva JT, Paschoalin VM (2002) Evidence for a modulation of neutral trehalase activity by Ca2+ and cAMP signaling pathways in *Saccharomyces cerevisiae*. Braz J Med Biol Res 35(1):11–16

Stambuk BU, de Araujo PS (2001) Kinetics of active alpha-glucoside transport in *Saccharomyces cerevisiae*. FEMS Yeast Res 1(1):73–78

Stambuk BU, De Araujo PS, Panek AD, Serrano R (1996) Kinetics and energetics of trehalose transport in *Saccharomyces cerevisiae*. Eur J Biochem 237(3):876–881

Stambuk BU, Panek AD, Crowe JH, Crowe LM, de Araujo PS (1998) Expression of high-affinity trehalose-H+ symport in *Saccharomyces cerevisiae*. Biochim Biophys Acta 1379(1):118–128

Stambuk BU, da Silva MA, Panek AD, de Araujo PS (1999) Active alpha-glucoside transport in *Saccharomyces cerevisiae*. FEMS Microbiol Lett 170(1):105–110

Stewart LC, Richtmeyer NK, Hudson CS (1950) The preparation of trehalose from yeast. J Am Chem Soc 72:2059–2061

Streeter JG, Bhagwat A (1999) Biosynthesis of trehalose from maltooligosaccharides in Rhizobia. Can J Microbiol 45(8):716–721

Strom AR (1998) Osmoregulation in the model organism *Escherichia coli*: genes governing the synthesis of glycine betaine and trehalose and their use in metabolic engineering of stress tolerance. J Biosci 23:437–445

Strøm AR, Falkenberg P, Landfald B (1986) Genetics of osmoregulation in *Escherichia coli* : uptake and biosynthesis of organic osmolytes. FEMS Microbiol Rev 39:79–86

Suarez R, Wong A, Ramirez M, Barraza A, Orozco Mdel C, Cevallos MA, Lara M, Hernandez G, Iturriaga G (2008) Improvement of drought tolerance and grain yield in common bean by overexpressing trehalose-6-phosphate synthase in rhizobia. Mol Plant Microbe Interact 21(7):958–966. doi:10.1094/mpmi-21-7-0958

Sugawara M, Cytryn EJ, Sadowsky MJ (2010) Functional role of *Bradyrhizobium japonicum* trehalose biosynthesis and metabolism genes during physiological stress and nodulation. Appl Environ Microbiol 76(4):1071–1081. doi:10.1128/aem.02483-09

Sugihara TF, L Kline (1968) Factors affecting the stability of frozen bread doughs. II. Prepared by the sponge and dough method. Bakers Dig 42:51–54, 69

Sun Y, Miao Y, Yamane Y, Zhang C, Shokat KM, Takematsu H, Kozutsumi Y, Drubin DG (2012) Orm protein phosphoregulation mediates transient sphingolipid biosynthesis response to heat stress via the Pkh-Ypk and Cdc55-PP2A pathways. Mol Biol Cell 23(12):2388–2398. doi:10.1091/mbc.E12-03-0209

Suomalainen H, Pfäffli S (1961) Changes in the carbohydrate reserves of baker's yeast during growth and on standing. J Inst Brew 67:249–254

Sussich F, Skopec C, Brady J, Cesaro A (2001) Reversible dehydration of trehalose and anhydrobiosis: from solution state to an exotic crystal? Carbohydr Res 334(3):165–176

Sussman AS (1954) Changes in the permeability of ascospores of *Neurospora tetrasperma* during germination. J Gen Physiol 38:59–77

Takemori Y, Sakaguchi A, Matsuda S, Mizukami Y, Sakurai H (2006) Stress-induced transcription of the endoplasmic reticulum oxidoreductin gene *ERO1* in the yeast *Saccharomyces cerevisiae*. Mol Genet Genomics 275(1):89–96. doi:10.1007/s00438-005-0065-9

Talarek N, Cameroni E, Jaquenoud M, Luo X, Bontron S, Lippman S, Devgan G, Snyder M, Broach JR, De Virgilio C (2010) Initiation of the TORC1-regulated G0 program requires Igo1/2, which license specific mRNAs to evade degradation via the 5'-3' mRNA decay pathway. Mol Cell 38(3):345–355. doi:10.1016/j.molcel.2010.02.039

Tan H, Dong J, Wang G, Xu H, Zhang C, Xiao D (2014) Enhanced freeze tolerance of baker's yeast by overexpressed trehalose-6-phosphate synthase gene (TPS1) and deleted trehalase genes in frozen dough. J Ind Microbiol Biotechnol 41(8):1275–1285. doi:10.1007/s10295-014-1467-7

Tanaka K, Matsumoto K, Toh EA (1989) IRA1, an inhibitory regulator of the RAS-cyclic AMP pathway in *Saccharomyces cerevisiae.* Mol Cell Biol 9(2):757–768

Tanaka K, Nakafuku M, Satoh T, Marshall MS, Gibbs JB, Matsumoto K, Kaziro Y, Toh-e A (1990a) *S. cerevisiae* genes *IRA1* and *IRA2* encode proteins that may be functionally equivalent to mammalian ras GTPase activating protein. Cell 60(5):803–807

Tanaka K, Nakafuku M, Tamanoi F, Kaziro Y, Matsumoto K, Toh-e A (1990b) *IRA2*, a second gene of *Saccharomyces cerevisiae* that encodes a protein with a domain homologous to mammalian ras GTPase-activating protein. Mol Cell Biol 10(8):4303–4313

Tanaka M, Machida Y, Niu S, Ikeda T, Jana NR, Doi H, Kurosawa M, Nekooki M, Nukina N (2004) Trehalose alleviates polyglutamine-mediated pathology in a mouse model of Huntington disease. Nat Med 10(2):148–154. doi:10.1038/nm985

Tapia H, Morano KA (2010) Hsp90 nuclear accumulation in quiescence is linked to chaperone function and spore development in yeast. Mol Biol Cell 21(1):63–72. doi:10.1091/mbc.E09-05-0376

Tereshina VM, Polotebnova MV, Feofilova EP (1988) Trehalase activity of spores of the wild-type strain of *Cunninghamella japonica* and mutants with a reduced rate of trehalase synthesis. Microbiology 56:587–592

Thaller MC, Schippa S, Rossolini GM (1998) Conserved sequence motifs among bacterial, eukaryotic, and archaeal phosphatases that define a new phosphohydrolase superfamily. Protein Sci 7(7):1647–1652. doi:10.1002/pro.5560070722

Thevelein JM (1984a) Activation of trehalase by membrane-depolarizing agents in yeast vegetative cells and ascospores. J Bacteriol 158(1):337–339

Thevelein JM (1984b) Cyclic-AMP content and trehalase activation in vegetative cells and ascospores of yeast. Arch Microbiol 138:64–67

Thevelein JM (1984c) Regulation of trehalose mobilization in fungi. Microbiol Rev 48(1):42–59

Thevelein JM (1991) Fermentable sugars and intracellular acidification as specific activators of the RAS-adenylate cyclase signalling pathway in yeast: the relationship to nutrient-induced cell cycle control. Mol Microbiol 5(6):1301–1307

Thevelein JM (1994) Signal transduction in yeast. Yeast 10:1753–1790

Thevelein JM, Beullens M (1985) Cyclic AMP and the stimulation of trehalase activity in the yeast *Saccharomyces cerevisiae* by carbon sources, nitrogen sources and inhibitors of protein synthesis. J Gen Microbiol 131(12):3199–3209

Thevelein JM, Hohmann S (1995) Trehalose synthase: guard to the gate of glycolysis in yeast? Trends Biochem Sci 20:3–10

Thevelein JM, Jones KA (1983) Reversibility characteristics of glucose-induced trehalase activation associated with the breaking of dormancy in yeast ascospores. Eur J Biochem 136:583–587

Thevelein JM, den Hollander JA, Shulman RG (1982) Changes in the activity and properties of trehalase during early germination of yeast ascospores: correlation with trehalose breakdown as studied by in vivo 13C NMR. Proc Natl Acad Sci USA 79:3503–3507

Thevelein JM, Voordeckers K (2009) Functioning and evolutionary significance of nutrient transceptors. Mol Biol Evol 26:2407–2414

Thompson SN (2003) Trehalose - The insect 'blood' sugar. In: Simpson SJ (ed) Advances in insect physiology. Academic, London, pp 205–285

Thomsson E, Svensson M, Larsson C (2005) Rapamycin pre-treatment preserves viability, ATP level and catabolic capacity during carbon starvation of *Saccharomyces cerevisiae.* Yeast 22(8):615–623. doi:10.1002/yea.1219

Thorat LJ, Gaikwad SM, Nath BB (2012) Trehalose as an indicator of desiccation stress in *Drosophila melanogaster* larvae: a potential marker of anhydrobiosis. Biochem Biophys Res Commun 419(4):638–642. doi:10.1016/j.bbrc.2012.02.065

Toda T, Uno I, Ishikawa T, Powers S, Kataoka T, Broek D, Cameron S, Broach J, Matsumoto K, Wigler M (1985) In yeast, RAS proteins are controlling elements of adenylate cyclase. Cell 40:27–36

Toda T, Cameron S, Sass P, Wigler M (1988) *SCH9*, a gene of *Saccharomyces cerevisiae* that encodes a protein distinct from, but functionally and structurally related to, cAMP-dependent protein kinase catalytic subunits. Genes Dev 2:517–527

Tompa P, Bar J, Batke J (1986) Interaction of enzymes involved in triosephosphate metabolism. Comparison of yeast and rabbit muscle cytoplasmic systems. Eur J Biochem 159(1):117–124

Treger JM, Schmitt AP, Simon JR, McEntee K (1998) Transcriptional factor mutations reveal regulatory complexities of heat shock and newly identified stress genes in *Saccharomyces cerevisiae.* J Biol Chem 273(41):26875–26879

Trevisol ET, Panek AD, Mannarino SC, Eleutherio EC (2011) The effect of trehalose on the fermentation performance of aged cells of *Saccharomyces cerevisiae.* Appl Microbiol Biotechnol 90(2):697–704. doi:10.1007/s00253-010-3053-x

Trevisol ET, Panek AD, De Mesquita JF, Eleutherio EC (2013) Regulation of the yeast trehalose-synthase complex by cyclic AMP-dependent phosphorylation. Biochim Biophys Acta. doi:10.1016/j.bbagen.2013.12.010

Tripp ML, Paznokas JL (1982) Glucose-initiated germination of *Mucor racemosus* sporangiospores. J Gen Microbiol 128(3):477–483

Trivedi NB, Jacobson G (1986) Recent advances in baker's yeast. Prog Ind Microbiol 23:45–71

Trotter EW, Berenfeld L, Krause SA, Petsko GA, Gray JV (2001) Protein misfolding and temperature upshift cause G1 arrest via a common mechanism dependent on heat shock factor in *Saccharomyces cerevisiae*. Proc Natl Acad Sci USA 98(13):7313–7318. doi:10.1073/pnas.121172998

Uno I, Matsumoto K, Adachi K, Ishikawa T (1983) Genetic and biochemical evidence that trehalase is a substrate of cAMP-dependent protein kinase in yeast. J Biol Chem 258(18):10867–10872

Van Aelst L, Hohmann S, Bulaya B, de Koning W, Sierkstra L, Neves MJ, Luyten K, Alijo R, Ramos J, Coccetti P, Martegani E, de Magalhaes-Rocha NM, Brandao RL, Van Dijck P, Vanhalewyn M, Durnez P, Jans AWH, Thevelein JM (1993) Molecular cloning of a gene involved in glucose sensing in the yeast *Saccharomyces cerevisiae*. Mol Microbiol 8:927–943

van der Plaat JB (1974) Cyclic 3',5'-adenosine monophosphate stimulates trehalose degradation in baker's yeast. Biochem Biophys Res Commun 56(3):580–587

Van Dijck P, Colavizza D, Smet P, Thevelein JM (1995) Differential importance of trehalose in stress resistance in fermenting and nonfermenting *Saccharomyces cerevisiae* cells. Appl Environ Microbiol 61(1):109–115

Van Dijck P, De Rop L, Szlufcik K, Van Ael E, Thevelein JM (2002) Disruption of the *Candida albicans TPS2* gene encoding trehalose-6-phosphate phosphatase decreases infectivity without affecting hypha formation. Infect Immun 70(4):1772–1782

van Dijken AJ, Schluepmann H, Smeekens SC (2004) *Arabidopsis* trehalose-6-phosphate synthase 1 is essential for normal vegetative growth and transition to flowering. Plant Physiol 135(2):969–977. doi:10.1104/pp.104.039743

van Heusden GP (2009) 14-3-3 Proteins: insights from genome-wide studies in yeast. Genomics 94(5):287–293. doi:10.1016/j.ygeno.2009.07.004

Van Houtte H, Lopez-Galvis L, Vandesteene L, Beeckman T, Van Dijck P (2013) Redundant and non-redundant roles of the trehalose-6-phosphate phosphatases in leaf growth, root hair specification and energy-responses in *Arabidopsis*. Plant Signal Behav 8(3):e23209. doi:10.4161/psb.23209

Van Laere AJ (1986a) Biochemistry of spore germination in *Phycomyces*. FEMS Microbiol Rev 32:189–198

Van Laere AJ (1986b) Resistance of germinating *Phycomyces* spores to desiccation, freezing, and acids. FEMS Microbiol Ecol 38:251–256

Van Laere A (1989) Trehalose, reserve and/or stress metabolite ? FEMS Microbiol Rev 63:201–210

Van Laere A, Slegers LK (1987) Trehalose breakdown in germinating spores of *Mucor rouxii*. FEMS Microbiol Lett 41:247–252

Van Mulders RM, Van Laere AJ (1984) Cyclic AMP, trehalase and germination of *Phycomyces blakesleeanus* spores. J Gen Microbiol 130:541–547

Van Nuland A, Vandormael P, Donaton M, Alenquer M, Lourenco A, Quintino E, Versele M, Thevelein JM (2006) Ammonium permease-based sensing mechanism for rapid ammonium activation of the protein kinase A pathway in yeast. Mol Microbiol 59(5):1485–1505. doi:10.1111/j.1365-2958.2005.05043.x

Van Zeebroeck G, Bonini BM, Versele M, Thevelein JM (2009) Transport and signaling via the amino acid binding site of the yeast Gap1 amino acid transceptor. Nat Chem Biol 5(1):45–52. doi:10.1038/nchembio.132

Van Zeebroeck G, Rubio-Texeira M, Schothorst J, Thevelein JM (2014) Specific analogues uncouple transport, signalling, oligo-ubiquitination and endocytosis in the yeast Gap1 amino acid transceptor. Mol Microbiol 93(2):213–233. doi:10.1111/mmi.12654

Vandamme J, Castermans D, Thevelein JM (2012) Molecular mechanisms of feedback inhibition of protein kinase A on intracellular cAMP accumulation. Cell Signal 24(8):1610–1618. doi:10.1016/j.cellsig.2012.04.001

Vandercammen A, Francois J, Hers HG (1989) Characterization of trehalose-6-phosphate synthase and trehalose-6-phosphate phosphatase of *Saccharomyces cerevisiae*. Eur J Biochem 182(3):613–620

Vandesteene L, Ramon M, Le Roy K, Van Dijck P, Rolland F (2010) A single active trehalose-6-P synthase (TPS) and a family of putative regulatory TPS-like proteins in *Arabidopsis*. Mol Plant 3(2):406–419. doi:10.1093/mp/ssp114

Vandesteene L, Lopez-Galvis L, Vanneste K, Feil R, Maere S, Lammens W, Rolland F, Lunn JE, Avonce N, Beeckman T, Van Dijck P (2012) Expansive evolution of the trehalose-6-phosphate phosphatase gene family in *Arabidopsis*. Plant Physiol 160(2):884–896. doi:10.1104/pp.112.201400

Veisova D, Macakova E, Rezabkova L, Sulc M, Vacha P, Sychrova H, Obsil T, Obsilova V (2012) Role of individual phosphorylation sites for the 14-3-3-protein-dependent activation of yeast neutral trehalase Nth1. Biochem J 443(3):663–670. doi:10.1042/bj20111615

Venters BJ, Wachi S, Mavrich TN, Andersen BE, Jena P, Sinnamon AJ, Jain P, Rolleri NS, Jiang C, Hemeryck-Walsh C, Pugh BF (2011) A comprehensive genomic binding map of gene and chromatin regulatory proteins in *Saccharomyces*. Mol Cell 41(4):480–492. doi:10.1016/j.molcel.2011.01.015

Versele M, de Winde JH, Thevelein JM (1999) A novel regulator of G protein signalling in yeast, Rgs2, downregulates glucose-activation of the cAMP pathway through direct inhibition of Gpa2. EMBO J 18(20):5577–5591. doi:10.1093/emboj/18.20.5577

Vicente-Soler J, Arguelles JC, Gacto M (1989) Presence of two trehalose-6-phosphate synthase enzymes in Candida utilis. FEMS Microbiol Lett 52(3):273–277

Vicente-Soler J, Arguelles JC, Gacto M (1991) Proteolytic activation of alpha, alpha-trehalose 6-phosphate synthase in Candida utilis. FEMS Microbiol Lett 66(2):157–161

Vicente-Soler J, Soto T, Madrid M, Nunez A, Cansado J, Gacto M (2009) Solubilization and characterization of a cell wall-bound trehalase from ascospores of the fission yeast *Schizosaccharomyces pombe*. Microbiol Res 164(3):304–311. doi:10.1016/j.micres.2006.12.005

Vidgren V, Ruohonen L, Londesborough J (2005) Characterization and functional analysis of the MAL and MPH Loci for maltose utilization in some ale and lager yeast strains. Appl Environ Microbiol 71(12):7846–7857. doi:10.1128/aem.71.12.7846-7857.2005

Vidgren V, Multanen JP, Ruohonen L, Londesborough J (2010) The temperature dependence of maltose transport in ale and lager strains of brewer's yeast. FEMS Yeast Res 10(4):402–411. doi:10.1111/j.1567-1364.2010.00627.x

Vidgren V, Kankainen M, Londesborough J, Ruohonen L (2011) Identification of regulatory elements in the *AGT1* promoter of ale and lager strains of brewer's yeast. Yeast 28(8):579–594. doi:10.1002/yea.1888

Vogel G, Aeschbacher RA, Muller J, Boller T, Wiemken A (1998) Trehalose-6-phosphate phosphatases from *Arabidopsis thaliana*: identification by functional complementation of the yeast *tps2* mutant. Plant J 13(5):673–683

von Meyenburg HK (1969) Energetics of the budding cycle of *Saccharomyces cerevisiae* during glucose limited aerobic growth. Arch Mikrobiol 66:289–303

Voordeckers K, Kimpe M, Haesendonckx S, Louwet W, Versele M, Thevelein JM (2011) Yeast 3-phosphoinositide-dependent protein kinase-1 (PDK1) orthologs Pkh1-3 differentially regulate phosphorylation of protein kinase A (PKA) and the protein kinase B (PKB)/S6K ortholog Sch9. J Biol Chem 286(25):22017–22027. doi:10.1074/jbc.M110.200071

Vuorio OE, Kalkkinen N, Londesborough J (1993) Cloning of two related genes encoding the 56-kDa and 123-kDa subunits of trehalose synthase from the yeast *Saccharomyces cerevisiae*. Eur J Biochem 216:849–861

Walker GM (1998) Yeast physiology and biotechnology. Wiley, New York

Walther T, Mtimet N, Alkim C, Vax A, Loret MO, Ullah A, Gancedo C, Smits GJ, Francois JM (2013) Metabolic phenotypes of *Saccharomyces cerevisiae* mutants with altered trehalose 6-phosphate dynamics. Biochem J 454(2):227–237. doi:10.1042/bj20130587

Walton EF, Carter BLA, Pringle JR (1979) An enrichment method for temperature-sensitive and auxotrophic mutants of yeast. Mol Gen Genet 171:111–114

Wang L, Renault G, Garreau H, Jacquet M (2004) Stress induces depletion of Cdc25p and decreases the cAMP producing capability in *Saccharomyces cerevisiae*. Microbiology 150(Pt 10):3383–3391. doi:10.1099/mic.0.27162-0

Wang PM, Zheng DQ, Chi XQ, Li O, Qian CD, Liu TZ, Zhang XY, Du FG, Sun PY, Qu AM, Wu XC (2014a) Relationship of trehalose accumulation with ethanol fermentation in industrial *Saccharomyces cerevisiae* yeast strains. Bioresour Technol 152:371–376. doi:10.1016/j.biortech.2013.11.033

Wang X, Fan Y, Shi R, Li J, Zhao S (2014b) Quality assessment of platelets stored in a modified platelet additive solution with trehalose at low temperature (10 degrees C) and *in vivo* effects on rabbit model of thrombocytopenia. Platelets. doi:10.3109/09537104.2013.872772

Wannet WJ, Op den Camp HJ, Wisselink HW, van der Drift C, Van Griensven LJ, Vogels GD (1998) Purification and characterization of trehalose phosphorylase from the commercial mushroom *Agaricus bisporus*. Biochim Biophys Acta 1425(1):177–188

Wannet WJB, Aben EMJ, van der Drift C, Van Griensven LJLD, Vogels GD, den Camp JMO (1999) Trehalose phosphorylase activity and carbohydrate levels during axenic fruiting in three *Agaricus bisporus* strains. Curr Microbiol 39:205–210

Watanabe M, Kikawada T, Minagawa N, Yukuhiro F, Okuda T (2002) Mechanism allowing an insect to survive complete dehydration and extreme temperatures. J Exp Biol 205(Pt 18):2799–2802

Welch AZ, Gibney PA, Botstein D, Koshland DE (2013) TOR and RAS pathways regulate desiccation tolerance in *Saccharomyces cerevisiae*. Mol Biol Cell 24(2):115–128. doi:10.1091/mbc.E12-07-0524

Welton RM, Hoffman CS (2000) Glucose monitoring in fission yeast via the Gpa2 galpha, the git5 Gbeta and the git3 putative glucose receptor. Genetics 156:513–521

Wera S, De Schrijver E, Geyskens I, Nwaka S, Thevelein JM (1999) Opposite roles of trehalase activity in heat-shock recovery and heat-shock survival in *Saccharomyces cerevisiae*. Biochem J 343(Pt 3):621–626

Westh P, Ramløv H (1991) Trehalose accumulation in the tardigrade *Adorybiotus coronifer* during anhydrobiosis. J Exp Zool 258:303–311

Wiemken A (1990) Trehalose in yeast, stress protectant rather than reserve carbohydrate. Antonie Leeuwenhoek J Microbiol 58:209–217

Wieser R, Adam G, Wagner A, Schuller C, Marchler G, Ruis H, Krawiec Z, Bilinski T (1991) Heat Shock Factor-independent heat control of transcription of the *CTT1* Gene encoding the cytosolic Catalase-T of *Saccharomyces cerevisiae*. J Biol Chem 266:12406–12411

Wilson RA, Talbot NJ (2009) Under pressure: investigating the biology of plant infection by *Magnaporthe oryzae*. Nat Rev Microbiol 7(3):185–195. doi:10.1038/nrmicro2032

Wilson RA, Jenkinson JM, Gibson RP, Littlechild JA, Wang ZY, Talbot NJ (2007) Tps1 regulates the pentose phosphate pathway, nitrogen metabolism and fungal virulence. EMBO J 26(15):3673–3685. doi:10.1038/sj.emboj.7601795

Wilson RA, Gibson RP, Quispe CF, Littlechild JA, Talbot NJ (2010) An NADPH-dependent genetic switch regulates plant infection by the rice blast fungus. Proc Natl Acad Sci USA 107(50):21902–21907. doi:10.1073/pnas.1006839107

Winderickx J, de Winde JH, Crauwels M, Hino A, Hohmann S, Van Dijck P, Thevelein JM (1996) Regulation of genes encoding subunits of the trehalose synthase complex in *Saccharomyces cerevisiae*: novel variations of STRE-mediated transcription control? Mol Gen Genet 252(4):470–482

Winkler K, Kienle I, Burgert M, Wagner JC, Holzer H (1991) Metabolic regulation of the trehalose content of vegetative yeast. FEBS Lett 291(2):269–272

Wolkers WF, Walker NJ, Tablin F, Crowe JH (2001) Human platelets loaded with trehalose survive freeze-drying. Cryobiology 42:79–87

Wolkers WF, Tablin F, Crowe JH (2002) From anhydrobiosis to freeze-drying of eukaryotic cells. Comp Biochem Physiol A Mol Integr Physiol 131(3):535–543

Wolschek MF, Kubicek CP (1997) The filamentous fungus *Aspergillus niger* contains two 'differentially regulated' trehalose-6-phosphate synthase-encoding genes, *tpsA* and *tpsB*. J Biol Chem 272:2729–2735

Wyatt TT, Wosten HA, Dijksterhuis J (2013) Fungal spores for dispersion in space and time. Adv Appl Microbiol 85:43–91. doi:10.1016/b978-0-12-407672-3.00002-2

Wykoff DD, O'Shea EK (2001) Phosphate transport and sensing in *Saccharomyces cerevisiae*. Genetics 159(4):1491–1499

Xia Y, Gao M, Clarkson J, Charnley A (2002) Molecular cloning, characterisation, and expression of a neutral trehalase from the insect pathogenic fungus *Metarhizium anisopliae*. J Invertebr Pathol 80(2):127–137

Xiao Y, Mitchell AP (2000) Shared roles of yeast glycogen synthase kinase 3 family members in nitrogen-responsive phosphorylation of meiotic regulator Ume6p. Mol Cell Biol 20(15):5447–5453

Xu Z, Yaguchi S, Tsurugi K (2004) Gts1p stabilizes oscillations in energy metabolism by activating the transcription of *TPS1* encoding trehalose-6-phosphate synthase 1 in the yeast *Saccharomyces cerevisiae*. Biochem J 383(Pt 1):171–178. doi:10.1042/bj20040967

Xu YB, Li HP, Zhang JB, Song B, Chen FF, Duan XJ, Xu HQ, Liao YC (2010) Disruption of the chitin synthase gene *CHS1* from *Fusarium asiaticum* results in an altered structure of cell walls and reduced virulence. Fungal Genet Biol 47(3):205–215. doi:10.1016/j.fgb.2009.11.003

Xu J, Sheng Z, Palli SR (2013) Juvenile hormone and insulin regulate trehalose homeostasis in the red flour beetle, *Tribolium castaneum*. PLoS Genet 9(6), e1003535. doi:10.1371/journal.pgen.1003535

Xue Y, Batlle M, Hirsch JP (1998) *GPR1* encodes a putative G protein-coupled receptor that associates with the Gpa2p Galpha subunit and functions in a Ras-independent pathway. EMBO J 17(7):1996–2007. doi:10.1093/emboj/17.7.1996

Yadav UP, Ivakov A, Feil R, Duan GY, Walther D, Giavalisco P, Piques M, Carillo P, Hubberten HM, Stitt M, Lunn JE (2014) The sucrose-trehalose 6-phosphate (Tre6P) nexus: specificity and mechanisms of sucrose signalling by Tre6P. J Exp Bot 65(4):1051–1068. doi:10.1093/jxb/ert457

Yamamoto A, Mizukami Y, Sakurai H (2005) Identification of a novel class of target genes and a novel type of binding sequence of heat shock transcription factor in *Saccharomyces cerevisiae*. J Biol Chem 280(12):11911–11919. doi:10.1074/jbc.M411256200

Yamamoto N, Maeda Y, Ikeda A, Sakurai H (2008) Regulation of thermotolerance by stress-induced transcription factors in *Saccharomyces cerevisiae*. Eukaryot Cell 7(5):783–790. doi:10.1128/ec.00029-08

Yang F, Chen S, Dai ZM, Chen DF, Duan RB, Wang HL, Jia SN, Yang WJ (2013) Regulation of trehalase expression inhibits apoptosis in diapause cysts of *Artemia*. Biochem J 456(2):185–194. doi:10.1042/bj20131020

Yeo ET, Kwon HB, Han SE, Lee JT, Ryu JC, Byu MO (2000) Genetic engineering of drought resistant potato plants by introduction of the trehalose-6-phosphate synthase (*TPS1*) gene from *Saccharomyces cerevisiae*. Mol Cells 10(3):263–268

Yost HJ, Lindquist S (1991) Heat shock proteins affect RNA processing during the heat shock response of *Saccharomyces cerevisiae*. Mol Cell Biol 11:1062–1068

Zaman S, Lippman SI, Zhao X, Broach JR (2008) How *Saccharomyces* responds to nutrients. Annu Rev Genet 42:27–81. doi:10.1146/annurev.genet.41.110306.130206

Zang B, Li H, Li W, Deng XW, Wang X (2011) Analysis of trehalose-6-phosphate synthase (TPS) gene family suggests the formation of TPS complexes in rice. Plant Mol Biol 76(6):507–522. doi:10.1007/s11103-011-9781-1

Zaparty M, Hagemann A, Brasen C, Hensel R, Lupas AN, Brinkmann H, Siebers B (2013) The first prokaryotic trehalose synthase complex identified in the hyperthermophilic crenarchaeon *Thermoproteus tenax*. PLoS One 8(4), e61354. doi:10.1371/journal.pone.0061354

Zaragoza O, Blazquez MA, Gancedo C (1998) Disruption of the *Candida albicans TPS1* gene encoding trehalose-6-phosphate synthase impairs formation of hyphae and decreases infectivity. J Bacteriol 180(15):3809–3815

Zaragoza O, de Virgilio C, Ponton J, Gancedo C (2002) Disruption in *Candida albicans* of the *TPS2* gene encoding trehalose-6-phosphate phosphatase affects cell integrity and decreases infectivity. Microbiology 148(Pt 5):1281–1290

Zeller CE, Parnell SC, Dohlman HG (2007) The RACK1 ortholog Asc1 functions as a G-protein beta subunit coupled to glucose responsiveness in yeast. J Biol Chem 282(34):25168–25176. doi:10.1074/jbc.M702569200

Zentella R, Mascorro-Gallardo JO, Van Dijck P, Folch-Mallol J, Bonini BM, Van Vaeck C, Gaxiola R, Covarrubias AA, Nieto-Sotelo J, Thevelein JM, Iturriaga G (1999) A *Selaginella lepidophylla* trehalose-6-phosphate synthase complements growth and stress-tolerance defects in a yeast *tps1* mutant. Plant Physiol 119:1473–1482

Zevenhuizen LPTM (1992) Levels of trehalose and glycogen in *Arthrobacter globiformis* under conditions of nutrient starvation and osmotic stress. A Van Leeuwenhoek 61:61–68

Zhang F, Wang ZP, Chi Z, Raoufi Z, Abdollahi S, Chi ZM (2013) The changes in Tps1 activity, trehalose content and expression of *TPS1* gene in the psychrotolerant yeast *Guehomyces pullulans* 17-1 grown at different temperatures. Extremophiles 17(2):241–249. doi:10.1007/s00792-013-0511-2

Zhou X, Zhang H, Li G, Shaw B, Xu JR (2012) The Cyclase-associated protein Cap1 is important for proper regulation of infection-related morphogenesis in *Magnaporthe oryzae*. PLoS Pathog 8(9), e1002911. doi:10.1371/journal.ppat.1002911

Zhu Z, Wang H, Shang Q, Jiang Y, Cao Y, Chai Y (2013) Time course analysis of *Candida albicans* metabolites during biofilm development. J Proteome Res 12(6):2375–2385. doi:10.1021/pr300447k

Zikmanis PB, Laivenieks MG, Auzinya LP, Kulaev IS, Beker ME (1985) Relationship between the content of high-molecular-weight polyphosphates and trehalose and viability of populations following dehydratation of the yeast *Saccharomyces cerevisiae.* Microbiology 54:326–330

Zikmanis PB, Kruche RV, Auzinya LP, Margevicha MV, Becker E (1988) Distribution of trehalose between dehydrated *Saccharomyces cerevisiae* cells and the rehydratation medium. Microbiology 57:414–416

Zimmermann P, Hirsch-Hoffmann M, Hennig L, Gruissem W (2004) GENEVESTIGATOR. *Arabidopsis* microarray database and analysis toolbox. Plant Physiol 136(1):2621–2632. doi:10.1104/pp.104.046367

Molecular Aspects of Biochemical Pathways

11 Regulation of Fungal Nitrogen Metabolism

RICHARD B. TODD[1]

CONTENTS

I. Introduction

Filamentous fungi can adapt their metabolism to utilize a wide range of nitrogen nutrients. To ensure efficient nitrogen utilization, a complex regulatory system controls expression of the genes required for nitrogen source uptake and metabolic breakdown for assimilation into cellular nitrogen compounds via biosynthetic processes. The regulation of nitrogen metabolic gene expression in filamentous fungi has been studied intensively for more than half a century and has been reviewed previously (Caddick 2004; Chang and Todd 2004; Marzluf 1997; Wong et al. 2008). Generally, the genes for uptake and catabolism of specific nitrogen nutrients are induced in response to availability of the corresponding nutrient. However, when a preferred or more easily metabolized nutrient such as ammonium or glutamine is available, the genes for uptake and metabolism of alternative nitrogen nutrients are not expressed. This process is known as **nitrogen metabolite repression** (NMR) (Arst and Cove 1973). NMR is relieved by the action of orthologs of the global positive-acting *Aspergillus nidulans* GATA transcription factor AreA at nitrogen metabolic gene promoters. AreA also binds to promoters and/or activates expression of certain secondary metabolism genes in response to nitrogen availability (Chang et al. 2000; Kim and Woloshuk 2008; Mihlan et al. 2003; Tudzynski et al. 1999). Nitrogen regulation of secondary metabolic genes has recently been reviewed (Tudzynski 2014). A transcriptional response to changes in nitrogen nutrient availability has been described for several plant and human pathogens, and it is thought that adaptation of metabolism to altered nitrogen nutrient availability is important for virulence (Coleman et al. 1997; Donofrio et al. 2006; Lee et al. 2011; Lopez-Berges et al. 2010a, b). The last decade has seen many advances in our understanding of fungal nitrogen regulation. These advances have increased our understanding of gene regulation of nitrogen utilization genes by AreA. Furthermore, studies of several transcription factors known for some years have afforded new insights into their roles in this complex regulatory system. Mechanistic differences in nitrogen regulation in filamentous fungi compared with the well-studied nitrogen regulatory system in *Saccharomyces cerevisiae* have emerged (Wong et al. 2008). In this chapter recent advances will be discussed

[1]Department of Plant Pathology, Kansas State University, Manhattan, KS 66506, USA; e-mail: rbtodd@k-state.edu

Biochemistry and Molecular Biology, 3rd Edition
The Mycota III
D. Hoffmeister (Ed.)

and the major regulators reviewed with particular focus on their action in regulating primary metabolic genes in one of the most studied traditional models for nitrogen regulation studies *A. nidulans*. These studies have revealed promoter-specific differences that fine-tune regulation of nitrogen metabolic genes. Many advances have also been made in understanding nitrogen regulation in other filamentous fungi, particularly in *Fusarium fujikuroi* and *Fusarium oxysporum* (e.g., Lopez-Berges et al. 2010a, b; Michielse et al. 2014; Schönig et al. 2008; Teichert et al. 2006, 2008; Wagner et al. 2010). Importantly, both similarities and differences in nitrogen regulation between the various models have emerged, highlighting the necessity of continued complementary studies in a variety of fungi.

II. Regulation by Global and Pathway-Specific Transcription Factors

Transcriptional regulation of many fungal metabolic pathway genes is achieved by the combined actions of global transcription regulators and pathway-specific transcription factors. Nitrogen metabolic genes generally share in common a conserved global regulator, the **GATA transcription factor AreA**, which typically activates gene transcription in the absence of the preferred nitrogen nutrients ammonium or glutamine (Arst and Cove 1973; Kudla et al. 1990). For many nitrogen metabolic pathways, a pathway-specific transcription factor, usually belonging to the Zn(II)$_2$Cys$_6$ zinc binuclear cluster transcription factor family, activates expression and/or mediates induction of genes encoding permeases and enzymes for metabolism of a particular nutrient (MacPherson et al. 2006; Todd and Andrianopoulos 1997). This class of transcription factor has been considerably expanded in *Ascomycetes* (Todd et al. 2014). These regulators usually act on genes within a pathway, or in related pathways, in response to a specific inducer (Fig. 11.1). There are several well-characterized pathway-specific nitrogen utilization regulator examples

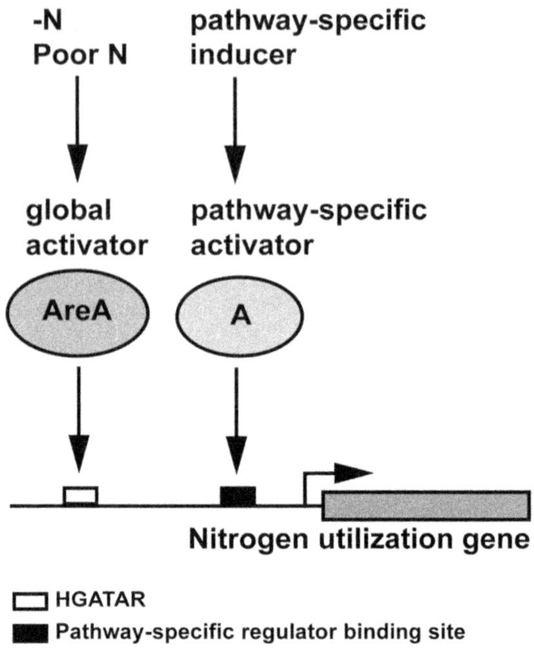

Fig. 11.1 Nitrogen regulation by global and pathway-specific regulators. The global transcription activator AreA relieves nitrogen metabolite repression by binding HGATAR sites in promoters of AreA-regulated nitrogen utilization genes and activating transcription during nitrogen limitation or nitrogen starvation. In many cases, a pathway-specific activator (A) responds to a specific induction signal and binds and activates a subset of nitrogen metabolic gene promoters to mediate induction. Gene expression levels are dictated by the combination of these controls

in *A. nidulans*, including AmdR (ω-amino acid and lactam catabolism), NirA (nitrate and nitrite utilization), ArcA (arginine utilization), PrnA (proline utilization), and UaY (uric acid utilization) (Andrianopoulos and Hynes 1990; Burger et al. 1991; Empel et al. 2001; Pokorska et al. 2000; Suarez et al. 1995). For most nitrogen metabolic genes, it is the combined action of the global regulator AreA and a pathway-specific regulator that dictates expression levels. However, for some nitrogen utilization genes such as *fmdS* (encoding formamidase for formamide utilization), there is no pathway-specific induction system, and regulation appears solely under global control (Fraser et al. 2001).

Table 11.1 Orthologous nitrogen transcription factors in filamentous fungi

Transcription factor					
Aspergillus nidulans	*Neurospora crassa*	*Fusarium fujikuroi*	*Magnaporthe oryzae*	Transcription factor class	Primary function
AreA	NIT2	AreA	NUT1	GATA zinc finger	DNA-binding activator
AreB	ASD4	AreB	Asd4	GATA zinc finger, leucine zipper	DNA-binding repressor or activator
NmrA	NMR1	Nmr1	Nmr1, Nmr2, Nmr3	NmrA-like family	Corepressor
MeaB	NC[a]	MeaB	NC	bZIP	DNA-binding activator
TamA	NC	NC	NC	$Zn(II)_2Cys_6$	Dual-function coactivator/ DNA-binding activator

NC not characterized

III. Global Regulation of Nitrogen Utilization Genes

A. Transcriptional Controls: The Key Players

The most important regulators of fungal nitrogen metabolic genes are the orthologs of the GATA transcription factors AreA and AreB, the transcription corepressor NmrA, the bZIP transcription factor MeaB, and the dual-function $Zn(II)_2Cys_6$ transcription factor TamA (Table 11.1). AreA orthologs are found throughout the fungal kingdom, and the other regulators are present in the *Eurotiomycetes* and *Sordariomycetes* and some other ascomycete clades; however, closely related orthologs are generally absent in the *Basidiomycetes* (Todd et al. 2014). Therefore, differences in nitrogen regulation are expected between the *Eurotiomycetes* and *Sordariomycetes* when compared with the *Saccharomycetes* and the *Basidiomycetes*.

1. The GATA Transcription Factor AreA

a) AreA Function

The major transcription activator of nitrogen metabolic genes AreA is a GATA transcription factor, which binds via a single four-cysteine Cys-X_2-Cys-X_{17}-Cys-X_2-Cys zinc finger DNA-binding domain to HGATAR sequences in the promoters of AreA-regulated target genes to activate transcription during nitrogen limitation or nitrogen starvation (Kudla et al. 1990; Ravagnani et al. 1997). AreA is required for growth on a wide range of nitrogen sources, and therefore AreA regulates expression of many nitrogen metabolic genes (Arst and Cove 1973; Tanzer et al. 2003). Detailed mutational analysis of the *A. nidulans* AreA DNA-binding domain coupled with structural information on DNA-contact residues from the AreA zinc finger-DNA target solution structure has identified the critical residues involved in DNA binding (Starich et al. 1998a, b; Wilson and Arst 1998). Some promoters of AreA-regulated genes contain multiple AreA binding sites. It is has been suggested that AreA may bind two closely spaced GATA sites with higher affinity than single GATA sites, like its *N. crassa* counterpart NIT2 where two sites were functionally important for DNA binding (Chiang and Marzluf 1994; Peters and Caddick 1994). However, at some AreA-regulated promoters such as *amdS* (acetamide utilization), AreA acts via a single major GATAA site (Davis and Hynes 2004), and in vitro DNA binding of AreA to DNA fragments containing a single GATA site is observed (Peters and Caddick 1994; Starich et al. 1998a). Several examples of AreA regulation of target genes have been characterized, with the most detailed DNA-binding and promoter analyses of the *prnD–prnB* (proline utilization) and *niiA–niaD* (nitrate utilization) divergent promoters. The *prnD–prnB* intergenic region contains 15 AreA binding sites, including canonical HGATAR sites and noncanonical GATA sites, but only two adjacent sites are involved in transcriptional regulation

(Gomez et al. 2003). The *niiA–niaD* intergenic region is also GATA rich and contains ten sites, but most of the AreA-dependent transcriptional activation occurs via four centrally clustered sites in a region of open chromatin (Muro-Pastor et al. 1999). One of these GATA sites is occupied by AreA constitutively, even under nitrogen-repressing conditions. Analysis of nucleosome positioning has demonstrated that AreA is required for chromatin remodeling by loss of at least five nucleosomes in this region (Muro-Pastor et al. 1999). Modification of histones by acetylation is important for chromatin remodeling, with acetylated histones correlated with open chromatin (Gacek and Strauss 2012; Tessarz and Kouzarides 2014). Analysis of histone acetylation in the *niiA–niaD* intergenic region showed low levels of histone H3 lysine 9 and lysine 14 (H3K9K14) acetylation in ammonium-grown cells, but H3K9K14 acetylation increased with AreA promoter occupancy during nitrogen starvation (Berger et al. 2008). Furthermore, this histone acetylation is lost in an *areA* loss-of-function mutant, suggesting that AreA recruits histone acetyltransferases (HATs) to the promoter for chromatin remodeling (Berger et al. 2008). Consistent with this hypothesis, AreA promoter occupancy on different nitrogen sources correlates with histone H3K9K14 acetylation levels and nucleosome accessibility (Berger et al. 2008). It is important to note, however, that chromatin remodeling is not sufficient for expression of *niiA* and *niaD* because activation is dependent on the pathway-specific nitrate activator NirA, which requires both induction by nitrate and AreA for promoter binding (Berger et al. 2006; Narendja et al. 2002). AreA also directly interacts with NirA, as is observed for their *N. crassa* orthologs (Feng and Marzluf 1998; Muro-Pastor et al. 2004). Therefore, AreA acts to recruit histone acetyltransferases for remodeling chromatin to an open conformation accessible for binding and pathway-specific activation by NirA in response to nitrate. Although it is clear that AreA is required for histone acetylation and chromatin remodeling for expression of the nitrate utilization genes, AreA may carry out this role at only a subset of AreA-dependent promoters as chromatin remodeling of the *prnD–prnB* intergenic region for activation is independent of AreA and the histone acetyltransferases GcnE and AdaB (Garcia et al. 2004; Reyes-Dominguez et al. 2008).

b) Conservation of AreA Function

AreA orthologs have now been characterized in a wide range of filamentous fungi including many ascomycetes: *Aspergillus nidulans* (Kudla et al. 1990), *A. oryzae* (Christensen et al. 1998), *A. niger* (MacCabe et al. 1998), *A. parasiticus* (Chang et al. 2000), *Neurospora crassa* (Fu and Marzluf 1990), *Penicillium chrysogenum* (Haas et al. 1995), *P. marneffei* (Bugeja et al. 2012), *Fusarium fujikuroi* (Tudzynski et al. 1999), *F. oxysporum* (Divon et al. 2005), *F. verticillioides* (Kim and Woloshuk 2008), *Gibberella zeae* (Min et al. 2012), *Colletotrichum lindemuthianum* (Pellier et al. 2003), *Magnaporthe grisea* (Froeliger and Carpenter 1996), and *M. oryzae* (Wilson et al. 2010). In all cases, AreA regulates nitrogen metabolic genes. AreA orthologs have also now been analyzed in the basidiomycete *Cryptococcus neoformans* (Kmetzsch et al. 2011; Lee et al. 2011) and *Ustilago maydis* (Horst et al. 2012). These analyses have revealed a conserved role for AreA orthologs in mediating relief from nitrogen metabolite repression for nitrogen uptake and catabolic genes in both *Ascomycetes* and *Basidiomycetes*. In the maize pathogen *U. maydis*, AreA also regulates filamentation in response to nitrogen limitation; however, filamentation observed in response to nitrogen limitation in the yeast growth phase of the dimorphic human pathogen *P. marneffei* is independent of AreA (Bugeja et al. 2012; Horst et al. 2012; Todd et al. 2003). Furthermore, AreA orthologs have been implicated as important for adaptation to the host nutritional environment during pathogenesis in several plant and human pathogens (Froeliger and Carpenter 1996; Lee et al. 2011; Pellier et al. 2003; Perez-Garcia et al. 2001; Snoeijers et al. 1999). In *C. neoformans*, the AreA ortholog GAT1/Are1, in addition to its conserved role in nitrogen metabolite repression of catabolic genes, also regulates several virulence traits including melanin production, capsule formation, and production of the infectious basidiospore (Lee et al. 2011).

c) Regulation of AreA Action

In *A. nidulans*, AreA activity is subject to many controls, including autogenous transcriptional regulation, regulation of *areA* mRNA stability, corepression by NmrA, and regulated AreA nuclear accumulation (Andrianopoulos et al. 1998; Langdon et al. 1995; Platt et al. 1996a, b; Todd et al. 2005) (Fig. 11.2). Together, these mechanisms modulate AreA activity in response to nitrogen nutrient quality and availability. The major controls mediating response to nitrogen quality are **differential mRNA stability** and corepression by interaction with the **transcription corepressor NmrA**. These controls were initially identified by mutations deleting the *areA* 3′ untranslated region (UTR) or truncating the AreA protein (Platt et al. 1996a, b). Either type of mutation separately leads to partial derepression, whereas when they are combined, full derepression is observed (Platt et al. 1996a). The role of NmrA will be described in Sect. III.A.2. AreA nuclear accumulation occurs in response to nitrogen starvation (i.e., nitrogen availability rather than nitrogen quality) and parallels further elevated AreA-dependent expression over derepressed levels for certain genes (Todd et al. 2005). Analysis of *glnA* glutamine synthase mutants (Margelis et al. 2001) and analysis of free glutamine pools (Berger et al. 2008; Schinko et al. 2010) showed that the activity of AreA is linked to intracellular glutamine levels, which are thought to signal nitrogen status.

d) Regulation of *areA* mRNA Stability via Caf1, CutA, and RrmA

Deletion of the *areA* 3′UTR leads to partial derepression of AreA activity due to increased transcript stability (Platt et al. 1996a). Quantitative Northern analysis of *areA* mRNA levels revealed that *areA* message stability is differentially regulated with a longer half-life during nitrogen limitation or nitrogen starvation and decreased stability during nitrogen sufficiency (Platt et al. 1996a). Furthermore, addition of the *areA* 3′UTR to a heterologous transcript stable under different nitrogen regimes conferred transcript instability in response to ammonium or glutamine compared with nitrogen starvation (Morozov et al. 2000). The molecular mechanism of differential *areA* mRNA stability occurs via shortening of the polyA tail as addition of ammonium to the media reduces polyA tail length, whereas the length of the polyA tail is maintained during nitrogen starvation (Morozov et al. 2000). Several nitrogen sources other than ammonium and glutamine also promote *areA* mRNA instability, including asparagine, glutamate, uric acid, nitrite, and aspartate (Morozov et al. 2001). Analysis of NADP-glutamate dehydrogenase or glutamine synthase mutants affected in ammonium metabolism indicated that ammonium does not directly signal transcript degradation and that glutamine or a derivative is the likely signal (Morozov et al. 2001). Systematic deletion of the *A. nidulans* deadenylation genes showed that loss of the Caf1 or Ccr4 members of the Caf1–Ccr4–Not deadenylation complex conferred increased *areA* mRNA stability, but only Caf1 is responsible for glutamine-regulated transcript instability (Morozov et al. 2010a). Caf1 prevents deadenylation-independent transcript decapping as the *caf1Δ* mutant shows decapped transcripts with abnormally long polyA tails (Morozov et al. 2010a, b). An important step in transcript degradation is modification of the shortened polyA tail. The nucleotidyltransferase CutA adds a CUCU modification to the 3′ end of *areA* transcripts prior to decapping, and disruption of *cutA* slows *areA* transcript degradation (Morozov et al. 2010a).

Krol et al. (2013) hypothesized that an RNA-binding protein specifically binds to the *areA* mRNA 3′UTR and regulates *areA* mRNA turnover. Pull down analysis of *A. nidulans* cell extracts with in vitro transcribed *areA* 3′UTR sequences and proteomics analysis identified an *areA* 3′UTR-interacting protein as the RNA recognition motif protein RrmA, which was already known as a regulator of arginine catabolism (Olszewska et al. 2007). Analysis of the *areA* mRNA transcript in the *rrmAΔ* mutant revealed that both glutamine-triggered mRNA instability and deadenylation were lost (Krol et al. 2013). Therefore RrmA is thought to accelerate deadenylation via the Ccr4–Caf1–Not complex.

e) Regulation of AreA Subcellular Localization

The subcellular distribution of hemagglutinin (HA)-epitope-tagged AreA is regulated in response to nitrogen nutrient availability

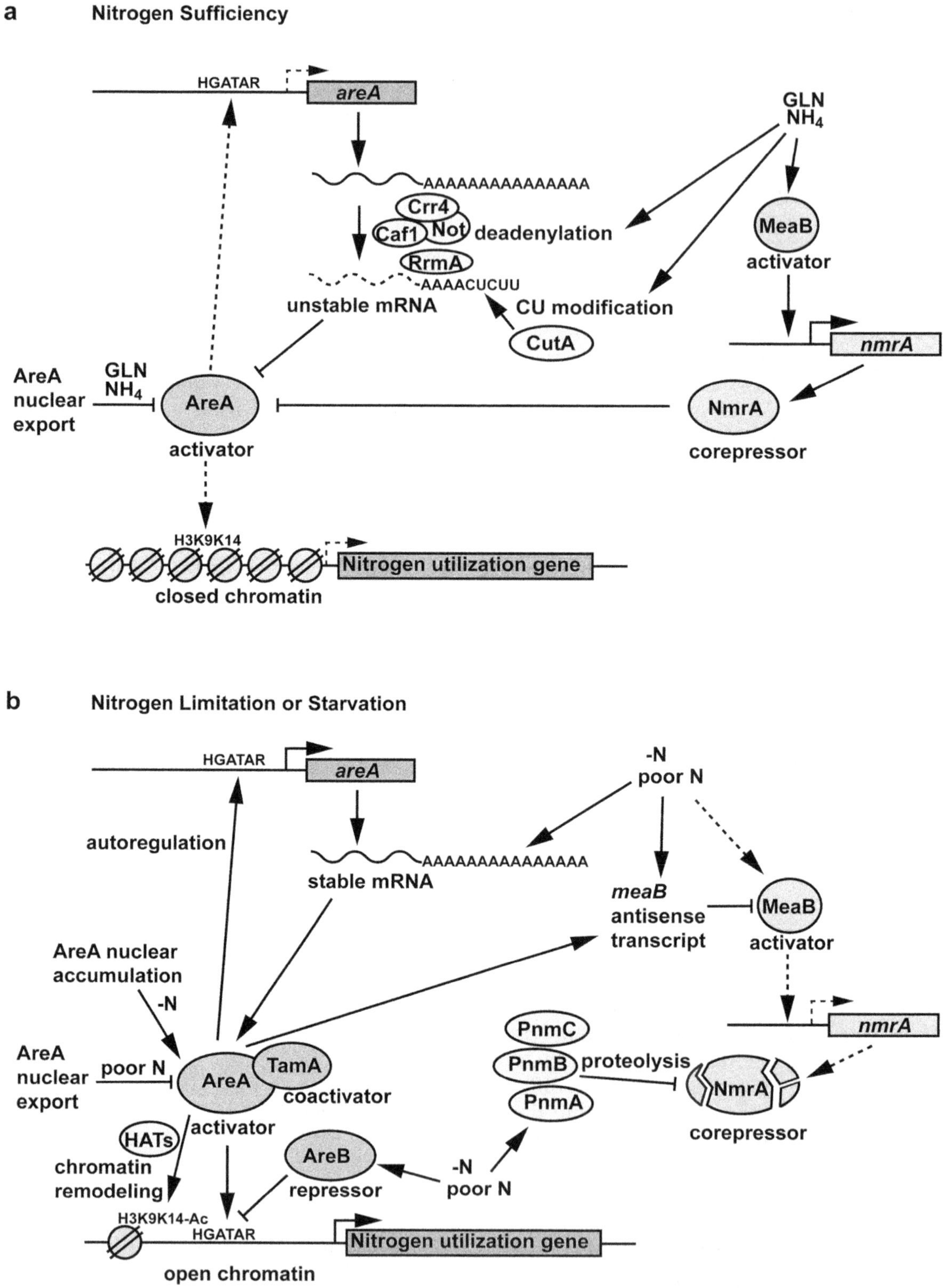

Fig. 11.2 Model for nitrogen regulation in *A. nidulans*. (a) During nitrogen sufficiency, AreA activity is low at most nitrogen metabolic gene promoters due to (1) *areA* mRNA instability mediated by CU modification

(Todd et al. 2005). AreA^{HA} shows nucleocytoplasmic distribution when nitrogen nutrients are available, whereas in the complete absence of nitrogen nutrients, AreA^{HA} accumulates in the nucleus. AreA^{HA} nuclear accumulation occurs slowly over the course of several hours of nitrogen starvation and is accompanied by elevated expression of certain AreA-dependent genes (*amdS* and *fmdS*) (Fraser et al. 2001; Todd et al. 2005) (Fig. 11.3). Furthermore, progressive AreA nuclear accumulation during nitrogen starvation correlates with increased AreA promoter occupancy in the *niiA–niaD* intergenic region (Berger et al. 2008). Nuclear accumulated AreA^{HA} is rapidly translocated to the cytoplasm within minutes of exogenous addition for all nitrogen nutrients tested, and the elevation in AreA-dependent gene expression is rapidly attenuated (Todd et al. 2005). The rapid kinetics of AreA^{HA} nuclear export compared with the slower nuclear accumulation of AreA^{HA} suggest that nuclear export is the key regulated step. Sequence analysis of the AreA protein revealed a CRM1 exportin-like nuclear export signal at residues 703–712. This sequence mediates AreA nuclear export as deletion of residues 703–712 confers AreA nuclear accumulation in the presence of nitrogen nutrients, and addition of this sequence to the constitutively nuclear protein PrnA (Pokorska et al. 2000) targets it to the cytoplasm (D.J. Downes, D.F. Clarke, M.A. Davis and R.B. Todd, unpublished data). The CRM1 exportin in some organisms is sensitive to the drug leptomycin B (LMB) due to specific effects on a cysteine residue [e.g., in *S. pombe* (Kudo et al. 1999)]. In *S. cerevisiae* CRM1 and in *A. nidulans* CrmA/KapK, the cysteine target residue for LMB is a threonine, and substitution of this residue for cysteine confers LMB sensitivity (Bernreiter et al. 2007; Neville and Rosbash 1999; Todd et al. 2005). Addition of LMB to the *crmA*^{T525C} mutant prevents ammonium-triggered AreA^{HA} nuclear export, indicating that AreA nuclear export occurs via CrmA (Todd et al. 2005). We have proposed that differential posttranslational modification of the AreA NES regulates AreA nuclear export, and we are using a genetic screen to identify the modifiers (Downes et al. 2014a; D.J. Downes and R.B. Todd, unpublished data).

In *F. fujikuroi*, AreA nuclear localization is also regulated as GFP-AreA is localized in the nucleus following 24 h nitrogen starvation, but is nucleocytoplasmic or predominantly cytoplasmic in mycelium grown on glutamine-rich media (Michielse et al. 2014). However, in *Gibberella zeae* AreA subcellular localization is not regulated as GFP-AreA was localized to nuclei in nitrogen rich complex media, during growth on nitrate as a sole nitrogen source, and during nitrogen starvation (Min et al. 2012).

Analysis of AreA nuclear import signals in *A. nidulans* revealed multiple nuclear localization signals (NLSs) that show redundancy and cooperation in mediating AreA nuclear localization (Hunter et al. 2014). AreA contains five classical NLSs conforming to the SV40 large T-antigen NLS. Most of the classical NLSs are conserved throughout fungal AreA orthologs with *S. cerevisiae* GLN3 and GAT1 being the notable exceptions. Deletion of the classical NLSs separately or in various combinations failed to prevent AreA^{HA} nuclear localization and did not affect AreA growth phenotypes or AreA-dependent gene expression. AreA also contains one noncanonical arginine-based bipartite NLS conserved in mammalian GATA4. This NLS type

Fig. 11.2 (continued) via CutA and deadenylation via Caf1–Crr4–Not and RrmA, (2) repression of AreA activity by the NmrA corepressor, and (3) AreA nuclear export. NmrA levels are increased at the transcription level by the action of MeaB. (**b**) During nitrogen limitation or nitrogen starvation, AreA activity is elevated due to (1) increased *areA* mRNA stability, (2) autoregulation, (3) relief of NmrA-mediated repression due to reduced NmrA levels and inactivation of NmrA by PnmA-/PnmB-/PnmC-mediated proteolysis, and (4) TamA action as a coactivator. Under these conditions, AreA is required for chromatin remodeling to an open conformation via histone acetyltransferases (HATs) and histone acetylation at some promoters, and the activity of AreA is tempered by AreB-mediated repression. During nitrogen starvation, AreA activity is further elevated, and this correlates with AreA nuclear accumulation due to a block in CrmA-/KapK-mediated AreA nuclear export

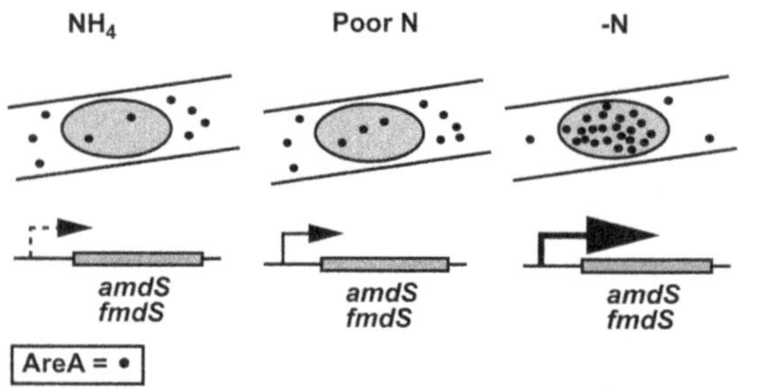

Fig. 11.3 Nuclear accumulation of AreA in *A. nidulans*. AreA shows nucleocytoplasmic distribution during growth on ammonium or during nitrogen limitation and accumulates in the nucleus during nitrogen starvation. AreA-dependent gene expression of the *amdS* and *fmdS* genes is elevated during nitrogen limitation and further elevated during nitrogen starvation

was discovered in GATA4, where mutation of all four arginines was necessary to prevent nuclear import via β-importin (Philips et al. 2007). The bipartite NLS is conserved in fungal AreA orthologs, but is absent in GLN3 and only partially conserved in GAT1, and is conserved in some other fungal GATA transcription factors (AreB and SreA) (Hunter et al. 2014). Point mutations of the four conserved arginine residues in the bipartite NLS did not prevent AreA^{HA} nuclear accumulation, but did confer a loss-of-function phenotype on nitrogen nutrient utilization and *fmdS–lacZ* reporter gene expression, presumably due to loss of DNA binding as these four arginine residues are DNA-contact residues. Several combinations of classical NLS deletions with the bipartite NLS point mutations, including mutation of all six NLSs, prevented AreA^{HA} nuclear localization. Therefore, the AreA NLSs show redundancy.

The functions of the AreA NLSs were also analyzed by fusion to Green Fluorescent Protein (GFP) (Hunter et al. 2014). Fusion of the bipartite NLS to GFP conferred nuclear localization. In contrast, fusion of the classical NLSs separately to GFP did not confer nuclear localization. However, certain combinations of classical NLSs fused to GFP direct the fusion protein to the nucleus suggesting that the AreA NLSs collaborate to mediate nuclear localization. The conservation of the NLSs suggests that each of these sequences is functionally important; and that the NLSs cannot be truly redundant. It will be interesting to determine whether there is a selective advantage of certain NLSs for *A. nidulans* growth under particular nutrient quality, availability or other growth conditions.

f) Nitrogen Regulation via TOR Signaling

In *S. cerevisiae*, GLN3 and GAT1 nuclear import and activation of target genes are regulated by the highly conserved rapamycin-responsive **TOR (target of rapamycin) kinase** signaling pathway (Beck and Hall 1999; Bertram et al. 2000; Cooper 2002). In *A. nidulans*, rapamycin does not cause rapid expression of nitrogen-repressed genes upon addition to nitrogen-rich media (Fitzgibbon et al. 2005). Furthermore, genetic analysis of the *A. nidulans* TOR pathway orthologs revealed only a minor role for TorA, FprA, JipA, and GstA (orthologs of *S. cerevisiae* TOR1/TOR2, FPR1, TIP41, and URE2) in regulation of nitrogen metabolic genes (Fitzgibbon et al. 2005; Fraser et al. 2002). In contrast to the rapid nuclear import for GLN3 and GAT1 in response to rapamycin in *S. cerevisiae*, rapamycin does not trigger AreA nuclear localization in *A. nidulans*, consistent with AreA nuclear export as the key regulated step (RB Todd, unpublished data). Proteomics analysis has now revealed that rapamycin induces autophagy genes in *A. nidulans* (Kim et al. 2011). In *F. fujikuroi*, macroarray hybridization experiments were used to identify genes differentially expressed in response to rapamycin treatment (Teichert et al. 2006). Some nitrogen metabolism genes as well as AreA-regulated secondary metabolism genes were identified as rapamycin targets. The rapamycin response was shown to be AreA dependent for the secondary metabolism genes, but AreA independent for other genes. In *F. oxysporum*, nitrogen availability regulates virulence functions via TOR; however, the response does not occur via inhibition of AreA function but instead operates via MeaB (Lopez-Berges et al. 2010a, b).

g) Unconventional Modes of AreA Action

While the majority of characterized AreA target genes are activated by AreA during nitrogen limitation and/or starvation, some genes are regulated by unconventional modes of AreA action including direct repression by AreA or AreA-dependent activation during nitrogen-repressing conditions. Direct repression by AreA is observed for the arginine utilization genes *agaA* and *otaA*. In contrast to other nitrogen-repressed genes where AreA activates gene expression to relieve repression, at the promoters of the arginine catabolism genes, AreA acts negatively to repress transcription on ammonium (Dzikowska et al. 2003; Macios et al. 2012). The action of AreA at these promoters occurs in conjunction with negative action of the GATA factor AreB (Macios et al. 2012). This may involve a direct interaction between AreA and AreB, as the orthologs of these two transcription factors directly interact in *F. fujikuroi* (Michielse et al. 2014).

Several nitrogen metabolism genes are activated by AreA during nitrogen sufficiency, despite low AreA levels due to autogenous control and reduced *areA* mRNA transcript stability, and repression by NmrA rendering AreA inactive at most promoters. These non-repressed promoters include those of genes for ammonium assimilation. The **major ammonium transporter** *meaA* is expressed during growth on ammonium, consistent with its role in ammonium uptake (Monahan et al. 2002, 2006). Semiquantitative reverse transcriptase-PCR showed that *meaA* expression is activated by AreA on ammonium, but the underlying molecular mechanism is not understood (Monahan et al. 2006). During growth on ammonium or glutamine, AreA activates transcription of the **NADP-glutamate dehydrogenase gene** *gdhA* (Christensen et al. 1998; Downes et al. 2014b; Polotnianka et al. 2004). AreA plays a lesser role than TamA in activating *gdhA* expression but is required for TamA DNA binding at the *gdhA* promoter suggesting that the interaction between AreA and TamA is important for the contribution of AreA to *gdhA* expression under these conditions (Downes et al. 2014b).

The glutamine synthase gene *glnA*, although not strongly regulated by nitrogen source, is activated by AreA under nitrogen sufficiency and nitrogen limitation (Margelis et al. 2001). Some other genes are induced by ammonium including the adenine deaminase *nadA* gene (Oestreicher et al. 2008) and the adenylo-succinate synthetase gene *adB* (Ribard et al. 2001). The molecular mechanism of ammonium induction for these two genes has not been reported. For *nadA*, AreA is also thought to act negatively during uric acid induction by the UaY pathway-specific activator via competitive binding of closely adjacent AreA and UaY DNA-binding sites. (Oestreicher et al. 2008)

2. The Corepressor NmrA

a) NmrA Function

The transcriptional activity of AreA is inhibited in the presence of the readily assimilated nitrogen sources ammonium or glutamine (i.e., nitrogen sufficiency) by interaction with the corepressor NmrA (Fig. 11.2). The *nmrA* gene was cloned by homology with the *Neurospora crassa nmr-1* gene (Andrianopoulos et al. 1998). In *N. crassa*, NMR1 represses the activity of the AreA ortholog NIT2, and mutations in *nmr-1* lead to derepression of NIT2-regulated genes (Fu et al. 1988; Xiao et al. 1995). NMR1 directly interacts with NIT2 via the NIT2 DNA-binding domain and C-terminus, and NMR1 prevents in vitro DNA binding by NIT2 (Pan et al. 1997; Xiao et al. 1995). In *A. nidulans*, the AreA extreme C-terminus is conserved with NIT2, and AreA C-terminal truncation mutants show partial derepression of AreA activity (Andrianopoulos et al. 1998; Platt et al. 1996a, b). Deletion of *nmrA* also leads to partial derepression of AreA-dependent nitrogen metabolic gene expression (Andrianopoulos et al. 1998; Todd et al. 2005). The NmrA sequence did not provide any evidence for a DNA-binding motif. The NmrA crystal structure revealed a Rossmann fold, structural similarity to short-chain dehydrogenase–reductase (SDR) enzymes, and that the dinucleotides NAD^+ or $NADP^+$ are bound by NmrA (Stammers et al. 2001). However, it is unlikely that NmrA has dehydrogenase–reductase activity as NmrA lacks an SDR catalytic motif (Stammers et al. 2001). SDR motifs are found in the NAD^+-dependent SIR2 silencing sirtuins involved in chromatin remodeling in *S. cerevisiae* (Denu 2003). The AreA DNA-binding domain and C-terminal region physically interact with NmrA (Lamb et al. 2003, 2004). However, in an in vitro mixture with NmrA and oligonucleotides containing GATA DNA-binding sites, AreA DNA binding was found to be preferential over binding to NmrA (Lamb et al. 2004). Furthermore, deletion of the C-terminal nine residues of AreA did not affect NmrA binding.

The role of NmrA binding to NAD^+ or $NADP^+$ dinucleotides is not yet understood. NmrA binds NAD^+ and $NADP^+$ with much stronger affinity than their reduced NADH/NADPH forms (Lamb et al. 2003). The NAD^+-/$NADP^+$-binding motif was defined as a conserved sequence NxxGxxA at residues 12–18, and substitution of a single residue T14V or a double substitution N12G/A18G within this site reduced NAD^+-/$NADP^+$-binding affinity (Lamb et al. 2004). The lower affinity for the reduced dinucleotides led to the proposal that NmrA may act as a redox sensor to link cellular metabolism to transcription regulation (Lamb et al. 2003; Stammers et al. 2001). However, on the basis of new evidence of the dissociation constant for NADPH compared with $NADP^+$ for the NmrA-like family member HSCARG, a redox sensor role is considered less likely (Zhao et al. 2010). In *Magnaporthe oryzae*, where there are three NMR corepressor genes, a T13V substitution within the $NADP^+$-binding site of NMR1 leads to a loss-of-function phenotype (Wilson et al. 2010), suggesting that $NADP^+$ binding is required for corepressor activity.

b) Conservation of NmrA

NmrA orthologs are found in *Eurotiomycetes*, *Sordariomycetes*, and *Dothidiomycetes*, but no close orthologs were identified by reciprocal BLAST searches in the *Saccharomycotina*, the *Schizosaccharomycotina*, or the *Basidiomycetes* (Todd et al. 2014). However, more distantly related NmrA-like proteins have been identified in the basidiomycetes *C. neoformans* and *U. maydis* (Horst et al. 2012; Lee et al. 2012). In *C. neoformans*, the NmrA homolog Tar1 represses Gat1/Are1 as the *tar1Δ* mutant shows derepression of non-preferred nitrogen utilization pathway genes under nitrogen-repressing conditions (Lee et al. 2012). The underlying molecular mechanism by which Tar1 acts is unclear, as the extreme C-terminal residues with which NmrA interacts in *A. nidulans* are not conserved in *C. neoformans* Gat1/Are1 and *TAR1* transcript levels are not regulated by nitrogen availability (Lee et al. 2012). NmrA orthologs have been studied in the sordariomycete rice pathogens *F. fujikuroi* and *M. oryzae*. In *F. fujikuroi*, Nmr1 interacts with AreA, and the *nmr1Δ* mutant shows chlorate resistance consistent with Nmr1 modulating AreA activity for nitrate utilization gene expression (Schönig et al. 2008). However, the molecular mechanism regulating Nmr1 action differs from *N. crassa* and *A. nidulans* as in *F. fujikuroi nmr1* expression is under AreA control and is repressed during nitrogen sufficiency (Schönig et al. 2008; Wagner et al. 2010). In *M. oryzae*, there are three NmrA transcriptional corepressors, Nmr1, Nmr2, and Nmr3, each of which binds $NADP^+$ and is regulated by the NADPH sensor protein trehalose-6-phosphate synthase (Tps1) (Wilson et al. 2007, 2010). Nmr1 and Nmr3 interact with the AreA ortholog NUT1 in yeast two-hybrid analysis suggesting functional conservation of these Nmr proteins in nitrogen regulation (Wilson et al. 2010). Interaction of Nmr1–Nmr3 with the GATA transcription factor Asd4 was also detected, and Nmr2 interacted with the light-sensing GATA factor Pas1. Furthermore, mutants in which either *nmr1*, *nmr2*, or *nmr3* was deleted showed partial derepression of nitrate reductase and also suppressed the reduced virulence of a *tps1Δ* mutant, indicating a role for these three Nmr proteins in both nitrogen regulation and pathogenesis (Wilson et al. 2010). These three Nmr proteins also regulate carbon catabolite repression in response to glucose-6-phosphate sensing by Tps1 independently of NUT1 (Fernandez et al. 2012).

c) Regulation of NmrA Function

In *A. nidulans*, the levels of NmrA under different conditions are important for controlling AreA activity. NmrA levels are adjusted both at the transcription and protein levels. RT-PCR analysis showed that *nmrA* mRNA levels are elevated under nitrogen-sufficient conditions (ammonium), intermediate during nitrogen limitation (alanine), and low under nitrogen starvation, and the elevated expression of *nmrA* was lost in an *meaBΔ* mutant lacking the bZIP transcription factor MeaB (Wong et al. 2007), suggesting that MeaB activates *nmrA* expression (see Sect. III.A.3). The same pattern of expression was observed for FLAG-epitope-tagged NmrA (Wong et al. 2007). Furthermore, transfer from ammonium media to nitrogen starvation media led to progressively reduced levels of $NmrA^{FLAG}$ over several hours. The levels of *nmrA* mRNA and NmrA protein showed an inverse correlation with *areA* mRNA and $AreA^{HA}$ protein levels (Wong et al. 2007). The levels of NmrA are important for function because overexpression of NmrA from the xylose-inducible *xylP* promoter inhibits AreA action and mimics an AreA loss-of-function phenotype, preventing growth on alternative nitrogen sources (Wong et al. 2007). Immunofluorescence microscopy analysis of NmrA using polyclonal antibodies revealed that the levels of nuclear NmrA are elevated in hyphae grown on ammonium compared with those grown on nitrate, and NmrA was not detected in nitrogen-starved hyphae (Zhao et al. 2010). The absence of detectable NmrA by immunofluorescence in hyphae (Zhao et al. 2010), along with the reduced levels of $NmrA^{FLAG}$ (Wong et al. 2007), after transfer to nitrogen starvation conditions suggested a proteolytic mechanism for turnover of NmrA. Three proteases—PNMA, PNMB, and PNMC—were purified by chromatographic fractionation from extracts of nitrogen-starved mycelium. One of these proteases, PNMB, was identified by a combination of mass spectrometry and LC/MS-MS followed by Mascot and BLAST searching and corresponds to a putative trypsin-like serine protease (Q5BAR4) encoded by AN2366 (Zhao et al. 2010). Together, these proteases mediate ordered proteolysis of NmrA in vitro. Initial cleavage occurs within a C-terminal loop, most likely between M306 and Q307 for PNMA and between K291 and G292 for PNMB. The cleavage product of PNMB is then digested at a specific site near the N-terminus by PNMA and/or PNMC. NmrA was protected from the subsequent cleavage steps by the presence of NAD^+ but not NADH. These proteases are absent in extracts from mycelia grown in nitrogen sufficiency, and therefore they are likely to specifically cleave NmrA in vivo during nitrogen starvation. The cleaved form of NmrA arising from proteolytic digestion within the C-terminal loop was unable to bind to the AreA zinc finger, indicating that proteolytic cleavage of NmrA may mediate relief from nitrogen metabolite repression.

Although the levels of NmrA are important for its corepressor activity, the molecular mechanism(s) underlying NmrA action remains unclear. It was suggested that NmrA may act as a cytoplasmic anchor to prevent AreA nuclear entry (Lamb et al. 2004). However, this appears unlikely because deletion of *nmrA* does not confer AreA nuclear accumulation (Todd et al. 2005), and NmrA is elevated within the nucleus under nitrogen-repressing conditions (Zhao et al. 2010). It has also been proposed that NmrA may prevent AreA DNA binding. In *N. crassa*, NMR1 prevents DNA binding by the AreA ortholog NIT2 in vitro (Xiao et al. 1995). NmrA was presumed to act similarly in preventing AreA DNA binding, and NmrA interacts with the zinc finger DNA-binding domain of AreA (Lamb et al. 2003). AreA DNA binding and interaction with NmrA were found to be mutually exclusive (Kotaka et al. 2008). However, AreA binds preferentially to its GATA DNA-binding site rather than to NmrA (Lamb et al. 2004), and it has been shown that AreA is bound at certain promoters under nitrogen-repressing conditions (Downes et al. 2014b; Muro-Pastor et al. 1999). One possibility is that NmrA prevents DNA binding to some sites, but not others, perhaps depending on the DNA-binding affinity of AreA for a particular site compared with its binding affinity for NmrA. Alternatively, NmrA may prevent activation and/or chromatin remodeling by AreA simply by NmrA–AreA interaction.

3. The bZIP Transcription Factor MeaB

a) MeaB as a DNA-Binding Protein

Mutations in the *A. nidulans meaB* gene confer derepression of several activities that are subject to nitrogen metabolite repression (Arst and Bailey 1980). The *meaB* gene was cloned by complementation, and sequence analysis revealed that MeaB was a novel protein with no significant sequence similarity to proteins in sequence databases (Polley and Caddick 1996). Northern analysis revealed a single transcript expressed under both repressing and derepressing conditions. Shortly after the MeaB sequence was published, the yeast YAP family of basic leucine-zipper (bZIP) transcription factors was identified by analysis of the *S. cerevisiae* genome sequence, and limited sequence similarity between the YAP bZIP domain and MeaB was noted (Fernandes et al. 1997). Subsequent sequence searches following database expansion revealed weak support for MeaB belonging to the bZIP family, and it was suggested that MeaB may act as a DNA-binding transcription factor (Chang and Todd 2004). In studies of *nmrA* regulation, semiquantitative RT-PCR showed elevated *nmrA* expression on ammonium compared with nitrogen starvation and that MeaB was required for this differential regulation (Wong et al. 2007). Therefore, *nmrA* was a potential target for MeaB. The *nmrA* promoter region required for differential expression of NmrA and for MeaB action was localized, using 5′ promoter truncations between −796 and −513 bp upstream of the translation start site. Sequence analysis identified within this region two related sequences separated by ten base pairs and conserved within *nmrA* promoters in aspergilli. Both of these sequences matched the known DNA-binding sites of the mammalian bZIP transcription factor C/EBPβ, and therefore these sites were tested as possible DNA-binding targets for MeaB. An *Escherichia coli*-expressed maltose-binding protein (MBP)–MeaB fusion protein was found to bind in vitro in electrophoretic mobility shift assay to a probe containing these DNA sequences (Wong et al. 2007). Furthermore, mutation of one site but not the other almost abolished DNA binding indicating that MeaB binds to the TTGCACCAT element. This DNA-binding site lacks palindromic half-sites that are often found in the optimal DNA-binding site for bZIP proteins. However, the bZIP transcription factor YAP1 is known to bind both palindromic and non-palindromic DNA-binding sites (Fernandes et al. 1997; Kuge and Jones 1994).

b) Conservation of MeaB

MeaB orthologs are present in the *Eurotiomycetes* and *Sordariomycetes* and in at least some *Dothidiomycetes* but are absent in *Basidiomycetes*, and no close orthologs are present in the *Saccharomycotina* or *Schizosaccharomycotina* (Todd et al. 2014). Protein sequence alignments of ascomycete MeaB orthologs revealed that the basic region of the putative bZIP domain was absolutely conserved in all MeaB orthologs examined, whereas the predicted leucine-zipper region was highly conserved within the aspergilli but has diverged considerably in other ascomycetes, including *N. crassa* and *F. fujikuroi* (Wong et al. 2007). Therefore, the role of MeaB in nitrogen regulation may be conserved in aspergilli, but may have diverged beyond this genus. However, in *F. oxysporum*, quantitative RT-PCR analysis showed that *Nmr1* transcripts were upregulated on ammonium-containing medium in wild type but not in the *meaB*Δ mutant, consistent with a conserved role with *A. nidulans* MeaB in elevating transcription of *Nmr1* under nitrogen-repressing conditions (Lopez-Berges et al. 2010a). In *F. fujikuroi*, *nmrA* also shows differential regulation by nitrogen, with higher expression on ammonium or glutamine than during nitrogen starvation, and this variation in transcript levels was reduced in the *meaB*Δ mutant consistent with MeaB contributing to this regulation (Wagner et al. 2010).

c) Regulation of MeaB

The activation of *nmrA* expression on ammonium by MeaB in *A. nidulans* suggested that MeaB levels and/or activity may respond to nitrogen nutrient quality or availability. Semi-quantitative RT-PCR analysis demonstrated that *meaB* expression was reduced after nitrogen starvation compared with when ammonium, glutamine, alanine, and glutamate are the sole nitrogen sources (Wong et al. 2007). Similarly, quantitative Northern analysis also showed *meaB* levels were reduced during nitrogen starvation compared with ammonium, glutamine, or alanine (Wagner et al. 2010). Furthermore, quantitative Northern analysis suggests that the levels of *meaB* transcript are distinguishably lower on alanine than ammonium (Sibthorp et al. 2013). Therefore, these data show that *meaB* levels are altered in response to the nitrogen source. Semiquantitative RT-PCR analysis also suggested that *meaB* levels were not substantially altered in ammonium-grown or nitrogen-starved cultures in *tamA*Δ, *areA*Δ, or *nmrA*Δ mutants compared with wild type (Wong et al. 2007). In contrast, quantitative Northern analysis revealed that the differential expression of *meaB* on ammonium, alanine, and nitrogen starvation was lost in the *areA49* loss-of-function mutant, demonstrating a role for AreA in *meaB* regulation (Sibthorp et al. 2013; Wagner et al. 2010).

Strand-specific whole transcriptome sequence analysis revealed a natural anti-sense transcript for *A. nidulans meaB* (Sibthorp et al. 2013). This anti-sense transcript begins in the first intron of the *meaB* gene, which contains potential GATA DNA-binding sites for AreA. Expression of the anti-sense transcript is AreA-dependent and different expression levels are observed on ammonium, alanine and during nitrogen starvation (Sibthorp et al. 2013). Deletion of the first intron, which removes the anti-sense transcript start-point, conferred similar high expression of the *meaB* sense transcript on ammonium and alanine, but down-regulation in response to nitrogen starvation was retained. This suggests that the *meaB* anti-sense transcript has a regulatory role in controlling *meaB* levels in response to nitrogen quality. The mechanism remains to be elucidated, however, as mutation of key components of the RNAi system do not confer altered *meaB* regulation. (Sibthorp et al. 2013)

4. AreB: A Second GATA Transcription Factor

a) AreB Is a Transcription Repressor of Nitrogen Metabolic Genes

A second GATA transcription factor, AreB, is also involved in nitrogen regulation in *A. nidulans*. AreB was first identified by mutation. *areB* gain-of-function mutants involving chromosomal rearrangements were identified in genetic screens as wide-domain suppressors of *areA* loss-of-function mutants (Tollervey and Arst 1982). A portion of the *areB* gene was recovered from degenerate PCR amplification of GATA zinc finger-encoding sequences and was used to isolate a complete gene, which was shown to span the chromosomal breakpoints in the *areB* translocation and inversion mutants (Conlon et al. 2001). Northern analysis revealed multiple *areB* transcripts, with the shortest transcript expressed at the highest levels on nitrate and the two longer transcripts expressed optimally on ammonium (Conlon et al. 2001). The *areB* mRNA shows differential stability, with a shorter half-life during growth in glutamine media than during nitrogen starvation (Caddick et al. 2006). *areB* encodes three protein isoforms AreBα (312 amino acids), AreBβ (320 amino acids), and AreBγ (436 amino acids), resulting from the combined effects of different transcription start points, alternative splicing, and the use of different translation start codons. All three AreB proteins contain a GATA zinc finger DNA-binding motif and a leucine-zipper motif. Computational analysis identified two classical and two bipartite nuclear localization signals (Wong et al. 2009), and an additional noncanonical $RRX_{33}RXR$ bipartite nuclear localization signal is found within the zinc finger (Hunter et al. 2014).

The *areB* gain-of-function mutants generate fusions of various sequences, thought to harbor cryptic transcription activation sequences, to AreB sequences N-terminal to the zinc finger (Conlon et al. 2001). *areB* loss-of-function mutants were selected as revertants of the *areB403* gain-of-function mutation, and molecular analysis indicated that these mutations truncate the predicted protein either N-terminal to the AreB sequences or within the AreB zinc finger (Conlon et al. 2001). Analysis of the *areB403/901* loss-of-function mutant suggested a negative role for AreB in regulation of arginine catabolic genes (Dzikowska et al. 2003). An additional third class of mutant, showing altered *areB* function, was identified by selection for partial restoration of nitrate utilization in a strain encoding fusion of AreA with the zinc finger replaced with two mouse zinc fingers (Conlon et al. 2001). This class of *areB* mutant encodes AreB proteins with an intact zinc finger truncated before or within the leucine zipper. Functional analysis by deletion of the zinc finger or leucine-zipper regions within AreBγ demonstrated that both of these domains are required for function (Wong et al. 2009).

Deletion of the *areB* gene revealed pleiotropic phenotypes, suggesting a broader role than in nitrogen regulation alone (Wong et al. 2009). The *areB*Δ mutant showed slightly reduced growth, reduced conidiation, and reduced and heterogeneous germination of both conidia and ascospores. There was no specific effect on growth on a range of nitrogen sources, and no derepression was observed in plate tests for derepression of nitrate, urea, or asparagine utilization genes on ammonium media containing chlorate, thiourea, or aspartate hydroxamate, respectively, or derepression of extracellular proteases as assessed by milk clearing assays in the presence of ammonium. Macios et al. (2012) also reported a lack of any specific growth effect for the *A. nidulans areB*Δ mutant on a range of nitrogen sources and observed a general slower growth in liquid culture due to a reduced growth rate at the start of log phase, consistent with the germination defects observed by Wong et al. (2009). However, the lack of a striking general growth reduction on solid media was considered at variance with the reduced growth rate reported by Wong et al. (2009). This reduced growth rate is slight and becomes more obvious when the colonies are grown over an extended growth period in the absence of contact inhibition (Wong et al. 2009). Assays of the AreA-dependent reporter gene *fmdS–lacZ* revealed elevation of gene expression during nitrogen limitation (alanine as the nitrogen source) or after transfer to media lacking a nitrogen source, suggesting a negative role for AreB during nitrogen limitation and nitrogen starvation (Wong et al. 2009). Analysis of the arginine catabolic genes *otaA* and *agaA* in the *areB*Δ and *areB403/901* mutants revealed derepression on ammonium plus arginine during carbon limitation (0.1 % fructose) but not in carbon-sufficient glucose medium, suggesting a negative role for AreB in regulating arginine catabolic genes during carbon limitation (Macios et al. 2012). The role of AreB was also analyzed by overexpression using the adjustable xylose-inducible promoter *xylP*(p) (Zadra et al. 2000). AreB overexpression using *xylP*(p)*areB*α or *xylP*(p)*areB*γ conferred growth inhibition and loss of AreA-dependent activation of the *fmdS–lacZ* reporter gene during nitrogen limitation or nitrogen starvation (Wong et al. 2009). Together, AreB overexpression and *areB* deletion suggest that AreB acts negatively in nitrogen regulation, and it has been proposed that AreB may compete with AreA for binding of GATA DNA-binding sites at the *fmdS* promoter.

AreB does not function similarly at all of its targets. AreB acts negatively to regulate the *fmdS* gene during nitrogen limitation and nitrogen starvation when the carbon supply is sufficient (glucose) (Wong et al. 2009). In contrast, AreB does not act under these same carbon conditions at *otaA* or *agaA*, but instead represses these two arginine catabolic genes under carbon-limiting conditions (Macios et al. 2012). It is notable that AreB acts at *fmdS* under conditions where AreA is most active (nitrogen limitation and nitrogen starvation) during carbon sufficiency, whereas at *otaA*/*agaA*, AreA and AreB both repress expression in the presence of ammonium, but AreA acts on glucose, whereas AreB fulfills this role during carbon limitation.

b) Conservation of AreB

AreB orthologs are found in the *Eurotiomycetes* and *Sordariomycetes* and some *Saccharomycetes* (Todd et al. 2014). Phylogenetic analysis revealed a single AreB ortholog in *A. nidulans*, *P. chrysogenum* (NreB), and *N. crassa* (ASD4), and sequence alignment showed that the AreB zinc finger and leucine zipper are highly conserved in the filamentous fungi, but there is an insertion in the leucine zipper of the *F. fujikuroi* and *N. crassa* AreB orthologs (Wong et al. 2009). The complex gene structure is also found in *F. fujikuroi areB*, with three transcripts encoding three AreB proteins (Michielse et al. 2014). In *F. fujikuroi*, AreB positively regulates some AreA-dependent nitrogen metabolic genes as well as gibberellic acid and fusaric acid biosynthetic genes, but represses certain other nitrogen metabolic genes (Michielse et al. 2014; Niehaus et al. 2014). Interestingly, in *F. fujikuroi*, AreB and AreA have been shown to interact in the nucleus via

bimolecular fluorescence complementation suggesting a direct interaction at least at some promoters (Michielse et al. 2014). In *N. crassa*, ASD4 has no apparent involvement in regulation of nitrogen metabolism genes, whereas in *P. chrysogenum*, overexpression of NreB leads to repression of the nitrate utilization genes (Feng et al. 2000; Haas et al. 1997).

5. The Dual-Function Transcription Factor TamA

a) TamA Is a Coactivator of AreA

TamA has been proposed to function as a transcription coactivator of AreA in *A. nidulans. tamA* mutants were identified by selection for simultaneous resistance to toxic nitrogen source analogs (Arst et al. 1982; Kinghorn and Pateman 1975) or suppression of GABA toxicity in an *areA102 amdR44* mutant (Davis et al. 1996). The *tamA* gene was cloned by complementation, and the TamA protein sequence revealed a predicted Zn(II)$_2$Cys$_6$ motif and similarity to the *S. cerevisiae* UGA35/DAL81 protein (Davis et al. 1996). However, two lines of evidence indicated that the TamA Zn(II)$_2$Cys$_6$ domain was dispensable for function. First, mutation of a critical zinc-coordinating cysteine residue C90 in the Zn(II)$_2$Cys$_6$ binuclear cluster did not abolish TamA function, as the *tamA*^{C90L} mutant gene fully complemented the *tamA24* mutant phenotypes of reduced growth on ammonium or acetamide, and methylammonium resistance (Davis et al. 1996). Second, a TamA in-frame deletion mutant construct lacking the Zn(II)$_2$Cys$_6$ binuclear cluster complemented the *tamA*Δ methylammonium resistance and growth phenotypes (Small et al. 1999, 2001). The dispensability of the Zn(II)$_2$Cys$_6$ motif suggested that TamA may interact with a DNA-binding protein to act as a coactivator of gene expression. The Zn(II)$_2$Cys$_6$ motif of DAL81 also is dispensable for known functions (Bricmont et al. 1991; Cardillo et al. 2012).

The *tamA*Δ mutant showed reduced *amdS-lacZ* reporter gene expression indicating that TamA activates *amdS* transcription (Small et al. 1999). TamA was shown to act as an activator by fusion of TamA residues 153–651, which lack the Zn(II)$_2$Cys$_6$ region, to the DNA-binding domains of FacB or AmdR (Small et al. 1999). These fusions complemented *facB* or *amdR* loss-of-function mutant phenotypes in growth tests and FacB-dependent or AmdR-dependent reporter gene assays, respectively. Importantly, the FacB-TamA and AmdR-TamA fusions complemented the *tamA*Δ methylammonium resistance phenotype. Sequence analysis of *tamA* mutant alleles identified sequences C-terminal to the Zn(II)$_2$Cys$_6$ motif as important for function (Small et al. 2001). The ability of in-frame deletion constructs to complement the poorer growth on ammonium phenotype and methylammonium resistance phenotype of the *tamA*Δ mutant was used to identify regions of TamA important for function (Small et al. 2001). Several lines of evidence indicated that TamA is a coactivator of AreA. First, activation of reporter gene expression by the FacB-TamA fusion requires AreA, as activation was lost in an *areA*Δ mutant (Small et al. 1999). Second, AreA deletion constructs were used to show that the extreme C-terminus of AreA was required for TamA function (Small et al. 1999). Furthermore, TamA and AreA physically interact in a yeast two-hybrid assay, and the AreA C-terminus was required for this interaction. Third, a single-copy gene-targeted *tamA*^{C90L} Zn(II)$_2$Cys$_6$ motif mutant gene complemented the *tamA*Δ mutant for resistance to thiourea, aspartate hydroxamate, methylammonium, and chlorate and also complemented the reduced expression levels in the *tamA*Δ of the AreA-dependent reporter genes *amdS-lacZ* and *fmdS-lacZ* (Downes et al. 2014b).

b) TamA Acts as a DNA-Binding Activator or as a Coactivator Depending on Promoter Context

TamA regulates expression of the NADP$^+$-dependent glutamate dehydrogenase (NADP-GDH) gene *gdhA*, as NADP-GDH enzyme activity and *gdhA* mRNA levels are reduced in a *tamA*Δ mutant (Polotnianka et al. 2004). Importantly, the effect of *tamA*Δ was greater than the effect of *areA*Δ on *gdhA* expression, suggesting that TamA has a role at *gdhA* in addition to its role as a coactivator of AreA. To identify additional partners for TamA, a yeast two-hybrid screen using TamA as bait was performed, and the leucine biosynthesis regulator LeuB was identified as an interactor (Polotnianka et al. 2004). LeuB also regulates *gdhA* expression as NADP-GDH activity, *gdhA* mRNA levels, and *gdhA-lacZ* expression were reduced in a *leuB*Δ mutant (Downes et al. 2013; Polotnianka et al. 2004). Reduction of NADP-GDH activity and *gdhA* mRNA levels is greater in the *tamA*Δ mutant than in the *leuB*Δ mutant (Polotnianka et al. 2004). Comparison of wild type, *tamA*Δ, and *tamA*^{C90L} for NADP-GDH activity on ammonium or *gdhA* mRNA levels on ammonium or glutamine revealed that, in contrast to the observations for *amdS* and *fmdS* expression, *tamA*^{C90L} only partially complements *gdhA* expression in the *tamA*Δ mutant (Downes et al. 2014b). This lack of full complementation indicates that the TamA Zn(II)2Cys6 motif functions at *gdhA*. A significant difference in NADP-GDH activity and *gdhA* levels between *tamA*Δ and *tamA*^{C90L} showed that—in addition to acting via its Zn(II)$_2$Cys$_6$ motif—*tamA* also acts at the full-length *gdhA* promoter independent of its Zn(II)$_2$Cys$_6$ motif. Bioinformatics searches for potential Zn(II)$_2$Cys$_6$ DNA-binding site triplets coupled with dissection of the *gdhA* promoter using promoter truncations and deletions of a *gdhA-lacZ* translational fusion were used to identify a site of action for the TamA Zn(II)$_2$Cys$_6$ motif (Downes et al. 2014b). Furthermore, Downes et al. (2014b) demonstrated binding by TamA to this site in vivo and in vitro. Chromatin immunoprecipitation (ChIP) assays showed that TamA^{FLAG} but not TamA$^{C90L.FLAG}$ bound to the *gdhA* promoter in vivo. Furthermore, using electrophoretic mobility shift assays, partially purified TamA^{FLAG} but not TamA$^{C90L.FLAG}$ bound to a double-stranded oligonucleotide sequence corresponding to this site of action. Therefore, TamA is a dual-function transcription factor, which acts by two distinct mechanisms depending on promoter context. TamA acts as a coactivator independent of its DNA-binding motif at most of its known targets, but acts as a DNA-binding transcription factor (DBTF) via its DNA-binding domain at *gdhA* (Fig. 11.4). It is important to note that both mechanisms act at *gdhA*, with TamA DNA binding via a proximal site and TamA coactivator action via a distal site (Downes et al. 2014b). Dual functionality may provide a means for the same transcription factor to mediate an additional level of combinatorial control.

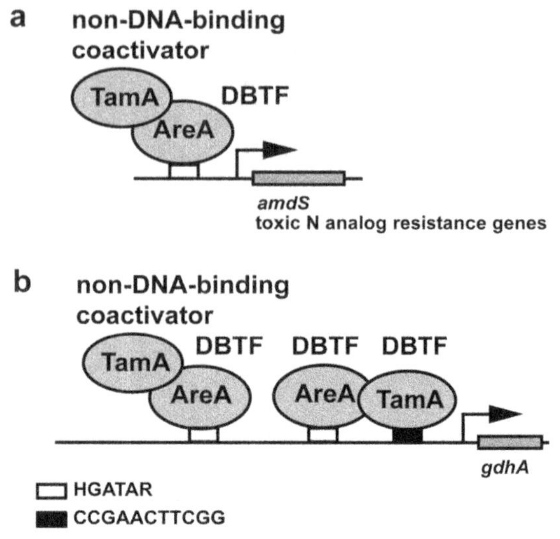

Fig. 11.4 Dual function of TamA. (**a**) TamA acts as a non-DNA-binding coactivator of AreA at *amdS* and at toxic nitrogen analog resistance genes. (**b**) TamA acts as a DNA-binding transcription factor (DBTF) via its Zn(II)$_2$Cys$_6$ DNA-binding motif at a proximal site and as a coactivator at a distal site in the *gdhA* promoter. TamA and AreA are reciprocally required for DNA binding at the proximal site

Downes et al. (2014b) further investigated TamA^{FLAG} DNA binding at *gdhA* using chromatin immunoprecipitation (ChIP). The levels of TamA^{FLAG} DNA binding varied for different nitrogen sources, and the level of DNA binding correlated with *gdhA-lacZ* expression. Furthermore, AreA was required for TamA DNA binding to the *gdhA* promoter, as TamA^{FLAG} did not exhibit ChIP in an *areA*Δ mutant. Interestingly, AreA^{HA} binding to the same region of the *gdhA* promoter was lost in a *tamA*Δ mutant, indicating that TamA is required for AreA binding. This suggests that AreA and TamA may bind at the *gdhA* promoter as a heteromeric DNA-binding complex, with both DNA-binding partners directly contacting the DNA. Therefore, TamA may facilitate AreA binding to the *gdhA* promoter under nitrogen-repressing conditions when AreA is not active at most nitrogen catabolism gene promoters.

c) Conservation of TamA

Phylogenetic and comparative sequence analyses of TamA revealed orthologs throughout the *Ascomycetes*, with high conservation within aspergilli (Downes et al. 2014b; Todd et al. 2014). Functional analysis demonstrated that the *Aspergillus oryzae* ortholog complements the *A. nidulans tamA*Δ mutant, but the *S. cerevisiae* ortholog DAL81 does not (Small et al. 2001). Interestingly, the pattern of conservation for TamA is unusual for a DNA-binding transcription factor. For most DNA-binding transcription factors, the DNA-binding motif is the most highly conserved part of the protein, presumably due to strong selection for binding to specific DNA-binding sites. The TamA Zn $(II)_2Cys_6$ motif shows lower levels of conservation than other parts of the protein when compared with orthologs across the *Ascomycetes* (Downes et al. 2014b). This suggests that the selective constraints are stronger for the non-DNA-binding functions of TamA compared with DNA-binding functions and suggests that there may be relatively few TamA DNA-binding targets. The conservation of the TamA Zn $(II)_2Cys_6$ DNA-binding motif within but not beyond aspergilli suggests that the TamA DNA-binding site sequence will likely differ for more distant orthologs (Downes et al. 2014b).

B. Posttranscriptional Controls

Regulation of transcript stability has emerged in *A. nidulans* as an important mechanism controlling expression of nitrogen metabolic genes. As discussed earlier [see Sect. III.A.1 (d)], mRNA stability is important for regulating AreA activity. Differential transcript degradation in response to nitrogen availability is not specific to *areA* mRNA, but also regulates a subset of transcripts for nitrogen metabolic genes. Ammonium or glutamine increased the instability of transcripts for the nitrate utilization genes *niiA*, *niaD*, *crnA*, and *nrtB*, an ammonium transporter *meaA*, the GABA transporter *gabA*, the proline transporter *prnB*, and the acetamidase *amdS* (Caddick et al. 2006). Transcript stability is also regulated in response to nitrogen conditions for the arginine metabolic genes *agaA* and *otaA* (Krol et al. 2013). However, transcript stability for *gdhA* (NADP-glutamate dehydrogenase) and *glnA* (glutamine synthase) was not regulated by nitrogen source (Caddick et al. 2006). As observed for glutamine-triggered instability of *areA* mRNA, regulated mRNA degradation occurred by deadenylation of *niaD* and *meaA* transcripts via the Caf1–Ccr4–Not deadenylation complex and CU modification of *meaA* mRNA by the nucleotidyltransferase CutA (Caddick et al. 2006; Morozov et al. 2010a). For *niaD*, *otaA*, and *agaA*, but not for *meaA*, the transcripts were stabilized by deletion of *rrmA*, which encodes an RNA-binding protein involved in deadenylation of *areA* mRNA (Krol et al. 2013). Therefore, RrmA appears to modulate the stability of specific transcripts. It is unclear how prevalent mRNA stability regulation of nitrogen metabolic genes by Caf1 and Rrm1 orthologs is in other species, but orthologs of these genes are found throughout the *Ascomycetes*, suggesting that nitrogen regulation of mRNA stability may be a conserved mechanism.

IV. Conclusions

Adaptation of fungal metabolism to changes in nitrogen nutrient quality and availability occurs by both transcriptional and posttranscriptional regulation of nitrogen metabolic genes and their regulators. The key global acting transcription factor AreA is conserved across the fungi and generally activates expression of metabolic target genes. The other main players NmrA, MeaB, AreB, and TamA are involved directly or indirectly in modulating AreA activity. These transcription factors are conserved, but not as widely as AreA, particularly less so in the *Basidiomycetes*. Therefore, while the general mechanisms underlying nitrogen regulation are emerging from analysis of these key players, differences in nitrogen regulation are expected to emerge in different fungal systems.

Acknowledgments Research in the author's laboratory has been supported by K-INBRE P20GM103416, Kansas NSF EPSCoR EPS-0903806, Kansas State University Johnson Center for Basic Cancer Research, Kansas State University Plant Biotechnology Center, Australian Research Council grant DP0558002, and the University of Melbourne. Thanks are due to Cameron Hunter and Damien Downes for comments on the manuscript and their contributions to the work. Collaborations with Meryl Davis, Koon Ho Wong, and Michael Hynes are gratefully acknowledged. This is contribution number 15-089-B from the Kansas Agricultural Experiment Station.

References

Andrianopoulos A, Hynes MJ (1990) Sequence and functional analysis of the positively acting regulatory gene *amdR* from *Aspergillus nidulans.* Mol Cell Biol 10:3194–3203

Andrianopoulos A, Kourambas S, Sharp JA, Davis MA, Hynes MJ (1998) Characterization of the *Aspergillus nidulans nmrA* gene involved in nitrogen metabolite repression. J Bacteriol 180:1973–1977

Arst HN Jr, Bailey CR (1980) Genetic evidence for a second asparaginase in *Aspergillus nidulans.* J Gen Microbiol 121:243–247

Arst HN Jr, Cove DJ (1973) Nitrogen metabolite repression in *Aspergillus nidulans.* Mol Gen Genet 126:111–141

Arst HN, Brownlee AG, Cousen SA (1982) Nitrogen metabolite repression in *Aspergillus nidulans* – a farewell to *tamA*? Curr Genet 6:245–257

Beck T, Hall MN (1999) The TOR signalling pathway controls nuclear localization of nutrient-regulated transcription factors. Nature 402:689–692

Berger H, Pachlinger R, Morozov I, Goller S, Narendja F, Caddick M, Strauss J (2006) The GATA factor AreA regulates localization and in vivo binding site occupancy of the nitrate activator NirA. Mol Microbiol 59:433–446

Berger H, Basheer A, Bock S, Reyes-Dominguez Y, Dalik T, Altmann F, Strauss J (2008) Dissecting individual steps of nitrogen transcription factor cooperation in the *Aspergillus nidulans* nitrate cluster. Mol Microbiol 69:1385–1398

Bernreiter A, Ramon A, Fernandez-Martinez J, Berger H, Araujo-Bazan L, Espeso EA, Pachlinger R, Gallmetzer A, Anderl I, Scazzocchio C, Strauss J (2007) Nuclear export of the transcription factor NirA is a regulatory checkpoint for nitrate induction in *Aspergillus nidulans.* Mol Cell Biol 27:791–802

Bertram PG, Choi JH, Carvalho J, Ai W, Zeng C, Chan TF, Zheng XF (2000) Tripartite regulation of Gln3p by TOR, Ure2p, and phosphatases. J Biol Chem 275:35727–35733

Bricmont PA, Daugherty JR, Cooper TG (1991) The *DAL81* gene product is required for induced expression of two differently regulated nitrogen catabolic genes in *Saccharomyces cerevisiae.* Mol Cell Biol 11:1161–1166

Bugeja HE, Hynes MJ, Andrianopoulos A (2012) AreA controls nitrogen source utilisation during both growth programs of the dimorphic fungus *Penicillium marneffei.* Fungal Biol 116:145–154

Burger G, Strauss J, Scazzocchio C, Lang BF (1991) *nirA*, the pathway-specific regulatory gene of nitrate assimilation in *Aspergillus nidulans*, encodes a putative GAL4-type zinc finger protein and contains four introns in highly conserved regions. Mol Cell Biol 11:5746–5755

Caddick MX (2004) Nitrogen regulation in mycelial fungi. In: Brambl R, Marzluf GA (eds) The Mycota: biochemistry and molecular biology, vol III, 2nd edn, The Mycota. Springer, Berlin, pp 349–368

Caddick MX, Jones MG, van Tonder JM, Le Cordier H, Narendja F, Strauss J, Morozov IY (2006) Opposing signals differentially regulate transcript stability in *Aspergillus nidulans.* Mol Microbiol 62:509–519

Cardillo SB, Levi CE, Bermudez Moretti M, Correa Garcia S (2012) Interplay between the transcription factors acting on the GATA- and GABA-responsive elements of *Saccharomyces cerevisiae* UGA promoters. Microbiology 158:925–935

Chang PK, Todd RB (2004) Metabolic pathway regulation. In: Arora DK (ed) Handbook of fungal biotechnology, vol 20, 2nd edn, Mycology series. Marcel Dekker, New York, pp 25–37

Chang PK, Yu J, Bhatnagar D, Cleveland TE (2000) Characterization of the *Aspergillus parasiticus* major nitrogen regulatory gene, *areA*. Biochim Biophys Acta 1491:263–266

Chiang TY, Marzluf GA (1994) DNA recognition by the NIT2 nitrogen regulatory protein: importance of the number, spacing, and orientation of GATA core elements and their flanking sequences upon NIT2 binding. Biochemistry 33:576–582

Christensen T, Hynes MJ, Davis MA (1998) Role of the regulatory gene *areA* of *Aspergillus oryzae* in nitrogen metabolism. Appl Environ Microbiol 64:3232–3237

Coleman M, Henricot B, Arnau J, Oliver RP (1997) Starvation-induced genes of the tomato pathogen *Cladosporium fulvum* are also induced during growth in planta. Mol Plant Microbe Interact 10:1106–1109

Conlon H, Zadra I, Haas H, Arst HN Jr, Jones MG, Caddick MX (2001) The *Aspergillus nidulans* GATA transcription factor gene *areB* encodes at least three proteins and features three classes of mutation. Mol Microbiol 40:361–375

Cooper TG (2002) Transmitting the signal of excess nitrogen in *Saccharomyces cerevisiae* from the

Tor proteins to the GATA factors: connecting the dots. FEMS Microbiol Rev 26:223–238

Davis MA, Hynes MJ (2004) Regulation of the *amdS* gene in *Aspergillus nidulans*. In: Brambl R, Marzluf GA (eds) The Mycota III. Biochemistry and molecular biology (2nd edn). Springer, Berlin

Davis MA, Small AJ, Kourambas S, Hynes MJ (1996) The *tamA* gene of *Aspergillus nidulans* contains a putative zinc cluster motif which is not required for gene function. J Bacteriol 178:3406–3409

Denu JM (2003) Linking chromatin function with metabolic networks: Sir2 family of NAD(+)-dependent deacetylases. Trends Biochem Sci 28:41–48

Divon HH, Rothan-Denoyes B, Davydov O, Dip A, Fluhr R (2005) Nitrogen-responsive genes are differentially regulated in planta during *Fusarium oxyspsorum f. sp. lycopersici* infection. Mol Plant Pathol 6:459–470

Donofrio NM, Oh Y, Lundy R, Pan H, Brown DE, Jeong JS, Coughlan S, Mitchell TK, Dean RA (2006) Global gene expression during nitrogen starvation in the rice blast fungus, *Magnaporthe grisea*. Fungal Genet Biol 43:605–617

Downes DJ, Davis MA, Kreutzberger SD, Taig BL, Todd RB (2013) Regulation of the NADP-glutamate dehydrogenase gene *gdhA* in *Aspergillus nidulans* by the $Zn(II)_2Cys_6$ transcription factor LeuB. Microbiology 159:2467–2480

Downes DJ, Chonofsky M, Tan K, Pfannenstiel BT, Reck-Peterson SL, Todd RB (2014a) Characterization of the mutagenic spectrum of 4-nitroquinoline 1-oxide (4-NQO) in *Aspergillus nidulans* by whole genome sequencing. G3 Bethesda 4:2483–2492

Downes DJ, Davis MA, Wong KH, Kreutzberger SD, Hynes MJ, Todd RB (2014b) Dual DNA binding and coactivator functions of *Aspergillus nidulans* TamA, a $Zn(II)_2Cys_6$ transcription factor. Mol Microbiol 92(6):1198–211

Dzikowska A, Kacprzak M, Tomecki R, Koper M, Scazzocchio C, Weglenski P (2003) Specific induction and carbon/nitrogen repression of arginine catabolism gene of *Aspergillus nidulans*-functional in vivo analysis of the *otaA* promoter. Fungal Genet Biol 38:175–186

Empel J, Sitkiewicz I, Andrukiewicz A, Lasocki K, Borsuk P, Weglenski P (2001) *arcA*, the regulatory gene for the arginine catabolic pathway in *Aspergillus nidulans*. Mol Genet Genomics 266:591–597

Feng B, Marzluf GA (1998) Interaction between major nitrogen regulatory protein NIT2 and pathway-specific regulatory factor NIT4 is required for their synergistic activation of gene expression in *Neurospora crassa*. Mol Cell Biol 18:3983–3990

Feng B, Haas H, Marzluf GA (2000) ASD4, a new GATA factor of *Neurospora crassa*, displays sequence-specific DNA binding and functions in ascus and ascospore development. Biochemistry 39:11065–11073

Fernandes L, Rodrigues-Pousada C, Struhl K (1997) Yap, a novel family of eight bZIP proteins in *Saccharomyces cerevisiae* with distinct biological functions. Mol Cell Biol 17:6982–6993

Fernandez J, Wright JD, Hartline D, Quispe CF, Madayiputhiya N, Wilson RA (2012) Principles of carbon catabolite repression in the rice blast fungus: Tps1, Nmr1-3, and a MATE-family pump regulate glucose metabolism during infection. PLoS Genet 8:e1002673

Fitzgibbon GJ, Morozov IY, Jones MG, Caddick MX (2005) Genetic analysis of the TOR pathway in *Aspergillus nidulans*. Eukaryot Cell 4:1595–1598

Fraser JA, Davis MA, Hynes MJ (2001) The formamidase gene of *Aspergillus nidulans*: regulation by nitrogen metabolite repression and transcriptional interference by an overlapping upstream gene. Genetics 157:119–131

Fraser JA, Davis MA, Hynes MJ (2002) A gene from *Aspergillus nidulans* with similarity to *URE2* of *Saccharomyces cerevisiae* encodes a glutathione S-transferase which contributes to heavy metal and xenobiotic resistance. Appl Environ Microbiol 68:2802–2808

Froeliger EH, Carpenter BE (1996) *NUT1*, a major nitrogen regulatory gene in *Magnaporthe grisea*, is dispensable for pathogenicity. Mol Gen Genet 251:647–656

Fu YH, Marzluf GA (1990) *nit-2*, the major nitrogen regulatory gene of *Neurospora crassa*, encodes a protein with a putative zinc finger DNA-binding domain. Mol Cell Biol 10:1056–1065

Fu YH, Young JL, Marzluf GA (1988) Molecular cloning and characterization of a negative-acting nitrogen regulatory gene of *Neurospora crassa*. Mol Gen Genet 214:74–79

Gacek A, Strauss J (2012) The chromatin code of fungal secondary metabolite gene clusters. Appl Microbiol Biotechnol 95:1389–1404

Garcia I, Gonzalez R, Gomez D, Scazzocchio C (2004) Chromatin rearrangements in the prnD-prnB bidirectional promoter: dependence on transcription factors. Eukaryot Cell 3:144–156

Gomez D, Garcia I, Scazzocchio C, Cubero B (2003) Multiple GATA sites: protein binding and physiological relevance for the regulation of the proline transporter gene of *Aspergillus nidulans*. Mol Microbiol 50:277–289

Haas H, Bauer B, Redl B, Stoffler G, Marzluf GA (1995) Molecular cloning and analysis of *nre*, the major nitrogen regulatory gene of *Penicillium chrysogenum*. Curr Genet 27:150–158

Haas H, Angermayr K, Zadra I, Stoffler G (1997) Overexpression of *nreB*, a new GATA factor-encoding gene of *Penicillium chrysogenum*, leads to repression of the nitrate assimilatory gene cluster. J Biol Chem 272:22576–22582

Horst RJ, Zeh C, Saur A, Sonnewald S, Sonnewald U, Voll LM (2012) The *Ustilago maydis* Nit2 homolog regulates nitrogen utilization and is required for efficient induction of filamentous growth. Eukaryot Cell 11:368–380

Hunter CC, Siebert KS, Downes DJ, Wong KH, Kreutzberger SD, Fraser JA, Clarke DF, Hynes MJ, Davis MA, Todd RB (2014) Multiple nuclear localization signals mediate nuclear localization of the GATA transcription factor AreA. Eukaryot Cell 13:527–538

Kim H, Woloshuk CP (2008) Role of AREA, a regulator of nitrogen metabolism, during colonization of maize kernels and fumonisin biosynthesis in *Fusarium verticillioides*. Fungal Genet Biol 45:947–953

Kim Y, Islam N, Moss BJ, Nandakumar MP, Marten MR (2011) Autophagy induced by rapamycin and carbon-starvation have distinct proteome profiles in *Aspergillus nidulans*. Biotechnol Bioeng 108:2705–2715

Kinghorn JR, Pateman JA (1975) Studies of partially repressed mutants at the *tamA* and *areA* loci in *Aspergillus nidulans*. Mol Gen Genet 140:137–147

Kmetzsch L, Staats CC, Simon E, Fonseca FL, Oliveira DL, Joffe LS, Rodrigues J, Lourenco RF, Gomes SL, Nimrichter L, Rodrigues ML, Schrank A, Vainstein MH (2011) The GATA-type transcriptional activator Gat1 regulates nitrogen uptake and metabolism in the human pathogen *Cryptococcus neoformans*. Fungal Genet Biol 48:192–199

Kotaka M, Johnson C, Lamb HK, Hawkins AR, Ren J, Stammers DK (2008) Structural analysis of the recognition of the negative regulator NmrA and DNA by the zinc finger from the GATA-type transcription factor AreA. J Mol Biol 381:373–382

Krol K, Morozov IY, Jones MG, Wyszomirski T, Weglenski P, Dzikowska A, Caddick MX (2013) RrmA regulates the stability of specific transcripts in response to both nitrogen source and oxidative stress. Mol Microbiol 89:975–988

Kudla B, Caddick MX, Langdon T, Martinez-Rossi NM, Bennett CF, Sibley S, Davies RW, Arst HN Jr (1990) The regulatory gene *areA* mediating nitrogen metabolite repression in *Aspergillus nidulans*. Mutations affecting specificity of gene activation alter a loop residue of a putative zinc finger. EMBO J 9:1355–1364

Kudo N, Matsumori N, Taoka H, Fujiwara D, Schreiner EP, Wolff B, Yoshida M, Horinouchi S (1999) Leptomycin B inactivates CRM1/exportin 1 by covalent modification at a cysteine residue in the central conserved region. Proc Natl Acad Sci USA 96:9112–9117

Kuge S, Jones N (1994) YAP1 dependent activation of *TRX2* is essential for the response of *Saccharomyces cerevisiae* to oxidative stress by hydroperoxides. EMBO J 13:655–664

Lamb HK, Leslie K, Dodds AL, Nutley M, Cooper A, Johnson C, Thompson P, Stammers DK, Hawkins AR (2003) The negative transcriptional regulator NmrA discriminates between oxidized and reduced dinucleotides. J Biol Chem 278:32107–32114

Lamb HK, Ren J, Park A, Johnson C, Leslie K, Cocklin S, Thompson P, Mee C, Cooper A, Stammers DK, Hawkins AR (2004) Modulation of the ligand binding properties of the transcription repressor NmrA by GATA-containing DNA and site-directed mutagenesis. Protein Sci 13:3127–3138

Langdon T, Sheerins A, Ravagnani A, Gielkens M, Caddick MX, Arst HN Jr (1995) Mutational analysis reveals dispensability of the N-terminal region of the Aspergillus transcription factor mediating nitrogen metabolite repression. Mol Microbiol 17:877–888

Lee IR, Chow EW, Morrow CA, Djordjevic JT, Fraser JA (2011) Nitrogen metabolite repression of metabolism and virulence in the human fungal pathogen *Cryptococcus neoformans*. Genetics 188:309–323

Lee IR, Lim JW, Ormerod KL, Morrow CA, Fraser JA (2012) Characterization of an Nmr homolog that modulates GATA factor-mediated nitrogen metabolite repression in *Cryptococcus neoformans*. PLoS One 7, e32585

Lopez-Berges MS, Rispail N, Prados-Rosales RC, Di Pietro A (2010a) A nitrogen response pathway regulates virulence functions in *Fusarium oxysporum* via the protein kinase TOR and the bZIP protein MeaB. Plant Cell 22:2459–2475

Lopez-Berges MS, Rispail N, Prados-Rosales RC, Di Pietro A (2010b) A nitrogen response pathway regulates virulence in plant pathogenic fungi: role of TOR and the bZIP protein MeaB. Plant Signal Behav 5:1623–1625

MacCabe AP, Vanhanen S, Sollewign Gelpke MD, van de Vondervoort PJ, Arst HN Jr, Visser J (1998) Identification, cloning and sequence of the *Aspergillus niger areA* wide domain regulatory gene controlling nitrogen utilisation. Biochim Biophys Acta 1396:163–168

Macios M, Caddick MX, Weglenski P, Scazzocchio C, Dzikowska A (2012) The GATA factors AREA and AREB together with the co-repressor NMRA, negatively regulate arginine catabolism in *Aspergillus nidulans* in response to nitrogen and carbon source. Fungal Genet Biol 49:189–198

MacPherson S, Larochelle M, Turcotte B (2006) A fungal family of transcriptional regulators: the zinc cluster proteins. Microbiol Mol Biol Rev 70:583–604

Margelis S, D'Souza C, Small AJ, Hynes MJ, Adams TH, Davis MA (2001) Role of glutamine synthetase in nitrogen metabolite repression in *Aspergillus nidulans*. J Bacteriol 183:5826–5833

Marzluf GA (1997) Genetic regulation of nitrogen metabolism in the fungi. Microbiol Mol Biol Rev 61:17–32

Michielse CB, Pfannmuller A, Macios M, Rengers P, Dzikowska A, Tudzynski B (2014) The interplay between the GATA transcription factors AreA, the global nitrogen regulator and AreB in *Fusarium fujikuroi*. Mol Microbiol 91:472–493

Mihlan M, Homann V, Liu TW, Tudzynski B (2003) AREA directly mediates nitrogen regulation of gibberellin biosynthesis in *Gibberella fujikuroi*, but its activity is not affected by NMR. Mol Microbiol 47:975–991

Min K, Shin Y, Son H, Lee J, Kim JC, Choi GJ, Lee YW (2012) Functional analyses of the nitrogen regulatory gene *areA* in *Gibberella zeae*. FEMS Microbiol Lett 334:66–73

Monahan BJ, Fraser JA, Hynes MJ, Davis MA (2002) Isolation and characterization of two ammonium permease genes, *meaA* and *mepA*, from *Aspergillus nidulans*. Eukaryot Cell 1:85–94

Monahan BJ, Askin MC, Hynes MJ, Davis MA (2006) Differential expression of *Aspergillus nidulans* ammonium permease genes is regulated by GATA transcription factor AreA. Eukaryot Cell 5:226–237

Morozov IY, Martinez MG, Jones MG, Caddick MX (2000) A defined sequence within the 3' UTR of the *areA* transcript is sufficient to mediate nitrogen metabolite signalling via accelerated deadenylation. Mol Microbiol 37:1248–1257

Morozov IY, Galbis-Martinez M, Jones MG, Caddick MX (2001) Characterization of nitrogen metabolite signalling in Aspergillus via the regulated degradation of *areA* mRNA. Mol Microbiol 42:269–277

Morozov IY, Jones MG, Razak AA, Rigden DJ, Caddick MX (2010a) CUCU modification of mRNA promotes decapping and transcript degradation in *Aspergillus nidulans*. Mol Cell Biol 30:460–469

Morozov IY, Jones MG, Spiller DG, Rigden DJ, Dattenbock C, Novotny R, Strauss J, Caddick MX (2010b) Distinct roles for Caf1, Ccr4, Edc3 and CutA in the co-ordination of transcript deadenylation, decapping and P-body formation in *Aspergillus nidulans*. Mol Microbiol 76:503–516

Muro-Pastor MI, Gonzalez R, Strauss J, Narendja F, Scazzocchio C (1999) The GATA factor AreA is essential for chromatin remodelling in a eukaryotic bidirectional promoter. EMBO J 18:1584–1597

Muro-Pastor MI, Strauss J, Ramon A, Scazzocchio C (2004) A paradoxical mutant GATA factor. Eukaryot Cell 3:393–405

Narendja F, Goller SP, Wolschek M, Strauss J (2002) Nitrate and the GATA factor AreA are necessary for in vivo binding of NirA, the pathway-specific transcriptional activator of *Aspergillus nidulans*. Mol Microbiol 44:573–583

Neville M, Rosbash M (1999) The NES-Crm1p export pathway is not a major mRNA export route in *Saccharomyces cerevisiae*. EMBO J 18:3746–3756

Niehaus EM, von Bargen KW, Espino JJ, Pfannmuller A, Humpf HU, Tudzynski B (2014) Characterization of the fusaric acid gene cluster in *Fusarium fujikuroi*. Appl Microbiol Biotechnol 98:1749–1762

Oestreicher N, Ribard C, Scazzocchio C (2008) The *nadA* gene of *Aspergillus nidulans*, encoding adenine deaminase, is subject to a unique regulatory pattern. Fungal Genet Biol 45:760–775

Olszewska A, Krol K, Weglenski P, Dzikowska A (2007) Arginine catabolism in *Aspergillus nidulans* is regulated by the *rrmA* gene coding for the RNA-binding protein. Fungal Genet Biol 44:1285–1297

Pan H, Feng B, Marzluf GA (1997) Two distinct protein-protein interactions between the NIT2 and NMR regulatory proteins are required to establish nitrogen metabolite repression in *Neurospora crassa*. Mol Microbiol 26:721–729

Pellier AL, Lauge R, Veneault-Fourrey C, Langin T (2003) CLNR1, the AREA/NIT2-like global nitrogen regulator of the plant fungal pathogen *Colletotrichum lindemuthianum* is required for the infection cycle. Mol Microbiol 48:639–655

Perez-Garcia A, Snoeijers SS, Joosten MH, Goosen T, De Wit PJ (2001) Expression of the Avirulence gene *Avr9* of the fungal tomato pathogen *Cladosporium fulvum* is regulated by the global nitrogen response factor NRF1. Mol Plant Microbe Interact 14:316–325

Peters DG, Caddick MX (1994) Direct analysis of native and chimeric GATA specific DNA binding proteins from *Aspergillus nidulans*. Nucleic Acids Res 22:5164–5172

Philips AS, Kwok JC, Chong BH (2007) Analysis of the signals and mechanisms mediating nuclear trafficking of GATA-4. Loss of DNA binding is associated with localization in intranuclear speckles. J Biol Chem 282:24915–24927

Platt A, Langdon T, Arst HN Jr, Kirk D, Tollervey D, Sanchez JM, Caddick MX (1996a) Nitrogen metabolite signalling involves the C-terminus and the GATA domain of the Aspergillus transcription factor AREA and the 3' untranslated region of its mRNA. EMBO J 15:2791–2801

Platt A, Ravagnani A, Arst H Jr, Kirk D, Langdon T, Caddick MX (1996b) Mutational analysis of the C-terminal region of AREA, the transcription factor mediating nitrogen metabolite repression in *Aspergillus nidulans*. Mol Gen Genet 250:106–114

Pokorska A, Drevet C, Scazzocchio C (2000) The analysis of the transcriptional activator PrnA reveals a tripartite nuclear localisation sequence. J Mol Biol 298:585–596

Polley SD, Caddick MX (1996) Molecular characterisation of *meaB*, a novel gene affecting nitrogen metabolite repression in *Aspergillus nidulans*. FEBS Lett 388:200–205

Polotnianka R, Monahan BJ, Hynes MJ, Davis MA (2004) TamA interacts with LeuB, the homologue of *Saccharomyces cerevisiae* Leu3p, to regulate *gdhA* expression in *Aspergillus nidulans*. Mol Genet Genomics 272:452–459

Ravagnani A, Gorfinkiel L, Langdon T, Diallinas G, Adjadj E, Demais S, Gorton D, Arst HN Jr, Scazzocchio C (1997) Subtle hydrophobic interactions between the seventh residue of the zinc finger loop and the first base of an HGATAR sequence determine promoter- specific recognition by the

Aspergillus nidulans GATA factor AreA. EMBO J 16:3974–3986

Reyes-Dominguez Y, Narendja F, Berger H, Gallmetzer A, Fernandez-Martin R, Garcia I, Scazzocchio C, Strauss J (2008) Nucleosome positioning and histone H3 acetylation are independent processes in the *Aspergillus nidulans prnD-prnB* bidirectional promoter. Eukaryot Cell 7:656–663

Ribard C, Scazzocchio C, Oestreicher N (2001) The *oxpA5* mutation of *Aspergillus nidulans* is an allele of *adB*, the gene encoding adenylosuccinate synthetase. Mol Genet Genomics 266:701–710

Schinko T, Berger H, Lee W, Gallmetzer A, Pirker K, Pachlinger R, Buchner I, Reichenauer T, Guldener U, Strauss J (2010) Transcriptome analysis of nitrate assimilation in *Aspergillus nidulans* reveals connections to nitric oxide metabolism. Mol Microbiol 78:720–738

Schönig B, Brown DW, Oeser B, Tudzynski B (2008) Cross-species hybridization with *Fusarium verticillioides* microarrays reveals new insights into *Fusarium fujikuroi* nitrogen regulation and the role of AreA and NMR. Eukaryot Cell 7:1831–1846

Sibthorp C, Wu H, Cowley G, Wong PW, Palaima P, Morozov IY, Weedall GD, Caddick MX (2013) Transcriptome analysis of the filamentous fungus *Aspergillus nidulans* directed to the global identification of promoters. BMC Genomics 14:847

Small AJ, Hynes MJ, Davis MA (1999) The TamA protein fused to a DNA-binding domain can recruit AreA, the major nitrogen regulatory protein, to activate gene expression in *Aspergillus nidulans*. Genetics 153:95–105

Small AJ, Todd RB, Zanker MC, Delimitrou S, Hynes MJ, Davis MA (2001) Functional analysis of TamA, a coactivator of nitrogen-regulated gene expression in *Aspergillus nidulans*. Mol Genet Genomics 265:636–646

Snoeijers SS, Vossen P, Goosen T, Van den Broek HW, De Wit PJ (1999) Transcription of the avirulence gene *Avr9* of the fungal tomato pathogen *Cladosporium fulvum* is regulated by a GATA-type transcription factor in *Aspergillus nidulans*. Mol Gen Genet 261:653–659

Stammers DK, Ren J, Leslie K, Nichols CE, Lamb HK, Cocklin S, Dodds A, Hawkins AR (2001) The structure of the negative transcriptional regulator NmrA reveals a structural superfamily which includes the short-chain dehydrogenase/reductases. EMBO J 20:6619–6626

Starich MR, Wikstrom M, Arst HN Jr, Clore GM, Gronenborn AM (1998a) The solution structure of a fungal AREA protein-DNA complex: an alternative binding mode for the basic carboxyl tail of GATA factors. J Mol Biol 277:605–620

Starich MR, Wikstrom M, Schumacher S, Arst HN Jr, Gronenborn AM, Clore GM (1998b) The solution structure of the Leu22->Val mutant AREA DNA binding domain complexed with a TGATAG core element defines a role for hydrophobic packing in the determination of specificity. J Mol Biol 277:621–634

Suarez T, de Queiroz MV, Oestreicher N, Scazzocchio C (1995) The sequence and binding specificity of UaY, the specific regulator of the purine utilization pathway in *Aspergillus nidulans*, suggest an evolutionary relationship with the PPR1 protein of *Saccharomyces cerevisiae*. EMBO J 14:1453–1467

Tanzer MM, Arst HN, Skalchunes AR, Coffin M, Darveaux BA, Heiniger RW, Shuster JR (2003) Global nutritional profiling for mutant and chemical mode-of-action analysis in filamentous fungi. Funct Integr Genomics 3:160–170

Teichert S, Wottawa M, Schonig B, Tudzynski B (2006) Role of the *Fusarium fujikuroi* TOR kinase in nitrogen regulation and secondary metabolism. Eukaryot Cell 5:1807–1819

Teichert S, Rutherford JC, Wottawa M, Heitman J, Tudzynski B (2008) Impact of ammonium permeases *mepA*, *mepB*, and *mepC* on nitrogen-regulated secondary metabolism in *Fusarium fujikuroi*. Eukaryot Cell 7:187–201

Tessarz P, Kouzarides T (2014) Histone core modifications regulating nucleosome structure and dynamics. Nat Rev Mol Cell Biol 15:703–708

Todd RB, Andrianopoulos A (1997) Evolution of a fungal regulatory gene family: the Zn(II)2Cys6 binuclear cluster DNA binding motif. Fungal Genet Biol 21:388–405

Todd RB, Greenhalgh JR, Hynes MJ, Andrianopoulos A (2003) TupA, the *Penicillium marneffei* Tup1p homologue, represses both yeast and spore development. Mol Microbiol 48:85–94

Todd RB, Fraser JA, Wong KH, Davis MA, Hynes MJ (2005) Nuclear accumulation of the GATA factor AreA in response to complete nitrogen starvation by regulation of nuclear export. Eukaryot Cell 4:1646–1653

Todd RB, Zhou M, Ohm RA, Leeggangers HA, Visser L, de Vries RP (2014) Prevalence of transcription factors in ascomycete and basidiomycete fungi. BMC Genomics 15:214

Tollervey DW, Arst HN Jr (1982) Domain-wide, locus-specific suppression of nitrogen metabolite repressed mutations in *Aspergillus nidulans*. Curr Genet 6:79–85

Tudzynski B (2014) Nitrogen regulation of fungal secondary metabolism in fungi. Front Microbiol 5:656

Tudzynski B, Homann V, Feng B, Marzluf GA (1999) Isolation, characterization and disruption of the *areA* nitrogen regulatory gene of *Gibberella fujikuroi*. Mol Gen Genet 261:106–114

Wagner D, Schmeinck A, Mos M, Morozov IY, Caddick MX, Tudzynski B (2010) The bZIP transcription factor MeaB mediates nitrogen metabolite repression at specific loci. Eukaryot Cell 9:1588–1601

Wilson RA, Arst HN Jr (1998) Mutational analysis of AREA, a transcriptional activator mediating nitrogen metabolite repression in *Aspergillus nidulans* and a member of the "streetwise" GATA family of transcription factors. Microbiol Mol Biol Rev 62:586–596

Wilson RA, Jenkinson JM, Gibson RP, Littlechild JA, Wang ZY, Talbot NJ (2007) Tps1 regulates the pentose phosphate pathway, nitrogen metabolism and fungal virulence. EMBO J 26:3673–3685

Wilson RA, Gibson RP, Quispe CF, Littlechild JA, Talbot NJ (2010) An NADPH-dependent genetic switch regulates plant infection by the rice blast fungus. Proc Natl Acad Sci USA 107:21902–21907

Wong KH, Hynes MJ, Todd RB, Davis MA (2007) Transcriptional control of *nmrA* by the bZIP transcription factor MeaB reveals a new level of nitrogen regulation in *Aspergillus nidulans*. Mol Microbiol 66:534–551

Wong KH, Hynes MJ, Davis MA (2008) Recent advances in nitrogen regulation: a comparison between *Saccharomyces cerevisiae* and filamentous fungi. Eukaryot Cell 7:917–925

Wong KH, Hynes MJ, Todd RB, Davis MA (2009) Deletion and overexpression of the *Aspergillus nidulans* GATA factor AreB reveals unexpected pleiotropy. Microbiology 155:3868–3880

Xiao X, Fu YH, Marzluf GA (1995) The negative-acting NMR regulatory protein of *Neurospora crassa* binds to and inhibits the DNA-binding activity of the positive-acting nitrogen regulatory protein NIT2. Biochemistry 34:8861–8868

Zadra I, Abt B, Parson W, Haas H (2000) *xylP* promoter-based expression system and its use for antisense downregulation of the *Penicillium chrysogenum* nitrogen regulator NRE. Appl Environ Microbiol 66:4810–4816

Zhao X, Hume SL, Johnson C, Thompson P, Huang J, Gray J, Lamb HK, Hawkins AR (2010) The transcription repressor NmrA is subject to proteolysis by three *Aspergillus nidulans* proteases. Protein Sci 19:1405–1419

12 Regulation of Sulfur Metabolism in Filamentous Fungi

John V. Paietta[1]

CONTENTS

I. Introduction

Filamentous fungi provide a valuable model for understanding the mechanisms of how organisms adjust to nutrient availability in order to avoid deficiencies of needed macro- and microelements. Sulfur represents an essential low-abundance element that is found in a wide variety of compounds and macromolecules (e.g., proteins, lipids, electron carriers, intermediary metabolites) that are necessary for routine cellular functions. Considered here will be how filamentous fungi use a complex regulatory network to monitor the cellular sulfur status and respond with appropriate acquisition and/or metabolic remodeling strategies. For example, when sulfur is limiting, filamentous fungi will typically respond with the induction of transport systems (e.g., methionine or sulfate permeases) and enzymes that can release sulfur from internal reserves or extracellular sulfur compounds (e.g., arylsulfatase or choline-*O*-sulfatase) (Marzluf 1997; Paietta 2010). These responses allow for efficient scavenging of the external and internal environment in order to obtain the limiting nutrient.

The emphasis here will be on the sulfur regulatory system in *Neurospora crassa* with consideration of other fungal species included throughout. Prior reviews have focused on various aspects of sulfur metabolism in the filamentous fungal model systems of *Aspergillus nidulans* (Paszewski et al. 2000) and *Neurospora crassa* (Marzluf 1997; Paietta 2004, 2010), as well as in yeast (Thomas and Surdin-Kerjan 1997). Our understanding of the molecular basis for the regulation of sulfur metabolism has greatly expanded and has changed our concept of the regulatory mechanisms involved in the control of sulfur metabolism, and these developments will serve as a point of focus. Also to be discussed are aspects of sulfur regulatory control which are still poorly understood, such as how the cellular sulfur status is monitored.

A number of studies provide an indication of the widespread applicability of the mechanistic findings derived from the analysis of model sulfur control systems to a variety of fungal species. Examples include (1) the human pathogen *Aspergillus fumigatus* in which the proper regulation of sulfur metabolism is essential for virulence (Amich et al. 2013), (2) the plant pathogen *Colletotrichum gloeosporioides* which has methionine-repressible arylsulfatase expression (Goodwin et al. 2000), (3) the penicillin producer *Penicillium chrysogenum* which has sulfur-regulated sulfate transporter genes (van

[1]Department of Biochemistry and Molecular Biology, Wright State University, Dayton, OH 45435, USA; e-mail: john.paietta@wright.edu

Biochemistry and Molecular Biology, 3rd Edition
The Mycota III
D. Hoffmeister (Ed.)

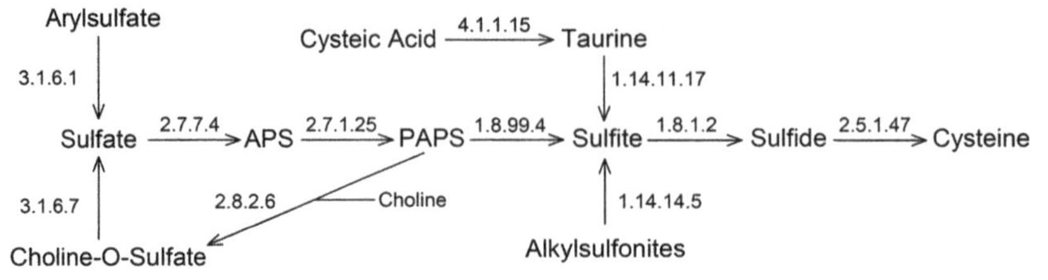

Fig. 12.1 The pathway in *Neurospora crassa* for sulfur acquisition routing to sulfur assimilation (Paietta 2010). Potential sources from the environment or internal stores (e.g., choline-*O*-sulfate) are shown. E.C. designations are given for each step. *APS* adenosine-5′-phosphosulfate, *PAPS* 3′-phosphoadenosine-5′-phosphosulfate

de Kamp et al. 2000), and (4) the dermatophytes *Trichophyton rubrum*, which has an extracellular protease released upon sulfur starvation (Apodaca and McKerrow 1989), and *Microsporum canis* in which sconC (a homologue of *scon-2*$^+$ in *N. crassa*) expression was upregulated and sulfur metabolism inhibited following fluconazole exposure (Uthman et al. 2005).

II. Acquisition of Sulfur

A. Sulfur Sources

Filamentous fungi are quite versatile in that they can accumulate sulfur atoms from a wide range of metabolically useful compounds. A major usage of the acquired sulfur is to transform it into a reduced state (as sulfide) which then can be used for the production of cysteine (Fig. 12.1). Transsulfuration allows the acquired sulfur in the form of cysteine to be routed through the intermediate cystathionine and then onto homocysteine and subsequently methionine (Fig. 12.2). Alternately, through reverse transsulfuration, homocysteine derived from methionine and the *S*-adenosylmethionine (AdoMet) cycle can be used to generate cysteine (Fig. 12.2).

In soil, inorganic sulfates, sulfate esters, sulfonates, and sulfur-containing amino acids are typically present but in variable proportions (e.g., as assayed by Autry and Fitzgerald 1990). A preferred and relatively abundant sulfur source for many fungi is inorganic sulfate which is transported into the cell and directly enters the **sulfur assimilation pathway** (Fig. 12.1). A variety of aliphatic and aromatic sulfate esters (e.g., choline-*O*-sulfate) can also directly provide sulfate using a variety of sulfatases (e.g., arylsulfatase, choline-*O*-sulfatase) (Marzluf 1997; Paietta 2004) (Fig. 12.1). Further, sulfite, sulfide, and thiosulfate can also be commonly used as sulfur sources by fungi (Paszewski et al. 2000; Thomas and Surdin-Kerjan 1997; Paietta 2004). Aliphatic sulfonates (e.g., ethanesulfonic acid, isethionic acid) can generate sulfite through the action of alkanesulfonate monooxygenases and are similarly used (Fig. 12.1). Cysteic acid and taurine provide another potential source of sulfite that can enter the sulfur assimilation pathway as shown (Fig. 12.1).

Exogenous protein can also serve as a direct source of cysteine and methionine for fungi. As will be discussed later, *N. crassa* produces an extracellular protease under sulfur-limiting conditions (Hanson and Marzluf 1975), and subsequent uptake of amino acids released from hydrolyzed protein can satisfy needed sulfur requirements. Dermatophytes, such as *Trichophyton rubrum*, also produce extracellular proteases in response to sulfur starvation (Apodaca and McKerrow 1989) and depend on the digested host proteins as a source of sulfur.

As might be predicted, there appears to be a variability in the range of alternative sulfur sources that can be used by different fungal species. There have been few methodical studies devoted to this topic, but an extensive survey of ascomycete yeasts for growth on sulfoxides, sulfones, sulfonates, sulfamates, and sulfate esters indicated a substantial range

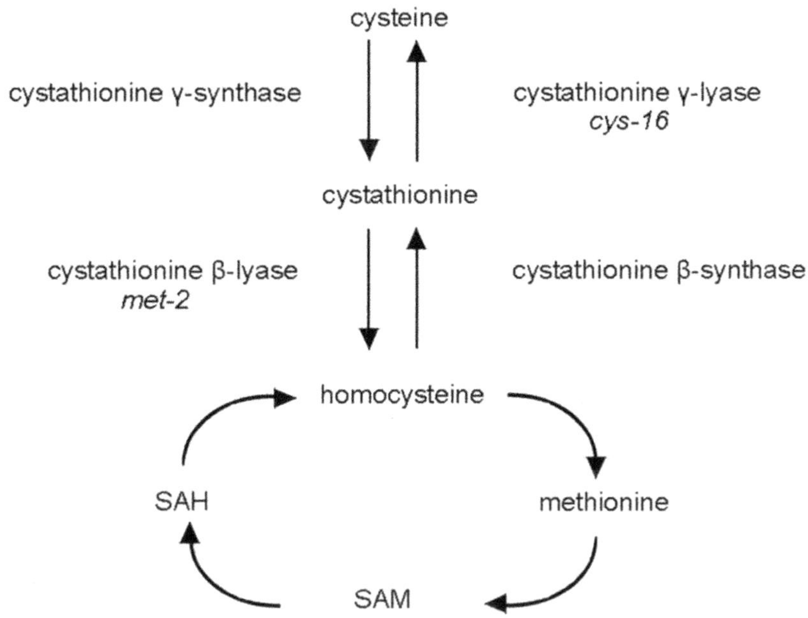

Fig. 12.2 Pathway of transsulfuration and methionine synthesis in *Neurospora crassa*. Two genes controlled by the sulfur regulatory system are shown designated as *cys-16* and *met-2* (Reveal and Paietta 2012, 2013). *SAM* *S*-adenosylmethionine (AdoMet), *SAH* *S*-adenosylhomocysteine

in capability to use different sulfur sources from highly versatile species (e.g., *Yarrowia lipolytica*) to those limited to using only sulfate esters (*Schizosaccharomyces pombe*) (Linder 2012). Genomic data also provides a means to make a preliminary assessment of sulfur compound utilization by identification of candidate genes as with *Neurospora crassa* (Borkovich et al. 2004).

B. Response to Sulfur Limitation

Neurospora crassa responds to growth under conditions of sulfur limitation (often termed **derepressing conditions**) through its sulfur regulatory system with the coordinate expression of a large set of genes related to sulfur acquisition and metabolism. The derepressed sulfur-related genes that have been studied the most thoroughly encode for a variety of proteins involved in the transport or hydrolysis of sulfur-containing compounds and include arylsulfatase, choline-*O*-sulfatase, choline sulfate permease, aromatic sulfate permease, methionine permease, sulfate permeases I and II, and an extracellular protease (reviewed by Marzluf 1997; Paietta 2004, 2010). Typically, derepressing conditions have been defined in a number of studies by growth on medium containing low levels of methionine as the sole sulfur source (i.e., 0.25 mM methionine). Conversely, repression of the entire set of genes is observed under high-sulfur (or sulfur-sufficient) conditions (i.e., 5.0 mM methionine). Since *Neurospora crassa* can use a variety of compounds as sulfur sources, the coordinated expression of a battery of such genes provides an efficient means of acquiring sulfur in various chemical forms from the environment.

Arylsulfatase (arylsulfate sulfohydrolase, EC 3.1.6.1) has been extensively studied in *Neurospora crassa*, and its regulation is representative of the set of proteins involved in obtaining environmental sulfur (Paietta 2004). Arylsulfatase expression is tightly regulated and appears upon derepression of the system following growth on low sulfur and depletion of internal sulfur stores. Figure 12.3a demonstrates the regulation of arylsulfatase production in wild-type

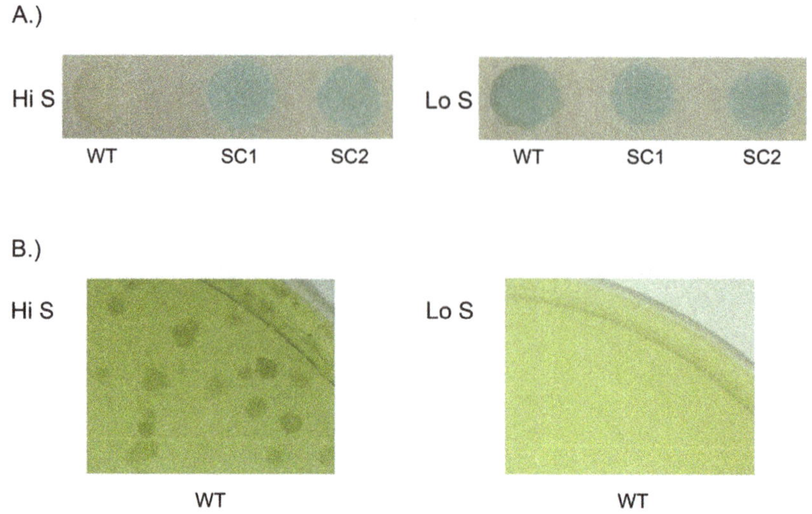

Fig. 12.3 Demonstration of the sulfur-regulated expression of arylsulfatase and sulfate permease in *Neurospora crassa*. (**a**) Wild type (WT), *scon-1* (SC1), and *scon-2* (SC2) were grown on either high sulfur (5 mM methionine) or low sulfur (0.25 mM methionine) and exposed to 10 mM X-sulfate for detection of arylsulfatase activity. Wild type shows a typical sulfur-regulated pattern of expression having arylsulfatase activity detectable under low-sulfur conditions but not high sulfur (i.e., colony remains colorless). The *scon-1* and *scon-2* negative regulatory mutants display constitutive expression of arylsulfatase under either sulfur growth conditions. (**b**) Wild type (WT) was plated on high-sulfur or low-sulfur (as above) medium followed by an overlay of 10 mM potassium chromate. If the sulfate permease is produced, then the chromate will be transported into the cells and no growth will result. Note that on high-sulfur medium, the sulfate permease is repressed and the cells grow normally, while on low-sulfur medium, the sulfate permease is derepressed and growth is completely inhibited by the chromate

and *scon* mutants (which produce arylsulfatase constitutively) plated on high or low sulfur and exposed to X-sulfate. Arylsulfatase activity is detectable in wild type only under sulfur limitation and constitutively in *scon* negative regulatory mutants (which reflects a loss of regulatory control, under either high or low sulfur) (Fig. 12.3a). Northern blot analysis and nuclear transcription analysis both indicate control of *ars-1*$^+$ is at the transcriptional level (Paietta 1989). Interestingly, arylsulfatase in *Neurospora crassa* has a complex localization in conidia and mycelia, with extracellular, intracellular, and bound activities detectable (Scott and Metzenberg 1970). The extracellular arylsulfatase can generate sulfate from a variety of organic compounds that naturally occur in the environment. Soil, for example, has a variety of accessible organic sulfur compounds (Autry and Fitzgerald 1990). The second crucial aspect to the acquisition of the arylsulfatase-liberated sulfate (i.e., extracellular) is the simultaneous production of sulfate permeases (transporters). *N. crassa* sulfate permeases, encoded by the *cys-13*$^+$ (primarily conidial expression) and *cys-14*$^+$ (primarily mycelial expression) genes, are similarly tightly regulated (Marzluf 1970). Figure 12.3b shows wild type grown on high and low sulfur in the presence of a toxic sulfate analog, chromate, which is transported into the cell by sulfate permeases. Under sulfur-sufficient conditions, sulfate permease production is repressed and the cells will grow normally in the presence of chromate, while low-sulfur conditions derepress sulfate permease production which results in no growth on chromate (Fig. 12.3b). Further, the production of an aromatic sulfate permease allows for direct uptake of ester sulfate compounds followed by hydrolysis by the intracellular arylsulfatase fraction. Acquired sulfate obtained by either route is primarily used through reduction to sulfide and subsequent biosynthesis of cysteine and methionine (Fig. 12.1).

Sulfur starvation in *N. crassa* also results in the derepression and secretion of an extracellular protease. Besides being under control of the sulfur regulatory system, the protease level is also derepressed under conditions of carbon and nitrogen starvation (Hanson and Marzluf 1975). To ensure efficient recovery of the free amino acids (e.g., methionine) resulting from hydrolyzed protein in the medium, the production of a methionine permease (Pall and Robertson 1988) is also coordinately derepressed. Taken together, the coordinate production of a battery of enzymes and transporters in *N. crassa* and other fungi allows for effective acquisition of needed sulfur (Paietta 2010).

Besides external sources of sulfur, fungi can hold reserves of sulfur in compounds such as choline-*O*-sulfate. Figure 12.1 shows how sulfur can be routed from PAPS (3′-phosphoadenosine-5′-phosphosulfate) using PAPS-choline sulfotransferase to generate choline-*O*-sulfate. Yet another coordinated derepression following sulfur limitation in *N. crassa* is the production of choline sulfatase. The choline sulfatase can act on the intracellular stores of choline-*O*-sulfate and release sulfate for entry into the sulfur assimilation pathway (Fig. 12.1) and help relieve a limiting level of sulfur. Additionally, *N. crassa* can take up exogenous choline-*O*-sulfate by a specific permease and either store it or obtain sulfate immediately using choline sulfatase. Specific examples of storage sulfur mobilization include the use of choline-*O*-sulfate during germination in *N. crassa* (McGuire and Marzluf 1974) and the upregulation of choline sulfatase expression in the mycelia to yeast transition in *Paracoccidioides brasiliensis* (Ferreira et al. 2006).

Figure 12.2 shows the central position of **cystathionine** in the transsulfuration (and reverse transsulfuration) pathway. Recent studies indicate that in *N. crassa* the available pool of cystathionine is accessed by derepressing both cystathionine β-lyase (which converts cystathionine to homocysteine) and cystathionine γ-lyase (which converts cystathionine to cysteine) when stressed by sulfur limitation (Reveal and Paietta 2012, 2013). Methionine synthase, which generates methionine from homocysteine, is also induced in *N. crassa* under the same sulfur-limiting culture conditions (Jensen and Paietta, unpublished data). These results provide a preliminary indication of the metabolic remodeling taking place under sulfur starvation. Thus, not only is there induction of the genes encoding the proteins needed to obtain sulfur from the environment, but components of pathways that will use the acquired sulfur in biosynthesis are similarly induced. A future direction will be to perform genome-wide microarray expression analysis to track the entire group of sulfur metabolism-related genes, most of which have been identified (Borkovich et al. 2004) following the sequencing of *N. crassa* and generate an overall model of the metabolic effect on cells of limiting sulfur:

Studies in yeast have revealed additional complexities to the effects of sulfur limitation. tRNA thiolation has been found to be down-regulated when sulfur is in short supply and allows for the organism to regulate translation based on sulfur availability (Laxman et al. 2013). In addition, Fauchon et al. (2002) found that there is a reduced production of sulfur-rich proteins, which are replaced by sulfur-depleted isozymes (particularly glycolytic enzymes), under sulfur limiting conditions. An overall sulfur amino acid savings of 30 % was extrapolated from the studies. An important part of the response was mediated through the action of the Met4p "master" regulator (a homolog of the *Neurospora crassa* CYS3 regulator and which controls a similar subset of genes) (Fauchon et al. 2002). In another study of sulfur-sparing at the proteome level, the work of Baudouin-Cornu et al. (2001) demonstrates that the sulfur metabolic pathway component proteins are sulfur depleted themselves. On an overall basis *Saccharomyces cerevisiae* may also contain significantly lower amounts of cysteine residues overall as compared to other eukaryotes (Scheibel et al. 1997). The availability of genomic data for numerous filamentous fungal species should allow for tests of how widespread this particular level of sulfur-related control occurs.

III. Detection of Sulfur Status

The more proximal steps in the sulfur regulatory system involving transcriptional activation, autoregulation, and regulated proteolysis have been defined (Paietta 2010). However, an important unresolved question is the nature of initial portion of the signal transduction pathway which regulates the response to sulfur limitation in fungi. The presumption is that a "sensor" protein detects the sulfur status of the cell and subsequently transmits a signal, perhaps a phosphorylation event, to a succeeding "partner" regulatory protein in the transduction pathway. In turn, the regulator (or chain of regulators) modifies, positively or negatively, the transcription of genes involved in the sulfur limitation response. Fungal cells, through this mechanism, would be able to detect changes in their internal or external environment and adjust to those changes by altering their gene expression profile and metabolism as discussed above.

What is the metabolic signal that links cellular sulfur status to gene expression? For filamentous fungi, there is limited indirect data from experiments in *Aspergillus nidulans* and *Neurospora crassa* that suggest cysteine or a related metabolite is the effector whose level is being monitored by the sulfur regulatory system in those organisms. Typically, limitation of methionine in the medium has been used as a means of derepressing the *Neurospora crassa* sulfur regulatory circuit. However, the experiments of Jacobson and Metzenberg (1977) suggest that the effector metabolite in the system is cysteine rather than methionine. In a *ser-l* mutant, high levels of methionine (5 mM) did not repress the system as typically observed in a wild-type strain. The *ser-l* strain, a leaky auxotroph, is starved for cysteine even if methionine is present. When the *ser-l* mutant was grown with supplemental serine, however, repression then occurred (i.e., with supplemented serine the mutant strain is no longer cysteine starved; Jacobson and Metzenberg 1977). A direct test using exogenous cysteine added to the medium was difficult because of the growth inhibitory effects of cysteine (Adiga et al. 1962; Jacobson and Metzenberg 1977). An approach around the cysteine toxicity problem uses cysteine derivatives which can be transported into cells which are then converted into cysteine. Compounds such as *N*-acetyl-L-cysteine and L-2-oxothiazolidine-4-carboxylate (Anderson and Meister 1987) function as an effective intracellular delivery system for L-cysteine and can be provided directly to the *N. crassa ser-l* mutant. Supplying cysteine in this alternate form results in repression of the sulfur regulatory system (Paietta, unpublished data) and strengthens the putative regulatory role of cysteine. In *Aspergillus nidulans*, cysteine (or glutathione) has been suggested as the likely effector (reviewed by Paszewski et al. 2000). In *A. nidulans*, for example, *cys* mutants blocked in the primary pathway of cysteine synthesis demonstrate derepression of a number of sulfur-related enzymes (e.g., in the sulfate assimilation pathway). In addition, starvation for cysteine in a *cysAmecB* strain also leads to derepression of the same group of enzymes:

In *Saccharomyces cerevisiae* the intracellular level of S-adenosylmethionine (AdoMet) has been proposed as the repression signal. Experiments with mutant strains which cannot synthesize AdoMet have been performed with growth in the presence of methionine. Such strains do not show repression of the system. The implication from the experiments is that intracellular AdoMet, rather than methionine is the signal for the repression of the MET genes (reviewed by Thomas and Surdin-Kerjan 1997). However, Hansen and Johannesen (2000) suggest a role for cysteine in the transcriptional regulation of sulfur assimilation genes in *S. cerevisiae*. The experiments carried out indicated a necessity for the formation of cysteine from homocysteine in order for methionine repression to take place for the MET14 and MET2S genes (Hansen and Johannesen 2000). A regulatory role is also indicated for cysteine in *S. cerevisiae* sulfur metabolism in studies of *O*-acetylserine and *O*-acetylhomoserine sulfhydrylase in mutant backgrounds (e.g., cystathionine γ-lyase defective; Ono et al. 1996, 1999). Recent studies with the yeast *Hansenula polymorpha* also suggest a central role for cysteine based on repression of a set of sulfur-related genes (Sohn et al. 2014).

Resolution of the identity of the metabolite whose level is monitored and connects the cell's sulfur status to the sulfur control system is an area crucial for future studies. An important set of eventual experiments will be to define the interactions of the effector molecule with the putative sulfur sensor.

A further unresolved question in fungi is the nature of the sulfur sensor. Relatively little is known about eukaryotic sulfur-nutrient sensors. Some data regarding a candidate sulfur sensor is available from studies on the effect of sulfur limitation in the green alga *Chlamydomonas reinhardtii*. In *C. reinhardtii*, several genes involved in the response to sulfur limitation have been studied (reviewed by Leustak et al. 2000). The *sac-l*, *sac-2*, and *sac-3* (sulfur acclimation) genes, however, do not seem to represent a parallel control system to that found in *A. nidulans*, *N. crassa*, and *S. cerevisiae*. *sac-1* encodes an integral membrane protein, and because of its substantial effect on the *Chlamydomonas*, sulfur regulatory system has been suggested as a regulator of the response (Davies et al. 1999). A close homologue of *sac-l* does not appear to exist in fungi (Paietta, unpublished data). Suggested roles, which may not be mutually exclusive, include the *sac-1* protein acting directly as a sensor of a sulfur metabolite that signals the intracellular sulfur level or by acting as transporter of a compound that signals the sulfur starvation response (Leustak et al. 2000). An additional regulatory component in *Chlamydomonas* is a SNRK2 (SNF1-related protein kinase 2) which is critical for sulfur-responsive gene expression (Gonzalez-Ballester et al. 2008).

However, fungal sulfur regulatory systems appear to be designed with a sulfur sensor whose output is relayed through a transduction pathway to mediate the changes in transcription observed when sulfur limitation occurs. For example, the action of the transcriptional activator protein CYS3 of *Neurospora crassa* appears to be controlled either directly by the hypothetical protein sensing the sulfur status of the cell or through an intermediate protein along the signaling pathway (Kumar and Paietta 1998). The CYS3 transcriptional activator itself does not appear to directly bind sulfur compounds (Paietta, unpublished data) with corresponding effects on promoter binding as observed in a number of sulfur-related bacterial regulatory proteins (Bykowski et al. 2002; Lochowska et al. 2001).

Among the filamentous fungi, the putative *sulfur controller-1* gene product of *N. crassa* appears be the best candidate for a sulfur sensor. The original mutant was isolated and characterized by Burton and Metzenberg (1972) and designated *scon*^{c}. Later, *scon*^{c} was designated as *scon-1* when other genes of the sulfur controller class were identified (Paietta 1990). The sulfur-related phenotype of *scon-1* is of constitutive expression of the sulfur structural genes (Burton and Metzenberg 1972). *scon-1* is also epistatic to other known elements in the sulfur regulatory system (Paietta 1990) and has an unusual "nuclear-limited" effect as demonstrated in heterokaryon studies (Burton and Metzenberg 1972). The "nuclear-limited" effect was that in a heterokaryon containing wild-type nuclei (i.e., *scon-1*$^+$) and *scon*-1 nuclei then the wild-type nuclei show normal sulfur regulation, while the *scon*-1 nuclei show constitutive sulfur gene expression as monitored by levels of electrophoretic variants of arylsulfatase that "represented" each nuclear type (Burton and Metzenberg 1972). Work is in progress to clone the unusual *scon-1* gene, which may lead to an understanding of how the cellular sulfur status is monitored and in what form the signal is transduced (Paietta, unpublished data).

IV. *Neurospora crassa* Sulfur Regulatory System

A. CYS3 Regulator

An important component of the *N. crassa* sulfur regulatory system involves the positive regulatory gene, *cys-3*$^+$ (Paietta et al. 1987; Paietta 1992, 1995; Marzluf 1997). *cys-3* mutants show a loss of expression of the entire set of sulfur-regulated enzymes (e.g., arylsulfatase, cystathionine γ-lyase) (see Fig. 12.4) and are sulfur auxotrophs. cys-3 mutants are also affected in a range of cellular properties such as nucleotide pools, energy charge, and ascospore viability (Paietta et al. 1987; Pall and Robertson 1988). The *cys-3*$^+$ gene is highly regulated and expression is induced under conditions of sulfur limitation (Paietta et al. 1987). The encoded CYS3 regulator is a basic region-leucine zipper (bZIP) DNAbinding protein that is functional as a homodimer in vivo and whose level is subject to autoregulation. Marzluf (1994, 1997) has reviewed earlier work on CYS3 in detail. More recently, Coulter and Marzluf (1998)

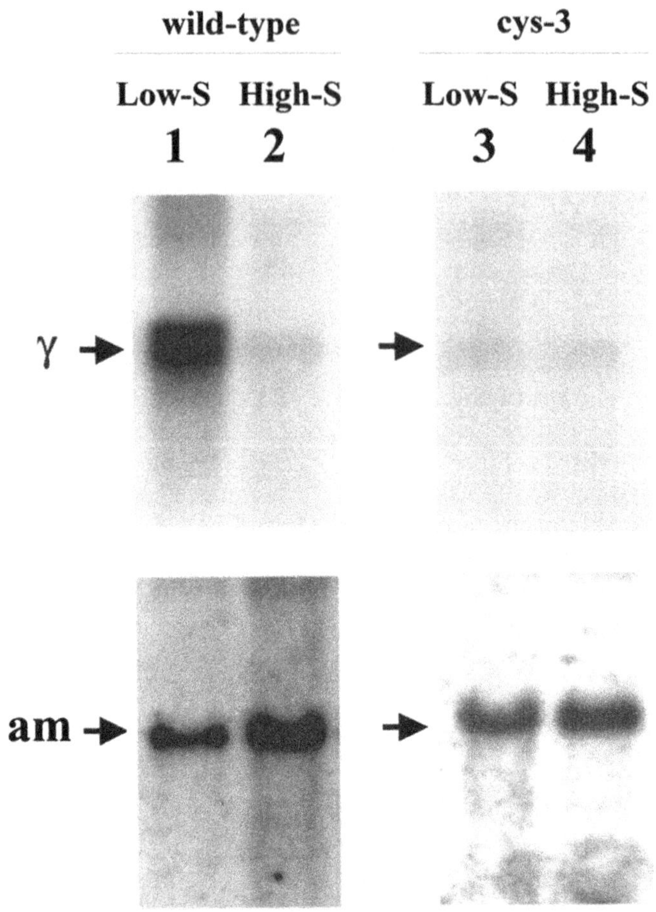

Fig. 12.4 Northern blot analysis showing the typical expression pattern for a gene under control of the *Neurospora crassa* sulfur regulatory system. Expression pattern of *cys-16*$^+$ (cystathionine γ-lyase) is shown (Reveal and Paietta 2012). Note the high level of *cys-16*$^+$ expression in wild type under low-sulfur conditions (*left panel*, band designated as γ). As expected in a mutant (Δ*cys-3*) with a deletion of the sulfur system transcriptional activator (*right panel*, displayed as cys-3), the induction does not occur. The *am* gene (glutamate dehydrogenase, constitutively expressed) serves as a control

have demonstrated that an amino-terminal serine/threonine-rich region acts as an activation domain. In addition, *cys-3*$^+$ transcripts and CYS3 protein appear to turn over rapidly under conditions of sulfur repression (Tao and Marzluf 1998). A key factor by which the sulfur regulatory system exerts its ultimate effect on gene expression is how the action and level of the CYS3 transcriptional activator is controlled (Kumar and Paietta 1998).

The coordinated expression of the entire set of sulfur-related genes is based on the binding of the CYS3 transcriptional activator to a promoter sequence that has been termed the **sulfur response element** or SuRE (Paietta 2000). A number of genes under CYS3-mediated sulfur-regulated control have been cloned and characterized with respect to regulation and the binding of CYS3 to promoter elements. The cystathionine γ-lyase gene (*cys-16*$^+$) shows a typical expression profile for genes under control of this system with high transcript levels only under sulfur-limited conditions and no expression in a *cys-3* mutant background (see Fig. 12.4; Reveal and Paietta 2012). The cystathionine γ-lyase gene also has a typical distribution of CYS3 binding sites within its promoter region as shown by gel mobility-shift analysis (see Fig. 12.5; Reveal and Paietta 2012). Shuler (1993) proposed a consensus-binding sequence

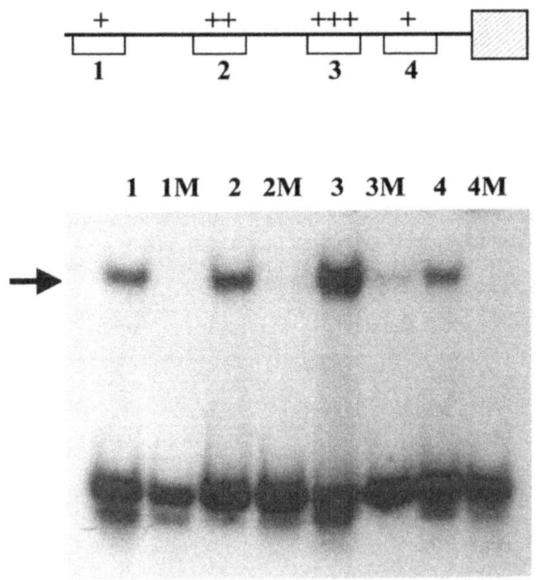

Fig. 12.5 Gel mobility-shift analysis of CYS3 binding to the *cys-16*$^{+}$ promoter in *Neurospora crassa*. *Top*, schematic of the *cys-16*$^{+}$ promoter showing the affinity of CYS3 for each binding site (*triple plus symbol*, strongest affinity; *plus symbol*, weakest affinity). *Bottom*, the *arrow* indicates DNA fragments exhibiting reduced electrophoretic mobility due to interaction with CYS3 for each corresponding site on the promoter (labeled 1, 2, 3, and 4). The specificity of the CYS3 interaction is shown by mutating a single nucleotide in the binding site which abolishes or reduces the shifted band (shown as 1, 2, 3, and 4 M) (Reveal and Paietta 2012)

of ATNPurPyrPurPyrCAT based on DNase I footprinting studies of the *ars-1*$^{+}$ (arylsulfatase) and *cys-3*$^{+}$ genes, while Li and Marzluf (1996) proposed a consensus binding site of ATGPurPyrPurPyrCAT based on mutational studies of several CYS3 binding sites within the *cys-14*$^{+}$(sulfate permease) and *cys-3*$^{+}$ promoters. Paietta (2008) used an approach of binding-site selection from random-sequence oligonucleotides to define a consensus sequence of ATGGCGCCAT for CYS3 binding. The CYS3 binding-site data will be of particular use for future genome-wide studies of *N. crassa* sulfur-regulated genes. Interestingly, the CYS3 binding site appears to be quite different, for example, compared to the binding site established (TCACGTG) for the MET4 bZIP protein which regulates sulfur gene expression in *Saccharomyces cerevisiae* (Thomas and Surdin-Kerjan 1997).

B. Sulfur Controller Regulators

Burton and Metzenberg (1972) used the designation "**sulfur controller**" for the original *scon*^{c} mutant (now renamed *scon-l*) that was characterized in *N. crassa*. *sulfur controller* mutants display a constitutive derepression of the entire set of sulfur-regulated enzymes (e.g., arylsulfatase, sulfate permease) and consequently defined as negative regulators in the sulfur regulatory system. To date, there are three genes that have been identified and given the "sulfur controller" designation: *scon-l* (*scon*^{c}) (Burton and Metzenberg 1972), *scon-2* (Paietta 1990; Kumar and Paietta 1998), and *scon-3* (Sizemore and Paietta 2002). *scon* is also used to designate a similar series of genes in *A. nidulans* (i.e., *sconA*, *sconB*, *sconC*, and *sconD*; Natorff et al. 1993; Paszewski et al. 2000). In *N. crassa* the constitutive *scon* phenotype provided a general approach for cloning the *scon* group of genes by transformation and restoration of wild-type regulation (Paietta 1990). The selection for transformants relies on the finding that chromate, a toxic sulfate analog (Holland and Avery 2011), is transported by sulfate permease into the cell (Marzluf 1970). Thus, a wild-type strain (or a *scon-2* strain transformed to *scon-2*$^{+}$) will be resistant to chromate when grown in high sulfur as a result of repression of sulfate permease. Because the *scon* mutants are not repressible for sulfate permease (or other sulfur enzymes), they are chromate sensitive under high-sulfur conditions. *scon*$^{+}$ transformants will have regained wild-type control of sulfate permease, with the permease now subject to sulfur repression. For example, *scon-2*$^{+}$ was cloned using this approach (Paietta 1990). The SCON2 and SCON3 proteins play key roles in the regulation of the system and are discussed in more detail below with regard to the F-box domain. The putative *scon-1*$^{+}$ gene product, as discussed in Section III, above, is the candidate sulfur sensor in the control circuit.

When the SCON2 protein of *N. crassa* was characterized, a strongly conserved domain aminoterminal to a WD-40 repeat region was discovered (Kumar and Paietta 1995). The "N-terminal domain" of SCON2 was characterized by several strongly conserved charged residues

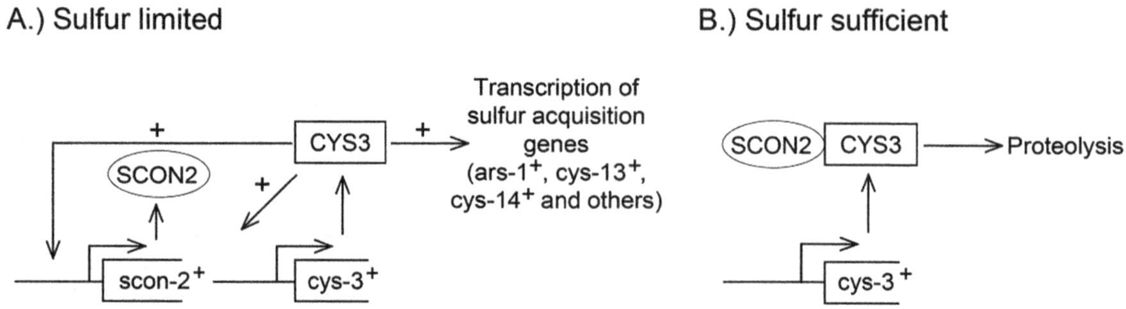

Fig. 12.6 Operation of the *Neurospora crassa* sulfur regulatory system showing primary regulatory interactions. (**a**) Sulfur-limited conditions. The CYS3 activator binds to the promoters of *cys-3*$^{+}$ (autoregulation), *scon-2*$^{+}$ (encodes F-box protein), as well as the large set of genes encoding sulfur-related enzymes and transporters. *Plus symbol* represents a positive effect. (**b**) Sulfur-sufficient conditions. In this case, binding of SCON2 to CYS3 and subsequent proteolysis represent the action of the SCF^{SCON2} complex which prevents CYS3-directed transcriptional activation (Paietta 2010)

(Asp, Glu, Lys, Arg) and an invariant Leu-Pro dipeptide. Later, Bai et al. (1996), working on yeast cell cycle regulation, designated the homologous domain the F-box (due to its presence within cyclin F). They proposed a model of control by regulated ubiquitin-mediated proteolysis with the Skp1p protein playing a key role by binding the F-box motif. Fbox proteins themselves were seen as adaptor proteins between the target protein to be regulated and the machinery necessary for proteolysis. The discovery of the F-box and its functional roles has led to substantial refinements in models for the regulation of sulfur metabolism (Kipreos and Pagano 2000; Jonkers and Rep 2009).

F-box motifs are found in a large number of regulatory proteins (Craig and Tyers 1999; Kipreos and Pagano 2000). In yeast, Fbox proteins have been shown to assemble with Skplp, Cdc53p, and Rbxlp to form a complex known as the SCF (Skplp/Cdc53p/F-box; Skowyra et al. 1997; Zheng et al. 2002). SCF complexes act as E3 ubiquitin ligases to target proteins for ubiquitin-mediated proteolysis. The F-box protein component is responsible for the target specificity of the SCF complex, with the F-box motif responsible for the interaction with Skplp and connection to the E2 component. The SCF complex, therefore, directs the ligation of ubiquitin to a particular targeted protein with subsequent proteasome destruction. In yeast, there are a number of major metabolic pathways that are regulated by SCF complexes which target either transcriptional activators or repressors for ubiquitination and degradation (Patton et al. 1998). For example, SCF^{Grrl} controls glucose induction and SCF^{Met30} regulates methionine repression.

The F-box domain of SCON2 has been demonstrated to be required for the role of SCON2 as a regulator within the *N. crassa* sulfur control system (Kumar and Paietta 1998). The ability of SCON2 to function as a negative regulator of sulfur-related gene expression appears due to the ability of SCON2 to form a functional SCF complex via protein-protein interactions between the F-box motif of SCON2 and the *N. crassa* homologue of yeast Skplp (which has been designated SCON3; Sizemore and Paietta 2002). A protein-protein interaction between SCON2 and SCON3 was confirmed using a two-hybrid approach (Sizemore and Paietta 2002). Presumably, an SCF^{SCON2} complex is regulating the level of the CYS3 transcriptional activator (Fig. 12.6).

C. Operation of the Control System

Overall, the *N. crassa* sulfur regulatory system monitors cellular sulfur status and expresses a set of sulfur-related genes under sulfur limitation to ensure an adequate internal supply of sulfur while subjecting the sulfur-related genes to repression under conditions of sulfur sufficiency (see Fig. 12.6). The *N. crassa* system can be outlined as follows: (1) the sulfur sensor, which is currently unidentified, but may involve the putative *scon-1*$^{+}$ "nuclear-limited" gene product, (2) the bZIP transcriptional activator CYS3 which autoregulates its own expression and binds to the promoter of the *scon-2*$^{+}$ gene and to the entire set of sulfur-regulated genes (Paietta 2004, 2010), and (3) the F-box protein *sulfur controller-2* (SCON2) (also containing a WD-40 repeat and subject to CYS3 transcriptional regulation) and is involved in the SCF^{SCON2}-directed control of CYS3 by targeted degradation (Kumar and Paietta 1998). The SCF complex also involves the Skp1p homologue SCON3 (among other needed components).

Presumably, the hypothetical *scon-1*$^{+}$ gene product acts as the sulfur sensor. Whether "SCON1" acts directly or transmits a signal to an intermediary protein (or proteins) is not known. Further, at this point, we do not know what the end result of the signal pathway involves on a molecular basis. A potential end result of the signal for cellular sulfur status would be to change the phosphorylation state of the CYS3 protein. A hypothetical model, based on the regulation seen in other SCF-controlled systems, would have the CYS3 protein existing in two states. For example, under sulfur-sufficient conditions, the CYS3 protein might be phosphorylated, and this modification would render binding by the SCON2 F-box protein possible. By this model CYS3 is a dual target of both the signal transduction chain and, subsequently, of the F-box receptor protein. When SCON2 binds CYS3, there would be an immediate sequestration of CYS3 so transcriptional activation is blocked. Subsequently, SCON2 (with bound CYS3) interacts with SCON3 to form the SCF^{SCON2} complex which targets CYS3 for degradation. The double tier of SCON2 regulation by sequestration and targeted degradation of CYS3 would effectively block transcriptional activation of the sulfur-related genes and is supported by SCON2 mutagenesis experiments (Kumar and Paietta 1998). Further, the CYS3 autoregulatory loop that is in operation under sulfur-limiting conditions (Paietta 1992) to upregulate *cys-3*$^{+}$ expression would be blocked.

In contrast, under sulfur-limiting conditions, CYS3 would be in an alternate state (perhaps dephosphorylated) that would not be favorable for binding to SCON2. SCON2 would not block transcriptional activation by CYS3 nor would there be targeting of CYS3 to the SCF^{SCON2} complex for degradation. The CYS3 autoregulatory loop would upregulate *cys-3*$^{+}$ expression as well as allow for sufficient levels of CYS3 to coordinately induce the expression of the entire set of sulfur-related genes. As the cells transition to sulfur sufficiency, CYS3 would be modified and subjected to the control mechanism outlined above:

A final note on the model is that SCON2 is present at its highest level when the system is fully derepressed (Paietta 1990), due to a feedback loop in which CYS3 activates *scon-2*$^{+}$ gene expression (Kumar and Paietta 1995). As a consequence, the system is poised for immediate repression by SCON2 when a state of cellular sufficiency is signaled. Although the model is speculative in some aspects, there are a number of resulting predictions that can be subjected to experimental testing.

V. Regulatory Comparison to *Aspergillus nidulans* and *Saccharomyces cerevisiae*

The *Aspergillus nidulans* sulfur regulatory system resembles that of *Neurospora crassa* in a number of respects, but there are also important differences. In *A. nidulans* MetR is a bZIP transcriptional activator (Natorff et al. 2003; Brzywczy et al. 2011) that corresponds to *N. crassa* CYS3 and *S. cerevisiae* Met4p. The most substantial homology between METR and CYS3 is in the DNA binding and dimerization domain (i.e., the bZIP region). Since a chimeric metR gene carrying a *cys-3* bZIP domain is functional in *A. nidulans*, this suggests that the promoter binding sites are likely similar (Natorff et al. 2003). However, in contrast to the tightly regulated expression of *cys-3*$^{+}$ where extremely low transcriptional levels are detectable under high-sulfur levels (Paietta et al. 1987), the *metR* transcript is apparently not regulated by sulfur source and not subject to autoregulation (Natorff et al. 2003). A seemingly fundamental difference in the respective control mechanisms used to regulate *metR* and *cys-3*$^{+}$ is unresolved at this point.

The *sconB* and *sconC* genes of *A. nidulans* have been cloned and characterized. *sconB* encodes a homologue of *N. crassa scon-2*$^{+}$ that is functionally interchangeable (Natorff et al. 1998). The *scon-2*$^{+}$ gene, as with *cys-3*$^{+}$, is tightly regulated and its transcript is present at low levels under repressing conditions (Paietta 1990). In contrast, the *A. nidulans sconB* transcript increases about twofold under sulfur limitation and is only slightly reduced in level under sulfur repression (Natorff et al. 1998). *sconC* encodes a Skp1 family protein that shares a high degree of homology to SCON3. For both *sconC* and *scon-3*$^{+}$, expression can be observed under sulfur-limiting or sulfur-sufficient conditions (Piotrowska et al. 2000; Sizemore and Paietta 2002).

At this point, a comparison of the extensively analyzed *S. cerevisiae* control system (reviewed by Thomas and Surdin-Kerjan 1997) to the *N. crassa* system is useful. As with *A. nidulans*, the yeast system demonstrates some similarities to *N. crassa* as well as some significant differences. Met4p, which contains transcriptional activation and bZIP domains, is probably most similar to CYS3, but the homology is weak and primarily confined to the leucine zipper. A second bZIP protein has been identified in yeast and designated Met28p; however, this protein lacks a transcriptional activation domain (Kuras et al. 1996). The yeast system appears to represent a case of combinatorial control by the activator Met4p, along with Met31p, Met32p, and Cbf1p (DNA-binding factors), along with a cofactor Met28p and a F-box protein Met30p (Carrillo et al. 2012; McIsaac et al. 2012). There are no data currently suggesting this level of complexity in either *N. crassa* or *A. nidulans* for this aspect of the sulfur regulatory system.

When yeast Met30p was characterized (Thomas et al. 1995), it turned out to be homologous to SCON2. Both proteins contain F-box and WD-40 domains. In contrast, deleting MET30 revealed that it was an essential gene (Thomas et al. 1995), while in *N. crassa* deletion of the *scon-2*$^{+}$ results in constitutive sulfur gene expression (Kumar and Paietta 1998). A substantial amount of data supports the involvement of Met30p and SCON2 in a SCF complex. SCON3 appears to play a similar role to Skp1p in the SCF control mechanism (Sizemore and Paietta 2002). Finally, there is currently no candidate gene in yeast or *A. nidulans* that appears equivalent to *scon-1*$^{+}$ of *N. crassa*.

VI. Conclusions

Studies in model organisms like *A. nidulans* and *N. crassa* have revealed significant insights into the regulation of sulfur metabolism in filamentous fungi. The *N. crassa* sulfur regulatory system can be divided up into three main areas: (1) the hypothetical detection of the effector metabolite by the sulfur sensor followed by transduction of a signal indicating the cellular sulfur status, (2) the transcriptional activation of the sulfur-regulated genes (including autoregulated and feedback-regulated genes), and (3) ultimate control of the transcriptional activation by the SCF complex through regulated proteolysis. Progress to date has been substantial concerning the more proximal steps needed for coordinate control of genes derepressed or repressed depending on available sulfur levels. However, major unanswered questions remain regarding the mechanism involved in monitoring the cellular sulfur level and the early steps in transmitting that signal. For *N. crassa*, the characterization and study of the *scon-1*$^{+}$ gene product, the putative sulfur sensor candidate, is a key research goal. An additional needed future research priority is to combine the traditional strengths of *Neurospora crassa* as an experimental system (Davis 2000; Davis and Perkins 2002) with microarray approaches to give us a system-level view of the entire regulatory system and components in operation.

References

Adiga PR, Sastry KS, Sarma PS (1962) Amino acid relationships in cysteine toxicity in *Neurospora crassa*. Gen Microbiol 29:149–155

Amich J, Schafferer L, Haas H, Krappmann S (2013) Regulation of sulphur assimilation is essential for virulence and affects iron homeostasis of the human-pathogenic mould *Aspergillus fumigatus*. PLoS Pathog 9, e1003573

Anderson ME, Meister A (1987) Intracellular delivery of cysteine. Methods Enzymol 143:313–325

Apodaca G, McKerrow JH (1989) Regulation of *Trichophyton rubrum* proteolytic activity. Infect Immun 57:3081–3090

Autry AR, Fitzgerald JW (1990) Sulfonate S: a major form of forest soil organic sulfur. Biol Fertil Soils 10:50–56

Bai C, Sen P, Hofmann K, Ma L, Goebl M, Harper JW, Elledge SJ (1996) SKP 1 connects cell cycle regulators to the ubiquitin proteolysis machinery through a novel motif, the F-box. Cell 86:263–274

Baudouin-Cornu P, Surdin-Kerjan Y, Marliere P, Thomas D (2001) Molecular evolution of protein atomic composition. Science 293:297–300

Borkovich KA, Alex LA, Yarden O, Freitag M, Turner GE, Read ND, Seiler S, Bell-Pederson D, Paietta JV, Plesofsky N, Plamann M, Goodrich-Tanrikulu M, Schulte U, Mannhaupt G, Nargang FE, Radford A, Selitrennikoff C, Galagan JE, Dunlap JC, Loros JJ, Catcheside D, Inoue H, Aramayo R, Polymenis M, Selker EU, Sachs MS, Marzluf GA, Paulsen I, Davis R, Ebbole DJ, Zelter A, Kalkman ER, O'Rourke R, Bowring F, Yeadon J, Ishii C, Suzuki K, Sakai W, Pratt R (2004) Lessons from the genome sequence of *Neurospora crassa*: tracing the path from genomic blueprint to multicellular organism. Microbiol Mol Biol Rev 68:1–108

Brzywczy J, Kacprzak MM, Paszewski A (2011) Novel mutations reveal two important regions in *Aspergillus nidulans* transcriptional activator MetR. Fungal Genet Biol 48:104–112

Burton EG, Metzenberg RL (1972) Novel mutation causing derepression of several enzymes of sulfur metabolism of several enzymes of sulfur metabolism in *Neurospora crassa*. J Bacteriol 109:140–151

Bykowski T, van der Ploeg JR, Iwanicka-Nowicka R, Hryniewicz MM (2002) The switch from inorganic sulphur assimilation in *Escherichia coli*: adenosine 5'-phosphosulphate (APS) as a signaling molecule for sulphate excess. Mol Microbiol 43:1347–1358

Carrillo E, Ben-Ari G, Wildenhain J, Tyers M, Grammentz D, Lee TA (2012) Characterizing the roles of Met31 and Met32 in coordinating Met4-activated transcription in the absence of Met30. Mol Biol Cell 23:1928–1942

Coulter KR, Marzluf GA (1998) Functional analysis of different regions of the positive-acting CYS3 regulatory protein of *Neurospora crassa*. Curr Genet 33:395–405

Craig KL, Tyers M (1999) The F-Box: a new motif for ubiquitin dependent proteolysis in cell cycle regulation and signal transduction. Prog Biophys Mol Biol 72:299–328

Davies JP, Yildiz FH, Grossman AR (1999) SACl, a putative regulator that is critical for survival of *Chlamydomonas reinhardtii* during sulfur deprivation. EMBO J 15:2150–2159

Davis RH (2000) *Neurospora*. Contributions of a model organism. Oxford Press, Oxford

Davis RH, Perkins DD (2002) *Neurospora*: a model of model microbes. Nat Rev Genet 3:397–403

Fauchon M, Lagniel G, Aude J-C, Lombardis L, Soularue P, Petat C, Marguerie G, Senenac A, Werner M, Labarre J (2002) Sulfur sparing in the yeast proteome in response to sulfur demand. Mol Cell 9:713–723

Ferreira MED, Marques ED, Malavazi I, Torres I, Restrepo A, Numes LR, deOliveira RC, Goldman MHS, Goldman GH (2006) Transcriptome analysis and molecular studies on sulfur metabolism in the human pathogenic fungus *Paracoccidioides brasiliensis*. Mol Gen Genomics 276:450–463

Gonzalez-Ballester D, Pollock SV, Pootakham W, Grossman AR (2008) The central role of a SNRK2 kinase in sulfur deprivation responses. Plant Physiol 147:216–227

Goodwin PH, Li J, Jin SM (2000) Evidence for sulfate derepression of an arylsulfatase gene of *Colletotrichum gloeosporioides* f. sp *malvae* during infection of round-leaved mallow, *Malva pusilla*. Physiol Mol Plant Pathol 57:169–176

Hansen J, Johannesen PF (2000) Cysteine is essential for transcriptional regulation of the sulfur assimilation genes in *Saccharomyces cerevisiae*. Mol Gen Genet 263:535–542

Hanson MA, Marzluf GA (1975) Control of the synthesis of a single enzyme by multiple regulatory circuits in *Neurospora crassa*. Proc Natl Acad Sci USA 72:1240–1244

Holland SL, Avery SV (2011) Chromate toxicity and the role of sulfur. Metallomics 3:1119–1123

Jacobson ES, Metzenberg RL (1977) Control of arylsulfatase in a serine auxotroph of *Neurospora*. J Bacteriol 130:1397–1398

Jonkers W, Rep M (2009) Lessons from fungal F-box proteins. Eukaryot Cell 8:677–695

Kipreos ET, Pagano M (2000) The F-box protein family. Genome Biol 1:3002.1–3002.7

Kumar A, Paietta JV (1995) The sulfur controller-2-negative regulatory gene of *Neurospora crassa* encodes a protein with β-transducin repeats. Proc Natl Acad Sci U S A 92:3343–3347

Kumar A, Paietta JV (1998) An additional role for the F-box motif: gene regulation within the *Neurospora crassa* sulfur control network. Proc Natl Acad Sci USA 95:2417–2422

Kuras L, Cherest H, Surdin-Kerjan Y, Thomas D (1996) A heteromeric complex containing the centromere binding factor 1 and two basic leucine zipper factors, met4 and me28, mediates the transcription activation of yeast sulfur metabolism. EMBO J 15:2519–2529

Laxman S, Sutter BM, Wu X, Kumar S, Guo X, Trudgian DC, Mirzaei H, Tu BP (2013) Sulfur amino acids regulated translational capacity and metabolic homeostasis through modulation of tRNA thiolation. Cell 154:416–429

Leustak T, Martin MN, Bick J-A, Davies JP (2000) Pathways and regulation of sulfur metabolism revealed through molecular and genetic studies. Annu Rev Plant Physiol Plant Mol Biol 51:141–146

Li Q, Marzluf GA (1996) Determination of the *Neurospora crassa* CYS3 sulfur regulatory protein consensus DNA-binding site: amino acid substitutions in the CYS3 bZIP domain that alter DNA-binding specificity. Curr Genet 30:298–304

Linder T (2012) Genomics of alternative sulfur utilization in ascomycetous yeasts. Microbiology 158:2585–2597

Lochowska A, Iwanicka-Nowicka R, Plochocka D, Hryniewica MM (2001) Functional dissection of the LysR-type CysB transcriptional regulator. Regions important for DNA binding, inducer response, oligomerization, and positive control. J Biol Chem 276:2098–2107

Marzluf GA (1970) Genetic and metabolic controls for sulfate metabolism in *Neurospora crassa*: isolation and study of chromate-resistant and sulfate transport negative mutants. J Bacteriol 102:716–721

Marzluf GA (1994) Genetics and molecular genetics of sulfur assimilation in the fungi. Adv Genet 31:187–206

Marzluf GA (1997) Molecular genetics of sulfur assimilation in filamentous fungi and yeast. Annu Rev Microbiol 51:73–96

McGuire WG, Marzluf GA (1974) Developmental regulation of choline sulfatase and arylsulfatase in *Neurospora crassa*. Arch Biochim Biophys 161:360–368

McIsaac RS, Petti AA, Bussemaker HJ, Botstein D (2012) Perturbation-based analysis and modeling of combinatorial regulation in the yeast sulfur assimilation pathway. Mol Biol Cell 23:2993–3007

Natorff R, Balinska M, Paszewski A (1993) At least four regulatory genes control sulphur metabolite repression in *Aspergillus nidulans*. Mol Gen Genet 238:185–192

Natorff R, Piotrowska M, Paszewski A (1998) The *Aspergillus nidulans* sulphur regulatory gene sconB encodes a protein with WD40 repeats and an F-box. Mol Gen Genet 257:255–263

Natorff R, Sienko M, Brzywczy J, Paszewski A (2003) The *Aspergillus nidulans metR* gene encodes a bZIP protein which activates transcription of sulphur metabolism genes. Mol Microbiol 49:1081–1094

Ono B, Kijima K, Ishii N, Kawato T, Matsuda A, Paszewski A, Shinoda S (1996) Regulation of sulphate assimilation in *Saccharomyces cerevisiae*. Yeast 12:1153–1162

Ono B, Hazu T, Yoshida S, Kawato T, Shinoda S, Brzvwczy J, Paszewski A (1999) Cysteine biosynthesis in *Saccharomyces cerevisiae*: a new outlook on pathway and regulation. Yeast 15:1365–1375

Paietta JV (1989) Molecular cloning and regulatory analysis of the arylsulfatase structural gene of *Neurospora crassa*. Mol Cell Biol 9:3630–3637

Paietta JV (1990) Molecular cloning and analysis of the *scon-2* negative regulatory gene of *Neurospora crassa*. Mol Cell Biol 10:5207–5214

Paietta JV (1992) Production of the CYS3 regulator, a bZIP DNA-binding protein is sufficient to induce sulfur gene expression in *Neurospora crassa*. Mol Cell Biol 12:1568–1577

Paietta JV (1995) Analysis of CYS3 regulator function in *Neurospora crassa* by modification of leucine zipper dimerization specificity. Nucleic Acids Res 23:1044–1049

Paietta JV (2000) Regulation of sulfur metabolism in *Neurospora crassa*. Fungal Genet Newslett 47S:8

Paietta JV (2004) Regulation of sulfur metabolism in mycelial fungi. In: Esser K (ed) The Mycota III. Biochemistry and molecular biology (2nd edn). [Brambl R, Marzluf GA (eds)]. Springer, Berlin, pp. 369–383

Paietta JV (2008) DNA-binding specificity of the CYS3 transcription factor of *Neurospora crassa* defined by binding-site selection. Fungal Genet Biol 45:1166–1171

Paietta JV (2010) Sulfur, phosphorus and iron metabolism. In: Borkovich KA, Ebbole DJ (eds) Cellular and molecular biology of filamentous fungi. ASM Press, Washington, DC, pp 359–375

Paietta JV, Akins RA, Lambowitz AM, Marzluf GA (1987) Molecular cloning and characterization of the *cys-3* regulatory gene of *Neurospora crassa*. Mol Cell Biol 7:2506–2511

Pall ML, Robertson CK (1988) Growth-regulation by GTP: regulation of nucleotide pools in *Neurospora* by nitrogen and sulfur control systems. J Biol Chem 263:11168–11174

Paszewski A, Natorff R, Piotrowska M, Brywczy J, Sienko M, Grynberg M, Pizzinini E, Turner G (2000) Regulation of sulfur amino acid biosynthesis in *Aspergillus nidulans*: physiological and genetic aspects. In: Brunhold C, Rennenberg H, De Kok LJ, Stulen I, Davidian J-C (eds) Sulfur nutrition and sulfur assimilation in higher plants. Paul Haupt, Bern, pp 93–105

Patton EE, Willems AR, Tyers M (1998) Combinatorial control in ubiquitin-dependent proteolysis: don't skp the F-box hypothesis. Trends Genet 14:236–243

Piotrowska M, Natorff R, Paszewski A (2000) sconC, a gene involved in the regulation of sulphur metabolism in *Aspergillus nidulans*, belongs to the *SKP1* gene family. Mol Gen Genet 264:276–282

Reveal BS, Paietta JV (2012) Analysis of the sulfur-regulated control of the cystathionine γ-lyase gene of *Neurospora crassa*. BMC Res Notes 5:339

Reveal BS, Paietta JV (2013) Sulfur-regulated control of the met-2+ gene of *Neurospora crassa* encoding cystathionine β-lyase. BMC Res Notes 6:259

Scheibel T, Bell S, Walke S (1997) *S. cerevisiae* and sulfur: a unique way to deal with the environment. FASEB J 11:917–921

Scott WA, Metzenberg RL (1970) Location of arylsulfatase in conidia and young mycelia of *Neurospora crassa*. J Bacteriol 104:1254–1265

Shuler JL (1993) Analysis of the DNA binding of the *Neurospora crassa* CYS3 sulfur regulatory protein to the *ars-1*$^{+}$ and *cys-3*$^{+}$ promoters and determination of the consensus binding site sequence. PhD Thesis. Wright State University, Dayton

Sizemore ST, Paietta JV (2002) Cloning and characterization of *scon-3*$^{+}$: a new member of the *Neurospora crassa* sulfur regulatory system. Eukaryot Cell 1:875–883

Skowyra D, Craig KL, Tyers M, Elledge SJ, Harper JW (1997) F-box proteins are receptors that recruit phosphorylated substrates to the SCF ubiquitin-ligase complex. Cell 91:209–219

Sohn MJ, Yoo SJ, Oh DB, Kwon O, Lee SY, Sibirny AA, Kang HA (2014) Novel cysteine-centered sulfur metabolic pathway in the thermotolerant methylotrophic yeast *Hansenula polymorpha*. PLoS One 24, e100725

Tao Y, Marzluf GA (1998) Synthesis and differential turnover of the CYS3 regulatory protein of *Neurospora crassa* are subject to sulfur control. J Bacteriol 180:478–482

Thomas D, Surdin-Kerjan Y (1997) Metabolism of sulfur amino acids in *Saccharomyces cerevisiae*. Microbiol Mol Biol Rev 61:503–532

Thomas D, Kuras L, Barbey R, Cherest H, Blaiseau P-L, Surdin-Kerjan Y (1995) Met30p, a yeast transcriptional inhibitor that responds to *S*-adenosylmethionine, is an essential protein with WD40 repeats. Mol Cell Biol 15:6526–6534

Uthman A, Dockal M, Soltz-Szots J, Tschachler E (2005) Fluconazole upregulates sconc expression and inhibits sulphur metabolism in *Microsporum canis*. Fungal Genet Biol 42:719–725

Van de Kamp M, Schuurs TA, Vos A, van der Lende TR, Konings WN, Driessen AJM (2000) Sulfur regulation of the sulfate transporter genes sutA and sutB in *Penicillium chrysogenum*. Appl Environ Microbiol 66:4536–4538

Zheng N, Schulman BA, Song L, Miller JJ, Jeffrey PD, Wang P, Chu C, Koepp DM, Elledge SM, Pagano M, Conaway RC, Conaway JW, Harper JW, Pavletich NP (2002) Structure of the Cul l-Rbx l-Skp l-F box^{skp2} SCF ubiquitin ligase complex. Nature 416:703–709

13 The Regulation of Carbon Metabolism in Filamentous Fungi

JOAN M. KELLY[1]

CONTENTS

I. Introduction

The unicellular and filamentous fungi grow in a wide range of environmental niches, and are able to use a diverse range of carbon sources to provide energy and metabolic intermediates. Faced with intense competition for resources, microorganisms have been selected to be metabolically efficient, and in most species, simple sugars such as glucose or sucrose are used in preference to more complex carbon sources, and the regulatory mechanisms required for achieving this in filamentous fungi form the focus of this chapter. The unicellular fungus *Saccharomyces cerevisiae* has been considered a model eukaryotic cell in gene regulation studies, and the regulation of carbon metabolism in this organism has been reviewed by others (Carlson 1999; Schuller 2003) and will only briefly be considered here for comparative purposes. Key aspects in the regulatory mechanisms, including environment sensing, transport, and carbon catabolite repression, are quite different between *S. cerevisiae* and the filamentous fungi that have been studied and indeed between filamentous fungi.

Carbon catabolite repression refers to the mechanism whereby energetically favourable repressing carbon sources are used preferentially to less readily metabolised carbon sources due to the reduced synthesis of enzymes for the utilisation of the latter in the presence of the former, and is the focus of this chapter. The best studied cases of carbon catabolite repression at the genetical, biochemical and molecular levels are in *Escherichia coli*, where, at the level of the gene promoter, there is a sophisticated molecular understanding of the mechanism, and of its interaction with the pathway-specific induction mechanism for several operons, in particular that for lactose utilisation, although new information is constantly emerging (see Wilson et al. 2007, for review). However, there is neither reason a priori nor evidence a posteriori to suggest that the mechanism should be conserved between prokaryotes and eukaryotes. In eukaryotes, work aimed at eliciting the mechanism of carbon catabolite repression has been undertaken in the yeast *S. cerevisiae* and in some filamentous fungi. In yeast, a

[1]Department of Genetics and Evolution, The University of Adelaide, Adelaide 5005, Australia; e-mail: joan.kelly@adelaide.edu.au

Biochemistry and Molecular Biology, 3rd Edition
The Mycota III
D. Hoffmeister (Ed.)

large number of genes have been implicated in the repression and derepressing mechanisms (see Gancedo 1998; Klein et al. 1998, for review). These genes have been identified by mutations that cause either an altered response in carbon catabolite repressing conditions or an altered response to the relief of carbon catabolite repression. *S. cerevisiae* is strongly adapted for single celled growth in sugar-rich environments, and thus studies in yeast will not reveal the complex mechanisms of carbon catabolite repression in the filamentous fungi, many of which can metabolise a wide range of complex carbon sources. Many filamentous fungi, including the model organisms for genetic analysis, *Aspergillus nidulans* and *Neurospora crassa*, are microbial eukaryotes that can grow on completely defined medium which facilitates studies of metabolism, and there are a number of very well-documented examples in the literature where the regulation of the production of an enzyme, or of all the enzymes in a particular biochemical pathway, has been studied in detail at the genetical, biochemical and molecular levels. In most cases, in addition to revealing gene- or pathway-specific controls, such as induction and repression, these studies have shown global regulatory controls operating at the transcriptional level.

Initially, genetic analysis was undertaken in these model fungi in order to identify genes that, when mutated, perturbe aspects of sensing, transport, induction and repression and inferences were made as to the role of the gene products in the regulatory mechanism. Molecular techniques allowed the identification and functional analysis of the roles of the gene products, and further forward and reverse genetics was used to test new hypotheses as they emerged. Genome-wide analysis, first at the genome level, and subsequently at the transcriptome, proteome, metabolome and network interactome levels, has led to a greater understanding and generated yet more hypotheses to be tested. Although much is known, this chapter underscores that we are still some way from a full understanding of the regulation of carbon metabolism in filamentous fungi.

II. Glucose Transport and Sensing

Glucose transport and sensing in *S. cerevisiae* have recently been reviewed (Kim et al. 2013), and a family of glucose transporters are involved. The Rgt2/Snf3 glucose induction pathway, the Snf1/Mig1 glucose repression pathway, and the cAMP-activated kinase pathway combine to allow expression of these transporters dependent on extracellular glucose concentration. In filamentous fungi, compared to *S. cerevisiae*, a much larger number of high- and low-affinity hexose transporters are present in the genomes, many with overlapping specificities (reviewed in Katz and Kelly 2010), and the functions of some of these have been analysed via complementation of yeast mutants. There are many examples that emphasise the added complexities due to the multicellular nature of filamentous fungi. In *Uromyces fabae*, Hxt1p is expressed only in the haustoria (Voegele et al. 2001), whereas in *A. nidulans* HxtA is expressed during sexual development (Wei et al. 2004). Also in *A. nidulans*, two high-affinity glucose transporter proteins, SorA and MstC, which show 90 % amino acid identity and apparently have the same function are expressed at different times and in different tissues; *sorA* is expressed early in conidial germination and is subject to carbon catabolite repression, whereas *mstC* is expressed in older mycelia in carbon starvation conditions (Forment et al. 2014).

In *A. nidulans*, mutations in *frA1*, which encodes a catalytic hexokinase, were selected due to their toxicity on D-fructose medium (McCullough et al. 1977), and mutations in *glkA*, which encodes a glucokinase, were selected due to their resistance to 2-deoxy glucose in glycerol medium (Flipphi et al. 2003). Mutations in the individual hexose phosphorylation genes do not affect growth on glucose, nor do they lead to carbon catabolite derepression (Arst et al. 1990; Roberts 1963; Ruijter et al. 1996); however, strains containing both *glkA* and *frA* mutations, which completely lack both glucose and fructose phosphorylation, show derepression, indicating that GlkA and FrA activities compensate for each other in carbon catabolite repression signalling (Flipphi et al. 2003). Filamentous fungi also contain atypical hexokinases which have a regulatory but not a catalytic role and are involved in carbon starvation (Bernardo et al. 2007; Katz et al. 1996, 2000, 2006).

III. Induction and Repression

Most fungi have evolved to use the most metabolically efficient carbon source when presented with a range of carbon sources, and there are a number of very well-documented examples in the literature where the regulation of the production of an enzyme, or of all the enzymes in a particular biochemical pathway, has been studied in detail at the genetical, biochemical and molecular levels. In most cases, in addition to revealing global regulatory controls, these studies have shown gene- or pathway-specific controls, such as induction operating primarily at the transcriptional level. Studies of the regulation of expression of enzymes that are involved only in carbon catabolism, and thus are subject only to specific induction controls and carbon catabolite repression, have been of great value as in these systems other global regulators do not interfere with carbon catabolite repression. These studies have been at the level of growth, enzyme activity and amount and synthesis of mRNA, and only two examples in model fungi are included here.

Induction involves the action of pathway-specific regulatory proteins. The induction of the alcohol and aldehyde dehydrogenases involved in ethanol utilisation in *A. nidulans*, *alcA* and *aldA*, has been studied in detail (Bailey and Arst 1975; Arst and Bailey 1977; Pateman et al. 1983; Lockington et al. 1985; Kulmberg et al. 1992a, b, 1993; Nikolaev et al. 1999; Flipphi et al. 2001; reviewed by Felenbok et al. 2001). The genes encoding alcohol dehydrogenase I (*alcA*) and aldehyde dehydrogenase (*aldA*), and the transcriptional regulator required for induction AlcR (*alcR*), were identified by mutation, and studies showed that induction is achieved via the intracellular accumulation of the physiological inducer, acetaldehyde. Expression depends on the binding of AlcR to a number of *cis*-acting elements 5′ of the *alcA*, *aldA* and *alcR* genes. This system provides such tight regulation that it has been used in a number of heterologous systems to drive gene expression, including in plants. The induction of the enzymes required for growth on quinate has also been studied in both *N. crassa* and *A. nidulans* (Bailey and Arst 1975; Giles et al. 1985; Geever et al. 1989; Lamb et al. 1996; Levett et al. 2000). In *N. crassa*, *qa-1F* mutants showed a recessive noninducible phenotype, indicative of a positive regulatory protein, while *qa-1S* mutants showed a recessive constitutive phenotype, indicative of a repressor. The molecular mechanism was unraveled in *A. nidulans*, where the QutR repressor protein interacts with the QutA activator protein to regulate transcription of the quinic acid utilisation (*qut*) gene cluster. In the presence of quinate, production of mRNA from the eight genes of the *qut* pathway is stimulated by the QutA activator. The N-terminal 88 amino acids of the QutR repressor binds to a region of QutA implicated in transcription activation, and QutA and QutR are in dynamic equilibrium between bound (transcriptionally inactive) and unbound (transcriptionally active) states. Both the *alc* and *qut* regulons are also subject to the global regulatory mechanism of carbon catabolite repression, as discussed later.

Other systems that are regulated by induction provide both a carbon and a nitrogen source to the cell, and studies of the regulation of these pathways are valuable in that they potentially provide an opportunity to unravel the interaction between carbon catabolite repression and ammonium repression. In *A. nidulans*, these include the study of the regulation of acetamidase (Hynes and Davis 2004) and the enzymes for L-proline catabolism (Scazzocchio et al. 1995). The *amdS* gene encoding acetamidase is induced by acetate via two positively acting regulatory proteins, FacB and AmdA, and by omega amino acids via a third positively acting regulatory protein, AmdR. The binding sites for each activator were initially identified via the isolation of *cis*-acting mutations in *amdS* which specifically affect the interaction with the regulatory protein. The enzymes needed for proline metabolism are induced by the presence of an inducer, and this has been shown to be due to transcriptional activation via the PrnA pathway specific regulatory mechanism. Advances in genome-wide transcriptome analysis have shed deeper light on the transcriptional effects of growth in the presence of particular inducing carbon sources, and these studies are referred to later in the chapter.

IV. Carbon Catabolite Repression

Genetic dissection of carbon catabolite repression, involving selection or construction of mutations that result in inappropriate repression or derepression, has been undertaken in a range of filamentous fungi.

A. CreA

1. *A. nidulans* creA Mutations

The CreA repressor protein is the primary DNA-binding regulatory protein involved in carbon catabolite repression in filamentous fungi. It was first identified in *A. nidulans* selection screens for mutations that lead to derepression. Initially, alleles were uncovered in screens involving the selection of suppressors of *areA* loss of function phenotypes (Arst and Cove 1973). The AreA DNA-binding protein is required for ammonium derepression (Wong et al. 2008 for review). The regulation of nitrogen metabolism is reviewed in Chap. 11, and only aspects relevant to the selection of mutants affected in carbon regulation are explored here. Loss of function *areA* alleles lead to an inability to grow on a range of nitrogen sources in glucose media due to the failure of these strains to express a range of genes subject to ammonium repression in the absence of ammonium. Enzymes involved in metabolising compounds that provide the cell with both a carbon and a nitrogen source are regulated by both carbon catabolite repression and ammonium repression, and the relief of either repression mechanism leads to expression (Hynes 1970; Arst and MacDonald 1975). Thus, *areA* loss of function strains can grow in medium containing compounds that provide both carbon and nitrogen in medium lacking D-glucose, but not as the only source of nitrogen in glucose medium. This observation was exploited in the selection of mutations where the enzymes for L-proline or acetamide metabolism are no longer sensitive to carbon catabolite repression, as these result in the phenotypic suppression of the *areA* lack of function mutations on D-glucose media (Arst and Cove 1973; Arst and Bailey 1977; Hynes and Kelly 1977; Arst 1981). Another screen for mutations resulting in derepression was the selection of suppressors of *pycA* and *pdhA* loss of function phenotypes. Lack of function mutations in the pyruvate carboxylase and pyruvate dehydrogenase encoding genes lead to the inability of the strain to produce acetyl CoA from pyruvate, and thus these strains require an alternative source of acetyl CoA (Romano and Kornberg 1968, 1969). This requirement can be met by ethanol or acetamide when these are present as sole carbon sources, but not in the presence of D-glucose (Bailey and Arst 1975). Mutations in the *creA* gene were present among phenotypic suppressors of *pycA* or *pdhA* mutant strains selected on D-glucose media containing ethanol or acetamide (Bailey and Arst 1975; Arst and Bailey 1977). The *frA1* mutation leads to toxicity on D-fructose medium (McCullough et al. 1977), and *frA* encodes a hexokinase, but absence of this activity alone does not interfere with glucose repression (Ruijter et al. 1996). Mutations in *creA* were present among phenotypic suppressors of *fr*A1 (Arst et al. 1990).

Mutations in *creA* result in various degrees of deregulated expression of a wide range of genes that would normally be repressed in the presence of glucose, and the various mutant alleles studied show no clear hierarchy for this derepression (for examples see Arst and Cove 1973; Arst 1981; Scazzocchio et al. 1995; Shroff et al. 1996, 1997). Enzyme assay data and transcript analysis indicate that, in addition to their effects of derepression, many mutations in *creA* lead to elevated levels of gene expression in both carbon catabolite repressing and derepressing conditions, indicating roles for CreA both in growth conditions generally regarded as repressing and derepressing. However, the possibility that the effects in derepressing conditions are due to the failure to identify an experimental condition that is truly derepressing without leading to starvation cannot be ruled out, and studies on the regulation of the cellular responses to carbon starvation will help to clarify this point. Although a wide range of enzymes show a degree of deregulated expression in *creA* mutant strains, not all enzymes that are subject to carbon catabolite repression are affected by the absence of *creA*, indicating the presence of CreA independent mechanisms of carbon regulation. Examples of these include extracellular lipase (Kawasaki et al. 1995) and penicillin biosynthetic enzymes (Espeso and Penalva 1994; Martin et al. 1999) in *A. nidulans*, and D-xylulose kinase (van Kuyk et al. 2001) in *A. niger*. Mutations in *creA* also affect colony morphology on complete medium with extreme alleles, such as *creA*^{d}*30*, *creA306* and *creA*Δ leading to a very small and compact morphology (Arst et al. 1990; Shroff et al. 1997), and also to altered sensitivity to toxic compounds, such as acriflavine added to complete medium.

2. CreA Mutant Phenotypes in Other Fungi

The identification of CreA as a major repressor protein in *A. nidulans* led to the isolation or construction of *creA* mutations in a range of filamentous fungi, particularly in industrially useful but less genetically amenable fungi, and shared and overlapping phenotypes are also apparent in these fungi.

In *T. reesei*, a number of studies have shown that industrial strains selected for high cellulolytic activity contain *cre1* mutations. Rut-C30, which produced cellulase mRNAs on glucose-containing medium, was shown to be a *cre1* truncation mutant (Ilmen et al. 1996), and deletion of phosphoglucose isomerase in a Rut-C30 background further increased cellulase expression in repressing conditions over that seen in a deletion in a wild-type strain (Limon et al. 2011). Further analysis of production strains showed that strain PC-3-7 contained a mutation in *cre1*, which was shown to be responsible for increased cellulase production (Porciuncula et al. 2013). An *Acremonium chrysogenum* strain with enhanced production of cephalosporin C was also shown to have lower levels of *cre1* gene expression (Jekosch and Kueck 2000a, b).

Mutations have been selected in *A. niger* using a similar technique to that used in *A. nidulans*, and these mutants were tested and showed partially deregulated expression of arabinases and enzymes of L-arabinose catabolism in glucose medium (Ruijter et al. 1997).

Mutations have been constructed using reverse genetics in a range of fungi. In *Aspergillus oryzae*, a *creA* deletion showed higher α-amylase activity in high concentrations of sugars (Ichinose et al. 2014). In *Penicillium canescens*, *creA* mutants showed derepressed β-galactosidase and endo-1,4-β-xylanase (Chulkin et al. 2011), and in *N. crassa*, inactivation of the *cre-1* gene resulted in reduced growth rate, abnormal hyphal morphology and altered carbon catabolite repression (Ziv et al. 2008).

In pathogenic fungi, the replacement of *CRE1* in *Fusarium oxysporum* resulted in derepressed expression of cell wall-degrading enzymes, and Cre1 controls the repression/derepression state of cell wall-degrading enzymes, and also affects pathogenicity and growth (Jonkers and Rep 2009). Deletion of *Beauveria bassiana BbcreA* resulted in pleiotropic effects, including derepression of secreted protease and lipase, which are enzymes critical in mediating pathogenesis, although insect bioassays indicated reduced virulence, implying that BbcreA functions in nutrient utilisation and virulence, as well as carbon catabolite repression (Luo et al. 2014).

3. Molecular Analysis of CreA in *A. nidulans*

CreA contains two DNA-binding zinc fingers of the Cys_2-His_2 class, an alanine rich region, and frequent S(T)PXX motifs which are commonly present in regulatory proteins (Dowzer and Kelly 1989, 1991). The zinc-finger region shows strong amino acid sequence similarity to the zinc finger DNA-binding regions of the yeast Mig1p repressor. Northern analysis revealed that the transcript is autoregulated (Arst et al. 1990; Dowzer and Kelly 1991; Shroff et al. 1996; Strauss et al. 1999). The CreA protein has subsequently been identified in a number of fungi, including, in order of sequence similarity, *A. oryzae* (EMBL AOR272151), *A. niger* (EMBL ANCREA; Drysdale et al. 1993), *Aspergillus aculeatus* (EMBL AB024314), *Gibberella fujikuroi* (EMBL GFY16626); Tudzynski et al. 2000), *Sclerotinia sclerotiorum* (EMBL SSCRES; Vautard et al. 1999), *Botrytis cinerea* (EMBL BCY16625; Tudzynski et al. 2000), *Acremonium chrysogenum* (EMBL ACH245727; Jekosch and Kueck 2000a, b), *Humicola grisea* (EMBL AB003106; Takashima et al. 1998), *T. reesei* (EMBL TR27356; Strauss et al. 1995; Ilmen et al. 1996), *Cochliobolus carbonum* (EMBL AF306571), *Metarhizium anisopliae* (EMBL MACRR1; Screen et al. 1997), *Neurospora crassa* (EMBL AF055464; de la Serna et al. 1999), and *Trichoderma harzianum* (EMBL THCRE1; Ilmen et al. 1996). Similar, but more distantly related proteins are encoded in the genomes of *Schizosaccharomyces pombe* (EMBL SPBC1D7), *Candida albicans* (EMBL CAL238242), *Kluyveromyces lactis* (EMBL KLDNAMIG1), *S. cerevisiae* (EMBL SCMIG1), *Kluyveromyces marxianus* (EMBL KMDNA-MIG1) and *Magnaporthe grisea* (EMBL BM865739), but there is no evidence that this latter group are functional homologues that can complement the phenotype of *A. nidulans creA* mutations.

Conserved domains and regions were identified using comparative analysis of the CreA sequence of 14 CreA amino acid sequences to identify conserved physiochemical profiles (Shroff 1997). Two of these regions are shared with Mig1p in both sequence and position, and theoretical and functional analyses of the Mig1p protein of *S. cerevisiae* have shown that these two domains, the N-terminal zinc finger DNA-binding domain and the C-terminal Mig1-effector domain, are required for repression (Ostling et al. 1996, 1998). Two other conserved regions identified in Mig1p, the R1 and R2 regulatory elements which inhibit Mig1p in the absence of glucose (Ostling et al. 1996), are not identifiable at the level of sequence conservation in CreA, but the sequence contains an acidic region containing potential serine/threonine phosphorylation sites, and two "conserved regions". A region of 42 amino acids shows extremely high sequence conservation between all of the CreA proteins identified to date, and it shares limited sequence similarity with a region of Rgr1p from yeast (Dowzer and Kelly 1991; Sakai et al. 1988, 1990). The sequence similarity with Rgr1p is unlikely to be significant since a construct containing a hybrid *creA* gene in which the *creA* Rgr1-similar region was replaced in-frame by the Rgr1p similar region from *S. cerevisiae* failed to complement the *A. nidulans creA204* or *creA20* mutations (Shroff 1997). However, the conservation of the sequence in this region between *creA* homologues is likely to be significant. There is a conserved region immediately N-terminal to the Rgr1-similar region that has a concentration of acidic residues.

The molecular nature of 19 in vivo generated mutations in *creA* has been determined (Arst et al. 1990; Shroff et al. 1996, 1997), and the predicted effects on the CreA protein fall into two clear classes. Seven mutant alleles contain a missense mutation in the zinc-finger region of the protein, and these mutations affect amino acids that are conserved among zinc-finger proteins of this class, and are predicted to alter (e.g. *creA204*) or prevent (e.g. *creA306*) binding to the *cis*-acting regulatory regions of genes that are regulated by CreA. Twelve mutant alleles are nonsense or frameshift mutations that result in a truncated protein, with *creA303* leading to the shortest peptide due to truncation at amino acid 68, and *creA322* having the longest peptide due to truncation at amino acid 336, and the fact that these mutations lead to different phenotypes implies that the truncated polypeptides are produced and in some cases retain some activity. Other than those in the DNA-binding domain, there were no *creA* missense mutations that lead to a derepressed phenotype, indicating that the repression domain probably acts as a region, and that alterations to individual amino acids do not lead to derepression. Thus, the in vivo isolated *A. nidulans* mutations that lead to derepression either alter the DNA-binding region or result in the absence of the effector domain due to truncation, and affect the exact regions that are conserved in sequence and position with regions in Mig1p (Ostling et al. 1996, 1998), and thus the functions of the conserved sequences are also conserved (Shroff et al. 1996, 1997).

An *A. nidulans* strain that is null for CreA was constructed using gene replacement techniques (Shroff et al. 1997), and the phenotype of this strain was similar to that of strains containing the mutations leading to the most severely truncated proteins, *creA303*, *creA304*, *creA305* and *creA30*, that is, the strain showed extreme morphological alteration on complete medium and a high level of derepression for a range of enzymes. However, a strain containing the *creA306* mutation had a more extreme phenotype than the null strain, in that the effect on colony morphology was more extreme, and the degree of derepression was higher based on phenotypic analysis and enzyme activity determinations. The *creA306* mutation is an A to T transition which results in a leucine to proline substitution at amino acid 109 within the recognition α-helix of the second zinc finger thus disrupting the highly conserved hydrophobic core which acts to stabilise the finger. Thus, the *creA306* mutation is the only mutation that is predicted to produce full length protein with no affinity for DNA, and the very extreme phenotype may be due to titration of proteins that interact with CreA.

4. Functional Analysis of CreA

The crystal structure of the Zif268–oligonucleotide interaction has provided a theoretical framework for the Cys_2-His_2 zinc finger class of regulators, and identified the important amino acid positions for interaction with DNA (Pavletich and Pabo 1991). CreA has been shown to bind DNA targets with a recognition sequence 5′ SYGGRG 3′, and the core recognition sequence is usually present in pairs (Kulmberg et al. 1993; Cubero and Scazzocchio 1994; Panozzo et al. 1998; Mathieu et al. 2000). This target recognition sequence is the same core recognition sequence that is bound by Mig1p of *S. cerevisiae* (Nehlin and Ronne 1990; Nehlin et al. 1991). Modeling studies on the interaction of CreA with an oligonucleotide sequence of 5′ GCGGGGGGCGT 3′ produced findings that are consistent with the interpretation that CreA interacts with DNA in a similar way to Zif268 (Paulino et al. 2002). Whilst core recognition sequences are required for binding, the mere presence of core sequences does not necessarily indicate functional in vivo binding, and potential recognition sequences need to be confirmed experimentally. For example, in the *alcA* and *alcR* promoters only a subset of potential consensus binding sites are functional, indicating the importance of residues outside the core binding consensus sequence for functionality (Panozzo et al. 1998; Mathieu et al. 2000).

For some systems, such as *alcA* encoding alcohol dehydrogenase I and *alcR* encoding a positively acting regulator, a "double lock" mechanism is in place, where both the pathway-specific regulatory gene and the enzyme-encoding gene are regulated directly by CreA binding (Kulmberg et al. 1993; Mathieu and Felenbok 1994; Panozzo et al. 1998; Mathieu et al. 2000). In other cases, such as for *aldA* encoding aldehyde dehydrogenase and *alcR* encoding a positively acting regulator, the carbon catabolite repression of *aldA* is achieved solely via the repression of *alcR*, and *aldA* is not directly regulated by CreA (Flipphi et al. 2001), and this may also be the case in other systems where the role of CreA is indirect, such as for *xlnA* and *xlnB* (Orejas et al. 1999, 2001). In the case of the L-proline utilisation regulon, only *prnB*, which encodes the permease, is directly regulated by CreA, and the deregulation of *prnD* and *prnC* in *creA* mutant strains is likely to be due to the reversal of inducer exclusion (Cubero et al. 2000). It is apparent that there are a number of possible mechanisms for CreA to effect repression once it is recruited to the promoter; in some cases a direct competition for overlapping DNA-binding sites with a pathway-specific regulatory protein can completely explain the repression, but in other cases a competition model cannot explain repression and some repression by CreA, either directly or by recruiting other proteins to the complex at the promoter, must be postulated.

5. Comparison Between Yeast and *A. nidulans*, and Within Filamentous Fungi

Although the DNA-binding zinc-finger regions of CreA and Mig1p are highly conserved, the mechanism of glucose repression in *S. cerevisiae* and *A. nidulans* is quite different. In yeast, Mig1p recruits the corepressor complex, Ssn6p–Tup1p, to promoters that are under carbon catabolite repression control (Nehlin et al. 1991; Treitel and Carlson 1995; Tzamarias and Struhl 1995), and Mig1p is translocated between the nucleus and cytoplasm depending on its phosphorylation by Snf1 kinase (De Vit and Johnston 1999; De Vit et al. 1997). However, glucose still regulates Mig1p-dependent repression in a mutant that no longer has nuclear export activity, indicating that Mig1p is regulated both by nuclear localisation and by modification of its ability to repress transcription (De Vit and Johnston 1999). Further, the Ssn6p–Tup1p complex is tethered to the promoter DNA of *GAL1* under both repressing and activating conditions, and Mig1p is not required for the Ssn6p–Tup1p complex to be tethered to the *GAL1* promoter (Papamichos-Chronakis et al. 2002). Although Mig1p is not essential for the binding of the Ssn6p–Tup1p complex to the promoter, it is important for Ssn6p–Tup1p-mediated glucose repression (Papamichos-Chronakis et al. 2004). Snf1p-dependent phosphorylation of Mig1p abolishes the interaction with Ssn6p–Tup1p and controls transcriptional repression or derepression, and it is the regulation of this interaction, not the Mig1p cytoplasmic localisation, that is the molecular switch that controls transcriptional repression and derepression of *GAL1* (Papamichos-Chronakis et al. 2002).

In *A. nidulans*, strains lacking the Tup1p homologue, RcoA, are not greatly affected for carbon catabolite repression (Hicks et al. 2001; Garcia et al. 2008), and it is unlikely that CreA acts by recruiting RcoA as part of a general repressor complex. For some promoters, deletion of *rcoA* leads to disruption of the chromatin default structure and repression occurs in the absence of nucleosome positioning, whereas for other promoters repression occurs with a nucleosome pattern identical to that found under conditions of full expression, whereas at a third class of promoter, deletion of *rcoA* leads to a novel chromatin pattern. Deletion of the putative Ssn6p homologue is lethal in *A. nidulans* (Garcia et al. 2008).

Carbon catabolite repression is, therefore, different between yeast and filamentous fungi, and Mig1p and CreA are not interchangeable. Even among filamentous fungi, where those CreA homologues that have been tested confer at least partial function when expressed heterologously, it is apparent that aspects of the molecular mechanisms are different. Functional regulation of CreA/Cre1 in filamentous fungi is not yet fully understood. In *A. nidulans*, the transcript of *creA* is present in higher amounts in strains grown in relatively derepressing carbon sources, such as arabinose than in strains grown in repressing levels of glucose (1 %) (Arst et al. 1990) which seems anomalous for a carbon catabolite repressor encoding gene. However, this seems to be because the *creA* gene is autoregulated, and the higher amounts seen in arabinose grown wild-type strains are also found in glucose grown *creA* mutant strains (Arst et al. 1990), and consistent with this, a number of CreA-binding consensus sequences are present upstream of the *creA* gene of *A. nidulans* (Shroff et al. 1996). Strauss and coworkers showed that there was a rapid transient increase in *creA* transcript amounts on the addition of either glucose or a nonrepressing monosaccharide to carbon starved mycelium, but that the high levels were subsequently downregulated on repressing carbon sources (Strauss et al. 1999). These findings indicate that steady state *creA* transcript levels do not directly correlate with the amount of CreA-mediated repression, and thus translational control or posttranslational modification and/or stability are likely to be important components of CreA function. Roy and colleagues showed that the autoregulated transcription of *creA* was not essential for repression or derepression, as a strain containing *creA* expressed from a constitutive promoter was phenotypically wild type (Roy et al. 2008).

Transcriptional autoregulation of *creA/cre1* was also present in *S. sclerotiorum*, *Trichoderma reesei*, *Hypocrea jecorina* and *P. canescens* (Ilmen et al. 1996; Strauss et al. 1999; Vautard-Mey and Fevre 2000; Chulkin et al. 2010). In contrast, expression of *G. fujikuroi* and *B. cinerea creA* was demonstrated to be continuously high in the presence of all carbon sources tested (Tudzynski et al. 2000); however, the transcript-abundance pattern of the *creA/cre1* gene did not correlate with the repressing activity of the protein (Strauss et al. 1999; Vautard-Mey et al. 1999; Tudzynski et al. 2000). In *Acremonium chrysogenum*, glucose dependent transcriptional expression patterns of the *cre1* gene were quite different to the gene expression patterns described in *T. reesei* and *A. nidulans* where the genes are autoregulated (Arst et al. 1990; Shroff et al. 1996; Ilmen et al. 1996; Strauss et al. 1999; Jekosch and Kueck 2000a, b). Further, this glucose-dependent transcriptional upregulation was absent from a strain of *A. chrysogenum*, which displayed enhanced production of the beta-lactam antibiotic cephalosporin C, consistent with the interpretation that the deregulation of *cre1* is connected with the increased production rate in this strain (Jekosch and Kueck 2000a, b). Studies of the *cre1* transcript in *S. sclerotiorum* also apparently revealed transcript levels that are positively correlated with the exocellular glucose concentration, being higher in glucose medium compared with glycerol or pectin media (cited in Vautard-Mey et al. 1999).

Thus, a posttranslational mechanism is almost certainly involved in the regulation of Cre-mediated repression activity. In fact, CreA/Cre1 has been implied to be undergoing several modes of posttranslational modification, including phosphorylation (Vautard-Mey and Fevre 2000; Cziferszsky et al. 2002) and ubiquitination (Lockington and Kelly 2001; Kamlangdee, 2008; Boase and Kelly 2004), which were found to affect transcription factor abundance and activity. The role of phosphorylation in the repressor activity of Cre1 has been investigated in *Hypocrea jecorina* (*T. reesei*; Cre1) (Cziferszsky et al. 2002, 2003). Phosphorylation of the Ser^{241} residue of *T. reesei* Cre1 was found to be required for DNA binding. Ser^{241} replaced by glutamic acid mimics phosphorylation, and Ser^{241} replaced by alanine allows binding without phosphorylation, and both these changes result in permanent carbon catabolite repression for cellobiohydrolyase I expression. This phenotype is similar to that of an *A. nidulans* strain that is deleted for the Rgr1-similar region and the adjacent acidic region (which includes the equivalent Serine residue) of CreA, which also fails to grow on a range of sole carbon sources (Shroff 1997; Shroff RA, Lockington RA, Kelly JM, Unpublished). On the other hand, a Glu^{244}Val substitution leads to no phosphorylation, binding or repression (Cziferszsky et al. 2002). Snf1 kinase is not involved in the phosphorylation of Cre1 in *T. reesei* (Cziferszsky et al. 2002), and a casein kinase II target consensus surrounds Ser^{241}. However, in *S. sclerotiorum*, substitution of Ser^{266} in CRE1 (equivalent to Ser^{241} in *T. reesei*) by alanine leads to derepression (Vautard-Mey and Fevre 2000). The *A. nidulans* CreA amino acid region SHED$^{262-265}$ corresponds to amino acids SNDE$^{241-244}$ in *T. reesei* and SHEE$^{266-269}$ in *S. sclerotiorum*. The observation that the *A. nidulans* strains containing deletions of this region are almost identical to the wild type with respect to repression and derepression indicates that this region is not essential for DNA binding or carbon catabolite repression in *A. nidulans* (Roy et al. 2008).

Functional studies of the CreA homologue from *S. sclerotiorum* (CRE1) have been undertaken (Vautard et al. 1999; Vautard-Mey et al. 1999; Vautard-Mey and Fevre 2000). Antibodies were raised to a CRE1::GST fusion protein, and they detected a protein that was present at higher levels in glucose and glycerol grown mycelia than in pectin grown mycelia, although there were no apparent differences in stability in the two media (Vautard-Mey et al. 1999). The subcellular localisation of the protein varied with glucose concentration, and although the presence or absence of a potential serine phosphorylation site was correlated with repressor activity, it did not affect subcellular localisation (Vautard-Mey et al. 1999; Vautard-Mey and Fevre 2000). In *A. nidulans*, Brown and colleagues showed that nuclear localisation of CreA:GFP varied according to carbon source, with high nuclear localisation in mono or disaccharides, intermediate in cellobiose and xylose, lower in glycerol, lower again in complex carbohydrates and absent in carbon starvation (Brown et al. 2013). However, in strains over expressing CreA:green fluorescent protein and exhibiting normal repression and derepression phenotypes, CreA:green fluorescent protein was present in the nucleus at high levels under both carbon catabolite repressing and derepressing conditions. Thus, subcellular localisation of CreA in *A. nidulans* is not a critical regulatory determinant (Roy et al. 2008). Chulkin and coworkers also observed that *P. canescens* CreA protein is nuclear localised regardless of the carbon source and glucose concentration in the medium (Chulkin et al. 2010), and in *Fusarium oxysporum*, a GFP-Cre1 fusion protein is not degraded nor exported out of the nucleus during growth on ethanol, a derepressing carbon source (Jonkers and Rep 2009).

SAGA complexes have been studied in yeast (reviewed by Baker and Grant 2007), and they act through the coordination of multiple histone posttranslational modifications, including acetylation, methylation, ubiquitination and phosphorylation. The Gcn5p subunit has histone acetyltransferase activity and acetylates histones. In *A. nidulans*, the relationship between transcriptional activation and nucleosome positioning has been investigated using the proline and ethanol regulons as a model (Garcia et al. 2004; Mathieu et al. 2005; Reyes-Dominguez et al. 2008). In the proline regulon, eight nucleosomes in the intergenic regulatory region are absent under inducing conditions, and they are partially restored under carbon- and nitrogen-repressing conditions when inducer is also present. CreA is required for partial nucleosome repositioning under repressing conditions, but not to establish the initial pattern. Trichostatin A, an inhibitor of deacetylases, results in total loss of nucleosome positioning under induced-repressing conditions (proline plus glucose), despite only a small derepression of *prnB*, indicating that CreA can repress independently of nucleosomes (Garcia et al. 2004). Transcriptional activation under derepressing conditions and chromatin remodelling were not affected in strains lacking GcnE and AdaB (essential components of the histone acetyltransferase), but the expression under induced-repressing conditions was affected along with the partial repositioning of histones under these conditions (Reyes-Dominguez et al. 2008). The effect of CreA/1 on nucleosome positioning has also been investigated in *T. reesei*, and CRE1 is important for correct nucleosome positioning. Under repressing conditions, CRE1 binds to several consensus recognition sequences in the *cbh1* promoter region in vitro, and the promoter and coding regions are occupied by several nucleosomes. In induced conditions, nucleosomes are lost from the coding region and nucleosomes in the promoter are repositioned preventing CRE1 binding. Cre1 mutant strains lose nucleosomes in the *cbh1* coding region in repressing conditions indicating a role for CRE1 in nucleosome positioning (Ries et al. 2014).

The effects of mutations affecting the Snf1 kinase on carbon catabolite repression and derepression have been studied in a range of filamentous fungi, and varied levels of involvement was apparent. In *A. nidulans*, null alleles of *snfA* kinase resulted in reduced cellulose-induced transcriptional responses, and *snfA*-deficient mutants were not able to derepress CreA-mediated carbon catabolite repression under derepressing conditions, such as growth on cellulose (Brown et al. 2013). In *C. carbonum*, mutation of ccSNF1 affected growth on simple sugars (Tonukari et al. 2000), and in *F. oxysporum*, *fosnf1* mutants showed reduced expression of several cell wall-degrading enzymes and grew poorly on certain carbon sources (Ospina-Giraldo et al. 2003). However, in *Ustilago maydis*, although expression of endoglucanase and pectinase was lower in the *Δsnf1* mutant strain in glucose-depleted conditions, the expression of two xylanase genes was higher in the *Δsnf1* strain, and Snf1 kinase was not required for the utilisation of alternative carbon sources (Nadal et al. 2010). Unlike the phosphorylation of Mig1p by Snf1p in yeast, in *T. reesei*, Snf1 kinase does not phosphorylate Cre1 (Cziferszsky et al. 2003).

B. A Role for Regulatory Ubiquitination

1. A Role for Deubiquitination?

Mutations affecting a deubiquitination system were identified in the same screens that identified CreA, and they showed pleiotropic effects on carbon metabolism. In the presence of repressing concentrations of glucose, these *creB* and *creC* mutations result in a degree of deregulated expression of some enzymes that would normally be subject to carbon catabolite repression, but the systems so affected are only a subset of those affected by mutations in *creA*, and in the absence of repressing concentrations of glucose, these same alleles result in failure to express enzymes for the utilisation of some carbon sources, such as quinate or L-proline (Hynes and Kelly 1977; Kelly and Hynes 1977; Arst 1981; Hunter et al. 2013), indicating a role or roles in conditions that are carbon catabolite repressing and in conditions that are carbon catabolite derepressing. Mutations in either gene also lead to altered sensitivity to some compounds in complete medium, such as increased sensitivity to acriflavine and decreased sensitivity to molybdate (Arst 1981), and they show minor alterations to colony morphology on complete medium (Hynes and Kelly, 1977). The *creB15* and *creC27* mutations greatly reduced proline and glutamate uptake, but no effects of the *creB15* and *creC27* mutations could be measured on the uptake of glucose (Hynes and Kelly 1997; Kelly 1980; Arst 1981). Biochemical studies have shown that mutations in *creB* and *creC* reduce external pH acidification, but the basis of this is not known except that it is not due to defects in the plasma-membrane-bound H^+-ATPase (Espeso et al. 1995; Abdallah et al. 2000).

CreB is a functional member of the ubiquitin processing protease (ubp) family (Hansen-Hagge et al. 1998; Lockington and Kelly 2001), and contains the six DUB (deubiquitination) homology domains common to this class of protein (D'Andrea and Pellman 1998), and a carboxy terminal extension that is predicted to form a coiled/coil involved in substrate recognition (Lockington and Kelly 2001). There are four high-scoring PEST (proline, glutamic acid, serine and threonine rich) sequences in the CreB protein (Lockington and Kelly 2002), indicative of ubiquitination and rapid degradation (Rechsteiner and Rogers 1996). There is no close homologue in *S. cerevisiae*. CreC contains a proline-rich region, a putative nuclear localisation region, and five WD40-repeat motifs likely to be involved in protein–protein interactions (Todd et al. 2000; van der Voorn and Ploegh 1992; Neer et al. 1994). The *A. nidulans* CreC protein shows weak sequence similarity to the *S. cerevisiae* Tup1p protein within the WD40-repeat regions of each protein (Todd et al. 2000). However, Tup1p and CreC are not homologous proteins, and there is at least one protein in *A. nidulans*, RcoA, that is significantly more similar to Tup1p than is CreC (Hicks et al. 2001).

Protein sequences conserved with CreB and CreC exist in higher eukaryotes, including humans, mice and *Drosophila*, and thus it is probable that these proteins are involved in a conserved regulatory pathway, but mutations in the genes leading to clear phenotypes have only been identified in filamentous fungi.

Some substrates of CreB have been identified. Strains lacking CreB grow poorly in media containing quinic acid as a carbon source, a phenotype unrelated to carbon catabolite repression. The quinate ion transporter QutD is regulated through deubiquitination by CreB. The quinate permease QutD is a ubiquitinated protein, and coimmunoprecipitation experiments have shown that and QutD and CreB can be coimmunoprecipated from protein extracts from quinic acid-induced mycelia, and thus the CreB deubiquitinating enzyme is present in a complex with QutD in vivo (Kamlangdee 2008). The failure of *creB* mutant strains to grow on quinic acid medium is not due to a failure of transcriptional induction of *qutD*, as the amount of mRNA was not lower in the *creB1937* mutant strain compared to a wild-type strain. Further, the concentration of QutD protein is lower in a *creB* null mutant background, indicating that deubiquitination is required to prevent protein turnover (Kamlangdee 2008). It is probable that a failure to deubiquitinate ubiquitinated permeases is likely to be the underlying cause of the phenotypes of *creB* and *creC* mutations that are found under carbon-derepressing conditions. It is, however, not likely that these effects on permeases account for the derepression phenotypes due to these mutations since, for example, the enzymes for ethanol utilisation are partially derepressed in the mutants, and ethanol does not require a permease to enter the cell. CreB was shown to have an effect on *alcA* (alcohol dehydrogenase I) transcription. Transcript levels of *alcA* were measured in a *creB1937* mutant using quantitative real-time PCR. In uninduced cultures, there was no evidence of elevation of *alcA* transcription due to *creB1937*. Induction with ethyl methyl ketone led to greater induction in the *creB1937* strain than in the wild-type strain, and significantly, while addition of a repressor reduced expression in the wild type to uninduced levels there was a 20-fold increase in *alcA* transcript levels compared to that in the uninduced culture (Hunter et al. 2013). Thus, *creB1937* leads to both elevation and partial derepression of *alcA* transcription.

There is some evidence that CreA is a direct target of CreB, in that a double round of purification of CreA-HA-His, first with a His-trap column followed by anti-HA matrix, also purified CreB-FLAG, indicating that CreA or a protein in a complex with CreA is a potential substrate of CreB. However, this evidence is from overexpressing strains, and thus requires confirmation. At least some of the pool of CreA has also been found among proteins purified as ubiquitinated proteins; however, since there were no great differences in the total amount of CreA in cells grown under repressing or derepressing conditions, CreA activity is not regulated by wholesale degradation (Kamlangdee 2008). Despite this, overexpression of CreA overcomes the defects in carbon catabolite repression seen in *creB* and *creC* mutant strains, perhaps suggesting that an activated form of CreA is the substrate for CreB.

Mutations in *creC* result in an identical range of phenotypes to mutations in *creB*, implicating both genes in the same regulatory network, either as partners in a protein complex, or as sequential steps in a regulatory pathway. Overexpression of CreB can partially compensate for a lack of CreC, but overexpression of the CreC cannot compensate for a lack of CreB, and thus the CreB deubiquitinating enzyme is the crucial active partner (Lockington and Kelly 2002). Coimmunoprecipitation experiments show that CreB and CreC are present in a high molecular weight complex in vivo as they can be coimmunoprecipitated from mycelia grown in either carbon catabolite repressing or carbon catabolite derepressing conditions (Lockington and Kelly 2002). However, each is not reliant on the other for their presence in the complex.

Mutations in *creB* have also been created in *T. reesei*, *Penicillium decumbens* and *A. oryzae* (Denton and Kelly 2011; Zhou et al. 2012; Hunter et al. 2013; Ichinose et al. 2014). The *T. reesei* and *A. oryzae* mutants exhibited phenotypes similar to the *A. nidulans creB* mutant strain in both the carbon catabolite repressing and carbon catabolite derepressing conditions tested (Denton and Kelly 2011; Hunter et al. 2013). The *A. oryzae* null strain showed increased activity levels of industrially important secreted enzymes, including cellulases, xylanases, amylases, and proteases, as well as alleviated inhibition of spore germination on glucose medium. Importantly, reverse transcription-quantitative PCR showed that the increased levels of amylase in *A. oryzae* are reflected at the *amyA*, *amyB* and *amyC* transcript level, indicating transcriptional regulation, and these findings mirror those for *alcA* in *A. nidulans* (Hunter et al. 2013). Ichinose and colleagues also investigated α-amylase in single and double deletion *creA* and *creB* mutants in *A. oryzae* (Ichinose et al. 2014), and showed that activity in the double deletion mutant was up to tenfold higher than that of the wild-type strain, suggesting genetic interaction resulting in dramatic improvements in the production of secretory glycoside hydrolases. In *P. decumbens*, endoglucanase, xylanase and exoglucanase activities were elevated, indicating that the deletion of *creB* also results in deregulation of some cellulases in this organism (Zhou et al. 2012). The *creC* gene has also been deleted in *A. oryzae*, and the strain showed improved protease amylase and xylanase activities, and has been assessed for use in the brewing and bioethanol industries (Watabe 2010).

2. A Role for Ubiquitination?

Suppressors of *creB* and *creC* mutations have been isolated that lead to tighter repression of some systems regulated by carbon catabolite repression. The effects of one suppressor, *creD34*, were pleiotropic in that in addition to effects of *creC27* on *alcA* as selected there were effects on other enzymes subject to carbon catabolite repression. However, the *creD34* mutation does not suppress the effects of *creC27* that are apparent in derepressing conditions, such as the poor growth on D-quinate medium (Kelly and Hynes 1977; Kelly 1980). The *creD34* mutation also suppresses some of the phenotypic effects of the *creB15* mutation, and, weakly, of the *creA204* mutation (Kelly and Hynes 1977; Kelly 1980). In an otherwise wild-type background, the *creD34* mutant strain is more resistant than wild type to the presence of acriflavine in complete medium, and less sensitive to the presence of molybdate, the reverse of the phenotype conferred by the *creA*, *creB* and *creC* mutations. CreD contains arrestin domains and PY motifs and is highly similar to the Rod1p and Rog3p proteins of *S. cerevisiae* which interact with the ubiquitin ligase Rsp5p. The *A. nidulans* homologue of Rsp5p was identified (HulA) and interaction with CreD was shown in the bacterial two-hybrid system (Boase and Kelly 2004).

Like the *creD34* mutation, mutations in *acrB* suppress the effects of *creB* and *creC* mutations on alcohol dehydrogenase expression in the presence of glucose and result in tighter repression of enzymes subject to carbon catabolite repression (Boase and Kelly 2004). *acrB* mutant strains have a pleiotropic phenotype with respect to carbon source utilisation. They show decreased ability to utilise a number of different sugars as sole carbon sources, including 1 % fructose, 1 % cellobiose, 1 % raffinose and 1 % starch, in comparison to both the wild-type and the *creD34* mutant strain, which may indicate a failure to derepress the enzymes required for their utilisation (Boase et al. 2003). The effects on acriflavine resistance of the *acrB2* and *creD34* mutations are additive, and both genes encode proteins that, when mutated, can lead to tighter carbon catabolite repression (Boase et al. 2003).

Mutations in the *acrB* gene, although originally selected through their resistance to acriflavine, also result in reduced growth on a range of sole carbon sources, including fructose, cellobiose, raffinose and starch, and reduced utilisation of omega-amino acids, including GABA and β-alanine, as sole carbon and nitrogen sources. AcrB is a novel protein that contains three putative transmembrane domains and a coiled-coil region, which also interacts with HulA in the bacterial two-hybrid system (Boase et al. 2003).

Further support for a role of ubiquitination and deubiquitination in carbon repression comes from the work on xylanase production in *A. nidulans* (Colabardini et al. 2012). The genes involved are repressed by CreA, and Colabardini and colleagues investigated a panel of *A. nidulans* F-box deletion mutants and identified *fbxA* that leads to decreased xylanase activity and reduced *xlnA* and *xlnD* mRNA. The mutant interacts genetically with *creA^{d}30*, *creB15* and *creC27* mutants. FbxA contains a functional F-box domain that binds to Skp1 ligase.

V. Genome-Wide Studies

Since carbon catabolite repression is of importance in industrial fermentations, many physiological studies have been performed on strains containing mutant alleles of *creA*. In *A. nidulans*, a study by Vanderveen et al. (1995) showed that in glucose grown mycelia of an extreme *creA* mutant, steady-state levels of some glycolytic enzyme activities and certain metabolite concentrations were different to those in wild-type glucose grown mycelia, and the flux of catabolism was diverted away from glycolysis towards increased synthesis and secretion of polyols. Other work described the use of antisense *creA* constructs in industrial contexts (Bautista et al. 2000) and the altered growth characteristics of a *creA* null allele in fermenters (Agger et al. 2002).

Genome data from a range of fungal species and strains continues to come available at an almost exponential rate, with close to 100 genomes from the *Ascomycota*, over 30 from the *Basidiomycota*, and smaller but increasing numbers from the *Chytridiomycota*, the Microsporidia, and the zygomycetes (Fungal Genome Initiative). Insightful studies using integrated approaches have shed light on metabolic pathways and networks, their regulation and metabolic flux through the cycles. An early example involved the stoichiometric modelling of central carbon metabolism using available information on *A. niger* metabolism, even before the fully annotated genome was available, by the integration of genomic, biochemical and physiological data, as well as extrapolation from closely related species (David et al. 2003). David and colleagues (David et al. 2005, 2006) have reconstructed complete metabolic networks for *A. nidulans* growing on glucose, glycerol and ethanol and shown that the shift from glucose to ethanol results in increased gluconeogenesis and decreased activity in the glycolytic and pentose phosphate pathways. Further, using ^{13}C-metabolic flux analysis, they have assessed the effects of the absence of CreA on metabolic flux (David et al. 2005). Expressed sequence tag and cDNA microarrays were used in early studies, such as those that investigated the fate of glucose in *T. reesei* and *A. oryzae* (Chambergo et al. 2002; Maeda et al. 2004).

Technical developments in ultra-high-throughput sequencing have changed the scale of investigations, with DNA sequencing and resequencing, transcriptome profiling (RNA-seq), and DNA–protein interactions (Chip-seq) now achievable on a genome wide scale. These advances have evolved alongside advances in data storage and bioinformatics to allow interpretation. Only a small number of recent examples will be discussed here, showing the power of the approach and the value of these studies both in advanced understanding of the particular biological process being studied, but also the value in making the data fully available for comparative purposes. A survey of the transcriptome in *A. oryzae* grown in several conditions uncovered unexpected complexity, with previously unidentified exons, untranslated regions, alternative upstream initiation codons and upstream open reading frames, and a large number of genes that showed alternative splicing (Wang et al. 2010). In studies of the use of *T. reesei* for second-generation biofuel production, genome-wide transcription on wheat straw was elucidated using RNA-seq and compared to published RNA-seq data on how *A. niger* senses and responds to wheat straw (Ries et al. 2013; Delmas et al. 2012). Also in *T. reesei*, a comparative analysis of metabolism on cellulose, sophorose and glucose was undertaken, aimed at understanding the molecular basis for lignocellulose-degrading enzyme production, and over 2000 genes were found to show alterations in their expression, many of which were previously not known to play any role (Castro et al. 2014). A genome-wide transcriptome analysis of the *N. crassa* response to glucose, xylose and arabinose (Li et al. 2014) uncovered novel metabolic responses, as well as novel transcription factors potentially involved in the process. In *A. nidulans*, further complexity of the transcriptome was revealed in studies using high-throughput RNA-sequencing from a strain grown on a number of different conditions (Sibthorp et al. 2013), and novel transcripts, anti-sense transcripts and alternative splice forms were apparent in a wide variety of systems, and these will no doubt be the focus of further investigation.

VI. Concluding Remarks

Seminal studies in *S. cerevisiae* have uncovered the molecular mechanism of numerous cellular processes, and have provided the foundation for studies in other organisms, from filamentous fungi to plants and animals. However, *S. cerevisiae* is a unicellular organism that has evolved to fill a very specific niche, and is one of the few yeasts that can grow rapidly in anaerobic conditions by fermentation of sugars. Strains have evolved to grow on sugar-rich substrates, but they are not particularly metabolically versatile and cannot grow on a range of alternative carbon sources. On the other hand, filamentous fungi are metabolically versatile, and have evolved to occupy a vast range of environmental niches, and their multicellularity allows further specialisation. The involvement of pathway-specific regulatory proteins in turning genes on in response to the presence of inducers appears similar at the mechanistic level in *S. cerevisiae* and in the filamentous fungi. Also similar is the involvement of a single major DNA-binding regulatory repressor protein in carbon catabolite repression in each organism, Mig1p and CreA/1 proteins, with similar zinc-finger domains and consensus recognition sequences. However, the similarities do not extend beyond binding, as the molecular mechanisms that result in repression by Mig1p/CreA/Cre1 are very different. Both Mig1p and the CreA/1 of some filamentous fungi have regulated subcellular localisation, with shuttling between the nucleus and the cytoplasm, but in neither organism is the export essential for derepression. In *S. cerevisiae*, the Tup1p/Ssn6p corepressor complex is a key component of the repression, but there is no strong evidence that this complex is a key component in filamentous fungi. Phosphorylation by Snf1p plays an important role in derepression in yeast, but a variable role in filamentous fungi, where *T. reesei* Cre1 is phosphorylated, but not by a Snf1 like kinase, *A. nidulans snfA* mutants fail to derepress some genes tested that are subject to carbon catabolite repression, and in *U. maydis Δsnf1* was unaffected for the use of alternative carbon sources, although it reduced the level of some genes. There is no key role for a regulatory ubiquitination mechanism in carbon catabolite repression in *S. cerevisiae*, whereas there is some evidence for such an involvement in filamentous fungi. Thus, it is apparent that although there are commonalities, the molecular mechanisms involved are quite diverse, and there is unlikely to be a single unifying model for induction and carbon catabolite repression in the filamentous fungi.

The recently developed and emerging technologies of systems biology will advance our understanding of carbon metabolism and its regulation in the near future, but it is apparent that even within the restricted framework of carbon utilisation research needs to be undertaken in a wide range of organisms to gain a complete picture. Even in a relatively simple microorganism, there are many regulatory circuits, including those for carbon starvation, nitrogen regulation, pH regulation, response to hypoxia and other stressors, cell cycle and developmental regulation, secondary metabolism, circadian cycling and many more; not only do we need to completely understand each circuit, but as all the circuits interact, we need to understand the complex interactions and networks. The events involved from environment signalling to integrated gene expression outcomes are still some way from being achieved.

References

Abdallah BM, Simoes T, Fernandes AR, Strauss J, Seiboth B, Sa-Correia I, Kubicek CP (2000) Glucose does not activate the plasma-membrane-bound H +-ATPase but affects *pmaA* transcript abundance in *Aspergillus nidulans*. Arch Microbiol 174:340–345

Agger T, Petersen JB, O'Connor SM, Murphy RL, Kelly JM, Nielsen J (2002) Physiological characterisation of recombinant *Aspergillus nidulans* strains with different *creA* genotypes expressing *A. oryzae* α-amylase. J Biotechnol 92:279–285

Arst HN (1981) Aspects of the control of gene expression in fungi. Symp Soc Gen Microbiol 31:131–160

Arst HN, Bailey CR (1977) The regulation of carbon metabolism in *Aspergillus nidulans*. In: Smith JE, Pateman JA (eds) Genetics and physiology of *Aspergillus nidulans*. Academic, London, pp 131–146

Arst HN, Cove DJ (1973) Nitrogen metabolite repression in *Aspergillus nidulans*. Mol Gen Genet 126:111–141

Arst HN, MacDonald DW (1975) A gene cluster in *Aspergillus nidulans* with an internally located cis-acting regulatory region. Nature 254:26–34

Arst HN, Tollervey D, Dowzer CEA, Kelly JM (1990) An inversion truncating the *creA* gene of *Aspergillus nidulans* results in carbon catabolite repression. Mol Microbiol 4:851–854

Bailey CR, Arst HN (1975) Carbon catabolite repression in *Aspergillus nidulans*. Eur J Biochem 51:573–577

Baker SP, Grant PA (2007) The SAGA continues: expanding the cellular role of a transcriptional co-activator complex. Oncogene 26:5329–5340

Bautista LF, Aleksenko A, Hentzer M, Santerre-Henriksen A, Nielsen J (2000) Antisense silencing of the *creA* gene in *Aspergillus nidulans*. Appl Environ Microbiol 66:4579–4581

Bernardo SMH, Gray K-A, Todd RB, Cheetham BF, Katz ME (2007) Characterization of regulatory non-catalytic hexokinases in *Aspergillus nidulans*. Mol Genet Genomics 277:519–532

Boase NA, Kelly JM (2004) A role for *creD*, a carbon catabolite repression gene from *Aspergillus nidulans*, in ubiquitination. Mol Microbiol 53:929–940

Boase NA, Lockington RA, Adams JRJ, Rodbourn L, Kelly JM (2003) Molecular characterization and analysis of the *acrB* gene of *Aspergillus nidulans*: a gene identified by genetic interaction as a component of the regulatory network that includes the CreB deubiquitination enzyme. Genetics 164:95–104

Brown NA, de Gouvea PF, Krohn NG, Savoldi M, Goldman GH (2013) Functional characterisation of the non-essential protein kinases and phosphatases regulating *Aspergillus nidulans* hydrolytic enzyme production. Biotechnol Biofuels 6:91

Carlson M (1999) Glucose repression in yeast. Curr Opin Microbiol 2:202–207

Castro LDS, Pedersoli WR, Antonio ACC, Steindorff AS, Silva-Rocha R, Martinez-Rossi NM, Rossi A, Brown NA, Goldman GH, Faa VM, Persinoti GF, Silva RN (2014) Comparative metabolism of cellulose, sophorose and glucose in *Trichoderma reesei* using high-throughput genomic and proteomic analyses. Biotechnol Biofuels 7:41

Chambergo FS, Bonaccorsi ED, Ferreira AJS, Ramos ASP, Junior JRF, Abrahao-Neto J, Farah JPS, El-Dorry H (2002) Elucidation of the metabolic fate of glucose in the filamentous fungus *Trichoderma reesei* using expressed sequence tag (EST) analysis and cDNA microarrays. J Biol Chem 277:13983–13988

Chulkin AM, Vavilova EA, Benevolenskij SV (2010) Transcriptional regulator of carbon catabolite repression CreA of filamentous fungus *Penicillium canescens*. Mol Biol 44:596–605

Chulkin AM, Vavilova EA, Benevolenskii SV (2011) Mutational analysis of carbon catabolite repression in filamentous fungus *Penicillium canescens*. Mol Biol 45:804–810

Colabardini AC, Humanes AC, Gouvea PF, Savoldi M, Goldman MHS, von Zeska Kress MR, Bayram O, de Castro Oliveira JV, Gomes MD, Braus GH, Goldman GH (2012) Molecular characterization of the *Aspergillus nidulans fbxA* encoding an F-box protein involved in xylanase induction. Fungal Genet Biol 49:130–140

Cubero B, Scazzocchio C (1994) Two different, adjacent and divergent zinc finger binding sites are necessary for CreA mediated carbon catabolite repression in the proline gene cluster of *Aspergillus nidulans*. EMBO J 13:407–415

Cubero B, Gomez D, Scazzocchio C (2000) Metabolite repression and inducer exclusion in the proline utilization gene cluster of *Aspergillus nidulans*. J Bacteriol 182:233–235

Cziferszsky A, Mach R, Cubicek CP (2002) Phosphorylation positively regulates DNA binding by the carbon catabolite repressor Crel of *Hypocrea jecorina* (*Trichoderma reesei*). J Biol Chem 277:14688–14694

Cziferszsky A, Mach R, Cubicek CP (2003) The Snfl kinase of the filamentous fungus *Hypocrea jecorina* phosphorylates regulation-relevant serine residues in the yeast carbon catabolite repressor Migl but not in the filamentous fungal counterpart Crel. Fungal Genet Biol 40:166–175

D'Andrea A, Pellman D (1998) Deubiquitinating enzymes: a new class of biological regulators. Crit Rev Biochem Mol Biol 33:337–352

David H, Akesson M, Nielsen J (2003) Reconstruction of the central carbon metabolism of *Aspergillus niger*. Eur J Biochem 270:4243–4253

David H, Krogh AM, Roca C, Akesson M, Nielsen J (2005) CreA influences the metabolic fluxes of *Aspergillus nidulans* during growth on glucose and xylose. Microbiology 151:2209–2221

David H, Hofmann G, Oliveira AP, Jarmer H, Nielsen J (2006) Metabolic network driven analysis of genome-wide transcription data from *Aspergillus nidulans*. Genome Biol 7:R108

de la Serna I, Ng D, Tyler BM (1999) Carbon regulation of ribosomal genes in *Neurospora crassa* occurs by a mechanism which does not require Cre-1, the homologue of the Aspergillus carbon catabolite repressor. CreA Fungal Genet Biol 26:253–269

De Vit MJ, Johnston M (1999) The nuclear exportin Msn5 is required for nuclear export of the Migl glucose repressor of *Saccharomyces cerevisiae*. Curr Biol 9:1231–1241

De Vit MJ, Waddle JA, Johnston M (1997) Regulated nuclear translocation of the Migl glucose repressor. Mol Biol Cell 8:1603–1618

Delmas S, Pullan S, Gaddipati S, Kokolski M, Malla S, Blythe MJ, Ibbett R, Campbell M, Liddell S, Aboobaker A, Tucker GA, Archer DB (2012) Uncovering the genome-wide transcriptional responses of the filamentous fungus *Aspergillus niger* to lignocellulose using RNA sequencing. PLoS Genet 8:e1002875

Denton J, Kelly JM (2011) Disruption of *Trichoderma reesei cre2*, encoding an ubiquitin C-terminal hydrolase, results in increased cellulase activity. BMC Biotechnol 11:9

Dowzer CEA, Kelly JM (1989) Cloning of *creA* from *Aspergillus nidulans*: a gene involved in carbon catabolite repression. Curr Genet 15:457–459

Dowzer CEA, Kelly JM (1991) Analysis of the *creA* gene, a regulator of carbon catabolite repression in *Aspergillus nidulans*. Mol Cell Biol 11:5701–5709

Drysdale MR, Kolze SE, Kelly JM (1993) The *Aspergillus niger* carbon catabolite repressor encoding gene, *creA*. Gene 130:241–245

Espeso EA, Penalva MA (1994) In vitro binding of the two-finger repressor CREA to several consensus and non-consensus sites at the *ipnA* upstream region is context dependent. FEBS Lett 342:43–48

Espeso EA, Fernandez-canon JM, Penalva MA (1995) Carbon regulation of penicillin biosynthesis in *Aspergillus nidulans* – a minor effect of mutations in *creB* and *creC*. FEMS Microbiol Lett 126:63–67

Felenbok B, Flipphi M, Nikolaev I (2001) Ethanol catabolism in *Aspergillus nidulans*: a model system for studying gene regulation. Prog Nucleic Acid Res Mol Biol 69:149–204

Flipphi M, Mathieu M, Cirpus I, Panozzo C, Felenbok B (2001) Regulation of the aldehyde dehydrogenase gene (*aldA*) and its role in the control of the coinducer level necessary for induction of the ethanol utilization pathway in *Aspergillus nidulans*. J Biol Chem 276:6950–6958

Flipphi M, van de Vondervoort PJ, Ruijter GJ, Visser J, Arst HN, Felenbok B (2003) Onset of carbon catabolite repression in *Aspergillus nidulans*. Parallel involvement of hexokinase and glucokinase in sugar signaling. J Biol Chem 278:11849–11857

Forment JV, Flipphi M, Ventura L, Gonzalez R, Ramon D, MacCabe AP (2014) High affinity glucose transport in *Aspergillus nidulans* is regulated by the products of two related but differentially expressed genes. PLoS One 9, e94662

Gancedo JM (1998) Yeast carbon catabolite repression. Microbiol Mol Biol Rev 62(334):1998

Garcia I, Gonzalez R, Gomez D, Scazzocchio C (2004) Chromatin rearrangements in the *prnD-prnB* bidirectional promoter: dependence on transcription factors. Eukaryot Cell 3:144–156

Garcia I, Mathieu M, Nikolaev I, Felenbok B, Scazzocchio C (2008) Roles of the *Aspergillus nidulans* homologues of Tupl and Ssn6 in chromatin structure and cell viability. FEMS Microbiol Lett 289:146–154

Geever RF, Huiet L, Baum JA, Tyler BM, Patel VB, Rutledge BJ, Case ME, Giles NH (1989) DNA sequence, organization and regulation of the *qa* gene cluster in *Neurospora crassa*. J Mol Biol 207:15–34

Giles NH, Case ME, Baum J, Geever R, Huiet L, Patel V, Tyler B (1985) Gene organisation and regulation in the *qa* (qunic acid) cluster of *Neurospora crassa*. Microbiol Rev 49:338–358

Hansen-Hagge TE, Janssen JWG, Hameister H, Papa FR, Zechner U, Seriu T, Jauch A, Becke D, Hochstrasser M, Bartram CR (1998) An evolutionarily conserved gene on chromosome 5q33-q34, UBH1, encodes a novel deubiquitinating enzyme. Genomics 49:411–418

Hicks J, Lockington RA, Strauss J, Dieringer D, Kubicek CP, Kelly J, Keller N (2001) RcoA has pleiotropic effects on *Aspergillus nidulans* cellular development. Mol Microbiol 39:1482–1493

Hunter AJ, Morris TA, Jin B, Kelly JM (2013) Deletion of *creB* in *Aspergillus oryzae* increases secreted hydrolytic enzyme activity. Appl Environ Microbiol 79:5480–5487

Hynes MJ (1970) Induction and repression of amidase enzymes in *Aspergillus nidulans.* J Bacteriol 103:482–487

Hynes MJ, Davis MA (2004) Regulation of the *amdS* gene in *Aspergillus nidulans.* In: Esser K (ed) The Mycota III: Biochemistry and molecular biology, 2nd edn. Springer, Berlin, pp 421–435

Hynes MJ, Kelly JM (1977) Pleiotropic mutants of *Aspergillus nidulans* altered in carbon metabolism. Mol Gen Genet 150:193–204

Ichinose S, Tanaka M, Shintani T, Gomi K (2014) Improved α-amylase production by Aspergillus oryzae after a double deletion of genes involved in carbon catabolite repression. Appl Microbiol Biotechnol 98:335–343

Ilmen M, Thrane C, Pentilla M (1996) The glucose repressor gene, *cre1* of *Trichoderma reesei* - isolation of a full length and truncated mutant form. Mol Gen Genet 251:451–460

Jekosch K, Kueck U (2000a) Glucose dependent transcriptional expression of the *cre1* gene in *Acremonium chrysogenum* strains showing different levels of cephalosporin C production. Curr Genet 37:388–395

Jekosch K, Kueck U (2000b) Loss of glucose repression in an *Acremonium chrysogenum* beta-lactam producer strain and its restoration by multiple copies of the *cre1* gene. Appl Microbiol Biotechnol 54:556–563

Jonkers W, Rep M (2009) Mutation of *CRE1* in *Fusarium oxysporum* reverts the pathogenicity defects of the *FRP1* deletion mutant. Mol Microbiol 74:1100–1113

Kamlangdee N (2008) Identifying target proteins of the CreB deubiquitination enzyme in the fungus, *Aspergillus nidulans.* University of Adelaide, Adelaide

Katz ME, Kelly JM (2010) Glucose. In: Borkovich K, Ebbole D (eds) Cellular and molecular biology of filamentous fungi. American Society for Microbiology Press, Washington, DC

Katz ME, Flynn PK, van Kuyk P, Cheetham BF (1996) Mutations affecting extracellular protease production in the filamentous fungus, *Aspergillus nidulans.* Mol Gen Genet 250:715–724

Katz ME, Masoumi A, Burrows SR, Shirtliff CG, Cheetham BF (2000) The *Aspergillus nidulans xprF* gene encodes a hexokinase-like protein involved in the regulation of the extracellular proteases. Genetics 156:1559–1571

Katz ME, Gray K-A, Cheetham BF (2006) The *Aspergillus nidulans xprG* (*phoG*) gene encodes a putative transcriptional activator involved in the response to nutrient limitation. Fungal Genet Biol 43:190–199

Kawasaki L, Farres A, Aguirre J (1995) *Aspergillus nidulans* mutants affected in acetate metabolism isolated as lipid nonutilizers. Exp Mycol 19:81–85

Kelly JM (1980) Pleiotropic mutants of *Aspergillus nidulans* affected in carbon metabolism. PhD Thesis, La Trobe University, Australia

Kelly JM, Hynes MJ (1977) Increased and decreased sensitivity to carbon catabolite repression of enzymes of acetate metabolism in mutants of *Aspergillus nidulans.* Mol Gen Genet 156:87–92

Kim J-H, Roy A, Jouandot D, Cho KH (2013) The glucose signaling network in yeast. Biochim Biophys Acta 1830:5204–5210

Klein CJL, Olsson L, Nielsen J (1998) Glucose control in *Saccharomyces cerevisiae* – the role of MIG1 in metabolic functions. Microbiology 144:13–24

Kulmberg P, Sequeval D, Lenouvel F, Mathieu M, Felenbok B (1992a) Identification of the promoter region involved in the autoregulation of the transcriptional activator ALCR in *Aspergillus nidulans.* Mol Cell Biol 12:1932–1939

Kulmberg P, Sequeval D, Lenouval F, Mathieu M, Felenbok B (1992b) Specific binding sites for the activator protein, ALCR, in the *alcA* promoter of the ethanol regulon of *Aspergillus nidulans.* J Biol Chem 267:21146–21153

Kulmberg P, Mathieu M, Dowzer C, Kelly J, Felenbok B (1993) Specific binding sites in the *alcR* and *alcA* promotors of the ethanol regulon for the CREA repressor mediating carbon catabolite repression in *Aspergillus nidulans.* Mol Microbiol 7:847–857

Lamb HK, Newton GH, Levett LJ, Cairns E, Roberts CF, Hawkins AR (1996) The QutA activator and QutR repressor proteins of *Aspergillus nidulans* interact to regulate transcription of the quinate utilization genes. Microbiology 142:1477–1490

Levett LJ, Si-Hoe SM, Liddle S, Wheeler K, Smith D, Lamb HK, Newton GH, Coggins JR, Hawkins AR (2000) Identification of domains responsible for signal recognition and transduction within the QUTR transcription repressor protein. Biochem J 350:189–197

Li J, Lin L, Li H, Tian C, Ma Y (2014) Transcriptional comparison of the filamentous fungus *Neurospora crassa* growing on three major monosaccharides D-glucose, D-xylose and L-arabinose. Biotechnol Biofuels 7:31

Limon MC, Pakula T, Saloheimo M, Penttila M (2011) The effects of disruption of phosphoglucose isomerase gene on carbon utilisation and cellulase production in *Trichoderma reesei* Rut-C30. Microb Cell Factories 10:40

Lockington RA, Kelly JM (2001) Carbon catabolite repression in *Aspergillus nidulans* involves deubiquitination. Mol Microbiol 40:1311–1321

Lockington RA, Kelly JM (2002) The WD40-repeat protein CreC interacts with and stabilizes the deubiquitinating enzyme CreB in vivo in *Aspergillus nidulans*. Mol Microbiol 43:1173–1182

Lockington RA, Sealy-Lewis HM, Scazzocchio C, Davies RW (1985) Cloning and characterization of the ethanol utilization regulon in *Aspergillus nidulans*. Gene 33:137–149

Luo Z, Qin Y, Pei Y, Keyhani NO (2014) Ablation of the *creA* regulator results in amino acid toxicity, temperature sensitivity, pleiotropic effects on cellular development and loss of virulence in the filamentous fungus *Beauveria bassiana*. Environ Microbiol 16:1122–1136

Maeda H, Sano M, Maruyama Y, Tanno T, Akao T, Totsuka Y, Endo M, Sakurada R, Yamagata Y, Machida M, Akita O, Hasegawa F, Abe K, Gomi K, Nakajima T, Iguchi Y (2004) Transcriptional analysis of genes for energy catabolism and hydrolytic enzymes in the filamentous fungus *Aspergillus oryzae* using cDNA microarrays and expressed sequence tags. Appl Microbiol Biotechnol 65:74–83

Martin JE, Casqueiro J, Kosalkova K, Marcos AT, Gutierrez S (1999) Penicillin and cephalosporin biosynthesis: Mechanism of carbon catabolite regulation of penicillin production, Antonie van Leeuwenhoek. Int J Gen Mol Microbiol 75:21–31

Mathieu M, Felenbok B (1994) The *Aspergillus nidulans* CREA protein mediates glucose repression of the ethanol regulon at various levels through competition with the ALCR-specific transactivator. EMBO J 13:4022–4027

Mathieu M, Fillinger S, Felenbok B (2000) In vivo studies of upstream regulatory cis-acting elements of the *alcR* gene encoding the transactivator of the ethanol regulon in *Aspergillus nidulans*. Mol Microbiol 36:123–131

Mathieu M, Nikolaev I, Scazzocchio C, Felenbok B (2005) Patterns of nucleosomal organization in the *alc* regulon of *Aspergillus nidulans*: roles of the AlcR transcriptional activator and the CreA global repressor. Mol Microbiol 56:535–548

McCullough W, Payton MA, Roberts CF (1977) Carbon metabolism in *Aspergillus nidulans*. In: Smith JE, Pateman JA (eds) Genetics and physiology of *Aspergillus*. Academic, London, pp 97–129

Nadal M, Garcia-Pedrajas MD, Gold SE (2010) The *snf1* gene of *Ustilago maydis* acts as a dual regulator of cell wall degrading enzymes. Phytopathology 100:1364–1372

Neer EJ, Schmidt CJ, Nambudripad R, Smith TF (1994) The ancient regulatory-protein family of WD-repeat proteins. Nature 371:297–300

Nehlin J, Ronne H (1990) Yeast MIG1 repressor is related to the mammalian early growth response and Wilms' tumor finger proteins. EMBO J 9:2891–2898

Nehlin JO, Carlberg M, Ronne H (1991) The control of yeast GAL genes by MIG1 repressor: a transcriptional cascade in the glucose response. EMBO J 10:3373–3377

Nikolaev I, Lenouvel F, Felenbok B (1999) Unique DNA binding specificity of the binuclear zinc AlcR activator of the ethanol utilization pathway in *Aspergillus nidulans*. J Biol Chem 274:9795–9802

Orejas M, MacCabe AP, Gonzalez JAP, Kumar S, Ramon D (1999) Carbon catabolite repression of the *Aspergillus nidulans xlnA* gene. Mol Microbiol 31:177–184

Orejas M, MacCabe AP, Perez-Gonzalez JA, Kumar S, Ramon D (2001) The wide-domain carbon catabolite repressor CreA indirectly controls expression of the *Aspergillus nidulans xlnB* gene, encoding the acidic endo-beta-(1,4)-xylanase X-24. J Bacteriol 183:1517–1523

Ospina-Giraldo MD, Mullins E, Kang S (2003) Loss of function of the *Fusarium oxysporum SNF1* gene reduces virulence on cabbage and Arabidopsis. Curr Genet 44:49–57

Ostling J, Carlberg M, Ronne H (1996) Functional domains in the Mig1 repressor. Mol Cell Biol 16:753–761

Ostling J, Cassart JP, Vandenhaute J, Ronne H (1998) Four hydrophobic amino acid residues in the C-terminal effector domain of the yeast Mig1p repressor are important for its in vivo activity. Mol Gen Genet 260:269–279

Panozzo C, Cornillot E, Felenbok B (1998) The CreA repressor is the sole DNA-binding protein responsible for carbon catabolite repression of the *alcA* gene in *Aspergillus nidulans* via its binding to a couple of sites. J Biol Chem 273:6367–6372

Papamichos-Chronakis M, Petrakis T, Ktistaki E, Topalidou I, Tzamarias D (2002) Cti6, a PHD domain protein, bridges the Cyc8-Tup1 corepressor and the SAGA coactivator to overcome repression at GAL1. Mol Cell 9:1297–1305

Papamichos-Chronakis M, Gligoris T, Tzamarias D (2004) The Snf1 kinase controls glucose repression in yeast by modulating interactions between the Mig1 repressor and the Cyc8-Tup1 co-repressor. EMBO Rep 5:368–372

Pateman JA, Doy CH, Olsen JE, Norris U, Creaser EH, Hynes MJ (1983) Regulation of alcohol dehydrogenase and aldehyde dehydrogenase in *Aspergillus nidulans*. Proc R Soc Lond Ser B 217: 243–264

Paulino M, Esperon P, Vega M, Scazzocchio C, Tapia O (2002) Modelling CreA protein-DNA recognition determinants. A molecular dynamics study of fully charged CreA-DNA model in water. J Mol Struct (Theochem) 580:225–242

Pavletich NP, Pabo CO (1991) Zinc finger-DNA recognition: crystal structure of a Zif268-DNA complex at 2.1 Å. Science 252:809–817

Porciuncula J de O, Furukawa T, Mori K, Shida Y, Hirakawa H, Tashiro K, Kuhara S, Nakagawa S, Morikawa Y, Ogasawara W (2013) Single nucleotide polymorphism analysis of a *Trichoderma reesei* hyper-cellulolytic mutant developed in Japan. Biosci Biotechnol Biochem 77:534–543

Rechsteiner M, Rogers S (1996) PEST sequences and regulation by proteolysis. TIBS 21:267–271

Reyes-Dominguez Y, Narendja F, Berger H, Gallmetzer H, Fernandez-Martin R, Garcia I, Scazzocchio C, Strauss J (2008) Nucleosome positioning and histone H3 acetylation are independent processes in the *Aspergillus nidulans prnD-prnB* bidirectional promoter. Eukaryot Cell 7:656–663

Ries L, Pullan ST, Delmas S, Sunir Malla S, Martin J, Blythe MJ, Archer DB (2013) Genome-wide transcriptional response of *Trichoderma reesei* to lignocellulose using RNA sequencing and comparison with *Aspergillus niger*. BMC Genomics 14:541

Ries L, Belshaw NJ, Ilmen M, Penttila ME, Alapuranen M, Archer DB (2014) The role of CRE1 in nucleosome positioning within the *cbh1* promoter and coding regions of *Trichoderma reesei*. Appl Microbiol Biotechnol 98:749–762

Roberts CF (1963) The genetic analysis of carbohydrate utilization in *Aspergillus nidulans*. J Gen Microbiol 31:45–58

Romano AH, Kornberg HL (1968) Regulation of sugar utilisation by *Aspergillus nidulans*. Biochem Biophys Acta 158:491–493

Romano AH, Kornberg HL (1969) Regulation of sugar uptake by *Aspergillus nidulans*. Proc R Soc Lond Ser B 173:475–490

Roy P, Lockington RA, Kelly JM (2008) CreA-mediated repression in *Aspergillus nidulans* does not require transcriptional auto-regulation, regulated intracellular localisation or degradation of CreA. Fungal Genet Biol 45:657–670

Ruijter GJG, Panneman H, Vandenbroeck HC, Bennett JM, Visser J (1996) Characterisation of *Aspergillus frA1* mutant – hexose phosphorylation and apparent lack of involvement of hexokinase in glucose repression. FEMS Microbiol Lett 139:223–228

Ruijter GJG, Vanhanen SA, Gielkens MMC, Vandevondervoort PJI, Visser J (1997) Isolation of *Aspergillus niger creA* mutants and effects of the mutations on expression of arabinases and L-arabinose catabolic enzymes. Microbiology 143:2991–2998

Sakai A, Shimizu Y, Hishinuma F (1988) Isolation and characterization of mutants which show an oversecretion phenotype in *Saccharomyces cerevisiae*. Genetics 119:499–506

Sakai A, Shimizu Y, Kondou S, Chibazakura T, Hishinuma F (1990) Structure and molecular analysis of RGR1, a gene required for glucose repression in *Saccharomyces cerevisiae*. Mol Cell Biol 10:4130–4138

Scazzocchio C, Gavrias V, Cubero B, Panozzo C, Mathieu M, Felenbok B (1995) Carbon catabolite repression in *Aspergillus nidulans* – A review. Can J Bot 73:S160–S166

Schuller HJ (2003) Transcriptional control of nonfermentative metabolism in the yeast *Saccharomyces cerevisiae*. Curr Genet 43:139–160

Screen S, Bailey A, Charnley K, Cooper R, Clarkson J (1997) Carbon regulation of the cuticle-degrading enzyme PR1 from *Metarhizium anisopliae* may involve a trans-acting DNA-binding protein CRR1, a functional equivalent of the *Aspergillus nidulans* CREA protein. Curr Genet 31:511–518

Shroff, RA (1997) PhD thesis, University of Adelaide, Australia

Shroff RA, Lockington RA, Kelly JM (1996) Analysis of mutations in the *creA* gene involved in carbon catabolite repression in *Aspergillus nidulans*. Can J Microbiol 42:950–959

Shroff RA, O'Connor SM, Hynes MJ, Lockington RA, Kelly JM (1997) Null alleles of *creA*, the regulator of carbon catabolite repression in *Aspergillus nidulans*. Fungal Genet Biol 22:28–38

Sibthorp C, Wu HH, Cowley G, Wong PWH, Palaima P, Morozov IY, Weedall GD, Caddick MX (2013) Transcriptome analysis of the filamentous fungus *Aspergillus nidulans* directed to the global identification of promoters. BMC Genomics 14:847

Strauss J, Mach RL, Zeilinger S, Hartler G, Stoffler G, Wolschek M, Kubicek CP (1995) Cre1, the carbon catabolite repressor protein from *Trichoderma reesei*. FEBS Lett 376:103–107

Strauss J, Horvath HK, Abdallah BM, Kindermann J, Mach RL, Kubicek CP (1999) The function of CreA, the carbon catabolite repressor of *Aspergillus nidulans*, is regulated at the transcriptional and post-transcriptional level. Mol Microbiol 32:168–178

Takashima S, Nakamura A, Hidaka M, Masaki H, Uozumi T (1998) Isolation of the *creA* gene from the cellulolytic fungus *Humicola grisea* and analysis of CreA binding sites upstream from the cellulase genes. Biosci Biotechnol Biochem 62:2364–2370

Todd RB, Lockington RA, Kelly JM (2000) The *Aspergillus nidulans creC* gene involved in carbon catabolite repression encodes a WD40 repeat protein. Mol Gen Genet 263:561–570

Tonukari NJ, Scott-Craig JS, Walton JD (2000) The *Cochliobolus carbonum SNF1* gene is required for cell wall-degrading enzyme expression and virulence on maize. Plant Cell 12:237–247

Treitel MA, Carlson M (1995) Repression by Ssn6p-Tup1p is directed by Mig1p, a repressor/activator protein. Proc Natl Acad Sci USA 92:3132–3136

Tudzynski B, Liu S, Kelly JM (2000) Carbon catabolite repression in plant pathogenic fungi: isolation and characterization of the *Gibberella fujikuroi* and

Botrytis cinerea creA genes. FEMS Microbiol Lett 184:9–15

Tzamarias D, Struhl K (1995) Distinct TPR motifs of Cyc8 are involved in recruiting the Cyc8-Tup1 corepressor to differentially regulated promoters. Genes Dev 9:821–831

van der Voorn L, Ploegh HL (1992) The WD-40 repeat. FEBS Lett 307:131–134

van Kuyk PA, de Groot MJL, Ruijter GJG, de Vries RP, Visser J (2001) The *Aspergillus niger* D-xylulose kinase gene is co-expressed with genes encoding arabinan degrading enzymes, and is essential for growth on D-xylose and L-arabinose. Eur J Biochem 268:5414–5423

Vanderveen P, Ruijter GJG, Visser J (1995) An extreme *creA* mutation in *Aspergillus nidulans* has severe effects on D-glucose utilization. Microbiology 141:2301–2306

Vautard G, Cotton P, Fevre M (1999) The glucose repressor CRE1 of *Sclerotinia sclerotiorum* is functionally related to CreA from *Aspergillus nidulans* but not to the Mig proteins from *Saccharomyces cerevisiae*. FEBS Lett 453:54–58

Vautard-Mey G, Fevre M (2000) Mutation of a putative AMPK phosphorylation site abolishes the repressor activity but not the nuclear targeting of the fungal glucose regulator CRE1. Curr Genet 37:328–332

Vautard-Mey G, Cotton P, Fevre M (1999) Expression and compartmentation of the glucose repressor CRE1 from the phytopathogenic fungus *Sclerotinia sclerotiorum*. Eur J Biochem 266:252–259

Voegele RT, Struck C, Hahn M, Mendgen K (2001) The role of haustori in sugar supply during infection of broad bean by the rust fungus *Uromyces fabae*. Proc Natl Acad Sci USA 98:8133–8138

Wang B, Guo G, Wang C, Lin Y, Wang X, Zhao M, Guo Y, He M, Zhang Y, Pan L (2010) Survey of the transcriptome of *Aspergillus oryzae* via massively parallel mRNA sequencing. Nucleic Acids Res 38:5075–5087

Watabe J (2010) *Aspergillus oryzae* in which function of gene encoding *creC* is deficient and which can be utilized for production of seasoning, and utilization thereof. Japanese Patent JP2010075131

Wei HK, Vienken R, Weber R, Bunting S, Requena N, Fischer R (2004) A putative high affinity hexose transporter, *hxtA*, of *Aspergillus nidulans* is induced in vegetative hyphae upon starvation and in ascogenous hyphae during cleistothecium formation. Fungal Genet Biol 41:148–156

Wilson CJ, Zhan H, Swint-Kruse L, Matthews KS (2007) The lactose repressor system: paradigms for regulation, allosteric behavior and protein folding. Cell Mol Life Sci 64:3–16

Wong KH, Hynes MJ, Davis MA (2008) Recent advances in nitrogen regulation: a comparison between *Saccharomyces cerevisiae* and filamentous fungi. Eukaryot Cell 7:917–925

Zhou G, Lü J, Li Z, Li J, Wang M, Qu Y, Xiao L, Qin S, Zhao H, Xia R, Fang X (2012) Enhanced cellulase production of *Penicillium decumbens* by knocking out *creB* encoding a deubiquitination enzyme. Chin J Biotechnol 28:959–972

Ziv C, Gorovits R, Yarden O (2008) Carbon source affects PKA-dependent polarity of *Neurospora crassa* in a CRE-1-dependent and independent manner. Fungal Genet Biol 45:103–116

14 Special Aspects of Fungal Catabolic and Anabolic Pathways

MATTHIAS BROCK[1,2,3], ELENA GEIB[1,3]

CONTENTS

I. Introduction

Sensing, uptake and metabolism of available nutrients is a prerequisite for all living organisms in order to proliferate, compete and maintain in the environment. Besides bacteria, fungi play a major role in carbon and nitrogen cycling by utilising plant-derived organic matter for growth. A large number of fungi are found as free-living saprophytes in the soil, and these species are generally able to synthesise all 20 proteinogenic amino acids de novo and do not depend on the external supply of vitamins. On the other hand, some fungi have specifically adapted to a pathogenic lifestyle infecting plants, insects or mammals. Although pathogenic adaptations may have led to the development of specific infection strategies that are unique to the pathogenic species, these adaptations can also lead to the loss of certain metabolic capacities that must be compensated by nutrient acquisition from the infected host. Extreme examples for such an adaptation are the *Microsporidia*, which are obligate intracellular pathogens that can infect a wide variety of animals from different phyla (Cuomo et al. 2012). *Microsporidia* seem to form the earliest branching clade within the sequenced fungi and have adapted to modulate the infected host cells to provide essential nutrients. These pathogens express specific nucleotide transporters to acquire these essential components from the host cells, because they have lost the capacity of de novo nucleotide synthesis. Additionally, they export hexokinases to the host cell, which most likely leads to activation of host storage sugars and increases the production of building blocks that are used by the pathogen for rapid proliferation (Cuomo et al. 2012). Another, but less dramatic form of adaptation that impacts cata- and anabolism is the synergism of fungi with phototrophic organisms such as algae or plants. In case of a synergism with algae - green algae and/or cyanobacteria - the resulting species are called lichens, which are able to colonise extreme environments (Stocker-Wörgötter 2008). However, in contrast to the *Microsporidia*, several fungal species identified from lichens can still

[1]Microbial Biochemistry and Physiology, Leibniz Institute for Natural Product Research and Infection Biology e. V., -Hans Knöll Institute, Beutenbergstraße 11a, 07745 Jena, Germany; e-mail: Matthias.brock@nottingham.ac.uk
[2]Institute for Microbiology, Friedrich Schiller University Jena, 07743 Jena, Germany; e-mail: Matthias.brock@nottingham.ac.uk
[3]Fungal Genetics and Biology Group, School of Life Sciences, University of Nottingham, University Park, NG7 2RD, Nottingham, UK; e-mail: Matthias.brock@nottingham.ac.uk

Biochemistry and Molecular Biology, 3rd Edition
The Mycota III
D. Hoffmeister (Ed.)

be cultivated without their interacting partner (Crittenden et al. 1995). Specialised fungi that grow in synergistic association with plant roots form a so-called mycorrhiza. Depending on the mode of interaction with plant roots, mycorrhizae can be subdivided into endo- or arbuscular mycorrhizae and the ectomycorrhiza (Balestrini and Bonfante 2014). However, regardless of the detailed interaction, the common feature is that the fungus obtains sugars from plant photosynthesis and delivers inorganic substrates such as nitrogen and phosphorous to the plant cell (Nehls et al. 2007; Fellbaum et al. 2014). In this respect, especially those fungi forming an endomycorrhiza are frequently unable to proliferate without the direct interaction with their host or synergistic partner, which also results in adapted regulation of fungal metabolism that changes in contact with the host (Trepanier et al. 2005).

The way of nutrient acquisition and metabolic properties makes fungi also interesting for a wide variety of technologic applications. Small nutritive molecules such as sugars can easily diffuse through the fungal cell wall and penetrate the membrane by specific transporters. In contrast, large and insoluble polymers cannot be consumed by this uptake mechanism. Thus, fungi that degrade plant material need to secrete enzymes that decompose polymers like cellulose, pectin or lignin outside the cell (Hori et al. 2013). The resulting monomers from these hydrolyses are then taken up and metabolised. Therefore, fungi are interesting organisms for the production of cellulases, xylanase, pectinases, amylases and other hydrolysing enzymes (Chi et al. 2009; Abdeljalil et al. 2013; Tiwari et al. 2013; Tu et al. 2013). Furthermore, besides secreted enzymes some of the metabolic intermediates that are secreted by a variety of fungi are of major importance in our daily life. Yeasts, with the prominent example of baker's yeast *Saccharomyces cerevisiae*, tend to rapidly metabolise glucose by fermentation accompanied by the secretion of ethanol, acetate and glycerol (Woo et al. 2014). In contrast to yeasts, filamentous fungi tend to prefer complete substrate oxidation to carbon dioxide. However, under certain conditions some filamentous fungi secrete large amounts of organic acids such as the citric acid producer *Aspergillus niger* (Papagianni 2007) or the itaconic acid producer *Aspergillus terreus* (Klement and Buchs 2013).

In conclusion, despite a sometimes very close phylogenetic relationship, the metabolic capacities of fungi show great diversity. Even more, some fungi have developed different metabolic pathways or strategies to make a given nutrient available. In this chapter selected metabolic pathways will be introduced that provide examples for metabolic diversity in fungi.

II. Propionyl-CoA, A Common Metabolic Intermediate

Propionyl-CoA is an activated short-chain fatty acid that can derive from the direct activation of propionate, the degradation of odd-chain fatty acids, the side chain of cholesterol and some amino acids such as isoleucine, valine and methionine (Maerker et al. 2005; Ballhausen et al. 2009). Interestingly, propionic acid, its potassium, calcium and sodium salts are used as preservatives (Coblentz et al. 2013). Especially feed of cattle is frequently enriched with propionate. This has two reasons. First, due to its ability to reduce growth of fungi, the addition of propionate prevents feed from getting mouldy and, thus, reduces the amount of toxins that is produced by several fungal species (Marin et al. 1999; Alam et al. 2010). Second, addition of propionate prevents or reduces metabolic ketoacidosis in cattle and sheep (Emmanuel and Kennelly 1984; Bigner et al. 1997). The reason for this is the ability of mammals to convert activated propionate via the methylmalonyl-CoA pathway into succinyl-CoA, an intermediate of the citric acid cycle with its end product oxaloacetate (Aschenbach et al. 2010). Oxaloacetate is an essential acceptor molecule for acetyl-CoA units that derive from β-oxidation of fatty acids. Additionally, oxaloacetate is required for gluconeogenesis to produce sugar molecules and is, thus, removed from the citric acid cycle and needs to be regen-

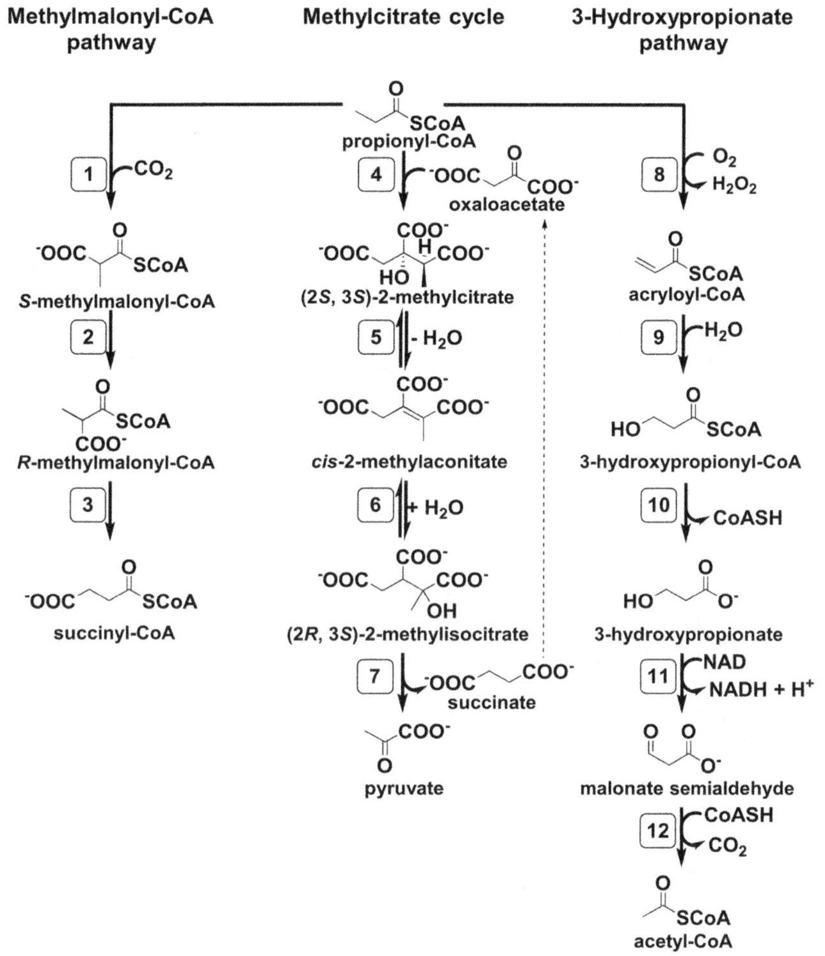

Fig. 14.1 Scheme of propionyl-CoA degradation pathways. *Left*: the methylmalonyl-CoA pathway leading to the formation of succinyl-CoA. This pathway is not present in fungi due to its dependence on coenzyme B_{12} as a cofactor in methylmalonyl-CoA mutase. *1* propionyl-CoA carboxylase, *2* methylmalonyl-CoA racemase, *3* methylmalonyl-CoA mutase. *Centre*: the methylcitrate cycle leading to the formation of pyruvate. This cycle is the main pathway for propionyl-CoA degradation in *Ascomycetes* and *Basidiomycetes*. *4* methylcitrate synthase, *5* methylcitrate dehydratase, *6* aconitase (methylisocitrate dehydratase), *7* methylisocitrate lyase. For details on the de- and rehydration reactions of (5) and (6), refer to the main text. *Right:* the 3-hydroxypropionate pathway leading to acetyl-CoA. This pathway is present in yeasts of the CUG clade. *8* acyl-CoA oxidase, *9* enoyl-CoA hydratase, *10* 3-hydroxypropionyl-CoA hydrolase, *11* 3-hydroxypropionate dehydrogenase, *12* malonate semialdehyde dehydrogenase

erated. Since mammals cannot produce oxaloacetate de novo from acetyl-CoA units, a lack of oxaloacetate can lead to severe ketoacidosis due to an arrest of fatty acid oxidation (Aschenbach et al. 2010). Thus, feeding of propionate can replenish the pool of oxaloacetate. In the methylmalonyl-CoA pathway, propionate is first carboxylated to yield *S*-methylmalonyl-CoA, which is subsequently converted into *R*-methylmalonyl-CoA via a specific racemase (Fig. 14.1). Then, a rearrangement of the carbon skeleton is required to convert the methylmalonyl-CoA into succinyl-CoA. This reaction is performed via a radical mechanism that is catalysed by the methylmalonyl-CoA mutase in the presence of the essential cofactor 5′deoxyadenosylcobalamin (coenzyme B_{12}) (Takahashi-Iniguez et al. 2012). Interestingly, despite the importance of propionyl-CoA degradation pathways, mammals do not possess genes to produce the cofactor and depend on the uptake of vitamin B_{12} from the diet (Jiang et al. 2013). Similarly, fungi are unable to produce coenzyme B_{12}, and a sufficient uptake from the environment is unlikely. Moreover, enzymatic activity determinations (Ledley et al. 1991) and genome analyses show that no gene for a methylmalonyl-CoA mutase is present (Otzen et al. 2014). Thus, since propionate is an abundant carbon source in the soil environment and propionyl-CoA is a common metabolic intermediate, alternative pathways for propionyl-CoA degradation need to exist.

The first report for such an alternative pathway derived from investigations on the yeast *Yarrowia lipolytica* that was grown on *n*-paraffins (Tabuchi and Hara 1974). A mutant of this yeast accumulated methylisocitrate in the growth medium, and the so-called methylcitrate cycle was postulated. Although enzymatic activities of some of the enzymes required for a methylcitrate cycle had been identified (Tabuchi and Uchiyama 1975), the genome sequence of *Y. lipolytica* was not available, and several questions on this pathway remained unaddressed. However, since this pathway was also detected in some bacteria, subsequent analyses on fungi and bacteria led to a more complete picture on the degradation of propionyl-CoA via this cycle.

A. Degradation of Propionyl-CoA via the Methylcitrate Cycle

First, we will start with a short overview of the reactions of the methylcitrate cycle (Fig. 14.1). It is now well established that the methylcitrate cycle starts with the condensation of propionyl-CoA with oxaloacetate to yield methylcitrate (Brock et al. 2000; Domin et al. 2009; Chittori et al. 2010; Kobayashi et al. 2013). This first intermediate undergoes a two-step de- and rehydration reaction to form methylisocitrate (Horswill and Escalante-Semerena 2001; Brock et al. 2002), which is subsequently cleaved into succinate and pyruvate and, thus, resembles an α-oxidation of pyruvate (Brock et al. 2001).

In bacteria, genes of the methylcitrate cycle are organised in a so-called *prp* operon from which at least two different main types exist. The first type is found in bacteria such as *Escherichia coli* or *Salmonella enterica* (Textor et al. 1997; Tsang et al. 1998). It contains genes for the regulator PrpR, the methylisocitrate lyase PrpB, the methylcitrate synthase PrpC, the methylcitrate dehydratase PrpD and the propionyl-CoA synthetase PrpE. The second class of bacterial *prp* operons contains a gene substitution in which the *prpD* gene is replaced by two other genes called *prpF* and *acnD*. This type is present, for example, in *Vibrio cholerae* or *Shewanella oneidensis* (Grimek and Escalante-Semerena 2004).

This gene replacement is of interest, because it indicates a special problem in the dehydration reaction of methylcitrate. Methylcitrate synthase produces (2*S*, 3*S*)-2-methylcitrate (Brock et al. 2000) and the product of an PrpD-catalysed dehydration reaction is *cis*-2-methylaconitate (Brock et al. 2002). Thus, a dehydrating enzyme must perform a *syn*-elimination of water for the direct generation of *cis*-2-methylaconitate (Fig. 14.1). However, β-eliminations of water, in which a proton is removed adjacent to a carboxylate anion, generally follow an antistereochemical mechanism (Creighton and Murthy 1990; Brock et al. 2000). Therefore, the methylcitrate dehydratase might catalyse two reactions: (1) the elimination of water from the C3–C4 bond of (2*S*, 3*S*)-2-methylcitrate, leading to γ-methylaconitate, which is (2) subsequently isomerised via a Δ-isomerase activity into *cis*-2-methylaconitate. Although experimental evidence for such a mechanism is lacking, the replacement of the *prp*D gene by *prpF* and *acnD* in some bacterial *prp*-operons strongly points to a common dehydration/isomerisation mechanism. AcnD is an aconitase like enzyme that performs the dehydration of (2*S*, 3*S*)-2-methylcitrate leading to a methylaconitate molecule with yet unknown stereochemistry (Grimek and Escalante-Semerena 2004). However, it is unlikely that *cis*-2-methylaconitate is formed by this reaction, since AcnD alone cannot replace PrpD in *S. enterica* and does not generate a product that can be used by a citric acid cycle aconitase for rehydration. Since PrpF was shown to isomerise *trans*-aconitate into *cis*-aconitate, it can be assumed that PrpF acts as an isomerase on a *trans*-methylaconitate isomer that might be produced by AcnD to yield *cis*-2-methylaconitate (Garvey et al. 2007).

In fungi the methylcitrate cycle seems to follow the *E. coli* and *S. enterica* reaction mechanisms. A **methylcitrate synthase**

(McsA) that is specifically induced in the presence of propionyl-CoA-generating nutrient sources initially forms methylcitrate from propionyl-CoA and oxaloacetate as shown for several fungi such as *Aspergillus nidulans* (Brock et al. 2000), *Aspergillus fumigatus* (Maerker et al. 2005), *Fusarium* species (Domin et al. 2009) and *Saccharomyces cerevisiae* (Graybill et al. 2007). The methylcitrate stereoisomer formed in this reaction is (2*S*, 3*S*)-2-methylcitrate as determined for the bacterial pathway. Since a deletion of the *mcsA* gene in *Aspergillus spp.* leads to accumulation of propionyl-CoA with severe disturbance of primary and secondary metabolism (Brock and Buckel 2004; Zhang et al. 2004), no alternative pathways for propionyl-CoA degradation seem to exist in these fungi. From an evolutionary point of view, methylcitrate synthases may have evolved from the citric acid cycle citrate synthase, because gene and protein alignments reveal > 50 % sequence identity (Brock et al. 2000).

The formation of methylcitrate is followed by de- and rehydration reactions. Although hardly investigated in fungi, preliminary analyses indicate that the dehydration is performed by a **methylcitrate dehydratase** (McdA) that possesses 50–60 % sequence identity to *PrpD* from *E. coli* (unpublished). Interestingly, except for a few itaconate-producing fungi such as *Aspergillus terreus* and *Aspergillus itaconicus* that possess a *cis*-aconitate decarboxylase with sequence similarity to methylcitrate dehydratases (Kanamasa et al. 2008; Steiger et al. 2013), no other enzymes similar to McdA are encoded in fungal genomes. This indicates that fungal methylcitrate dehydratases were directly acquired from a bacterial source. The subsequent rehydration of *cis*-2-methylaconitate is most likely performed by a citric acid cycle **aconitase**, as shown for *E. coli* and *S. enterica* (Horswill and Escalante-Semerena 2001; Brock et al. 2002). This is possible, because *cis*-2-methylaconitate fits into the active site of aconitases when bound in the isocitrate mode resulting in (2*R*, 3S)-2-methylisocitrate (Lauble and Stout 1995).

The final pathway-specific reaction is the cleavage of (2*R*, 3*S*)-2-methylisocitrate into succinate and pyruvate by a **methylisocitrate lyase**. Succinate can be used to replenish the oxaloacetate pool, whereas pyruvate can be used for gluconeogenesis or energy metabolism. Phylogenetic analyses of fungal methylisocitrate lyases have shown that these enzymes likely evolved from the isocitrate lyase of the glyoxylate cycle (Müller et al. 2011) that cleaves isocitrate into succinate and glyoxylate (for further details of the glyoxylate cycle, see below and Fig. 14.2). However, the bacterial methylisocitrate lyases only share low sequence identity to isocitrate lyases and fungal methylisocitrate lyases, indicating an independent evolution of bacterial and fungal methylisocitrate lyases (Müller et al. 2011). Nevertheless, the structural fold of methylisocitrate lyases appears to be conserved (Liu et al. 2005), and a specific active site motif required to harbour the methyl group of methylisocitrate is conserved in these independently evolved enzymes (Müller et al. 2011). This shows that nature developed the same enzymatic solution for a specific metabolic problem.

B. Alternative Propionyl-CoA Degradation Pathways in Fungi

The methylcitrate cycle appears as the main metabolic pathway in fungi to cope with the metabolic intermediate propionyl-CoA. Key enzymes such as the methylcitrate synthase can be found encoded in genomes of *Basidiomycetes* and most *Ascomycetes* (Müller et al. 2011). However, by searching the currently available genomes of *Zygomycetes*, no homologous genes are detected. This indicates either that the fungal methylcitrate cycle has specifically evolved in a common ancestor of *Basidiomycetes* and *Ascomycetes* after the branch from the *Zygomycetes* or that the genes have been completely lost in the latter groups. Thus, it has remained unclear by which mechanisms *Zygomycetes* metabolise propionyl-CoA. However, even more surprisingly, the so-called CUG clade of the *Saccharomycetales*, among them the important human pathogen *Candida albicans*, is also lacking methylcitrate cycle genes, and no activities for methylcitrate cycle enzymes can be detected from cell-free extracts (Otzen et al. 2014).

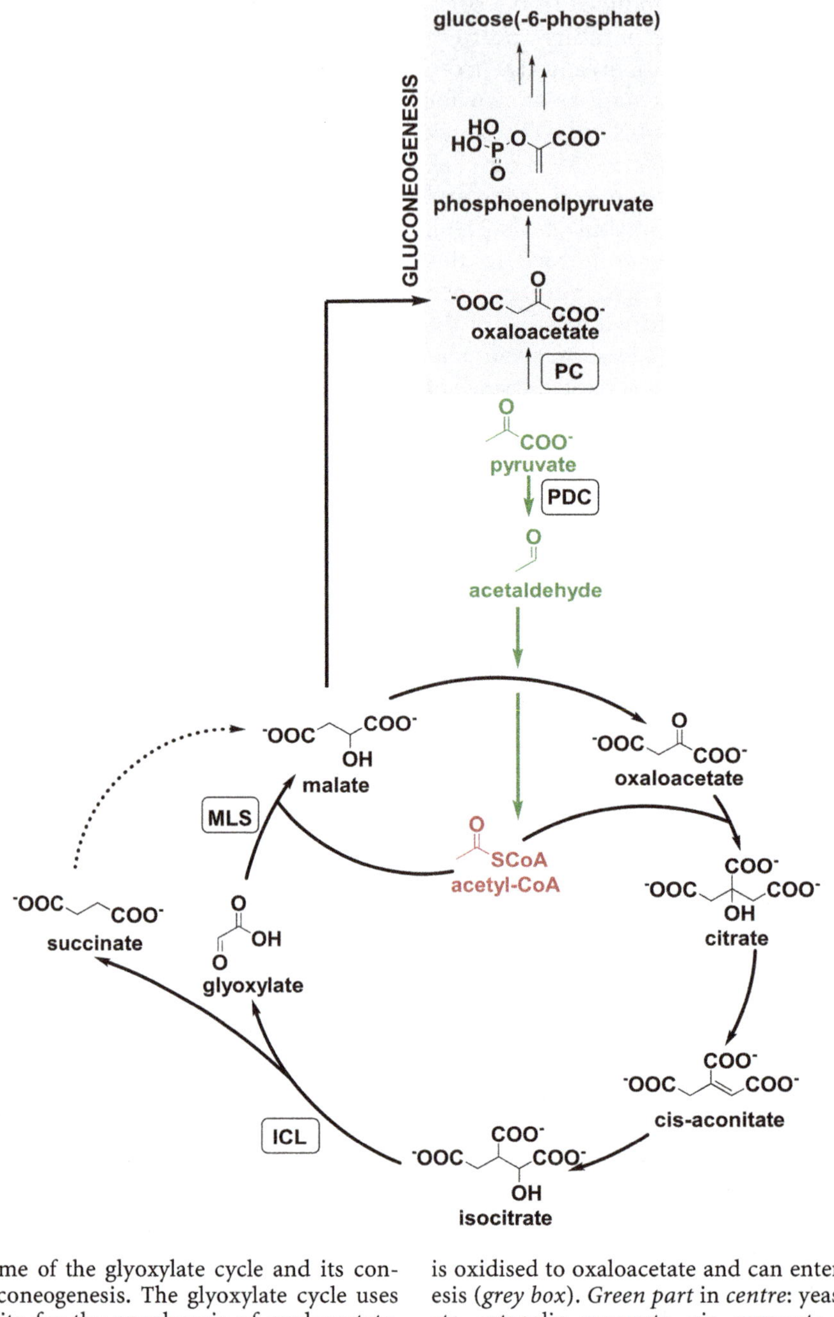

Fig. 14.2 Scheme of the glyoxylate cycle and its connection to gluconeogenesis. The glyoxylate cycle uses acetyl-CoA units for the anaplerosis of oxaloacetate. Acetyl-CoA condenses with oxaloacetate to form citrate, which is isomerised to isocitrate. The glyoxylate cycle-specific isocitrate lyase (**ICL**) cleaves isocitrate into succinate and glyoxylate. Succinate is regenerated by citric acid cycle enzymes to malate. Glyoxylate is condensed via malate synthase (**MLS**) to malate, which is oxidised to oxaloacetate and can enter gluconeogenesis (*grey box*). *Green part* in *centre*: yeasts decarboxylate cytosolic pyruvate via pyruvate decarboxylase (**PDC**) to acetyl-CoA. To enter gluconeogenesis, acetyl-CoA must be converted to oxaloacetate via the glyoxylate cycle. In filamentous fungi, pyruvate carboxylase (**PC**) carboxylates pyruvate to oxaloacetate, which can directly enter gluconeogenesis without the need for the glyoxylate cycle

Nevertheless, *C. albicans* is able to grow, though at low rates, on propionate and its growth is well supported on the odd-chain fatty acid valerate that is composed of five carbon units (Otzen et al. 2014). A valerate degradation via β-oxidation leads to equal amounts of acetyl- and propionyl-CoA. Since an intracellular accumulation of propionyl-CoA would interfere with other enzymatic reactions dealing with CoA-ester substrates such as the pyruvate dehydrogenase complex or the succinyl-CoA synthetase, an efficient removal of propionyl-CoA is also essentially required (Brock and Buckel 2004).

An alternative pathway for propionyl-CoA degradation had been suggested for some insects and plants that proceeds via the key intermediate 3-hydroxypropionate and finally ends in the formation of acetyl-CoA (Halarnkar and Blomquist 1989). However, experimental evidence for the existence of the respective enzymes was mostly lacking. This has recently changed by analysing the proteome and transcriptome of *C. albicans* cells grown on propionate or valerate medium (Otzen et al. 2014). As depicted in Fig. 14.1, it is assumed that propionyl-CoA initially enters the normal pathway of β-oxidation where a **fatty acid oxidase** generates acryloyl-CoA. In the next step acryloyl-CoA is hydrated by the bifunctional **enoyl-CoA hydratase/dehydrogenase** yielding 3-hydroxypropionyl-CoA. This intermediate is transported from peroxisomes to mitochondria, where the specific enzymes of the 3-hydroxypropionate pathway are located. A **hydroxypropionyl-CoA hydrolase** cleaves the CoA-ester, which leads to the intermediate 3-hydroxypropionate. This key intermediate is oxidised to malonate semialdehyde as shown by enzymatic characterisation of the essential **3-hydroxypropionate dehydrogenase.** Additionally, a 3-hydroxypropionate dehydrogenase deletion mutant accumulates significant amounts of 3-hydroxypropionate when grown on acetate/propionate or valerate containing medium (Otzen et al. 2014). Finally, malonate semialdehyde is oxidatively decarboxylated to yield acetyl-CoA via an **aldehyde dehydrogenase.**

Although the 3-hydroxypropionate pathway is very elegantly used to degrade propionyl-CoA and to provide mitochondrial acetyl-CoA, it also causes a severe problem for *C. albicans*. While filamentous *Ascomycetes* are able to transport mitochondrial acetyl-CoA into the cytosol by using an export of citrate with subsequent cytosolic acetyl-CoA formation via ATP citrate lyase, this latter enzyme is not present in *C. albicans* (Hynes and Murray 2010). Therefore, acetyl-CoA is trapped within mitochondria and can only be used for energy metabolism but not for anabolic processes. This may explain the very low growth rate of *C. albicans* on propionate. On the other hand, growth on valerate is well supported.

Valerate first undergoes one round of β-oxidation, which leads to the formation of peroxisomal acetyl-CoA, whereas the resulting propionyl-CoA forms mitochondrial acetyl-CoA. The peroxisomal acetyl-CoA can now be used for biosynthetic purposes via conversion by the malate forming glyoxylate cycle, whereas the mitochondrial acetyl-CoA is available for energy metabolism. Thus, the modified β-oxidation with a branch into the 3-hydroxypropionate pathway can be efficiently used for the utilisation of propionyl-CoA from fatty acids, but appears less suitable for the utilisation of propionate. However, the yeast *S. cerevisiae* appears, from a phylogenetic and metabolic point of view, more closely related to *C. albicans* than to filamentous *Ascomycetes* but metabolises propionate via the methylcitrate cycle. Nevertheless, it should be mentioned that in contrast to filamentous fungi, *S. cerevisiae* can only metabolise propionate in co-metabolism with limited amounts of glucose and cannot use propionate as sole carbon source (Pronk et al. 1994). Thus, at least two questions remain open: (1) Is there a specific advantage of using either the methylcitrate cycle or the 3-hydroxypropionate pathway? (2) At which time did the fungi from the CUG clade lose the genes of a functional methylcitrate cycle? An answer to this question might derive from studies on *Zygomycetes*. Analyses of currently available genomes imply that genes for a methylcitrate cycle are lacking, but there is no experimental proof for the existence of a modified β-oxidation pathway via 3-hydroxypropionate. A survey on currently available zygomycete genomes is provided in Chap. 15. Therefore, *Zygomycetes* either developed a third fungal, yet unknown, pathway for propionyl-CoA degradation or they are using the *C. albicans* pathway. If so, the 3-hydroxypropionate pathway would be more ancient than the methylcitrate cycle. This means that modern *Basidiomycetes* and *Ascomycetes* lost this ancient pathway and developed the methylcitrate cycle, which might make them more independent of fatty acid β-oxidation and produces the more versatile intermediate pyruvate rather than acetyl-CoA.

III. Growth on Gluconeogenic Substrates

Sugars are of major importance for fungal growth and development. The fungal cell wall is mainly composed of a mixture of glucans (mainly β-1,3- and β-1,6-glucan), chitin, chitosan and glycosylated proteins (Free 2013). For more details on the fungal cell wall please refer to Chaps. 8 and 9. Additionally, sugars are required for nucleotide synthesis and for the provision of building blocks for amino acid synthesis. Therefore, when cultivated on sugar-free media, fungi require the de novo synthesis of glucose via gluconeogenesis. The common intermediate to initiate gluconeogenesis is pyruvate, which is carboxylated to the energy-rich intermediate oxaloacetate that is required to circumvent the irreversible pyruvate kinase reaction from glycolysis (Hers and Hue 1983).

As outlined above, propionyl-CoA degradation via the methylcitrate cycle generates pyruvate, which is directly suitable for gluconeogenesis. In contrast, the 3-hydroxypropionate pathway forms acetyl-CoA, which cannot serve as a direct gluconeogenic precursor. Similarly, degradation of fatty acids via β-oxidation or direct activation of acetate yields acetyl-CoA. The ability of using acetyl-CoA as a gluconeogenic precursor requires the action of a specific metabolic pathway, the so-called **glyoxylate cycle** (Fig. 14.2). In this cycle two acetyl units are condensed to yield malate, which is suitable for gluconeogenesis (Peraza-Reyes and Berteaux-Lecellier 2013). In humans genes for a functional glyoxylate cycle are lacking. Therefore, they are unable to survive for a longer period with fatty acids as a sole carbon and energy source although fatty acids are degraded to acetyl-CoA units that serve for energy metabolism. Thus, additional nutrients such as proteins are required to replenish the pool of biosynthetic precursors during fat-burning starvation periods frequently resulting in the degradation of muscle-derived proteins.

In contrast to humans, the glyoxylate cycle is common to most fungi, which allows them to live at the expense of fatty acids as sole carbon and energy source (Dunn et al. 2009). However, a prominent exception is the model yeast *Schizosaccharomyces pombe*. Although a gene for isocitrate lyase, which is one of the two key enzymes of the cycle, can be identified by BLAST analyses, its sequence seems to have accumulated several point mutations that might have resulted in a loss of function (Müller et al. 2011). In agreement, no glyoxylate cycle enzyme activity is detected in *S. pombe* and acetate is only utilised in the co-metabolism with glucose (Tsai et al. 1989). However, isocitrate lyase is detected in *Chytridiomycota*, *Zygomycetes*, *Basidiomycetes* and *Ascomycetes* showing that this cycle is very ancient in the fungal lineage. However, even among *Ascomycetes* the regulation, importance and use of the glyoxylate cycle differ as will be outlined below.

A. Utilisation of the Glyoxylate Cycle in *Ascomycetes*

The glyoxylate cycle branches from the citric acid cycle and is characterised mainly by two key enzymes: **isocitrate lyase** (ICL) and **malate synthase** (MLS) (Dunn et al. 2009). Similar to the citric acid cycle, acetyl-CoA initially condenses with oxaloacetate to form citrate, which is subsequently converted into isocitrate via aconitase de- and rehydration reactions. While the citric acid cycle would normally continue with the first decarboxylation reaction that yields α-ketoglutarate, isocitrate lyase mediates a C–C bond cleavage on isocitrate resulting in the products glyoxylate and succinate (Fig. 14.2). While succinate is regenerated to oxaloacetate in order to replenish the substrate for the initial citrate synthase reaction, glyoxylate undergoes a condensation with a second molecule of acetyl-CoA. This reaction is mediated by MLS and leads to malate, which is easily oxidised to oxaloacetate. Thus, from two acetyl units, a new molecule of oxaloacetate is formed that can be used for biosynthetic purposes. It should be mentioned that, in contrast to the mitochondrial citric acid cycle, the fungal glyoxylate cycle is generally located within peroxisomes (Peraza-Reyes and Berteaux-Lecellier 2013). This requires additional shuttling of molecules over compartmental membranes, which will not be discussed in detail.

In general, activation of expression of the glyoxylate cycle genes is under control of carbon catabolite repression, which means that glyoxylate cycle genes are not transcribed or rapidly degraded in the presence of glucose (De Lucas et al. 1994a, b). In *Aspergillus spp.* the main carbon catabolite repressor CreA binds to the promoter region of the target genes and interferes with initiation of gene transcription (Strauss et al. 1999). Similarly, in the yeast *S. cerevisiae*, glyoxylate cycle genes are under negative control of the carbon catabolite repressor Mig1p (Turcotte et al. 2010).

In the absence of glucose, the carbon catabolite repressor does no longer bind to target promoter sequences. This enables transcription factors such as the glyoxylate cycle-specific transcription factor FacB in *A. nidulans* (Todd et al. 1998) or the more global gluconeogenic transcription factors Cat8p, Sip4p, Adr1p and Rds2p in *S. cerevisiae* (Soontorngun et al. 2012) to activate gene expression. Besides this rather general carbon control mechanism of gene regulation, the dimorphic pathogenic fungus *Paracoccidioides brasiliensis* follows a different strategy to regulate activation and inactivation of glyoxylate cycle genes.

P. brasiliensis grows as a filamentous fungus at temperatures below 26 °C, but switches into yeast growth at elevated temperatures. Regulation of the glyoxylate cycle in the filamentous form follows the regulation mechanisms as seen for *Aspergillus spp.* However, regardless of the available nutrient source, glyoxylate cycle genes are constitutively transcribed when cells are growing in the yeast form.

Interestingly, although ICL, as a marker for the glyoxylate cycle, can be detected by Western blot on all nutrient sources, ICL activity is only detected on acetyl-CoA generating nutrient sources, but not on glucose (Cruz et al. 2011). Two-dimensional gel electrophoresis revealed an extensive phosphorylation of ICL on glucose, which was strongly reduced when yeast cells were shifted to acetate medium. In addition, dephosphorylation of ICL that derived from glucose medium results in strongly increased activity, indicating that a phosphorylation/dephosphorylation mechanism is responsible for glyoxylate cycle regulation in *P. brasiliensis* yeast cells (Cruz et al. 2011).

Besides these different strategies to regulate the activity of the glyoxylate cycle, the impact of this cycle on various nutrient sources also differs among fungi. In *Aspergillus spp.*, the glyoxylate cycle is essential when cells are growing on nutrients exclusively producing acetyl-CoA such as ethanol, acetate, acetamide or fatty acids, and the cycle is specifically induced by FacB (Todd et al. 1997). However, the pathway is dispensable when growing on peptone or most amino acids since sufficient pyruvate is formed that can be carboxylated to oxaloacetate for gluconeogenic purposes (Brock 2009). However, this situation differs in yeasts as shown for *C. albicans*. Here, isocitrate lyase is induced in the presence of all gluconeogenic nutrient sources and during carbon starvation. Additionally, a *C. albicans* ICL mutant is unable to grow on most gluconeogenic nutrient sources, implying that the glyoxylate cycle is directly linked with gluconeogenesis (Brock 2009). How is this explained?

First, transcription of the glyoxylate cycle genes in yeasts seems to be tightly linked with transcription of genes from gluconeogenesis (Turcotte et al. 2010). In contrast, this regulation is separated in *Aspergillus spp.* Here, a heterodimer of transcription factors AcuM and AcuK is responsible for activation of gluconeogenesis. AcuM and AcuK are always activated in the absence of glucose, independent of an acetyl-CoA-generating nutrient source (Suzuki et al. 2012), and these transcription factors are insufficient to induce ICL and MLS. This requires FacB, the glyoxylate cycle-specific transcription factor that is only induced in the presence of nutrients generating acetyl-CoA.

Another difference between yeasts and filamentous fungi is the completeness of substrate oxidation. Filamentous fungi generally rely on complete substrate oxidation for gain of energy. This requires the transport of pyruvate and acetyl-CoA to mitochondria where oxidative phosphorylation takes place. In contrast, yeasts prefer fermentation (Tylicki et al. 2008). Thus, on glucose yeasts produce ethanol from pyruvate. Pyruvate is decarboxylated in the cytoplasm via the **pyruvate decarboxylase** to acetaldehyde, which is either reduced to ethanol or oxidised to acetate (Fig. 14.2). The key enzyme pyruvate decarboxylase is highly active on glucose, but expression analyses with a luciferase reporter in *C. albicans* revealed that the gene is also strongly transcribed on gluconeogenic substrates (Jacobsen et al. 2014). Thus, cytosolic pyruvate, deriving both from glycolysis and also from the degradation of gluconeogenic substrates, is rapidly decarboxylated to acetaldehyde and might not be amenable to carboxylation by a mitochondrial **pyruvate carboxylase.** By this strategy large amounts of cytosolic acetyl-CoA can be generated, which makes an ATP citrate lyase dispensable. However, in turn, it requires the glyoxylate cycle to generate oxaloacetate from acetyl-CoA units for gluconeogenesis.

In conclusion, the glyoxylate cycle appears well conserved among fungi, but its regulation and use differ among fungal species and are dependent on the set of enzymatic activities from accompanying metabolic pathways.

IV. Amino Acid Biosynthesis and Utilisation as Nutrient Sources

Most fungi have specifically adapted to sense, transport and utilise amino acids, as de novo synthesis of amino acids is very energy consuming. Additionally, amino acids can provide the cell with important macroelements such as nitrogen and sulphur and can serve as carbon sources. Since environmental amino acids are frequently bound in proteins that are too large for direct import, fungi are rich in extracellular proteases and peptidases that cleave proteins in smaller subunits that can be taken up by various peptide and amino acid transporters (Yike 2011). Besides that, proteases and peptidases can support the virulence of pathogenic fungi, because these enzymes not only provide nutrients from the host environment but also ease tissue penetration, inactivate components of the immune system or cause inflammation and allergies (Gropp et al. 2009; Yike 2011).

The uptake of amino acids as nitrogen sources in fungi follows a specific preference. The nitrogen-rich amino acids glutamine and arginine as well as ammonia generally constitute preferred nitrogen sources in fungi, whereas other amino acids, urea, purines, nitrate and nitrite are less preferred. Thus, similar to carbon catabolite repression, a nitrogen catabolite repression system is involved in sensing and utilisation of the available nitrogen sources. A key player in this regulatory path is a GATA-type transcription factor. One of the best studied members is AreA from *Aspergillus spp.* (Gonzalez et al. 1997), which has close relatives in other fungi such as NIT-2 from *Neurospora crassa* (Tao and Marzluf 1999) or Gat1 in the basidiomycete *Cryptococcus neoformans* (Kmetzsch et al. 2011). Fungal nitrogen metabolism is reviewed in great depth in Chap. 11. The AreA-type GATA factors sense the intracellular glutamine levels and interact with other transcription factors that are specific for a given nitrogen source as exemplified for nitrate assimilation (Berger et al. 2008) or proline utilisation (Garcia et al. 2004).

Interestingly, some amino acids cannot be used as carbon sources and a few even do not serve as a nitrogen sources in selected fungi as will be exemplified by the degradation of histidine. However, regardless the ability to utilise external amino acids as carbon, nitrogen or sulphur sources, fungi are able to import all proteinogenic amino acids to replenish the intracellular amino acid pool. Even more, amino acids can be stored in vacuoles and mobilised from these compartments under conditions of starvation (Sekito et al. 2014).

Although fungi prefer to procure amino acids from the environment, there is a strong competition for nutrients in microbial communities and some or all amino acids may become limited. To cope with this limitation, fungi are able to synthesise all amino acids de novo. This contrasts the anabolic capacity of mammals, which have lost pathways for the biosynthesis of several amino acids and therefore must obtain amino acids from the diet. Examples are the biosynthesis of the basic amino acids histidine, lysine and arginine, the aromatic amino acids tryptophane and phenylalanine or the aliphatic branched-chain amino acids isoleucine, leucine and valine. Thus, these amino acids are called "essential" amino acids, although it should be mentioned that also the uptake of other amino acids for which biosynthetic pathways still exist is required by humans (Wu 2014). Therefore, it is of great interest to modify microorganisms for amino acid production in order to use the amino acids as food and feed additives. While most amino acids are produced from bacterial sources such as *Corynebacterium glutamicum* and *E. coli* (Wendisch 2014), especially the synthesis of the amino acid lysine has been studied in detail in several fungal species. This amino acid not only is important as a food additive but also provides precursors for penicillins and cephalosporins. Therefore, knowledge on its biosynthesis is indispensable for industrial production of β-lactam antibiotics (Fazius et al. 2013). Fungal lysine biosynthesis seems to have evolved independently of bacterial and plant lysine biosynthesis, and at least the use of different isoforms of one enzyme is dependent on the fungal lifestyle as will be outlined later.

A. Histidine Degradation in Fungi

Histidine is an important amino acid frequently found in the catalytic centre of enzymes because it is the only amino acid from which a protonation and deprotonation of the side chain at neutral pH is possible (Kulis-Horn et al. 2014). Examples for enzymes with essential functional histidine residues are the bisphosphoglycerate mutase from glycolysis/gluconeogenesis (White et al. 1993), the catalytic triad of serine proteases (Betzel et al. 2001) or transcription factors like the carbon catabolite repressor Mig1 from *S. cerevisiae* that contains zinc coordinating Cys_2–His_2 zinc fingers (De Vit et al. 1997). Histidine has been assumed as a very ancient amino acid (Kulis-Horn et al. 2014) since its synthesis starts from phosphoribosyl pyrophosphate and ATP.

All genes identified in bacteria to contribute to histidine biosynthesis are also found in yeast (Alifano et al. 1996).

Although histidine can be synthesised by *S. cerevisiae* de novo, this yeast is unable to proliferate with histidine as sole nitrogen source (Brunke et al. 2014). However, *HIS3* codes for the **imidazole glycerol-phosphate dehydratase** and deletions are frequently used as a selection marker in yeast transformations (Baudin et al. 1993). Importantly, *his3* mutants are viable in the presence of externally supplied histidine, indicating that *S. cerevisiae* can utilise external histidine to replenish the intracellular histidine pool. This is in agreement with the presence of high- and low-affinity histidine transport systems (Bajmoczi et al. 1998).

However, in contrast to the inability of *S. cerevisiae* to utilise histidine as a nitrogen source, studies on the filamentous fungus *A. nidulans* showed that histidine serves at least for some fungi as nitrogen source, although it cannot be used as a sole source of carbon (Polkinghorne and Hynes 1982). When grown with histidine as sole nitrogen source, *A. nidulans* most likely releases ammonium by a **histidase** (histidine ammonia-lyase) since the reaction product urocanate accumulated in the culture broth (Fig. 14.3). Urocanate is a well-known intermediate from bacterial and most likely also from the mammalian histidine degradation pathway that leads to the end product glutamate. Urocanate is subsequently hydrated by an **urocanase** yielding 4-imidazolone-5-propionate. Due to the accumulation of urocanate in the medium and the lack of urocanase activity in *A. nidulans*, the enzymatic set for glutamate synthesis appears to be lost (Polkinghorne and Hynes 1982) explaining the inability of *A. nidu-*

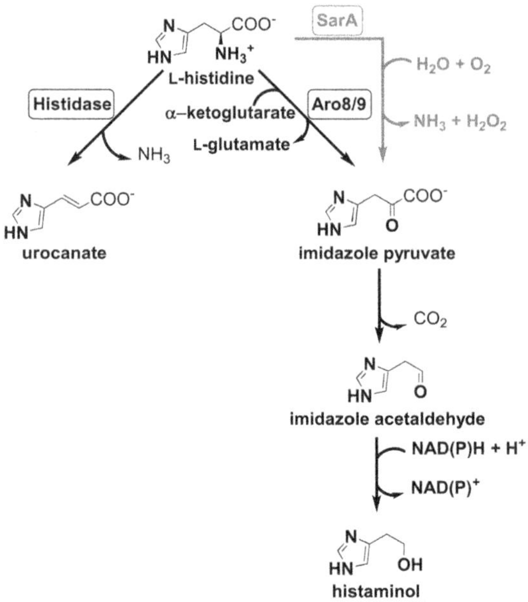

Fig. 14.3 Postulated fungal pathways for the use of histidine as nitrogen source. A histidase releases ammonia from histidine resulting in urocanate, which is secreted to the medium. Alternatively, an L-amino acid oxidase such as SarA from *A. nidulans* releases ammonia from histidine by producing hydrogen peroxide and imidazole pyruvate. The fate of imidazole pyruvate in filamentous fungi and the impact of this pathway are unclear, since only low L-amino acid oxidase activity is detected in wild-type cells. Finally, yeasts use aromatic amino acid aminotransferases such as Aro8 or Aro9 to transfer the amino group from histidine to an α-keto acid resulting in imidazole pyruvate. Via reactions of the Ehrlich pathway, imidazole pyruvate is further converted to the fusel alcohol histaminol

lans to use histidine as carbon source. Furthermore, a BLAST search on fungal genomes against urocanase from bacteria or mammals revealed no positive hit from any fungal species. This implies that (1) either the amino acid sequence for urocanases is not highly conserved and therefore not identified by sequence analyses, (2) the degradation of urocanate follows a different pathway or (3) histidine can generally not be used as a carbon source by fungi. However, due to the presence of histidase-like enzymes in many fungal species (Brunke et al. 2014), degradation to urocanate might provide a general mechanism for utilisation of histidine as a nitrogen source. Nevertheless, even for *A. nidulans* an alternative degradation pathway might exist (Fig. 14.3).

A point mutation in the *areA* gene of *A. nidulans*, denoted *areA102*, causes increased growth rates on several amino acids, among them histidine (Davis et al. 2005). It has been shown that this mutation leads to increased expression of *sarA*, which encodes an L-amino acid oxidase with broad substrate specificity. Thus, histidine might be converted to imidazole pyruvate under the formation of H_2O_2 and the release of ammonia. However, in wild-type cells this oxidase activity is hardly detectable (Davis et al. 2005), leaving room for speculation on the normal in vivo function of this reaction, since it would not lead to the formation of urocanate as detected in the culture broth of *A. nidulans* wild-type cells (Polkinghorne and Hynes 1982).

However, a third pathway might account for histidine utilisation in *Candida glabrata*, a yeast that is closely related to *S. cerevisiae*. Growth of *C. glabrata* on histidine was well supported, although no histidase was found in its genome. In contrast, two genes coding for **aromatic amino acid aminotransferases** can be detected (Fig. 14.3). One of these aromatic aminotransferases, called Aro8, was essentially required for growth on aromatic amino acids, whereas the second aromatic aminotransferase, Aro9, was dispensable. Unexpectedly, deletion of *C. glabrata ARO8* also resulted in the inability to utilise histidine as nitrogen source (Brunke et al. 2014). In agreement with a requirement of Aro8 for growth on histidine, purified Aro8 protein catalysed the conversion of histidine to imidazole pyruvate when α-ketoglutarate was added as acceptor for the amino group. Thus, *C. glabrata* most likely acquires nitrogen from histidine via an aminotransferase reaction.

Surprisingly, Aro8 and Aro9 are also conserved in *S. cerevisiae* and Sc*ARO8* shows a similar expression pattern as Cg*ARO8*. However, while in *C. glabrata* Cg*ARO9* cannot be induced -most likely due to a genome inversion upstream Cg*ARO9* - *S. cerevisiae* Sc*ARO9* is strongly induced by aromatic amino acids and also by histidine. Due to the simultaneous activation of Sc*ARO8* and Sc*ARO9* this might result in aminotransferase levels that produce toxic amounts of intermediates from histidine degradation. In agreement, when *S. cerevisiae* was cultivated with aromatic amino acids that strongly induce Sc*ARO9* expression, histidine supplementation completely abolished growth. In contrast, histidine only had a moderate growth inhibitory effect when added in combination with ammonium as nitrogen source, which does not induce Sc*ARO8* or Sc*ARO9* expression (Brunke et al. 2014). Thus, *S. cerevisiae* and *C. glabrata*, can, in principle, acquire ammonium from histidine via an aminotransferase mechanism. However, the co-expression of Sc*ARO9* might result in accumulation of toxic amounts of intermediates from imidazole pyruvate, which might be circumvented by *C. glabrata* either due to lower aromatic aminotransferase levels or due to an optimised detoxification of intermediates. It can be assumed that the final product of histidine degradation from imidazole pyruvate in yeasts are fusel alcohols that are generated via the Ehrlich pathway (Hazelwood et al. 2008). This is supported by studies on the fate of histidine during alcoholic fermentation of grape must. Added histidine was slowly consumed and resulted in the formation of imidazole ethanol (histaminol) (Lopez-Rituerto et al. 2013). Histaminol can be assumed to derive from a decarboxylation of imidazole pyruvate yielding imidazole acetaldehyde, which is subsequently reduced to histaminol (Fig. 14.3). Especially the aldehyde intermediate might cause toxic effects.

In conclusion, histidine does not seem to serve as a carbon source for any fungus. Moreover, utilisation of histidine as a nitrogen source differs among fungal species in respect of the mechanism by which nitrogen is released. In addition, enzyme levels might require a strict control to avoid the production of toxic levels of intermediates as seen from the comparison of *C. glabrata* and *S. cerevisiae*.

B. Synthesis of the Amino Acid Lysine

While the degradation of amino acids in fungi has hardly been studied, more information is available on the biosynthesis of amino acids. Of special interest are those amino acids for which no biosynthetic pathway is present in humans. These "essential" amino acids are not only valuable food and feed additives but also interesting targets for new antibiotics, because interruption of their synthesis is assumed to cause only minor side effects on human metabolism (Hutton et al. 2007).

Of special interest is the biosynthesis of the amino acid lysine, for two reasons. First, it has important functions in several enzymatic reaction mechanisms, in binding proteins such as histones and transcription factors to DNA and in signal sequences for subcellular localisation of proteins. Second, lysine is essential for the structure of the bacterial cell wall and the production of β-lactam antibiotics (Fazius et al. 2013). Although lysine can be synthesised by bacteria, plants and fungi, different pathways are involved, whereby plant lysine biosynthesis is highly related to the bacterial pathway and the fungal pathway seems to have evolved independently, most likely from the citric acid cycle and leucine biosynthesis (Irvin and Bhattacharjee 1998; Xu et al. 2006).

With the exception of some thermophilic bacteria from the *Deinococcus-Thermus* phylum (Nishida and Nishiyama 2012), bacteria and plants generally synthesise lysine via the *meso*-diaminopimelate pathway, whereby different isoforms of the pathway exist (Hudson et al. 2005). In brief, synthesis starts by phosphorylation of aspartate that is converted to aspartate semialdehyde. The semialdehyde condenses with pyruvate resulting in 2,3-dihydrodipicolinate and subsequently in tetrahydrodipicolinate. This is the common intermediate in all variants of the pathway. Then, either a succinylase, an acetylase, a dehydrogenase or an aminotransferase branch leads to the formation of *meso*-diaminopimelate that is finally decarboxylated to lysine (Watanabe and James 2011).

In contrast to bacteria and plants, fungi and bacteria from the *Deinococcus-Thermus* phylum synthesise lysine via the **α-aminoadipate pathway** (Fig. 14.4). This pathway can be divided in two parts. The first part leads to the intermediate α-aminoadipate, which is a precursor molecule for the synthesis of β-lactam antibiotics, whereas the second part leads to the formation of the final product lysine (Fazius et al. 2013). Synthesis starts by condensation of α-ketoglutarate with acetyl-CoA via **homocitrate synthase** leading to the product homocitrate (Schöbel et al. 2010). While filamentous fungi only seem to possess one enzyme encoding homocitrate synthase with moderate sensitivity to feedback inhibition by lysine, *S. cerevisiae* possesses two isozymes of homocitrate synthase that show different expression

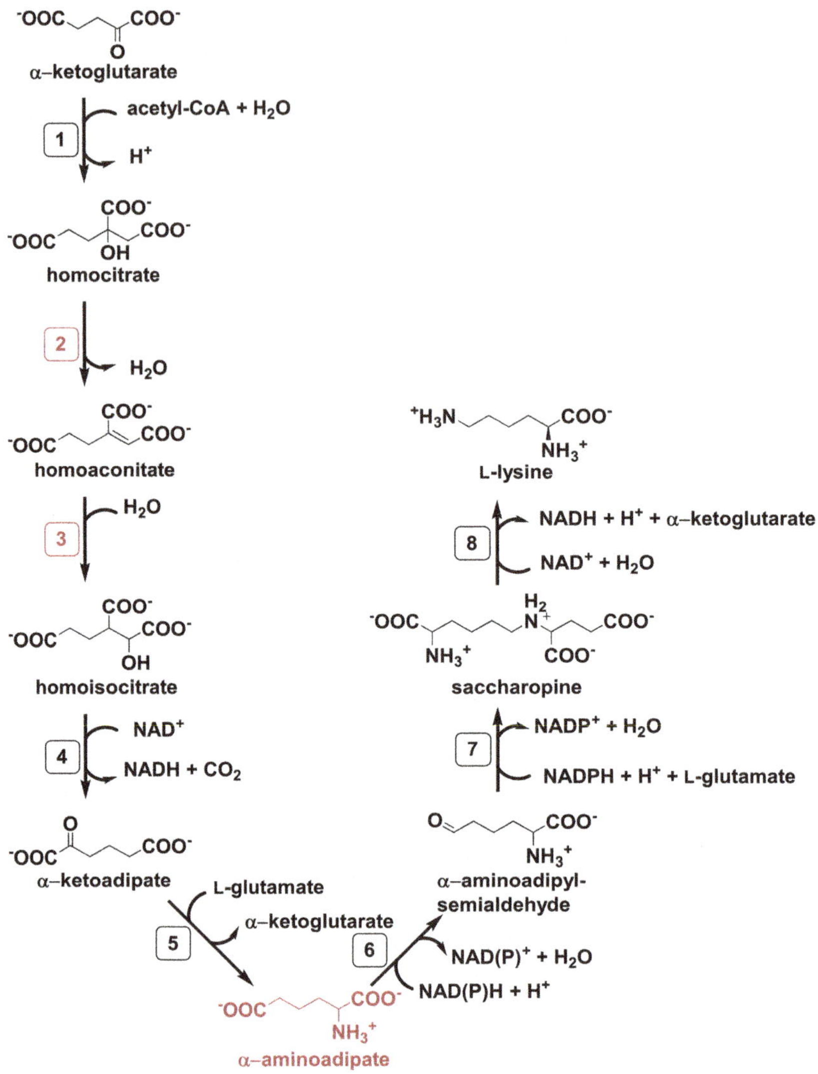

Fig. 14.4 Scheme of the fungal α-aminoadipate pathway. *1* homocitrate synthase, *2* aconitase, *3* homoaconitase, *4* homoisocitrate dehydrogenase, *5* α-aminoadipate aminotransferase, *6* α-aminoadipate reductase, *7* saccharopine reductase, *8* saccharopine dehydrogenase. Details on reactions (2) and (3) (shown in *red*) and the differences between yeasts and filamentous fungi are described in the text

levels depending on respiratory or fermentative growth and display different degrees of feedback inhibition by lysine (Quezada et al. 2011). In *S. cerevisiae* expression of lysine, biosynthesis is under control of the pathway-specific transcriptional activator Lys14, which is induced by the downstream intermediate α-aminoadipate semialdehyde (Feller et al. 1999). Thus, due to lysine feedback inhibition of homocitrate synthases, *S. cerevisiae* can counteract this activation by measuring the internal lysine pool. Interestingly, such regulatory circuit and the existence of a transcription factor specifically contributing to induction of lysine biosynthesis have not been described for filamentous fungi. This might be explained by the fact that yeasts are unable to produce β-lactam antibiotics and exclusively use the α-aminoadipate pathway for lysine biosynthesis. In contrast, in β-lactam-producing fungi, a significant proportion of α-aminoadipate may be required and withdrawn for antibiotic production even in the presence of high internal lysine levels. Thus, a feedback mechanism by lysine or the activation by a downstream intermediate would interfere with the dual function of this pathway.

Homocitrate undergoes a de-and rehydration reaction to form homoisocitrate (Fazius et al. 2012). Biochemical and genetic analyses have shown that this reaction is split on two separate enzymes, similar to the de- and rehydration of methylcitrate to methylisocitrate in the methylcitrate cycle described above. The enzymes involved in these reactions are an **aconitase** and a pathway-specific **homoaconitase** (Fazius et al. 2012). Since a striking difference between yeasts and filamentous fungi in the utilisation of the aconitase has been observed, this isomerisation will be explained in more detail below.

Homoisocitrate is oxidatively decarboxylated by a **homoisocitrate dehydrogenase** to α-ketoadipate (Lin et al. 2009) that is subsequently transaminated to α-aminoadipate by an **α-aminoadipate aminotransferase.** Interestingly, in *S. cerevisiae* this aminotransferase reaction may be carried out by Aro8 (Bulfer et al. 2013), which can be assumed to be involved in the transamination of histidine as described above. This reaction completes the first part of the α-aminoadipate pathway. The second half of the pathway mainly constitutes reversible reactions that may also be used for the degradation of lysine, in which lysine is degraded via α-aminoadipate and α-ketoadipate to acetoacetyl-CoA (Pink et al. 2011).

Interestingly, while the initial reactions of the α-aminoadipate pathway are shared between the thermophilic bacteria and fungi, the subsequent synthesis of lysine branches between those groups of organisms. In the thermophilic bacteria, the α-aminoadipate becomes *N*-protected by addition of the molecule to a small 54-amino acid protein, at which all intermediates maintain attached until the final release of lysine (Horie et al. 2009). In contrast, no *N*-protection has been observed in fungal lysine biosynthesis, and subsequent reactions involve the reduction of α-aminoadipate to the semialdehyde by an **aminoadipate reductase** (Yan et al. 2007). The α-aminoadipate semialdehyde is subsequently condensed with a glutamate via **saccharopine reductase** yielding saccharopine (Vashishtha et al. 2009). The final step in the synthesis is performed by a **saccharopine dehydrogenase** catalysing an oxidative deamination of saccharopine, which results in the products α-ketoglutarate and lysine (Kumar et al. 2012).

While the α-aminoadipate pathway is well conserved among fungi, it involves at least one enzyme that is shared by several metabolic pathways. This enzyme is the aconitase that is involved in the citric acid cycle, in the methylcitrate cycle and in lysine biosynthesis.

In the citric acid cycle the enzyme performs the de- and rehydration of citrate to isocitrate by binding the substrate *cis*-aconitate either in citrate mode or 180° rotated in the isocitrate mode. In the methylcitrate cycle, *cis*-methylaconitate can only be bound in the isocitrate mode due to sterical hindrance of the methyl-group with an aspartate residue (Lauble and Stout 1995). Finally, in lysine biosynthesis *cis*-homoaconitate can only bind in the citrate mode, because the enhanced aliphatic chain in *cis*-homoaconitate does not allow correct positioning in the isocitrate mode (Fazius et al. 2012).

Both yeasts and filamentous fungi use an aconitase for the dehydration of homocitrate to homoaconitate. Genome analyses revealed that fungi contain at least two aconitase-like enzymes (AcoA and AcoB in *Aspergillus spp.*, Aco1 and Aco2 in *S. cerevisiae*). In *Aspergillus spp.* AcoB seems to have lost its function, since no phenotype is observed when the gene is deleted, and AcoB cannot complement aconitase-negative mutants from *S. cerevisiae.* However, deletion of AcoA is lethal in *A. fumigatus*, confirming its essential function, and AcoA is the only aconitase contributing to lysine biosynthesis (Fazius et al. 2012). In contrast, *S. cerevisiae* has specifically adapted Aco2 to serve for lysine biosynthesis. Aco1 is the main citric acid cycle aconitase in *S. cerevisiae* and a deletion leads to glutamate auxotrophy

(Gangloff et al. 1990). However, although a lysine auxotrophy would be simultaneously expected, growth in the absence of external lysine is only slightly retarded. When Aco2 is deleted, no glutamate auxotrophy is observed, but, again, growth is retarded in the absence of external lysine. Thus, a deletion of both *ACO1* and *ACO2* is required to cause a lysine auxotrophic phenotype in *S. cerevisiae* (Fazius et al. 2012). Biochemical characterisation additionally revealed that Aco2 is highly specific for homocitrate as substrate and hardly active with citrate. This confirms that *S. cerevisiae* specifically adapted Aco2 to serve in lysine biosynthesis, which is not the case for AcoB in *Aspergillus spp.*

The reason for the difference between filamentous fungi and yeast might be due to the different mode of glucose utilisation. While filamentous fungi rely on a functional citric acid cycle for substrate oxidation and energy gain, yeasts tend to fermentation and can grow independent from a citric acid cycle when glutamate is supplied. Thus, due to the adaptation of Aco2 to serve lysine biosynthesis, yeast can grow by fermentation without disturbing accompanying metabolic pathways (Fazius et al. 2012).

In conclusion, lysine biosynthesis is highly conserved among fungi. However, differences exist in the regulation of the pathway. This is caused by the fact that yeasts use the α-aminoadipate solely for lysine biosynthesis, whereas several filamentous fungi also require the intermediate α-aminoadipate for production of β-lactam antibiotics. Additionally, the adaptation of an aconitase isozyme in *S. cerevisiae* appears due to its ability of decoupling energy metabolism and anaplerosis, a mechanism that is not possible for oxidatively growing filamentous fungi.

V. Conclusions

Investigation of fungal metabolism has mainly been performed on *Ascomycetes* such as the yeast *S. cerevisiae* or the filamentous fungus *A. nidulans*. In case of bakers' yeast, this is mainly due to the early use of this organism in food and ethanol production and its easy cultivation (Nevoigt 2008). Thus, early investigations of cellular metabolism based on *S. cerevisiae* from which important metabolic pathways such as glycolysis had been identified (Buchholz and Collins 2013). Besides yeast, *A. nidulans* established as a model due to its homothallic sexual mating and the early development of sexual crossing systems that allowed the identification of complementation groups in developmental and metabolic mutants (Nierman et al. 2005). In contrast, due to the lack of transformation systems or problems in laboratory cultivation, studies on the metabolism of *Basidiomycetes* or *Zygomycetes* are limited. However, as exemplified in this chapter, even relatively closely related *Ascomycetes* show a great diversity in their metabolic properties. These differences appear related to fungal morphology (yeast against hyphae), nutritional niches, gene loss and acquisition and organismal interaction. Taking into account that even closely related fungal species already display a great variety in use and occurrence of metabolic pathways, it can be expected that future analyses of metabolism and physiology of *Basidiomycota*, *Zygomycota* and *Chytridiomycota* provide a treasure chest of new pathways that might become important for several biotechnological processes.

References

Abdeljalil S et al (2013) Improvement of cellulase and xylanase production by solid-state fermentation of *Stachybotrys microspora*. Biotechnol Appl Biochem. doi: 10.1002/bab.1195

Alam S, Shah HU, Magan N (2010) Effect of calcium propionate and water activity on growth and aflatoxins production by *Aspergillus flavus*. J Food Sci 75:M61–M64

Alifano P et al (1996) Histidine biosynthetic pathway and genes: structure, regulation, and evolution. Microbiol Rev 60:44–69

Aschenbach JR, Kristensen NB, Donkin SS, Hammon HM, Penner GB (2010) Gluconeogenesis in dairy cows: the secret of making sweet milk from sour dough. IUBMB Life 62:869–877

Bajmoczi M, Sneve M, Eide DJ, Drewes LR (1998) *TAT1* encodes a low-affinity histidine transporter in *Saccharomyces cerevisiae*. Biochem Biophys Res Commun 243:205–209

Balestrini R, Bonfante P (2014) Cell wall remodeling in mycorrhizal symbiosis: a way towards biotrophism. Front Plant Sci 5:237

Ballhausen D, Mittaz L, Boulat O, Bonafe L, Braissant O (2009) Evidence for catabolic pathway of propionate metabolism in CNS: expression pattern of methylmalonyl-CoA mutase and propionyl-CoA carboxylase alpha-subunit in developing and adult rat brain. Neuroscience 164:578–587

Baudin A, Ozier-Kalogeropoulos O, Denouel A, Lacroute F, Cullin C (1993) A simple and efficient method for direct gene deletion in *Saccharomyces cerevisiae*. Nucleic Acids Res 21:3329–3330

Berger H et al (2008) Dissecting individual steps of nitrogen transcription factor cooperation in the *Aspergillus nidulans* nitrate cluster. Mol Microbiol 69:1385–1398

Betzel C et al (2001) Structure of a serine protease proteinase K from *Tritirachium album limber* at 0.98 A resolution. Biochemistry 40:3080–3088

Bigner DR, Goff JP, Faust MA, Tyler HD, Horst RL (1997) Comparison of oral sodium compounds for the correction of acidosis. J Dairy Sci 80:2162–2166

Brock M (2009) Fungal metabolism in host niches. Curr Opin Microbiol 12:371–376

Brock M, Buckel W (2004) On the mechanism of action of the antifungal agent propionate. Eur J Biochem 271:3227–3241

Brock M, Fischer R, Linder D, Buckel W (2000) Methylcitrate synthase from *Aspergillus nidulans*: implications for propionate as an antifungal agent. Mol Microbiol 35:961–973

Brock M, Darley D, Textor S, Buckel W (2001) 2-Methylisocitrate lyases from the bacterium *Escherichia coli* and the filamentous fungus *Aspergillus nidulans*: characterization and comparison of both enzymes. Eur J Biochem 268:3577–3586

Brock M, Maerker C, Schutz A, Volker U, Buckel W (2002) Oxidation of propionate to pyruvate in *Escherichia coli*. Involvement of methylcitrate dehydratase and aconitase. Eur J Biochem 269:6184–6194

Brunke S et al (2014) Histidine degradation via an aminotransferase increases the nutritional flexibility of *Candida glabrata*. Eukaryot Cell 13:758–765

Buchholz K, Collins J (2013) The roots–a short history of industrial microbiology and biotechnology. Appl Microbiol Biotechnol 97:3747–3762

Bulfer SL, Brunzelle JS, Trievel RC (2013) Crystal structure of *Saccharomyces cerevisiae* Aro8, a putative alpha-aminoadipate aminotransferase. Protein Sci 22:1417–1424

Chi Z, Chi Z, Liu G, Wang F, Ju L, Zhang T (2009) *Saccharomycopsis fibuligera* and its applications in biotechnology. Biotechnol Adv 27:423–431

Chittori S, Simanshu DK, Savithri HS, Murthy MR (2010) Preliminary X-ray crystallographic analysis of 2-methylcitrate synthase from *Salmonella typhimurium*. Acta Crystallogr Sect F: Struct Biol Cryst Commun 66:467–470

Coblentz WK, Coffey KP, Young AN, Bertram MG (2013) Storage characteristics, nutritive value, energy content, and in vivo digestibility of moist, large rectangular bales of alfalfa-orchardgrass hay treated with a propionic acid-based preservative. J Dairy Sci 96:2521–2535

Creighton DJ, Murthy NSRK (1990) Stereospecificity in enzymatic reactions. In: Sigman DS, Boyer PD (eds) The enzymes, 3rd edn. Academic, San Diego

Crittenden PD, Davis JC, Hawksworth DL, Campbell FS (1995) Attempted isolation and success in culturing of a broad spectrum of lichen forming and lichenicolous fungi. New Phytol 130:267–297

Cruz AH et al (2011) Phosphorylation is the major mechanism regulating isocitrate lyase activity in *Paracoccidioides brasiliensis* yeast cells. FEBS J 278:2318–2332

Cuomo CA et al (2012) Microsporidian genome analysis reveals evolutionary strategies for obligate intracellular growth. Genome Res 22:2478–2488

Davis MA, Askin MC, Hynes MJ (2005) Amino acid catabolism by an *areA*-regulated gene encoding an L-amino acid oxidase with broad substrate specificity in *Aspergillus nidulans*. Appl Environ Microbiol 71:3551–3555

De Lucas JR, Gregory S, Turner G (1994a) Analysis of the regulation of the *Aspergillus nidulans acuD* gene, encoding isocitrate lyase, by construction of a hybrid promoter. Mol Gen Genet 243:654–659

De Lucas JR, Valenciano S, Laborda F, Turner G (1994b) Glucose-induced inactivation of isocitrate lyase in *Aspergillus nidulans*. Arch Microbiol 162:409–413

De Vit MJ, Waddle JA, Johnston M (1997) Regulated nuclear translocation of the Mig1 glucose repressor. Mol Biol Cell 8:1603–1618

Domin N, Wilson D, Brock M (2009) Methylcitrate cycle activation during adaptation of *Fusarium solani* and *Fusarium verticillioides* to propionyl-CoA-generating carbon sources. Microbiology 155:3903–3912

Dunn MF, Ramirez-Trujillo JA, Hernandez-Lucas I (2009) Major roles of isocitrate lyase and malate synthase in bacterial and fungal pathogenesis. Microbiology 155:3166–3175

Emmanuel B, Kennelly JJ (1984) Effect of propionic acid on ketogenesis in lactating sheep fed restricted rations or deprived of food. J Dairy Sci 67:344–350

Fazius F, Shelest E, Gebhardt P, Brock M (2012) The fungal alpha-aminoadipate pathway for lysine biosynthesis requires two enzymes of the aconitase family for the isomerization of homocitrate to homoisocitrate. Mol Microbiol 86:1508–1530

Fazius F, Zaehle C, Brock M (2013) Lysine biosynthesis in microbes: relevance as drug target and prospects for beta-lactam antibiotics production. Appl Microbiol Biotechnol 97:3763–3772

Fellbaum CR et al (2014) Fungal nutrient allocation in common mycorrhizal networks is regulated by the carbon source strength of individual host plants. New Phytol 203:646–656

Feller A, Ramos F, Pierard A, Dubois E (1999) In *Saccharomyces cerevisae*, feedback inhibition of homocitrate synthase isoenzymes by lysine modulates the activation of *LYS* gene expression by Lys14p. Eur J Biochem 261:163–170

Free SJ (2013) Fungal cell wall organization and biosynthesis. Adv Genet 81:33–82

Gangloff SP, Marguet D, Lauquin GJ (1990) Molecular cloning of the yeast mitochondrial aconitase gene (*ACO1*) and evidence of a synergistic regulation of expression by glucose plus glutamate. Mol Cell Biol 10:3551–3561

Garcia I, Gonzalez R, Gomez D, Scazzocchio C (2004) Chromatin rearrangements in the *prnD-prnB* bidirectional promoter: dependence on transcription factors. Eukaryot Cell 3:144–156

Garvey GS, Rocco CJ, Escalante-Semerena JC, Rayment I (2007) The three-dimensional crystal structure of the PrpF protein of *Shewanella oneidensis* complexed with *trans*-aconitate: insights into its biological function. Protein Sci 16:1274–1284

Gonzalez R, Gavrias V, Gomez D, Scazzocchio C, Cubero B (1997) The integration of nitrogen and carbon catabolite repression in *Aspergillus nidulans* requires the GATA factor AreA and an additional positive-acting element, ADA. EMBO J 16:2937–2944

Graybill ER, Rouhier MF, Kirby CE, Hawes JW (2007) Functional comparison of citrate synthase isoforms from *S. cerevisiae*. Arch Biochem Biophys 465:26–37

Grimek TL, Escalante-Semerena JC (2004) The *acnD* genes of *Shewenella oneidensis* and *Vibrio cholerae* encode a new Fe/S-dependent 2-methylcitrate dehydratase enzyme that requires *prpF* function in vivo. J Bacteriol 186:454–462

Gropp K, Schild L, Schindler S, Hube B, Zipfel PF, Skerka C (2009) The yeast *Candida albicans* evades human complement attack by secretion of aspartic proteases. Mol Immunol 47:465–475

Halarnkar PP, Blomquist GJ (1989) Comparative aspects of propionate metabolism. Comp Biochem Physiol B 92:227–231

Hazelwood LA, Daran JM, van Maris AJ, Pronk JT, Dickinson JR (2008) The Ehrlich pathway for fusel alcohol production: a century of research on *Saccharomyces cerevisiae* metabolism. Appl Environ Microbiol 74:2259–2266

Hers HG, Hue L (1983) Gluconeogenesis and related aspects of glycolysis. Annu Rev Biochem 52:617–653

Hori C et al (2013) Genomewide analysis of polysaccharides degrading enzymes in 11 white- and brown-rot *Polyporales* provides insight into mechanisms of wood decay. Mycologia 105:1412–1427

Horie A et al (2009) Discovery of proteinaceous *N*-modification in lysine biosynthesis of *Thermus thermophilus*. Nat Chem Biol 5:673–679

Horswill AR, Escalante-Semerena JC (2001) In vitro conversion of propionate to pyruvate by *Salmonella enterica* enzymes: 2-methylcitrate dehydratase (PrpD) and aconitase enzymes catalyze the conversion of 2-methylcitrate to 2-methylisocitrate. Biochemistry 40:4703–4713

Hudson AO et al (2005) Biosynthesis of lysine in plants: evidence for a variant of the known bacterial pathways. Biochim Biophys Acta 1721:27–36

Hutton CA, Perugini MA, Gerrard JA (2007) Inhibition of lysine biosynthesis: an evolving antibiotic strategy. Mol Biosyst 3:458–465

Hynes MJ, Murray SL (2010) ATP-citrate lyase is required for production of cytosolic acetyl coenzyme A and development in *Aspergillus nidulans*. Eukaryot Cell 9:1039–1048

Irvin SD, Bhattacharjee JK (1998) A unique fungal lysine biosynthesis enzyme shares a common ancestor with tricarboxylic acid cycle and leucine biosynthetic enzymes found in diverse organisms. J Mol Evol 46:401–408

Jacobsen ID, Lüttich A, Kurzai O, Hube B, Brock M (2014) In vivo imaging of disseminated murine *Candida albicans* infection reveals unexpected host sites of fungal persistence during antifungal therapy. J Antimicrob Chemother 69(10):2785–96

Jiang W, Nakayama Y, Sequeira JM, Quadros EV (2013) Mapping the functional domains of TCblR/CD320, the receptor for cellular uptake of transcobalamin-bound cobalamin. FASEB J 27:2988–2994

Kanamasa S, Dwiarti L, Okabe M, Park EY (2008) Cloning and functional characterization of the cis-aconitic acid decarboxylase (CAD) gene from *Aspergillus terreus*. Appl Microbiol Biotechnol 80:223–229

Klement T, Buchs J (2013) Itaconic acid – a biotechnological process in change. Bioresour Technol 135:422–431

Kmetzsch L et al (2011) The GATA-type transcriptional activator Gat1 regulates nitrogen uptake and metabolism in the human pathogen *Cryptococcus neoformans*. Fungal Genet Biol 48:192–199

Kobayashi K, Hattori T, Honda Y, Kirimura K (2013) Gene identification and functional analysis of methylcitrate synthase in citric acid-producing *Aspergillus niger* WU-2223L. Biosci Biotechnol Biochem 77:1492–1498

Kulis-Horn RK, Persicke M, Kalinowski J (2014) Histidine biosynthesis, its regulation and biotechnological application in *Corynebacterium glutamicum*. Microb Biotechnol 7:5–25

Kumar VP, West AH, Cook PF (2012) Supporting role of lysine 13 and glutamate 16 in the acid-base mechanism of saccharopine dehydrogenase from *Saccharomyces cerevisiae*. Arch Biochem Biophys 522:57–61

Lauble H, Stout CD (1995) Steric and conformational features of the aconitase mechanism. Proteins 22:1–11

Ledley FD, Crane AM, Klish KT, May GS (1991) Is there methylmalonyl CoA mutase in *Aspergillus nidulans*? Biochem Biophys Res Commun 177:1076–1081

Lin Y, West AH, Cook PF (2009) Site-directed mutagenesis as a probe of the acid-base catalytic mechanism of homoisocitrate dehydrogenase from *Saccharomyces cerevisiae*. Biochemistry 48:7305–7312

Liu S, Lu Z, Han Y, Melamud E, Dunaway-Mariano D, Herzberg O (2005) Crystal structures of 2-methylisocitrate lyase in complex with product and with isocitrate inhibitor provide insight into lyase substrate specificity, catalysis and evolution. Biochemistry 44:2949–2962

Lopez-Rituerto E, Avenoza A, Busto JH, Peregrina JM (2013) NMR study of histidine metabolism during alcoholic and malolactic fermentations of wine and their influence on histamine production. J Agric Food Chem 61:9464–9469

Maerker C, Rohde M, Brakhage AA, Brock M (2005) Methylcitrate synthase from *Aspergillus fumigatus*. Propionyl-CoA affects polyketide synthesis, growth and morphology of conidia. FEBS J 272:3615–3630

Marin S et al (1999) Control of growth and fumonisin B1 production by *Fusarium verticillioides* and *Fusarium proliferatum* isolates in moist maize with propionate preservatives. Food Addit Contam 16:555–563

Müller S, Fleck CB, Wilson D, Hummert C, Hube B, Brock M (2011) Gene acquisition, duplication and metabolic specification: the evolution of fungal methylisocitrate lyases. Environ Microbiol 13:1534–1548

Nehls U, Grunze N, Willmann M, Reich M, Kuster H (2007) Sugar for my honey: carbohydrate partitioning in ectomycorrhizal symbiosis. Phytochemistry 68:82–91

Nevoigt E (2008) Progress in metabolic engineering of *Saccharomyces cerevisiae*. Microbiol Mol Biol Rev 72:379–412

Nierman WC, May G, Kim HS, Anderson MJ, Chen D, Denning DW (2005) What the *Aspergillus* genomes have told us. Med Mycol 43(Suppl 1):S3–S5

Nishida H, Nishiyama M (2012) Evolution of lysine biosynthesis in the phylum *Deinococcus-Thermus*. Int J Evol Biol 2012:745931

Otzen C, Bardl B, Jacobsen ID, Nett M, Brock M (2014) *Candida albicans* utilizes a modified beta-oxidation pathway for the degradation of toxic propionyl-CoA. J Biol Chem 289:8151–8169

Papagianni M (2007) Advances in citric acid fermentation by *Aspergillus niger*: biochemical aspects, membrane transport and modeling. Biotechnol Adv 25:244–263

Peraza-Reyes L, Berteaux-Lecellier V (2013) Peroxisomes and sexual development in fungi. Front Physiol 4:244

Pink DB et al (2011) Lysine alpha-ketoglutarate reductase, but not saccharopine dehydrogenase, is subject to substrate inhibition in pig liver. Nutr Res 31:544–554

Polkinghorne MA, Hynes MJ (1982) L-histidine utilization in *Aspergillus nidulans*. J Bacteriol 149:931–940

Pronk JT, van der Linden-Beuman A, Verduyn C, Scheffers WA, van Dijken JP (1994) Propionate metabolism in *Saccharomyces cerevisiae*: implications for the metabolon hypothesis. Microbiology 140(Pt 4):717–722

Quezada H et al (2011) The Lys20 homocitrate synthase isoform exerts most of the flux control over the lysine synthesis pathway in *Saccharomyces cerevisiae*. Mol Microbiol 82:578–590

Schöbel F, Jacobsen ID, Brock M (2010) Evaluation of lysine biosynthesis as an antifungal drug target: biochemical characterization of *Aspergillus fumigatus* homocitrate synthase and virulence studies. Eukaryot Cell 9:878–893

Sekito T, Chardwiriyapreecha S, Sugimoto N, Ishimoto M, Kawano-Kawada M, Kakinuma Y (2014) Vacuolar transporter Avt4 is involved in excretion of basic amino acids from the vacuoles of *Saccharomyces cerevisiae*. Biosci Biotechnol Biochem 78:969–975

Soontorngun N et al (2012) Genome-wide location analysis reveals an important overlap between the targets of the yeast transcriptional regulators Rds2 and Adr1. Biochem Biophys Res Commun 423:632–637

Steiger MG, Blumhoff ML, Mattanovich D, Sauer M (2013) Biochemistry of microbial itaconic acid production. Front Microbiol 4:23

Stocker-Wörgötter E (2008) Metabolic diversity of lichen-forming ascomycetous fungi: culturing, polyketide and shikimate metabolite production, and PKS genes. Nat Prod Rep 25:188–200

Strauss J, Horvath HK, Abdallah BM, Kindermann J, Mach RL, Kubicek CP (1999) The function of CreA, the carbon catabolite repressor of *Aspergillus nidulans*, is regulated at the transcriptional and post-transcriptional level. Mol Microbiol 32:169–178

Suzuki Y, Murray SL, Wong KH, Davis MA, Hynes MJ (2012) Reprogramming of carbon metabolism by the transcriptional activators AcuK and AcuM in *Aspergillus nidulans*. Mol Microbiol 84:942–964

Tabuchi T, Hara S (1974) Production of 2-methylisocitric acid from *n*-paraffins by mutants of *Candida lipolytica*. Agric Biol Chem 38:1105–1106

Tabuchi T, Uchiyama H (1975) Methylcitrate condensing and methylisocitrate cleaving enzymes; evidence for the pathway of oxidation of propionyl-CoA to pyruvate via C_7-tricarboxylic acids. Agric Biol Chem 39:2035–2042

Takahashi-Iniguez T, Garcia-Hernandez E, Arreguin-Espinosa R, Flores ME (2012) Role of vitamin B_{12}

on methylmalonyl-CoA mutase activity. J Zhejiang Univ Sci B 13:423–437

Tao Y, Marzluf GA (1999) The NIT2 nitrogen regulatory protein of Neurospora: expression and stability of *nit-2* mRNA and protein. Curr Genet 36:153–158

Textor S et al (1997) Propionate oxidation in *Escherichia coli*: evidence for operation of a methylcitrate cycle in bacteria. Arch Microbiol 168:428–436

Tiwari P, Misra BN, Sangwan NS (2013) beta-Glucosidases from the fungus *Trichoderma*: an efficient cellulase machinery in biotechnological applications. Biomed Res Int 2013:203735

Todd RB et al (1997) The acetate regulatory gene *facB* of *Aspergillus nidulans* encodes a $Zn(II)_2Cys_6$ transcriptional activator. Mol Gen Genet 254:495–504

Todd RB, Andrianopoulos A, Davis MA, Hynes MJ (1998) FacB, the *Aspergillus nidulans* activator of acetate utilization genes, binds dissimilar DNA sequences. EMBO J 17:2042–2054

Trepanier M et al (2005) Dependence of arbuscular-mycorrhizal fungi on their plant host for palmitic acid synthesis. Appl Environ Microbiol 71:5341–5347

Tsai CS, Mitton KP, Johnson BF (1989) Acetate assimilation by the fission yeast, *Schizosaccharomyces pombe*. Biochem Cell Biol 67:464–467

Tsang AW, Horswill AR, Escalante-Semerena JC (1998) Studies of regulation of expression of the propionate (prpBCDE) operon provide insights into how *Salmonella typhimurium* LT2 integrates its 1,2-propanediol and propionate catabolic pathways. J Bacteriol 180:6511–6518

Tu T et al (2013) High-yield production of a low-temperature-active polygalacturonase for papaya juice clarification. Food Chem 141:2974–2981

Turcotte B, Liang XB, Robert F, Soontorngun N (2010) Transcriptional regulation of nonfermentable carbon utilization in budding yeast. FEMS Yeast Res 10:2–13

Tylicki A, Ziolkowska G, Bolkun A, Siemieniuk M, Czerniecki J, Nowakiewicz A (2008) Comparative study of the activity and kinetic properties of malate dehydrogenase and pyruvate decarboxylase from *Candida albicans*, *Malassezia pachydermatis*, and *Saccharomyces cerevisiae*. Can J Microbiol 54:734–741

Vashishtha AK, West AH, Cook PF (2009) Chemical mechanism of saccharopine reductase from *Saccharomyces cerevisiae*. Biochemistry 48:5899–5907

Watanabe N, James MN (2011) Structural insights for the substrate recognition mechanism of LL-diaminopimelate aminotransferase. Biochim Biophys Acta 1814:1528–1533

Wendisch VF (2014) Microbial production of amino acids and derived chemicals: synthetic biology approaches to strain development. Curr Opin Biotechnol 30C:51–58

White MF, Fothergill-Gilmore LA, Kelly SM, Price NC (1993) Substitution of His-181 by alanine in yeast phosphoglycerate mutase leads to cofactor-induced dissociation of the tetrameric structure. Biochem J 291(Pt 2):479–483

Woo JM, Yang KM, Kim SU, Blank LM, Park JB (2014) High temperature stimulates acetic acid accumulation and enhances the growth inhibition and ethanol production by *Saccharomyces cerevisiae* under fermenting conditions. Appl Microbiol Biotechnol 98:6085–6094

Wu G (2014) Dietary requirements of synthesizable amino acids by animals: a paradigm shift in protein nutrition. J Anim Sci Biotechnol 5:34

Xu H, Andi B, Qian J, West AH, Cook PF (2006) The alpha-aminoadipate pathway for lysine biosynthesis in fungi. Cell Biochem Biophys 46:43–64

Yan H, He P, Cheng HR, Shen A, Jiang N (2007) Cloning, sequencing and characterization of the alpha-aminoadipate reductase gene (*LYS2*) from *Saccharomycopsis fibuligera*. Yeast 24:189–199

Yike I (2011) Fungal proteases and their pathophysiological effects. Mycopathologia 171:299–323

Zhang YQ, Brock M, Keller NP (2004) Connection of propionyl-CoA metabolism to polyketide biosynthesis in *Aspergillus nidulans*. Genetics 168:785–794

15 Genetic and Metabolic Aspects of Primary and Secondary Metabolism of the Zygomycetes

Kerstin Voigt[1,2], Thomas Wolf[3], Katrin Ochsenreiter[4], Gábor Nagy[5], Kerstin Kaerger[6], Ekaterina Shelest[3], Tamás Papp[5]

CONTENTS

[1]Jena Microbial Resource Collection, Leibniz Institute for Natural Product Research and Infection Biology - Hans-Knöll-Institute, Adolf-Reichwein-Strasse 23, 07745 Jena, Germany; e-mail: kerstin.voigt@leibniz-hki.de

[2]Department of Microbiology and Molecular Biology, Institute of Microbiology, University of Jena, Neugasse 25, 07743 Jena, Germany; e-mail: kerstin.voigt@leibniz-hki.de

[3]Systems Biology/Bioinformatics, Leibniz Institute for Natural Product Research and Infection Biology - Hans-Knöll-Institute, Adolf-Reichwein-Strasse 23, 07745 Jena, Germany

[4]Karlsruhe Institute of Technology (KIT), Institute of Process Engineering in Life Sciences, Section II: Technical Biology, Engler-Bunte-Ring 1, 76131 Karlsruhe, Germany

[5]Faculty of Science and Informatics, Department of Microbiology, University of Szeged, Közép fasor 52, 6726 Szeged, Hungary

[6]National Reference Center for Invasive Mycoses, Leibniz Institute for Natural Product Research and Infection Biology – Hans-Knöll-Institute, Adolf-Reichwein-Strasse 23, 07745 Jena, Germany

I. Introduction

Zygomycetes constitute a remarkable group of microscopic fungi (formerly classified into the phylum *Zygomycota*) basal to *Ascomycota* and *Basidiomycota* (for review see Voigt 2012; Voigt and Kirk 2014). These fungi are mainly soil inhabitants living as saprobes and decomposers of organic matter and herbivorous feces (coprophiles). Some taxa are parasitic or predacious, in which case developing mycelium is immersed in the host tissue:

Traditionally, the Zygomycota, represent the most basal terrestrial phylum of the kingdom of Fungi. The Zygomycota are not accepted as a valid phylum (as "Phylum des Zygomycètes"; Whittaker 1969; Cavalier-Smith 1981 because of a lacking compliance to the International Code of Botanical Nomenclature/International Code of Nomenclature for algae, fungi and plants (Hawksworth 2011) and lacking resolution of the basal fungal clades (James et al. 2006). Molecular phylogenetic analyses based on informal phylogenetic trees where molecular phylogenies are substituted with traditional taxonomic information revealed dispersal into five subphyla containing one to four orders (Hibbett et al. 2007; Hoffmann et al. 2011, for review see: Benny et al. 2014). The phylogenetic relationships between these subphyla and their orders is still not well resolved. However, based on the potential of all five subphyla to produce zygospores during conjugation of two yoke-shaped gametangia it is referred to a phylogenetically coherent group named zygosporic fungi as a whole group, which share morphological features but consists of phylogenetically unrelated subphyla. Therefore, the

Biochemistry and Molecular Biology, 3rd Edition
The Mycota III
D. Hoffmeister (Ed.)

phylum referred to as "Zygomycota" is employed to make clear the term is being used in a colloquial sense, for instance the inclusion of all basal lineages of terrestrial fungi with the potential to form zygospores or sharing any other of the plesiomorphic morphological characters of the former phylum.

Cavalier-Smith (1981, 1998) introduced the name as "cl. nov." and comments that it does not appear to have been validly published elsewhere. Likewise, this class *Zygomycetes* does not appear to be monophyletic (James et al. 2006, for review and comprehensive phylogeny see Voigt and de Hoog 2013). Here the term "zygomycetes" is printed in lower-case letters and used in a colloquial sense for ecological groupings sharing soil as their main habitat. Zygomycetes are common and cosmopolitan components of the mycoflora of dung, soil, and other substrates that support their growth and sporulation.

A. Zygomycetes: Evolution, Systematics, and Ecology

Zygomycetes are ubiquitously distributed. The occurrence of zygomycetes dates back to the Precambrian era, 800–1400 million years ago (Heckman et al. 2001; Mendoza et al. 2014). During the course of their long evolutions, which has its roots in the Precambrian, they have learned to interact with many other microorganisms in a wide variety of interplays, e.g., symbiosis with endo- and ectosymbionts; commensalism and parasitism during zygomycetes become successful pathogens of plants, animals, and human. Interaction pattern appears to be instrumental for fitness reasons as shown in aphid–bacterium–fungus alliances lowering the rate of transmission diseases (Scarborough et al. 2005).

Zygomycetes encompass nine orders: *Asellariales*, *Dimargaritales*, *Endogonales*, *Entomophthorales*, *Harpellales*, *Kickxellales*, *Mortierellales*, *Mucorales*, and *Zoopagales* (for review see Voigt and Kirk 2014). Members of the *Asellariales* (18 species) have filamentous, branched thalli and reproduce asexually by arthrospore-like cells that disarticulate from their corresponding thallus. They inhabit the digestive tract of terrestrial, aquatic, and marine isopods as well as springtails by attachment to the cuticle or digestive tract via a holdfast. They are not immersed in the host tissue (Moss 1975; Lichtwardt and Manier 1978). The *Dimargaritales* (18 species) is comprised by **obligate haustorial mycoparasites of the** *Mucorales* (rarely, species of *Chaetomium* [*Ascomycota*: *Sordariales*]) which are saprobic, or coprophilous, and share the same habitat.

The *Endogonales* (15 species) is an order of **mainly ectomycorrhizal fungi**, in addition to some saprobes. Endogonalean fungi are widely associated with the earliest branching land plants. During their evolution, they give way to arbuscular mycorrhizal glomeromycotan fungi in later lineages. It has been hypothesized that *Endogone*-like fungi rather than (as previously proposed, Simon et al. 1993; Parniske 2008) the *Glomeromycota* enabled the establishment and growth of early land colonists and thus facilitated terrestrialization (Bidartondo et al. 2011).

The *Entomophthorales* (250 species) consists of **mainly entomogenous/entomopathogenic fungi** producing one of the most spectacular insect-killing mechanisms. They are occasionally saprobic and found in soil, but mainly parasites of insects ("insect destroyers"), and other arthropods, rarely of nematodes and tardigrades as hosts. Most species are obligate parasites, and, therefore, these are so highly adapted to their hosts that their lifestyle obligately relies on the close relation to the host insect throughout the entire fungal ontogeny making a fungal cultivation in axenic cultures impossible. Few recipes of pure cultures have been reported which are highly complex media often containing natural products or biopolymers (Grundschober et al. 1998; Delalibera et al. 2003). Even under these conditions, it is unlikely that growth will be typical, and certainly sporulation will rarely be present. The exceptions to this general property are species of the genus *Conidiobolus* which are saprobes from the soil and are of widespread distribution. They are frequently isolated from the soil and are easy to grow in culture. *Conidiobolus coronatus* is found to be associated with medical and veterinary cases of mainly local, chronically lapsed, entomophthoromycotic infections (for review see Rothhardt et al.

2011; Mendoza et al. 2014). Zygospores, where known, are formed on differentiated hyphae.

The ecology of the *Harpellales* (252 species) is similar to that of the *Asellariales* by endocommensal association with the aquatic larvae of arthropods (incl. crustaceans and diplopods, rarely isopods), found attached to the gut lining of the aquatic larvae.

The *Kickxellales* (37 species) comprise mainly **saprobes from soil or coprophilous in dung**, rarely as mycoparasites. The *Kickxellales* are of widespread occurrence apparently favoring somewhat dry climates rather than the wet tropics but are relatively under-recorded, so their true distribution, like that for many of the fungi, is unclear.

The order *Mortierellales* (79 species) possesses an extremely high ecological and physiological diversity enabling them to be distributed worldwide (for comprehensive phylogeny, see Nagy et al. 2011; Wagner et al. 2013). Most species are polyunsaturated fatty acid-based lipid-accumulating organisms (e.g., *Mortierella alpina*, for overview see Münchberg et al. 2012, 2015). One thermotolerant species, *M. wolfii*, has clinical relevance and appears as a causative agent of bovine abortion (Papp et al. 2011). Zygospores are mostly thin walled, not ornamented, and nonpigmented.

The order *Mucorales* (237 species) is the most prominent and the most studied group among the zygomycetes (Voigt and Kirk 2014). Members of the *Mucorales* constitute a remarkable group which encompass a wide variety of morphological appearances, ecological niches, and lifestyles (saprobic, facultative parasitic, opportunistic pathogenic) facilitating extensive evolutionary studies (Voigt and Wöstemeyer 2001; Voigt et al. 2009, 2013; Hoffmann et al. 2009, 2013).

Mucoralean species are **predominantly saprotrophic, soil inhabitants**, rarely mycoparasites (biotrophic fusion parasites) on other mucoralean hosts. Due to airborne spores, high germination, and growth rates, mucoralean species belong to the primary colonizers of organic substrates. As typical indoor contaminants and post-harvest pathogens on fruits and food causing food spoilage, mucoralean fungi are the most successful and most abundant zygosporic fungi encountering permanent presence in the human environment. Some mucoralean species are able to develop life-threatening infections within immunocompromised patients (mucormycosis) (de Hoog et al. 2000, 2014; Mendoza et al. 2014; Ibrahim 2011; Chayakulkeeree et al. 2006; Greenberg et al. 2004; Bitar et al. 2009; Chakrabarti et al. 2008, 2009; Morace and Borghi 2012; Casadevall and Pirofski 2001). On the other hand, mucoralean species are used for fermentation of soy-based food in Asia since centuries and for the application of *Rhizopus* species in biotechnological production of enzymes for decades.

The *Zoopagales* (208 species) is, even though species rich, relatively unknown concerning its frequency and distribution. They appear to be cosmopolitan as obligate haustorial parasites of fungi and animals (nematodes, Amoeba, and other small terrestrial invertebrates).

B. The Cooperative Nature of Zygomycetes: Bacterial–Fungal Alliances

The observation that progressive coupling of fungal host and bacterial endosymbiont metabolic and reproductive interests leads to an acceleration of studies reporting the cooperative nature of bacterial-fungal alliances in zygomycetes (Partida-Martinez and Hertweck 2005). The macrocyclic polyketide metabolite rhizoxin has been frequently isolated from cultures of *Rhizopus microsporus*, which is infamous for causing rice seedling blight (Tsuruo et al. 1986; White et al. 2002: *R. chinensis* synonym of *R. microsporus*; Dolatabadi et al. 2014a). Among other antimicrotubule agents, **rhizoxin** was proven to be particularly effective in small-cell lung cancer cell lines with a potential application in the salvage treatment of refractory or relapsed patients suffering small-cell lung cancer to overcome drug-resistance (Ikubo et al. 1999). **Rhizoxin is not biosynthesized by the fungus itself** but by an endosymbiotic, that is, intracellular living, bacterium of the genus *Burkholderia* (Partida-Martinez and Hertweck 2005). The remarkably complex symbiotic–pathogenic relationship that extends the fungus–plant interaction to a third, bacterial, key player unveils new perspectives for pest control. This finding appeared to be initially unexpected and unique, but the cases of endosymbiotic bacterial alliances with zygomycetes have increased during the following time (*Mortierella elongata*: Sato et al. 2010; Bonito et al. 2013, *Rhizopus chinensis*: White et al. 2002). All bacterial endosymbionts discovered so far in the zygomycetes belong to the family *Burkholderiaceae* (class *Betaproteobacteria*, Sato et al. 2010) and are closely related to *Glomeribacter gigasporarum*, which is an obligate endosymbiotic bacterium of the arbuscular mycorrhizal fungus *Gigaspora margarita* (Bianciotto et al. 2003). *Glomeribacter gigasporarum* reveals an interphylum network of nutritional interactions (Ghignone et al. 2012). On the other hand, the ~2.6 MB endosymbiont genome of *M. elongata* is larger than that of *Glomeribacter* but reduced compared to free-living *Burkholderia* (Bonito et al. 2013; Fujimura et al. 2014). Thus, intimate coevolution seems to be more recent than that of the alliance between *Glomeribacter gigasporarum* and *Gigaspora margarita.* Although many genes have been lost (e.g., genes encoding starch- or sucrose-degradation enzymes, phosphofructokinase leading to an incomplete glycolysis pathway, enzymes involved in the synthesis of the essential amino acids arginine, isoleucine, leucine, methionine, phenylalanine, tryptophan, histidine, and valine), some gene families have expanded including those involved in protein metabolism and electron transport (e.g., genes encoding amino acid transporters such as proteins involved in phosphate, zinc, and putrescine uptake; Ghignone et al. 2012). A gene cluster coding for a dipeptide/heme/δ-aminolevulinic acid transporters (*dpp* operon) contains the *dppA* gene, the product of which is responsible for the specificity of the imported oligopeptides and is present in at least 20 copies, suggesting that peptide uptake is crucial for bacterial cell function (Ghignone et al. 2012):

Rhizopus species appear to be trans-kingdom pathogens causing soil-, air and foodborne diseases in plants and humans (de Hoog et al. 2000; Dolatabadi et al. 2014b). The frequency of opportunistic mycoses in human began to rise since the mid 1990s (Ribes et al. 2000; Kauffman 2004; Chamilos et al. 2007). With regard to human pathogens, endosymbiotic toxin-producing bacteria in clinical *Rhizopus* isolates appear to be rather an exception then a general feature (Partida-Martinez et al. 2008). No evidence was found that bacterial endosymbionts and rhizoxin contribute to the pathogenesis of mucormycosis (Ibrahim et al. 2008). Consequently, it remains unclear if the paradigm of modulation of virulence of opportunistic fungi by widespread use of antibacterials can be applied (Chamilos et al. 2007).

II. Key Aspects in the Metabolism of Zygomycetes: Biotechnological Implications

Since centuries zygomycetes are traditionally used for the fermentative production of food in China and Southeast Asia, e.g., for tempeh or tofu (Wikandari et al. 2012; Hesseltine 1983). However, recent studies of the last few years have shown that mucoralean species can also produce a large amount of interesting and biotechnological relevant metabolites, including organic acid, e.g., lactic acid and fumaric acid, biofuels, e.g., ethanol and biodiesel, polyunsaturated fatty acids, carotenoids, chitosan, and various enzymes, e.g., amylases, cellulases, steroid 11α-hydroxylases, phytases, proteases, and lipases. Additionally, biomass can be used as animal and fish feed due to its high nutritional value (Ferreira et al. 2013; Karimi and Zamani 2013; Meussen et al. 2012).

Zygomycetous fungi show several characteristics which are advantageous in biotechnological applications: (1) one of the **highest fungal growth rates**, enabling fast biomass accumulation; (2) ability of growing at **higher temperature** for many of the species; (3) **dimorphism** of various genera, transition from filamentous growth to yeastlike growth under oxygen limitation or at high glucose concentrations (e.g., Orlowsky 1991); (4) simple demands on culture conditions; and (5) ability to produce a high diversity of enzymes enabling growth on diverse substrates, like starch or starch-containing residual materials, lignocellulosic substrates or whey, within wide temperature and pH ranges (Zhang et al. 2007; Dyal et al. 2005; Millati et al. 2005; Sautour et al. 2002; Nahas 1988; Sajbidor et al. 1988). Mucoralean species are amylase positive and are able to use pentoses; therefore, they can be directly applied to ethanol production from starch-containing or lignocellulosic substrates (SSF—simultaneous saccharification and fermentation) (Deng et al. 2012; Zhang et al. 2007; Jin et al. 2005). Nevertheless, the number of mucoralean species applied to established biotechnological processes is scarce:

Currently, only few species have been fully characterized regarding their potential to produce metabolites and enzymes. Research is primarily focussed on three genera: *Rhizopus* species for the production of organic acids, *Mucor* species for the production of ethanol and single cell oil (Ferreira et al. 2013) and *Cunninghamella* species for single cell oil production (Fakas et al. 2009). Interestingly, the transition from filamentous growth to yeast-like growth in dimorphic species is triggered by similar conditions favourable for organic acid and ethanol production, particularly high glucose concentrations along with elevated CO_2 contents (Lennartsson et al. 2009; Sharifia et al. 2008; Wolff and Arnau 2002; Serrano et al. 2001; Orlowsky 1991; Bartnicki-Garcia 1968). Therefore, due to the highly beneficial characteristics and promising biotechnological potential of mucoralean species, research exploring metabolite and enzyme production is urgently needed.

Biotechnologically relevant metabolites produced by zygomycetes are ethanol, carotenoids, fatty acids, organic acids, and single cell oils (SCOs) rich in polyunsaturated fatty acids (PUFAs) which are also named storage lipids. SCOs are known for their bifunction as a supplier of functional oils, and feedstock for biodiesel production (Huang et al. 2013). Especially organic acids and single cell oils containing PUFAs have a high market value, but suitable production strains and economically efficient processes are not available. Therefore, research on these substances would imply a high impact on white biotechnological issues.

A. Carotene Biosynthesis and Degradation: Primary Meets Secondary Metabolism

Despite their negative impact on humans and agriculture, zygomycetes could also be used in a positive way, like fermentations of food and sterols or the production of additives for food, feed, or pharmaceuticals with major interest in biological production of carotenoids, e.g. zeaxanthin, lycopene, or carotene (Hesseltine 1991; Nout and Kiers 2005; Liu et al. 2012; Rodríquez-Sáiz et al. 2012; Voigt and Kirk 2014). Since animals are not able to produce carotenoids by themselves, they depend on external sources and producers like plants, microorganisms, or fungi. Carotenoids are important pigments in animals and plants serving light protection and other physiological functions, e.g., as antioxidants, chromophores in photosynthesis or photoprotection, membrane stabilizers, and precursors for vitamin A. Carotenoid biosynthesis is known to be light stimulated (Rodríguez-Ortiz et al. 2012). Within the fungi, precursors of the carotenoids originate from the mevalonate pathway and are processed via geranylgeranyl diphosphate, phytoene, lycopene to, e.g., β-carotene. Known enzymes involved in **β-carotene synthesis** (Fig. 15.1) comprise CarRA (CarRP in *Mucor circinelloides*, containing the two domains CarR (lycopene cyclase) and CarA (phytoene synthase) (Fig. 15.1; Torres-Martínez et al. 1980; Arrach et al. 2001), CarB (phytoene dehydrogenase) (Ruiz-Hidalgo et al. 1997), CarI (Roncero and Cerdá-Olmedo 1982), CarF (Mehta et al. 1997), CarC (Revuelta and Eslava 1983), and CarD (Salgado et al. 1989).

β-carotene itself is processed differently in different fungal phyla, e.g., to neurosporaxanthin in *Ascomycota* or dihydroactinidiolide and β-ionone serving as flavor compounds in the *Basidiomycota* (Zorn et al. 2003). In the basal fungal lineages, however, carotene is cleaved to pheromones facilitating sexual recognition between mating partners—the sesquiterpene sirene in the *Chytridiomycota* (*Blastocladiomycetes*, *Neocallimastigomycetes*, *Monoblepharidomycetes*, and *Chytridiomycetes* for review see Voigt et al. 2013) and trisporoids in the zygomycetes (for review see Wöstemeyer et al. 2005). Zygomycetes recruit their recognition molecules from β-carotene biosynthesis (Fig. 15.1) and degradation (Fig. 15.1) pathways bridging primary and secondary metabolism, a feature comparable to abscisic acid in plants (Schwartz et al. 1997). Trisporoids are a rather

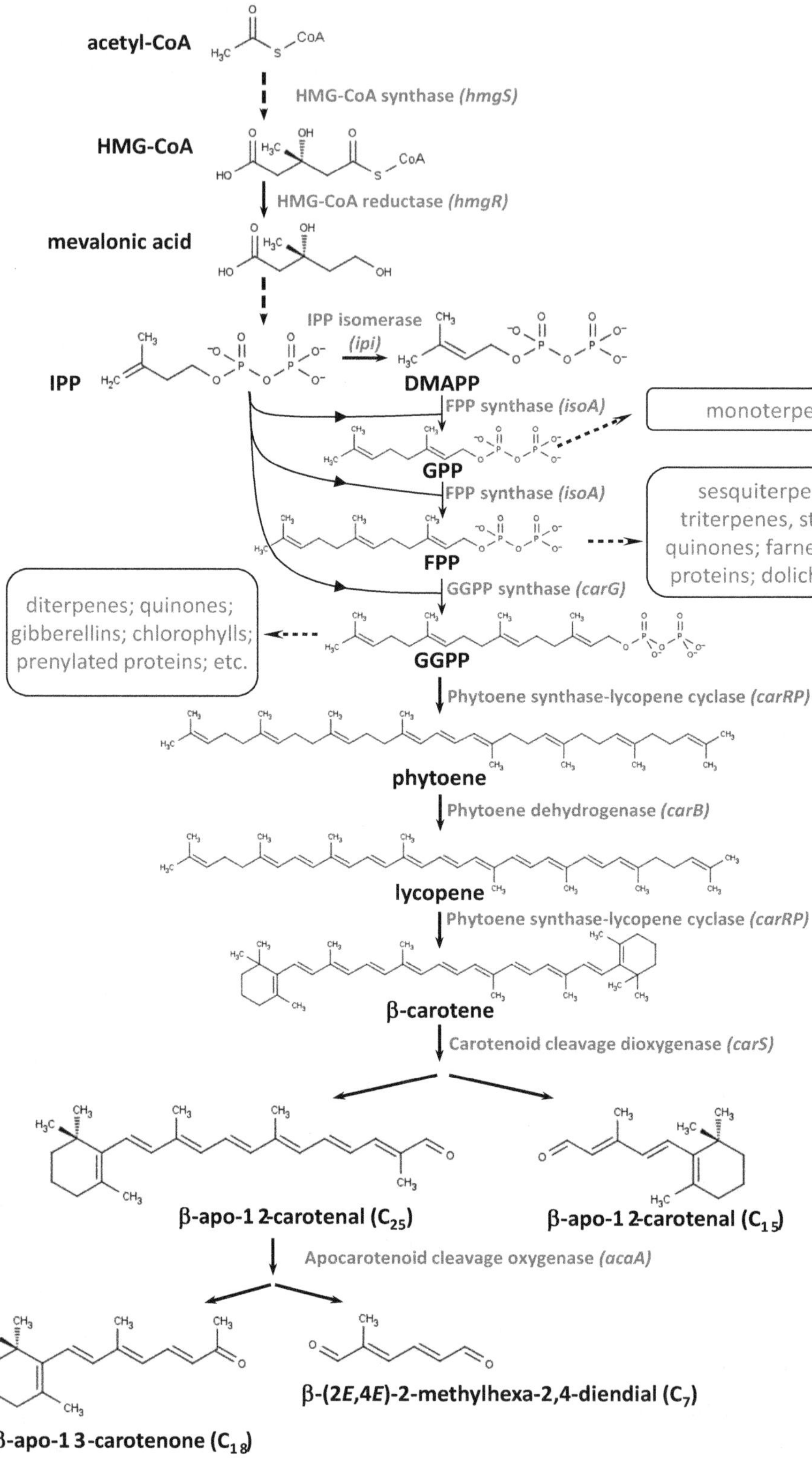

Fig. 15.1 Carotene biosynthesis and degradation pathways: Main steps of the acetate–mevalonate pathway, the specific β-carotene biosynthesis in *M. circinelloides* and presumable cleavage of β-carotene (C40)

unusual degradation product since it is not involved in cell-supporting or cell-protective functions but in pheromone action in sexual communication of those fungi if compared to higher fungi which rely on modified peptides (Gooday 1974; Jones and Bennett 2011). Since a multitude of structural diverse carotenoids is involved in a very broad and diverse spectrum of applications in all organisms, an even much more diversity of enzymes involved in carotenogenesis and processing should be reasonable. Enzymes cleaving specific double bonds are termed **carotenoid cleavage oxygenases** (CCOs) or more specified mono- (CMOs) or dioxygenases (CCDs):

Elucidation and clarification of enzymatic mechanisms started only several years ago with the first crystal structure of a CCO from *Synechocystis* sp. converting β-apo-carotenals as sole substrates (Kloer et al. 2005). Since then, only few amino acid residues are believed to be essential for enzyme activity, namely four histidine and three glutamate/aspartate residues (Poliakov et al. 2005; Takahashi et al. 2005). According to the first structurally described CCO, these amino acids correspond to sequence positions Glu150, His183, His238, His304, Glu370, Glu426 and His484 of *Synechocystis*.

All representative carotenoid cleavage enzymes (shown in Fig. 15.1) possess these conserved amino acids. In accordance with their first description by Medina et al. 2011, few sequences (clustering within the zygomycetous order *Mucorales* and termed "unknown") possess similar sequences but lack some of the conserved residues (essentially His183 and His238). The gene coding for the CCO from *Phycomyces blakesleeanus* (acaA) seems to have been duplicated recently. A functional characterization remains to be done for the duplicated *aca*A and the presumed genes of unknown function (Medina et al. 2011). The phylogeny of the CCOs (Fig. 15.2) shows that each clade has evolved its more or less specific carotenoid cleavage enzymes with similar cleaving sites, but with different natural substrates, which is presumably due to their wide variety within carotenogenesis and specific organismal requirements. The carotenoid-cleaving enzymes in the *Mucorales* are unique for this group of fungi with no similar cleaving enzymes in other fungal groups (Sahadevan et al. 2013):

Mucorales seem to possess also only one enzyme capable to cleave β-carotene, an enzyme crucial for all subsequent utilizations of β-carotene. This gene, called *carS*, is an 11′–12′ carotenoid cleavage dioxygenase (Fig. 15.1; Medina et al. 2011; Tagua et al. 2012; Rodríquez-Sáiz et al. 2012; Rodríguez-Ortiz et al. 2012), which cleaves β-carotene (C_{40}) into β-apo-12-carotenal (C_{25}) and β-apo-12-carotenal (C_{15}, Fig. 15.1, Sahadevan et al. 2013). CarS should not be misapplied as an orthologue of the *carS* in the ascomycete *Fusarium* sp., which codes for a regulatory protein, most likely corresponding to CrgA from *Mucor circinelloides* (Navarro et al. 2001). After cleavage by CarS, β-apo-12-carotenal (C_{25}) is further processed by the apocarotenoid cleavage oxygenase AcaA, presumably cleaved at its 13–14 position, resulting finally in two more fragments, namely β-apo-13-carotenone (C_{18}, also named d'orenone, Sahadevan et al. 2013), and probably (2*E*,4*E*)-2-methylhexa-2,4-diendial (C_7) (Fig. 15.1; Polaino et al. 2010; Medina et al. 2011).

1. Regulation, Genetic Manipulation: What Have We Learned from the Major Model Organisms *Mucor circinelloides*, *Phycomyces blakesleeanus*, and *Blakeslea trispora*?

Members of the order *Mucorales* are known as β-carotene-producing fungi. Among them, *Blakeslea trispora*, *Mucor circinelloides*, and *Phycomyces blakesleeanus* are involved in the study of the carotenoid biosynthesis as model organisms. *B. trispora* is already an industrial

Fig. 15.1 (continued) at position C11–C12 by the carotenoid cleavage dioxygenase CarS, resulting in two fragments of β-apo-12-carotenal, (C25) and (C15) in *P. blakesleeanus*, the final cleavage of β-apo-12-carotenal (C25) to β-apo-13-carotenone (C18) and probably (2*E*, 4*E*)-2-methylhexa-2,4-diendial (C7) by the apocarotenoid cleavage oxygenase AcaA. The most important enzymes and the encoding genes are indicated with *gray*. *HMG* hydroxymethylglutaryl, *IPP* isopentenyl pyrophosphate, *DMAPP* dimethylallyl pyrophosphate, *GPP* geranyl pyrophosphate, *FPP* farnesyl pyrophosphate, *GGPP* geranylgeranyl pyrophosphate

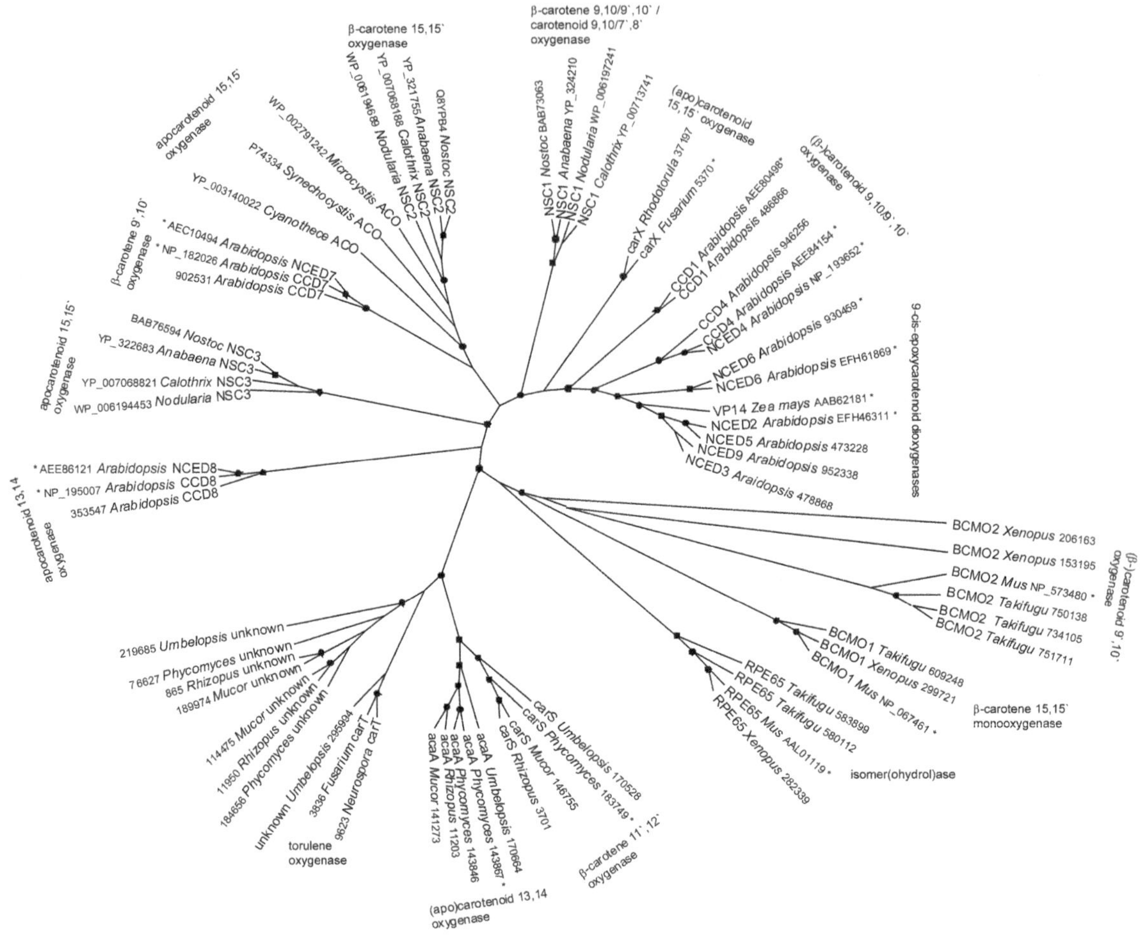

Fig. 15.2 Evolutionary relationships of representative genes involved in β-carotene degradation and their representative cleavage sites. The tree can be roughly divided into three sub-trees, each following species phylogeny. Carotenoid cleavage oxygenases from *Mucorales* comprise CarS and AcaA as well as several so far uncharacterized sequences. Bootstrap values greater or equal to 90 % are indicated by *black dots*. *Names on branches* indicate prominent cleavage sites

source of β-carotene, while the application of *M. circinelloides* and *P. blakesleeanus* is in a developmental phase (Dufossé 2006, 2008). However, improvement and study of the carotenogenesis in *B. trispora* and *P. blakesleeanus* are hampered by the lack of efficient methods for genetic manipulation; i.e., their genetic transformation has still been unsuccessful (Obraztsova et al. 2004; Sanz et al. 2011; Garre et al. 2015). *M. circinelloides* seems to be more amenable to molecular techniques as well-developed transformation systems including vectors, promoters, recipient strains, and methods (i.e., PEG-mediated protoplast transformation and electroporation) are available (van Heeswijck and Roncero 1984; Wolff and Arnau 2002; Appel et al. 2004; Papp et al. 2010; Gutiérrez et al. 2011). Moreover, this fungus has an ability to maintain and express exogenous genes from related fungi (e.g., *P. blakesleeanus*, *B. trispora*, or *Rhizomucor miehei*) and other organisms (e.g., *Xanthophyllomyces dendrorhous* or *Paracoccus* sp. N81106) (Iturriaga et al. 1992; Ruiz-Hidalgo et al. 1999; Quiles-Rosillo et al. 2003; Rodríguez-Sáiz et al. 2004; Lukács et al. 2009; Papp et al. 2006, 2013; Csernetics et al. 2015).

Carotenoids are terpenoid compounds, and their biosynthesis can be regarded as a side route of the general acetate–mevalonate (AMV) pathway, in which precursors of the different terpene derivatives are synthesized from acetyl CoA. Several genes encoding the enzymes, which catalyze the main steps of the AMV pathway and carotenoid biosynthesis, have been isolated and characterized in *M. circinelloides* (Velayos et al. 2000a, b, 2003; Csernetics et al. 2011; Nagy et al. 2014). Carotenogenic genes of *B. trispora* and *P. blakesleeanus* were also identified, and their functions were analyzed by expressing them in *M. circinelloides* (Rodríguez-Sáiz et al. 2004; Sanz et al. 2011).

One of the key enzymes of the general AMV pathway is 3-hydroxy-3-methylglutaryl-CoA (HMG-CoA) reductase, which catalyzes the formation of HMG-CoA from mevalonic acid. As HMG-CoA is a common intermediate of numerous different terpenoid compounds, such as carotenoids, ergosterol, prenyl groups of certain proteins, and ubiquinone, its formation is considered to be rate limiting for the carotenoid synthesis (Wang and Keasling 2002). *M. circinelloides* has three HMG-CoA reductase genes (*hmgR*), which respond differently to temperature and the oxygen level of the environment (Nagy et al. 2014). Among them, *hmgR2* and *hmgR3* seem to be especially involved in the carotenoid biosynthesis. Overexpression of these genes by changing their own promoter to that of the endogenous glyceraldehyde-3-phosphate dehydrogenase 1 gene (*gpd1*) and elevating their copy numbers increased the whole carotenoid content of the fungus 1.5–1.7-fold (Nagy et al. 2014).

Another important section of this pathway is the synthesis of the prenyl-chain intermediate compounds, which serve as precursors in the different specific side routes. The most important steps of this process are the isomerization of dimethylallyl pyrophosphate (DMAPP) and isopentenyl pyrophosphate (IPP) catalyzed by the IPP isomerase, the condensation of IPP and DMAPP to form geranyl pyrophosphate (GPP), and the extension of the prenyl chain by the addition of further IPP units to the carbon chain forming farnesyl and geranylgeranyl pyrophosphate (FPP and GGPP, respectively). Synthesis of GPP and FPP is catalyzed by the FPP synthase, while formation of GGPP is managed by the GGPP synthase. GGPP is the direct precursor of carotenoids as their specific biosynthesis starts with the condensation of two 20-carbon GGPP units leading to the synthesis of carotenoid phytoene (Iturriaga et al. 2000). In *M. circinelloides*, the IPP isomerase and the FPP and GGPP synthases are encoded by the *ipi*, *isoA*, and *carG* genes (Velayos et al. 2003, 2004; Csernetics et al. 2011). Overexpression of these genes significantly enhanced the carotenoid biosynthesis. In this study, the step determined by the *carG* gene proved to be the first bottleneck for carotenoid production, placing it under the control of the *Mucor gpd1* promoter resulted in a fourfold increase in the carotenoid content (Csernetics et al. 2011). Total carotenoid content of these strains was more than 2 mg/g (dry weight). Similarly, the expression of the *ipi* and the *carG* genes of *B. trispora* in an engineered, carotenoid-producing *E. coli* strain led to a twofold increase in the carotenoid production of the bacterium (Sun et al. 2012). These studies indicated that *ipi* and *carG* genes can be applied to improve the carotenoid production of mucoralean fungi. *M. circinelloides* requires light for carotenoid biosynthesis and transcription of *carG*, and the carotenoid-specific genes (i.e., *carB*-encoding phytoene dehydrogenase and *carRP*-encoding phytoene synthase–lycopene cyclase) are induced by blue light (Velayos et al. 2000a, b, 2003). White collar-1-like proteins, Mcwc-1b, and Mcwc-1c were found to be involved in the activation of the carotenogenic genes of *M. circinelloides* (Silva et al. 2006, 2008), while the protein CrgA proved to be a repressor of the carotenoid biosynthesis in *Mucor* (Navarro et al. 2001). Deletion of the *crgA* gene resulted in enhanced accumulation of carotenoids under both dark and light conditions (Navarro et al. 2001). Moreover, deletion of *crgA* could be used to increase the lycopene production of a *mutant M. circinelloides* strain achieving a lycopene content of 54 g/L (Nicolás-Molina et al. 2008). Recently,

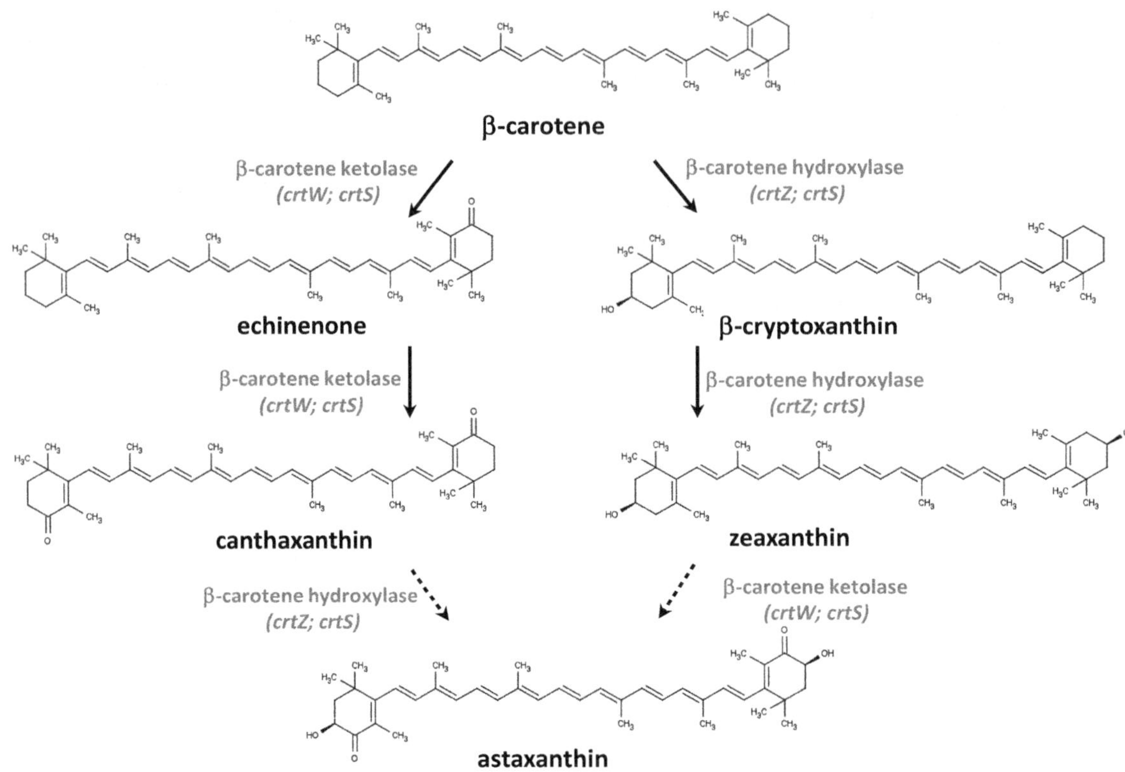

Fig. 15.3 Conversion of β-carotene to its oxygenated derivatives, the carotenoids (xanthophylls)

it has been supposed that CrgA may be an ubiquitin ligase, and one of its functions may include preventing Mcwc-1b to activate the transcription of the carotenoid biosynthesis genes (Silva et al. 2008; Navarro et al. 2013):

By expressing exogenous carotenoid biosynthesis genes, production of new carotenoid compounds, such as oxygenated derivatives of β-carotene, can be achieved. *Paracoccus* sp. N81106 is a marine, astaxanthin producing bacterium, in which the conversion of β-carotene to astaxanthin is catalyzed by the enzymes β-carotene ketolase (CrtW) and hydroxylase (CrtZ). Production of xanthophylls (i.e. β-cryptoxanthin, zeaxanthin, canthaxanthin and astaxanthin, Fig. 15.3) could be carried out by transforming *M. circinelloides* with autonomously replicating vectors, which harboured the *crtW* and the *crtZ* genes fused with the regulatory sequences of *Mucor gpd1* (Papp et al. 2006). Multiple integration of the *crtW* gene into the *Mucor* genome resulted in strains accumulating canthaxanthin as the main carotenoid instead of β-carotene (Papp et al. 2013). The astaxanthin biosynthesis gene (*crtS*) of *Xanthophyllomyces dendrorhous* also could be used to obtain xanthophyll-producing *M. circinelloides* strains (Álvarez et al. 2006; Csernetics et al. 2015; Rodríquez-Sáiz et al. 2012). In these experiments, the *crtS* gene was driven by the promoter of the *Blakeslea carRA* or the *Mucor gpd1*.

2. Carotene Degradation Is Linked to Sexual Interactions

All zygomycetes are coherently united by the potential to form the chemotactic pheromone **trisporic acid** (Fig. 15.4; for review see Wöstemeyer et al. 2002, 2005). This compound is morphogenic by its ability to induce the genesis of zygophores subsequently followed by zygospores during conjugation of two yoke-shaped gametangia (**gametangiogamy**) in compatible mating interactions. Trisporic acid is the universal gamone, which is cooperatively formed between both mating partners (Schimek et al. 2003; Schachtschabel et al. 2008). Trisporoid acid has a multitude of derivatives (trisporoids), which possess deviating biological activ-

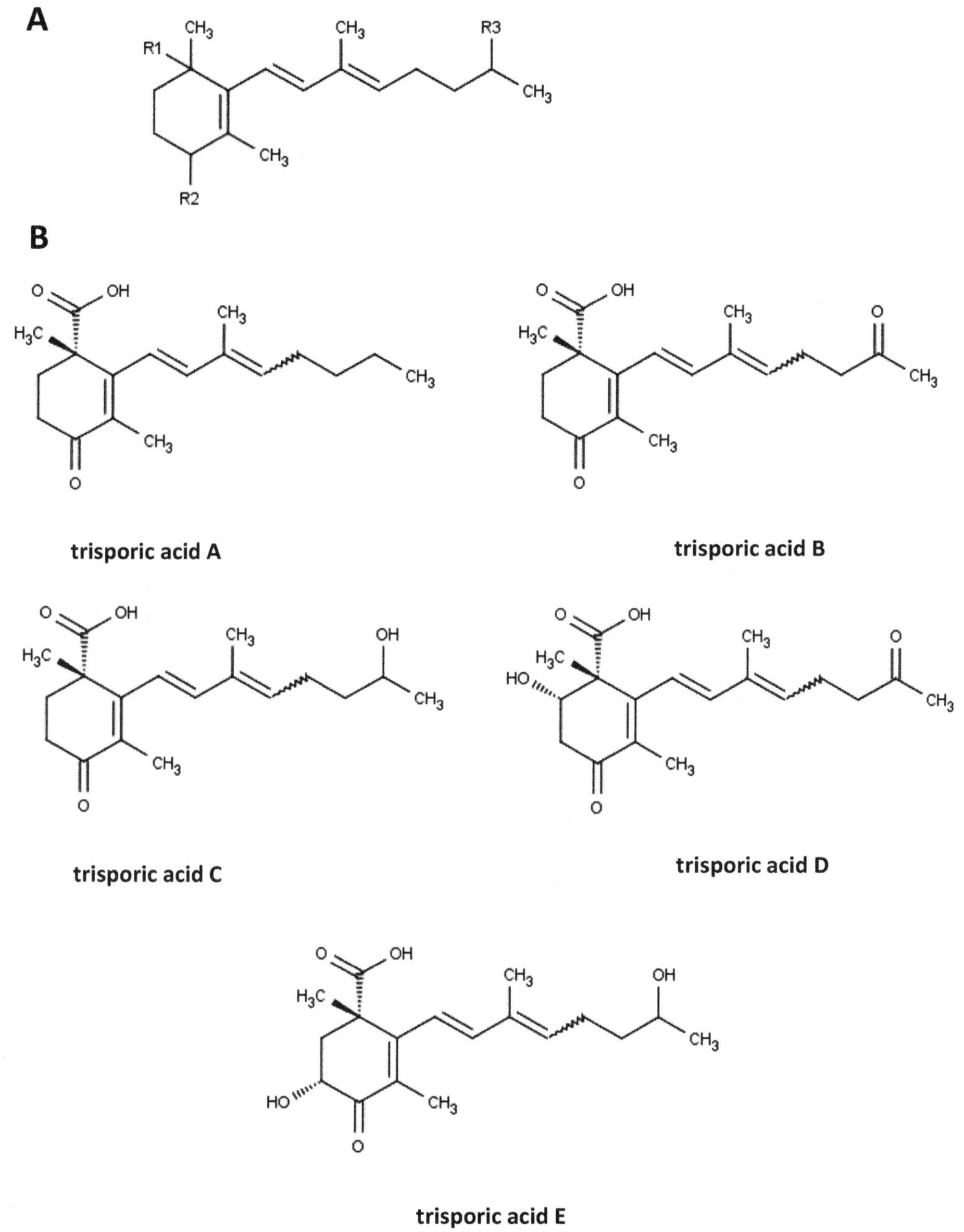

Fig. 15.4 Chemical structures of trisporic acid, the universal sexual pheromone of the zygomycetes. (**a**) Basic chemical structure. (**b**) Structures of trisporic acids A–D. Trisporic acid D was postulated but was never experimentally proven

ity (Schachtschabel et al. 2005). The biosynthesis of trisporoids starts with the cleavage of β-carotene which is mediated by a trisporic acid-regulated β-carotene oxygenases *tsp3* and *tsp4* aiming a C_{18} compound (Burmester et al. 2007). The sexual pheromones of the *Mucorales* are processed from the C_{18} compound resulting finally in trisporic acids. Yet many of the enzymatic steps remain unknown. The only enzymes known so far in the processing of the

C_{18} compound are two enzymes belonging to two different families of oxidoreductases: *tsp2*, a short-chain dehydrogenase involved in the processing of 4-dihydrotrisporin, and *tsp1*, an aldo-keto reductase involved in the processing of 4-dihydromethyltrisporate (Czempinski et al. 1996; Wetzel et al. 2009).

B. Fatty Acids

Fungi of the order *Mortierellales* and *Mucorales* have attracted considerable interest as **industrial lipid producers**. They are easily cultivated in solid or liquid culture and have been shown to grow on various different carbon sources (Dyal and Narine 2005; Gao et al. 2013; Zeng et al. 2013), on numerous different agricultural waste products (Chaudhuri et al. 1998; Jang et al. 2000; Zeng et al. 2013), and on glycerol, a by-product of biodiesel production (Hou 2008; Dedyukhina et al. 2011; Chatzifragkou et al. 2011). Hence, industrial and agricultural waste products can be converted as low-cost substrate into valuable products, providing an excellent biotechnological application for the zygomycetes. For example, lignocellulosic biomass, which is the most available and renewable source in nature, might be an ideal raw material for single cell oils (see Sect. D) production (Huang et al. 2013). Especially *Mortierella* spp. can accumulate large amounts of unusual lipids containing polyunsaturated fatty acids depending on species, strain, and growth conditions (Münchberg et al. 2012, 2015). The characterization of the genomes from oleaginous fungi like *Mortierella alpina* (Wang et al. 2011) and *M. elongata* (Bonito et al. 2013) provides insights into the genomic basis of fatty acid production. First insights into the *M. elongata* genome reveal preliminary enrichments of genes related to lipid metabolism (e.g., sphingolipids, ether lipids, and glycerophospholipids), tryptophan metabolism, siderophore group nonribosomal peptides, and glucan 1,4-alpha glucosidases compared to genome sequences of other basal fungi (Bonito et al. 2013).

C. Organic Acids

The production of relevant organic acids, namely, L-**lactic acid** and **fumaric acid**, is based on pyruvate, the end product of the glycolysis. Whereas L-lactic acid is formed directly from pyruvate by lactate dehydrogenase (Skory 2000; Pritchard 1971, 1973), fumaric acid is formed via the oxidative branch of the TCA cycle located in the cytoplasm (Fig. 15.5; Goldberg et al. 2006).

Both organic acids can be diversely applied in food industry, textile sector, cosmetic industry, and chemical and pharmaceutical industry. Lactic acid is the most abundantly produced organic acid in nature. Therefore, lactic acid production by *Rhizopus* species is a subject of intensive research and has the potential to replace the established lactic acid production processes using chemical methods or lactobacilli fermentation. When producing lactic acid by *Rhizopus* species, low-cost substrates (e.g., agricultural waste products containing any kind of plant fibers) and a wide variety of carbon sources ranging from monosaccharides to polysaccharides can be used (Guo et al. 2010; Vially et al. 2010; Yen and Lee 2010; Bulut et al. 2009; Bai et al. 2008), resulting in very high yields ranging near the theoretical maximum (Ferreira et al. 2013; Meussen et al. 2012).

Fumaric acid, a C_4-dicarboxylic acid, was identified by the US Department of Energy as one of 12 promising platform chemicals from biomass with high added value (Werpy and Petersen 2004). Presently, fumaric acid is chemically produced from crude oil and is applied in food industry as acidulant, food preservative, and flavor enhancer. Due to its bifunctionality and the double-bond fumaric acid, it is also suitable to act as polymerization starter unit for plastics or resins (Anonymus 2007; Willke and Vorlop 2004). As for lactic acid, high yields near the theoretical maximum can be achieved by microbial fermentation when using glucose as carbon source (Meussen et al. 2012; Roa Engel et al. 2008; Cao et al. 1996). Noteworthily, for each molecule of formed fumaric acid, one molecule CO_2 is fixated (Osmani and Scrutton 1985; Overman and Romano 1969).

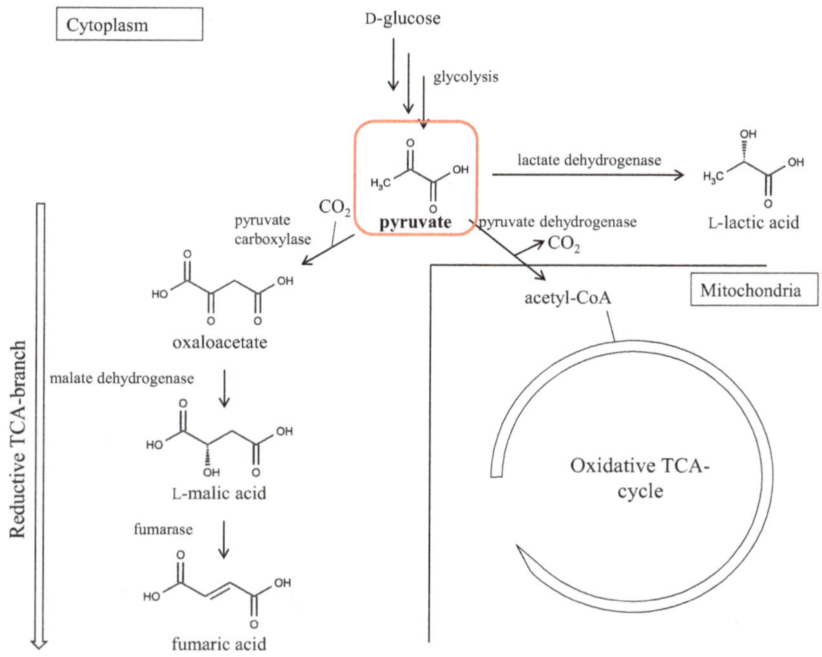

Fig. 15.5 Formation of lactic and fumaric acid is based on pyruvate, the end product of the glycolysis. Whereas L-lactic acid is formed directly from pyruvate by lactate dehydrogenase (Skory 2000; Pritchard 1971, 1973), fumaric acid is formed via the oxidative branch of the TCA cycle located in the cytoplasm

Whether fumaric or lactic acid is produced from pyruvate depends on activity and substrate affinity of the respective enzymes and seemed to be strain dependent. However, Saito et al. (2004) proved that *Rhizopus oryzae* strains with two genes for lactate dehydrogenase produce mainly lactic acid, whereas strains with only one gene produce mainly fumaric acid.

Phylogenetic studies using further independent DNA markers by Abe et al. (2007) revealed that fumaric acid and lactic acid producers can be separated into two sibling species, *R. oryzae* sensu stricto (also known as *R. arrhizus*) and *R. delemar* correlating with the lactic acid and fumaric–malic acid producers, respectively. Reclassification of strains in the fumaric–malic acid group as *R. delemar* and therefore reclassification of the genome strain *R. oryzae* 99-880 into *R. delemar* were proposed (Gryganskyi et al. 2010), which was later converted into *Rhizopus arrhizus* var. *delemar* (Dolatabadi et al. 2014a).

D. Storage Lipids and Single Cell Oils

All living organisms have to synthesize a minimum amount of lipids to build up membranes. However, only few organisms are able to accumulate more than 20 % of their dry biomass in form of storage lipids. The term "oleaginous" refers to microorganisms, including yeasts, fungi, and microalgae, which meet this criterion and store lipids in form of triacylglycerols (Ratledge and Wynn 2002). Storage lipids, which are also known as "single cell oils" (SCPs), are rich in polyunsaturated fatty acids and are of special interest due to their bifunction as a supplier of functional oils and feedstock for biodiesel production (Huang et al. 2013). Especially **γ-linoleic acid** (GLA, C18:3n−6) is biotechnologically relevant (Fig. 15.6). It is commercially applied in pharmaceutical industry and is currently obtained by extraction of selected plant oils. However, higher amounts of GLA are also produced by some mucoralean genera, like *Cunninghamella*, *Mucor* (including *Zygorhynchus*), *Rhizopus*

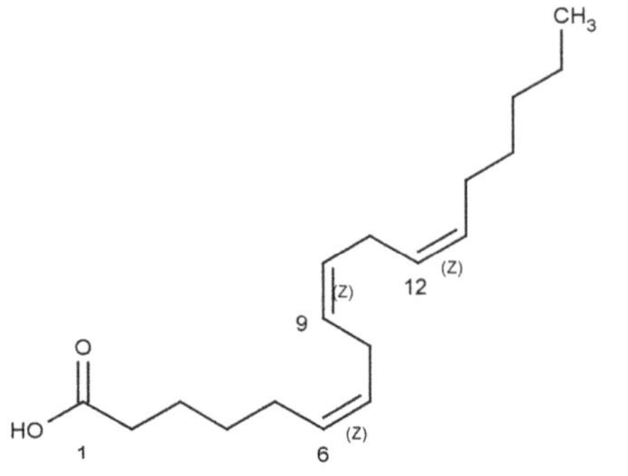

Fig. 15.6 Chemical structure of γ-linoleic acid (GLA, C18:3n−6)

(Kavadia et al. 2001; van der Westhuizen et al. 1994), *Choanephora*, *Phycomyces* (van der Westhuizen et al. 1994), and *Thamnidium* (Stredansky et al. 2000). Oleaginous microorganisms start the accumulation of single cell oil when grown in a medium with excess of carbon source but with a limitation of another nutrient. Oleagincity is characterized by the ability to produce a continuous supply of both acetyl CoA and NADPH as necessary precursors and reduction equivalent in fatty acid biosynthesis and is realized by the key enzymes ATP citrate lyase and AMP deaminase (Ratledge 2004). *Mortierellales* have great biotechnological importance as industrial producers of polyunsaturated fatty acids, such as arachidonic acid or eicosapentaenoic acid. Both the content of fatty acids and their rate of saturation are known to be dependent on the temperature during production and also vary due to utilization of different nutrients in the cultivation media (Münchberg et al. 2012, 2015).

E. Enzymes

Zygomycetes are known to produce a vast variety of enzymes, e.g., amylases, cellulases, xylanases, steroid 11α-hydroxylases, phytases, proteases, and lipases which have a multitude of applications in industrial and pharmaceutical applications (for review see Krisch et al. 2010; Voigt and Kirk 2014).

Amylases are one of the main enzymes used in industry. They hydrolyze starch molecules into polymers composed of glucose units or oligosaccharides. Amylases have potential application in industrial processes such as food, textile, paper, and detergent industries as well as fermentation and pharmaceutical industries. As starch is an important constituent of the human diet and is a major storage product of many economically important crops such as wheat, rice, maize, tapioca, and potato, starch-converting enzymes are used in the production of maltodextrin, modified starches, or glucose and fructose syrups. For the production α-amylases using submerged and solid-state fermentation systems, distribution, structural–functional aspects, physical and chemical parameters, and the use of these enzymes in industrial applications, see the review by Monteiro de Souza and de Oliveira Magalhães (2010).

Cellulases catalyze cellulolysis, the decomposition of cellulose, which is the most abundant organic source of feed/food, fuel, and chemicals (Spano et al. 1976). Cellulase breaks down the cellulose molecule into mono- and oligosaccharides by hydrolysis of the 1,4-beta-D-glycosidic linkages in cellulose in its derivative hemicellulose, lichenin, and cereal beta-D-glucans. Cellulases represent a naturally occurring mixture of various enzymes acting serially or synergistically to decompose cellulosic material. Zygomycetes (e.g., *Mucor circinelloides*) were frequently found as straw-colonizing fungi producing total cellulases, endo-beta-1,4 glucanase, and endo-beta-1,4 xylanase in solid-state fermentation (Lee et al. 2011).

Lipases are water-soluble enzymes that act on insoluble substrates and catalyze the hydrolysis of long-chain triglycerides. They play a vital role in the food, detergent, chemical, and pharmaceutical industries and have gained significant attention in the industries due to their substrate specificity and stability under varied chemical and physical conditions (for review see Gopinath et al. 2013).

Phytases are myo-inositol hexakisphosphate phosphohydrolases and represent any type of phosphatase enzyme that catalyzes the hydrolysis of phytic acid (myo-inositol hexakisphosphate)—an indigestible, organic form of phosphorus that is found in grains and oil seeds—and releases a usable form of inorganic phosphorus (Mullaney et al. 2000). Phytases have been most commonly detected and characterized from fungi (Mullaney and Ullah 2003), specifically in the zygomycete *Rhizopus oligosporus* (DSMZ 1964), which is commonly used for tempeh production (Azeke et al. 2011). The phytases from *R. oligosporus* exhibit a broad affinity for various phosphorylated compounds. Practical interest in phytases has been stimulated by the fact that phytase supplements increase the availability of phosphorus in pig and poultry feed and thereby reduce environmental pollution due to excess phosphate excretion in areas where there is intensive livestock production.

Proteases produced by zygomycetes are rennin-like proteases secreted by several mucoralean species that are used in cheese production. In particular, mucoralean fungi (*Rhizopus oryzae, Circinella muscae, Mucor subtilissimus, Mucor hiemalis* f. *hiemalis, Syncephalastrum racemosum, Rhizopus microsporus* var. *chinensis*, and *Absidia cylindrospora*) were frequently isolated from maize flour, corn meal, and cooked cornflakes using surface and depth plate methods with subsequent measurement of significant proteolytic activities (de Azevedo Santiago and de Souza Motta 2008).

Steroid 11α-hydroxylases are encoded by genes of the cytochrome P450 superfamily of enzymes containing a heme cofactor (hemoproteins, Sigel et al. 2007). The cytochrome P450 proteins are monooxygenases that catalyze many reactions involved in drug metabolism and synthesis of cholesterol, steroids, and other lipids. They are, in general, the terminal oxidase enzymes in electron transfer chains, broadly categorized as P450-containing systems. The term *P450* is derived from the spectrophotometric peak at the wavelength of the absorption maximum of the enzyme (450 nm) when it is in the reduced state and complexed with CO (Sigel et al. 2007). To overcome the chemically laborious stereo- and regioselective hydroxylation steps in the pharmaceutical production of corticosteroids and progestogens, zygomycetes, e.g. *Rhizopus* spp., are employed to perform the 11α-hydroxylation of the steroid skeleton, thereby significantly simplifying steroid drug production (Petrič et al. 2010).

Xylanases degrade the linear polysaccharide beta-1,4-xylan into xylose, thus breaking down hemicellulose, one of the major components of plant cell walls (Beg et al. 2001). Zygomycetes (e.g. *Mucor circinelloides*) are straw-colonizing fungi producing xylanolytic enzymes such as endo-beta-1,4 xylanase in solid-state fermentation (Lee et al. 2011).

III. The Dogma of the Unability of Zygomycetes to Produce Natural Products

It has been a widespread dogma that zygomycetes are not capable to produce own secondary metabolites, despite of those produced by endosymbiotic bacteria (see Sect. I.B.) (for examples, see Jennessen et al. 2005). However, it has been shown that zygomycetes react on other fungal secondary metabolites by morphogenic changes as shown by the sesterterpene-type phytotoxin ophiobolin produced by fungi belonging mainly to the ascomycetous genus *Bipolaris* (Krizsán et al. 2010). Ophiobolin A caused morphological changes in *Mucor circinelloides*; the fungus formed degenerated, thick or swollen cells with septa and cytoplasm effusions from the damaged cells. Here we explore the potential of zygomycetes to produce secondary metabolites together with other microorganisms in a cooperative manner.

To estimate the genomic potential of the zygomycetes to produce secondary metabolites, all final, publicly available drafts of zygomycete genomes were scanned for the presence of genes encoding **polyketide synthases** (PKSs), **nonribosomal peptide synthetases** (NRPSs), and **L-tryptophan dimethylallyl transferases** (DMATs). For more information on regulation of secondary metabolism, see also Chap. 2. A total of eight genomes were screened and analyzed: one entomophthoralean (*Conidiobolus coronatus*), one kickxellalean (*Coemansia reversa*), two mortierellalean, and four mucoralean genomes (Table 15.1). All species possess the genomic prerequisite for the production of natural products. On average, each species encodes two DMATs and 1–2 NRPSs, with a

Table 15.1 The presence of gene families encoding polyketide synthases (PKSs), nonribosomal peptide synthetases (NRPSs), dimethylallyl pyrophosphate: L-tryptophan dimethylallyl transferase (DMAT synthase) which were predicted in the genomes of 15 species comprising five subphyla of the *Zygomycota*, as of 6th of July, 2015; genome resources (if not stated elsewhere): Joint Genome Institute Broad Institute of Harvard and MIT, Origins of Multicellularity Sequencing Project *Mortierella verticillata*

		Genome resource	(PKS)/FAS[a]	NRPS	DMAT
Mucorales	*Lichtheimia hyalospora*	JGI, unpublished	1	1	2
	Mucor circinelloides	Lee et al. (2014)	2	3	2
	Rhizopus microsporus var. chinensis	Wang et al. (2013)	1(+1)	2	6
	Rhizopus arrhizus var. *delemar* (syn. *R. oryzae*)[b]	Ma et al. (2009)	1	1	3
Kickxellales	*Coemansia reversa*	Chang et al. (2015)	4	1	2
Entomophthorales	*Conidiobolus coronatus*	Chang et al. (2015)	″(1)	3	2
Mortierellales	*Mortierella verticillata*	Bonito et al. (2013)	0	1	0
	Mortierella alpina	Wang et al. (2011)[c]	1	21	0

For an overview of genome projects on basal fungi incl. *Zygomycota*, see Shelest and Voigt (2014)
[a]The genes predicted as PKS-like have the typical structure of the FAS alpha subunit
[b]Nomenclature: Gryganskyi et al. (2010) and Dolatabadi et al. (2014a)
[c]Proteome of *M. alpina* available from phylomeDB: ftp://phylomedb.org/phylomedb/proteomes/ ftp://phylomedb.org/phylomedb/proteomes/685557.1.fa.gz

noteworthy exception detecting 21 NRPSs in the genome of *Mortierella alpina*. The genes encoding typical PKS/fatty acid synthase (FAS) ketosynthase domains are most likely FAS alpha subunits: they reveal a characteristic domain order. BLAST searches using the tools at the specific home pages of the genomes confirm their annotation as FASs. Most of the discovered NRPS genes encode monomodular enzymes, except for the one in *Mortierella verticillata*, where we find a five-module protein encoded. Many NRPS-like genes do not possess the minimal set of domains necessary for the full enzymatic activity. These genes are therefore characterized as NRPS-like. Three genes which putatively encode for NRPSs were found in *Mucor circinelloides* f. *circinelloides* (Table 15.1) and can also be considered NRPS-like (Lee et al. 2014). Within the genus *Lichtheimia*, genes encoding PKSs, NRPSs, and DMATs are present in *L. hyalospora*, but absent in *L. corymbifera* (Schwartze et al. 2014). In some cases it can be supposed that the genes predicted actually represent the full enzymes but cannot be correctly annotated due to erroneous gene prediction. Another problem connected with genome assembly is the high AT content, which renders the bioinformation content low and prevents motif-based cluster prediction. Transcription regulators of secondary metabolism have not been yet systematically characterized, as reliable data is scarce. At this stage, we are mostly aware of pathway-specific regulators of clusters, but it is premature to draw general picture yet. In fact, known clusters in the *Ascomycetes* build the main basis of such genome-mining analyses, whereas information on proven clusters in other fungal phyla is lacking. In the *Ascomycetes*, about 60 % of PKS- and NRPS-encoding gene clusters include an embedded transcription regulator gene, which encodes in majority of cases a **zinc cluster transcription factor** (TF) (Brakhage 2013). The neighboring sequence regions of the genes encoding PKSs, NRPSs, and DMATs in zygomycetes were analyzed in order to confirm this preference for Zn cluster TFs. For this, we predicted that all TFs genome-wide extracted the TF annotations in regions of ± 10 genes around those encoding secondary metabolite enzymes. We assigned these TFs to families of DNA-binding domains based on InterProScan predictions as described by Shelest (2008). Interestingly, Zn clusters are very modestly represented among these SM-accompanying TFs, leaving the first place to C_2H_2 Zn finger TFs and TFs of the homeodomain-like class. This observation becomes less surprising, however, if we think about the overall predominance of C_2H_2 Zn fingers and especially of homeodomain-like TFs observed in zygomycetes. Our analysis suggests that every second, NRPS has a TF in the vicinity of 10 genes (8 of 15 NRPSs). For DMATs this number is higher (20 of 34, ~60 %). This corresponds to the number of the TFs in known ascomycete clusters (60 % for PKSs and NRPSs). The number for zygomycetes can be lower because we consider only the vicinity of 10 genes, whereas the cluster can be longer; on the other hand, considering longer region can give more false-positive predictions. The most frequent TFs in the vicinity of NRPSs and DMATs are homeodomain-like DNA-binding domain family TFs and C_2H_2 Zn finger TFs, comprising in sum nearly half of the total number of TFs that can potentially be the regulators of secondary metabolism in zygomycetes. This is not very surprising since these two families are the most abundant in at least *Mucorales* and *Entomophthorales* genomes (Schwartze et al. 2014). It is interesting to notice, however, that the most numerous families take over the regulation of the secondary metabolite clusters in fungi: in *Ascomycetes*, where Zn cluster is the dominating TF family, Zn cluster TFs are most frequently embedded in SM clusters, and in a similar picture we observe now for C_2H_2 and homeobox TFs in zygomycetes.

One promising strategy to explore and to broaden the biotechnological potential of the zygomycetes could be the investigation of zygomycetes in co-cultures with other microorganisms sharing the same ecological niche. This procedure has been shown successful in *Aspergillus* spp. for activation of silent gene clusters (Schroeckh et al. 2009; Nützmann et al. 2011, 2012) and is consistent with the cooperative nature of the zygomycetes as shown at the cellular and molecular level (Schachtschabel et al. 2008; Schimek and Wöstemeyer 2009; Krizsán et al. 2010; Voigt and Kirk 2014). Cocultivation of zygomycetes with other microorganisms sharing the same habitat under nature-close cultivation conditions has a high potential to increase the metabolic activity of zygomycetes, which are commonly known to be low producers of secondary compounds.

To sum it up, we show that all considered zygomycetes have a pronounced potential to produce natural products.

IV. Conclusions

- Zygomycetes have an integral role in the development of microbial ecosystems, a property which has the potential to be converted for biotechnological and industrial applications ranging from food technology to drug development.
- Understanding the biology, ecology, and biotrophic interactions of the zygomycetes with other microorganisms can help to explore novel secondary metabolites which are central to ecological functions and are useful for effective and innovative biotechnological utilizations,
- The genomic architecture and transcription factor repertoire of the zygomycetes largely differs from that of more recent fungal lineages. C_2H_2 transcription factors are predominant transcription factors.
- The degradation of β-carotene to pheromones has been extensively studied. The impact on biotechnological importance and applications has been neglected so far. Systematic–phylogenetic approaches may help with the screening for suitable production strains for biotechnological applications.

References

Abe A, Oda Y, Asano K, Sone T (2007) *Rhizopus delemar* is the proper name for Rhizopus oryzae fumaric-malic acid producers. Mycologia 99:714–722

Álvarez V, Rodríguez-Sáiz M, de la Fuente JL, Gudiña EJ, Godio RP, Martín JF, Barredo JL (2006) The crtS gene of *Xanthophyllomyces dendrorhous* encodes a novel cytochrome-P450 hydroxylase involved in the conversion of β-carotene into astaxanthin and other xanthophylls. Fungal Genet Biol 43:261–272

Anonymus (2007) Product focus: maleic anhydride. Chem Week 39

Appel KF, Wolff AM, Arnau J (2004) A multicopy vector system for genetic studies in Mucor circinelloides and other zygomycetes. Mol Genet Genomics 271:595–602

Arrach N, Fernández-Martín R, Cerdá-Olmedo E, Avalos J (2001) A single gene for lycopene cyclase, phytoene synthase, and regulation of carotene biosynthesis in *Phycomyces*. Proc Natl Acad Sci USA 98:1687–1692

Azeke MA, Greiner R, Jany KD (2011) Purification and characterization of two intracellular phytases from the tempeh fungus *Rhizopus oligosporus*. J Food Biochem 35:213–227

Bai DM, Li SZ, Liu ZL, Cui ZF (2008) Enhanced L-(+)-lactic acid production by an adapted strain of *Rhizopus oryzae* using corncob hydrolysate. Appl Biochem Biotechnol 144:79–85

Bartnicki-Garcia S (1968) Control of dimorphism in Mucor by hexoses: inhibition of hyphal morphogenesis. J Bacteriol 96:1586–1594

Beg QK, Kapoor M, Mahajan L, Hoondal GS (2001) Microbial xylanases and their industrial applications: a review. Appl Microbiol Biotechnol 56:326–338

Benny GL, Humber RA, Voigt K (2014) Zygomycetous Fungi: phylum Entomophthoromycota and subphyla Kickxellomycotina, Mortierellomycotina, Mucoromycotina, and Zoopagomycotina. In: McLaughlin DJ, Spatafora JW (eds) The Mycota, vol VIIA, 2nd edn, Systematics and evolution. Springer, Berlin, pp 209–250, Chapter 8

Bianciotto V, Lumini E, Bonfante P, Vandamme P (2003) "*Candidatus Glomeribacter gigasporarum*" gen. nov, sp. nov., an endosymbiont of arbuscular mycorrhizal fungi. Int J Syst Evol Microbiol 53:121–124

Bidartondo MI, Read DJ, Trappe JM, Merckx V, Ligrone R, Duckett JG (2011) The dawn of symbiosis between plants and fungi. Biol Lett 7:574–577

Bitar D, Van Cauteren D, Lanternier F, Dannaoui E, Che D, Dromer F, Desenclos JC, Lortholary O (2009) Increasing incidence of zygomycosis (mucormycosis), France, 1997-2006. Emerg Infect Dis 15:1395–1401

Bonito G, Gryganskyi A, Hameed K, Schadt C, Pelletier D, Schaefer A, Tuskan G, Labbe J, Martin F, Doktycz M, LaButti K, Ohm R, Grigoriev I, Vilgalys R (2013) Co-evolution of *Mortierella elongata* and its endosymbiotic bacterium. In: Abstracts submitted for presentation at the 2013 APS-MSA joint meeting, Austin, TX (2013-08-10–2013-08-14). St. Paul. American Phytopathological Society. Phytopathology 103 (6, Suppl 2):18–19 Presented at APS-MSA Joint Meeting

Brakhage AA (2013) Regulation of fungal secondary metabolism. Nat Rev Microbiol 11:21–32

Bulut S, Elibol M, Ozer D (2009) Optimization of process parameters and culture medium for L-(+)-lactic acid production by *Rhizopus oryzae*. J Chem Eng Jpn 42:589–595

Burmester A, Richter M, Schultze K, Voelz K, Schachtschabel D, Boland W, Wöstemeyer J, Schimek C (2007) Cleavage of β-carotene as the first step in sexual hormone synthesis in zygomycetes is mediated by a trisporic acid regulated β-carotene oxygenase. Fungal Genet Biol 44:1096–1108

Cao NJ, Du JX, Gong CS, Tsao GT (1996) Simultaneous production and recovery of fumaric acid from immobilized *Rhizopus oryzae* with a rotary biofilm contractor and adsorption column. Appl Environ Microbiol 62:2926–2931

Casadevall A, Pirofski L (2001) Host-pathogen interactions. The attributes of virulence. J Infect Dis 184 (3):337–344

Cavalier-Smith T (1981) Eukaryote kingdoms: seven or nine? BioSystems 14:461–481

Cavalier-Smith T (1998) A revised six-kingdom system of life. Biol Rev 73:203–266

Chakrabarti A, Chatterjee SS, Shivaprakash MR (2008) Overview of opportunistic fungal infections in India. Nihon Ishinkin Gakkai Zasshi 49:165–172

Chakrabarti A, Chatterjee SS, Das A, Panda N, Shivaprakash MR, Kaur A, Varma SC, Singhi S, Bhansali A, Sakhuja V (2009) Invasive zygomycosis in India: experience in a tertiary care hospital. Postgrad Med J 85:573–581

Chamilos G, Lewis RE, Kontoyiannis DP (2007) Multidrug-resistant endosymbiotic bacteria account for the emergence of zygomycosis: a hypothesis. Fungal Genet Biol 44(2):88–92

Chang Y, Wang S, Sekimoto S, Aerts AL, Choi C, Clum A, LaButti KM, Lindquist EA, Yee Ngan C, Ohm RA, Salamov AA, Grigoriev IV, Spatafora JW, Berbee ML (2015) Phylogenomic analyses indicate that early fungi evolved digesting cell walls of algal ancestors of land plants. Genome Biol Evol 7(6):1590–1601

Chatzifragkou A, Makri A, Belka A, Bellou S, Mavrou M, Mastoridou M, Mystrioti P, Onjaro G, Aggelis G, Papanikolaou S (2011) Biotechnological conversions of biodiesel derived waste glycerol by yeast and fungal species. Energy 36:1097–1108

Chaudhuri S, Ghosh S, Bhattacharyya DK, Bandyopadhyay S (1998) Effect of mustard meal on the production of arachidonic acid by *Mortierella elongata* SC-208. J Am Oil Chem Soc 75:1053–1055

Chayakulkeeree M, Ghannoum MA, Perfect JR (2006) Zygomycosis: the re-emerging fungal infection. Eur J Clin Microbiol Infect Dis 25:215–229

Csernetics Á, Nagy G, Iturriaga EA, Szekeres A, Eslava AP, Vágvölgyi CS, Papp T (2011) Expression of three isoprenoid biosynthesis genes and their effects on the carotenoid production of the zygomycete *Mucor circinelloides*. Fungal Genet Biol 48:696–703

Csernetics Á, Tóth E, Farkas A, Nagy G, Bencsik O, Vágvölgyi C, Papp T (2015) Expression of *Xanthophyllomyces dendrorhous* cytochrome-P450 hydroxylase and reductase in *Mucor circinelloides*. World J Microbiol Biotechnol 31:321–336

Czempinski K, Kruft V, Wöstemeyer J, Burmester A (1996) 4-dihydromethyltrisporate dehydrogenase from *Mucor mucedo*, an enzyme of the sexual hormone pathway:purification, and cloning of the corresponding gene. Microbiology 142:2647–2654

de Azevedo Santiago AL, de Souza Motta CM (2008) Isolation of Mucorales from processed maize (*Zea mays* L.) and screening for protease activity. Braz J Microbiol 39:698–700

de Hoog GS, Guarro J, Genè J, Figueras MJ (2000) Zygomycota. In: Atlas of clinical fungi, pp 58–124, Centraalbureau voor Schimmelcultures, Universitat Rovira I Virgili, Utrecht, The Netherlands

de Hoog GS, Ibrahim AS, Voigt K (2014) Emerging zygomycetes: an emerging problem in the clinical laboratory. Mycoses 57(Suppl 3):1

Dedyukhina E, Chistyakova T, Vainshtein M (2011) Biosynthesis of arachidonic acid by micromycetes (review). Appl Biochem Microbiol 47:109–117

Delalibera I Jr, Hajek AE, Humber RA (2003) Use of cell culture media for cultivation of the mite pathogenic fungi *Neozygites tanajoae* and *Neozygites floridana*. J Invertebr Pathol 84:119–127

Deng Y, Li S, Xu Q, Gao M, Huang H (2012) Production of fumaric acid by simultaneous saccharification and fermentation of starchy materials with 2-deoxyglucose-resistant mutant strains of *Rhizopus oryzae*. Bioresour Technol 107:363–367

Dolatabadi S, de Hoog GS, Meis JF, Walther G (2014a) Species boundaries and nomenclature of *Rhizopus arrhizus* (syn. *R. oryzae*). Mycoses 57:108–127

Dolatabadi S, Walther G, van den Ende AHGG, de Hoog GS (2014b) Diversity and delimitation of *Rhizopus microsporus*. Fungal Div 64:145–163

Dufossé L (2006) Microbial production of food grade pigments. Food Technol Biotechnol 44:313–321

Dufossé L (2008) Pigments from microalgae and microorganisms: sources of food colorants. In: Sociaciu C (ed) Food colorants, chemical and functional properties. CRC Press, Boca Raton, FL, pp 399–427

Dyal SD, Narine SS (2005) Implications for the use of *Mortierella* fungi in the industrial production of essential fatty acids. Food Res Int 38:445–467

Dyal SD, Bouzidi L, Narine SS (2005) Maximizing the production of γ-linolenic acid in *Mortierella ramanniana* var. *ramanniana* as a function of pH, temperature and carbon source, nitrogen source, metal ions and oil supplementation. Food Res Int 38:815–829

Fakas S, Papanikolaou S, Batsos A, Galiotou-Panayotou M, Mallouchos A, Aggelis G (2009) Evaluating renewable carbon sources as substrates for single cell oil production by *Cunninghamella echinulata* and *Mortierella isabellina*. Biomass Bioenergy 33:573–580

Ferreira JA, Lennartsson PR, Edebo L, Taherzadeh MJ (2013) Zygomycetes-based biorefinery: Present status and future prospects. Bioresour Technol 135:523–532

Fujimura R, Nishimura A, Ohshima S, Sato Y, Nishizawa T, Oshima K, Hattori M, Narisawa K, Ohta H (2014) Draft Genome Sequence of the Betaproteobacterial Endosymbiont Associated with the Fungus *Mortierella elongata* FMR23-6. Genome Announc 2(6):e01272-14

Gao D, Zeng J, Zheng Y, Yu X, Chen S (2013) Microbial lipid production from xylose by *Mortierella isabellina*. Bioresour Technol 133:315–321

Garre V, Barredo JL, Iturriaga EA (2015) Transformation of *Mucor circinelloides* f. *lusitanicus* protoplasts. In: van den Berg MA, Maruthachalam K (eds) Genetic transformation systems in fungi, Volume 1. Springer International Publishing, pp 49–59

Ghignone S, Salvioli A, Anca I, Lumini E, Ortu G, Petiti L, Cruveiller S, Bianciotto V, Piffanelli P, Lanfranco L, Bonfante P (2012) The genome of the obligate endobacterium of an AM fungus reveals an interphylum network of nutritional interactions. ISME J 6(1):136–145

Goldberg I, Rokem JS, Pines O (2006) Organic acids: old metabolites, new themes. J Chem Tech Biotechnol 81:1601–1611

Gooday GW (1974) Fungal sex hormones. Annu Rev Biochem 43:35–49

Gopinath SCB, Anbu P, Lakshmipriya T, Hilda A (2013) Strategies to characterize fungal lipases for applications in medicine and dairy industry. Biomed Res Int 2013:1545–1549

Greenberg RN, Scott LJ, Vaughn HH, Ribes JA (2004) Zygomycosis (mucormycosis): emerging clinical importance and new treatments. Curr Opin Infect Dis 17:517–525

Grundschober A, Tuor U, Aebi M (1998) *In vitro* cultivation and sporulation of *Neozygites parvispora* (Zygomycetes: Entomophthorales). Syst Appl Microbiol 21:461–469

Gryganskyi AP, Lee SC, Litvintseva AP, Smith ME, Bonito G, Porter T, Anishchenko IM, Heitman J, Vilgalys R (2010) Structure, function, and phylogeny of the mating locus in the *Rhizopus oryzae* complex. PLoS One 5(12), e15273

Guo Y, Yan Q, Jiang Z, Teng C, Wang X (2010) Efficient production of lactic acid from sucrose and corncob hydrolysate by a newly isolated Rhizopus oryzae GY18. J Ind Microbiol Biotechnol 37:1137–1143

Gutiérrez A, López-García S, Garre V (2011) High reliability transformation of the basal fungus *Mucor circinelloides* by electroporation. J Microbiol Met 84:442–446

Hawksworth DL (2011) A new dawn for the naming of fungi: impacts of decisions made in Melbourne in July 2011 on the future publication and regulation of fungal names. MycoKeys 1:7–20

Heckman DS, Geiser DM, Eidell BR, Stauffer RL, Kardos NL, Hedges SB (2001) Molecular evidence for the early colonization of land by fungi and plants. Science 293:1129–1133

Hesseltine CW (1983) Microbiology of oriental fermented foods. Annu Rev Microbiol 37:575–601

Hesseltine CW (1991) Zygomycetes in food fermentations. Mycologist 5:162–169

Hibbett DS, Binder M, Bischoff JF, Blackwell M, Cannon PF, Eriksson O, Huhndorf S, James T, Kirk PM, Lücking R, Lumbsch T, Lutzoni F, Matheny PB, McLaughlin DJ, Powell MJ, Redhead S, Schoch CL, Spatafora JW, Stalpers JA, Vilgalys R, Aime MC, Aptroot A, Bauer R, Begerow D, Benny GL, Castlebury LA, Crous PW, Dai Y-C, Gams W, Geiser DM, Griffith GW, Gueidan C, Hawksworth DL, Hestmark G, Hosaka K, Humber RA, Hyde K, Köljalb U, Kurtzman CP, Larsson K-H, Lichtwardt R, Longcore J, Miadlikowska J, Miller A, Moncalvo J-M, Mozley-Standridge S, Oberwinkler F, Parmasto R, Reeb V, Rogers JD, Roux C, Ryvarden L, Sampaio JP, Schuessler A, Sugiyama J, Thorn RG, Tibell L, Untereiner WA, Walker C, Wang A, Weir A, Weiss M, White M, Winka K, Yao Y-J, Zhang N (2007) A higher-level phylogenetic classification of the Fungi. Mycol Res 111:509–547

Hoffmann K, Telle S, Walther G, Eckart M, Kirchmair M, Prillinger HJ, Prazenica A, Newcombe G, Dölz F, Papp T, Vágvölgyi C, de Hoog S, Olsson L, Voigt K (2009) Diversity, genotypic identification, ultrastructural and phylogenetic characterization of zygomycetes from different ecological habitats and climatic regions: limitations and utility of nuclear ribosomal DNA barcode markers. In: Gherbawy Y, Mach RL, Rai M (eds) Current advances in molecular mycology. Nova, New York, pp 263–312

Hoffmann K, Voigt K, Kirk PM (2011) *Mortierellomycotina* subphyl. nov. based on multi-gene genealogies. Mycotaxon 115:353–363

Hoffmann K, Pawlowska J, Walther G, Wrzosek M, de Hoog GS, Benny GL, Kirk PM, Voigt K (2013) The family structure of the Mucorales: a synoptic revision based on comprehensive multigene-genealogies. Persoonia 30:57–76

Hou C (2008) Production of arachidonic acid and dihomo-linolenic acid from glycerol by oil-producing filamentous fungi, *Mortierella* in the ARS culture collection. J Ind Microbiol Biotechnol 35:501–506

Huang C, Chen XF, Xiong L, Chen XD, Ma LL, Chen Y (2013) Single cell oil production from low-cost substrates: the possibility and potential of its industrialization. Biotechnol Adv 31(2):129–139

Ibrahim AS (2011) Host cell invasion in mucormycosis: role of iron. Curr Opin Microbiol 14:406–411

Ibrahim AS, Gebremariam T, Liu M, Chamilos G, Kontoyiannis DP, Mink R, Kwon-Chung KJ, Fu Y, Skory CD, Edwards JE Jr, Spellberg B (2008) Bacterial Endosymbiosis is widely present among Zygomycetes but does not contribute to the pathogenesis of mucormycosis. J Infect Dis 198:1083–1090

Ikubo S, Takigawa N, Ueoka H, Kiura K, Tabata M, Shibayama T, Chikamori M, Aoe K, Matsushita A, Harada M (1999) *In vitro* evaluation of antimicrotubule agents in human small-cell lung cancer cell lines. Anticancer Res 19:3985–3988

Iturriaga EA, Díaz-Mínguez JM, Benito EP, Álvarez MI, Eslava AP (1992) Heterologous transformation of *Mucor circinelloides* with the *Phycomyces blakesleeanus leu1* gene. Curr Genet 21:215–223

Iturriaga EA, Velayos A, Eslava AP (2000) The structure and function of the genes involved in the biosynthesis of carotenoids in the Mucorales. Biotechnol Bioprocess Eng 5:263–274

James TY, Kauff F, Schoch CL, Matheny PB, Hofstetter V, Cox CJ, Celio G, Gueidan C, Fraker E, Miadlikowska J, Lumbsch HT, Rauhut A, Reeb V, Arnold AE, Amtoft A, Stajich JE, Hosaka K, Sung GH, Johnson D, O'Rourke B, Crockett M, Binder M, Curtis JM, Slot JC, Wang Z, Wilson AW, Schüler A, Longcore JE, O'Donnell K, Mozley-Standridge S, Porter D, Letcher PM, Powell MJ, Taylor JW, White MM, Griffith GW, Davies DR, Humber RA, Morton JB, Sugiyama J, Rossman AY, Rogers JD, Pfister DH, Hewitt D, Hansen K, Hambleton S, Shoemaker RA, Kohlmeyer J, Volkmann-Kohlmeyer B, Spotts RA, Serdani M, Crous PW, Hughes KW, Matsuura K, Langer E, Langer G, Untereiner WA, Lücking R, Büdel B, Geiser DM, Aptroot A, Diederich P, Schmitt I, Schultz M, Yahr R, Hibbett DS, Lutzoni F, McLaughlin DJ, Spatafora JW, Vilgalys R (2006) Reconstructing the early evolution of the fungi using a six-gene phylogeny. Nature 443:818–822

Jang HD, Lin YY, Yang SS (2000) Polyunsaturated fatty acid production with *Mortierella alpina* by solid substrate fermentation. Bot Bull Acad Sin 41:41–48

Jennessen J, Nielsen KF, Houbraken J, Lyhne EK, Schnürer J, Frisvad JC, Samson RA (2005) Secondary metabolite and mycotoxin production by the *Rhizopus microsporus* group. J Agric Food Chem 53(5):1833–1840

Jin B, Yin P, Ma Y, Zhao L (2005) Production of lactic acid and fungal biomass by *Rhizopus* fungi from food processing waste water. J Ind Microbiol Biotechnol 32:678–686

Jones SK Jr, Bennett RJ (2011) Fungal mating pheromones: choreographing the dating game. Fungal Genet Biol 48:668–676

Karimi K, Zamani A (2013) *Mucor indicus*: biology and industrial application perspectives: a review. Biotechnol Adv 31:466–481

Kauffman CA (2004) Zygomycosis: reemergence of an old pathogen. Clin Infect Dis 39:588–590

Kavadia A, Komaitis M, Chevalot I, Blanchard F, Marc I, Aggelis G (2001) Lipid and gamma-linolenic acid accumulation in strains of zygomycetes growing on glucose. J Am Oil Chem Soc 78:341–346

Kloer DP, Ruch S, Al-Babili S, Beyer P, Schulz GE (2005) The structure of a retinal-forming carotenoid oxygenase. Science 8:267–269

Krisch J, Takó M, Papp T, Vágvölyi C (2010) Characteristics and potential use of β-glucosidases from Zygomycetes. In: Méndez-Vilas a (ed.) Current research, technology and education topics in applied microbiology and microbial biotechnology, Formatex Research Center, pp 891-896.

Krizsán K, Bencsik O, Nyilasi I, Galgóczy L, Vágvölgyi C, Papp T (2010) Effect of the sesterterpene-type metabolites, ophiobolins A and B, on zygomycetes fungi. FEMS Microbiol Lett 313:135–140

Lee S, Jang Y, Lee YM, Lee J, Lee H, Kim GH, Kim JJ (2011) Rice straw-decomposing fungi and their cellulolytic and xylanolytic enzymes. J Microbiol Biotechnol 21:1322–1329

Lee SC, Billmyre RB, Li A, Carson S, Sykes SM, Huh EY, Mieczkowski P, Ko DC, Cuomo CA, Heitman J (2014) Analysis of a food-borne fungal pathogen outbreak: virulence and genome of a *Mucor circinelloides* isolate from yogurt. mBio 5(4):e01390-14

Lennartsson PR, Karimi K, Edebo L, Taherzadeh MJ (2009) Effects of different growth forms of *Mucor indicus* on cultivation on dilute-acid lignocellulosic hydrolyzate, inhibitor tolerance, and cell wall composition. J Biotechnol 143:255–261

Lichtwardt RW, Manier JF (1978) Validation of the Harpellales and Asellariales. Mycotaxon 7:441–442

Liu XJ, Liu RS, Li HM, Tang YJ (2012) Lycopene production from synthetic medium by *Blakeslea trispora* NRRL2895(+) and 2896(-) in a stirred-tank fermenter. Bioprocess Biosyst Eng 35:739–749

Lukács GY, Papp T, Somogyvári F, Csernetics Á, Nyilasi I, Vágvölgyi CS (2009) Cloning of the *Rhizomucor miehei* 3-hydroxy-3-methylglutaryl-coenzyme A reductase gene and its heterologous expression in *Mucor circinelloides*. Ant Leeuwenhoek 95:55–64

Ma LJ, Ibrahim AS, Skory C, Grabherr MG, Burger G, Butler M, Elias M, Idnurm A, Lang BF, Sone T, Abe A, Calvo SE, Corrochano LM, Engels R, Fu J, Hansberg W, Kim JM, Kodira CD, Koehrsen MJ, Liu B, Miranda-Saavedra D, O'Leary S, Ortiz-Castellanos L, Poulter R, Rodriguez-Romero J, Ruiz-Herrera J, Shen YQ, Zeng Q, Galagan J, Birren BW, Cuomo CA, Wickes BL (2009) Genomic analysis of the basal lineage fungus *Rhizopus oryzae* reveals a whole-genome duplication. PLoS Genet 5, e1000549

Medina HR, Cerdá-Olmedo E, Al-Babili S (2011) Cleavage oxygenases for the biosynthesis of trisporoids and other apocarotenoids in *Phycomyces*. Mol Microbiol 82(1):199–208

Mehta BJ, Salgado LM, Bejarano ER, Cerdá-Olmedo E (1997) New mutants of *Phycomyces blakesleeanus* for beta-carotene production. Appl Environ Microbiol 63:3657–3661

Mendoza L, Vilela R, Voelz K, Ibrahim AS, Voigt K, Lee SC (2014) Human fungal pathogens of Mucorales and Entomophthorales, Chapter 27. In: Casadevall A, Mitchell AP, Berman J, Kwon-Chung KJ, Perfect JR, Heitman J (eds) Cold spring harbor perspectives: fungal pathogens. Cold Spring Harbor Press, Cold Spring Harb Perspect Med 5(4) pii: a019562

Meussen BJ, de Graaff LH, Sanders JP, Weusthuis RA (2012) Metabolic engineering of *Rhizopus oryzae* for the production of platform chemicals. Appl Microbiol Biotechnol 94:875–886

Millati R, Edebo L, Taherzadeh MJ (2005) Performance of *Rhizopus*, *Rhizomucor*, and *Mucor* in ethanol production from glucose, xylose, and wood hydrolyzates. Enzyme Microb Technol 36:294–300

Monteiro de Souza P, de Oliveira Magalhães P (2010) Application of microbial α-amylase in industry – a review. Braz J Microbiol 41(4):850–861

Morace G, Borghi E (2012) invasive mold infections: virulence and pathogenesis of mucorales. Int J Microbiol 2012, Article ID 349278, 5 p

Moss ST (1975) Commensalism of the trichomycetes. In: Batra LR (ed) Insect-fungus symbiosis: nutrition, mutualism and commensalism. Allanheld, Osmun & Co., Montclair, pp 175–227

Mullaney EJ, Ullah AH (2003) The term phytase comprises several different classes of enzymes. Biochem Biophys Res Commun 312:179–184

Mullaney EJ, Daly CB, Ullah AH (2000) Advances in phytase research. Adv Appl Microbiol 47:157–199

Münchberg U, Wagner L, Spielberg ET, Voigt K, Rösch P, Popp J (2012) Spatially resolved investigation of the oil composition in single intact hyphae of *Mortierella* spp. with micro-Raman spectroscopy. Biochim Biophys Acta 1831:341–349

Münchberg U, Wagner L, Rohrer C, Voigt K, Jahreis G, Rösch P, Popp J (2015) Quantitative assessment of the degree of lipid unsaturation in intact *Mortierella* by Raman microspectroscopy. Anal Bioanal Chem 407:3303–3311

Nagy LG, Petkovits T, Kovács GM, Voigt K, Vágvölgyi C, Papp T (2011) Where is the unseen fungal diversity hidden? A study of *Mortierella* reveals a large contribution of reference collections to the identification of fungal environmental sequences. New Phytol 191:789–794

Nagy G, Farkas A, Csernetics Á, Bencsik O, Szekeres A, Nyilasi I, Vágvölgyi CS, Papp T (2014) Transcription analysis of the three HMG-CoA reductase genes of *Mucor circinelloides*. BMC Microbiol 14:93

Nahas E (1988) Control of lipase production by *Rhizopus oligosporus* under various growth conditions. J Gen Microbiol 134:227–233

Navarro E, Lorca-Pascual JM, Quiles-Rosillo MD, Nicolas FE, Garre V, Torres-Martinez S, Ruiz-Vazquez RM (2001) A negative regulator of light-inducible carotenogenesis in *Mucor circinelloides*. Mol Gen Genomics 266:463–470

Navarro E, Peñaranda A, Hansberg W, Torres-Martínez S, Garre V (2013) A white collar 1-like protein mediates opposite regulatory functions in *Mucor circinelloides*. Fungal Genet Biol 52:42–52

Nicolás-Molina FE, Navarro E, Ruiz-Vázquez RM (2008) Lycopene over-accumulation by disruption of the negative regulator gene crgA in *Mucor circinelloides*. Appl Microbiol Biotechnol 78:131–137

Nout MJR, Kiers JL (2005) Tempeh fermentation, innovation and functionality: update into the third millennium. J Appl Microbiol 98:789–805

Nützmann HW, Reyes-Dominguez Y, Scherlach K, Schroeckh V, Horn F, Gacek A, Schümann J, Hertweck C, Strauss J, Brakhage AA (2011) Bacteria-induced natural product formation in the fungus *Aspergillus nidulans* requires Saga/Ada-mediated histone acetylation. Proc Natl Acad Sci USA 108 (34):14282–14287

Nützmann HW, Schroeckh V, Brakhage AA (2012) Regulatory cross talk and microbial induction of fungal secondary metabolite gene clusters. Methods Enzymol 517:325–341

Obraztsova IN, Prados N, Holzmann K, Avalos J, Cerda-Olmedo E (2004) Genetic damage following introduction of DNA in *Phycomyces*. Fungal Genet Biol 41:68–180

Orlowsky M (1991) *Mucor* dimorphism. Microbiol Rev 55:234–258

Osmani SA, Scrutton MC (1985) The sub-cellular localisation and regulatory properties of pyruvate carboxylase from *Rhizopus arrhizus*. Eur J Biochem 147:119–128

Overman SA, Romano AH (1969) Pyruvate carboxylase of Rhizopus nigricans and its role in fumaric acid production. Biochem Biophys Res Commun 37:457–463

Papp T, Velayos A, Bartók T, Eslava AP, Vágvölgyi CS, Iturriaga EA (2006) Heterologous expression of astaxanthin biosynthesis genes in *Mucor circinelloides*. Appl Microbiol Biotechnol 69:526–531

Papp T, Csernetics Á, Nyilasi I, Ábrók M, Vágvölgyi CS (2010) Genetic transformation of Zygomycetes fungi. In: Rai MK, Kövics GJ (eds) Progress in mycology. Springer, Netherlands, pp 75–94

Papp T, Hoffmann K, Nyilasi I, Petkovits T, Wagner L, Vágvölgyi C, Voigt K (2011) *Mortierella*. In: Liu D (ed) Molecular detection of human fungal pathogens. Taylor & Francis CRC Press, pp 749–758

Papp T, Csernetics Á, Nagy G, Bencsik O, Iturriaga EA, Eslava AP, Vágvölgyi CS (2013) Canthaxanthin production with modified *Mucor circinelloides* strains. Appl Microbiol Biotechnol 97:4937–4950

Parniske M (2008) Arbuscular mycorrhiza: the mother of plant root endosymbioses. Nat Rev Microbiol 10:763–775

Partida-Martinez LP, Hertweck C (2005) Pathogenic fungus harbours endosymbiotic bacteria for toxin production. Nature 437:884–888

Partida-Martinez LP, Bandemer S, Rüchel R, Dannaoui E, Hertweck C (2008) Lack of evidence of endosymbiotic toxin-producing bacteria in clinical *Rhizopus* isolates. Mycoses 51(3):266–269

Petrič S, Hakki T, Bernhardt R, Zigon D, Crešnar B (2010) Discovery of a steroid 11α-hydroxylase from *Rhizopus oryzae* and its biotechnological application. J Biotechnol 150:428–437

Polaino S, Herrador MM, Cerdá-Olmedo E, Barrero AF (2010) Splitting of beta-carotene in the sexual interaction of Phycomyces. Org Biomol Chem 8 (19):4229–4231

Poliakov E, Gentleman S, Cunningham FX Jr, Miller-Ihli NJ, Redmond TM (2005) Key role of conserved histidines in recombinant mouse beta-carotene 15′15-monooxygenase-1 activity. J Biol Chem 280:29217–29223

Pritchard GG (1971) An NAD+-independent L-lactate dehydrogenase from *Rhizopus oryzae*. Biochim Biophys Acta 250:25–34

Pritchard GG (1973) Factors affecting the activity and synthesis of NAD dependent lactate dehydrogenase in *Rhizopus oryzae*. J Gen Microbiol 78:125–137

Quiles-Rosillo MD, Ruiz-Vázquez RM, Torres-Martínez S, Garre V (2003) Cloning, characterization and heterologous expression of the *Blakeslea trispora* gene encoding orotidine-5P-monophosphate decarboxylase. FEMS Microbiol Lett 222:229–236

Ratledge C (2004) Fatty acid biosynthesis in microorganisms being used for Single Cell Oil production. Biochimie 86:807–815

Ratledge C, Wynn JP (2002) The biochemistry and molecular biology of lipid accumulation in oleaginous microorganisms. Adv Appl Microbiol 51:1–51

Revuelta JL, Eslava AP (1983) A new gene (carC) involved in the regulation of carotenogenesis in *Phycomyces*. Mol Gen Genet 192:225–229

Ribes JA, Vanover-Sams CL, Baker DJ (2000) Zygomycetes in human disease. Clin Microbiol Rev 13 (2):236–301

Roa Engel CA, Straathof AJ, Zijlmans TW, van Gulik WM, van der Wielen LA (2008) Fumaric acid production by fermentation. Appl Microbiol Biotechnol 78:379–389

Rodríguez-Ortiz R, Michielse C, Rep M, Limón MC, Avalos J (2012) Genetic basis of carotenoid overproduction in *Fusarium oxysporum*. Fungal Genet Biol 49:684–696

Rodríguez-Sáiz M, Paz B, de la Fuente JL, López-Nieto MJ, Cabri W, Barredo JL (2004) *Blakeslea trispora* genes for carotene biosynthesis. Appl Environ Microbiol 70:5589–5594

Rodríquez-Sáiz M, de la Fuente JL, Barredo JL (2012) Metabolic engineering of *Mucor circinelloides* for zeaxanthin production. Methods Mol Biol 898:133–151

Roncero MIG, Cerdá-Olmedo E (1982) Genetics of carotene biosynthesis in *Phycomyces*. Curr Genet 5:5–8

Rothhardt J, Schwartze V, Voigt K (2011) Entomophthorales. In: Liu D (ed) Molecular detection of human fungal pathogens. Taylor & Francis CRC Press, pp 723–734

Ruiz-Hidalgo MJ, Benito EP, Sandmann G, Eslava AP (1997) The phytoene dehydrogenase gene of *Phycomyces*: regulation of its expression by blue light and vitamin A. Mol Gen Genet 253:734–744

Ruiz-Hidalgo MJ, Eslava AP, Alvarez MI, Benito EP (1999) Heterologous expression of the *Phycomyces blakesleeanus* phytoene dehydrogenase gene (*carB*) in *Mucor circinelloides*. Curr Microbiol 39:259–264

Sahadevan Y, Richter-Fecken M, Kaerger K, Voigt K, Boland W (2013) Early and late trisporoids differentially regulate β-carotene production and gene transcript Levels in the mucoralean fungi *Blakeslea trispora* and *Mucor mucedo*. Appl Environ Microbiol 79(23):7466–7475

Saito K, Saito A, Ohnishi M, Oda Y (2004) Genetic diversity in *Rhizopus oryzae* strains as revealed by the sequence of lactate dehydrogenase genes. Arch Microbiol 182:30–36

Sajbidor J, Certik M, Dobronova S (1988) Influence of different carbon sources on growth, lipid content and fatty acid composition in four strains belonging to Mucorales. Biotechnol Lett 10:347–350

Salgado LM, Bejarano ER, Cerdá-Olmedo E (1989) Carotene superproducing mutants of *Phycomyces blakesleeanus*. Exp Mycol 13:332–336

Sanz C, Velayos A, Álvarez MI, Benito EP, Eslava AP (2011) Functional analysis of the *Phycomyces carRA* gene encoding the enzymes phytoene synthase and lycopene cyclase. PLos One 6:e23102

Sato Y, Narisawa K, Tsuruta K, Umezu M, Nishizawa T, Tanaka K, Yamaguchi K, Komatsuzaki M, Ohta H (2010) Detection of *Betaproteobacteria* inside the mycelium of the fungus *Mortierella elongata*. Microbes Environ 25(4):321–324

Sautour M, Soares Mansur C, Divies C, Bensoussan M, Dantigny P (2002) Comparison of the effects of temperature and water activity on growth rate of food spoilage moulds. J Ind Microbiol Biotechnol 28:311–315

Scarborough CL, Ferrari J, Godfray HC (2005) Aphid protected from pathogen by endosymbiont. Science 310(5755):1781

Schachtschabel D, Schimek C, Wöstemeyer J, Boland W (2005) Biological activity of trisporoids and trisporoid analogues in *Mucor mucedo* (-). Phytochemistry 66(11):1358–1365

Schachtschabel D, David A, Menzel KD, Schimek C, Wöstemeyer J, Boland W (2008) Cooperative biosynthesis of trisporoids by the (+) and (-) mating types of the zygomycete *Blakeslea trispora*. ChemBioChem 9(18):3004–3012

Schimek C, Wöstemeyer J (2009) Carotene derivatives in sexual communication of zygomycete fungi. Phytochemistry 70:1867–1875

Schimek C, Kleppe K, Saleem AR, Voigt K, Burmester A, Wöstemeyer J (2003) Sexual reactions in Mortierellales are mediated by the trisporic acid system. Mycol Res 107:736–747

Schroeckh V, Scherlach K, Nützmann HW, Shelest E, Schmidt-Heck W, Schuemann J, Martin K, Hertweck C, Brakhage AA (2009) Intimate bacterial-fungal interaction triggers biosynthesis of archetypal polyketides in *Aspergillus nidulans*. Proc Natl Acad Sci USA 106(34):14558–14563

Schwartz SH, Tan BC, Gage DA, Zeevaart JA, McCarty DR (1997) Specific oxidative cleavage of carotenoids by VP14 of maize. Science 276:1872

Schwartze VU, Winter S, Shelest E, Marcet-Houben M, Horn F, Wehner S, Linde J, Valiante V, Sammeth M, Riege K, Nowrousian M, Kaerger K, Jacobsen ID, Marz M, Brakhage AA, Gabaldón T, Böcker S, Voigt K (2014) Gene expansion shapes genome architecture in the human pathogen *Lichtheimia corymbifera*: an evolutionary genomics analysis in the ancient terrestrial Mucorales (Mucoromycotina). PLoS Genet 10(8):e1004496. doi:10.1371/journal.pgen.1004496

Serrano I, Lopes da Silva T, Carlos Roseiro J (2001) Ethanol-induced dimorphism and lipid composition changes in Mucor fragilis CCMI 142. Lett Appl Microbiol 33:89–93

Sharifia M, Karimi K, Taherzadeh MJ (2008) Production of ethanol by filamentous and yeast-like forms of Mucor indicus from fructose, glucose, sucrose, and molasses. J Ind Microbiol Biotechnol 35:1253–1259

Shelest E (2008) Transcription factors in fungi. FEMS Microbiol Lett 286:145–151

Shelest E, Voigt K (2014) Genomics to study basal lineage fungal biology: phylogenomics suggests a common origin. In: Nowrousian M (ed) The Mycota, vol XIII, 2nd edn, Fungal genomics. Springer, Berlin, pp 31–60

Sigel R, Sigel A, Sigel H (2007) The ubiquitous roles of cytochrome P450 proteins: metal ions in life sciences. Wiley, New York. ISBN 0-470-01672-8

Silva F, Torres-Martinez S, Garre V (2006) Distinct white collar-1 genes control specific light responses in *Mucor circinelloides*. Mol Microbiol 61:1023–1037

Silva F, Navarro E, Peñaranda A, Murcia-Flores L, Torres-Martínez S, Garre V (2008) A RING-finger protein regulates carotenogenesis via proteolysis-independent ubiquitylation of a white collar-1-like activator. Mol Microbiol 70:1026–1036

Simon L, Bousquet J, Lévesque RC, Lalonde M (1993) Origin and diversification of endomycorrhizal fungi and co-incidence with vascular land plants. Nature 363:67–69

Skory CD (2000) Isolation and expression of lactate dehydrogenase genes from *Rhizopus oryzae*. Appl Environ Microbiol 66:2343–2348

Spano LA, Medeiros J, Mandels M (1976) Enzymatic hydrolysis of cellulose wastes to glucose. Resour Recover Conserv 1:279–294

Stredansky M, Conti E, Stredanska S, Zanetti F (2000) γ-Linolenic acid production with *Thamnidium elegans* by solid-state fermentation on apple pomace. Bioresour Technol 73:41–45

Sun J, Sun XX, Tang PW, Yuan QP (2012) Molecular cloning and functional expression of two key carotene synthetic genes derived from *Blakeslea trispora* into *E. coli* for increased β-carotene production. Biotechnol Lett 34:2077–2082

Tagua VG, Medina HR, Martín-Domínguez R, Eslava AP, Corrochano LM, Cerdá-Olmedo E, Idnurm A (2012) A gene for carotene cleavage required for pheromone biosynthesis and carotene regulation in the fungus *Phycomyces blakesleeanus*. Fungal Genet Biol 49(5):398–404

Takahashi Y, Moiseyev G, Chen Y, Ma JX (2005) Identification of conserved histidines and glutamic acid as key residues for isomerohydrolase activity of RPE65, an enzyme of the visual cycle in the retinal pigment epithelium. FEBS Lett 579:5414–5418

Torres-Martínez S, Murillo FJ, Cerdá-Olmedo E (1980) Genetics of locopene cyclization and substrate transfer in beta-carotene biosynthesis in *Phycomyces*. Genet Res 36:299–946

Tsuruo T, Oh-hara T, Iida H, Tsukagoshi S, Sato Z, Matsuda I, Iwasaki S, Okuda S, Shimizu F, Sasagawa K, Fukami M, Fukada K, Arakawa M (1986) Rhizoxin, a macrocyclic lactone antibiotic, as a new antitumor agent against human and murine tumor cells and their vincristine-resistant sublines. Cancer Res 46:381–385

van der Westhuizen JPJ, Kock JLF, Botha A, Botes PJ (1994) The distribution of the ω3- and ω 6-series of cellular long-chain fatty acids in fungi. Sys Appl Microbiol 17:327–345

van Heeswijck R, Roncero MIG (1984) High frequency transformation of *Mucor* with recombinant plasmid DNA. Carlsberg Res Commun 49:691–702

Velayos A, Blasco JL, Alvarez MI, Iturriaga EA, Eslava AP (2000a) Blue-light regulation of the phytoene dehydrogenase (carB) gene expression in Mucor circinelloides. Planta 210:938–946

Velayos A, Eslava AP, Iturriaga EA (2000b) A bifunctional enzyme with lycopene cyclase and phytoene synthase activities is encoded by the carRP gene of Mucor circinelloides. Eur J Biochem 267:1–12

Velayos A, Papp T, Aguilar-Elena R, Fuentes-Vicente M, Eslava AP, Iturriaga EA, Álvarez MI (2003) Expression of the *carG* gene, encoding geranylgeranyl pyrophosphate synthase, is up-regulated by blue light in *Mucor circinelloides*. Curr Genet 43:112–120

Velayos A, Fuentes-Vicente M, Aguilar-Elena R, Eslava AP, Iturriaga EA (2004) A novel fungal prenyl diphosphate synthase in the dimorphic zygomycete *Mucor circinelloides*. Curr Genet 45:371–377

Vially G, Marchal R, Guilbert N (2010) L(+) Lactate production from carbohydrates and lignocellulosic materials by *Rhizopus oryzae* UMIP 4.77. World J Microb Biot 26:607–614

Voigt K (2012) Zygomycota. In: Frey W (ed) Syllabus of plant families - A. Engler's Syllabus der Pflanzenfamilien. Part 1/1: Blue-green algae, Myxomycetes and Myxomycete-like organisms, Phytoparasitic protists, Heterotrophic Heterokontobionta and Fungi p.p. Borntraeger Verlag, Stuttgart, pp 130–162

Voigt K, de Hoog GS (2013) The zygomycetes in a phylogenetic perspective. Special issue Persoonia: molecular phylogeny and evolution of fungi, vol. 30. Naturalis Biodiversity Center Leiden and Centraalbureau voor Schimmelcultures Utrecht, The Netherlands, p 125

Voigt K, Kirk PM (2014) 136. FUNGI | Classification of the zygomycetes: reappraisal as coherent class based on a comparison between traditional *versus* molecular systematics. In: Batt CA, Tortorello ML (eds) Encyclopedia of food microbiology, Vol 2. Elsevier, Academic, pp 54–67

Voigt K, Wöstemeyer J (2001) Phylogeny and origin of 82 zygomycetes from all 54 genera of the Mucorales and Mortierellales based on combined analysis of actin and translation elongation factor EF-1α genes. Gene 270:113–120

Voigt K, Hoffmann K, Einax E, Eckart M, Papp T, Vágvölgyi C, Olsson L (2009) Revision of the family structure of the Mucorales (Mucoromycotina, Zygomycetes) based on multigene-genealogies: phylogenetic analyses suggest a bigeneric Phycomycetaceae with *Spinellus* as sister group to *Phycomyces*. In: Gherbawy Y, Mach RL, Rai M (eds) Current advances in molecular mycology. Nova, New York, pp 313–332

Voigt K, Marano AV, Gleason F (2013) Ecological and economical importance of parasitic and zoosporic true fungi. In: Kempken F (ed) The Mycota, vol XI, 2nd edn, Agricultural applications. Springer, Berlin, pp 243–270

Wagner L, Stielow B, Hoffmann K, Petkovits T, Papp T, Vágvölgyi C, de Hoog GS, Verkley G, Voigt K (2013) A comprehensive molecular phylogeny of the Mortierellales (Mortierellomycotina) based on nuclear ribosomal DNA. Persoonia 30:77–93

Wang GY, Keasling JD (2002) Amplification of HMG-CoA reductase production enhances carotenoid accumulation in *Neurospora crassa*. Metab Eng 4:193–201

Wang L, Chen W, Feng Y, Ren Y, Gu Z et al (2011) Genome characterization of the oleaginous fungus *Mortierella alpina*. PLoS One 6:e28319

Wang D, Wu R, Xu Y, Li M (2013) Draft genome sequence of *Rhizopus chinensis* CCTCCM201021, used for brewing traditional Chinese alcoholic beverages. Genome Announc 1(2), e0019512

Werpy T, Petersen G (2004) Top value added chemicals from biomass, vol 1, Results of screening for potential candidates from sugars and synthesis gas. US Department of Energy, Washington, DC

Wetzel J, Scheibner O, Burmester A, Schimek C, Wöstemeyer J (2009) 4-Dihydrotrisporin-dehydrogenase, an enzyme of the sex hormone pathway of *Mucor mucedo*: purification, cloning of the corresponding gene, and developmental expression. Eukaryotic Cell 8:88–95

White JD, Blakemore PR, Green NJ et al (2002) Total synthesis of rhizoxin D, a potent antimitotic agent from the fungus *Rhizopus chinensis*. J Org Chem 67:7750–7760

Whittaker RH (1969) New concepts of kingdoms of organisms. Science 163:150–160

Wikandari R, Millati R, Lennartsson PR, Harmayani E, Taherzadeh MJ (2012) Isolation and characterization of zygomycetes fungi from tempeh for ethanol production and biomass applications. Appl Biochem Biotechnol 167:1501–1512

Willke T, Vorlop KD (2004) Industrial bioconversion of renewable resources as an alternative to conventional chemistry. Appl Microbiol Biotechnol 66:131–142

Wolff AM, Arnau J (2002) Cloning of glyceraldehyde-3-phosphate dehydrogenase-encoding genes in *Mucor circinelloides* (Syn. *racemosus*) and use of the *gpd1* promoter for recombinant protein production. Fungal Genet Biol 35:21–29

Wöstemeyer J, Burmester A, Wöstemeyer A, Schultze K, Voigt K (2002) Gene transfer in the fungal host-parasite system *Absidia glauca-Parasitella parasitica* depends on infection. In: Syvanen M, Kado CI (eds) Horizontal gene transfer, 2nd edn. Academic, San Diego, CA, pp 241–247, Chapter 21

Wöstemeyer J, Grünler A, Schimek C, Voigt K (2005) Genetic regulation of carotenoid biosynthesis in fungi. In: Applied mycology and biotechnology Vol 5: Genes and genomics. Elsevier, pp 257–274

Yen HW, Lee YC (2010) Production of lactic acid from raw sweet potato powders by rhizopus oryzae immobilized in sodium alginate capsules. Appl Biochem Biotechnol 162:607–615

Zeng J, Zheng Y, Yu X, Yu L, Gao D, Chen S (2013) Lignocellulosic biomass as a carbohydrate source for lipid production by *Mortierella isabellina*. Bioresour Technol 128:385–391

Zhang ZY, Jin B, Kelly JM (2007) Production of lactic acid from renewable materials by *Rhizopus* fungi. Biochem Eng J 35:251–263

Zorn H, Langhoff S, Scheibner M, Berger RG (2003) Cleavage of β-carotene to flavor compounds by fungi. Appl Microbiol Biotechnol 62(4):331–336

Biosystematic Index

Biochemistry and Molecular Biology, 3rd Edition
The Mycota III
D. Hoffmeister (Ed.)

Subject Index

Biochemistry and Molecular Biology, 3rd Edition
The Mycota III
D. Hoffmeister (Ed.)

www.ingramcontent.com/pod-product-compliance
Lightning Source LLC
Chambersburg PA
CBHW081926120926
51338CB00004B/25
* 9 7 8 3 3 1 9 2 7 7 8 8 2 *